STATA TIME-SERIES
REFERENCE MANUAL
RELEASE 12

A Stata Press Publication
StataCorp LP
College Station, Texas

Table of contents

Cross-referencing the documentation

When reading this manual, you will find references to other Stata manuals. For example,

[U] **26 Overview of Stata estimation commands**

[R] **regress**

[D] **reshape**

The first example is a reference to chapter 26, *Overview of Stata estimation commands*, in the *User's Guide*; the second is a reference to the regress entry in the *Base Reference Manual*; and the third is a reference to the reshape entry in the *Data-Management Reference Manual*.

All the manuals in the Stata Documentation have a shorthand notation:

[GSM]	*Getting Started with Stata for Mac*
[GSU]	*Getting Started with Stata for Unix*
[GSW]	*Getting Started with Stata for Windows*
[U]	*Stata User's Guide*
[R]	*Stata Base Reference Manual*
[D]	*Stata Data-Management Reference Manual*
[G]	*Stata Graphics Reference Manual*
[XT]	*Stata Longitudinal-Data/Panel-Data Reference Manual*
[MI]	*Stata Multiple-Imputation Reference Manual*
[MV]	*Stata Multivariate Statistics Reference Manual*
[P]	*Stata Programming Reference Manual*
[SEM]	*Stata Structural Equation Modeling Reference Manual*
[SVY]	*Stata Survey Data Reference Manual*
[ST]	*Stata Survival Analysis and Epidemiological Tables Reference Manual*
[TS]	*Stata Time-Series Reference Manual*
[I]	*Stata Quick Reference and Index*
[M]	*Mata Reference Manual*

Detailed information about each of these manuals may be found online at

http://www.stata-press.com/manuals/

Title

> **intro** — Introduction to time-series manual

Description

This entry describes this manual and what has changed since Stata 11.

Remarks

This manual documents Stata's time-series commands and is referred to as [TS] in cross-references.

After this entry, [TS] **time series** provides an overview of the ts commands. The other parts of this manual are arranged alphabetically. If you are new to Stata's time-series features, we recommend that you read the following sections first:

[TS] **time series**	Introduction to time-series commands
[TS] **tsset**	Declare a dataset to be time-series data

Stata is continually being updated, and Stata users are always writing new commands. To ensure that you have the latest features, you should install the most recent official update; see [R] **update**.

What's new

1. **MGARCH**, which is to say, multivariate GARCH, which is to say, estimation of multivariate generalized autoregressive conditional heteroskedasticity models of volatility, and this includes constant, dynamic, and varying conditional correlations, also known as the CCC, DCC, and VCC models. Innovations in these models may follow multivariate normal or Student's t distributions. See [TS] **mgarch**.

2. **UCM**, which is to say, unobserved-components models, also known as structural time-series models that decompose a series into trend, seasonal, and cyclical components, and which were popularized by Harvey (1989). See [TS] **ucm**.

3. **ARFIMA**, which is to say, autoregressive fractionally integrated moving-average models, useful for long-memory processes. See [TS] **arfima**.

4. **Filters for extracting business and seasonal cycles**. Four popular time-series filters are provided: the Baxter–King and the Christiano–Fitzgerald band-pass filters, and the Butterworth and the Hodrick–Prescott high-pass filters. See [TS] **tsfilter**.

5. **Business dates** allow you to define your own calendars so that they display correctly and lags and leads work as they should. You could create file lse.stbcal that recorded the days the London Stock Exchange is open (or closed) and then Stata would understand format %tblse just as it understands the usual date format %td. Once you define a calendar, Stata deeply understands it. You can, for instance, easily convert between %tblse and %td values. See [D] **datetime business calendars**.

6. **Improved documentation for date and time variables**. Anyone who has ever been puzzled by Stata's date and time variables, which is to say, anyone who uses them, should see [D] **datetime**, [D] **datetime translation**, and [D] **datetime display formats**.

1

7. **Contrasts**, which is to say, tests of linear hypotheses involving factor variables and their interactions from the most recently fit model. Tests include ANOVA-style tests of main effects, simple effects, interactions, and nested effects. Effects can be decomposed into comparisons with reference categories, comparisons of adjacent levels, comparisons with the grand mean, and more. New commands `contrast` and `margins, contrast` are available after many time-series estimation commands. See [R] **contrast** and [R] **margins, contrast**.

8. **Pairwise comparisons** available after many time-series estimation commands. See [R] **pwcompare** and [R] **margins, pwcompare**.

9. **Graphs of margins, marginal effects, contrasts, and pairwise comparisons** available after most time-series estimation commands. See [R] **marginsplot**.

10. **Estimation output improved.**

    a. **Implied zero coefficients now shown.** When a coefficient is omitted, it is now shown as being zero and the reason it was omitted—collinearity, base, empty—is shown in the standard-error column. (The word "omitted" is shown if the coefficient was omitted because of collinearity.)

    b. **You can set displayed precision for all values in coefficient tables** using `set cformat`, `set pformat`, and `set sformat`. Or you may use options `cformat()`, `pformat()`, and `sformat()` now allowed on all estimation commands. See [R] **set cformat** and [R] **estimation options**.

    c. **Estimation commands now respect the width of the Results window.** This feature may be turned off by new display option `nolstretch`. See [R] **estimation options**.

    d. **You can now set whether base levels, empty cells, and omitted are shown** using `set showbaselevels`, `set showemptycells`, and `set showomitted`. See [R] **set showbaselevels**.

11. **Spectral densities from parametric models** via new postestimation command `psdensity` lets you estimate using `arfima`, `arima`, and `ucm` and then obtain the implied spectral density. See [TS] **psdensity**.

12. **dvech renamed mgarch dvech.** The command for fitting the diagonal VECH model is now named `mgarch dvech`, and innovations may follow multivariate normal or Student's t distributions. See [TS] **mgarch**.

13. **Loading data from Haver Analytics supported on all 64-bit Windows.** See [TS] **haver**.

14. **Option addplot() now places added graphs above or below.** Graph commands that allow option `addplot()` can now place the added plots above or below the command's plots. Affected by this are the commands `corrgram`, `cumsp`, `pergram`, `varstable`, `vecstable`, `wntestb`, and `xcorr`.

For a complete list of all the new features in Stata 12, see [U] **1.3 What's new**.

Reference

Harvey, A. C. 1989. *Forecasting, Structural Time Series Models and the Kalman Filter*. Cambridge: Cambridge University Press.

Also see

[U] **1.3 What's new**

[R] **intro** — Introduction to base reference manual

Title

> **time series** — Introduction to time-series commands

Description

The *Time-Series Reference Manual* organizes the commands alphabetically, making it easy to find individual command entries if you know the name of the command. This overview organizes and presents the commands conceptually, that is, according to the similarities in the functions that they perform. The table below lists the manual entries that you should see for additional information.

Data-management tools and time-series operators.
   These commands help you prepare your data for further analysis.

Univariate time series.
   These commands are grouped together because they are either estimators or filters designed for univariate time series or preestimation or postestimation commands that are conceptually related to one or more univariate time-series estimators.

Multivariate time series.
   These commands are similarly grouped together because they are either estimators designed for use with multivariate time series or preestimation or postestimation commands conceptually related to one or more multivariate time-series estimators.

Within these three broad categories, similar commands have been grouped together.

Data-management tools and time-series operators

[TS] **tsset**	Declare data to be time-series data
[TS] **tsfill**	Fill in gaps in time variable
[TS] **tsappend**	Add observations to a time-series dataset
[TS] **tsreport**	Report time-series aspects of a dataset or estimation sample
[TS] **tsrevar**	Time-series operator programming command
[TS] **haver**	Load data from Haver Analytics database
[TS] **rolling**	Rolling-window and recursive estimation
[D] **datetime business calendars**	User-definable business calendars

Univariate time series

Estimators

[TS] **arfima**	Autoregressive fractionally integrated moving-average models
[TS] **arfima postestimation**	Postestimation tools for arfima
[TS] **arima**	ARIMA, ARMAX, and other dynamic regression models
[TS] **arima postestimation**	Postestimation tools for arima
[TS] **arch**	Autoregressive conditional heteroskedasticity (ARCH) family of estimators
[TS] **arch postestimation**	Postestimation tools for arch
[TS] **newey**	Regression with Newey–West standard errors
[TS] **newey postestimation**	Postestimation tools for newey
[TS] **prais**	Prais–Winsten and Cochrane–Orcutt regression
[TS] **prais postestimation**	Postestimation tools for prais
[TS] **ucm**	Unobserved-components model
[TS] **ucm postestimation**	Postestimation tools for ucm

Time-series smoothers and filters

[TS] **tsfilter bk**	Baxter–King time-series filter
[TS] **tsfilter bw**	Butterworth time-series filter
[TS] **tsfilter cf**	Christiano–Fitzgerald time-series filter
[TS] **tsfilter hp**	Hodrick–Prescott time-series filter
[TS] **tssmooth ma**	Moving-average filter
[TS] **tssmooth dexponential**	Double-exponential smoothing
[TS] **tssmooth exponential**	Single-exponential smoothing
[TS] **tssmooth hwinters**	Holt–Winters nonseasonal smoothing
[TS] **tssmooth shwinters**	Holt–Winters seasonal smoothing
[TS] **tssmooth nl**	Nonlinear filter

Diagnostic tools

[TS] **corrgram**	Tabulate and graph autocorrelations
[TS] **xcorr**	Cross-correlogram for bivariate time series
[TS] **cumsp**	Cumulative spectral distribution
[TS] **pergram**	Periodogram
[TS] **psdensity**	Parametric spectral density estimation
[TS] **dfgls**	DF-GLS unit-root test
[TS] **dfuller**	Augmented Dickey–Fuller unit-root test
[TS] **pperron**	Phillips–Perron unit-root test
[R] **regress postestimation time series**	Postestimation tools for regress with time series
[TS] **wntestb**	Bartlett's periodogram-based test for white noise
[TS] **wntestq**	Portmanteau (Q) test for white noise

Multivariate time series

Estimators

[TS]	**dfactor**	Dynamic-factor models
[TS]	**dfactor postestimation**	Postestimation tools for dfactor
[TS]	**mgarch ccc**	Constant conditional correlation multivariate GARCH models
[TS]	**mgarch ccc postestimation**	Postestimation tools for mgarch ccc
[TS]	**mgarch dcc**	Dynamic conditional correlation multivariate GARCH models
[TS]	**mgarch dcc postestimation**	Postestimation tools for mgarch dcc
[TS]	**mgarch dvech**	Diagonal vech multivariate GARCH models
[TS]	**mgarch dvech postestimation**	Postestimation tools for mgarch dvech
[TS]	**mgarch vcc**	Varying conditional correlation multivariate GARCH models
[TS]	**mgarch vcc postestimation**	Postestimation tools for mgarch vcc
[TS]	**sspace**	State-space models
[TS]	**sspace postestimation**	Postestimation tools for sspace
[TS]	**var**	Vector autoregressive models
[TS]	**var postestimation**	Postestimation tools for var
[TS]	**var svar**	Structural vector autoregressive models
[TS]	**var svar postestimation**	Postestimation tools for svar
[TS]	**varbasic**	Fit a simple VAR and graph IRFs or FEVDs
[TS]	**varbasic postestimation**	Postestimation tools for varbasic
[TS]	**vec**	Vector error-correction models
[TS]	**vec postestimation**	Postestimation tools for vec

Diagnostic tools

[TS]	**varlmar**	Perform LM test for residual autocorrelation
[TS]	**varnorm**	Test for normally distributed disturbances
[TS]	**varsoc**	Obtain lag-order selection statistics for VARs and VECMs
[TS]	**varstable**	Check the stability condition of VAR or SVAR estimates
[TS]	**varwle**	Obtain Wald lag-exclusion statistics
[TS]	**veclmar**	Perform LM test for residual autocorrelation
[TS]	**vecnorm**	Test for normally distributed disturbances
[TS]	**vecrank**	Estimate the cointegrating rank of a VECM
[TS]	**vecstable**	Check the stability condition of VECM estimates

Forecasting, inference, and interpretation

[TS]	**irf create**	Obtain IRFs, dynamic-multiplier functions, and FEVDs
[TS]	**fcast compute**	Compute dynamic forecasts of dependent variables
[TS]	**vargranger**	Perform pairwise Granger causality tests

Graphs and tables

[TS] **corrgram**	Tabulate and graph autocorrelations
[TS] **xcorr**	Cross-correlogram for bivariate time series
[TS] **pergram**	Periodogram
[TS] **irf graph**	Graph IRFs, dynamic-multiplier functions, and FEVDs
[TS] **irf cgraph**	Combine graphs of IRFs, dynamic-multiplier functions, and FEVDs
[TS] **irf ograph**	Graph overlaid IRFs, dynamic-multiplier functions, and FEVDs
[TS] **irf table**	Create tables of IRFs, dynamic-multiplier functions, and FEVDs
[TS] **irf ctable**	Combine tables of IRFs, dynamic-multiplier functions, and FEVDs
[TS] **fcast graph**	Graph forecasts of variables computed by fcast compute
[TS] **tsline**	Plot time-series data
[TS] **varstable**	Check the stability condition of VAR or SVAR estimates
[TS] **vecstable**	Check the stability condition of VECM estimates
[TS] **wntestb**	Bartlett's periodogram-based test for white noise

Results management tools

[TS] **irf add**	Add results from an IRF file to the active IRF file
[TS] **irf describe**	Describe an IRF file
[TS] **irf drop**	Drop IRF results from the active IRF file
[TS] **irf rename**	Rename an IRF result in an IRF file
[TS] **irf set**	Set the active IRF file

Remarks

Remarks are presented under the following headings:

> *Data-management tools and time-series operators*
> *Univariate time series*
> > *Estimators*
> > *Time-series smoothers and filters*
> > *Diagnostic tools*
> *Multivariate time series*
> > *Estimators*
> > *Diagnostic tools*

We also offer a NetCourse on Stata's time-series capabilities; see http://www.stata.com/netcourse/nc461.html.

Data-management tools and time-series operators

Because time-series estimators are, by definition, a function of the temporal ordering of the observations in the estimation sample, Stata's time-series commands require the data to be sorted and indexed by time, using the `tsset` command, before they can be used. `tsset` is simply a way for you to tell Stata which variable in your dataset represents time; `tsset` then sorts and indexes the data appropriately for use with the time-series commands. Once your dataset has been `tsset`, you can use Stata's time-series operators in data manipulation or programming using that dataset and when specifying the syntax for most time-series commands. Stata has time-series operators for representing the lags, leads, differences, and seasonal differences of a variable. The time-series operators are documented in [TS] **tsset**.

You can also define a business-day calendar so that Stata's time-series operators respect the structure of missing observations in your data. The most common example is having Monday come after Friday in market data. [D] **datetime business calendars** provides a discussion and examples.

tsset can also be used to declare that your dataset contains cross-sectional time-series data, often referred to as panel data. When you use tsset to declare your dataset to contain panel data, you specify a variable that identifies the panels and a variable that identifies the time periods. Once your dataset has been tsset as panel data, the time-series operators work appropriately for the data.

tsfill, which is documented in [TS] **tsfill**, can be used after tsset to fill in missing times with missing observations. tsset will report any gaps in your data, and tsreport will provide more details about the gaps. tsappend adds observations to a time-series dataset by using the information set by tsset. This function can be particularly useful when you wish to predict out of sample after fitting a model with a time-series estimator. tsrevar is a programmer's command that provides a way to use *varlist*s that contain time-series operators with commands that do not otherwise support time-series operators.

The haver commands documented in [TS] **haver** allow you to load and describe the contents of a Haver Analytics (http://www.haver.com) file.

rolling performs rolling regressions, recursive regressions, and reverse recursive regressions. Any command that saves results in e() or r() can be used with rolling.

Univariate time series

Estimators

The six univariate time-series estimators currently available in Stata are arfima, arima, arch, newey, prais, and ucm. newey and prais are really just extensions to ordinary linear regression. When you fit a linear regression on time-series data via ordinary least squares (OLS), if the disturbances are autocorrelated, the parameter estimates are usually consistent, but the estimated standard errors tend to be underestimated. Several estimators have been developed to deal with this problem. One strategy is to use OLS for estimating the regression parameters and use a different estimator for the variances, one that is consistent in the presence of autocorrelated disturbances, such as the Newey–West estimator implemented in newey. Another strategy is to model the dynamics of the disturbances. The estimators found in prais, arima, arch, arfima, and ucm are based on such a strategy.

prais implements two such estimators: the Prais–Winsten and the Cochrane–Orcutt generalized least-squares (GLS) estimators. These estimators are GLS estimators, but they are fairly restrictive in that they permit only first-order autocorrelation in the disturbances. Although they have certain pedagogical and historical value, they are somewhat obsolete. Faster computers with more memory have made it possible to implement full information maximum likelihood (FIML) estimators, such as Stata's arima command. These estimators permit much greater flexibility when modeling the disturbances and are more efficient estimators.

arima provides the means to fit linear models with autoregressive moving-average (ARMA) disturbances, or in the absence of linear predictors, autoregressive integrated moving-average (ARIMA) models. This means that, whether you think that your data are best represented as a distributed-lag model, a transfer-function model, or a stochastic difference equation, or you simply wish to apply a Box–Jenkins filter to your data, the model can be fit using arima. arch, a conditional maximum likelihood estimator, has similar modeling capabilities for the mean of the time series but can also model autoregressive conditional heteroskedasticity in the disturbances with a wide variety of specifications for the variance equation.

arfima estimates the parameters of autoregressive fractionally integrated moving-average (ARFIMA) models, which handle higher degrees of dependence than ARIMA models. ARFIMA models allow the autocorrelations to decay at the slower hyperbolic rate, whereas ARIMA models handle processes whose autocorrelations decay at an exponential rate.

Unobserved-components models (UCMs) decompose a time series into trend, seasonal, cyclical, and idiosyncratic components and allow for exogenous variables. ucm estimates the parameters of UCMs by maximum likelihood. UCMs can also model the stationary cyclical component using the stochastic-cycle parameterization that has an intuitive frequency-domain interpretation.

Time-series smoothers and filters

In addition to the estimators mentioned above, Stata also provides time-series filters and smoothers. The Baxter–King and Christiano–Fitzgerald band-pass filters and the Butterworth and Hodrick–Prescott high-pass filters are implemented in tsfilter; see [TS] **tsfilter** for an overview.

Also included are a simple, uniformly weighted, moving-average filter with unit weights; a weighted moving-average filter in which you can specify the weights; single- and double-exponential smoothers; Holt–Winters seasonal and nonseasonal smoothers; and a nonlinear smoother. Most of these smoothers were originally developed as ad hoc procedures and are used for reducing the noise in a time series (smoothing) or forecasting. Although they have limited application for signal extraction, these smoothers have all been found to be optimal for some underlying modern time-series models; see [TS] **tssmooth**.

Diagnostic tools

Stata's time-series commands also include several preestimation and postestimation diagnostic and interpretation commands. corrgram estimates the autocorrelation function and partial autocorrelation function of a univariate time series, as well as Q statistics. These functions and statistics are often used to determine the appropriate model specification before fitting ARIMA models. corrgram can also be used with wntestb and wntestq to examine the residuals after fitting a model for evidence of model misspecification. Stata's time-series commands also include the commands pergram and cumsp, which provide the log-standardized periodogram and the cumulative-sample spectral distribution, respectively, for time-series analysts who prefer to estimate in the frequency domain rather than the time domain.

psdensity computes the spectral density implied by the parameters estimated by arfima, arima, or ucm. The estimated spectral density shows the relative importance of components at different frequencies.

xcorr estimates the cross-correlogram for bivariate time series and can similarly be used both for preestimation and postestimation. For example, the cross-correlogram can be used before fitting a transfer-function model to produce initial estimates of the IRF. This estimate can then be used to determine the optimal lag length of the input series to include in the model specification. It can also be used as a postestimation tool after fitting a transfer function. The cross-correlogram between the residual from a transfer-function model and the prewhitened input series of the model can be examined for evidence of model misspecification.

When you fit ARMA or ARIMA models, the dependent variable being modeled must be covariance stationary (ARMA models), or the order of integration must be known (ARIMA models). Stata has three commands that can test for the presence of a unit root in a time-series variable: dfuller performs the augmented Dickey–Fuller test, pperron performs the Phillips–Perron test, and dfgls performs a modified Dickey–Fuller test. arfima can also be used to investigate the order of integration.

The remaining diagnostic tools for univariate time series are for use after fitting a linear model via OLS with Stata's `regress` command. They are documented collectively in [R] **regress postestimation time series**. They include `estat dwatson`, `estat durbinalt`, `estat bgodfrey`, and `estat archlm`. `estat dwatson` computes the Durbin–Watson d statistic to test for the presence of first-order autocorrelation in the OLS residuals. `estat durbinalt` likewise tests for the presence of autocorrelation in the residuals. By comparison, however, Durbin's alternative test is more general and easier to use than the Durbin–Watson test. With `estat durbinalt`, you can test for higher orders of autocorrelation, the assumption that the covariates in the model are strictly exogenous is relaxed, and there is no need to consult tables to compute rejection regions, as you must with the Durbin–Watson test. `estat bgodfrey` computes the Breusch–Godfrey test for autocorrelation in the residuals, and although the computations are different, the test in `estat bgodfrey` is asymptotically equivalent to the test in `estat durbinalt`. Finally, `estat archlm` performs Engle's LM test for the presence of autoregressive conditional heteroskedasticity.

Multivariate time series

Estimators

Stata provides commands for fitting the most widely applied multivariate time-series models. `var` and `svar` fit vector autoregressive and structural vector autoregressive models to stationary data. `vec` fits cointegrating vector error-correction models. `dfactor` fits dynamic-factor models. `mgarch ccc`, `mgarch dcc`, `mgarch dvech`, and `mgarch vcc` fit multivariate GARCH models. `sspace` fits state-space models. Many linear time-series models, including vector autoregressive moving-average (VARMA) models and structural time-series models, can be cast as state-space models and fit by `sspace`.

Diagnostic tools

Before fitting a multivariate time-series model, you must specify the number of lags of the dependent variable to include. `varsoc` produces statistics for determining the order of a VAR or VECM.

Several postestimation commands perform the most common specification analysis on a previously fitted VAR or SVAR. You can use `varlmar` to check for serial correlation in the residuals, `varnorm` to test the null hypothesis that the disturbances come from a multivariate normal distribution, and `varstable` to see if the fitted VAR or SVAR is stable. Two common types of inference about VAR models are whether one variable Granger-causes another and whether a set of lags can be excluded from the model. `vargranger` reports Wald tests of Granger causation, and `varwle` reports Wald lag exclusion tests.

Similarly, several postestimation commands perform the most common specification analysis on a previously fitted VECM. You can use `veclmar` to check for serial correlation in the residuals, `vecnorm` to test the null hypothesis that the disturbances come from a multivariate normal distribution, and `vecstable` to analyze the stability of the previously fitted VECM.

VARs and VECMs are often fit to produce baseline forecasts. `fcast` produces dynamic forecasts from previously fitted VARs and VECMs.

Many researchers fit VARs, SVARs, and VECMs because they want to analyze how unexpected shocks affect the dynamic paths of the variables. Stata has a suite of `irf` commands for estimating IRF functions and interpreting, presenting, and managing these estimates; see [TS] **irf**.

References

Baum, C. F. 2005. Stata: The language of choice for time-series analysis? *Stata Journal* 5: 46–63.

Hamilton, J. D. 1994. *Time Series Analysis.* Princeton: Princeton University Press.

Lütkepohl, H. 1993. *Introduction to Multiple Time Series Analysis.* 2nd ed. New York: Springer.

——. 2005. *New Introduction to Multiple Time Series Analysis.* New York: Springer.

Pisati, M. 2001. sg162: Tools for spatial data analysis. *Stata Technical Bulletin* 60: 21–37. Reprinted in *Stata Technical Bulletin Reprints*, vol. 10, pp. 277–298. College Station, TX: Stata Press.

Stock, J. H., and M. W. Watson. 2001. Vector autoregressions. *Journal of Economic Perspectives* 15: 101–115.

Also see

[U] **1.3 What's new**

[R] **intro** — Introduction to base reference manual

Title

> **arch** — Autoregressive conditional heteroskedasticity (ARCH) family of estimators

Syntax

arch *depvar* $\left[\,indepvars\,\right]$ $\left[\,if\,\right]$ $\left[\,in\,\right]$ $\left[\,weight\,\right]$ $\left[\,,\ options\,\right]$

options	Description
Model	
<u>no</u>constant	suppress constant term
arch(*numlist*)	ARCH terms
garch(*numlist*)	GARCH terms
<u>sa</u>arch(*numlist*)	simple asymmetric ARCH terms
<u>t</u>arch(*numlist*)	threshold ARCH terms
<u>aa</u>rch(*numlist*)	asymmetric ARCH terms
<u>n</u>arch(*numlist*)	nonlinear ARCH terms
narchk(*numlist*)	nonlinear ARCH terms with single shift
<u>ab</u>arch(*numlist*)	absolute value ARCH terms
<u>at</u>arch(*numlist*)	absolute threshold ARCH terms
<u>sd</u>garch(*numlist*)	lags of σ_t
<u>e</u>arch(*numlist*)	news terms in Nelson's (1991) EGARCH model
egarch(*numlist*)	lags of $\ln(\sigma_t^2)$
<u>pa</u>rch(*numlist*)	power ARCH terms
tparch(*numlist*)	threshold power ARCH terms
aparch(*numlist*)	asymmetric power ARCH terms
nparch(*numlist*)	nonlinear power ARCH terms
nparchk(*numlist*)	nonlinear power ARCH terms with single shift
pgarch(*numlist*)	power GARCH terms
<u>c</u>onstraints(*constraints*)	apply specified linear constraints
<u>coll</u>inear	keep collinear variables
Model 2	
archm	include ARCH-in-mean term in the mean-equation specification
<u>archml</u>ags(*numlist*)	include specified lags of conditional variance in mean equation
<u>archme</u>xp(*exp*)	apply transformation in *exp* to any ARCH-in-mean terms
arima(#$_p$,#$_d$,#$_q$)	specify ARIMA(p, d, q) model for dependent variable
ar(*numlist*)	autoregressive terms of the structural model disturbance
ma(*numlist*)	moving-average terms of the structural model disturbances
Model 3	
<u>d</u>istribution(*dist* $\left[\,\#\,\right]$)	use *dist* distribution for errors (may be <u>gau</u>ssian, <u>norm</u>al, t, or ged; default is gaussian)
het(*varlist*)	include *varlist* in the specification of the conditional variance
<u>save</u>space	conserve memory during estimation

Priming

arch0(xb)	compute priming values on the basis of the expected unconditional variance; the default
arch0(xb0)	compute priming values on the basis of the estimated variance of the residuals from OLS
arch0(xbwt)	compute priming values on the basis of the weighted sum of squares from OLS residuals
arch0(xb0wt)	compute priming values on the basis of the weighted sum of squares from OLS residuals, with more weight at earlier times
arch0(zero)	set priming values of ARCH terms to zero
arch0(#)	set priming values of ARCH terms to #
arma0(zero)	set all priming values of ARMA terms to zero; the default
arma0(p)	begin estimation after observation p, where p is the maximum AR lag in model
arma0(q)	begin estimation after observation q, where q is the maximum MA lag in model
arma0(pq)	begin estimation after observation $(p + q)$
arma0(#)	set priming values of ARMA terms to #
condobs(#)	set conditioning observations at the start of the sample to #

SE/Robust

vce(vcetype)	vcetype may be opg, robust, or oim

Reporting

level(#)	set confidence level; default is level(95)
detail	report list of gaps in time series
nocnsreport	do not display constraints
display_options	control column formats, row spacing, and line width

Maximization

maximize_options	control the maximization process; seldom used
coeflegend	display legend instead of statistics

You must tsset your data before using arch; see [TS] **tsset**.

depvar and *varlist* may contain time-series operators; see [U] **11.4.4 Time-series varlists**.

by, rolling, statsby, and xi are allowed; see [U] **11.1.10 Prefix commands**.

iweights are allowed; see [U] **11.1.6 weight**.

coeflegend does not appear in the dialog box.

See [U] **20 Estimation and postestimation commands** for more capabilities of estimation commands.

To fit an ARCH($\#_m$) model with Gaussian errors, type

 . arch *depvar* ... , arch(1/$\#_m$)

To fit a GARCH($\#_m$, $\#_k$) model assuming that the errors follow Student's t distribution with 7 degrees of freedom, type

 . arch *depvar* ... , arch(1/$\#_m$) garch(1/$\#_k$) distribution(t 7)

You can also fit many other models.

Details of syntax

The basic model `arch` fits is

$$y_t = \mathbf{x}_t\boldsymbol{\beta} + \epsilon_t$$
$$\mathrm{Var}(\epsilon_t) = \sigma_t^2 = \gamma_0 + A(\boldsymbol{\sigma}, \boldsymbol{\epsilon}) + B(\boldsymbol{\sigma}, \boldsymbol{\epsilon})^2 \tag{1}$$

The y_t equation may optionally include ARCH-in-mean and ARMA terms:

$$y_t = \mathbf{x}_t\boldsymbol{\beta} + \sum_i \psi_i g(\sigma_{t-i}^2) + \mathrm{ARMA}(p, q) + \epsilon_t$$

If no options are specified, $A() = B() = 0$, and the model collapses to linear regression. The following options add to $A()$ (α, γ, and κ represent parameters to be estimated):

Option	Terms added to $A()$
`arch()`	$A() = A() + \alpha_{1,1}\epsilon_{t-1}^2 + \alpha_{1,2}\epsilon_{t-2}^2 + \cdots$
`garch()`	$A() = A() + \alpha_{2,1}\sigma_{t-1}^2 + \alpha_{2,2}\sigma_{t-2}^2 + \cdots$
`saarch()`	$A() = A() + \alpha_{3,1}\epsilon_{t-1} + \alpha_{3,2}\epsilon_{t-2} + \cdots$
`tarch()`	$A() = A() + \alpha_{4,1}\epsilon_{t-1}^2(\epsilon_{t-1} > 0) + \alpha_{4,2}\epsilon_{t-2}^2(\epsilon_{t-2} > 0) + \cdots$
`aarch()`	$A() = A() + \alpha_{5,1}(|\epsilon_{t-1}| + \gamma_{5,1}\epsilon_{t-1})^2 + \alpha_{5,2}(|\epsilon_{t-2}| + \gamma_{5,2}\epsilon_{t-2})^2 + \cdots$
`narch()`	$A() = A() + \alpha_{6,1}(\epsilon_{t-1} - \kappa_{6,1})^2 + \alpha_{6,2}(\epsilon_{t-2} - \kappa_{6,2})^2 + \cdots$
`narchk()`	$A() = A() + \alpha_{7,1}(\epsilon_{t-1} - \kappa_7)^2 + \alpha_{7,2}(\epsilon_{t-2} - \kappa_7)^2 + \cdots$

The following options add to $B()$:

Option	Terms added to $B()$
`abarch()`	$B() = B() + \alpha_{8,1}|\epsilon_{t-1}| + \alpha_{8,2}|\epsilon_{t-2}| + \cdots$
`atarch()`	$B() = B() + \alpha_{9,1}|\epsilon_{t-1}|(\epsilon_{t-1} > 0) + \alpha_{9,2}|\epsilon_{t-2}|(\epsilon_{t-2} > 0) + \cdots$
`sdgarch()`	$B() = B() + \alpha_{10,1}\sigma_{t-1} + \alpha_{10,2}\sigma_{t-2} + \cdots$

Each option requires a *numlist* argument (see [U] **11.1.8 numlist**), which determines the lagged terms included. `arch(1)` specifies $\alpha_{1,1}\epsilon_{t-1}^2$, `arch(2)` specifies $\alpha_{1,2}\epsilon_{t-2}^2$, `arch(1,2)` specifies $\alpha_{1,1}\epsilon_{t-1}^2 + \alpha_{1,2}\epsilon_{t-2}^2$, `arch(1/3)` specifies $\alpha_{1,1}\epsilon_{t-1}^2 + \alpha_{1,2}\epsilon_{t-2}^2 + \alpha_{1,3}\epsilon_{t-3}^2$, etc.

If the `earch()` or `egarch()` option is specified, the basic model fit is

$$y_t = \mathbf{x}_t\boldsymbol{\beta} + \sum_i \psi_i g(\sigma_{t-i}^2) + \mathrm{ARMA}(p, q) + \epsilon_t$$
$$\ln\mathrm{Var}(\epsilon_t) = \ln\sigma_t^2 = \gamma_0 + C(\ln\boldsymbol{\sigma}, \mathbf{z}) + A(\boldsymbol{\sigma}, \boldsymbol{\epsilon}) + B(\boldsymbol{\sigma}, \boldsymbol{\epsilon})^2 \tag{2}$$

where $z_t = \epsilon_t/\sigma_t$. $A()$ and $B()$ are given as above, but $A()$ and $B()$ now add to $\ln\sigma_t^2$ rather than σ_t^2. (The options corresponding to $A()$ and $B()$ are rarely specified here.) $C()$ is given by

Option	Terms added to $C()$
`earch()`	$C() = C() + \alpha_{11,1}z_{t-1} + \gamma_{11,1}(|z_{t-1}| - \sqrt{2/\pi})$ $+ \alpha_{11,2}z_{t-2} + \gamma_{11,2}(|z_{t-2}| - \sqrt{2/\pi}) + \cdots$
`egarch()`	$C() = C() + \alpha_{12,1}\ln\sigma_{t-1}^2 + \alpha_{12,2}\ln\sigma_{t-2}^2 + \cdots$

Instead, if the `parch()`, `tparch()`, `aparch()`, `nparch()`, `nparchk()`, or `pgarch()` options are specified, the basic model fit is

$$y_t = \mathbf{x}_t\boldsymbol{\beta} + \sum_i \psi_i g(\sigma_{t-i}^2) + \mathrm{ARMA}(p, q) + \epsilon_t \tag{3}$$

$$\{\mathrm{Var}(\epsilon_t)\}^{\varphi/2} = \sigma_t^{\varphi} = \gamma_0 + D(\boldsymbol{\sigma}, \boldsymbol{\epsilon}) + A(\boldsymbol{\sigma}, \boldsymbol{\epsilon}) + B(\boldsymbol{\sigma}, \boldsymbol{\epsilon})^2$$

where φ is a parameter to be estimated. $A()$ and $B()$ are given as above, but $A()$ and $B()$ now add to σ_t^{φ}. (The options corresponding to $A()$ and $B()$ are rarely specified here.) $D()$ is given by

Option	Terms added to $D()$
`parch()`	$D() = D() + \alpha_{13,1}\epsilon_{t-1}^{\varphi} + \alpha_{13,2}\epsilon_{t-2}^{\varphi} + \cdots$
`tparch()`	$D() = D() + \alpha_{14,1}\epsilon_{t-1}^{\varphi}(\epsilon_{t-1} > 0) + \alpha_{14,2}\epsilon_{t-2}^{\varphi}(\epsilon_{t-2} > 0) + \cdots$
`aparch()`	$D() = D() + \alpha_{15,1}(|\epsilon_{t-1}| + \gamma_{15,1}\epsilon_{t-1})^{\varphi} + \alpha_{15,2}(|\epsilon_{t-2}| + \gamma_{15,2}\epsilon_{t-2})^{\varphi} + \cdots$
`nparch()`	$D() = D() + \alpha_{16,1}|\epsilon_{t-1} - \kappa_{16,1}|^{\varphi} + \alpha_{16,2}|\epsilon_{t-2} - \kappa_{16,2}|^{\varphi} + \cdots$
`nparchk()`	$D() = D() + \alpha_{17,1}|\epsilon_{t-1} - \kappa_{17}|^{\varphi} + \alpha_{17,2}|\epsilon_{t-2} - \kappa_{17}|^{\varphi} + \cdots$
`pgarch()`	$D() = D() + \alpha_{18,1}\sigma_{t-1}^{\varphi} + \alpha_{18,2}\sigma_{t-2}^{\varphi} + \cdots$

Common models

Common term	Options to specify
ARCH (Engle 1982)	`arch()`
GARCH (Bollerslev 1986)	`arch() garch()`
ARCH-in-mean (Engle, Lilien, and Robins 1987)	`archm arch()` $\big[$`garch()`$\big]$
GARCH with ARMA terms	`arch() garch() ar() ma()`
EGARCH (Nelson 1991)	`earch() egarch()`
TARCH, threshold ARCH (Zakoian 1994)	`abarch() atarch() sdgarch()`
GJR, form of threshold ARCH (Glosten, Jagannathan, and Runkle 1993)	`arch() tarch()` $\big[$`garch()`$\big]$
SAARCH, simple asymmetric ARCH (Engle 1990)	`arch() saarch()` $\big[$`garch()`$\big]$
PARCH, power ARCH (Higgins and Bera 1992)	`parch()` $\big[$`pgarch()`$\big]$
NARCH, nonlinear ARCH	`narch()` $\big[$`garch()`$\big]$
NARCHK, nonlinear ARCH with one shift	`narchk()` $\big[$`garch()`$\big]$
A-PARCH, asymmetric power ARCH (Ding, Granger, and Engle 1993)	`aparch()` $\big[$`pgarch()`$\big]$
NPARCH, nonlinear power ARCH	`nparch()` $\big[$`pgarch()`$\big]$

In all cases, you type

$$\texttt{arch } \textit{depvar } \big[\textit{indepvars}\big] \texttt{, } \textit{options}$$

where *options* are chosen from the table above. Each option requires that you specify as its argument a *numlist* that specifies the lags to be included. For most ARCH models, that value will be 1. For instance, to fit the classic first-order GARCH model on cpi, you would type

```
. arch cpi, arch(1) garch(1)
```

If you wanted to fit a first-order GARCH model of cpi on wage, you would type

```
. arch cpi wage, arch(1) garch(1)
```

If, for any of the options, you want first- and second-order terms, specify *optionname*(1/2). Specifying garch(1) arch(1/2) would fit a GARCH model with first- and second-order ARCH terms. If you specified arch(2), only the lag 2 term would be included.

Reading arch output

The regression table reported by `arch` when using the normal distribution for the errors will appear as

op.depvar	Coef.	Std. Err.	z	P>|z|	[95% Conf. Interval]
depvar					
x1	# ...				
x2					
L1.	# ...				
L2.	# ...				
_cons	# ...				
ARCHM					
sigma2	# ...				
ARMA					
ar					
L1.	# ...				
ma					
L1.	# ...				
HET					
z1	# ...				
z2					
L1.	# ...				
L2.	# ...				
ARCH					
arch					
L1.	# ...				
garch					
L1.	# ...				
aparch					
L1.	# ...				
etc.					
_cons	# ...				
POWER					
power	# ...				

Dividing lines separate "equations".

The first one, two, or three equations report the mean model:

$$y_t = \mathbf{x}_t\boldsymbol{\beta} + \sum_i \psi_i g(\sigma^2_{t-i}) + \mathrm{ARMA}(p, q) + \epsilon_t$$

The first equation reports β, and the equation will be named [*depvar*]; if you fit a model on d.cpi, the first equation would be named [cpi]. In Stata, the coefficient on x1 in the above example could be referred to as [*depvar*]_b[x1]. The coefficient on the lag 2 value of x2 would be referred to as [*depvar*]_b[L2.x2]. Such notation would be used, for instance, in a later test command; see [R] **test**.

The [ARCHM] equation reports the ψ coefficients if your model includes ARCH-in-mean terms; see options discussed under the **Model 2** tab below. Most ARCH-in-mean models include only a contemporaneous variance term, so the term $\sum_i \psi_i g(\sigma_{t-i}^2)$ becomes $\psi \sigma_t^2$. The coefficient ψ will be [ARCHM]_b[sigma2]. If your model includes lags of σ_t^2, the additional coefficients will be [ARCHM]_b[L1.sigma2], and so on. If you specify a transformation $g()$ (option archmexp()), the coefficients will be [ARCHM]_b[sigma2ex], [ARCHM]_b[L1.sigma2ex], and so on. sigma2ex refers to $g(\sigma_t^2)$, the transformed value of the conditional variance.

The [ARMA] equation reports the ARMA coefficients if your model includes them; see options discussed under the **Model 2** tab below. This equation includes one or two "variables" named ar and ma. In later test statements, you could refer to the coefficient on the first lag of the autoregressive term by typing [ARMA]_b[L1.ar] or simply [ARMA]_b[L.ar] (the L operator is assumed to be lag 1 if you do not specify otherwise). The second lag on the moving-average term, if there were one, could be referred to by typing [ARMA]_b[L2.ma].

The next one, two, or three equations report the variance model.

The [HET] equation reports the multiplicative heteroskedasticity if the model includes it. When you fit such a model, you specify the variables (and their lags), determining the multiplicative heteroskedasticity; after estimation, their coefficients are simply [HET]_b[op.varname].

The [ARCH] equation reports the ARCH, GARCH, etc., terms by referring to "variables" arch, garch, and so on. For instance, if you specified arch(1) garch(1) when you fit the model, the conditional variance is given by $\sigma_t^2 = \gamma_0 + \alpha_{1,1} \epsilon_{t-1}^2 + \alpha_{2,1} \sigma_{t-1}^2$. The coefficients would be named [ARCH]_b[_cons] (γ_0), [ARCH]_b[L.arch] ($\alpha_{1,1}$), and [ARCH]_b[L.garch] ($\alpha_{2,1}$).

The [POWER] equation appears only if you are fitting a variance model in the form of (3) above; the estimated φ is the coefficient [POWER]_b[power].

Also, if you use the distribution() option and specify either Student's t or the generalized error distribution but do not specify the degree-of-freedom or shape parameter, then you will see two additional rows in the table. The final row contains the estimated degree-of-freedom or shape parameter. Immediately preceding the final row is a transformed version of the parameter that arch used during estimation to ensure that the degree-of-freedom parameter is greater than two or that the shape parameter is positive.

The naming convention for estimated ARCH, GARCH, etc., parameters is as follows (definitions for parameters α_i, γ_i, and κ_i can be found in the tables for $A()$, $B()$, $C()$, and $D()$ above):

Option	1st parameter	2nd parameter	Common parameter
arch()	α_1 = [ARCH]_b[arch]		
garch()	α_2 = [ARCH]_b[garch]		
saarch()	α_3 = [ARCH]_b[saarch]		
tarch()	α_4 = [ARCH]_b[tarch]		
aarch()	α_5 = [ARCH]_b[aarch]	γ_5 = [ARCH]_b[aarch_e]	
narch()	α_6 = [ARCH]_b[narch]	κ_6 = [ARCH]_b[narch_k]	
narchk()	α_7 = [ARCH]_b[narch]	κ_7 = [ARCH]_b[narch_k]	
abarch()	α_8 = [ARCH]_b[abarch]		
atarch()	α_9 = [ARCH]_b[atarch]		
sdgarch()	α_{10} = [ARCH]_b[sdgarch]		
earch()	α_{11} = [ARCH]_b[earch]	γ_{11} = [ARCH]_b[earch_a]	
egarch()	α_{12} = [ARCH]_b[egarch]		
parch()	α_{13} = [ARCH]_b[parch]		φ = [POWER]_b[power]
tparch()	α_{14} = [ARCH]_b[tparch]		φ = [POWER]_b[power]
aparch()	α_{15} = [ARCH]_b[aparch]	γ_{15} = [ARCH]_b[aparch_e]	φ = [POWER]_b[power]
nparch()	α_{16} = [ARCH]_b[nparch]	κ_{16} = [ARCH]_b[nparch_k]	φ = [POWER]_b[power]
nparchk()	α_{17} = [ARCH]_b[nparch]	κ_{17} = [ARCH]_b[nparch_k]	φ = [POWER]_b[power]
pgarch()	α_{18} = [ARCH]_b[pgarch]		φ = [POWER]_b[power]

Menu

ARCH/GARCH

Statistics > Time series > ARCH/GARCH > ARCH and GARCH models

EARCH/EGARCH

Statistics > Time series > ARCH/GARCH > Nelson's EGARCH model

ABARCH/ATARCH/SDGARCH

Statistics > Time series > ARCH/GARCH > Threshold ARCH model

ARCH/TARCH/GARCH

Statistics > Time series > ARCH/GARCH > GJR form of threshold ARCH model

ARCH/SAARCH/GARCH

Statistics > Time series > ARCH/GARCH > Simple asymmetric ARCH model

PARCH/PGARCH

Statistics > Time series > ARCH/GARCH > Power ARCH model

NARCH/GARCH

Statistics > Time series > ARCH/GARCH > Nonlinear ARCH model

NARCHK/GARCH

Statistics > Time series > ARCH/GARCH > Nonlinear ARCH model with one shift

APARCH/PGARCH

Statistics > Time series > ARCH/GARCH > Asymmetric power ARCH model

NPARCH/PGARCH

Statistics > Time series > ARCH/GARCH > Nonlinear power ARCH model

Description

arch fits regression models in which the volatility of a series varies through time. Usually, periods of high and low volatility are grouped together. ARCH models estimate future volatility as a function of prior volatility. To accomplish this, arch fits models of autoregressive conditional heteroskedasticity (ARCH) by using conditional maximum likelihood. In addition to ARCH terms, models may include multiplicative heteroskedasticity. Gaussian (normal), Student's t, and generalized error distributions are supported.

Concerning the regression equation itself, models may also contain ARCH-in-mean and ARMA terms.

Options

```
          ┌ Model ┐
```

noconstant; see [R] **estimation options**.

arch(*numlist*) specifies the ARCH terms (lags of ϵ_t^2).

> Specify arch(1) to include first-order terms, arch(1/2) to specify first- and second-order terms, arch(1/3) to specify first-, second-, and third-order terms, etc. Terms may be omitted. Specify arch(1/3 5) to specify terms with lags 1, 2, 3, and 5. All the options work this way.

> arch() may not be specified with aarch(), narch(), narchk(), nparchk(), or nparch(), as this would result in collinear terms.

garch(*numlist*) specifies the GARCH terms (lags of σ_t^2).

saarch(*numlist*) specifies the simple asymmetric ARCH terms. Adding these terms is one way to make the standard ARCH and GARCH models respond asymmetrically to positive and negative innovations. Specifying saarch() with arch() and garch() corresponds to the SAARCH model of Engle (1990).

> saarch() may not be specified with narch(), narchk(), nparchk(), or nparch(), as this would result in collinear terms.

tarch(*numlist*) specifies the threshold ARCH terms. Adding these is another way to make the standard ARCH and GARCH models respond asymmetrically to positive and negative innovations. Specifying tarch() with arch() and garch() corresponds to one form of the GJR model (Glosten, Jagannathan, and Runkle 1993).

> tarch() may not be specified with tparch() or aarch(), as this would result in collinear terms.

aarch(*numlist*) specifies the lags of the two-parameter term $\alpha_i(|\epsilon_t| + \gamma_i\epsilon_t)^2$. This term provides the same underlying form of asymmetry as including arch() and tarch(), but it is expressed in a different way.

> aarch() may not be specified with arch() or tarch(), as this would result in collinear terms.

narch(*numlist*) specifies the lags of the two-parameter term $\alpha_i(\epsilon_t - \kappa_i)^2$. This term allows the minimum conditional variance to occur at a value of lagged innovations other than zero. For any term specified at lag L, the minimum contribution to conditional variance of that lag occurs when $\epsilon_{t-L}^2 = \kappa_L$—the squared innovations at that lag are equal to the estimated constant κ_L.

narch() may not be specified with arch(), saarch(), narchk(), nparchk(), or nparch(), as this would result in collinear terms.

narchk(*numlist*) specifies the lags of the two-parameter term $\alpha_i(\epsilon_t - \kappa)^2$; this is a variation of narch() with κ held constant for all lags.

  narchk() may not be specified with arch(), saarch(), narch(), nparchk(), or nparch(), as this would result in collinear terms.

abarch(*numlist*) specifies lags of the term $|\epsilon_t|$.

atarch(*numlist*) specifies lags of $|\epsilon_t|(\epsilon_t > 0)$, where $(\epsilon_t > 0)$ represents the indicator function returning 1 when true and 0 when false. Like the TARCH terms, these ATARCH terms allow the effect of unanticipated innovations to be asymmetric about zero.

sdgarch(*numlist*) specifies lags of σ_t. Combining atarch(), abarch(), and sdgarch() produces the model by Zakoian (1994) that the author called the TARCH model. The acronym TARCH, however, refers to any model using thresholding to obtain asymmetry.

earch(*numlist*) specifies lags of the two-parameter term $\alpha z_t + \gamma(|z_t| - \sqrt{2/\pi})$. These terms represent the influence of news—lagged innovations—in Nelson's (1991) EGARCH model. For these terms, $z_t = \epsilon_t/\sigma_t$, and arch assumes $z_t \sim N(0, 1)$. Nelson derived the general form of an EGARCH model for any assumed distribution and performed estimation assuming a generalized error distribution (GED). See Hamilton (1994) for a derivation where z_t is assumed normal. The z_t terms can be parameterized in either of these two equivalent ways. arch uses Nelson's original parameterization; see Hamilton (1994) for an equivalent alternative.

egarch(*numlist*) specifies lags of $\ln(\sigma_t^2)$.

For the following options, the model is parameterized in terms of $h(\epsilon_t)^\varphi$ and σ_t^φ. One φ is estimated, even when more than one option is specified.

parch(*numlist*) specifies lags of $|\epsilon_t|^\varphi$. parch() combined with pgarch() corresponds to the class of nonlinear models of conditional variance suggested by Higgins and Bera (1992).

tparch(*numlist*) specifies lags of $(\epsilon_t > 0)|\epsilon_t|^\varphi$, where $(\epsilon_t > 0)$ represents the indicator function returning 1 when true and 0 when false. As with tarch(), tparch() specifies terms that allow for a differential impact of "good" (positive innovations) and "bad" (negative innovations) news for lags specified by *numlist*.

  tparch() may not be specified with tarch(), as this would result in collinear terms.

aparch(*numlist*) specifies lags of the two-parameter term $\alpha(|\epsilon_t| + \gamma\epsilon_t)^\varphi$. This asymmetric power ARCH model, A-PARCH, was proposed by Ding, Granger, and Engle (1993) and corresponds to a Box–Cox function in the lagged innovations. The authors fit the original A-PARCH model on more than 16,000 daily observations of the Standard and Poor's 500, and for good reason. As the number of parameters and the flexibility of the specification increase, more data are required to estimate the parameters of the conditional heteroskedasticity. See Ding, Granger, and Engle (1993) for a discussion of how seven popular ARCH models nest within the A-PARCH model.

  When γ goes to 1, the full term goes to zero for many observations and can then be numerically unstable.

nparch(*numlist*) specifies lags of the two-parameter term $\alpha|\epsilon_t - \kappa_i|^\varphi$.

  nparch() may not be specified with arch(), saarch(), narch(), narchk(), or nparchk(), as this would result in collinear terms.

nparchk(*numlist*) specifies lags of the two-parameter term $\alpha|\epsilon_t - \kappa|^\varphi$; this is a variation of nparch() with κ held constant for all lags. This is the direct analog of narchk(), except for the power of φ. nparchk() corresponds to an extended form of the model of Higgins and Bera (1992) as presented by Bollerslev, Engle, and Nelson (1994). nparchk() would typically be combined with the pgarch() option.

nparchk() may not be specified with arch(), saarch(), narch(), narchk(), or nparch(), as this would result in collinear terms.

pgarch(*numlist*) specifies lags of σ_t^φ.

constraints(*constraints*), collinear; see [R] **estimation options**.

    ⌐‾‾ Model 2 ⌐‾‾

archm specifies that an ARCH-in-mean term be included in the specification of the mean equation. This term allows the expected value of *depvar* to depend on the conditional variance. ARCH-in-mean is most commonly used in evaluating financial time series when a theory supports a tradeoff between asset risk and return. By default, no ARCH-in-mean terms are included in the model.

archm specifies that the contemporaneous expected conditional variance be included in the mean equation. For example, typing

    . arch y x, archm arch(1)

specifies the model

$$y_t = \beta_0 + \beta_1 x_t + \psi\sigma_t^2 + \epsilon_t$$
$$\sigma_t^2 = \gamma_0 + \gamma\epsilon_{t-1}^2$$

archmlags(*numlist*) is an expansion of archm that includes lags of the conditional variance σ_t^2 in the mean equation. To specify a contemporaneous and once-lagged variance, specify either archm archmlags(1) or archmlags(0/1).

archmexp(*exp*) applies the transformation in *exp* to any ARCH-in-mean terms in the model. The expression should contain an X wherever a value of the conditional variance is to enter the expression. This option can be used to produce the commonly used ARCH-in-mean of the conditional standard deviation. With the example from archm, typing

    . arch y x, archm arch(1) archmexp(sqrt(X))

specifies the mean equation $y_t = \beta_0 + \beta_1 x_t + \psi\sigma_t + \epsilon_t$. Alternatively, typing

    . arch y x, archm arch(1) archmexp(1/sqrt(X))

specifies $y_t = \beta_0 + \beta_1 x_t + \psi/\sigma_t + \epsilon_t$.

arima($\#_p$,$\#_d$,$\#_q$) is an alternative, shorthand notation for specifying autoregressive models in the dependent variable. The dependent variable and any independent variables are differenced $\#_d$ times, 1 through $\#_p$ lags of autocorrelations are included, and 1 through $\#_q$ lags of moving averages are included. For example, the specification

    . arch y, arima(2,1,3)

is equivalent to

    . arch D.y, ar(1/2) ma(1/3)

The former is easier to write for classic ARIMA models of the mean equation, but it is not nearly as expressive as the latter. If gaps in the AR or MA lags are to be modeled, or if different operators are to be applied to independent variables, the latter syntax is required.

ar(*numlist*) specifies the autoregressive terms of the structural model disturbance to be included in the model. For example, ar(1/3) specifies that lags 1, 2, and 3 of the structural disturbance be included in the model. ar(1,4) specifies that lags 1 and 4 be included, possibly to account for quarterly effects.

If the model does not contain regressors, these terms can also be considered autoregressive terms for the dependent variable; see [TS] **arima**.

ma(*numlist*) specifies the moving-average terms to be included in the model. These are the terms for the lagged innovations or white-noise disturbances.

⌐‾‾| Model 3 |‾‾‾

distribution(*dist* [#]) specifies the distribution to assume for the error term. *dist* may be gaussian, normal, t, or ged. gaussian and normal are synonyms, and # cannot be specified with them.

If distribution(t) is specified, arch assumes that the errors follow Student's t distribution, and the degree-of-freedom parameter is estimated along with the other parameters of the model. If distribution(t #) is specified, then arch uses Student's t distribution with # degrees of freedom. # must be greater than 2.

If distribution(ged) is specified, arch assumes that the errors have a generalized error distribution, and the shape parameter is estimated along with the other parameters of the model. If distribution(ged #) is specified, then arch uses the generalized error distribution with shape parameter #. # must be positive. The generalized error distribution is identical to the normal distribution when the shape parameter equals 2.

het(*varlist*) specifies that *varlist* be included in the specification of the conditional variance. *varlist* may contain time-series operators. This varlist enters the variance specification collectively as multiplicative heteroskedasticity; see Judge et al. (1985). If het() is not specified, the model will not contain multiplicative heteroskedasticity.

Assume that the conditional variance depends on variables x and w and has an ARCH(1) component. We request this specification by using the het(x w) arch(1) options, and this corresponds to the conditional-variance model

$$\sigma_t^2 = \exp(\lambda_0 + \lambda_1 x_t + \lambda_2 w_t) + \alpha \epsilon_{t-1}^2$$

Multiplicative heteroskedasticity enters differently with an EGARCH model because the variance is already specified in logs. For the het(x w) earch(1) egarch(1) options, the variance model is

$$\ln(\sigma_t^2) = \lambda_0 + \lambda_1 x_t + \lambda_2 w_t + \alpha z_{t-1} + \gamma(|z_{t-1}| - \sqrt{2/\pi}) + \delta \ln(\sigma_{t-1}^2)$$

savespace conserves memory by retaining only those variables required for estimation. The original dataset is restored after estimation. This option is rarely used and should be specified only if there is insufficient memory to fit a model without the option. arch requires considerably more temporary storage during estimation than most estimation commands in Stata.

⌐‾‾| Priming |‾‾‾

arch0(*cond_method*) is a rarely used option that specifies how to compute the conditioning (presample or priming) values for σ_t^2 and ϵ_t^2. In the presample period, it is assumed that $\sigma_t^2 = \epsilon_t^2$ and that this value is constant. If arch0() is not specified, the priming values are computed as the expected unconditional variance given the current estimates of the β coefficients and any ARMA parameters.

arch0(xb), the default, specifies that the priming values are the expected unconditional variance of the model, which is $\sum_1^T \widehat{\epsilon}_t^2 / T$, where $\widehat{\epsilon}_t$ is computed from the mean equation and any ARMA terms.

arch0(xb0) specifies that the priming values are the estimated variance of the residuals from an OLS estimate of the mean equation.

arch0(xbwt) specifies that the priming values are the weighted sum of the $\widehat{\epsilon}_t^2$ from the current conditional mean equation (and ARMA terms) that places more weight on estimates of ϵ_t^2 at the beginning of the sample.

arch0(xb0wt) specifies that the priming values are the weighted sum of the $\widehat{\epsilon}_t^2$ from an OLS estimate of the mean equation (and ARMA terms) that places more weight on estimates of ϵ_t^2 at the beginning of the sample.

arch0(zero) specifies that the priming values are 0. Unlike the priming values for ARIMA models, 0 is generally not a consistent estimate of the presample conditional variance or squared innovations.

arch0(#) specifies that $\sigma_t^2 = \epsilon_t^2 = \#$ for any specified nonnegative #. Thus arch0(0) is equivalent to arch0(zero).

arma0(*cond_method*) is a rarely used option that specifies how the ϵ_t values are initialized at the beginning of the sample for the ARMA component, if the model has one. This option has an effect only when AR or MA terms are included in the model (the ar(), ma(), or arima() options specified).

arma0(zero), the default, specifies that all priming values of ϵ_t be taken as 0. This fits the model over the entire requested sample and takes ϵ_t as its expected value of 0 for all lags required by the ARMA terms; see Judge et al. (1985).

arma0(p), arma0(q), and arma0(pq) specify that estimation begin after priming the recursions for a certain number of observations. p specifies that estimation begin after the pth observation in the sample, where p is the maximum AR lag in the model; q specifies that estimation begin after the qth observation in the sample, where q is the maximum MA lag in the model; and pq specifies that estimation begin after the $(p+q)$th observation in the sample.

During the priming period, the recursions necessary to generate predicted disturbances are performed, but results are used only to initialize preestimation values of ϵ_t. To understand the definition of preestimation, say that you fit a model in 10/100. If the model is specified with ar(1,2), preestimation refers to observations 10 and 11.

The ARCH terms σ_t^2 and ϵ_t^2 are also updated over these observations. Any required lags of ϵ_t before the priming period are taken to be their expected value of 0, and ϵ_t^2 and σ_t^2 take the values specified in arch0().

arma0(#) specifies that the presample values of ϵ_t are to be taken as # for all lags required by the ARMA terms. Thus arma0(0) is equivalent to arma0(zero).

condobs(#) is a rarely used option that specifies a fixed number of conditioning observations at the start of the sample. Over these priming observations, the recursions necessary to generate predicted disturbances are performed, but only to initialize preestimation values of ϵ_t, ϵ_t^2, and σ_t^2. Any required lags of ϵ_t before the initialization period are taken to be their expected value of 0 (or the value specified in arma0()), and required values of ϵ_t^2 and σ_t^2 assume the values specified by arch0(). condobs() can be used if conditioning observations are desired for the lags in the ARCH terms of the model. If arma() is also specified, the maximum number of conditioning observations required by arma() and condobs(#) is used.

⌈ SE/Robust ⌊

vce(*vcetype*) specifies the type of standard error reported, which includes types that are robust to some kinds of misspecification and that are derived from asymptotic theory; see [R] *vce_option*.

For ARCH models, the robust or quasi–maximum likelihood estimates (QMLE) of variance are robust to symmetric nonnormality in the disturbances. The robust variance estimates generally are not robust to functional misspecification of the mean equation; see Bollerslev and Wooldridge (1992).

The robust variance estimates computed by arch are based on the full Huber/White/sandwich formulation, as discussed in [P] **_robust**. Many other software packages report robust estimates that set some terms to their expectations of zero (Bollerslev and Wooldridge 1992), which saves them from calculating second derivatives of the log-likelihood function.

⌈ Reporting ⌊

level(*#*); see [R] **estimation options**.

detail specifies that a detailed list of any gaps in the series be reported, including gaps due to missing observations or missing data for the dependent variable or independent variables.

nocnsreport; see [R] **estimation options**.

display_options: vsquish, cformat(*% fmt*), pformat(*% fmt*), sformat(*% fmt*), and nolstretch; see [R] **estimation options**.

⌈ Maximization ⌊

maximize_options: <u>difficult</u>, <u>tech</u>nique(*algorithm_spec*), <u>iter</u>ate(*#*), [<u>no</u>]log, <u>trace</u>, gradient, showstep, <u>hess</u>ian, <u>showtol</u>erance, <u>tol</u>erance(*#*), <u>ltol</u>erance(*#*), gtolerance(*#*), <u>nrtol</u>erance(*#*), <u>nonrtol</u>erance, and from(*init_specs*); see [R] **maximize** for all options except gtolerance(), and see below for information on gtolerance().

These options are often more important for ARCH models than for other maximum likelihood models because of convergence problems associated with ARCH models—ARCH model likelihoods are notoriously difficult to maximize.

Setting technique() to something other than the default or BHHH changes the *vcetype* to vce(oim).

The following options are all related to maximization and are either particularly important in fitting ARCH models or not available for most other estimators.

gtolerance(*#*) specifies the tolerance for the gradient relative to the coefficients. When $|g_i b_i| \leq$ gtolerance() for all parameters b_i and the corresponding elements of the gradient g_i, the gradient tolerance criterion is met. The default gradient tolerance for arch is gtolerance(.05).

gtolerance(999) may be specified to disable the gradient criterion. If the optimizer becomes stuck with repeated "(backed up)" messages, the gradient probably still contains substantial values, but an uphill direction cannot be found for the likelihood. With this option, results can often be obtained, but whether the global maximum likelihood has been found is unclear.

When the maximization is not going well, it is also possible to set the maximum number of iterations (see [R] **maximize**) to the point where the optimizer appears to be stuck and to inspect the estimation results at that point.

from(*init_specs*) specifies the initial values of the coefficients. ARCH models may be sensitive to initial values and may have coefficient values that correspond to local maximums. The default starting values are obtained via a series of regressions, producing results that, on the basis of asymptotic theory, are consistent for the β and ARMA parameters and generally reasonable for the rest. Nevertheless, these values may not always be feasible in that the likelihood function cannot be evaluated at the initial values `arch` first chooses. In such cases, the estimation is restarted with ARCH and ARMA parameters initialized to zero. It is possible, but unlikely, that even these values will be infeasible and that you will have to supply initial values yourself.

The standard syntax for `from()` accepts a matrix, a list of values, or coefficient name value pairs; see [R] **maximize**. `arch` also allows the following:

`from(archb0)` sets the starting value for all the ARCH/GARCH/... parameters in the conditional-variance equation to 0.

`from(armab0)` sets the starting value for all ARMA parameters in the model to 0.

`from(archb0 armab0)` sets the starting value for all ARCH/GARCH/... and ARMA parameters to 0.

The following option is available with `arch` but is not shown in the dialog box:

`coeflegend`; see [R] **estimation options**.

Remarks

The volatility of a series is not constant through time; periods of relatively low volatility and periods of relatively high volatility tend to be grouped together. This is a commonly observed characteristic of economic time series and is even more pronounced in many frequently sampled financial series. ARCH models seek to estimate this time-dependent volatility as a function of observed prior volatility. Sometimes the model of volatility is of more interest than the model of the conditional mean. As implemented in `arch`, the volatility model may also include regressors to account for a structural component in the volatility—usually referred to as multiplicative heteroskedasticity.

ARCH models were introduced by Engle (1982) in a study of inflation rates, and there has since been a barrage of proposed parametric and nonparametric specifications of autoregressive conditional heteroskedasticity. Overviews of the literature can found in Bollerslev, Engle, and Nelson (1994) and Bollerslev, Chou, and Kroner (1992). Introductions to basic ARCH models appear in many general econometrics texts, including Davidson and MacKinnon (1993, 2004), Greene (2012), Kmenta (1997), Stock and Watson (2011), and Wooldridge (2009). Harvey (1989) and Enders (2004) provide introductions to ARCH in the larger context of econometric time-series modeling, and Hamilton (1994) gives considerably more detail in the same context.

`arch` fits models of autoregressive conditional heteroskedasticity (ARCH, GARCH, etc.) using conditional maximum likelihood. By "conditional", we mean that the likelihood is computed based on an assumed or estimated set of priming values for the squared innovations ϵ_t^2 and variances σ_t^2 prior to the estimation sample; see Hamilton (1994) or Bollerslev (1986). Sometimes more conditioning is done on the first a, g, or $a + g$ observations in the sample, where a is the maximum ARCH term lag and g is the maximum GARCH term lag (or the maximum lags from the other ARCH family terms).

The original ARCH model proposed by Engle (1982) modeled the variance of a regression model's disturbances as a linear function of lagged values of the squared regression disturbances. We can write an ARCH(m) model as

$$y_t = \mathbf{x}_t \boldsymbol{\beta} + \epsilon_t \qquad \text{(conditional mean)}$$
$$\sigma_t^2 = \gamma_0 + \gamma_1 \epsilon_{t-1}^2 + \gamma_2 \epsilon_{t-2}^2 + \cdots + \gamma_m \epsilon_{t-m}^2 \qquad \text{(conditional variance)}$$

where
$$\epsilon_t^2 \text{ is the squared residuals (or innovations)}$$
$$\gamma_i \text{ are the ARCH parameters}$$

The ARCH model has a specification for both the conditional mean and the conditional variance, and the variance is a function of the size of prior unanticipated innovations—ϵ_t^2. This model was generalized by Bollerslev (1986) to include lagged values of the conditional variance—a GARCH model. The GARCH(m, k) model is written as

$$y_t = \mathbf{x}_t \boldsymbol{\beta} + \epsilon_t$$
$$\sigma_t^2 = \gamma_0 + \gamma_1 \epsilon_{t-1}^2 + \gamma_2 \epsilon_{t-2}^2 + \cdots + \gamma_m \epsilon_{t-m}^2 + \delta_1 \sigma_{t-1}^2 + \delta_2 \sigma_{t-2}^2 + \cdots + \delta_k \sigma_{t-k}^2$$

where
$$\gamma_i \text{ are the ARCH parameters}$$
$$\delta_i \text{ are the GARCH parameters}$$

In his pioneering work, Engle (1982) assumed that the error term, ϵ_t, followed a Gaussian (normal) distribution: $\epsilon_t \sim N(0, \sigma_t^2)$. However, as Mandelbrot (1963) and many others have noted, the distribution of stock returns appears to be leptokurtotic, meaning that extreme stock returns are more frequent than would be expected if the returns were normally distributed. Researchers have therefore assumed other distributions that can have fatter tails than the normal distribution; arch allows you to fit models assuming the errors follow Student's t distribution or the generalized error distribution. The t distribution has fatter tails than the normal distribution; as the degree-of-freedom parameter approaches infinity, the t distribution converges to the normal distribution. The generalized error distribution's tails are fatter than the normal distribution's when the shape parameter is less than two and are thinner than the normal distribution's when the shape parameter is greater than two.

The GARCH model of conditional variance can be considered an ARMA process in the squared innovations, although not in the variances as the equations might seem to suggest; see Hamilton (1994). Specifically, the standard GARCH model implies that the squared innovations result from

$$\epsilon_t^2 = \gamma_0 + (\gamma_1 + \delta_1)\epsilon_{t-1}^2 + (\gamma_2 + \delta_2)\epsilon_{t-2}^2 + \cdots + (\gamma_k + \delta_k)\epsilon_{t-k}^2 + w_t - \delta_1 w_{t-1} - \delta_2 w_{t-2} - \delta_3 w_{t-3}$$

where
$$w_t = \epsilon_t^2 - \sigma_t^2$$
$$w_t \text{ is a white-noise process that is fundamental for } \epsilon_t^2$$

One of the primary benefits of the GARCH specification is its parsimony in identifying the conditional variance. As with ARIMA models, the ARMA specification in GARCH allows the conditional variance to be modeled with fewer parameters than with an ARCH specification alone. Empirically, many series with a conditionally heteroskedastic disturbance have been adequately modeled with a GARCH(1,1) specification.

An ARMA process in the disturbances can easily be added to the mean equation. For example, the mean equation can be written with an ARMA$(1, 1)$ disturbance as

$$y_t = \mathbf{x}_t \boldsymbol{\beta} + \rho(y_{t-1} - \mathbf{x}_{t-1} \boldsymbol{\beta}) + \theta \epsilon_{t-1} + \epsilon_t$$

with an obvious generalization to ARMA(p, q) by adding terms; see [TS] **arima** for more discussion of this specification. This change affects only the conditional-variance specification in that ϵ_t^2 now results from a different specification of the conditional mean.

Much of the literature on ARCH models focuses on alternative specifications of the variance equation. arch allows many of these specifications to be requested using the saarch() through pgarch() options, which imply that one or more terms may be changed or added to the specification of the variance equation.

These alternative specifications also address asymmetry. Both the ARCH and GARCH specifications imply a symmetric impact of innovations. Whether an innovation ϵ_t^2 is positive or negative makes no difference to the expected variance σ_t^2 in the ensuing periods; only the size of the innovation matters—good news and bad news have the same effect. Many theories, however, suggest that positive and negative innovations should vary in their impact. For risk-averse investors, a large unanticipated drop in the market is more likely to lead to higher volatility than a large unanticipated increase (see Black [1976], Nelson [1991]). saarch(), tarch(), aarch(), abarch(), earch(), aparch(), and tparch() allow various specifications of asymmetric effects.

narch(), narchk(), nparch(), and nparchk() imply an asymmetric impact of a specific form. All the models considered so far have a minimum conditional variance when the lagged innovations are all zero. "No news is good news" when it comes to keeping the conditional variance small. narch(), narchk(), nparch(), and nparchk() also have a symmetric response to innovations, but they are not centered at zero. The entire news-response function (response to innovations) is shifted horizontally so that minimum variance lies at some specific positive or negative value for prior innovations.

ARCH-in-mean models allow the conditional variance of the series to influence the conditional mean. This is particularly convenient for modeling the risk–return relationship in financial series; the riskier an investment, with all else equal, the lower its expected return. ARCH-in-mean models modify the specification of the conditional mean equation to be

$$y_t = \mathbf{x_t}\beta + \psi\sigma_t^2 + \epsilon_t \qquad \text{(ARCH-in-mean)}$$

Although this linear form in the current conditional variance has dominated the literature, arch allows the conditional variance to enter the mean equation through a nonlinear transformation $g()$ and for this transformed term to be included contemporaneously or lagged.

$$y_t = \mathbf{x}_t\beta + \psi_0 g(\sigma_t^2) + \psi_1 g(\sigma_{t-1}^2) + \psi_2 g(\sigma_{t-2}^2) + \cdots + \epsilon_t$$

Square root is the most commonly used $g()$ transformation because researchers want to include a linear term for the conditional standard deviation, but any transform $g()$ is allowed.

▷ Example 1: ARCH model

Consider a simple model of the U.S. Wholesale Price Index (WPI) (Enders 2004, 87–93), which we also consider in [TS] **arima**. The data are quarterly over the period 1960q1 through 1990q4.

In [TS] **arima**, we fit a model of the continuously compounded rate of change in the WPI, $\ln(\text{WPI}_t) - \ln(\text{WPI}_{t-1})$. The graph of the differenced series—see [TS] **arima**—clearly shows periods of high volatility and other periods of relative tranquility. This makes the series a good candidate for ARCH modeling. Indeed, price indices have been a common target of ARCH models. Engle (1982) presented the original ARCH formulation in an analysis of U.K. inflation rates.

First, we fit a constant-only model by OLS and test ARCH effects by using Engle's Lagrange-multiplier test (estat archlm).

```
. use http://www.stata-press.com/data/r12/wpi1
. regress D.ln_wpi
```

Source	SS	df	MS		
Model	0	0	.	Number of obs =	123
Residual	.02521709	122	.000206697	F(0, 122) =	0.00
				Prob > F =	.
				R-squared =	0.0000
				Adj R-squared =	0.0000
Total	.02521709	122	.000206697	Root MSE =	.01438

D.ln_wpi	Coef.	Std. Err.	t	P>|t|	[95% Conf. Interval]	
_cons	.0108215	.0012963	8.35	0.000	.0082553	.0133878

```
. estat archlm, lags(1)
LM test for autoregressive conditional heteroskedasticity (ARCH)
```

lags(p)	chi2	df	Prob > chi2
1	8.366	1	0.0038

H0: no ARCH effects   vs.   H1: ARCH(p) disturbance

Because the LM test shows a p-value of 0.0038, which is well below 0.05, we reject the null hypothesis of no ARCH(1) effects. Thus we can further estimate the ARCH(1) parameter by specifying arch(1). See [R] **regress postestimation time series** for more information on Engle's LM test.

The first-order generalized ARCH model (GARCH, Bollerslev 1986) is the most commonly used specification for the conditional variance in empirical work and is typically written GARCH(1, 1). We can estimate a GARCH(1, 1) process for the log-differenced series by typing

```
. arch D.ln_wpi, arch(1) garch(1)
(setting optimization to BHHH)
Iteration 0:   log likelihood =  355.23458
Iteration 1:   log likelihood =  365.64586
 (output omitted )
Iteration 10:  log likelihood =   373.1894
ARCH family regression
```

Sample: 1960q2 - 1990q4	Number of obs =	123
Distribution: Gaussian	Wald chi2(.) =	.
Log likelihood = 373.234	Prob > chi2 =	.

D.ln_wpi	Coef.	OPG Std. Err.	z	P>|z|	[95% Conf. Interval]	
ln_wpi						
_cons	.0061167	.0010616	5.76	0.000	.0040361	.0081974
ARCH						
arch						
L1.	.4364123	.2437428	1.79	0.073	-.0413147	.9141394
garch						
L1.	.4544606	.1866606	2.43	0.015	.0886127	.8203086
_cons	.0000269	.0000122	2.20	0.028	2.97e-06	.0000508

We have estimated the ARCH(1) parameter to be 0.436 and the GARCH(1) parameter to be 0.454, so our fitted GARCH(1, 1) model is

$$y_t = 0.0061 + \epsilon_t$$
$$\sigma_t^2 = 0.436\,\epsilon_{t-1}^2 + 0.454\,\sigma_{t-1}^2$$

where $y_t = \ln(\mathtt{wpi}_t) - \ln(\mathtt{wpi}_{t-1})$.

The model Wald test and probability are both reported as missing (.). By convention, Stata reports the model test for the mean equation. Here and fairly often for ARCH models, the mean equation consists only of a constant, and there is nothing to test.   ◁

▷ Example 2: ARCH model with ARMA process

We can retain the GARCH$(1,1)$ specification for the conditional variance and model the mean as an ARMA process with AR(1) and MA(1) terms as well as a fourth-lag MA term to control for quarterly seasonal effects by typing

```
. arch D.ln_wpi, ar(1) ma(1 4) arch(1) garch(1)

(setting optimization to BHHH)
Iteration 0:   log likelihood =   380.9997
Iteration 1:   log likelihood =  388.57823
Iteration 2:   log likelihood =  391.34143
Iteration 3:   log likelihood =  396.36991
Iteration 4:   log likelihood =  398.01098
(switching optimization to BFGS)
Iteration 5:   log likelihood =  398.23668
BFGS stepping has contracted, resetting BFGS Hessian (0)
Iteration 6:   log likelihood =  399.21497
Iteration 7:   log likelihood =  399.21537   (backed up)
  (output omitted )
(switching optimization to BHHH)
Iteration 15:   log likelihood =  399.51441
Iteration 16:   log likelihood =  399.51443
Iteration 17:   log likelihood =  399.51443

ARCH family regression -- ARMA disturbances
Sample: 1960q2 - 1990q4                        Number of obs   =        123
Distribution: Gaussian                         Wald chi2(3)    =     153.56
Log likelihood =  399.5144                      Prob > chi2     =     0.0000
```

D.ln_wpi	Coef.	OPG Std. Err.	z	P>|z|	[95% Conf. Interval]	
ln_wpi						
_cons	.0069541	.0039517	1.76	0.078	−.000791	.0146992
ARMA						
ar						
L1.	.7922674	.1072225	7.39	0.000	.5821153	1.00242
ma						
L1.	−.341774	.1499943	−2.28	0.023	−.6357575	−.0477905
L4.	.2451724	.1251131	1.96	0.050	−.0000447	.4903896
ARCH						
arch						
L1.	.2040449	.1244991	1.64	0.101	−.0399688	.4480587
garch						
L1.	.6949687	.1892176	3.67	0.000	.3241091	1.065828
_cons	.0000119	.0000104	1.14	0.253	−8.52e−06	.0000324

To clarify exactly what we have estimated, we could write our model as

$$y_t = 0.007 + 0.792\,(y_{t-1} - 0.007) - 0.342\,\epsilon_{t-1} + 0.245\,\epsilon_{t-4} + \epsilon_t$$
$$\sigma_t^2 = 0.204\,\epsilon_{t-1}^2 + .695\,\sigma_{t-1}^2$$

where $y_t = \ln(\mathtt{wpi}_t) - \ln(\mathtt{wpi}_{t-1})$.

The ARCH(1) coefficient, 0.204, is not significantly different from zero, but the ARCH(1) and GARCH(1) coefficients are significant collectively. If you doubt this, you can check with `test`.

```
. test [ARCH]L1.arch [ARCH]L1.garch

 ( 1)   [ARCH]L.arch = 0
 ( 2)   [ARCH]L.garch = 0

           chi2(  2) =     84.92
         Prob > chi2 =     0.0000
```

(For comparison, we fit the model over the same sample used in the example in [TS] **arima**; Enders fits this GARCH model but over a slightly different sample.)

◁

❑ Technical note

The rather ugly iteration log on the previous result is typical, as difficulty in converging is common in ARCH models. This is actually a fairly well-behaved likelihood for an ARCH model. The "switching optimization to ..." messages are standard messages from the default optimization method for `arch`. The "backed up" messages are typical of BFGS stepping as the BFGS Hessian is often overoptimistic, particularly during early iterations. These messages are nothing to be concerned about.

Nevertheless, watch out for the messages "BFGS stepping has contracted, resetting BFGS Hessian" and "backed up", which can flag problems that may result in an iteration log that goes on and on. Stata will never report convergence and will never report final results. The question is, when do you give up and press *Break*, and if you do, what then?

If the "BFGS stepping has contracted" message occurs repeatedly (more than, say, five times), it often indicates that convergence will never be achieved. Literally, it means that the BFGS algorithm was stuck and reset its Hessian and take a steepest-descent step.

The "backed up" message, if it occurs repeatedly, also indicates problems, but only if the likelihood value is simultaneously not changing. If the message occurs repeatedly but the likelihood value is changing, as it did above, all is going well; it is just going slowly.

If you have convergence problems, you can specify options to assist the current maximization method or try a different method. Or, your model specification and data may simply lead to a likelihood that is not concave in the allowable region and thus cannot be maximized.

If you see the "backed up" message with no change in the likelihood, you can reset the gradient tolerance to a larger value. Specifying the `gtolerance(999)` option disables gradient checking, allowing convergence to be declared more easily. This does not guarantee that convergence will be declared, and even if it is, the global maximum likelihood may not have been found.

You can also try to specify initial values.

Finally, you can try a different maximization method; see options discussed under the **Maximization** tab above.

ARCH models are notorious for having convergence difficulties. Unlike in most estimators in Stata, it is common for convergence to require many steps or even to fail. This is particularly true of the explicitly nonlinear terms such as aarch(), narch(), aparch(), or archm (ARCH-in-mean), and of any model with several lags in the ARCH terms. There is not always a solution. You can try other maximization methods or different starting values, but if your data do not support your assumed ARCH structure, convergence simply may not be possible.

ARCH models can be susceptible to irrelevant regressors or unnecessary lags, whether in the specification of the conditional mean or in the conditional variance. In these situations, arch will often continue to iterate, making little to no improvement in the likelihood. We view this conservative approach as better than declaring convergence prematurely when the likelihood has not been fully maximized. arch is estimating the conditional form of second sample moments, often with flexible functions, and that is asking much of the data.

❑

❑ Technical note

if *exp* and in *range* are interpreted differently with commands accepting time-series operators. The time-series operators are resolved *before* the conditions are tested, which may lead to some confusion. Note the results of the following list commands:

```
. use http://www.stata-press.com/data/r12/archxmpl
. list t y l.y in 5/10
```

	t	y	L.y
5.	1961q1	30.8	30.7
6.	1961q2	30.5	30.8
7.	1961q3	30.5	30.5
8.	1961q4	30.6	30.5
9.	1962q1	30.7	30.6
10.	1962q2	30.6	30.7

```
. keep in 5/10
(118 observations deleted)
. list t y l.y
```

	t	y	L.y
1.	1961q1	30.8	.
2.	1961q2	30.5	30.8
3.	1961q3	30.5	30.5
4.	1961q4	30.6	30.5
5.	1962q1	30.7	30.6
6.	1962q2	30.6	30.7

We have one more lagged observation for y in the first case: l.y was resolved before the in restriction was applied. In the second case, the dataset no longer contains the value of y to compute the first lag. This means that

```
. use http://www.stata-press.com/data/r12/archxmpl, clear
. arch y l.x if twithin(1962q2, 1990q3), arch(1)
```

is not the same as

```
. keep if twithin(1962q2, 1990q3)
. arch y l.x, arch(1)
```

❑

➢ Example 3: Asymmetric effects—EGARCH model

Continuing with the WPI data, we might be concerned that the economy as a whole responds differently to unanticipated increases in wholesale prices than it does to unanticipated decreases. Perhaps unanticipated increases lead to cash flow issues that affect inventories and lead to more volatility. We can see if the data support this supposition by specifying an ARCH model that allows an asymmetric effect of "news"—innovations or unanticipated changes. One of the most popular such models is EGARCH (Nelson 1991). The full first-order EGARCH model for the WPI can be specified as follows:

```
. use http://www.stata-press.com/data/r12/wpi1, clear
. arch D.ln_wpi, ar(1) ma(1 4) earch(1) egarch(1)
(setting optimization to BHHH)
Iteration 0:   log likelihood =   227.5251
Iteration 1:   log likelihood =  381.68426
 (output omitted )
Iteration 23:  log likelihood =  405.31453
ARCH family regression -- ARMA disturbances
```

```
Sample: 1960q2 - 1990q4                      Number of obs   =        123
Distribution: Gaussian                       Wald chi2(3)    =     156.02
Log likelihood =  405.3145                    Prob > chi2     =     0.0000
```

D.ln_wpi	Coef.	OPG Std. Err.	z	P>|z|	[95% Conf. Interval]	
ln_wpi						
_cons	.0087342	.0034004	2.57	0.010	.0020696	.0153989
ARMA						
ar						
L1.	.7692121	.0968395	7.94	0.000	.5794101	.9590141
ma						
L1.	-.3554618	.1265725	-2.81	0.005	-.6035393	-.1073842
L4.	.241463	.0863832	2.80	0.005	.072155	.4107711
ARCH						
earch						
L1.	.4064002	.1163509	3.49	0.000	.1783566	.6344438
earch_a						
L1.	.2467349	.1233365	2.00	0.045	.0049999	.4884699
egarch						
L1.	.8417294	.0704079	11.96	0.000	.7037325	.9797264
_cons	-1.4884	.6604394	-2.25	0.024	-2.782837	-.1939622

Our result for the variance is

$$\ln(\sigma_t^2) = -1.49 + .406\, z_{t-1} + .247\left(\left|z_{t-1}\right| - \sqrt{2/\pi}\,\right) + .842\ln(\sigma_{t-1}^2)$$

where $z_t = \epsilon_t/\sigma_t$, which is distributed as $N(0,1)$.

This is a strong indication for a leverage effect. The positive `L1.earch` coefficient implies that positive innovations (unanticipated price increases) are more destabilizing than negative innovations. The effect appears strong (0.406) and is substantially larger than the symmetric effect (0.247). In fact, the relative scales of the two coefficients imply that the positive leverage completely dominates the symmetric effect.

This can readily be seen if we plot what is often referred to as the news-response or news-impact function. This curve shows the resulting conditional variance as a function of unanticipated news, in the form of innovations, that is, the conditional variance σ_t^2 as a function of ϵ_t. Thus we must evaluate σ_t^2 for various values of ϵ_t—say, -4 to 4—and then graph the result.

◁

▷ Example 4: Asymmetric power ARCH model

As an example of a frequently sampled, long-run series, consider the daily closing indices of the Dow Jones Industrial Average, variable `dowclose`. To avoid the first half of the century, when the New York Stock Exchange was open for Saturday trading, only data after 1jan1953 are used. The compound return of the series is used as the dependent variable and is graphed below.

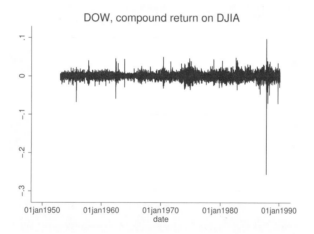

We formed this difference by referring to `D.ln_dow`, but only after playing a trick. The series is daily, and each observation represents the Dow closing index for the day. Our data included a time variable recorded as a daily date. We wanted, however, to model the log differences in the series, and we wanted the span from Friday to Monday to appear as a single-period difference. That is, the day before Monday is Friday. Because our dataset was `tsset` with `date`, the span from Friday to Monday was 3 days. The solution was to create a second variable that sequentially numbered the observations. By `tsset`ing the data with this new variable, we obtained the desired differences.

```
. generate t = _n
. tsset t
```

Now our data look like this:

```
. use http://www.stata-press.com/data/r12/dow1

. generate dayofwk = dow(date)

. list date dayofwk t ln_dow D.ln_dow in 1/8
```

	date	dayofwk	t	ln_dow	D. ln_dow
1.	02jan1953	5	1	5.677096	.
2.	05jan1953	1	2	5.682899	.0058026
3.	06jan1953	2	3	5.677439	-.0054603
4.	07jan1953	3	4	5.672636	-.0048032
5.	08jan1953	4	5	5.671259	-.0013762
6.	09jan1953	5	6	5.661223	-.0100365
7.	12jan1953	1	7	5.653191	-.0080323
8.	13jan1953	2	8	5.659134	.0059433

```
. list date dayofwk t ln_dow D.ln_dow in -8/1
```

	date	dayofwk	t	ln_dow	D. ln_dow
9334.	08feb1990	4	9334	7.880188	.0016198
9335.	09feb1990	5	9335	7.881635	.0014472
9336.	12feb1990	1	9336	7.870601	-.011034
9337.	13feb1990	2	9337	7.872665	.0020638
9338.	14feb1990	3	9338	7.872577	-.0000877
9339.	15feb1990	4	9339	7.88213	.009553
9340.	16feb1990	5	9340	7.876863	-.0052676
9341.	20feb1990	2	9341	7.862054	-.0148082

The difference operator D spans weekends because the specified time variable, t, is not a true date and has a difference of 1 for all observations. We must leave this contrived time variable in place during estimation, or arch will be convinced that our dataset has gaps. If we were using calendar dates, we would indeed have gaps.

Ding, Granger, and Engle (1993) fit an A-PARCH model of daily returns of the Standard and Poor's 500 (S&P 500) for 3jan1928–30aug1991. We will fit the same model for the Dow data shown above. The model includes an AR(1) term as well as the A-PARCH specification of conditional variance.

```
. arch D.ln_dow, ar(1) aparch(1) pgarch(1)

(setting optimization to BHHH)
Iteration 0:   log likelihood =  31139.547
Iteration 1:   log likelihood =  31350.751
 (output omitted )
Iteration 68:  log likelihood =  32273.555  (backed up)
Iteration 69:  log likelihood =  32273.555

ARCH family regression -- AR disturbances
```

Sample: 2 - 9341   Number of obs   =   9340
Distribution: Gaussian   Wald chi2(1)   =   175.46
Log likelihood =   32273.56   Prob > chi2   =   0.0000

D.ln_dow	Coef.	OPG Std. Err.	z	P>|z|	[95% Conf. Interval]	
ln_dow						
_cons	.0001786	.0000875	2.04	0.041	7.15e-06	.00035
ARMA						
ar L1.	.1410944	.0106519	13.25	0.000	.1202171	.1619716
ARCH						
aparch L1.	.0626323	.0034307	18.26	0.000	.0559082	.0693564
aparch_e L1.	-.3645093	.0378485	-9.63	0.000	-.4386909	-.2903277
pgarch L1.	.9299015	.0030998	299.99	0.000	.923826	.935977
_cons	7.19e-06	2.53e-06	2.84	0.004	2.23e-06	.0000121
POWER						
power	1.585187	.0629186	25.19	0.000	1.461869	1.708505

In the iteration log, the final iteration reports the message "backed up". For most estimators, ending on a "backed up" message would be a cause for great concern, but not with `arch` or, for that matter, `arima`, as long as you do not specify the `gtolerance()` option. `arch` and `arima`, by default, monitor the gradient and declare convergence only if, in addition to everything else, the gradient is small enough.

The fitted model demonstrates substantial asymmetry, with the large negative `L1.aparch_e` coefficient indicating that the market responds with much more volatility to unexpected drops in returns (bad news) than it does to increases in returns (good news).

◁

▷ Example 5: ARCH model with nonnormal errors

Stock returns tend to be leptokurtotic, meaning that large returns (either positive or negative) occur more frequently than one would expect if returns were in fact normally distributed. Here we refit the previous A-PARCH model assuming the errors follow the generalized error distribution, and we let `arch` estimate the shape parameter of the distribution.

```
. use http://www.stata-press.com/data/r12/dow1, clear

. arch D.ln_dow, ar(1) aparch(1) pgarch(1) distribution(ged)

(setting optimization to BHHH)
Iteration 0:   log likelihood =  31139.547
Iteration 1:   log likelihood =   31348.13
  (output omitted)
Iteration 60:  log likelihood = 32486.461

ARCH family regression -- AR disturbances
```

Sample: 2 - 9341	Number of obs	=	9340
Distribution: GED	Wald chi2(1)	=	178.22
Log likelihood = 32486.46	Prob > chi2	=	0.0000

D.ln_dow	Coef.	OPG Std. Err.	z	P>|z|	[95% Conf. Interval]	
ln_dow						
_cons	.0002735	.000078	3.51	0.000	.0001207	.0004264
ARMA						
ar						
L1.	.1337478	.0100188	13.35	0.000	.1141114	.1533842
ARCH						
aparch						
L1.	.0641785	.0049402	12.99	0.000	.0544958	.0738612
aparch_e						
L1.	-.4051993	.0573039	-7.07	0.000	-.5175128	-.2928857
pgarch						
L1.	.9341717	.004567	204.55	0.000	.9252206	.9431228
_cons	.0000216	.0000117	1.84	0.066	-1.39e-06	.0000446
POWER						
power	1.325311	.1030753	12.86	0.000	1.123287	1.527335
/lnshape	.3527036	.009482	37.20	0.000	.3341193	.3712879
shape	1.422909	.013492			1.39671	1.4496

The ARMA and ARCH coefficients are similar to those we obtained when we assumed normally distributed errors, though we do note that the power term is now closer to 1. The estimated shape parameter for the generalized error distribution is shown at the bottom of the output. Here the shape parameter is 1.42; because it is less than 2, the distribution of the errors has tails that are fatter than they would be if the errors were normally distributed.

◁

▷ Example 6: ARCH model with constraints

Engle's (1982) original model, which sparked the interest in ARCH, provides an example requiring constraints. Most current ARCH specifications use GARCH terms to provide flexible dynamic properties without estimating an excessive number of parameters. The original model was limited to ARCH terms, and to help cope with the collinearity of the terms, a declining lag structure was imposed in the parameters. The conditional variance equation was specified as

$$\sigma_t^2 = \alpha_0 + \alpha(.4\,\epsilon_{t-1} + .3\,\epsilon_{t-2} + .2\,\epsilon_{t-3} + .1\,\epsilon_{t-4})$$
$$= \alpha_0 + .4\,\alpha\epsilon_{t-1} + .3\,\alpha\epsilon_{t-2} + .2\,\alpha\epsilon_{t-3} + .1\,\alpha\epsilon_{t-4}$$

From the earlier arch output, we know how the coefficients will be named. In Stata, the formula is

$$\sigma_t^2 = \text{[ARCH]\_cons} + .4\,\text{[ARCH]L1.arch}\,\epsilon_{t-1} + .3\,\text{[ARCH]L2.arch}\,\epsilon_{t-2}$$
$$+ .2\,\text{[ARCH]L3.arch}\,\epsilon_{t-3} + .1\,\text{[ARCH]L4.arch}\,\epsilon_{t-4}$$

We could specify these linear constraints many ways, but the following seems fairly intuitive; see [R] **constraint** for syntax.

```
. use http://www.stata-press.com/data/r12/wpi1, clear
. constraint 1 (3/4)*[ARCH]l1.arch = [ARCH]l2.arch
. constraint 2 (2/4)*[ARCH]l1.arch = [ARCH]l3.arch
. constraint 3 (1/4)*[ARCH]l1.arch = [ARCH]l4.arch
```

The original model was fit on U.K. inflation; we will again use the WPI data and retain our earlier specification of the mean equation, which differs from Engle's U.K. inflation model. With our constraints, we type

```
. arch D.ln_wpi, ar(1) ma(1 4) arch(1/4) constraints(1/3)
(setting optimization to BHHH)
Iteration 0:   log likelihood =  396.80198
Iteration 1:   log likelihood =  399.07809
 (output omitted )
Iteration 9:   log likelihood =  399.46243
ARCH family regression -- ARMA disturbances
Sample: 1960q2 - 1990q4                          Number of obs    =        123
Distribution: Gaussian                           Wald chi2(3)     =     123.32
Log likelihood =  399.4624                        Prob > chi2      =     0.0000
 ( 1)  .75*[ARCH]L.arch - [ARCH]L2.arch = 0
 ( 2)  .5*[ARCH]L.arch - [ARCH]L3.arch = 0
 ( 3)  .25*[ARCH]L.arch - [ARCH]L4.arch = 0
```

D.ln_wpi	Coef.	OPG Std. Err.	z	P>|z|	[95% Conf. Interval]	
ln_wpi						
_cons	.0077204	.0034531	2.24	0.025	.0009525	.0144883
ARMA						
ar						
L1.	.7388168	.1126811	6.56	0.000	.5179659	.9596676
ma						
L1.	-.2559691	.1442861	-1.77	0.076	-.5387646	.0268264
L4.	.2528923	.1140185	2.22	0.027	.02942	.4763645
ARCH						
arch						
L1.	.2180138	.0737787	2.95	0.003	.0734101	.3626174
L2.	.1635103	.055334	2.95	0.003	.0550576	.2719631
L3.	.1090069	.0368894	2.95	0.003	.0367051	.1813087
L4.	.0545034	.0184447	2.95	0.003	.0183525	.0906544
_cons	.0000483	7.66e-06	6.30	0.000	.0000333	.0000633

L1.arch, L2.arch, L3.arch, and L4.arch coefficients have the constrained relative sizes.

◁

Saved results

arch saves the following in e():

Scalars

e(N)	number of observations
e(N_gaps)	number of gaps
e(condobs)	number of conditioning observations
e(k)	number of parameters
e(k_eq)	number of equations in e(b)
e(k_eq_model)	number of equations in overall model test
e(k_dv)	number of dependent variables
e(k_aux)	number of auxiliary parameters
e(df_m)	model degrees of freedom
e(ll)	log likelihood
e(chi2)	χ^2
e(p)	significance
e(archi)	$\sigma_0^2 = \epsilon_0^2$, priming values
e(archany)	1 if model contains ARCH terms, 0 otherwise
e(tdf)	degrees of freedom for Student's t distribution
e(shape)	shape parameter for generalized error distribution
e(tmin)	minimum time
e(tmax)	maximum time
e(power)	φ for power ARCH terms
e(rank)	rank of e(V)
e(ic)	number of iterations
e(rc)	return code
e(converged)	1 if converged, 0 otherwise

Macros

e(cmd)	arch
e(cmdline)	command as typed
e(depvar)	name of dependent variable
e(covariates)	list of covariates
e(eqnames)	names of equations
e(wtype)	weight type
e(wexp)	weight expression
e(title)	title in estimation output
e(tmins)	formatted minimum time
e(tmaxs)	formatted maximum time
e(dist)	distribution for error term: gaussian, t, or ged
e(mhet)	1 if multiplicative heteroskedasticity
e(dfopt)	yes if degrees of freedom for t distribution or shape parameter for GED distribution was estimated; no otherwise
e(chi2type)	Wald; type of model χ^2 test
e(vce)	*vcetype* specified in vce()
e(vcetype)	title used to label Std. Err.
e(ma)	lags for moving-average terms
e(ar)	lags for autoregressive terms
e(arch)	lags for ARCH terms
e(archm)	ARCH-in-mean lags
e(archmexp)	ARCH-in-mean exp
e(earch)	lags for EARCH terms
e(egarch)	lags for EGARCH terms
e(aarch)	lags for AARCH terms
e(narch)	lags for NARCH terms
e(aparch)	lags for A-PARCH terms
e(nparch)	lags for NPARCH terms
e(saarch)	lags for SAARCH terms
e(parch)	lags for PARCH terms
e(tparch)	lags for TPARCH terms
e(abarch)	lags for ABARCH terms
e(tarch)	lags for TARCH terms
e(atarch)	lags for ATARCH terms
e(sdgarch)	lags for SDGARCH terms
e(pgarch)	lags for PGARCH terms
e(garch)	lags for GARCH terms
e(opt)	type of optimization
e(ml_method)	type of ml method
e(user)	name of likelihood-evaluator program
e(technique)	maximization technique
e(tech)	maximization technique, including number of iterations
e(tech_steps)	number of iterations performed before switching techniques
e(properties)	b V
e(estat_cmd)	program used to implement estat
e(predict)	program used to implement predict
e(marginsok)	predictions allowed by margins
e(marginsnotok)	predictions disallowed by margins

Matrices
    e(b)   coefficient vector
    e(Cns)   constraints matrix
    e(ilog)   iteration log (up to 20 iterations)
    e(gradient)   gradient vector
    e(V)   variance–covariance matrix of the estimators
    e(V_modelbased)   model-based variance

Functions
    e(sample)   marks estimation sample

Methods and formulas

arch is implemented as an ado-file using the `ml` commands; see [R] **ml**.

The mean equation for the model fit by arch and with ARMA terms can be written as

$$
y_t = \mathbf{x_t}\boldsymbol{\beta} + \sum_{i=1}^{p} \psi_i g(\sigma_{t-i}^2) + \sum_{j=1}^{p} \rho_j \left\{ y_{t-j} - x_{t-j}\boldsymbol{\beta} - \sum_{i=1}^{p} \psi_i g(\sigma_{t-j-i}^2) \right\}
$$
$$
+ \sum_{k=1}^{q} \theta_k \epsilon_{t-k} + \epsilon_t \qquad \text{(conditional mean)}
$$

where

$\boldsymbol{\beta}$ are the regression parameters,

$\boldsymbol{\psi}$ are the ARCH-in-mean parameters,

ρ are the autoregression parameters,

$\boldsymbol{\theta}$ are the moving-average parameters, and

$g()$ is a general function, see the `archmexp()` option.

Any of the parameters in this full specification of the conditional mean may be zero. For example, the model need not have moving-average parameters ($\boldsymbol{\theta} = 0$) or ARCH-in-mean parameters ($\boldsymbol{\psi} = 0$). The variance equation will be one of the following:

$$
\sigma^2 = \gamma_0 + A(\boldsymbol{\sigma}, \boldsymbol{\epsilon}) + B(\boldsymbol{\sigma}, \boldsymbol{\epsilon})^2 \tag{1}
$$
$$
\ln \sigma_t^2 = \gamma_0 + C(\ln\boldsymbol{\sigma}, \mathbf{z}) + A(\boldsymbol{\sigma}, \boldsymbol{\epsilon}) + B(\boldsymbol{\sigma}, \boldsymbol{\epsilon})^2 \tag{2}
$$
$$
\sigma_t^{\varphi} = \gamma_0 + D(\boldsymbol{\sigma}, \boldsymbol{\epsilon}) + A(\boldsymbol{\sigma}, \boldsymbol{\epsilon}) + B(\boldsymbol{\sigma}, \boldsymbol{\epsilon})^2 \tag{3}
$$

where $A(\boldsymbol{\sigma}, \boldsymbol{\epsilon})$, $B(\boldsymbol{\sigma}, \boldsymbol{\epsilon})$, $C(\ln\boldsymbol{\sigma}, \mathbf{z})$, and $D(\boldsymbol{\sigma}, \boldsymbol{\epsilon})$ are linear sums of the appropriate ARCH terms; see *Details of syntax* for more information. Equation (1) is used if no EGARCH or power ARCH terms are included in the model, (2) if EGARCH terms are included, and (3) if any power ARCH terms are included; see *Details of syntax*.

Methods and formulas are presented under the following headings:

    Priming values
    Likelihood from prediction error decomposition
    Missing data

Priming values

The above model is recursive with potentially long memory. It is necessary to assume preestimation sample values for ϵ_t, ϵ_t^2, and σ_t^2 to begin the recursions, and the remaining computations are therefore conditioned on these priming values, which can be controlled using the arch0() and arma0() options. See options discussed under the **Priming** tab above.

The arch0(xb0wt) and arch0(xbwt) options compute a weighted sum of estimated disturbances with more weight on the early observations. With either of these options,

$$\sigma_{t_0-i}^2 = \epsilon_{t_0-i}^2 = (1 - .7) \sum_{t=0}^{T-1} .7^{T-t-1} \epsilon_{T-t}^2 \qquad \forall i$$

where t_0 is the first observation for which the likelihood is computed; see options discussed under the **Priming** tab above. The ϵ_t^2 are all computed from the conditional mean equation. If arch0(xb0wt) is specified, β, ψ_i, ρ_j, and θ_k are taken from initial regression estimates and held constant during optimization. If arch0(xbwt) is specified, the current estimates of β, ψ_i, ρ_j, and θ_k are used to compute ϵ_t^2 on every iteration. If any ψ_i is in the mean equation (ARCH-in-mean is specified), the estimates of ϵ_t^2 from the initial regression estimates are not consistent.

Likelihood from prediction error decomposition

The likelihood function for ARCH has a particularly simple form. Given priming (or conditioning) values of ϵ_t, ϵ_t^2, and σ_t^2, the mean equation above can be solved recursively for every ϵ_t (prediction error decomposition). Likewise, the conditional variance can be computed recursively for each observation by using the variance equation. Using these predicted errors, their associated variances, and the assumption that $\epsilon_t \sim N(0, \sigma_t^2)$, we find that the log likelihood for each observation t is

$$\ln L_t = -\frac{1}{2} \left\{ \ln(2\pi\sigma_t^2) + \frac{\epsilon_t^2}{\sigma_t^2} \right\}$$

If we assume that $\epsilon_t \sim t(\mathrm{df})$, then as given in Hamilton (1994, 662),

$$\ln L_t = \ln \Gamma \left(\frac{\mathrm{df} + 1}{2} \right) - \ln \Gamma \left(\frac{\mathrm{df}}{2} \right) - \frac{1}{2} \left[\ln \left\{ (\mathrm{df} - 2)\pi\sigma_t^2 \right\} + (\mathrm{df} + 1) \ln \left\{ 1 + \frac{\epsilon_t^2}{(\mathrm{df} - 2)\sigma_t^2} \right\} \right]$$

The likelihood is not defined for $\mathrm{df} \leq 2$, so instead of estimating df directly, we estimate $m = \ln(\mathrm{df} - 2)$. Then $\mathrm{df} = \exp(m) + 2 > 2$ for any m.

Following Bollerslev, Engle, and Nelson (1994, 2978), the log likelihood for the tth observation, assuming $\epsilon_t \sim \mathrm{GED}(s)$, is

$$\ln L_t = \ln s - \ln \lambda - \frac{s+1}{s} \ln 2 - \ln \Gamma \left(s^{-1} \right) - \frac{1}{2} \left| \frac{\epsilon_t}{\lambda\sigma_t} \right|^s$$

where

$$\lambda = \left\{ \frac{\Gamma \left(s^{-1} \right)}{2^{2/s} \Gamma \left(3s^{-1} \right)} \right\}^{1/2}$$

To enforce the restriction that $s > 0$, we estimate $r = \ln s$.

This command supports the Huber/White/sandwich estimator of the variance using vce(robust). See [P] _robust, particularly *Maximum likelihood estimators* and *Methods and formulas*.

Missing data

ARCH allows missing data or missing observations but does not attempt to condition on the surrounding data. If a dynamic component cannot be computed—ϵ_t, ϵ_t^2, and/or σ_t^2—its priming value is substituted. If a covariate, the dependent variable, or the entire observation is missing, the observation does not enter the likelihood, and its dynamic components are set to their priming values for that observation. This is acceptable only asymptotically and should not be used with a great deal of missing data.

Robert Fry Engle (1942–) was born in Syracuse, New York. He earned degrees in physics and economics at Williams College and Cornell and then worked at MIT and the University of California, San Diego, before moving to New York University Stern School of Business in 2000. He was awarded the 2003 Nobel Prize in Economics for research on autoregressive conditional heteroskedasticity and is a leading expert in time-series analysis, especially the analysis of financial markets.

References

Adkins, L. C., and R. C. Hill. 2008. *Using Stata for Principles of Econometrics*. 3rd ed. Hoboken, NJ: Wiley.

Baum, C. F. 2000. sts15: Tests for stationarity of a time series. *Stata Technical Bulletin* 57: 36–39. Reprinted in *Stata Technical Bulletin Reprints*, vol. 10, pp. 356–360. College Station, TX: Stata Press.

Baum, C. F., and R. Sperling. 2000. sts15.1: Tests for stationarity of a time series: Update. *Stata Technical Bulletin* 58: 35–36. Reprinted in *Stata Technical Bulletin Reprints*, vol. 10, pp. 360–362. College Station, TX: Stata Press.

Baum, C. F., and V. L. Wiggins. 2000. sts16: Tests for long memory in a time series. *Stata Technical Bulletin* 57: 39–44. Reprinted in *Stata Technical Bulletin Reprints*, vol. 10, pp. 362–368. College Station, TX: Stata Press.

Berndt, E. K., B. H. Hall, R. E. Hall, and J. A. Hausman. 1974. Estimation and inference in nonlinear structural models. *Annals of Economic and Social Measurement* 3/4: 653–665.

Black, F. 1976. Studies of stock price volatility changes. *Proceedings of the American Statistical Association, Business and Economics Statistics* 177–181.

Bollerslev, T. 1986. Generalized autoregressive conditional heteroskedasticity. *Journal of Econometrics* 31: 307–327.

Bollerslev, T., R. Y. Chou, and K. F. Kroner. 1992. ARCH modeling in finance. *Journal of Econometrics* 52: 5–59.

Bollerslev, T., R. F. Engle, and D. B. Nelson. 1994. ARCH models. In *Handbook of Econometrics, Volume IV*, ed. R. F. Engle and D. L. McFadden. New York: Elsevier.

Bollerslev, T., and J. M. Wooldridge. 1992. Quasi-maximum likelihood estimation and inference in dynamic models with time-varying covariances. *Econometric Reviews* 11: 143–172.

Davidson, R., and J. G. MacKinnon. 1993. *Estimation and Inference in Econometrics*. New York: Oxford University Press.

———. 2004. *Econometric Theory and Methods*. New York: Oxford University Press.

Diebold, F. X. 2003. The *ET* Interview: Professor Robert F. Engle. *Econometric Theory* 19: 1159–1193.

Ding, Z., C. W. J. Granger, and R. F. Engle. 1993. A long memory property of stock market returns and a new model. *Journal of Empirical Finance* 1: 83–106.

Enders, W. 2004. *Applied Econometric Time Series*. 2nd ed. New York: Wiley.

Engle, R. F. 1982. Autoregressive conditional heteroscedasticity with estimates of the variance of United Kingdom inflation. *Econometrica* 50: 987–1007.

——. 1990. Discussion: Stock volatility and the crash of '87. *Review of Financial Studies* 3: 103–106.

Engle, R. F., D. M. Lilien, and R. P. Robins. 1987. Estimating time varying risk premia in the term structure: The ARCH-M model. *Econometrica* 55: 391–407.

Glosten, L. R., R. Jagannathan, and D. E. Runkle. 1993. On the relation between the expected value and the volatility of the nominal excess return on stocks. *Journal of Finance* 48: 1779–1801.

Greene, W. H. 2012. *Econometric Analysis.* 7th ed. Upper Saddle River, NJ: Prentice Hall.

Hamilton, J. D. 1994. *Time Series Analysis.* Princeton: Princeton University Press.

Harvey, A. C. 1989. *Forecasting, Structural Time Series Models and the Kalman Filter.* Cambridge: Cambridge University Press.

——. 1990. *The Econometric Analysis of Time Series.* 2nd ed. Cambridge, MA: MIT Press.

Higgins, M. L., and A. K. Bera. 1992. A class of nonlinear ARCH models. *International Economic Review* 33: 137–158.

Hill, R. C., W. E. Griffiths, and G. C. Lim. 2011. *Principles of Econometrics.* 4th ed. Hoboken, NJ: Wiley.

Judge, G. G., W. E. Griffiths, R. C. Hill, H. Lütkepohl, and T.-C. Lee. 1985. *The Theory and Practice of Econometrics.* 2nd ed. New York: Wiley.

Kmenta, J. 1997. *Elements of Econometrics.* 2nd ed. Ann Arbor: University of Michigan Press.

Mandelbrot, B. 1963. The variation of certain speculative prices. *Journal of Business* 36: 394–419.

Nelson, D. B. 1991. Conditional heteroskedasticity in asset returns: A new approach. *Econometrica* 59: 347–370.

Press, W. H., S. A. Teukolsky, W. T. Vetterling, and B. P. Flannery. 2007. *Numerical Recipes in C: The Art of Scientific Computing.* 3rd ed. Cambridge: Cambridge University Press.

Stock, J. H., and M. W. Watson. 2011. *Introduction to Econometrics.* 3rd ed. Boston: Addison–Wesley.

Wooldridge, J. M. 2009. *Introductory Econometrics: A Modern Approach.* 4th ed. Cincinnati, OH: South-Western.

Zakoian, J. M. 1994. Threshold heteroskedastic models. *Journal of Economic Dynamics and Control* 18: 931–955.

Also see

[TS] **arch postestimation** — Postestimation tools for arch

[TS] **tsset** — Declare data to be time-series data

[TS] **arima** — ARIMA, ARMAX, and other dynamic regression models

[TS] **mgarch** — Multivariate GARCH models

[R] **regress** — Linear regression

[U] **20 Estimation and postestimation commands**

Title

> **arch postestimation** — Postestimation tools for arch

Description

The following postestimation commands are available after `arch`:

Command	Description
estat	AIC, BIC, VCE, and estimation sample summary
estimates	cataloging estimation results
lincom	point estimates, standard errors, testing, and inference for linear combinations of coefficients
lrtest	likelihood-ratio test
margins	marginal means, predictive margins, marginal effects, and average marginal effects
marginsplot	graph the results from margins (profile plots, interaction plots, etc.)
nlcom	point estimates, standard errors, testing, and inference for nonlinear combinations of coefficients
predict	predictions, residuals, influence statistics, and other diagnostic measures
predictnl	point estimates, standard errors, testing, and inference for generalized predictions
test	Wald tests of simple and composite linear hypotheses
testnl	Wald tests of nonlinear hypotheses

See the corresponding entries in the *Base Reference Manual* for details.

Syntax for predict

> predict [*type*] *newvar* [*if*] [*in*] [, *statistic options*]

statistic	Description
Main	
xb	predicted values for mean equation—the differenced series; the default
y	predicted values for the mean equation in y—the undifferenced series
variance	predicted values for the conditional variance
het	predicted values of the variance, considering only the multiplicative heteroskedasticity
residuals	residuals or predicted innovations
yresiduals	residuals or predicted innovations in y—the undifferenced series

These statistics are available both in and out of sample; type `predict ... if e(sample) ...` if wanted only for the estimation sample.

options	Description
Options	
<u>dynamic</u>(*time_constant*)	how to handle the lags of y_t
<u>at</u>(*varname*$_\epsilon$ | #$_\epsilon$ *varname*$_{\sigma^2}$ | #$_{\sigma^2}$)	make static predictions
t0(*time_constant*)	set starting point for the recursions to *time_constant*
<u>structural</u>	calculate considering the structural component only

time_constant is a # or a time literal, such as td(1jan1995) or tq(1995q1), etc.; see
   Conveniently typing SIF values in [D] **datetime**.

Menu

Statistics > Postestimation > Predictions, residuals, etc.

Options for predict

Six statistics can be computed by using predict after arch: the predictions of the mean equation (option xb, the default), the undifferenced predictions of the mean equation (option y), the predictions of the conditional variance (option variance), the predictions of the multiplicative heteroskedasticity component of variance (option het), the predictions of residuals or innovations (option residuals), and the predictions of residuals or innovations in terms of y (option yresiduals). Given the dynamic nature of ARCH models and because the dependent variable might be differenced, there are other ways of computing each statistic. We can use all the data on the dependent variable available right up to the time of each prediction (the default, which is often called a one-step prediction), or we can use the data up to a particular time, after which the predicted value of the dependent variable is used recursively to make later predictions (option dynamic()). Either way, we can consider or ignore the ARMA disturbance component, which is considered by default and is ignored if you specify the structural option. We might also be interested in predictions at certain fixed points where we specify the prior values of ϵ_t and σ_t^2 (option at()).

⌐ Main ⌐
---

xb, the default, calculates the predictions from the mean equation. If D.*depvar* is the dependent variable, these predictions are of D.*depvar* and not of *depvar* itself.

y specifies that predictions of *depvar* are to be made even if the model was specified for, say, D.*depvar*.

variance calculates predictions of the conditional variance $\widehat{\sigma}_t^2$.

het calculates predictions of the multiplicative heteroskedasticity component of variance.

residuals calculates the residuals. If no other options are specified, these are the predicted innovations ϵ_t; that is, they include any ARMA component. If the structural option is specified, these are the residuals from the mean equation, ignoring any ARMA terms; see structural below. The residuals are always from the estimated equation, which may have a differenced dependent variable; if *depvar* is differenced, they are not the residuals of the undifferenced *depvar*.

yresiduals calculates the residuals for *depvar*, even if the model was specified for, say, D.*depvar*. As with residuals, the yresiduals are computed from the model, including any ARMA component. If the structural option is specified, any ARMA component is ignored and yresiduals are the residuals from the structural equation; see structural below.

dynamic(*time_constant*) specifies how lags of y_t in the model are to be handled. If dynamic() is not specified, actual values are used everywhere lagged values of y_t appear in the model to produce one-step-ahead forecasts.

dynamic(*time_constant*) produces dynamic (also known as recursive) forecasts. *time_constant* specifies when the forecast is to switch from one step ahead to dynamic. In dynamic forecasts, references to y_t evaluate to the prediction of y_t for all periods at or after *time_constant*; they evaluate to the actual value of y_t for all prior periods.

dynamic(10), for example, would calculate predictions where any reference to y_t with $t < 10$ evaluates to the actual value of y_t and any reference to y_t with $t \geq 10$ evaluates to the prediction of y_t. This means that one-step-ahead predictions would be calculated for $t < 10$ and dynamic predictions would be calculated thereafter. Depending on the lag structure of the model, the dynamic predictions might still refer to some actual values of y_t.

You may also specify dynamic(.) to have predict automatically switch from one-step-ahead to dynamic predictions at $p + q$, where p is the maximum AR lag and q is the maximum MA lag.

at(*varname*$_\epsilon$ | #$_\epsilon$ *varname*$_{\sigma^2}$ | #$_{\sigma^2}$) makes static predictions. at() and dynamic() may not be specified together.

Specifying at() allows static evaluation of results for a given set of disturbances. This is useful, for instance, in generating the news response function. at() specifies two sets of values to be used for ϵ_t and σ_t^2, the dynamic components in the model. These specified values are treated as given. Also, any lagged values of *depvar* in the model are obtained from the real values of the dependent variable. All computations are based on actual data and the given values.

at() requires that you specify two arguments, which can be either a variable name or a number. The first argument supplies the values to be used for ϵ_t; the second supplies the values to be used for σ_t^2. If σ_t^2 plays no role in your model, the second argument may be specified as '.' to indicate missing.

t0(*time_constant*) specifies the starting point for the recursions to compute the predicted statistics; disturbances are assumed to be 0 for $t < $ t0(). The default is to set t0() to the minimum t observed in the estimation sample, meaning that observations before that are assumed to have disturbances of 0.

t0() is irrelevant if structural is specified because then all observations are assumed to have disturbances of 0.

t0(5), for example, would begin recursions at $t = 5$. If your data were quarterly, you might instead type t0(tq(1961q2)) to obtain the same result.

Any ARMA component in the mean equation or GARCH term in the conditional-variance equation makes arch recursive and dependent on the starting point of the predictions. This includes one-step-ahead predictions.

structural makes the calculation considering the structural component only, ignoring any ARMA terms, and producing the steady-state equilibrium predictions.

Remarks

▷ Example 1

Continuing with our EGARCH model example (example 3) in [TS] **arch**, we can see that `predict`, `at()` calculates σ_t^2 given a set of specified innovations $(\epsilon_t, \epsilon_{t-1}, \ldots)$ and prior conditional variances $(\sigma_{t-1}^2, \sigma_{t-2}^2, \ldots)$. The syntax is

        . predict *newvar*, variance at(*epsilon sigma2*)

epsilon and *sigma2* are either variables or numbers. Using *sigma2* is a little tricky because you specify values of σ_t^2, which `predict` is supposed to predict. *predict* does not simply copy variable *sigma2* into *newvar* but uses the lagged values contained in *sigma2* to produce the predicted value of σ_t^2. It does this for all t, and those results are stored in *newvar*. (If you are interested in dynamic predictions of σ_t^2, see *Options for predict*.)

We will generate predictions for σ_t^2, assuming that the lagged values of σ_t^2 are 1, and we will vary ϵ_t from -4 to 4. First, we will create variable et containing ϵ_t, and then we will create and graph the predictions:

        . generate et = (_n-64)/15

        . predict sigma2, variance at(et 1)

        . line sigma2 et in 2/1, m(i) c(l) title(News response function)

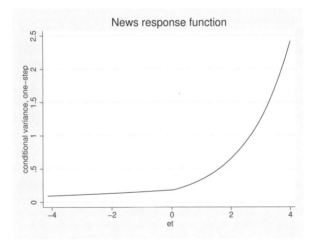

The positive asymmetry does indeed dominate the shape of the news response function. In fact, the response is a monotonically increasing function of news. The form of the response function shows that, for our simple model, only positive, unanticipated price increases have the destabilizing effect that we observe as larger conditional variances.

◁

▷ Example 2

Continuing with our ARCH model with constraints example (example 6) in [TS] **arch**, using lincom we can recover the α parameter from the original specification.

```
. lincom [ARCH]l1.arch/.4
 ( 1)  2.5 [ARCH]L.arch = 0
```

D.ln_wpi	Coef.	Std. Err.	z	P>|z|	[95% Conf. Interval]	
(1)	.5450344	.1844468	2.95	0.003	.1835253	.9065436

Any arch parameter could be used to produce an identical estimate.

◁

Methods and formulas

All postestimation commands listed above are implemented as ado-files.

Also see

[TS] **arch** — Autoregressive conditional heteroskedasticity (ARCH) family of estimators

[U] **20 Estimation and postestimation commands**

Title

arfima — Autoregressive fractionally integrated moving-average models

Syntax

arfima *depvar* [*indepvars*] [*if*] [*in*] [, *options*]

options	Description
Model	
noconstant	suppress constant term
ar(*numlist*)	autoregressive terms
ma(*numlist*)	moving-average terms
smemory	estimate short-memory model without fractional integration
mle	maximum likelihood estimates; the default
mpl	maximum modified-profile-likelihood estimates
constraints(*numlist*)	apply specified linear constraints
collinear	do not drop collinear variables
SE/Robust	
vce(*vcetype*)	*vcetype* may be oim or robust
Reporting	
level(#)	set confidence level; default is level(95)
nocnsreport	do not display constraints
display_options	control column formats, row spacing, line width, and display of omitted variables and base and empty cells
Maximization	
maximize_options	control the maximization process; seldom used
coeflegend	display legend instead of statistics

You must tsset your data before using arfima; see [TS] **tsset**.
depvar and *indepvars* may contain time-series operators; see [U] **11.4.4 Time-series varlists**.
indepvars may contain factor variables; see [U] **11.4.3 Factor variables**.
by, rolling, and statsby are allowed; see [U] **11.1.10 Prefix commands**.
coeflegend does not appear in the dialog box.
See [U] **20 Estimation and postestimation commands** for more capabilities of estimation commands.

Menu

Statistics > Time series > ARFIMA models

Description

arfima estimates the parameters of autoregressive fractionally integrated moving-average (ARFIMA) models.

Long-memory processes are stationary processes whose autocorrelation functions decay more slowly than short-memory processes. The ARFIMA model provides a parsimonious parameterization of long-memory processes that nests the autoregressive moving-average (ARMA) model, which is widely used for short-memory processes. By allowing for fractional degrees of integration, the ARFIMA model also generalizes the autoregressive integrated moving-average (ARIMA) model with integer degrees of integration. See [TS] **arima** for ARMA and ARIMA parameter estimation.

Options

⌐ Model ⌐

noconstant; see [R] **estimation options**.

ar(*numlist*) specifies the autoregressive (AR) terms to be included in the model. An AR(p), $p \geq 1$, specification would be ar(1/p). This model includes all lags from 1 to p, but not all lags need to be included. For example, the specification ar(1 p) would specify an AR(p) with only lags 1 and p included, setting all the other AR lag parameters to 0.

ma(*numlist*) specifies the moving-average terms to be included in the model. These are the terms for the lagged innovations (white-noise disturbances). ma(1/q), $q \geq 1$, specifies an MA(q) model, but like the ar() option, not all lags need to be included.

smemory causes arfima to fit a short-memory model with $d = 0$. This option causes arfima to estimate the parameters of an ARMA model by a method that is asymptotically equivalent to that produced by arima; see [TS] **arima**.

mle causes arfima to estimate the parameters by maximum likelihood. This method is the default.

mpl causes arfima to estimate the parameters by maximum modified profile likelihood (MPL). The MPL estimator of the fractional-difference parameter has less small-sample bias than the maximum likelihood estimator when there are covariates in the model. mpl may only be specified when there is a constant term or *indepvars* in the model, and it may not be combined with the mle option.

constraints(*numlist*), collinear; see [R] **estimation options**.

⌐ SE/Robust ⌐

vce(*vcetype*) specifies the type of standard error reported, which includes types that are robust to some kinds of misspecification and that are derived from asymptotic theory; see [R] ***vce_option***.

Options vce(robust) and mpl may not be combined.

⌐ Reporting ⌐

level(#), nocnsreport; see [R] **estimation options**.

display_options: noomitted, vsquish, noemptycells, baselevels, allbaselevels, cformat(% *fmt*), pformat(% *fmt*), sformat(% *fmt*), and nolstretch; see [R] **estimation options**.

⌐‾ Maximization ⌐‾

maximize_options: <u>diff</u>icult, <u>techni</u>que(*algorithm_spec*), <u>iter</u>ate(*#*), [<u>no</u>]<u>log</u>, <u>tr</u>ace, <u>grad</u>ient, showstep, <u>hess</u>ian, <u>showtol</u>erance, <u>tol</u>erance(*#*), <u>ltol</u>erance(*#*), <u>nrtol</u>erance(*#*), <u>gtol</u>erance(*#*), <u>nonrtol</u>erance(*#*), and from(*init_specs*); see [R] **maximize** for all options.

Some special points for `arfima`'s *maximize_options* are listed below.

`technique`(*algorithm_spec*) sets the optimization algorithm. The default algorithm is BFGS and BHHH is not allowed. See [R] **maximize** for a description of the available optimization algorithms.

You can specify multiple optimization methods. For example, `technique(bfgs 10 nr)` requests that the optimizer perform 10 BFGS iterations and then switch to Newton–Raphson until convergence.

`iterate`(*#*) sets the maximum number of iterations. When the maximization is not going well, set the maximum number of iterations to the point where the optimizer appears to be stuck and inspect the estimation results at that point.

`from`(*matname*) allows you to specify starting values for the model parameters in a row vector. We recommend that you use the `iterate(0)` option, retrieve the initial estimates from `e(b)`, and modify these elements.

The following option is available with `arfima` but is not shown in the dialog box:

`coeflegend`; see [R] **estimation options**.

Remarks

Long-memory processes are stationary processes whose autocorrelation functions decay more slowly than short-memory processes. Because the autocorrelations die out so slowly, long-memory processes display a type of long-run dependence. The autoregressive fractionally integrated moving-average (ARFIMA) model provides a parsimonious parameterization of long-memory processes. This parameterization nests the autoregressive moving-average (ARMA) model, which is widely used for short-memory processes.

The ARFIMA model also generalizes the autoregressive integrated moving-average (ARIMA) model with integer degrees of integration. ARFIMA models provide a solution for the tendency to overdifference stationary series that exhibit long-run dependence. In the ARIMA approach, a nonstationary time series is differenced d times until the differenced series is stationary, where d is an integer. Such series are said to be integrated of order d, denoted $I(d)$, with not differencing, $I(0)$, being the option for stationary series. Many series exhibit too much dependence to be $I(0)$ but are not $I(1)$, and ARFIMA models are designed to represent these series.

The ARFIMA model allows for a continuum of fractional differences, $-0.5 < d < 0.5$. The generalization to fractional differences allows the ARFIMA model to handle processes that are neither $I(0)$ nor $I(1)$, to test for overdifferencing, and to model long-run effects that only die out at long horizons.

❏ Technical note

An ARIMA model for the series y_t is given by

$$\rho(L)(1 - L)^d y_t = \theta(L)\epsilon_t \tag{1}$$

where $\rho(L) = (1 - \rho_1 L - \rho_2 L^2 - \cdots - \rho_p L^p)$ is the autoregressive (AR) polynomial in the lag operator L; $L y_t = y_{t-1}$; $\theta(L) = (1 + \theta_1 L + \theta_2 L^2 + \cdots + \theta_p L^p)$ is the moving-average (MA) lag polynomial; ϵ_t is the independent and identically distributed innovation term; and d is the integer number of differences required to make the y_t stationary. An ARFIMA model is also specified by (1) with the generalization that $-0.5 < d < 0.5$. Series with $d \geq 0.5$ are handled by differencing and subsequent ARFIMA modeling.

❏

Because long-memory processes are stationary, one might be tempted to approximate the processes with many terms in an ARMA model. But these approximate models are difficult to fit and to interpret because ARMA models with many terms are difficult to estimate and the ARMA parameterization has an inherent short-run nature. In contrast, the ARFIMA model has the d parameter for the long-run dependence and ARMA parameters for short-run dependence. Using different parameters for different types of dependence facilitates estimation and interpretation, as discussed by Sowell (1992a).

❏ Technical note

An ARFIMA model specifies a fractionally integrated ARMA process. Formally, the ARFIMA model specifies that

$$y_t = (1 - L)^{-d}\{\rho(L)\}^{-1}\theta(L)\epsilon_t$$

The short-run ARMA process $\rho(L)^{-1}\theta(L)\epsilon_t$ captures the short-run effects, and the long-run effects are captured by fractionally integrating the short-run ARMA process.

Essentially, the fractional-integration parameter d captures the long-run effects, and the ARMA parameters capture the short-run effects. Having separate parameters for short-run and long-run effects makes the ARFIMA model more flexible and easier to interpret than the ARMA model. After estimating the ARFIMA parameters, the short-run effects are obtained by setting $d = 0$, whereas the long-run effects use the estimated value for d. The short-run effects describe the behavior of the fractionally differenced process $(1 - L)^d y_t$, whereas the long-run effects describe the behavior of the fractionally integrated y_t.

❏

ARFIMA models have been useful in fields as diverse as hydrology and economics. Long-memory processes were first introduced in hydrology by Hurst (1951). Hosking (1981), in hydrology, and Granger and Joyeux (1980), in economics, independently discovered the ARFIMA representation of long-memory processes. Beran (1994), Baillie (1996), and Palma (2007) provide good introductions to long-memory processes and ARFIMA models.

▷ Example 1: Mount Campito tree ring data

Baillie (1996) discusses a time series of measurements of the widths of the annual rings of a Mount Campito Bristlecone pine. The series contains measurements on rings formed in the tree from 3436 BC to 1969 AD. Essentially, larger widths were good years for the tree and narrower widths were harsh years.

We begin by plotting the time series.

```
. use http://www.stata-press.com/data/r12/campito
(Campito Mnt. tree ring data from 3435BC to 1969AD)
. tsline width, xlabel(-3435(500)1969) ysize(2)
```

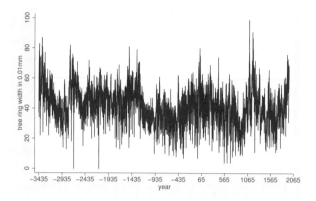

Good years and bad years seem to run together, causing the appearance of local trends. The local trends are evidence of dependence, but they are not as pronounced as those in a nonstationary series.

We plot the autocorrelations for another view:

```
. ac width, ysize(2)
```

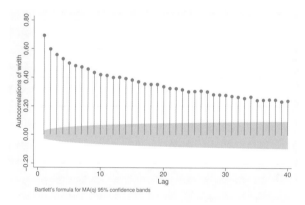

The autocorrelations do not approach 1 but decay very slowly.

Granger and Joyeux (1980) show that the autocorrelations from an ARMA model decay exponentially, whereas the autocorrelations from an ARFIMA process decay at the much slower hyperbolic rate. Box, Jenkins, and Reinsel (2008) define short-memory processes as those whose autocorrelations decay exponentially fast and long-memory processes as those whose autocorrelations decay at the hyperbolic rate. The above plot of autocorrelations looks closer to hyperbolic than exponential.

Together, the above plots make us suspect that the series was generated by a long-memory process. We see evidence that the series is stationary but that the autocorrelations die out much slower than a short-memory process would predict.

Given that we believe the data was generated by a stationary process, we begin by fitting the data to an ARMA model. We begin by using a short-memory model because a comparison of the results highlights the advantages of using an ARFIMA model for a long-memory process.

```
. arima width, ar(1/2) ma(1) technique(bhhh 4 nr)

(setting optimization to BHHH)
Iteration 0:   log likelihood = -18934.593
Iteration 1:   log likelihood = -18914.337
Iteration 2:   log likelihood = -18913.407
Iteration 3:   log likelihood =  -18913.24
(switching optimization to Newton-Raphson)
Iteration 4:   log likelihood = -18913.214
Iteration 5:   log likelihood = -18913.208
Iteration 6:   log likelihood = -18913.208

ARIMA regression

Sample:  -3435 - 1969                      Number of obs    =       5405
                                           Wald chi2(3)     = 133686.46
Log likelihood = -18913.21                 Prob > chi2      =     0.0000
```

width	Coef.	OIM Std. Err.	z	P>|z|	[95% Conf. Interval]
width					
_cons	42.45055	1.02142	41.56	0.000	40.44861 44.4525
ARMA					
ar					
L1.	1.264367	.0253199	49.94	0.000	1.214741 1.313994
L2.	-.2848827	.0227534	-12.52	0.000	-.3294785 -.2402869
ma					
L1.	-.8066007	.0189699	-42.52	0.000	-.8437811 -.7694204
/sigma	8.005814	.0770004	103.97	0.000	7.854896 8.156732

```
Note: The test of the variance against zero is one sided, and the two-sided
      confidence interval is truncated at zero.
```

The roots of the AR polynomial are 0.971 and 0.293, and the root of the MA polynomial is -0.807; all of these are less than one in magnitude, indicating that the series is stationary and invertible but has a high level of persistence. See Hamilton (1994, 59) for how to compute the roots of the polynomials from the estimated coefficients.

Below we estimate the parameters of an ARFIMA model with only the fractional difference parameter and a constant.

```
. arfima width
Iteration 0:    log likelihood = -18918.219
Iteration 1:    log likelihood =  -18916.84
Iteration 2:    log likelihood = -18908.508
Iteration 3:    log likelihood = -18908.508  (backed up)
Iteration 4:    log likelihood = -18907.379
Iteration 5:    log likelihood = -18907.318
Iteration 6:    log likelihood = -18907.279
Iteration 7:    log likelihood = -18907.279
Refining estimates:
Iteration 0:    log likelihood = -18907.279
Iteration 1:    log likelihood = -18907.279

ARFIMA regression

Sample: -3435 - 1969                          Number of obs   =        5405
                                              Wald chi2(1)    =     1864.43
Log likelihood = -18907.279                   Prob > chi2     =      0.0000
```

width	Coef.	OIM Std. Err.	z	P>|z|	[95% Conf. Interval]	
width						
_cons	44.01432	9.174318	4.80	0.000	26.03299	61.99565
ARFIMA						
d	.4468887	.0103497	43.18	0.000	.4266038	.4671737
/sigma2	63.92927	1.229754	51.99	0.000	61.519	66.33955

Note: The test of the variance against zero is one sided, and the two-sided
      confidence interval is truncated at zero.

The estimate of d is large and statistically significant. The relative parsimony of the ARFIMA model is illustrated by the fact that the estimates of the standard deviation of the idiosyncratic errors are about the same in the 5-parameter ARMA model and the 3-parameter ARFIMA model.

Let's add an AR parameter to the above ARFIMA model:

```
. arfima width, ar(1)
Iteration 0:   log likelihood = -18910.997
Iteration 1:   log likelihood = -18910.949  (backed up)
Iteration 2:   log likelihood = -18908.158  (backed up)
Iteration 3:   log likelihood = -18907.248
Iteration 4:   log likelihood = -18907.233
Iteration 5:   log likelihood = -18907.233
Iteration 6:   log likelihood = -18907.233
Refining estimates:
Iteration 0:   log likelihood = -18907.233
Iteration 1:   log likelihood = -18907.233

ARFIMA regression
Sample: -3435 - 1969                        Number of obs    =       5405
                                            Wald chi2(2)     =    1875.34
Log likelihood = -18907.233                 Prob > chi2      =     0.0000
```

width	Coef.	OIM Std. Err.	z	P>|z|	[95% Conf. Interval]
width						
_cons	43.98774	8.685319	5.06	0.000	26.96482	61.01065
ARFIMA						
ar						
L1.	.0063319	.0210492	0.30	0.764	-.0349238	.0475876
d	.4432473	.0159172	27.85	0.000	.4120502	.4744445
/sigma2	63.92915	1.229756	51.99	0.000	61.51887	66.33943

Note: The test of the variance against zero is one sided, and the two-sided
      confidence interval is truncated at zero.

That the estimated AR term is tiny and statistically insignificant indicates that the d parameter has accounted for all the dependence in the series.

◁

As mentioned above, there is a sense in which the main advantages of an ARFIMA model over an ARMA model for long-memory processes are the relative parsimony of the ARFIMA parameterization and the ability of the ARFIMA parameterization to separate out the long-run effects from the short-run effects. If the true process was generated from an ARFIMA model, an ARMA model with many terms can approximate the process, but the terms make estimation difficult and the lack of separate long-run and short-run parameters complicates interpretation.

The previous example highlights the relative parsimony of the ARFIMA model. In the examples below, we illustrate the advantages of having separate parameters for long-run and short-run effects.

❑ Technical note

You may be wondering what long-run effects can be produced by a model for stationary processes. Because the autocorrelations of a long-memory process die out so slowly, the spectral density becomes infinite as the frequency goes to 0 and the impulse–response functions die out at a much slower rate.

The spectral density of a process describes the relative contributions of random components at different frequencies to the variance of the process, with the low-frequency components corresponding to long-run effects. See [TS] **psdensity** for an introduction to estimating and interpreting spectral densities implied by the estimated parameters of parametric models.

Granger and Joyeux (1980) motivate ARFIMA models by noting that their implied spectral densities are finite except at frequency 0 with $0 < d < 0.5$, whereas stationary ARMA models have finite spectral densities at all frequencies. Granger and Joyeux (1980) argue that the ability of ARFIMA models to capture this long-range dependence, which cannot be captured by stationary ARMA models, is an important advantage of ARFIMA models over ARMA models when modeling long-memory processes.

Impulse–response functions are the coefficients on the infinite-order MA representation of a process, and they describe how a shock feeds though the dynamic system. If the process is stationary, the coefficients decay to 0 and they sum to a finite constant. As expected, the coefficients from an ARFIMA model die out at a slower rate than those from an ARMA model. Because the ARMA terms model the short-run effects and the d parameter models the long-run effects, an ARFIMA model specifies both a short-run impulse–response function and a long-run impulse–response function. When an ARMA model is used to approximate a long-memory model, the ARMA impulse–response-function coefficients confound the two effects.

❏

▷ Example 2

In this example, we model the log of the monthly levels of carbon dioxide above Mauna Loa, Hawaii. To remove the seasonality, we model the twelfth seasonal difference of the log of the series. This example illustrates that the ARFIMA model parameterizes long-run and short-run effects, whereas the ARMA model confounds the two effects. (Sowell [1992a] discusses this point in greater depth.)

We begin by fitting the series to an ARMA model with an AR(1) term and an MA(2).

```
. use http://www.stata-press.com/data/r12/mloa
. arima S12.log, ar(1) ma(2)
(setting optimization to BHHH)
Iteration 0:   log likelihood =  2000.9262
Iteration 1:   log likelihood =  2001.5484
Iteration 2:   log likelihood =  2001.5637
Iteration 3:   log likelihood =  2001.5641
Iteration 4:   log likelihood =  2001.5641
ARIMA regression
Sample:  1960m1 - 1990m12                    Number of obs     =        372
                                             Wald chi2(2)      =     500.41
Log likelihood =  2001.564                   Prob > chi2       =     0.0000
```

S12.log	Coef.	OPG Std. Err.	z	P>|z|	[95% Conf. Interval]
log					
_cons	.0036754	.0002475	14.85	0.000	.0031903 .0041605
ARMA					
ar					
L1.	.7354346	.0357715	20.56	0.000	.6653237 .8055456
ma					
L2.	.1353086	.0513156	2.64	0.008	.0347319 .2358853
/sigma	.0011129	.0000401	27.77	0.000	.0010344 .0011914

Note: The test of the variance against zero is one sided, and the two-sided confidence interval is truncated at zero.

All the parameters are statistically significant, and they indicate a high degree of dependence.

Below we nest the previously fit ARMA model into an ARFIMA model.

```
. arfima S12.log, ar(1) ma(2)
Iteration 0:   log likelihood =  2006.0757
Iteration 1:   log likelihood =  2006.0774  (backed up)
Iteration 2:   log likelihood =  2006.0775  (backed up)
Iteration 3:   log likelihood =  2006.0804
Iteration 4:   log likelihood =  2006.0805
Refining estimates:
Iteration 0:   log likelihood =  2006.0805
Iteration 1:   log likelihood =  2006.0805

ARFIMA regression

Sample: 1960m1 - 1990m12                    Number of obs   =        372
                                            Wald chi2(3)    =     248.88
Log likelihood =  2006.0805                 Prob > chi2     =     0.0000
```

S12.log	Coef.	OIM Std. Err.	z	P>|z|	[95% Conf. Interval]	
S12.log						
_cons	.003616	.0012968	2.79	0.005	.0010743	.0061578
ARFIMA						
ar						
L1.	.2160894	.101556	2.13	0.033	.0170433	.4151355
ma						
L2.	.1633916	.0516907	3.16	0.002	.0620796	.2647035
d	.4042573	.0805422	5.02	0.000	.2463974	.5621171
/sigma2	1.20e-06	8.84e-08	13.63	0.000	1.03e-06	1.38e-06

```
Note: The test of the variance against zero is one sided, and the two-sided
      confidence interval is truncated at zero.
```

All the parameters are statistically significant at the 5% level. That the confidence interval for the fractional-difference parameter d includes numbers greater than 0.5 is evidence that the series may be nonstationary. Alternatively, we proceed as if the series is stationary, and the wide confidence interval for d reflects the difficulty of fitting a complicated dynamic model with only 372 observations.

With the above caveat, we can now proceed to compare the interpretations of the ARMA and ARFIMA estimates. We compare these estimates in terms of their implied spectral densities. The spectral density of a stationary time series describes the relative importance of components at different frequencies. See [TS] **psdensity** for an introduction to spectral densities.

Below we quietly refit the ARMA model and use `psdensity` to estimate the parametric spectral density implied by the ARMA parameter estimates.

```
. quietly arima S12.log, ar(1) ma(2)
. psdensity d_arma omega1
```

The `psdensity` command above put the estimated ARMA spectral density into the new variable `d_arma` at the frequencies stored in the new variable `omega1`.

Below we quietly refit the ARFIMA model and use `psdensity` to estimate the long-run parametric spectral density and then the short-run parametric spectral density implied by the ARFIMA parameter estimates. The long-run estimates use the estimated d, and the short-run estimates set d to 0 (as is implied by specifying the `smemory` option). The long-run estimates describe the fractionally integrated series, and the short-run estimates describe the fractionally differenced series.

```
. quietly arfima S12.log, ar(1) ma(2)
. psdensity d_arfima omega2
. psdensity ds_arfima omega3, smemory
```

Now that we have the ARMA estimates, the long-run ARFIMA estimates, and the short-run ARFIMA estimates, we graph them below.

```
. line d_arma d_arfima  omega1, name(lmem) nodraw
. line d_arma ds_arfima omega1, name(smem) nodraw
. graph combine lmem smem, cols(1) xcommon
```

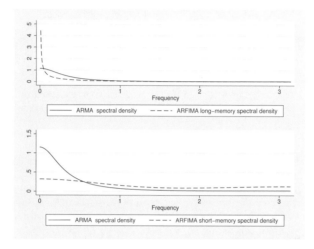

The top graph contains a plot of the spectral densities implied by the ARMA parameter estimates and by the long-run ARFIMA parameter estimates. As discussed by Granger and Joyeux (1980), the two models imply different spectral densities for frequencies close to 0 when $d > 0$. When $d > 0$, the spectral density implied by the ARFIMA estimates diverges to infinity, whereas the spectral density implied by the ARMA estimates remains finite at frequency 0 for stable ARMA processes. This difference reflects the ability of ARFIMA models to capture long-run effects that ARMA models only capture as the parameters approach those of an unstable model.

The bottom graph contains a plot of the spectral densities implied by the ARMA parameter estimates and by the short-run ARFIMA parameter estimates, which are the ARMA parameters for the fractionally differenced process. Comparing the two plots illustrates the ability of the short-run ARFIMA parameters to capture both low-frequency and high-frequency components in the fractionally differenced series. In contrast, the ARMA parameters captured only low-frequency components in the fractionally integrated series.

Comparing the ARFIMA and ARMA spectral densities in the two graphs illustrates that the additional fractional-difference parameter allows the ARFIMA model to identify both long-run and short-run effects, which the ARMA model confounds.

◁

❏ Technical note

As noted above, the spectral density of an ARFIMA process with $d > 0$ diverges to infinity as the frequency goes to 0. In contrast, the spectral density of an ARFIMA process with $d < 0$ is 0 at frequency 0.

The autocorrelation function of an ARFIMA process with $d < 0$ also decays at the slower hyperbolic rate. ARFIMA processes with $d < 0$ are sometimes called antipersistent because all the autocorrelations for lags greater than 0 are negative.

Hosking (1981), Baillie (1996), and others refer to ARFIMA processes with $d < 0$ as "intermediate memory" processes and ARFIMA processes with $d > 0$ as long-memory processes. Box, Jenkins, and Reinsel (2008, 429) define long-memory processes as those with the slower hyperbolic rate of decay, which includes ARFIMA processes with $d < 0$. We follow Box, Jenkins, and Reinsel (2008) and thus call ARFIMA processes for $-0.5 < d < 0$ and $0 < d < 0.5$ long-memory processes.

Sowell (1992a) uses the properties of ARFIMA processes with $d < 0$ to derive tests for whether a series was generated by an $I(1)$ process or an $I(d)$ process with $d < 1$.

❏

➢ Example 3

In this example, we use `arfima` to test whether a series is nonstationary. More specifically, we test whether the series was generated by an $I(1)$ process by testing whether the first difference of the series is overdifferenced.

We have monthly data on the log of the number of reported cases of mumps in New York City between January 1928 and December 1972. We believe that the series is stationary, after accounting for the monthly seasonal effects. We use an ARFIMA model for differenced series to test the null hypothesis of nonstationarity. We use the confidence interval for the d parameter from an ARFIMA model for the first difference of the log of the series to perform the test. If the right-hand end of the 95% CI is less than 0, we conclude that the differenced series was overdifferenced, which implies that the original series was not nonstationary.

More formally, if y_t is $I(1)$, then $\Delta y_t = y_t - y_{t-1}$ must be $I(0)$. If Δy_t is $I(d)$ with $d < 0$, then Δy_t is overdifferenced and y_t is $I(d)$ with $d < 1$.

We use seasonal indicators to account for the seasonal effects. In the output below, we specify the `mpl` option to use the MPL estimator that is less biased in the presence of covariates.

`arfima` computes the maximum likelihood estimates (MLE) for the parameters of this stationary and invertible Gaussian process. Alternatively, the maximum MPL estimates may be computed. See *Methods and formulas* for a description of these two estimation techniques, but suffice it to say that the MLE estimates for d are biased in the presence of exogenous variables, even the constant term, for small samples. The MPL estimator reduces this bias; see Hauser (1999) and Doornik and Ooms (2004).

```
. use http://www.stata-press.com/data/r12/mumps2, clear
(Hipel and Mcleod (1994), http://robjhyndman.com/tsdldata/epi/mumps.dat)

. arfima D.log i.month, ma(1 2) mpl
Iteration 0:   log modified profile likelihood =  53.766763
Iteration 1:   log modified profile likelihood =   54.38864
Iteration 2:   log modified profile likelihood =  54.934726  (backed up)
Iteration 3:   log modified profile likelihood =  54.937524  (backed up)
Iteration 4:   log modified profile likelihood =  55.002186
Iteration 5:   log modified profile likelihood =   55.20462
Iteration 6:   log modified profile likelihood =  55.205939
Iteration 7:   log modified profile likelihood =  55.205949
Iteration 8:   log modified profile likelihood =  55.205949
Refining estimates:
Iteration 0:   log modified profile likelihood =  55.205949
Iteration 1:   log modified profile likelihood =  55.205949
```

```
ARFIMA regression

Sample: 1928m2 - 1972m6                        Number of obs    =        533
                                               Wald chi2(14)    =    1360.28
Log modified profile likelihood =  55.205949   Prob > chi2      =     0.0000
```

D.log	Coef.	OIM Std. Err.	z	P>|z|	[95% Conf. Interval]	
D.log						
month						
2	-.220719	.0428112	-5.16	0.000	-.3046275	-.1368105
3	.0314683	.0424718	0.74	0.459	-.0517749	.1147115
4	-.2800296	.0460084	-6.09	0.000	-.3702043	-.1898548
5	-.3703179	.0449932	-8.23	0.000	-.4585029	-.2821329
6	-.4722035	.0446764	-10.57	0.000	-.5597676	-.3846394
7	-.9613239	.0448375	-21.44	0.000	-1.049204	-.873444
8	-1.063042	.0449272	-23.66	0.000	-1.151098	-.9749868
9	-.7577301	.0452529	-16.74	0.000	-.8464242	-.669036
10	-.3024251	.0462887	-6.53	0.000	-.3931494	-.2117009
11	-.0115317	.0426911	-0.27	0.787	-.0952046	.0721413
12	.0247135	.0430401	0.57	0.566	-.0596435	.1090705
_cons	.3656807	.0303215	12.06	0.000	.3062517	.4251096
ARFIMA						
ma						
L1.	.258056	.0684414	3.77	0.000	.1239134	.3921986
L2.	.1972011	.0506439	3.89	0.000	.0979409	.2964613
d	-.2329426	.067336	-3.46	0.001	-.3649187	-.1009664

We interpret the fact that the estimated 95% CI is strictly less than 0 to mean that the differenced series is overdifferenced, which implies that the original series is stationary.

◁

Saved results

arfima saves the following in e():

Scalars

e(N)	number of observations
e(k)	number of parameters
e(k_eq)	number of equations in e(b)
e(k_dv)	number of dependent variables
e(k_aux)	number of auxiliary parameters
e(df_m)	model degrees of freedom
e(ll)	log likelihood
e(chi2)	χ^2
e(p)	significance
e(s2)	idiosyncratic error variance estimate, if e(method) = mpl
e(tmin)	minimum time
e(tmax)	maximum time
e(ar_max)	maximum AR lag
e(ma_max)	maximum MA lag
e(rank)	rank of e(V)
e(ic)	number of iterations
e(rc)	return code
e(converged)	1 if converged, 0 otherwise
e(constant)	0 if noconstant, 1 otherwise

Macros
e(cmd)	arfima
e(cmdline)	command as typed
e(depvar)	name of dependent variable
e(covariates)	list of covariates
e(eqnames)	names of equations
e(title)	title in estimation output
e(tmins)	formatted minimum time
e(tmaxs)	formatted maximum time
e(chi2type)	Wald; type of model χ^2 test
e(vce)	*vcetype* specified in vce()
e(vcetype)	title used to label Std. Err.
e(ma)	lags for MA terms
e(ar)	lags for AR terms
e(technique)	maximization technique
e(tech_steps)	number of iterations performed before switching techniques
e(properties)	b V
e(estat_cmd)	program used to implement estat
e(predict)	program used to implement predict
e(marginsok)	predictions allowed by margins
e(marginsnotok)	predictions disallowed by margins

Matrices
e(b)	coefficient vector
e(Cns)	constraints matrix
e(ilog)	iteration log (up to 20 iterations)
e(gradient)	gradient vector
e(V)	variance–covariance matrix of the estimators
e(V_modelbased)	model-based variance

Functions
e(sample)	marks estimation sample

Methods and formulas

arfima is implemented as an ado-file.

Methods and formulas are presented under the following headings:

> *Introduction*
> *The likelihood function*
> *The autocovariance function*
> *The profile likelihood*
> *The MPL*

Introduction

We model an observed second-order stationary time-series y_t, $t = 1, \ldots, T$, using the ARFIMA(p, d, q) model defined as

$$\rho(L^p)(1 - L)^d (y_t - \mathbf{x}_t \boldsymbol{\beta}) = \boldsymbol{\theta}(L^q)\epsilon_t$$

where

$$\rho(L^p) = 1 - \rho_1 L - \rho_2 L^2 - \cdots - \rho_p L^p$$

$$\boldsymbol{\theta}(L^q) = 1 + \theta_1 L + \theta_2 L^2 + \cdots + \theta_q L^q$$

$$(1 - L)^d = \sum_{j=0}^{\infty} (-1)^j \frac{\Gamma(j + d)}{\Gamma(j + 1)\Gamma(d)} L^j$$

and the lag operator is defined as $L^j y_t = y_{t-j}$, $t = 1, \ldots, T$ and $j = 1, \ldots, t-1$; $\epsilon_t \sim N(0, \sigma^2)$; $\Gamma()$ is the gamma function; and $-0.5 < d < 0.5$, $d \neq 0$. The row vector $\mathbf{x}_t$ contains the exogenous variables specified as *indepvars* in the `arfima` syntax.

The process is stationary and invertible for $-0.5 < d < 0.5$; the roots of the AR polynomial, $\rho(z) = 1 - \rho_1 z - \rho_2 z^2 - \cdots - \rho_p z^p = 0$, and the MA polynomial, $\theta(z) = 1 + \theta_1 z + \theta_2 z^2 + \cdots + \theta_q z^q = 0$, lie outside the unit circle and there are no common roots. When $0 < d < 0.5$, the process has long memory in that the autocovariance function, γ_h, decays to 0 at a hyperbolic rate, such that $\sum_{h=-\infty}^{\infty} |\gamma_h| = \infty$. When $-0.5 < d < 0$, the process also has long memory in that the autocovariance function, γ_h, decays to 0 at a hyperbolic rate such that $\sum_{h=-\infty}^{\infty} |\gamma_h| < \infty$. (As discussed in the text, some authors refer to ARFIMA processes with $-0.5 < d < 0$ as having intermediate memory, but we follow Box, Jenkins, and Reinsel [2008] and refer to them as long-memory processes.)

Granger and Joyeux (1980), Hosking (1981), Sowell (1992b), Sowell (1992a), Baillie (1996), and Palma (2007) provide overviews of long-memory processes, fractional integration, and introductions to ARFIMA models.

The likelihood function

Estimation of the ARFIMA parameters ρ, θ, d, β and σ^2 is done by the method of maximum likelihood. The log Gaussian likelihood of $\mathbf{y}$ given parameter estimates $\widehat{\eta} = (\widehat{\rho}', \widehat{\theta}', \widehat{d}, \widehat{\beta}', \widehat{\sigma}^2)$ is

$$\ell(\mathbf{y}|\widehat{\eta}) = -\frac{1}{2}\left\{ T\log(2\pi) + \log|\widehat{\mathbf{V}}| + (\mathbf{y} - \mathbf{X}\widehat{\beta})'\widehat{\mathbf{V}}^{-1}(\mathbf{y} - \mathbf{X}\widehat{\beta}) \right\} \tag{2}$$

where the covariance matrix $\mathbf{V}$ has a Toeplitz structure

$$\mathbf{V} = \begin{pmatrix} \gamma_0 & \gamma_1 & \gamma_2 & \cdots & \gamma_{T-1} \\ \gamma_1 & \gamma_0 & \gamma_1 & \cdots & \gamma_{T-2} \\ \vdots & \vdots & \vdots & \ddots & \vdots \\ \gamma_{T-1} & \gamma_{T-2} & \gamma_{T-3} & \cdots & \gamma_0 \end{pmatrix}$$

$\text{Var}(y_t) = \gamma_0$, $\text{Cov}(y_t, y_{t-h}) = \gamma_h$ (for $h = 1, \ldots, t-1$), and $t = 1, \ldots, T$ (Sowell 1992b).

We use the Durbin–Levinson algorithm (Palma 2007; Golub and Van Loan 1996) to factor and invert $\mathbf{V}$. Using only the vector of autocovariances γ, the Durbin–Levinson algorithm will compute $\widehat{\epsilon} = \widehat{\mathbf{D}}^{-0.5}\widehat{\mathbf{L}}^{-1}(\mathbf{y} - \mathbf{X}\widehat{\beta})$, where $\mathbf{L}$ is lower triangular and $\mathbf{V} = \mathbf{L}\mathbf{D}\mathbf{L}'$ and $\mathbf{D} = \text{Diag}(\nu)$, $\nu_t = \text{Var}(y_t)$. The algorithm performs these computations without generating the $T \times T$ matrix $\mathbf{L}^{-1}$.

During optimization, we restrict the fractional-integration parameter to $(-0.5, 0.5)$ using a logistic transform, $d^* = \log\{(x + 0.5)/(0.5 - x)\}$, so that the range of d^* encompasses the real line. During the "Refining estimates" step, the fractional-integration parameter is transformed back to the restricted space, where we obtain its standard error from the observed information matrix.

The autocovariance function

Computation of the autocovariances γ_h is given by Sowell (1992b) with numerical enhancements by Doornik and Ooms (2003) and is reviewed by Palma (2007, sec. 3.2.4). We reproduce it here. The autocovariance of an ARFIMA$(0, d, 0)$ process is

$$\gamma_h^* = \sigma^2 \frac{\Gamma(1 - 2d)}{\Gamma(1 - d)\Gamma(d)} \frac{\Gamma(h + d)}{\Gamma(1 + h - d)}$$

where $h = 0, 1, \ldots$. For ARFIMA(p, d, q), we have

$$\gamma_h = \sigma^2 \sum_{i=-q}^{q} \sum_{j=1}^{p} \psi(i)\xi_j C(d, p + i - h, \rho_j) \tag{3}$$

where

$$\psi(i) = \sum_{k=\max(0,i)}^{\min(q,q+i)} \theta_k \theta_{k-i}$$

$$\xi_j = \left(\rho_j \prod_{i=1}^{p}(1 - \rho_i\rho_j) \prod_{m \neq j}(\rho_j - \rho_m) \right)^{-1}$$

and

$$C(d, h, \rho) = \frac{\gamma_h^*}{\sigma^2}\left\{ \rho^{2p} F(d + h, 1, 1 - d + h, \rho) + F(d - h, 1, 1 - d - h, \rho) - 1 \right\}$$

$F()$ is the hypergeometric series (Gradshteyn and Ryzhik 2007)

$$F(a, b, c, x) = 1 + \frac{ab}{c \cdot 1}x + \frac{a(a + 1)b(b + 1)}{c(c + 1) \cdot 1 \cdot 2}x^2 + \frac{a(a + 1)(a + 2)b(b + 1)(b + 2)}{c(c + 1)(c + 2) \cdot 1 \cdot 2 \cdot 3}x^3 + \ldots$$

The series recursions are evaluated backward as Doornik and Ooms (2003) emphasize. Doornik and Ooms (2003) also provide other computational enhancements, such as not dividing by ρ_j in (3).

The profile likelihood

Doornik and Ooms (2003) show that the parameters σ^2 and β can be concentrated out of the likelihood. Using (2), the MLE for σ^2 is

$$\widehat{\sigma}^2 = \frac{1}{T}(\mathbf{y} - \mathbf{X}\widehat{\beta})'\widehat{\mathbf{R}}^{-1}(\mathbf{y} - \mathbf{X}\widehat{\beta}) \tag{4}$$

where $\mathbf{R} = \frac{1}{\sigma^2}\mathbf{V}$ and

$$\widehat{\beta} = (\mathbf{X}'\widehat{\mathbf{R}}^{-1}\mathbf{X})^{-1}\mathbf{X}'\widehat{\mathbf{R}}^{-1}\mathbf{y} \tag{5}$$

is the weighted least-squares estimates for β. Substituting (4) into (2) results in the profile likelihood

$$\ell_p(\mathbf{y}|\widehat{\boldsymbol{\eta}}_r) = -\frac{T}{2}\left\{ 1 + \log(2\pi) + \frac{1}{T}\log|\widehat{\mathbf{R}}| + \log\widehat{\sigma}^2 \right\}$$

We compute the MLEs using the profile likelihood for the reduced parameter set $\boldsymbol{\eta}_r = (\boldsymbol{\rho}', \boldsymbol{\theta}', d)$. Equations (4) and (5) provide MLEs for σ^2 and β to create the full parameter vector $\boldsymbol{\eta} = (\boldsymbol{\beta}', \boldsymbol{\rho}', \boldsymbol{\theta}', d, \sigma^2)$. We follow with the "Refining estimates" step, optimizing on the log likelihood (1). The refining step does not change the estimates; it produces the coefficient variance–covariance matrix from the observed information matrix.

Using this profile likelihood prevents the use of the BHHH optimization method because there are no observation-level scores.

The MPL

The small-sample MLE for d can be biased when there are exogenous variables in the model. The MPL reduces this bias (Hauser 1999; Doornik and Ooms 2004). The `mpl` option will direct `arfima` to use this optimization criterion. The MPL is expressed as

$$\ell_m(\mathbf{y}|\widehat{\boldsymbol{\eta}}_r) = -\frac{T}{2}\left\{1 + \log(2\pi)\right\} - \left(\frac{1}{T} - \frac{1}{2}\right)\log|\widehat{\mathbf{R}}| - \left(\frac{T-k-2}{2}\right)\log\widehat{\sigma}^2 - \frac{1}{2}\log|\mathbf{X}'\widehat{\mathbf{R}}^{-1}\mathbf{X}|$$

where $k = \mathrm{rank}(\mathbf{X})$ (An and Bloomfield 1993).

There is no MPL estimator for σ^2, and you will notice its absence from the coefficient table. However, the unbiased estimate assuming ARFIMA$(0, 0, 0)$,

$$\widetilde{\sigma}^2 = \frac{(\mathbf{y} - \mathbf{X}\widehat{\boldsymbol{\beta}})'\widehat{\mathbf{R}}^{-1}(\mathbf{y} - \mathbf{X}\widehat{\boldsymbol{\beta}})}{T - k}$$

is stored in `e()` for postestimation computation of the forecast and residual root mean squared errors.

References

An, S., and P. Bloomfield. 1993. Cox and Reid's modification in regression models with correlated errors. Technical report, Department of Statistics, North Carolina State University, Raleigh, NC.

Baillie, R. T. 1996. Long memory processes and fractional integration in econometrics. *Journal of Econometrics* 73: 5–59.

Beran, J. 1994. *Statistics for Long-Memory Processes.* Boca Raton: Chapman & Hall/CRC.

Box, G. E. P., G. M. Jenkins, and G. C. Reinsel. 2008. *Time Series Analysis: Forecasting and Control.* 4th ed. Hoboken, NJ: Wiley.

Doornik, J. A., and M. Ooms. 2003. Computational aspects of maximum likelihood estimation of autoregressive fractionally integrated moving average models. *Computational Statistics & Data Analysis* 42: 333–348.

———. 2004. Inference and forecasting for ARFIMA models with an application to US and UK inflation. *Studies in Nonlinear Dynamics & Econometrics* 8: 1–23.

Golub, G. H., and C. F. Van Loan. 1996. *Matrix Computations.* 3rd ed. Baltimore: Johns Hopkins University Press.

Gradshteyn, I. S., and I. M. Ryzhik. 2007. *Table of Integrals, Series, and Products.* 7th ed. San Diego: Elsevier.

Granger, C. W. J., and R. Joyeux. 1980. An introduction to long-memory time series models and fractional differencing. *Journal of Time Series Analysis* 1: 15–29.

Hamilton, J. D. 1994. *Time Series Analysis.* Princeton: Princeton University Press.

Hauser, M. A. 1999. Maximum likelihood estimators for ARMA and ARFIMA models: a Monte Carlo study. *Journal of Statistical Planning and Inference* 80: 229–255.

Hosking, J. R. M. 1981. Fractional differencing. *Biometrika* 68: 165–176.

Hurst, H. E. 1951. Long-term storage capacity of reservoirs. *Transactions of the American Society of Civil Engineers* 116: 770–779.

Palma, W. 2007. *Long-Memory Time Series: Theory and Methods.* Hoboken, NJ: Wiley.

Sowell, F. 1992a. Modeling long-run behavior with the fractional ARIMA model. *Journal of Monetary Economics* 29: 277–302.

———. 1992b. Maximum likelihood estimation of stationary univariate fractionally integrated time series models. *Journal of Econometrics* 53: 165–188.

Also see

[TS] **arfima postestimation** — Postestimation tools for arfima

[TS] **tsset** — Declare data to be time-series data

[TS] **arima** — ARIMA, ARMAX, and other dynamic regression models

[TS] **sspace** — State-space models

[U] **20 Estimation and postestimation commands**

Title

> **arfima postestimation** — Postestimation tools for arfima

Description

The following postestimation command is of special interest after `arfima`:

Command	Description
psdensity	estimate the spectral density

For information about `psdensity`, see [TS] **psdensity**.

The following standard postestimation commands are also available:

Command	Description
contrast	contrasts and ANOVA-style joint tests of estimates
*estat	AIC, BIC, VCE, and estimation sample summary
estimates	cataloging estimation results
lincom	point estimates, standard errors, testing, and inference for linear combinations of coefficients
lrtest	likelihood-ratio test
*margins	marginal means, predictive margins, marginal effects, and average marginal effects
*marginsplot	graph the results from margins (profile plots, interaction plots, etc.)
*nlcom	point estimates, standard errors, testing, and inference for nonlinear combinations of coefficients
predict	predictions, residuals, influence statistics, and other diagnostic measures
*predictnl	point estimates, standard errors, testing, and inference for generalized predictions
pwcompare	pairwise comparisons of estimates
test	Wald tests of simple and composite linear hypotheses
testnl	Wald tests of nonlinear hypotheses

* `estat ic`, `margins`, `marginsplot`, `nlcom`, and `predictnl` are not appropriate after `arfima, mpl`.

See the corresponding entries in the *Base Reference Manual* for details.

Syntax for predict

> predict [*type*] *newvar* [*if*] [*in*] [, *statistic options*]

statistic	Description
Main	
xb	predicted values; the default
residuals	predicted innovations
rstandard	standardized innovations
fdifference	fractionally differenced series

These statistics are available both in and out of sample; type `predict ... if e(sample) ...` if wanted only for the estimation sample.

options	Description
Options	
<u>rmse</u>([*type*] *newvar*)	put the estimated root mean squared error of the predicted statistic in a new variable; only permitted with options xb and residuals
<u>dynamic</u>(*datetime*)	forecast the time series starting at *datetime*; only permitted with option xb

datetime is a # or a time literal, such as td(1jan1995) or tq(1995q1); see [D] **datetime**.

Menu

Statistics > Postestimation > Predictions, residuals, etc.

Options for predict

⎡ Main ⎤

xb, the default, calculates the predictions for the level of *depvar*.

residuals calculates the predicted innovations.

rstandard calculates the standardized innovations.

fdifference calculates the fractionally differenced predictions of *depvar*.

⎡ Options ⎤

rmse([*type*] *newvar*) puts the root mean squared errors of the predicted statistics into the specified new variables. The root mean squared errors measure the variances due to the disturbances but do not account for estimation error. rmse() is only permitted with the xb and residuals options.

dynamic(*datetime*) specifies when predict starts producing dynamic forecasts. The specified *datetime* must be in the scale of the time variable specified in tsset, and the *datetime* must be inside a sample for which observations on the dependent variables are available. For example, dynamic(tq(2008q4)) causes dynamic predictions to begin in the fourth quarter of 2008, assuming that your time variable is quarterly; see [D] **datetime**. If the model contains exogenous variables, they must be present for the whole predicted sample. dynamic() may only be specified with xb.

Remarks

We assume that you have already read [TS] **arfima**. In this entry, we illustrate some of the features of predict after fitting an ARFIMA model using arfima.

▷ Example 1

We have monthly data on the one-year Treasury bill secondary market rate imported from the Federal Reserve Bank (FRED) database using freduse; see Drukker (2006) for an introduction to freduse. Below we fit an ARFIMA model with two autoregressive terms and one moving-average term to the data.

```
. use http://www.stata-press.com/data/r12/tb1yr
(FRED,  1-year treasury bill; secondary market rate, monthy 1959-2001)

. arfima tb1yr, ar(1/2) ma(1)
Iteration 0:   log likelihood = -235.31856
Iteration 1:   log likelihood = -235.26104  (backed up)
Iteration 2:   log likelihood = -235.25974  (backed up)
Iteration 3:   log likelihood =  -235.2544  (backed up)
Iteration 4:   log likelihood = -235.13353
Iteration 5:   log likelihood = -235.13065
Iteration 6:   log likelihood = -235.12107
Iteration 7:   log likelihood = -235.11917
Iteration 8:   log likelihood = -235.11869
Iteration 9:   log likelihood = -235.11868
Refining estimates:
Iteration 0:   log likelihood = -235.11868
Iteration 1:   log likelihood = -235.11868

ARFIMA regression
```

Sample: 1959m7 - 2001m8

Log likelihood = -235.11868

Number of obs   =   506
Wald chi2(4)   =   1864.15
Prob > chi2   =   0.0000

tb1yr	Coef.	OIM Std. Err.	z	P>|z|	[95% Conf. Interval]	
tb1yr						
_cons	5.496708	2.92036	1.88	0.060	-.2270925	11.22051
ARFIMA						
ar						
L1.	.2326105	.1136696	2.05	0.041	.0098222	.4553989
L2.	.3885212	.0835657	4.65	0.000	.2247354	.5523071
ma						
L1.	.7755849	.0669566	11.58	0.000	.6443524	.9068174
d	.460649	.0646564	7.12	0.000	.3339247	.5873733
/sigma2	.1466495	.009232	15.88	0.000	.1285551	.1647439

Note: The test of the variance against zero is one sided, and the two-sided
      confidence interval is truncated at zero.

All the parameters are statistically significant at the 5% level, and they indicate a high degree of dependence in the series. In fact, the confidence interval for the fractional-difference parameter d indicates that the series may be nonstationary. We will proceed as if the series is stationary and suppose that it is fractionally integrated of order 0.46.

We begin our postestimation analysis by predicting the series in sample:

```
. predict ptb
(option xb assumed)
```

We continue by using the estimated fractional-difference parameter to fractionally difference the original series and by plotting the original series, the predicted series, and the fractionally differenced series. See [TS] **arfima** for a definition of the fractional-difference operator.

```
. predict fdtb, fdifference
. twoway tsline tb1yr ptb fdtb, legend(cols(1))
```

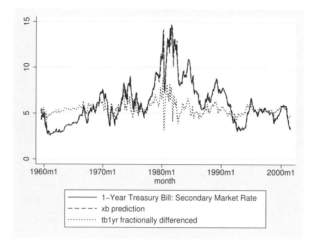

The above graph shows that the in-sample predictions appear to track the original series well and that the fractionally differenced series looks much more like a stationary series than does the original.

◁

▷ Example 2

In this example, we use the above estimates to produce a dynamic forecast and a confidence interval for the forecast for the one-year treasury bill rate and plot them.

We begin by extending the dataset and using `predict` to put the dynamic forecast in the new `ftb` variable and the root mean squared error of the forecast in the new `rtb` variable. (As discussed in *Methods and formulas*, the root mean squared error of the forecast accounts for the idiosyncratic error but not for the estimation error.)

```
. tsappend, add(12)
. predict ftb, xb dynamic(tm(2001m9)) rmse(rtb)
```

Now we compute a 90% confidence interval around the dynamic forecast and plot the original series, the in-sample forecast, the dynamic forecast, and the confidence interval of the dynamic forecast.

```
. scalar z = invnormal(0.95)

. generate lb = ftb - z*rtb if month>=tm(2001m9)
(506 missing values generated)

. generate ub = ftb + z*rtb if month>=tm(2001m9)
(506 missing values generated)

. twoway tsline tb1yr ftb if month>tm(1998m12) ||
>          tsrline lb ub if month>=tm(2001m9),
>          legend(cols(1) label(3 "90% prediction interval"))
```

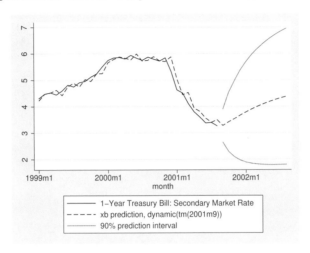

◁

Methods and formulas

All postestimation commands listed above are implemented as ado-files.

Denote γ_h, $h = 1, \ldots, t$, to be the autocovariance function of the ARFIMA(p, d, q) process for two observations, y_t and y_{t-h}, h time periods apart. The covariance matrix $\mathbf{V}$ of the process of length T has a Toeplitz structure of

$$
\mathbf{V} = \begin{pmatrix}
\gamma_0 & \gamma_1 & \gamma_2 & \cdots & \gamma_{T-1} \\
\gamma_1 & \gamma_0 & \gamma_1 & \cdots & \gamma_{T-2} \\
\vdots & \vdots & \vdots & \ddots & \vdots \\
\gamma_{T-1} & \gamma_{T-2} & \gamma_{T-3} & \cdots & \gamma_0
\end{pmatrix}
$$

where the process variance is $\gamma_0 = \mathrm{Var}(y_t)$. We factor $\mathbf{V} = \mathbf{LDL}'$, where $\mathbf{L}$ is lower triangular and $\mathbf{D} = \mathrm{Diag}(\nu_t)$. The structure of $\mathbf{L}^{-1}$ is of importance.

$$
\mathbf{L}^{-1} = \begin{pmatrix}
1 & 0 & 0 & \cdots & 0 & 0 \\
-\tau_{1,1} & 1 & 0 & \cdots & 0 & 0 \\
-\tau_{2,2} & -\tau_{2,1} & 1 & \cdots & 0 & 0 \\
\vdots & \vdots & \vdots & \ddots & \vdots & \vdots \\
-\tau_{T-1,T-1} & -\tau_{T-1,T-2} & -\tau_{T-1,T-2} & \cdots & -\tau_{T-1,1} & 1
\end{pmatrix}
$$

Let $z_t = y_t - \mathbf{x}_t\boldsymbol{\beta}$. The best linear predictor of z_{t+1} based on $z_1, z_2, \ldots, z_t$ is $\widehat{z}_{t+1} = \sum_{k=1}^{t} \tau_{t,k} z_{t-k+1}$. Define $-\boldsymbol{\tau}_t = (-\tau_{t,t}, -\tau_{t,t-1}, \ldots, -\tau_{t-1,1})$ to be the tth row of $\mathbf{L}^{-1}$ up to, but not including, the diagonal. Then $\boldsymbol{\tau}_t = \mathbf{V}_t^{-1}\boldsymbol{\gamma}_t$, where $\mathbf{V}_t$ is the $t \times t$ upper left submatrix of $\mathbf{V}$ and $\boldsymbol{\gamma}_t = (\gamma_1, \gamma_2, \ldots, \gamma_t)'$. Hence, the best linear predictor of the innovations is computed as $\widehat{\epsilon} = \mathbf{L}^{-1}\mathbf{z}$, and the one-step predictions are $\widehat{\mathbf{y}} = \widehat{\epsilon} + \mathbf{X}\widehat{\boldsymbol{\beta}}$. In practice, the computation is

$$\widehat{\mathbf{y}} = \widehat{\mathbf{L}}^{-1}\left(\mathbf{y} - \mathbf{X}\widehat{\boldsymbol{\beta}}\right) + \mathbf{X}\widehat{\boldsymbol{\beta}}$$

where $\widehat{\mathbf{L}}$ and $\widehat{\mathbf{V}}$ are computed from the maximum likelihood estimates. We use the Durbin–Levinson algorithm (Palma 2007; Golub and Van Loan 1996) to factor $\widehat{\mathbf{V}}$, invert $\widehat{\mathbf{L}}$, and scale $\mathbf{y} - \mathbf{X}\widehat{\boldsymbol{\beta}}$ using only the vector of estimated autocovariances $\widehat{\boldsymbol{\gamma}}$.

The prediction error variances of the one-step predictions are computed recursively in the Durbin–Levinson algorithm. They are the ν_t elements in the diagonal matrix $\mathbf{D}$ computed from the Cholesky factorization of $\mathbf{V}$. The recursive formula is $\nu_0 = \gamma_0$, and $\nu_t = \nu_{t-1}(1 - \tau_{t,t}^2)$.

Forecasting is carried out as described by Beran (1994, sec. 8.7), $\widehat{\mathbf{z}}_{T+k} = \widetilde{\boldsymbol{\gamma}}_k' \widehat{\mathbf{V}}^{-1}\widetilde{\mathbf{z}}$, where $\widetilde{\boldsymbol{\gamma}}_k' = (\widehat{\gamma}_{T+k-1}, \widehat{\gamma}_{T+k-2}, \ldots, \widehat{\gamma}_k)$. The forecast mean squared error is computed as $\text{MSE}(\widehat{\mathbf{z}}_{T+k}) = \widehat{\gamma}_0 - \widetilde{\boldsymbol{\gamma}}_k' \widehat{\mathbf{V}}^{-1}\widetilde{\boldsymbol{\gamma}}_k$. Computation of $\widehat{\mathbf{V}}^{-1}\widetilde{\boldsymbol{\gamma}}_k$ is carried out efficiently using algorithm 4.7.2 of Golub and Van Loan (1996).

References

Beran, J. 1994. *Statistics for Long-Memory Processes*. Boca Raton: Chapman & Hall/CRC.

Drukker, D. M. 2006. Importing Federal Reserve economic data. *Stata Journal* 6: 384–386.

Golub, G. H., and C. F. Van Loan. 1996. *Matrix Computations*. 3rd ed. Baltimore: Johns Hopkins University Press.

Palma, W. 2007. *Long-Memory Time Series: Theory and Methods*. Hoboken, NJ: Wiley.

Also see

[TS] **arfima** — Autoregressive fractionally integrated moving-average models

[TS] **psdensity** — Parametric spectral density estimation after arima, arfima, and ucm

[U] **20 Estimation and postestimation commands**

Title

> **arima** — ARIMA, ARMAX, and other dynamic regression models

Syntax

Basic syntax for a regression model with ARMA *disturbances*

> arima *depvar* [*indepvars*] , ar(*numlist*) ma(*numlist*)

Basic syntax for an ARIMA(p, d, q) *model*

> arima *depvar* , arima($\#_p, \#_d, \#_q$)

Basic syntax for a multiplicative seasonal ARIMA$(p, d, q) \times (P, D, Q)_s$ *model*

> arima *depvar* , arima($\#_p, \#_d, \#_q$) sarima($\#_P, \#_D, \#_Q, \#_s$)

Full syntax

> arima *depvar* [*indepvars*] [*if*] [*in*] [*weight*] [, *options*]

options	Description
Model	
<u>noc</u>onstant	suppress constant term
arima($\#_p, \#_d, \#_q$)	specify ARIMA(p, d, q) model for dependent variable
ar(*numlist*)	autoregressive terms of the structural model disturbance
ma(*numlist*)	moving-average terms of the structural model disturbance
<u>c</u>onstraints(*constraints*)	apply specified linear constraints
<u>col</u>linear	keep collinear variables
Model 2	
sarima($\#_P, \#_D, \#_Q, \#_s$)	specify period-$\#_s$ multiplicative seasonal ARIMA term
mar(*numlist*, $\#_s$)	multiplicative seasonal autoregressive term; may be repeated
mma(*numlist*, $\#_s$)	multiplicative seasonal moving-average term; may be repeated
Model 3	
<u>cond</u>ition	use conditional MLE instead of full MLE
<u>save</u>space	conserve memory during estimation
<u>diff</u>use	use diffuse prior for starting Kalman filter recursions
p0(# | *matname*)	use alternate prior for starting Kalman recursions; seldom used
state0(# | *matname*)	use alternate state vector for starting Kalman filter recursions
SE/Robust	
vce(*vcetype*)	*vcetype* may be opg, <u>r</u>obust, or oim

Reporting

level(#)	set confidence level; default is level(95)
detail	report list of gaps in time series
nocnsreport	do not display constraints
display_options	control column formats, row spacing, and line width

Maximization

maximize_options	control the maximization process; seldom used
coeflegend	display legend instead of statistics

You must tsset your data before using arima; see [TS] **tsset**.

depvar and *indepvars* may contain time-series operators; see [U] **11.4.4 Time-series varlists**.

by, rolling, statsby, and xi are allowed; see [U] **11.1.10 Prefix commands**.

iweights are allowed; see [U] **11.1.6 weight**.

coeflegend does not appear in the dialog box.

See [U] **20 Estimation and postestimation commands** for more capabilities of estimation commands.

Menu

Statistics > Time series > ARIMA and ARMAX models

Description

arima fits univariate models with time-dependent disturbances. arima fits a model of *depvar* on *indepvars* where the disturbances are allowed to follow a linear autoregressive moving-average (ARMA) specification. The dependent and independent variables may be differenced or seasonally differenced to any degree. When independent variables are included in the specification, such models are often called ARMAX models; and when independent variables are not specified, they reduce to Box–Jenkins autoregressive integrated moving-average (ARIMA) models in the dependent variable. Multiplicative seasonal ARMAX and ARIMA models can also be fit. Missing data are allowed and are handled using the Kalman filter and methods suggested by Harvey (1989 and 1993); see *Methods and formulas*.

In the full syntax, *depvar* is the variable being modeled, and the structural or regression part of the model is specified in *indepvars*. ar() and ma() specify the lags of autoregressive and moving-average terms, respectively; and mar() and mma() specify the multiplicative seasonal autoregressive and moving-average terms, respectively.

arima allows time-series operators in the dependent variable and independent variable lists, and making extensive use of these operators is often convenient; see [U] **11.4.4 Time-series varlists** and [U] **13.9 Time-series operators** for an extended discussion of time-series operators.

arima typed without arguments redisplays the previous estimates.

Options

    Model

noconstant; see [R] **estimation options**.

arima($\#_p$,$\#_d$,$\#_q$) is an alternative, shorthand notation for specifying models with ARMA disturbances. The dependent variable and any independent variables are differenced $\#_d$ times, and 1 through $\#_p$ lags of autocorrelations and 1 through $\#_q$ lags of moving averages are included in the model. For example, the specification

        . arima D.y, ar(1/2) ma(1/3)

is equivalent to

        . arima y, arima(2,1,3)

The latter is easier to write for simple ARMAX and ARIMA models, but if gaps in the AR or MA lags are to be modeled, or if different operators are to be applied to independent variables, the first syntax is required.

ar(*numlist*) specifies the autoregressive terms of the structural model disturbance to be included in the model. For example, ar(1/3) specifies that lags of 1, 2, and 3 of the structural disturbance be included in the model; ar(1 4) specifies that lags 1 and 4 be included, perhaps to account for additive quarterly effects.

If the model does not contain regressors, these terms can also be considered autoregressive terms for the dependent variable.

ma(*numlist*) specifies the moving-average terms to be included in the model. These are the terms for the lagged innovations (white-noise disturbances).

constraints(*constraints*), collinear; see [R] **estimation options**.

If constraints are placed between structural model parameters and ARMA terms, the first few iterations may attempt steps into nonstationary areas. This process can be ignored if the final solution is well within the bounds of stationary solutions.

⌐ Model 2 ⌐

sarima(#$_P$,#$_D$,#$_Q$,#$_s$) is an alternative, shorthand notation for specifying the multiplicative seasonal components of models with ARMA disturbances. The dependent variable and any independent variables are lag-#$_s$ seasonally differenced #$_D$ times, and 1 through #$_P$ seasonal lags of autoregressive terms and 1 through #$_Q$ seasonal lags of moving-average terms are included in the model. For example, the specification

        . arima DS12.y, ar(1/2) ma(1/3) mar(1/2,12) mma(1/2,12)

is equivalent to

        . arima y, arima(2,1,3) sarima(2,1,2,12)

mar(*numlist*, #$_s$) specifies the lag-#$_s$ multiplicative seasonal autoregressive terms. For example, mar(1/2,12) requests that the first two lag-12 multiplicative seasonal autoregressive terms be included in the model.

mma(*numlist*, #$_s$) specified the lag-#$_s$ multiplicative seasonal moving-average terms. For example, mma(1 3,12) requests that the first and third (but not the second) lag-12 multiplicative seasonal moving-average terms be included in the model.

⌐ Model 3 ⌐

condition specifies that conditional, rather than full, maximum likelihood estimates be produced. The presample values for ϵ_t and μ_t are taken to be their expected value of zero, and the estimate of the variance of ϵ_t is taken to be constant over the entire sample; see Hamilton (1994, 132). This estimation method is not appropriate for nonstationary series but may be preferable for long series or for models that have one or more long AR or MA lags. diffuse, p0(), and state0() have no meaning for models fit from the conditional likelihood and may not be specified with condition.

If the series is long and stationary and the underlying data-generating process does not have a long memory, estimates will be similar, whether estimated by unconditional maximum likelihood (the default), conditional maximum likelihood (`condition`), or maximum likelihood from a diffuse prior (`diffuse`).

In small samples, however, results of conditional and unconditional maximum likelihood may differ substantially; see Ansley and Newbold (1980). Whereas the default unconditional maximum likelihood estimates make the most use of sample information when all the assumptions of the model are met, Harvey (1989) and Ansley and Kohn (1985) argue for diffuse priors often, particularly in ARIMA models corresponding to an underlying structural model.

The `condition` or `diffuse` options may also be preferred when the model contains one or more long AR or MA lags; this avoids inverting potentially large matrices (see `diffuse` below).

When `condition` is specified, estimation is performed by the `arch` command (see [TS] **arch**), and more control of the estimation process can be obtained using `arch` directly.

`condition` cannot be specified if the model contains any multiplicative seasonal terms.

savespace specifies that memory use be conserved by retaining only those variables required for estimation. The original dataset is restored after estimation. This option is rarely used and should be used only if there is not enough space to fit a model without the option. However, `arima` requires considerably more temporary storage during estimation than most estimation commands in Stata.

diffuse specifies that a diffuse prior (see Harvey 1989 or 1993) be used as a starting point for the Kalman filter recursions. Using `diffuse`, nonstationary models may be fit with `arima` (see the p0() option below; `diffuse` is equivalent to specifying p0(1e9)).

By default, `arima` uses the unconditional expected value of the state vector ξ_t (see *Methods and formulas*) and the mean squared error (MSE) of the state vector to initialize the filter. When the process is stationary, this corresponds to the expected value and expected variance of a random draw from the state vector and produces unconditional maximum likelihood estimates of the parameters. When the process is not stationary, however, this default is not appropriate, and the unconditional MSE cannot be computed. For a nonstationary process, another starting point must be used for the recursions.

In the absence of nonsample or presample information, `diffuse` may be specified to start the recursions from a state vector of zero and a state MSE matrix corresponding to an effectively infinite variance on this initial state. This method amounts to an uninformative and improper prior that is updated to a proper MSE as data from the sample become available; see Harvey (1989).

Nonstationary models may also correspond to models with infinite variance given a particular specification. This and other problems with nonstationary series make convergence difficult and sometimes impossible.

diffuse can also be useful if a model contains one or more long AR or MA lags. Computation of the unconditional MSE of the state vector (see *Methods and formulas*) requires construction and inversion of a square matrix that is of dimension $\{\max(p, q + 1)\}^2$, where p and q are the maximum AR and MA lags, respectively. If $q = 27$, for example, we would require a 784-by-784 matrix. Estimation with `diffuse` does not require this matrix.

For large samples, there is little difference between using the default starting point and the `diffuse` starting point. Unless the series has a long memory, the initial conditions affect the likelihood of only the first few observations.

p0(*#* | *matname*) is a rarely specified option that can be used for nonstationary series or when an alternate prior for starting the Kalman recursions is desired (see diffuse above for a discussion of the default starting point and *Methods and formulas* for background).

matname specifies a matrix to be used as the MSE of the state vector for starting the Kalman filter recursions—$P_{1|0}$. Instead, one number, *#*, may be supplied, and the MSE of the initial state vector $P_{1|0}$ will have this number on its diagonal and all off-diagonal values set to zero.

This option may be used with nonstationary series to specify a larger or smaller diagonal for $P_{1|0}$ than that supplied by diffuse. It may also be used with state0() when you believe that you have a better prior for the initial state vector and its MSE.

state0(*#* | *matname*) is a rarely used option that specifies an alternate initial state vector, $\xi_{1|0}$ (see *Methods and formulas*), for starting the Kalman filter recursions. If *#* is specified, all elements of the vector are taken to be *#*. The default initial state vector is state0(0).

SE/Robust

vce(*vcetype*) specifies the type of standard error reported, which includes types that are robust to some kinds of misspecification and that are derived from asymptotic theory; see [R] **vce_option**.

For state-space models in general and ARMAX and ARIMA models in particular, the robust or quasi−maximum likelihood estimates (QMLEs) of variance are robust to symmetric nonnormality in the disturbances, including, as a special case, heteroskedasticity. The robust variance estimates are not generally robust to functional misspecification of the structural or ARMA components of the model; see Hamilton (1994, 389) for a brief discussion.

Reporting

level(*#*); see [R] **estimation options**.

detail specifies that a detailed list of any gaps in the series be reported, including gaps due to missing observations or missing data for the dependent variable or independent variables.

nocnsreport; see [R] **estimation options**.

display_options: vsquish, cformat(%*fmt*), pformat(%*fmt*), sformat(%*fmt*), and nolstretch; see [R] **estimation options**.

Maximization

maximize_options: difficult, technique(*algorithm_spec*), iterate(*#*), [no]log, trace, gradient, showstep, hessian, showtolerance, tolerance(*#*), ltolerance(*#*), nrtolerance(*#*), gtolerance(*#*), nonrtolerance(*#*), and from(*init_specs*); see [R] **maximize** for all options except gtolerance(), and see below for information on gtolerance().

These options are sometimes more important for ARIMA models than most maximum likelihood models because of potential convergence problems with ARIMA models, particularly if the specified model and the sample data imply a nonstationary model.

Several alternate optimization methods, such as Berndt–Hall–Hall–Hausman (BHHH) and Broyden– Fletcher–Goldfarb–Shanno (BFGS), are provided for ARIMA models. Although ARIMA models are not as difficult to optimize as ARCH models, their likelihoods are nevertheless generally not quadratic and often pose optimization difficulties; this is particularly true if a model is nonstationary or nearly nonstationary. Because each method approaches optimization differently, some problems can be successfully optimized by an alternate method when one method fails.

Setting technique() to something other than the default or BHHH changes the *vcetype* to vce(oim).

The following options are all related to maximization and are either particularly important in fitting ARIMA models or not available for most other estimators.

technique(*algorithm_spec*) specifies the optimization technique to use to maximize the likelihood function.

technique(bhhh) specifies the Berndt–Hall–Hall–Hausman (BHHH) algorithm.

technique(dfp) specifies the Davidon–Fletcher–Powell (DFP) algorithm.

technique(bfgs) specifies the Broyden–Fletcher–Goldfarb–Shanno (BFGS) algorithm.

technique(nr) specifies Stata's modified Newton–Raphson (NR) algorithm.

You can specify multiple optimization methods. For example,

    technique(bhhh 10 nr 20)

requests that the optimizer perform 10 BHHH iterations, switch to Newton–Raphson for 20 iterations, switch back to BHHH for 10 more iterations, and so on.

The default for arima is technique(bhhh 5 bfgs 10).

gtolerance(#) specifies the tolerance for the gradient relative to the coefficients. When $|g_i \, b_i| \leq$ gtolerance() for all parameters b_i and the corresponding elements of the gradient g_i, the gradient tolerance criterion is met. The default gradient tolerance for arima is gtolerance(.05).

gtolerance(999) may be specified to disable the gradient criterion. If the optimizer becomes stuck with repeated "(backed up)" messages, the gradient probably still contains substantial values, but an uphill direction cannot be found for the likelihood. With this option, results can often be obtained, but whether the global maximum likelihood has been found is unclear.

When the maximization is not going well, it is also possible to set the maximum number of iterations (see [R] **maximize**) to the point where the optimizer appears to be stuck and to inspect the estimation results at that point.

from(*init_specs*) allows you to set the starting values of the model coefficients; see [R] **maximize** for a general discussion and syntax options.

The standard syntax for from() accepts a matrix, a list of values, or coefficient name value pairs; see [R] **maximize**. arima also accepts from(armab0), which sets the starting value for all ARMA parameters in the model to zero prior to optimization.

ARIMA models may be sensitive to initial conditions and may have coefficient values that correspond to local maximums. The default starting values for arima are generally good, particularly in large samples for stationary series.

The following option is available with arima but is not shown in the dialog box:

coeflegend; see [R] **estimation options**.

Remarks

Remarks are presented under the following headings:

> *Introduction*
> *ARIMA models*
> *Multiplicative seasonal ARIMA models*
> *ARMAX models*
> *Dynamic forecasting*

Introduction

arima fits both standard ARIMA models that are autoregressive in the dependent variable and structural models with ARMA disturbances. Good introductions to the former models can be found in Box, Jenkins, and Reinsel (2008); Hamilton (1994); Harvey (1993); Newton (1988); Diggle (1990); and many others. The latter models are developed fully in Hamilton (1994) and Harvey (1989), both of which provide extensive treatment of the Kalman filter (Kalman 1960) and the state-space form used by arima to fit the models.

Consider a first-order autoregressive moving-average process. Then arima estimates all the parameters in the model

$$y_t = \mathbf{x}_t \boldsymbol{\beta} + \mu_t \qquad \qquad \textit{structural equation}$$
$$\mu_t = \rho \mu_{t-1} + \theta \epsilon_{t-1} + \epsilon_t \qquad \qquad \textit{disturbance}, \text{ARMA}(1,1)$$

where

ρ	is the first-order autocorrelation parameter
θ	is the first-order moving-average parameter
ϵ_t	$\sim$ *i.i.d.* $N(0, \sigma^2)$, meaning that ϵ_t is a white-noise disturbance

You can combine the two equations and write a general ARMA(p, q) in the disturbances process as

$$y_t = \mathbf{x_t}\boldsymbol{\beta} + \rho_1(y_{t-1} - \mathbf{x}_{t-1}\boldsymbol{\beta}) + \rho_2(y_{t-2} - \mathbf{x}_{t-2}\boldsymbol{\beta}) + \cdots + \rho_p(y_{t-p} - \mathbf{x}_{t-p}\boldsymbol{\beta})$$
$$+ \theta_1 \epsilon_{t-1} + \theta_2 \epsilon_{t-2} + \cdots + \theta_q \epsilon_{t-q} + \epsilon_t$$

It is also common to write the general form of the ARMA model more succinctly using lag operator notation as

$$\rho(L^p)(y_t - \mathbf{x}_t\boldsymbol{\beta}) = \boldsymbol{\theta}(L^q)\epsilon_t \qquad \qquad \text{ARMA}(p,q)$$

where

$$\rho(L^p) = 1 - \rho_1 L - \rho_2 L^2 - \cdots - \rho_p L^p$$
$$\boldsymbol{\theta}(L^q) = 1 + \theta_1 L + \theta_2 L^2 + \cdots + \theta_q L^q$$

and $L^j y_t = y_{t-j}$.

For stationary series, full or unconditional maximum likelihood estimates are obtained via the Kalman filter. For nonstationary series, if some prior information is available, you can specify initial values for the filter by using state0() and p0() as suggested by Hamilton (1994) or assume an uninformative prior by using the diffuse option as suggested by Harvey (1989).

ARIMA models

Pure ARIMA models without a structural component do not have regressors and are often written as autoregressions in the dependent variable, rather than autoregressions in the disturbances from a structural equation. For example, an ARMA$(1, 1)$ model can be written as

$$y_t = \alpha + \rho y_{t-1} + \theta \epsilon_{t-1} + \epsilon_t \qquad \qquad (1a)$$

Other than a scale factor for the constant term α, these models are equivalent to the ARMA in the disturbances formulation estimated by arima, though the latter are more flexible and allow a wider class of models.

To see this effect, replace $\mathbf{x}_t\beta$ in the structural equation above with a constant term β_0 so that

$$
\begin{aligned}
y_t &= \beta_0 + \mu_t \\
&= \beta_0 + \rho\mu_{t-1} + \theta\epsilon_{t-1} + \epsilon_t \\
&= \beta_0 + \rho(y_{t-1} - \beta_0) + \theta\epsilon_{t-1} + \epsilon_t \\
&= (1-\rho)\beta_0 + \rho y_{t-1} + \theta\epsilon_{t-1} + \epsilon_t
\end{aligned}
\tag{1b}
$$

Equations (1a) and (1b) are equivalent, with $\alpha = (1-\rho)\beta_0$, so whether we consider an ARIMA model as autoregressive in the dependent variable or disturbances is immaterial. Our illustration can easily be extended from the ARMA$(1,1)$ case to the general ARIMA(p,d,q) case.

▷ Example 1: ARIMA model

Enders (2004, 87–93) considers an ARIMA model of the U.S. Wholesale Price Index (WPI) using quarterly data over the period 1960q1 through 1990q4. The simplest ARIMA model that includes differencing and both autoregressive and moving-average components is the ARIMA(1,1,1) specification. We can fit this model with `arima` by typing

```
. use http://www.stata-press.com/data/r12/wpi1

. arima wpi, arima(1,1,1)
(setting optimization to BHHH)
Iteration 0:   log likelihood = -139.80133
Iteration 1:   log likelihood =  -135.6278
Iteration 2:   log likelihood = -135.41838
Iteration 3:   log likelihood = -135.36691
Iteration 4:   log likelihood = -135.35892
(switching optimization to BFGS)
Iteration 5:   log likelihood = -135.35471
Iteration 6:   log likelihood = -135.35135
Iteration 7:   log likelihood = -135.35132
Iteration 8:   log likelihood = -135.35131

ARIMA regression

Sample:  1960q2 - 1990q4                         Number of obs     =       123
                                                 Wald chi2(2)      =    310.64
Log likelihood = -135.3513                       Prob > chi2       =    0.0000
```

D.wpi	Coef.	OPG Std. Err.	z	P>|z|	[95% Conf. Interval]	
wpi						
_cons	.7498197	.3340968	2.24	0.025	.0950019	1.404637
ARMA						
ar						
L1.	.8742288	.0545435	16.03	0.000	.7673256	.981132
ma						
L1.	-.4120458	.1000284	-4.12	0.000	-.6080979	-.2159938
/sigma	.7250436	.0368065	19.70	0.000	.6529042	.7971829

```
Note: The test of the variance against zero is one sided, and the two-sided
      confidence interval is truncated at zero.
```

Examining the estimation results, we see that the AR(1) coefficient is 0.874, the MA(1) coefficient is -0.412, and both are highly significant. The estimated standard deviation of the white-noise disturbance ϵ is 0.725.

This model also could have been fit by typing

```
. arima D.wpi, ar(1) ma(1)
```

The D. placed in front of the dependent variable wpi is the Stata time-series operator for differencing. Thus we would be modeling the first difference in WPI from the second quarter of 1960 through the fourth quarter of 1990 because the first observation is lost because of differencing. This second syntax allows a richer choice of models.

◁

▷ Example 2: ARIMA model with additive seasonal effects

After examining first-differences of WPI, Enders chose a model of differences in the natural logarithms to stabilize the variance in the differenced series. The raw data and first-difference of the logarithms are graphed below.

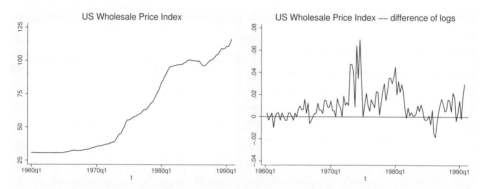

On the basis of the autocorrelations, partial autocorrelations (see graphs below), and the results of preliminary estimations, Enders identified an ARMA model in the log-differenced series.

```
. ac D.ln_wpi, ylabels(-.4(.2).6)
. pac D.ln_wpi, ylabels(-.4(.2).6)
```

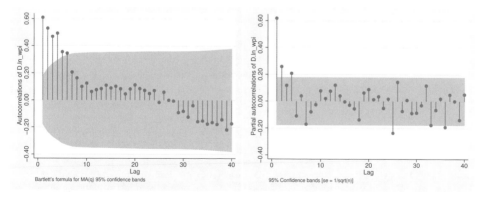

In addition to an autoregressive term and an MA(1) term, an MA(4) term is included to account for a remaining quarterly effect. Thus the model to be fit is

$$\Delta \ln(\mathrm{wpi}_t) = \beta_0 + \rho_1 \{\Delta \ln(\mathrm{wpi}_{t-1}) - \beta_0\} + \theta_1 \epsilon_{t-1} + \theta_4 \epsilon_{t-4} + \epsilon_t$$

We can fit this model with `arima` and Stata's standard difference operator:

```
. arima D.ln_wpi, ar(1) ma(1 4)

(setting optimization to BHHH)
Iteration 0:   log likelihood =  382.67447
Iteration 1:   log likelihood =  384.80754
Iteration 2:   log likelihood =  384.84749
Iteration 3:   log likelihood =  385.39213
Iteration 4:   log likelihood =  385.40983
(switching optimization to BFGS)
Iteration 5:   log likelihood =   385.9021
Iteration 6:   log likelihood =  385.95646
Iteration 7:   log likelihood =  386.02979
Iteration 8:   log likelihood =  386.03326
Iteration 9:   log likelihood =  386.03354
Iteration 10:  log likelihood =  386.03357

ARIMA regression

Sample:  1960q2 - 1990q4                         Number of obs     =        123
                                                 Wald chi2(3)      =     333.60
Log likelihood =  386.0336                        Prob > chi2       =     0.0000
```

D.ln_wpi	Coef.	OPG Std. Err.	z	P>|z|	[95% Conf. Interval]	
ln_wpi						
_cons	.0110493	.0048349	2.29	0.022	.0015731	.0205255
ARMA						
ar						
L1.	.7806991	.0944946	8.26	0.000	.5954931	.965905
ma						
L1.	-.3990039	.1258753	-3.17	0.002	-.6457149	-.1522928
L4.	.3090813	.1200945	2.57	0.010	.0737003	.5444622
/sigma	.0104394	.0004702	22.20	0.000	.0095178	.0113609

```
Note: The test of the variance against zero is one sided, and the two-sided
      confidence interval is truncated at zero.
```

In this final specification, the log-differenced series is still highly autocorrelated at a level of 0.781, though innovations have a negative impact in the ensuing quarter (-0.399) and a positive seasonal impact of 0.309 in the following year.

◁

❑ Technical note

In one way, the results differ from most of Stata's estimation commands: the standard error of the coefficients is reported as OPG Std. Err. As noted in *Options*, the default standard errors and covariance matrix for `arima` estimates are derived from the outer product of gradients (OPG). This is one of three asymptotically equivalent methods of estimating the covariance matrix of the coefficients (only two of which are usually tractable to derive). Discussions and derivations of all three estimates can be found in Davidson and MacKinnon (1993), Greene (2012), and Hamilton (1994). Bollerslev, Engle, and Nelson (1994) suggest that the OPG estimates are more numerically stable in time-series regressions when the likelihood and its derivatives depend on recursive computations, which is certainly the case for the Kalman filter. To date, we have found no numerical instabilities in either estimate of the covariance matrix—subject to the stability and convergence of the overall model.

Most of Stata's estimation commands provide covariance estimates derived from the Hessian of the likelihood function. These alternate estimates can also be obtained from `arima` by specifying the `vce(oim)` option.

❏

Multiplicative seasonal ARIMA models

Many time series exhibit a periodic seasonal component, and a seasonal ARIMA model, often abbreviated SARIMA, can then be used. For example, monthly sales data for air conditioners have a strong seasonal component, with sales high in the summer months and low in the winter months.

In the previous example, we accounted for quarterly effects by fitting the model

$$(1 - \rho_1 L)\{\Delta \ln(\text{wpi}_t) - \beta_0\} = (1 + \theta_1 L + \theta_4 L^4)\epsilon_t$$

This is an *additive* seasonal ARIMA model, in the sense that the first- and fourth-order MA terms work additively: $(1 + \theta_1 L + \theta_4 L^4)$.

Another way to handle the quarterly effect would be to fit a multiplicative seasonal ARIMA model. A multiplicative SARIMA model of order $(1, 1, 1) \times (0, 0, 1)_4$ for the $\ln(\text{wpi}_t)$ series is

$$(1 - \rho_1 L)\{\Delta \ln(\text{wpi}_t) - \beta_0\} = (1 + \theta_1 L)(1 + \theta_{4,1} L^4)\epsilon_t$$

or, upon expanding terms,

$$\Delta \ln(\text{wpi}_t) = \beta_0 + \rho_1\{\Delta \ln(\text{wpi}_t) - \beta_0\} + \theta_1 \epsilon_{t-1} + \theta_{4,1}\epsilon_{t-4} + \theta_1\theta_{4,1}\epsilon_{t-5} + \epsilon_t \qquad (2)$$

In the notation $(1, 1, 1) \times (0, 0, 1)_4$, the $(1, 1, 1)$ means that there is one nonseasonal autoregressive term $(1 - \rho_1 L)$ and one nonseasonal moving-average term $(1 + \theta_1 L)$ and that the time series is first-differenced one time. The $(0, 0, 1)_4$ indicates that there is no lag-4 seasonal autoregressive term, that there is one lag-4 seasonal moving-average term $(1 + \theta_{4,1} L^4)$, and that the series is seasonally differenced zero times. This is known as a *multiplicative* SARIMA model because the nonseasonal and seasonal factors work multiplicatively: $(1 + \theta_1 L)(1 + \theta_{4,1} L^4)$. Multiplying the terms imposes nonlinear constraints on the parameters of the fifth-order lagged values; `arima` imposes these constraints automatically.

To further clarify the notation, consider a $(2, 1, 1) \times (1, 1, 2)_4$ multiplicative SARIMA model:

$$(1 - \rho_1 L - \rho_2 L^2)(1 - \rho_{4,1} L^4)\Delta\Delta_4 z_t = (1 + \theta_1 L)(1 + \theta_{4,1} L^4 + \theta_{4,2} L^8)\epsilon_t \qquad (3)$$

where Δ denotes the difference operator $\Delta y_t = y_t - y_{t-1}$ and Δ_s denotes the lag-s seasonal difference operator $\Delta_s y_t = y_t - y_{t-s}$. Expanding (3), we have

$$\widetilde{z}_t = \rho_1 \widetilde{z}_{t-1} + \rho_2 \widetilde{z}_{t-2} + \rho_{4,1} \widetilde{z}_{t-4} - \rho_1\rho_{4,1} \widetilde{z}_{t-5} - \rho_2\rho_{4,1} \widetilde{z}_{t-6}$$
$$+ \theta_1 \epsilon_{t-1} + \theta_{4,1}\epsilon_{t-4} + \theta_1\theta_{4,1}\epsilon_{t-5} + \theta_{4,2}\epsilon_{t-8} + \theta_1\theta_{4,2}\epsilon_{t-9}$$

where

$$\widetilde{z}_t = \Delta\Delta_4 z_t = \Delta(z_t - z_{t-4}) = z_t - z_{t-1} - (z_{t-4} - z_{t-5})$$

and $z_t = y_t - \mathbf{x}_t\boldsymbol{\beta}$ if regressors are included in the model, $z_t = y_t - \beta_0$ if just a constant term is included, and $z_t = y_t$ otherwise.

More generally, a $(p, d, q) \times (P, D, Q)_s$ multiplicative SARIMA model is

$$\rho(L^p)\boldsymbol{\rho}_s(L^P)\Delta^d\Delta_s^D z_t = \theta(L^q)\boldsymbol{\theta}_s(L^Q)$$

where

$$\boldsymbol{\rho}_s(L^P) = (1 - \rho_{s,1}L^s - \rho_{s,2}L^{2s} - \cdots - \rho_{s,P}L^{Ps})$$
$$\boldsymbol{\theta}_s(L^Q) = (1 + \theta_{s,1}L^s + \theta_{s,2}L^{2s} + \cdots + \theta_{s,Q}L^{Qs})$$

$\rho(L^p)$ and $\theta(L^q)$ were defined previously, Δ^d means apply the Δ operator d times, and similarly for Δ_s^D. Typically, d and D will be 0 or 1; and p, q, P, and Q will seldom be more than 2 or 3. s will typically be 4 for quarterly data and 12 for monthly data. In fact, the model can be extended to include both monthly and quarterly seasonal factors, as we explain below.

If a plot of the data suggests that the seasonal effect is proportional to the mean of the series, then the seasonal effect is probably multiplicative and a multiplicative SARIMA model may be appropriate. Box, Jenkins, and Reinsel (2008, sec. 9.3.1) suggest starting with a multiplicative SARIMA model with any data that exhibit seasonal patterns and then exploring nonmultiplicative SARIMA models if the multiplicative models do not fit the data well. On the other hand, Chatfield (2004, 14) suggests that taking the logarithm of the series will make the seasonal effect additive, in which case an additive SARIMA model as fit in the previous example would be appropriate. In short, the analyst should probably try both additive and multiplicative SARIMA models to see which provides better fits and forecasts.

Unless the `diffuse` option is used, `arima` must create square matrices of dimension $\{\max(p, q + 1)\}^2$, where p and q are the maximum AR and MA lags, respectively; and the inclusion of long seasonal terms can make this dimension rather large. For example, with monthly data, you might fit a $(0, 1, 1) \times (0, 1, 2)_{12}$ SARIMA model. The maximum MA lag is $2 \times 12 + 1 = 25$, requiring a matrix with $26^2 = 676$ rows and columns.

▷ Example 3: Multiplicative SARIMA model

One of the most common multiplicative SARIMA specifications is the $(0, 1, 1) \times (0, 1, 1)_{12}$ "airline" model of Box, Jenkins, and Reinsel (2008, sec. 9.2). The dataset `airline.dta` contains monthly international airline passenger data from January 1949 through December 1960. After first- and seasonally differencing the data, we do not suspect the presence of a trend component, so we use the `noconstant` option with `arima`:

```
. use http://www.stata-press.com/data/r12/air2
(TIMESLAB: Airline passengers)

. generate lnair = ln(air)

. arima lnair, arima(0,1,1) sarima(0,1,1,12) noconstant
(setting optimization to BHHH)
Iteration 0:   log likelihood =   223.8437
Iteration 1:   log likelihood =  239.80405
  (output omitted)
Iteration 8:   log likelihood =  244.69651

ARIMA regression

Sample:  14 - 144                       Number of obs    =      131
                                        Wald chi2(2)     =    84.53
Log likelihood =  244.6965              Prob > chi2      =   0.0000
```

DS12.lnair	Coef.	OPG Std. Err.	z	P>|z|	[95% Conf. Interval]
ARMA					
ma					
L1.	-.4018324	.0730307	-5.50	0.000	-.5449698 -.2586949
ARMA12					
ma					
L1.	-.5569342	.0963129	-5.78	0.000	-.745704 -.3681644
/sigma	.0367167	.0020132	18.24	0.000	.0327708 .0406625

Note: The test of the variance against zero is one sided, and the two-sided
       confidence interval is truncated at zero.

Thus our model of the monthly number of international airline passengers is

$$\Delta\Delta_{12}\text{lnair}_t = -0.402\epsilon_{t-1} - 0.557\epsilon_{t-12} + 0.224\epsilon_{t-13} + \epsilon_t$$

$$\hat{\sigma} = 0.037$$

In (2), for example, the coefficient on ϵ_{t-13} is the product of the coefficients on the ϵ_{t-1} and ϵ_{t-12} terms $(0.224 \approx -0.402 \times -0.557)$. arima labeled the dependent variable DS12.lnair to indicate that it has applied the difference operator Δ and the lag-12 seasonal difference operator Δ_{12} to lnair; see [U] **11.4.4 Time-series varlists** for more information.

We could have fit this model by typing

```
. arima DS12.lnair, ma(1) mma(1, 12) noconstant
```

For simple multiplicative models, using the sarima() option is easier, though this second syntax allows us to incorporate more complicated seasonal terms.

◁

The mar() and mma() options can be repeated, allowing us to control for multiple seasonal patterns. For example, we may have monthly sales data that exhibit a quarterly pattern as businesses purchase our product at the beginning of calendar quarters when new funds are budgeted, and our product is purchased more frequently in a few months of the year than in most others, even after we control for quarterly fluctuations. Thus we might choose to fit the model

$$(1-\rho L)(1-\rho_{4,1}L^4)(1-\rho_{12,1}L^{12})(\Delta\Delta_4\Delta_{12}\text{sales}_t - \beta_0) = (1+\theta L)(1+\theta_{4,1}L^4)(1+\theta_{12,1}L^{12})\epsilon_t$$

Although this model looks rather complicated, estimating it using `arima` is straightforward:

```
. arima DS4S12.sales, ar(1) mar(1, 4) mar(1, 12) ma(1) mma(1, 4) mma(1, 12)
```

If we instead wanted to include two lags in the lag-4 seasonal AR term and the first and third (but not the second) term in the lag-12 seasonal MA term, we would type

```
. arima DS4S12.sales, ar(1) mar(1 2, 4) mar(1, 12) ma(1) mma(1, 4) mma(1 3, 12)
```

However, models with multiple seasonal terms can be difficult to fit. Usually, one seasonal factor with just one or two AR or MA terms is adequate.

ARMAX models

Thus far all our examples have been pure ARIMA models in which the dependent variable was modeled solely as a function of its past values and disturbances. Also, `arima` can fit ARMAX models, which model the dependent variable in terms of a linear combination of independent variables, as well as an ARMA disturbance process. The `prais` command, for example, allows you to control for only AR(1) disturbances, whereas `arima` allows you to control for a much richer dynamic error structure. `arima` allows for both nonseasonal and seasonal ARMA components in the disturbances.

▷ Example 4: ARMAX model

For a simple example of a model including covariates, we can estimate an update of Friedman and Meiselman's (1963) equation representing the quantity theory of money. They postulate a straight-forward relationship between personal-consumption expenditures (`consump`) and the money supply as measured by M2 (`m2`).

$$\text{consump}_t = \beta_0 + \beta_1 \text{m2}_t + \mu_t$$

Friedman and Meiselman fit the model over a period ending in 1956; we will refit the model over the period 1959q1 through 1981q4. We restrict our attention to the period prior to 1982 because the Federal Reserve manipulated the money supply extensively in the later 1980s to control inflation, and the relationship between consumption and the money supply becomes much more complex during the later part of the decade.

To demonstrate `arima`, we will include both an autoregressive term and a moving-average term for the disturbances in the model; the original estimates included neither. Thus we model the disturbance of the structural equation as

$$\mu_t = \rho\mu_{t-1} + \theta\epsilon_{t-1} + \epsilon_t$$

As per the original authors, the relationship is estimated on seasonally adjusted data, so there is no need to include seasonal effects explicitly. Obtaining seasonally unadjusted data and simultaneously modeling the structural and seasonal effects might be preferable.

We will restrict the estimation to the desired sample by using the `tin()` function in an `if` expression; see [D] **functions**. By leaving the first argument of `tin()` blank, we are including all available data through the second date (1981q4). We fit the model by typing

```
. use http://www.stata-press.com/data/r12/friedman2, clear

. arima consump m2 if tin(, 1981q4), ar(1) ma(1)

(setting optimization to BHHH)
Iteration 0:   log likelihood = -344.67575
Iteration 1:   log likelihood = -341.57248
 (output omitted )
Iteration 10:  log likelihood = -340.50774

ARIMA regression

Sample:  1959q1 - 1981q4                    Number of obs    =        92
                                           Wald chi2(3)     =   4394.80
Log likelihood = -340.5077                 Prob > chi2      =    0.0000
```

consump	Coef.	OPG Std. Err.	z	P>|z|	[95% Conf. Interval]
consump						
m2	1.122029	.0363563	30.86	0.000	1.050772	1.193286
_cons	-36.09872	56.56703	-0.64	0.523	-146.9681	74.77062
ARMA						
ar						
L1.	.9348486	.0411323	22.73	0.000	.8542308	1.015467
ma						
L1.	.3090592	.0885883	3.49	0.000	.1354293	.4826891
/sigma	9.655308	.5635157	17.13	0.000	8.550837	10.75978

```
Note: The test of the variance against zero is one sided, and the two-sided
      confidence interval is truncated at zero.
```

We find a relatively small money velocity with respect to consumption (1.122) over this period, although consumption is only one facet of the income velocity. We also note a very large first-order autocorrelation in the disturbances, as well as a statistically significant first-order moving average.

We might be concerned that our specification has led to disturbances that are heteroskedastic or non-Gaussian. We refit the model by using the vce(robust) option.

```
. arima consump m2 if tin(, 1981q4), ar(1) ma(1) vce(robust)
(setting optimization to BHHH)
Iteration 0:   log pseudolikelihood = -344.67575
Iteration 1:   log pseudolikelihood = -341.57248
 (output omitted )
Iteration 10:  log pseudolikelihood = -340.50774

ARIMA regression

Sample:  1959q1 - 1981q4                    Number of obs     =        92
                                           Wald chi2(3)      =   1176.26
Log pseudolikelihood = -340.5077           Prob > chi2       =    0.0000
```

consump	Coef.	Semirobust Std. Err.	z	P>|z|	[95% Conf. Interval]
consump					
m2	1.122029	.0433302	25.89	0.000	1.037103 1.206954
_cons	-36.09872	28.10478	-1.28	0.199	-91.18308 18.98563
ARMA					
ar					
L1.	.9348486	.0493428	18.95	0.000	.8381385 1.031559
ma					
L1.	.3090592	.1605359	1.93	0.054	-.0055854 .6237038
/sigma	9.655308	1.082639	8.92	0.000	7.533375 11.77724

```
Note: The test of the variance against zero is one sided, and the two-sided
      confidence interval is truncated at zero.
```

We note a substantial increase in the estimated standard errors, and our once clearly significant moving-average term is now only marginally significant.

◁

Dynamic forecasting

Another feature of the arima command is the ability to use predict afterward to make dynamic forecasts. Suppose that we wish to fit the regression model

$$y_t = \beta_0 + \beta_1 x_t + \rho y_{t-1} + \epsilon_t$$

by using a sample of data from $t = 1 \ldots T$ and make forecasts beginning at time f.

If we use regress or prais to fit the model, then we can use predict to make one-step-ahead forecasts. That is, predict will compute

$$\widehat{y_f} = \widehat{\beta_0} + \widehat{\beta_1} x_f + \widehat{\rho} y_{f-1}$$

Most importantly, here predict will use the actual value of y at period $f - 1$ in computing the forecast for time f. Thus, if we use regress or prais, we cannot make forecasts for any periods beyond $f = T + 1$ unless we have observed values for y for those periods.

If we instead fit our model with arima, then predict can produce dynamic forecasts by using the Kalman filter. If we use the dynamic(f) option, then for period f predict will compute

$$\widehat{y_f} = \widehat{\beta_0} + \widehat{\beta_1} x_f + \widehat{\rho} y_{f-1}$$

by using the observed value of y_{f-1} just as predict after regress or prais. However, for period $f + 1$ predict *newvar*, dynamic(f) will compute

$$\widehat{y}_{f+1} = \widehat{\beta_0} + \widehat{\beta_1} x_{f+1} + \widehat{\rho}\widetilde{y}_f$$

using the *predicted* value of y_f instead of the observed value. Similarly, the period $f + 2$ forecast will be

$$\widehat{y}_{f+2} = \widehat{\beta_0} + \widehat{\beta_1} x_{f+2} + \widehat{\rho}\widetilde{y}_{f+1}$$

Of course, because our model includes the regressor x_t, we can make forecasts only through periods for which we have observations on x_t. However, for pure ARIMA models, we can compute dynamic forecasts as far beyond the final period of our dataset as desired.

For more information on predict after arima, see [TS] **arima postestimation**.

Saved results

arima saves the following in e():

Scalars

e(N)	number of observations
e(N_gaps)	number of gaps
e(k)	number of parameters
e(k_eq)	number of equations in e(b)
e(k_eq_model)	number of equations in overall model test
e(k_dv)	number of dependent variables
e(k1)	number of variables in first equation
e(df_m)	model degrees of freedom
e(ll)	log likelihood
e(sigma)	sigma
e(chi2)	χ^2
e(p)	significance
e(tmin)	minimum time
e(tmax)	maximum time
e(ar_max)	maximum AR lag
e(ma_max)	maximum MA lag
e(rank)	rank of e(V)
e(ic)	number of iterations
e(rc)	return code
e(converged)	1 if converged, 0 otherwise

Macros
    e(cmd)   arima
    e(cmdline)   command as typed
    e(depvar)   name of dependent variable
    e(covariates)   list of covariates
    e(eqnames)   names of equations
    e(wtype)   weight type
    e(wexp)   weight expression
    e(title)   title in estimation output
    e(tmins)   formatted minimum time
    e(tmaxs)   formatted maximum time
    e(chi2type)   Wald; type of model χ^2 test
    e(vce)   *vcetype* specified in vce()
    e(vcetype)   title used to label Std. Err.
    e(ma)   lags for moving-average terms
    e(ar)   lags for autoregressive terms
    e(mar*i*)   multiplicative AR terms and lag $i=1...$ (*#* seasonal AR terms)
    e(mma*i*)   multiplicative MA terms and lag $i=1...$ (*#* seasonal MA terms)
    e(seasons)   seasonal lags in model
    e(unsta)   unstationary or blank
    e(opt)   type of optimization
    e(ml_method)   type of ml method
    e(user)   name of likelihood-evaluator program
    e(technique)   maximization technique
    e(tech_steps)   number of iterations performed before switching techniques
    e(properties)   b V
    e(estat_cmd)   program used to implement estat
    e(predict)   program used to implement predict
    e(marginsok)   predictions allowed by margins
    e(marginsnotok)   predictions disallowed by margins

Matrices
    e(b)   coefficient vector
    e(Cns)   constraints matrix
    e(ilog)   iteration log (up to 20 iterations)
    e(gradient)   gradient vector
    e(V)   variance–covariance matrix of the estimators
    e(V_modelbased)   model-based variance

Functions
    e(sample)   marks estimation sample

Methods and formulas

arima is implemented as an ado-file.

Estimation is by maximum likelihood using the Kalman filter via the prediction error decomposition; see Hamilton (1994), Gourieroux and Monfort (1997), or, in particular, Harvey (1989). Any of these sources will serve as excellent background for the fitting of these models with the state-space form; each source also provides considerable detail on the method outlined below.

Methods and formulas are presented under the following headings:

> *ARIMA model*
> *Kalman filter equations*
> *Kalman filter or state-space representation of the ARIMA model*
> *Kalman filter recursions*
> *Kalman filter initial conditions*
> *Likelihood from prediction error decomposition*
> *Missing data*

ARIMA model

The model to be fit is

$$y_t = \mathbf{x}_t \boldsymbol{\beta} + \mu_t$$

$$\mu_t = \sum_{i=1}^{p} \rho_i \mu_{t-i} + \sum_{j=1}^{q} \theta_j \epsilon_{t-j} + \epsilon_t$$

which can be written as the single equation

$$y_t = \mathbf{x}_t \boldsymbol{\beta} + \sum_{i=1}^{p} \rho_i (y_{t-i} - x_{t-i} \boldsymbol{\beta}) + \sum_{j=1}^{q} \theta_j \epsilon_{t-j} + \epsilon_t$$

Some of the ρs and θs may be constrained to zero or, for multiplicative seasonal models, the products of other parameters.

Kalman filter equations

We will roughly follow Hamilton's (1994) notation and write the Kalman filter

$$\boldsymbol{\xi}_t = \mathbf{F} \boldsymbol{\xi}_{t-1} + \mathbf{v}_t \qquad \qquad (state\ equation)$$

$$\mathbf{y}_t = \mathbf{A}' \mathbf{x}_t + \mathbf{H}' \boldsymbol{\xi}_t + \mathbf{w}_t \qquad \qquad (observation\ equation)$$

and

$$\begin{pmatrix} \mathbf{v}_t \\ \mathbf{w}_t \end{pmatrix} \sim N \left\{ \mathbf{0}, \begin{pmatrix} \mathbf{Q} & \mathbf{0} \\ \mathbf{0} & \mathbf{R} \end{pmatrix} \right\}$$

We maintain the standard Kalman filter matrix and vector notation, although for univariate models $\mathbf{y}_t$, $\mathbf{w}_t$, and $\mathbf{R}$ are scalars.

Kalman filter or state-space representation of the ARIMA model

A univariate ARIMA model can be cast in state-space form by defining the Kalman filter matrices as follows (see Hamilton [1994], or Gourieroux and Monfort [1997], for details):

$$F = \begin{bmatrix} \rho_1 & \rho_2 & \cdots & \rho_{p-1} & \rho_p \\ 1 & 0 & \cdots & 0 & 0 \\ 0 & 1 & \cdots & 0 & 0 \\ 0 & 0 & \cdots & 1 & 0 \end{bmatrix}$$

$$\mathbf{v}_t = \begin{bmatrix} \epsilon_{t-1} \\ 0 \\ \cdots \\ \cdots \\ \cdots \\ 0 \end{bmatrix}$$

$$\mathbf{A}' = \beta$$

$$\mathbf{H}' = \begin{bmatrix} 1 & \theta_1 & \theta_2 & \cdots & \theta_q \end{bmatrix}$$

$$\mathbf{w}_t = 0$$

The Kalman filter representation does not require the moving-average terms to be invertible.

Kalman filter recursions

To demonstrate how missing data are handled, the updating recursions for the Kalman filter will be left in two steps. Writing the updating equations as one step using the gain matrix $\mathbf{K}$ is common. We will provide the updating equations with little justification; see the sources listed above for details.

As a linear combination of a vector of random variables, the state $\boldsymbol{\xi}_t$ can be updated to its expected value on the basis of the prior state as

$$\boldsymbol{\xi}_{t|t-1} = \mathbf{F}\boldsymbol{\xi}_{t-1} + \mathbf{v}_{t-1} \tag{1}$$

This state is a quadratic form that has the covariance matrix

$$\mathbf{P}_{t|t-1} = \mathbf{F}\mathbf{P}_{t-1}\mathbf{F}' + \mathbf{Q} \tag{2}$$

The estimator of $\mathbf{y}_t$ is

$$\widehat{\mathbf{y}}_{t|t-1} = \mathbf{x}_t\beta + \mathbf{H}'\boldsymbol{\xi}_{t|t-1}$$

which implies an innovation or prediction error

$$\widehat{\iota}_t = \mathbf{y}_t - \widehat{\mathbf{y}}_{t|t-1}$$

This value or vector has mean squared error (MSE)

$$\mathbf{M}_t = \mathbf{H}'\mathbf{P}_{t|t-1}\mathbf{H} + \mathbf{R}$$

Now the expected value of $\boldsymbol{\xi}_t$ conditional on a realization of $\mathbf{y}_t$ is

$$\boldsymbol{\xi}_t = \boldsymbol{\xi}_{t|t-1} + \mathbf{P}_{t|t-1}\mathbf{H}\mathbf{M}_t^{-1}\widehat{\iota}_t \tag{3}$$

with MSE

$$\mathbf{P}_t = \mathbf{P}_{t|t-1} - \mathbf{P}_{t|t-1}\mathbf{H}\mathbf{M}_t^{-1}\mathbf{H}'\mathbf{P}_{t|t-1} \tag{4}$$

This expression gives the full set of Kalman filter recursions.

Kalman filter initial conditions

When the series is stationary, conditional on $\mathbf{x}_t\boldsymbol{\beta}$, the initial conditions for the filter can be considered a random draw from the stationary distribution of the state equation. The initial values of the state and the state MSE are the expected values from this stationary distribution. For an ARIMA model, these can be written as

$$\xi_{1|0} = \mathbf{0}$$

and

$$\text{vec}(\mathbf{P}_{1|0}) = (\mathbf{I}_{r^2} - \mathbf{F} \otimes \mathbf{F})^{-1}\text{vec}(\mathbf{Q})$$

where vec() is an operator representing the column matrix resulting from stacking each successive column of the target matrix.

If the series is not stationary, the initial state conditions do not constitute a random draw from a stationary distribution, and some other values must be chosen. Hamilton (1994) suggests that they be chosen based on prior expectations, whereas Harvey suggests a diffuse and improper prior having a state vector of $\mathbf{0}$ and an infinite variance. This method corresponds to $\mathbf{P}_{1|0}$ with diagonal elements of ∞. Stata allows either approach to be taken for nonstationary series—initial priors may be specified with state0() and p0(), and a diffuse prior may be specified with diffuse.

Likelihood from prediction error decomposition

Given the outputs from the Kalman filter recursions and assuming that the state and observation vectors are Gaussian, the likelihood for the state-space model follows directly from the resulting multivariate normal in the predicted innovations. The log likelihood for observation t is

$$\ln L_t = -\frac{1}{2}\left\{ \ln(2\pi) + \ln(|\mathbf{M}_t|) - \widehat{\boldsymbol{\iota}}_t'\mathbf{M}_t^{-1}\widehat{\boldsymbol{\iota}}_t \right\}$$

This command supports the Huber/White/sandwich estimator of the variance using vce(robust). See [P] _robust, particularly *Maximum likelihood estimators* and *Methods and formulas*.

Missing data

Missing data, whether a missing dependent variable y_t, one or more missing covariates $\mathbf{x}_t$, or completely missing observations, are handled by continuing the state-updating equations without any contribution from the data; see Harvey (1989 and 1993). That is, (1) and (2) are iterated for every missing observation, whereas (3) and (4) are ignored. Thus, for observations with missing data, $\boldsymbol{\xi}_t = \boldsymbol{\xi}_{t|t-1}$ and $\mathbf{P}_t = \mathbf{P}_{t|t-1}$. Without any information from the sample, this effectively assumes that the prediction error for the missing observations is 0. Other methods of handling missing data on the basis of the EM algorithm have been suggested, for example, Shumway (1984, 1988).

George Edward Pelham Box (1919–) was born in Kent, England, and earned degrees in statistics at the University of London. After work in the chemical industry, he taught and researched at Princeton and the University of Wisconsin. His many major contributions to statistics include papers and books in Bayesian inference, robustness (a term he introduced to statistics), modeling strategy, experimental design and response surfaces, time-series analysis, distribution theory, transformations, and nonlinear estimation.

Gwilym Meirion Jenkins (1933–1982) was a British mathematician and statistician who spent his career in industry and academia, working for extended periods at Imperial College London and the University of Lancaster before running his own company. His interests were centered on time series and he collaborated with G. E. P. Box on what are often called Box–Jenkins models. The last years of Jenkins' life were marked by a slowly losing battle against Hodgkin's disease.

References

Ansley, C. F., and R. Kohn. 1985. Estimation, filtering, and smoothing in state space models with incompletely specified initial conditions. *Annals of Statistics* 13: 1286–1316.

Ansley, C. F., and P. Newbold. 1980. Finite sample properties of estimators for autoregressive moving average models. *Journal of Econometrics* 13: 159–183.

Baum, C. F. 2000. sts15: Tests for stationarity of a time series. *Stata Technical Bulletin* 57: 36–39. Reprinted in *Stata Technical Bulletin Reprints*, vol. 10, pp. 356–360. College Station, TX: Stata Press.

Baum, C. F., and T. Room. 2001. sts18: A test for long-range dependence in a time series. *Stata Technical Bulletin* 60: 37–39. Reprinted in *Stata Technical Bulletin Reprints*, vol. 10, pp. 370–373. College Station, TX: Stata Press.

Baum, C. F., and R. Sperling. 2000. sts15.1: Tests for stationarity of a time series: Update. *Stata Technical Bulletin* 58: 35–36. Reprinted in *Stata Technical Bulletin Reprints*, vol. 10, pp. 360–362. College Station, TX: Stata Press.

Baum, C. F., and V. L. Wiggins. 2000. sts16: Tests for long memory in a time series. *Stata Technical Bulletin* 57: 39–44. Reprinted in *Stata Technical Bulletin Reprints*, vol. 10, pp. 362–368. College Station, TX: Stata Press.

Berndt, E. K., B. H. Hall, R. E. Hall, and J. A. Hausman. 1974. Estimation and inference in nonlinear structural models. *Annals of Economic and Social Measurement* 3/4: 653–665.

Bollerslev, T., R. F. Engle, and D. B. Nelson. 1994. ARCH models. In *Handbook of Econometrics, Volume IV*, ed. R. F. Engle and D. L. McFadden. New York: Elsevier.

Box, G. E. P. 1983. Obituary: G. M. Jenkins, 1933–1982. *Journal of the Royal Statistical Society, Series A* 146: 205–206.

Box, G. E. P., G. M. Jenkins, and G. C. Reinsel. 2008. *Time Series Analysis: Forecasting and Control*. 4th ed. Hoboken, NJ: Wiley.

Chatfield, C. 2004. *The Analysis of Time Series: An Introduction*. 6th ed. Boca Raton, FL: Chapman & Hall/CRC.

David, J. S. 1999. sts14: Bivariate Granger causality test. *Stata Technical Bulletin* 51: 40–41. Reprinted in *Stata Technical Bulletin Reprints*, vol. 9, pp. 350–351. College Station, TX: Stata Press.

Davidson, R., and J. G. MacKinnon. 1993. *Estimation and Inference in Econometrics*. New York: Oxford University Press.

DeGroot, M. H. 1987. A conversation with George Box. *Statistical Science* 2: 239–258.

Diggle, P. J. 1990. *Time Series: A Biostatistical Introduction*. Oxford: Oxford University Press.

Enders, W. 2004. *Applied Econometric Time Series*. 2nd ed. New York: Wiley.

Friedman, M., and D. Meiselman. 1963. The relative stability of monetary velocity and the investment multiplier in the United States, 1897–1958. In *Stabilization Policies*, Commission on Money and Credit, 123–126. Englewood Cliffs, NJ: Prentice Hall.

Gourieroux, C., and A. Monfort. 1997. *Time Series and Dynamic Models*. Trans. ed. G. M. Gallo. Cambridge: Cambridge University Press.

Greene, W. H. 2012. *Econometric Analysis*. 7th ed. Upper Saddle River, NJ: Prentice Hall.

Hamilton, J. D. 1994. *Time Series Analysis*. Princeton: Princeton University Press.

Harvey, A. C. 1989. *Forecasting, Structural Time Series Models and the Kalman Filter*. Cambridge: Cambridge University Press.

———. 1993. *Time Series Models*. 2nd ed. Cambridge, MA: MIT Press.

Hipel, K. W., and A. I. McLeod. 1994. *Time Series Modelling of Water Resources and Environmental Systems*. Amsterdam: Elsevier.

Kalman, R. E. 1960. A new approach to linear filtering and prediction problems. *Transactions of the ASME–Journal of Basic Engineering, Series D* 82: 35–45.

McDowell, A. W. 2002. From the help desk: Transfer functions. *Stata Journal* 2: 71–85.

———. 2004. From the help desk: Polynomial distributed lag models. *Stata Journal* 4: 180–189.

Newton, H. J. 1988. *TIMESLAB: A Time Series Analysis Laboratory*. Belmont, CA: Wadsworth.

Press, W. H., S. A. Teukolsky, W. T. Vetterling, and B. P. Flannery. 2007. *Numerical Recipes in C: The Art of Scientific Computing*. 3rd ed. Cambridge: Cambridge University Press.

Shumway, R. H. 1984. Some applications of the EM algorithm to analyzing incomplete time series data. In *Time Series Analysis of Irregularly Observed Data*, ed. E. Parzen, 290–324. New York: Springer.

———. 1988. *Applied Statistical Time Series Analysis*. Upper Saddle River, NJ: Prentice Hall.

Also see

Title

arima postestimation — Postestimation tools for arima

Description

The following postestimation command is of special interest after `arima`:

Command	Description
psdensity	estimate the spectral density

For information about `psdensity`, see [TS] **psdensity**.

The following standard postestimation commands are also available:

Command	Description
estat	AIC, BIC, VCE, and estimation sample summary
estimates	cataloging estimation results
lincom	point estimates, standard errors, testing, and inference for linear combinations of coefficients
lrtest	likelihood-ratio test
margins	marginal means, predictive margins, marginal effects, and average marginal effects
marginsplot	graph the results from margins (profile plots, interaction plots, etc.)
nlcom	point estimates, standard errors, testing, and inference for nonlinear combinations of coefficients
predict	predictions, residuals, influence statistics, and other diagnostic measures
predictnl	point estimates, standard errors, testing, and inference for generalized predictions
test	Wald tests of simple and composite linear hypotheses
testnl	Wald tests of nonlinear hypotheses

See the corresponding entries in the *Base Reference Manual* for details.

Syntax for predict

> predict [*type*] *newvar* [*if*] [*in*] [, *statistic options*]

statistic	Description
Main	
xb	predicted values for mean equation—the differenced series; the default
stdp	standard error of the linear prediction
y	predicted values for the mean equation in y—the undifferenced series
mse	mean squared error of the predicted values
residuals	residuals or predicted innovations
yresiduals	residuals or predicted innovations in y, reversing any time-series operators

These statistics are available both in and out of sample; type `predict ... if e(sample) ...` if wanted only for the estimation sample.

Predictions are not available for conditional ARIMA models fit to panel data.

options	Description
Options	
<u>dynamic</u>(*time_constant*)	how to handle the lags of y_t
t0(*time_constant*)	set starting point for the recursions to *time_constant*
<u>structural</u>	calculate considering the structural component only

time_constant is a # or a time literal, such as td(1jan1995) or tq(1995q1); see
  Conveniently typing SIF values in [D] **datetime**.

Menu

Statistics > Postestimation > Predictions, residuals, etc.

Options for predict

Five statistics can be computed using predict after arima: the predictions from the model (the default also given by xb), the predictions after reversing any time-series operators applied to the dependent variable (y), the MSE of xb (mse), the predictions of residuals or innovations (residual), and the predicted residuals or innovations in terms of y (yresiduals). Given the dynamic nature of the ARMA component and because the dependent variable might be differenced, there are other ways of computing each. We can use all the data on the dependent variable that is available right up to the time of each prediction (the default, which is often called a one-step prediction), or we can use the data up to a particular time, after which the predicted value of the dependent variable is used recursively to make later predictions (dynamic()). Either way, we can consider or ignore the ARMA disturbance component (the component is considered by default and is ignored if you specify structural).

All calculations can be made in or out of sample.

<u>Main</u>

xb, the default, calculates the predictions from the model. If D.*depvar* is the dependent variable, these predictions are of D.*depvar* and not of *depvar* itself.

stdp calculates the standard error of the linear prediction xb. stdp does not include the variation arising from the disturbance equation; use mse to calculate standard errors and confidence bands around the predicted values.

y specifies that predictions of *depvar* be made, even if the model was specified in terms of, say, D.*depvar*.

mse calculates the MSE of the predictions.

residuals calculates the residuals. If no other options are specified, these are the predicted innovations ϵ_t; that is, they include the ARMA component. If structural is specified, these are the residuals μ_t from the structural equation; see structural below.

yresiduals calculates the residuals in terms of *depvar*, even if the model was specified in terms of, say, D.*depvar*. As with residuals, the yresiduals are computed from the model, including any ARMA component. If structural is specified, any ARMA component is ignored, and yresiduals are the residuals from the structural equation; see structural below.

dynamic(*time_constant*) specifies how lags of y_t in the model are to be handled. If dynamic() is not specified, actual values are used everywhere that lagged values of y_t appear in the model to produce one-step-ahead forecasts.

dynamic(*time_constant*) produces dynamic (also known as recursive) forecasts. *time_constant* specifies when the forecast is to switch from one step ahead to dynamic. In dynamic forecasts, references to y_t evaluate to the prediction of y_t for all periods at or after *time_constant*; they evaluate to the actual value of y_t for all prior periods.

For example, dynamic(10) would calculate predictions in which any reference to y_t with $t < 10$ evaluates to the actual value of y_t and any reference to y_t with $t \geq 10$ evaluates to the prediction of y_t. This means that one-step-ahead predictions are calculated for $t < 10$ and dynamic predictions thereafter. Depending on the lag structure of the model, the dynamic predictions might still refer some actual values of y_t.

You may also specify dynamic(.) to have predict automatically switch from one-step-ahead to dynamic predictions at $p + q$, where p is the maximum AR lag and q is the maximum MA lag.

t0(*time_constant*) specifies the starting point for the recursions to compute the predicted statistics; disturbances are assumed to be 0 for $t <$ t0(). The default is to set t0() to the minimum t observed in the estimation sample, meaning that observations before that are assumed to have disturbances of 0.

t0() is irrelevant if structural is specified because then all observations are assumed to have disturbances of 0.

t0(5) would begin recursions at $t = 5$. If the data were quarterly, you might instead type t0(tq(1961q2)) to obtain the same result.

The ARMA component of ARIMA models is recursive and depends on the starting point of the predictions. This includes one-step-ahead predictions.

structural specifies that the calculation be made considering the structural component only, ignoring the ARMA terms, producing the steady-state equilibrium predictions.

Remarks

We assume that you have already read [TS] **arima**. See [TS] **psdensity** for an introduction to estimating spectral densities using the parameters estimated by arima. In this entry, we illustrate some of the features of predict after fitting ARIMA, ARMAX, and other dynamic models by using arima. In example 2 of [TS] **arima**, we fit the model

$$\Delta \ln(wpi_t) = \beta_0 + \rho_1\{\Delta \ln(wpi_{t-1}) - \beta_0\} + \theta_1\epsilon_{t-1} + \theta_4\epsilon_{t-4} + \epsilon_t$$

by typing

```
. use http://www.stata-press.com/data/r12/wpi1
. arima D.ln_wpi, ar(1) ma(1 4)
  (output omitted )
```

If we use the command

```
. predict xb, xb
```

then Stata computes xb_t as

$$xb_t = \widehat{\beta}_0 + \widehat{\rho}_1\{\Delta \ln(wpi_{t-1}) - \widehat{\beta}_0\} + \widehat{\theta}_1\widehat{\epsilon}_{t-1} + \widehat{\theta}_4\widehat{\epsilon}_{t-4}$$

where

$$\widehat{\epsilon}_{t-j} = \begin{cases} \Delta \ln(wpi_{t-j}) - \mathrm{xb}_{t-j} & t - j > 0 \\ 0 & \text{otherwise} \end{cases}$$

meaning that predict *newvar*, xb calculates predictions by using the metric of the dependent variable. In this example, the dependent variable represented *changes* in $\ln(wpi_t)$, and so the predictions are likewise for *changes* in that variable.

If we instead use

```
. predict y, y
```

Stata computes y_t as $y_t = \mathrm{xb}_t + \ln(wpi_{t-1})$ so that y_t represents the predicted *levels* of $\ln(wpi_t)$. In general, predict *newvar*, y will reverse any time-series operators applied to the dependent variable during estimation.

If we want to ignore the ARMA error components when making predictions, we use the structural option,

```
. predict xbs, xb structural
```

which generates $\mathrm{xbs}_t = \widehat{\beta}_0$ because there are no regressors in this model, and

```
. predict ys, y structural
```

generates $\mathrm{ys}_t = \widehat{\beta}_0 + \ln(wpi_{t-1})$

▷ Example 1: Dynamic forecasts

An attractive feature of the arima command is the ability to make dynamic forecasts. In example 4 of [TS] **arima**, we fit the model

$$\text{consump}_t = \beta_0 + \beta_1 \text{m2}_t + \mu_t$$

$$\mu_t = \rho \mu_{t-1} + \theta \epsilon_{t-1} + \epsilon_t$$

First, we refit the model by using data up through the first quarter of 1978, and then we will evaluate the one-step-ahead and dynamic forecasts.

```
. use http://www.stata-press.com/data/r12/friedman2
. keep if time<=tq(1981q4)
(67 observations deleted)
. arima consump m2 if tin(, 1978q1), ar(1) ma(1)
  (output omitted )
```

To make one-step-ahead forecasts, we type

```
. predict chat, y
(52 missing values generated)
```

(Because our dependent variable contained no time-series operators, we could have instead used predict chat, xb and accomplished the same thing.) We will also make dynamic forecasts, switching from observed values of consump to forecasted values at the first quarter of 1978:

```
. predict chatdy, dynamic(tq(1978q1)) y
(52 missing values generated)
```

The following graph compares the forecasted values to the observed values for the first few years following the estimation sample:

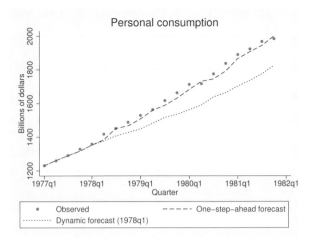

The one-step-ahead forecasts never deviate far from the observed values, though over time the dynamic forecasts have larger errors. To understand why that is the case, rewrite the model as

$$\texttt{consump}_t = \beta_0 + \beta_1 \texttt{m2}_t + \rho \mu_{t-1} + \theta \epsilon_{t-1} + \epsilon_t$$
$$= \beta_0 + \beta_1 \texttt{m2}_t + \rho \left(\texttt{consump}_{t-1} - \beta_0 - \beta_1 \texttt{m2}_{t-1} \right) + \theta \epsilon_{t-1} + \epsilon_t$$

This form shows that the forecasted value of consumption at time t depends on the value of consumption at time $t-1$. When making the one-step-ahead forecast for period t, we know the actual value of consumption at time $t-1$. On the other hand, with the `dynamic(tq(1978q1))` option, the forecasted value of consumption for period 1978q1 is based on the observed value of consumption in period 1977q4, but the forecast for 1978q2 is based on the forecast value for 1978q1, the forecast for 1978q3 is based on the forecast value for 1978q2, and so on. Thus, with dynamic forecasts, prior forecast errors accumulate over time. The following graph illustrates this effect.

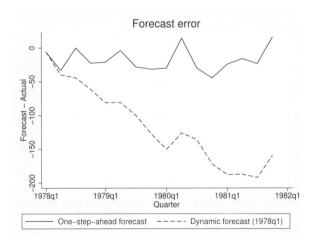

◁

Methods and formulas

All postestimation commands listed above are implemented as ado-files.

Also see

[TS] **arima** — ARIMA, ARMAX, and other dynamic regression models

[TS] **psdensity** — Parametric spectral density estimation after arima, arfima, and ucm

[U] **20 Estimation and postestimation commands**

Title

> **corrgram** — Tabulate and graph autocorrelations

Syntax

Autocorrelations, partial autocorrelations, and portmanteau (Q) statistics

> corrgram *varname* [*if*] [*in*] [, *corrgram_options*]

Graph autocorrelations with confidence intervals

> ac *varname* [*if*] [*in*] [, *ac_options*]

Graph partial autocorrelations with confidence intervals

> pac *varname* [*if*] [*in*] [, *pac_options*]

corrgram_options	Description
Main	
<u>lags</u>(#)	calculate # autocorrelations
<u>noplot</u>	suppress character-based plots
yw	calculate partial autocorrelations by using Yule–Walker equations

ac_options	Description
Main	
<u>lags</u>(#)	calculate # autocorrelations
<u>generate</u>(*newvar*)	generate a variable to hold the autocorrelations
<u>level</u>(#)	set confidence level; default is level(95)
fft	calculate autocorrelation by using Fourier transforms
Plot	
line_options	change look of dropped lines
marker_options	change look of markers (color, size, etc.)
marker_label_options	add marker labels; change look or position
CI plot	
<u>ciopts</u>(*area_options*)	affect rendition of the confidence bands
Add plots	
<u>addplot</u>(*plot*)	add other plots to the generated graph
Y axis, X axis, Titles, Legend, Overall	
twoway_options	any options other than by() documented in [G-3] *twoway_options*

pac_options	Description
Main	
lags(#)	calculate # partial autocorrelations
generate(newvar)	generate a variable to hold the partial autocorrelations
yw	calculate partial autocorrelations by using Yule–Walker equations
level(#)	set confidence level; default is level(95)
Plot	
line_options	change look of dropped lines
marker_options	change look of markers (color, size, etc.)
marker_label_options	add marker labels; change look or position
CI plot	
ciopts(area_options)	affect rendition of the confidence bands
SRV plot	
srv	include standardized residual variances in graph
srvopts(marker_options)	affect rendition of the plotted standardized residual variances (SRVs)
Add plots	
addplot(plot)	add other plots to the generated graph
Y axis, X axis, Titles, Legend, Overall	
twoway_options	any options other than by() documented in [G-3] *twoway_options*

You must tsset your data before using corrgram, ac, or pac; see [TS] **tsset**. Also, the time series
   must be dense (nonmissing and no gaps in the time variable) in the sample if you specify the fft option.
varname may contain time-series operators; see [U] **11.4.4 Time-series varlists**.

Menu

corrgram

Statistics > Time series > Graphs > Autocorrelations & partial autocorrelations

ac

Statistics > Time series > Graphs > Correlogram (ac)

pac

Statistics > Time series > Graphs > Partial correlogram (pac)

Description

corrgram produces a table of the autocorrelations, partial autocorrelations, and portmanteau (Q)
statistics. It also displays a character-based plot of the autocorrelations and partial autocorrelations.
See [TS] **wntestq** for more information on the Q statistic.

ac produces a correlogram (a graph of autocorrelations) with pointwise confidence intervals that
is based on Bartlett's formula for MA(q) processes.

pac produces a partial correlogram (a graph of partial autocorrelations) with confidence intervals calculated using a standard error of $1/\sqrt{n}$. The residual variances for each lag may optionally be included on the graph.

Options for corrgram

‖ Main ‖

lags(#) specifies the number of autocorrelations to calculate. The default is to use $\min(\lfloor n/2 \rfloor - 2, 40)$, where $\lfloor n/2 \rfloor$ is the greatest integer less than or equal to $n/2$.

noplot prevents the character-based plots from being in the listed table of autocorrelations and partial autocorrelations.

yw specifies that the partial autocorrelations be calculated using the Yule–Walker equations instead of using the default regression-based technique. yw cannot be used if srv is used.

Options for ac and pac

‖ Main ‖

lags(#) specifies the number of autocorrelations to calculate. The default is to use $\min(\lfloor n/2 \rfloor - 2, 40)$, where $\lfloor n/2 \rfloor$ is the greatest integer less than or equal to $n/2$.

generate(*newvar*) specifies a new variable to contain the autocorrelation (ac command) or partial autocorrelation (pac command) values. This option is required if the nograph option is used.

nograph (implied when using generate() in the dialog box) prevents ac and pac from constructing a graph. This option requires the generate() option.

yw (pac only) specifies that the partial autocorrelations be calculated using the Yule–Walker equations instead of using the default regression-based technique. yw cannot be used if srv is used.

level(#) specifies the confidence level, as a percentage, for the confidence bands in the ac or pac graph. The default is level(95) or as set by set level; see [R] **level**.

fft (ac only) specifies that the autocorrelations be calculated using two Fourier transforms. This technique can be faster than simply iterating over the requested number of lags.

‖ Plot ‖

line_options, *marker_options*, and *marker_label_options* affect the rendition of the plotted autocorrelations (with ac) or partial autocorrelations (with pac).

line_options specify the look of the dropped lines, including pattern, width, and color; see [G-3] **line_options**.

marker_options specify the look of markers. This look includes the marker symbol, the marker size, and its color and outline; see [G-3] **marker_options**.

marker_label_options specify if and how the markers are to be labeled; see [G-3] **marker_label_options**.

‖ CI plot ‖

ciopts(*area_options*) affects the rendition of the confidence bands; see [G-3] **area_options**.

⌐ SRV plot ⌐

srv (pac only) specifies that the standardized residual variances be plotted with the partial autocorrelations. srv cannot be used if yw is used.

srvopts(*marker_options*) (pac only) affects the rendition of the plotted standardized residual variances; see [G-3] *marker_options*. This option implies the srv option.

⌐ Add plots ⌐

addplot(*plot*) adds specified plots to the generated graph; see [G-3] *addplot_option*.

⌐ Y axis, X axis, Titles, Legend, Overall ⌐

twoway_options are any of the options documented in [G-3] *twoway_options*, excluding by(). These include options for titling the graph (see [G-3] *title_options*) and for saving the graph to disk (see [G-3] *saving_option*).

Remarks

corrgram tabulates autocorrelations, partial autocorrelations, and portmanteau (Q) statistics and plots the autocorrelations and partial autocorrelations. The Q statistics are the same as those produced by [TS] **wntestq**. ac produces graphs of the autocorrelations, and pac produces graphs of the partial autocorrelations.

> Example 1

Here we use the international airline passengers dataset (Box, Jenkins, and Reinsel 2008, Series G). This dataset has 144 observations on the monthly number of international airline passengers from 1949 through 1960. We can list the autocorrelations and partial autocorrelations by using corrgram.

```
. use http://www.stata-press.com/data/r12/air2
(TIMESLAB: Airline passengers)

. corrgram air, lags(20)
```

| | | | | | -1   0   1 | -1   0   1 |
LAG	AC	PAC	Q	Prob>Q	[Autocorrelation]	[Partial Autocor]
1	0.9480	0.9589	132.14	0.0000		
2	0.8756	-0.3298	245.65	0.0000		
3	0.8067	0.2018	342.67	0.0000		
4	0.7526	0.1450	427.74	0.0000		
5	0.7138	0.2585	504.8	0.0000		
6	0.6817	-0.0269	575.6	0.0000		
7	0.6629	0.2043	643.04	0.0000		
8	0.6556	0.1561	709.48	0.0000		
9	0.6709	0.5686	779.59	0.0000		
10	0.7027	0.2926	857.07	0.0000		
11	0.7432	0.8402	944.39	0.0000		
12	0.7604	0.6127	1036.5	0.0000		
13	0.7127	-0.6660	1118	0.0000		
14	0.6463	-0.3846	1185.6	0.0000		
15	0.5859	0.0787	1241.5	0.0000		
16	0.5380	-0.0266	1289	0.0000		
17	0.4997	-0.0581	1330.4	0.0000		
18	0.4687	-0.0435	1367	0.0000		
19	0.4499	0.2773	1401.1	0.0000		
20	0.4416	-0.0405	1434.1	0.0000		

We can use `ac` to produce a graph of the autocorrelations.

```
. ac air, lags(20)
```

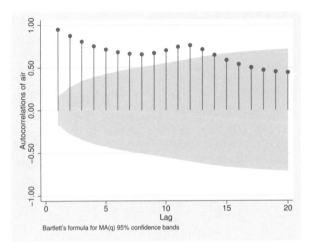

The data probably have a trend component as well as a seasonal component. First-differencing will mitigate the effects of the trend, and seasonal differencing will help control for seasonality. To accomplish this goal, we can use Stata's time-series operators. Here we graph the partial autocorrelations after controlling for trends and seasonality. We also use `srv` to include the standardized residual variances.

```
. pac DS12.air, lags(20) srv
```

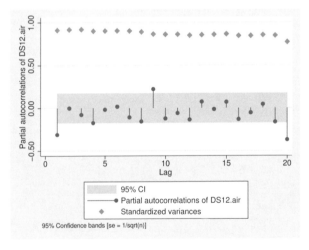

See [U] **11.4.4 Time-series varlists** for more information about time-series operators.

◁

Saved results

corrgram saves the following in r():

Scalars
> r(lags)   number of lags
> r(ac#)   AC for lag #
> r(pac#)   PAC for lag #
> r(q#)   Q for lag #

Matrices
> r(AC)   vector of autocorrelations
> r(PAC)   vector of partial autocorrelations
> r(Q)   vector of Q statistics

Methods and formulas

corrgram, ac, and pac are implemented as ado-files.

Box, Jenkins, and Reinsel (2008, sec. 2.1.4); Newton (1988); Chatfield (2004); and Hamilton (1994) provide excellent descriptions of correlograms. Newton (1988) also discusses the calculation of the various quantities.

The autocovariance function for a time series $x_1, x_2, \ldots, x_n$ is defined for $|v| < n$ as

$$\widehat{R}(v) = \frac{1}{n} \sum_{i=1}^{n-|v|} (x_i - \overline{x})(x_{i+v} - \overline{x})$$

where $\overline{x}$ is the sample mean, and the autocorrelation function is then defined as

$$\widehat{\rho}_v = \frac{\widehat{R}(v)}{\widehat{R}(0)}$$

The variance of $\widehat{\rho}_v$ is given by Bartlett's formula for MA(q) processes. From Brockwell and Davis (2002, 94), we have

$$\mathrm{Var}(\widehat{\rho}_v) = \begin{cases} 1/n & v = 1 \\ \frac{1}{n}\left\{1 + 2\sum_{i=1}^{v-1} \widehat{\rho}^2(i)\right\} & v > 1 \end{cases}$$

The partial autocorrelation at lag v measures the correlation between x_t and x_{t+v} after the effects of $x_{t+1}, \ldots, x_{t+v-1}$ have been removed. By default, corrgram and pac use a regression-based method to estimate it. We run an OLS regression of x_t on $x_{t-1}, \ldots, x_{t-v}$ and a constant term. The estimated coefficient on x_{t-v} is our estimate of the vth partial autocorrelation. The residual variance is the estimated variance of that regression, which we then standardize by dividing by $\widehat{R}(0)$.

If the yw option is specified, corrgram and pac use the Yule–Walker equations to estimate the partial autocorrelations. Per Enders (2010, 66–67), let ϕ_{vv} denote the vth partial autocorrelation coefficient. We then have

$$\widehat{\phi}_{11} = \widehat{\rho}_1$$

and for $v > 1$

$$\widehat{\phi}_{vv} = \frac{\widehat{\rho}_v - \sum_{j=1}^{v-1} \widehat{\phi}_{v-1,j}\widehat{\rho}_{v-j}}{1 - \sum_{j=1}^{v-1} \widehat{\phi}_{v-1,j}\widehat{\rho}_j}$$

and

$$\widehat{\phi}_{vj} = \widehat{\phi}_{v-1,j} - \widehat{\phi}_{vv}\widehat{\phi}_{v-1,v-j} \qquad j = 1, 2, \ldots, v - 1$$

Unlike the regression-based method, the Yule–Walker equations-based method ensures that the first-sample partial autocorrelation equal the first-sample autocorrelation coefficient, as must be true in the population; see Greene (2008, 725).

McCullough (1998) discusses other methods of estimating ϕ_{vv}; he finds that relative to other methods, such as linear regression, the Yule–Walker equations-based method performs poorly, in part because it is susceptible to numerical error. Box, Jenkins, and Reinsel (2008, 69) also caution against using the Yule–Walker equations-based method, especially with data that are nearly nonstationary.

Acknowledgment

The ac and pac commands are based on the ac and pac commands written by Sean Becketti (1992), a past editor of the *Stata Technical Bulletin*.

References

Becketti, S. 1992. sts1: Autocorrelation and partial autocorrelation graphs. *Stata Technical Bulletin* 5: 27–28. Reprinted in *Stata Technical Bulletin Reprints*, vol. 1, pp. 221–223. College Station, TX: Stata Press.

Box, G. E. P., G. M. Jenkins, and G. C. Reinsel. 2008. *Time Series Analysis: Forecasting and Control*. 4th ed. Hoboken, NJ: Wiley.

Brockwell, P. J., and R. A. Davis. 2002. *Introduction to Times Series and Forecasting*. 2nd ed. New York: Springer.

Chatfield, C. 2004. *The Analysis of Time Series: An Introduction*. 6th ed. Boca Raton, FL: Chapman & Hall/CRC.

Enders, W. 2010. *Applied Econometric Time Series*. 3rd ed. New York: Wiley.

Greene, W. H. 2008. *Econometric Analysis*. 6th ed. Upper Saddle River, NJ: Prentice Hall.

Hamilton, J. D. 1994. *Time Series Analysis*. Princeton: Princeton University Press.

McCullough, B. D. 1998. Algorithm choice for (partial) autocorrelation functions. *Journal of Economic and Social Measurement* 24: 265–278.

Newton, H. J. 1988. *TIMESLAB: A Time Series Analysis Laboratory*. Belmont, CA: Wadsworth.

Also see

Title

> **cumsp** — Cumulative spectral distribution

Syntax

cumsp *varname* [*if*] [*in*] [, *options*]

options	Description
Main	
generate(*newvar*)	create *newvar* holding distribution values
Plot	
cline_options	affect rendition of the plotted points connected by lines
marker_options	change look of markers (color, size, etc.)
marker_label_options	add marker labels; change look or position
Add plots	
addplot(*plot*)	add other plots to the generated graph
Y axis, X axis, Titles, Legend, Overall	
twoway_options	any options other than by() documented in [G-3] *twoway_options*

You must tsset your data before using cumsp; see [TS] **tsset**. Also, the time series must be dense
(nonmissing with no gaps in the time variable) in the sample specified.

varname may contain time-series operators; see [U] **11.4.4 Time-series varlists**.

Menu

Statistics > Time series > Graphs > Cumulative spectral distribution

Description

cumsp plots the cumulative sample spectral-distribution function evaluated at the natural frequencies
for a (dense) time series.

Options

> Main

generate(*newvar*) specifies a new variable to contain the estimated cumulative spectral-distribution
values.

> Plot

cline_options affect the rendition of the plotted points connected by lines; see [G-3] ***cline_options***.

marker_options specify the look of markers. This look includes the marker symbol, the marker size,
and its color and outline; see [G-3] ***marker_options***.

109

marker_label_options specify if and how the markers are to be labeled; see [G-3] ***marker_label_options***.

_____
⌐ Add plots ⌐
_____

addplot(*plot*) provides a way to add other plots to the generated graph; see [G-3] ***addplot_option***.

_____
⌐ Y axis, X axis, Titles, Legend, Overall ⌐
_____

twoway_options are any of the options documented in [G-3] ***twoway_options***, excluding by(). These include options for titling the graph (see [G-3] ***title_options***) and for saving the graph to disk (see [G-3] ***saving_option***).

Remarks

▷ Example 1

Here we use the international airline passengers dataset (Box, Jenkins, and Reinsel 2008, Series G). This dataset has 144 observations on the monthly number of international airline passengers from 1949 through 1960. In the cumulative sample spectral distribution function for these data, we also request a vertical line at frequency 1/12. Because the data are monthly, there will be a pronounced jump in the cumulative sample spectral-distribution plot at the 1/12 value if there is an annual cycle in the data.

```
. use http://www.stata-press.com/data/r12/air2
(TIMESLAB: Airline passengers)
. cumsp air, xline(.083333333)
```

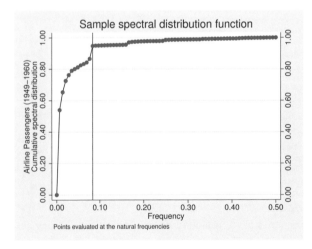

The cumulative sample spectral-distribution function clearly illustrates the annual cycle.

◁

Methods and formulas

cumsp is implemented as an ado-file.

A time series of interest is decomposed into a unique set of sinusoids of various frequencies and amplitudes.

A plot of the sinusoidal amplitudes versus the frequencies for the sinusoidal decomposition of a time series gives us the spectral density of the time series. If we calculate the sinusoidal amplitudes for a discrete set of "natural" frequencies $(1/n, 2/n, \ldots, q/n)$, we obtain the periodogram.

Let $x(1), \ldots, x(n)$ be a time series, and let $\omega_k = (k-1)/n$ denote the natural frequencies for $k = 1, \ldots, \lfloor n/2 \rfloor + 1$ where $\lfloor\ \rfloor$ indicates the greatest integer function. Define

$$
C_k^2 = \frac{1}{n^2} \left| \sum_{t=1}^{n} x(t) e^{2\pi i(t-1)\omega_k} \right|^2
$$

A plot of nC_k^2 versus ω_k is then called the periodogram.

The sample spectral density may then be defined as $\widehat{f}(\omega_k) = nC_k^2$.

If we let $\widehat{f}(\omega_1), \ldots, \widehat{f}(\omega_Q)$ be the sample spectral density function of the time series evaluated at the frequencies $\omega_j = (j-1)/Q$ for $j = 1, \ldots, Q$ and we let $q = \lfloor Q/2 \rfloor + 1$, then

$$
\widehat{F}(\omega_k) = \frac{\displaystyle\sum_{i=1}^{k} \widehat{f}(\omega_j)}{\displaystyle\sum_{i=1}^{q} \widehat{f}(\omega_j)}
$$

is the sample spectral-distribution function of the time series.

References

Box, G. E. P., G. M. Jenkins, and G. C. Reinsel. 2008. *Time Series Analysis: Forecasting and Control*. 4th ed. Hoboken, NJ: Wiley.

Newton, H. J. 1988. *TIMESLAB: A Time Series Analysis Laboratory*. Belmont, CA: Wadsworth.

Also see

[TS] **tsset** — Declare data to be time-series data

[TS] **corrgram** — Tabulate and graph autocorrelations

[TS] **pergram** — Periodogram

Title

> **dfactor** — Dynamic-factor models

Syntax

dfactor *obs_eq* [*fac_eq*] [*if*] [*in*] [, *options*]

obs_eq specifies the equation for the observed dependent variables, and it has the form

(*depvars* = [*exog_d*] [, *sopts*])

fac_eq specifies the equation for the unobserved factors, and it has the form

(*facvars* = [*exog_f*] [, *sopts*])

depvars are the observed dependent variables. *exog_d* are the exogenous variables that enter into the equations for the observed dependent variables. (All factors are automatically entered into the equations for the observed dependent variables.) *facvars* are the names for the unobserved factors in the model. You may specify the names of existing variables in *facvars*, but dfactor treats them only as names and takes no notice that they are also variables. *exog_f* are the exogenous variables that enter into the equations for the factors.

options	Description
Model	
constraints(*constraints*)	apply specified linear constraints
SE/Robust	
vce(*vcetype*)	*vcetype* may be oim or robust
Reporting	
level(#)	set confidence level; default is level(95)
nocnsreport	do not display constraints
display_options	control column formats, row spacing, and display of omitted variables and base and empty cells
Maximization	
maximize_options	control the maximization process; seldom used
from(*matname*)	specify initial values for the maximization process; seldom used
Advanced	
method(*method*)	specify the method for calculating the log likelihood; seldom used
coeflegend	display legend instead of statistics

sopts	Description
Model	
<u>nocons</u>tant	suppress constant term from the equation; allowed only in *obs_eq*
ar(*numlist*)	autoregressive terms
<u>ars</u>tructure(*arstructure*)	structure of autoregressive coefficient matrices
<u>cov</u>structure(*covstructure*)	covariance structure

arstructure	Description
<u>di</u>agonal	diagonal matrix; the default
ltriangular	lower triangular matrix
<u>g</u>eneral	general matrix

covstructure	Description
<u>i</u>dentity	identity matrix
<u>ds</u>calar	diagonal scalar matrix
<u>di</u>agonal	diagonal matrix
<u>un</u>structured	symmetric, positive-definite matrix

method	Description
<u>h</u>ybrid	use the stationary Kalman filter and the De Jong diffuse Kalman filter; the default
<u>de</u>jong	use the stationary De Jong method and the De Jong diffuse Kalman filter

You must tsset your data before using dfactor; see [TS] **tsset**.

exog_d and *exog_f* may contain factor variables; see [U] **11.4.3 Factor variables**.

depvars, *exog_d*, and *exog_f* may contain time-series operators; see [U] **11.4.4 Time-series varlists**.

by, statsby, and rolling are allowed; see [U] **11.1.10 Prefix commands**.

coeflegend does not appear in the dialog box.

See [U] **20 Estimation and postestimation commands** for more capabilities of estimation commands.

Menu

Statistics > Multivariate time series > Dynamic-factor models

Description

dfactor estimates the parameters of dynamic-factor models by maximum likelihood. Dynamic-factor models are flexible models for multivariate time series in which unobserved factors have a vector autoregressive structure, exogenous covariates are permitted in both the equations for the latent factors and the equations for observable dependent variables, and the disturbances in the equations for the dependent variables may be autocorrelated.

Options

```
                    Model
```

constraints(*constraints*) apply linear constraints. Some specifications require linear constraints for parameter identification.

noconstant suppresses the constant term.

ar(*numlist*) specifies the vector autoregressive lag structure in the equation. By default, no lags are included in either the observable or the factor equations.

arstructure(diagonal|ltriangular|general) specifies the structure of the matrices in the vector autoregressive lag structure.

> arstructure(diagonal) specifies the matrices to be diagonal—separate parameters for each lag, but no cross-equation autocorrelations. arstructure(diagonal) is the default for both the observable and the factor equations.

> arstructure(ltriangular) specifies the matrices to be lower triangular—parameterizes a recursive, or Wold causal, structure.

> arstructure(general) specifies the matrices to be general matrices—separate parameters for each possible autocorrelation and cross-correlation.

covstructure(identity | dscalar | diagonal | unstructured) specifies the covariance structure of the errors.

> covstructure(identity) specifies a covariance matrix equal to an identity matrix, and it is the default for the errors in the factor equations.

> covstructure(dscalar) specifies a covariance matrix equal to σ^2 times an identity matrix.

> covstructure(diagonal) specifies a diagonal covariance matrix, and it is the default for the errors in the observable variables.

> covstructure(unstructured) specifies a symmetric, positive-definite covariance matrix with parameters for all variances and covariances.

```
                    SE/Robust
```

vce(*vcetype*) specifies the estimator for the variance–covariance matrix of the estimator.

> vce(oim), the default, causes dfactor to use the observed information matrix estimator.

> vce(robust) causes dfactor to use the Huber/White/sandwich estimator.

```
                    Reporting
```

level(*#*); see [R] **estimation options**.

nocnsreport; see [R] **estimation options**.

display_options: noomitted, vsquish, noemptycells, baselevels, allbaselevels, cformat(%*fmt*), pformat(%*fmt*), and sformat(%*fmt*); see [R] **estimation options**.

```
                    Maximization
```

maximize_options: difficult, technique(*algorithm_spec*), iterate(*#*), [no]log, trace, gradient, showstep, hessian, showtolerance, tolerance(*#*), ltolerance(*#*), nrtolerance(*#*), and from(*matname*); see [R] **maximize** for all options except from(), and see below for information on from(). These options are seldom used.

from(*matname*) specifies initial values for the maximization process. from(b0) causes dfactor to begin the maximization algorithm with the values in b0. b0 must be a row vector; the number of columns must equal the number of parameters in the model; and the values in b0 must be in the same order as the parameters in e(b). This option is seldom used.

⌐ Advanced ⌐

method(*method*) specifies how to compute the log likelihood. dfactor writes the model in state-space form and uses sspace to estimate the parameters; see [TS] **sspace**. method() offers two methods for dealing with some of the technical aspects of the state-space likelihood. This option is seldom used.

method(hybrid), the default, uses the Kalman filter with model-based initial values when the model is stationary and uses the De Jong (1988, 1991) diffuse Kalman filter when the model is nonstationary.

method(dejong) uses the De Jong (1988) method for estimating the initial values for the Kalman filter when the model is stationary and uses the De Jong (1988, 1991) diffuse Kalman filter when the model is nonstationary.

The following option is available with dfactor but is not shown in the dialog box:

coeflegend; see [R] **estimation options**.

Remarks

Remarks are presented under the following headings:

An introduction to dynamic-factor models
Some examples

An introduction to dynamic-factor models

dfactor estimates the parameters of dynamic-factor models by maximum likelihood (ML). Dynamic-factor models represent a vector of k endogenous variables as linear functions of $n_f < k$ unobserved factors and some exogenous covariates. The unobserved factors and the disturbances in the equations for the observed variables may follow vector autoregressive structures.

Dynamic-factor models have been developed and applied in macroeconomics; see Geweke (1977), Sargent and Sims (1977), Stock and Watson (1989, 1991), and Watson and Engle (1983).

Dynamic-factor models are very flexible; in a sense, they are too flexible. Constraints must be imposed to identify the parameters of dynamic-factor and static-factor models. The parameters in the default specifications in dfactor are identified, but other specifications require additional restrictions. The factors are identified only up to a sign, which means that the coefficients on the unobserved factors can flip signs and still produce the same predictions and the same log likelihood. The flexibility of the model sometimes produces convergence problems.

dfactor is designed to handle cases in which the number of modeled endogenous variables, k, is small. The ML estimator is implemented by writing the model in state-space form and by using the Kalman filter to derive and implement the log likelihood. As k grows, the number of parameters quickly exceeds the number that can be estimated.

A dynamic-factor model has the form

$$\mathbf{y}_t = \mathbf{P}\mathbf{f}_t + \mathbf{Q}\mathbf{x}_t + \mathbf{u}_t$$

$$\mathbf{f}_t = \mathbf{R}\mathbf{w}_t + \mathbf{A}_1\mathbf{f}_{t-1} + \mathbf{A}_2\mathbf{f}_{t-2} + \cdots + \mathbf{A}_{t-p}\mathbf{f}_{t-p} + \boldsymbol{\nu}_t$$

$$\mathbf{u}_t = \mathbf{C}_1\mathbf{u}_{t-1} + \mathbf{C}_2\mathbf{u}_{t-2} + \cdots + \mathbf{C}_{t-q}\mathbf{u}_{t-q} + \boldsymbol{\epsilon}_t$$

where the definitions are given in the following table:

Item	Dimension	Definition
$\mathbf{y}_t$	$k \times 1$	vector of dependent variables
$\mathbf{P}$	$k \times n_f$	matrix of parameters
$\mathbf{f}_t$	$n_f \times 1$	vector of unobservable factors
$\mathbf{Q}$	$k \times n_x$	matrix of parameters
$\mathbf{x}_t$	$n_x \times 1$	vector of exogenous variables
$\mathbf{u}_t$	$k \times 1$	vector of disturbances
$\mathbf{R}$	$n_f \times n_w$	matrix of parameters
$\mathbf{w}_t$	$n_w \times 1$	vector of exogenous variables
$\mathbf{A}_i$	$n_f \times n_f$	matrix of autocorrelation parameters for $i \in \{1, 2, \ldots, p\}$
$\boldsymbol{\nu}_t$	$n_f \times 1$	vector of disturbances
$\mathbf{C}_i$	$k \times k$	matrix of autocorrelation parameters for $i \in \{1, 2, \ldots, q\}$
$\boldsymbol{\epsilon}_t$	$k \times 1$	vector of disturbances

By selecting different numbers of factors and lags, the dynamic-factor model encompasses the six models in the table below:

Dynamic factors with vector autoregressive errors	(DFAR)	$n_f > 0$	$p > 0$	$q > 0$
Dynamic factors	(DF)	$n_f > 0$	$p > 0$	$q = 0$
Static factors with vector autoregressive errors	(SFAR)	$n_f > 0$	$p = 0$	$q > 0$
Static factors	(SF)	$n_f > 0$	$p = 0$	$q = 0$
Vector autoregressive errors	(VAR)	$n_f = 0$	$p = 0$	$q > 0$
Seemingly unrelated regression	(SUR)	$n_f = 0$	$p = 0$	$q = 0$

In addition to the time-series models, `dfactor` can estimate the parameters of SF models and SUR models. `dfactor` can place equality constraints on the disturbance covariances, which `sureg` and `var` do not allow.

Some examples

▷ Example 1

Stock and Watson (1989, 1991) wrote a simple macroeconomic model as a DF model, estimated the parameters by ML, and extracted an economic indicator. In this example, we estimate the parameters of a DF model. In [TS] **dfactor postestimation**, we extend this example and extract an economic indicator for the differenced series.

We have data on an industrial-production index, `ipman`; real disposable income, `income`; an aggregate weakly hours index, `hours`; and aggregate unemployment, `unemp`. We believe that these variables are first-difference stationary. We model their first-differences as linear functions of an unobserved factor that follows a second-order autoregressive process.

```
. use http://www.stata-press.com/data/r12/dfex
(St. Louis Fed (FRED) macro data)
. dfactor (D.(ipman income hours unemp) = , noconstant) (f = , ar(1/2))
searching for initial values ...................
(setting technique to bhhh)
Iteration 0:   log likelihood = -675.18934
Iteration 1:   log likelihood = -667.47825
  (output omitted )
Refining estimates:
Iteration 0:   log likelihood = -662.09507
Iteration 1:   log likelihood = -662.09507

Dynamic-factor model
Sample: 1972m2 - 2008m11                   Number of obs    =        442
                                           Wald chi2(6)     =     751.95
Log likelihood = -662.09507                Prob > chi2      =     0.0000
```

	Coef.	OIM Std. Err.	z	P>|z|	[95% Conf. Interval]	
f						
f						
L1.	.2651932	.0568663	4.66	0.000	.1537372	.3766491
L2.	.4820398	.0624635	7.72	0.000	.3596136	.604466
D.ipman						
f	.3502249	.0287389	12.19	0.000	.2938976	.4065522
D.income						
f	.0746338	.0217319	3.43	0.001	.0320401	.1172276
D.hours						
f	.2177469	.0186769	11.66	0.000	.1811407	.254353
D.unemp						
f	-.0676016	.0071022	-9.52	0.000	-.0815217	-.0536816
Variance						
De.ipman	.1383158	.0167086	8.28	0.000	.1055675	.1710641
De.income	.2773808	.0188302	14.73	0.000	.2404743	.3142873
De.hours	.0911446	.0080847	11.27	0.000	.0752988	.1069903
De.unemp	.0237232	.0017932	13.23	0.000	.0202086	.0272378

```
Note: Tests of variances against zero are one sided, and the two-sided
      confidence intervals are truncated at zero.
```

For a discussion of the atypical iteration log, see example 1 in [TS] **sspace**.

The header in the output describes the estimation sample, reports the log-likelihood function at the maximum, and gives the results of a Wald test against the null hypothesis that the coefficients on the independent variables, the factors, and the autoregressive components are all zero. In this example, the null hypothesis that all parameters except for the variance parameters are zero is rejected at all conventional levels.

The results in the estimation table indicate that the unobserved factor is quite persistent and that it is a significant predictor for each of the observed variables.

dfactor writes the DF model as a state-space model and uses the same methods as sspace to estimate the parameters. Example 5 in [TS] **sspace** writes the model considered here in state-space form and uses sspace to estimate the parameters.

◁

❑ Technical note

The signs of the coefficients on the unobserved factors are not identified. They are not identified because we can multiply the unobserved factors and the coefficients on the unobserved factors by negative one without changing the log likelihood or any of the model predictions.

Altering either the starting values for the maximization process, the maximization technique() used, or the platform on which the command is run can cause the signs of the estimated coefficients on the unobserved factors to change.

Changes in the signs of the estimated coefficients on the unobserved factors do not alter the implications of the model or the model predictions.

❑

▷ Example 2

Here we extend the previous example by allowing the errors in the equations for the observables to be autocorrelated. This extension yields a constrained VAR model with an unobserved autocorrelated factor.

We estimate the parameters by typing

```
. dfactor (D.(ipman income hours unemp) = , noconstant ar(1)) (f = , ar(1/2))
searching for initial values ..............
(setting technique to bhhh)
Iteration 0:   log likelihood = -654.19377
Iteration 1:   log likelihood = -627.46986
  (output omitted )
Refining estimates:
Iteration 0:   log likelihood = -610.28846
Iteration 1:   log likelihood = -610.28846
```

Dynamic-factor model

Sample: 1972m2 - 2008m11

Log likelihood = -610.28846

	Number of obs	=	442
	Wald chi2(10)	=	990.91
	Prob > chi2	=	0.0000

	Coef.	OIM Std. Err.	z	P>|z|	[95% Conf. Interval]	
f						
f						
L1.	.4058457	.0906183	4.48	0.000	.2282371	.5834544
L2.	.3663499	.0849584	4.31	0.000	.1998344	.5328654
De.ipman						
e.ipman						
LD.	-.2772149	.068808	-4.03	0.000	-.4120761	-.1423538
De.income						
e.income						
LD.	-.2213824	.0470578	-4.70	0.000	-.3136141	-.1291508
De.hours						
e.hours						
LD.	-.3969317	.0504256	-7.87	0.000	-.495764	-.2980994
De.unemp						
e.unemp						
LD.	-.1736835	.0532071	-3.26	0.001	-.2779675	-.0693995
D.ipman						
f	.3214972	.027982	11.49	0.000	.2666535	.3763408
D.income						
f	.0760412	.0173844	4.37	0.000	.0419684	.110114
D.hours						
f	.1933165	.0172969	11.18	0.000	.1594151	.2272179
D.unemp						
f	-.0711994	.0066553	-10.70	0.000	-.0842435	-.0581553
Variance						
De.ipman	.1387909	.0154558	8.98	0.000	.1084981	.1690837
De.income	.2636239	.0179043	14.72	0.000	.2285322	.2987157
De.hours	.0822919	.0071096	11.57	0.000	.0683574	.0962265
De.unemp	.0218056	.0016658	13.09	0.000	.0185407	.0250704

Note: Tests of variances against zero are one sided, and the two-sided
    confidence intervals are truncated at zero.

The autoregressive (AR) terms are displayed in error notation. e.*varname* stands for the error in the equation for *varname*. The estimate of the pth AR term from *y1* on *y2* is reported as Lpe.*y1* in equation e.*y2*. In the above output, the estimated first-order AR term of D.ipman on D.ipman is -0.277 and is labeled as LDe.ipman in equation De.ipman.

◁

The previous two examples illustrate how to use dfactor to estimate the parameters of DF models. Although the previous example indicates that the more general DFAR model fits the data well, we use these data to illustrate how to estimate the parameters of more restrictive models.

▷ Example 3

In this example, we use dfactor to estimate the parameters of a SUR model with constraints on the error-covariance matrix. The model is also a constrained VAR with constraints on the error-covariance matrix, because we include the lags of two dependent variables as exogenous variables to model the dynamic structure of the data. Previous exploratory work suggested that we should drop the lag of D.unemp from the model.

```
. constraint 1 [cov(De.unemp,De.income)]_cons  = 0

. dfactor (D.(ipman income unemp) = LD.(ipman income), noconstant
> covstructure(unstructured)), constraints(1)
searching for initial values ...........
(setting technique to bhhh)
Iteration 0:   log likelihood = -569.3512
Iteration 1:   log likelihood = -548.76963
  (output omitted )
Refining estimates:
Iteration 0:   log likelihood = -535.12973
Iteration 1:   log likelihood = -535.12973

Dynamic-factor model

Sample: 1972m3 - 2008m11                    Number of obs   =       441
                                            Wald chi2(6)    =     88.32
Log likelihood = -535.12973                 Prob > chi2     =    0.0000
  ( 1)  [cov(De.income,De.unemp)]_cons = 0
```

	Coef.	OIM Std. Err.	z	P>|z|	[95% Conf. Interval]	
D.ipman						
ipman LD.	.206276	.0471654	4.37	0.000	.1138335	.2987185
income LD.	.1867384	.0512139	3.65	0.000	.086361	.2871158
D.income						
ipman LD.	.1043733	.0434048	2.40	0.016	.0193015	.1894451
income LD.	-.1957893	.0471305	-4.15	0.000	-.2881634	-.1034153
D.unemp						
ipman LD.	-.0865823	.0140747	-6.15	0.000	-.1141681	-.0589964
income LD.	-.0200749	.0152828	-1.31	0.189	-.0500285	.0098788
Variance De.ipman	.3243902	.0218533	14.84	0.000	.2815584	.3672219
Covariance De.ipman						
De.income	.0445794	.013696	3.25	0.001	.0177358	.071423
De.unemp	-.0298076	.0047755	-6.24	0.000	-.0391674	-.0204478
Variance De.income	.2747234	.0185008	14.85	0.000	.2384624	.3109844
Covariance De.income						
De.unemp	0	(omitted)				
Variance De.unemp	.0288866	.0019453	14.85	0.000	.0250738	.0326994

Note: Tests of variances against zero are one sided, and the two-sided
   confidence intervals are truncated at zero.

The output indicates that the model fits well, except that the lag of first-differenced income is not a significant predictor of first-differenced unemployment.

◁

❏ Technical note

The previous example shows how to use dfactor to estimate the parameters of a SUR model with constraints on the error-covariance matrix. Neither sureg nor var allow for constraints on the error-covariance matrix. Without the constraints on the error-covariance matrix and including the lag of D.unemp,

```
. dfactor (D.(ipman income unemp) = LD.(ipman income unemp),
> noconstant covstructure(unstructured))
  (output omitted )

. var D.(ipman income unemp), lags(1) noconstant
  (output omitted )
```

and

```
. sureg (D.ipman  LD.(ipman income unemp), noconstant)
>       (D.income LD.(ipman income unemp), noconstant)
>       (D.unemp  LD.(ipman income unemp), noconstant)
  (output omitted )
```

produce the same estimates after allowing for small numerical differences.

❏

▷ Example 4

The previous example estimated the parameters of a constrained VAR model with a constraint on the error-covariance matrix. This example makes two refinements on the previous one: we use an unconditional estimator instead of a conditional estimator, and we constrain the AR parameters to have a lower triangular structure. (See the next technical note for a discussion of conditional and unconditional estimators.) The results are

```
. constraint 1 [cov(De.unemp,De.income)]_cons  = 0

. dfactor (D.(ipman income unemp) = , ar(1) arstructure(ltriangular) noconstant
> covstructure(unstructured)), constraints(1)
searching for initial values ............
(setting technique to bhhh)
Iteration 0:   log likelihood = -543.89836
Iteration 1:   log likelihood = -541.47455
 (output omitted )
Refining estimates:
Iteration 0:   log likelihood = -540.36159
Iteration 1:   log likelihood = -540.36159

Dynamic-factor model

Sample: 1972m2 - 2008m11                    Number of obs    =       442
                                            Wald chi2(6)     =     75.48
Log likelihood = -540.36159                 Prob > chi2      =    0.0000
 ( 1)  [cov(De.income,De.unemp)]_cons = 0
```

	Coef.	OIM Std. Err.	z	P>|z|	[95% Conf. Interval]	
De.ipman						
e.ipman LD.	.2297308	.0473147	4.86	0.000	.1369957	.3224659
De.income						
e.ipman LD.	.1075441	.0433357	2.48	0.013	.0226077	.1924805
e.income LD.	-.2209485	.047116	-4.69	0.000	-.3132943	-.1286028
De.unemp						
e.ipman LD.	-.0975759	.0151301	-6.45	0.000	-.1272304	-.0679215
e.income LD.	-.0000467	.0147848	-0.00	0.997	-.0290244	.0289309
e.unemp LD.	-.0795348	.0482213	-1.65	0.099	-.1740469	.0149773
Variance De.ipman	.3335286	.0224282	14.87	0.000	.2895702	.377487
Covariance De.ipman						
De.income	.0457804	.0139123	3.29	0.001	.0185127	.0730481
De.unemp	-.0329438	.0051423	-6.41	0.000	-.0430226	-.022865
Variance De.income	.2743375	.0184657	14.86	0.000	.2381454	.3105296
Covariance De.income De.unemp	0	(omitted)				
Variance De.unemp	.0292088	.00199	14.68	0.000	.0253083	.0331092

Note: Tests of variances against zero are one sided, and the two-sided
    confidence intervals are truncated at zero.

The estimated AR terms of D.income and D.unemp on D.unemp are -0.000047 and -0.079535, and they are not significant at the 1% or 5% levels. The estimated AR term of D.ipman on D.income is 0.107544 and is significant at the 5% level but not at the 1% level.

◁

❏ Technical note

We obtained the unconditional estimator in example 4 by specifying the ar() option instead of including the lags of the endogenous variables as exogenous variables, as we did in example 3. The unconditional estimator has an additional observation and is more efficient. This change is analogous to estimating an AR coefficient by arima instead of using regress on the lagged endogenous variable. For example, to obtain the unconditional estimator in a univariate model, typing

```
. arima D.ipman, ar(1) noconstant technique(nr)
(output omitted )
```

will produce the same estimated AR coefficient as

```
. dfactor (D.ipman, ar(1) noconstant)
(output omitted )
```

We obtain the conditional estimator by typing either

```
. regress D.ipman LD.ipman, noconstant
(output omitted )
```

or

```
. dfactor (D.ipman = LD.ipman, noconstant)
(output omitted )
```

❏

⊳ Example 5

In this example, we fit regional unemployment data to an SF model. We have data on the unemployment levels for the four regions in the U.S. census: west for the West, south for the South, ne for the Northeast, and midwest for the Midwest. We treat the variables as first-difference stationary and model the first-differences of these variables. Using dfactor yields

```
. use http://www.stata-press.com/data/r12/urate
(Monthly unemployment rates in US Census regions)
. dfactor (D.(west south ne midwest) = , noconstant) (z = )
searching for initial values ............
(setting technique to bhhh)
Iteration 0:   log likelihood =  872.72029
Iteration 1:   log likelihood =  873.04781
  (output omitted )
Refining estimates:
Iteration 0:   log likelihood =   873.0755
Iteration 1:   log likelihood =   873.0755

Dynamic-factor model
Sample: 1990m2 - 2008m12                      Number of obs    =        227
                                              Wald chi2(4)     =     342.56
Log likelihood =    873.0755                  Prob > chi2      =     0.0000
```

		Coef.	OIM Std. Err.	z	P>|z|	[95% Conf. Interval]	
D.west							
	z	.0978324	.0065644	14.90	0.000	.0849664	.1106983
D.south							
	z	.0859494	.0061762	13.92	0.000	.0738442	.0980546
D.ne							
	z	.0918607	.0072814	12.62	0.000	.0775893	.106132
D.midwest							
	z	.0861102	.0074652	11.53	0.000	.0714787	.1007417
Variance							
De.west		.0036887	.0005834	6.32	0.000	.0025453	.0048322
De.south		.0038902	.0005228	7.44	0.000	.0028656	.0049149
De.ne		.0064074	.0007558	8.48	0.000	.0049261	.0078887
De.midwest		.0074749	.0008271	9.04	0.000	.0058538	.009096

```
Note: Tests of variances against zero are one sided, and the two-sided
      confidence intervals are truncated at zero.
```

The estimates indicate that we could reasonably suppose that the unobserved factor has the same effect on the changes in unemployment in all four regions. The output below shows that we cannot reject the null hypothesis that these coefficients are the same.

```
. test [D.west]z = [D.south]z = [D.ne]z = [D.midwest]z
 ( 1)  [D.west]z - [D.south]z = 0
 ( 2)  [D.west]z - [D.ne]z = 0
 ( 3)  [D.west]z - [D.midwest]z = 0
           chi2(  3) =      3.58
         Prob > chi2 =     0.3109
```

◁

▷ Example 6

In this example, we impose the constraint that the unobserved factor has the same impact on changes in unemployment in all four regions. This constraint was suggested by the results of the previous example. The previous example did not allow for any dynamics in the variables, a problem we alleviate by allowing the disturbances in the equation for each observable to follow an AR(1) process.

```
. constraint 2 [D.west]z = [D.south]z
. constraint 3 [D.west]z = [D.ne]z
. constraint 4 [D.west]z = [D.midwest]z
. dfactor (D.(west south ne midwest) = , noconstant ar(1)) (z = ),
> constraints(2/4)
searching for initial values .............
(setting technique to bhhh)
Iteration 0:   log likelihood =  828.22533
Iteration 1:   log likelihood =  874.84221
 (output omitted )
Refining estimates:
Iteration 0:   log likelihood =  880.97488
Iteration 1:   log likelihood =  880.97488
```

Dynamic-factor model

Sample: 1990m2 - 2008m12

	Number of obs	=	227
	Wald chi2(5)	=	363.34
	Prob > chi2	=	0.0000

Log likelihood = 880.97488
```
 ( 1)  [D.west]z - [D.south]z = 0
 ( 2)  [D.west]z - [D.ne]z = 0
 ( 3)  [D.west]z - [D.midwest]z = 0
```

		Coef.	OIM Std. Err.	z	P>|z|	[95% Conf. Interval]
De.west						
e.west						
	LD.	.1297198	.0992663	1.31	0.191	-.0648386 .3242781
De.south						
e.south						
	LD.	-.2829014	.0909205	-3.11	0.002	-.4611023 -.1047004
De.ne						
e.ne						
	LD.	.2866958	.0847851	3.38	0.001	.12052 .4528715
De.midwest						
e.midwest						
	LD.	.0049427	.0782188	0.06	0.950	-.1483634 .1582488
D.west						
	z	.0904724	.0049326	18.34	0.000	.0808047 .1001401
D.south						
	z	.0904724	.0049326	18.34	0.000	.0808047 .1001401
D.ne						
	z	.0904724	.0049326	18.34	0.000	.0808047 .1001401
D.midwest						
	z	.0904724	.0049326	18.34	0.000	.0808047 .1001401
Variance						
De.west		.0038959	.0005111	7.62	0.000	.0028941 .0048977
De.south		.0035518	.0005097	6.97	0.000	.0025528 .0045507
De.ne		.0058173	.0006983	8.33	0.000	.0044488 .0071859
De.midwest		.0075444	.0008268	9.12	0.000	.0059239 .009165

Note: Tests of variances against zero are one sided, and the two-sided
    confidence intervals are truncated at zero.

The results indicate that the model might not fit well. Two of the four AR coefficients are statistically insignificant, while the two significant coefficients have opposite signs and sum to about zero. We suspect that a DF model might fit these data better than an SF model with autocorrelated disturbances.

◁

Saved results

dfactor saves the following in e():

Scalars

e(N)	number of observations
e(k)	number of parameters
e(k_aux)	number of auxiliary parameters
e(k_eq)	number of equations in e(b)
e(k_eq_model)	number of equations in overall model test
e(k_dv)	number of dependent variables
e(k_obser)	number of observation equations
e(k_factor)	number of factors specified
e(o_ar_max)	number of AR terms for the disturbances
e(f_ar_max)	number of AR terms for the factors
e(df_m)	model degrees of freedom
e(ll)	log likelihood
e(chi2)	χ^2
e(p)	significance
e(tmin)	minimum time in sample
e(tmax)	maximum time in sample
e(stationary)	1 if the estimated parameters indicate a stationary model, 0 otherwise
e(rank)	rank of VCE
e(ic)	number of iterations
e(rc)	return code
e(converged)	1 if converged, 0 otherwise

Macros

e(cmd)	dfactor
e(cmdline)	command as typed
e(depvar)	unoperated names of dependent variables in observation equations
e(obser_deps)	names of dependent variables in observation equations
e(covariates)	list of covariates
e(indeps)	independent variables
e(factor_deps)	names of unobserved factors in model
e(tvar)	variable denoting time within groups
e(eqnames)	names of equations
e(model)	type of dynamic-factor model specified
e(title)	title in estimation output
e(tmins)	formatted minimum time
e(tmaxs)	formatted maximum time
e(o_ar)	list of AR terms for disturbances
e(f_ar)	list of AR terms for factors
e(observ_cov)	structure of observation-error covariance matrix
e(factor_cov)	structure of factor-error covariance matrix
e(chi2type)	Wald; type of model χ^2 test
e(vce)	*vcetype* specified in vce()
e(vcetype)	title used to label Std. Err.
e(opt)	type of optimization
e(method)	likelihood method
e(initial_values)	type of initial values
e(technique)	maximization technique
e(tech_steps)	iterations taken in maximization technique(s)
e(datasignature)	the checksum
e(datasignaturevars)	variables used in calculation of checksum
e(properties)	b V

e(estat_cmd)	program used to implement estat
e(predict)	program used to implement predict
e(marginsok)	predictions allowed by margins
e(marginsnotok)	predictions disallowed by margins

Matrices

e(b)	coefficient vector
e(Cns)	constraints matrix
e(ilog)	iteration log (up to 20 iterations)
e(gradient)	gradient vector
e(V)	variance–covariance matrix of the estimators
e(V_modelbased)	model-based variance

Functions

e(sample)	marks estimation sample

Methods and formulas

dfactor is implemented as an ado-file.

dfactor writes the specified model as a state-space model and uses sspace to estimate the parameters by maximum likelihood. See Lütkepohl (2005, 619–621) for how to write the DF model in state-space form. See [TS] **sspace** for the technical details.

References

De Jong, P. 1988. The likelihood for a state space model. *Biometrika* 75: 165–169.

———. 1991. The diffuse Kalman filter. *Annals of Statistics* 19: 1073–1083.

Geweke, J. 1977. The dynamic factor analysis of economic time series models. In *Latent Variables in Socioeconomic Models*, ed. D. J. Aigner and A. S. Goldberger, 365–383. Amsterdam: North-Holland.

Lütkepohl, H. 2005. *New Introduction to Multiple Time Series Analysis*. New York: Springer.

Sargent, T. J., and C. A. Sims. 1977. Business cycle modeling without pretending to have too much a priori economic theory. In *New Methods in Business Cycle Research: Proceedings from a Conference*, ed. C. A. Sims, 45–109. Minneapolis: Federal Reserve Bank of Minneapolis.

Stock, J. H., and M. W. Watson. 1989. New indexes of coincident and leading economic indicators. In *NBER Macroeconomics Annual 1989*, ed. O. J. Blanchard and S. Fischer, vol. 4, 351–394. Cambridge, MA: MIT Press.

———. 1991. A probability model of the coincident economic indicators. In *Leading Economic Indicators: New Approaches and Forecasting Records*, ed. K. Lahiri and G. H. Moore, 63–89. Cambridge: Cambridge University Press.

Watson, M. W., and R. F. Engle. 1983. Alternative algorithms for the estimation of dymanic factor, MIMIC and varying coefficient regression models. *Journal of Econometrics* 23: 385–400.

Also see

[TS] **dfactor postestimation** — Postestimation tools for dfactor

[TS] **tsset** — Declare data to be time-series data

[TS] **sspace** — State-space models

[TS] **var** — Vector autoregressive models

[TS] **arima** — ARIMA, ARMAX, and other dynamic regression models

[R] **regress** — Linear regression

[R] **sureg** — Zellner's seemingly unrelated regression

[U] **20 Estimation and postestimation commands**

Title

dfactor postestimation — Postestimation tools for dfactor

Description

The following standard postestimation commands are available after `dfactor`:

Command	Description
estat	AIC, BIC, VCE, and estimation sample summary
estimates	cataloging estimation results
lincom	point estimates, standard errors, testing, and inference for linear combinations of coefficients
lrtest	likelihood-ratio test
nlcom	point estimates, standard errors, testing, and inference for nonlinear combinations of coefficients
predict	predictions, residuals, influence statistics, and other diagnostic measures
predictnl	point estimates, standard errors, testing, and inference for generalized predictions
test	Wald tests of simple and composite linear hypotheses
testnl	Wald tests of nonlinear hypotheses

See the corresponding entries in the *Base Reference Manual* for details.

Syntax for predict

predict [*type*] { *stub** | *newvarlist* } [*if*] [*in*] [, *statistic options*]

statistic	Description
Main	
y	dependent variable, which is xbf + residuals
xb	linear predictions using the observable independent variables
xbf	linear predictions using the observable independent variables plus the factor contributions
<u>factor</u>s	unobserved factor variables
<u>resi</u>duals	autocorrelated disturbances
<u>innov</u>ations	innovations, the observed dependent variable minus the predicted y

These statistics are available both in and out of sample; type predict ... if e(sample) ... if wanted only for the estimation sample.

129

options	Description
Options	
equation(*eqnames*)	specify name(s) of equation(s) for which predictions are to be made
rmse(*stub** | *newvarlist*)	put estimated root mean squared errors of predicted objects in new variables
dynamic(*time_constant*)	begin dynamic forecast at specified time
Advanced	
smethod(*method*)	method for predicting unobserved states

method	Description
onestep	predict using past information
smooth	predict using all sample information
filter	predict using past and contemporaneous information

Menu

Statistics > Postestimation > Predictions, residuals, etc.

Options for predict

The mathematical notation used in this section is defined in *Description* of [TS] **dfactor**.

    ⌐ Main ⌐

y, xb, xbf, factors, residuals, and innovations specify the statistic to be predicted.

    y, the default, predicts the dependent variables. The predictions include the contributions of the unobserved factors, the linear predictions by using the observable independent variables, and any autocorrelation, $\widehat{\mathbf{P}}\widehat{\mathbf{f}}_t + \widehat{\mathbf{Q}}\mathbf{x}_t + \widehat{\mathbf{u}}_t$.

    xb calculates the linear prediction by using the observable independent variables, $\widehat{\mathbf{Q}}\mathbf{x}_t$.

    xbf calculates the contributions of the unobserved factors plus the linear prediction by using the observable independent variables, $\widehat{\mathbf{P}}\widehat{\mathbf{f}}_t + \widehat{\mathbf{Q}}\mathbf{x}_t$.

    factors estimates the unobserved factors, $\widehat{\mathbf{f}}_t = \widehat{\mathbf{R}}\mathbf{w}_t + \widehat{\mathbf{A}}_1\widehat{\mathbf{f}}_{t-1} + \widehat{\mathbf{A}}_2\widehat{\mathbf{f}}_{t-2} + \cdots + \widehat{\mathbf{A}}_{t-p}\widehat{\mathbf{f}}_{t-p}$.

    residuals calculates the autocorrelated residuals, $\widehat{\mathbf{u}}_t = \widehat{\mathbf{C}}_1\widehat{\mathbf{u}}_{t-1} + \widehat{\mathbf{C}}_2\widehat{\mathbf{u}}_{t-2} + \cdots + \widehat{\mathbf{C}}_{t-q}\widehat{\mathbf{u}}_{t-q}$.

    innovations calculates the innovations, $\widehat{\boldsymbol{\epsilon}}_t = \mathbf{y}_t - \widehat{\mathbf{P}}\widehat{\mathbf{f}}_t + \widehat{\mathbf{Q}}\mathbf{x}_t - \widehat{\mathbf{u}}_t$.

    ⌐ Options ⌐

equation(*eqnames*) specifies the equation(s) for which the predictions are to be calculated.

    You specify equation names, such as equation(income consumption) or equation(factor1 factor2), to identify the equations. For the factors statistic, you must specify names of equations for factors; for all other statistics, you must specify names of equations for observable variables.

If you do not specify equation() and do not specify *stub**, the results are the same as if you had specified the name of the first equation for the predicted statistic.

equation() may not be specified with *stub**.

rmse(*stub* | newvarlist*) puts the root mean squared errors of the predicted objects into the specified new variables. The root mean squared errors measure the variances due to the disturbances but do not account for estimation error.

dynamic(*time_constant*) specifies when predict starts producing dynamic forecasts. The specified *time_constant* must be in the scale of the time variable specified in tsset, and the *time_constant* must be inside a sample for which observations on the dependent variables are available. For example, dynamic(tq(2008q4)) causes dynamic predictions to begin in the fourth quarter of 2008, assuming that your time variable is quarterly, see [D] **datetime**. If the model contains exogenous variables, they must be present for the whole predicted sample. dynamic() may not be specified with xb, xbf, innovations, smethod(filter), or smethod(smooth).

_____⌐ Advanced ⌐_____

smethod(*method*) specifies the method used to predict the unobserved states in the model. smethod() may not be specified with xb.

smethod(onestep), the default, causes predict to use previous information on the dependent variables. The Kalman filter is performed on previous periods, but only the one-step predictions are made for the current period.

smethod(smooth) causes predict to estimate the states at each time period using all the sample data by the Kalman smoother.

smethod(filter) causes predict to estimate the states at each time period using previous and contemporaneous data by the Kalman filter. The Kalman filter is performed on previous periods and the current period. smethod(filter) may be specified only with factors and residuals.

Remarks

We assume that you have already read [TS] **dfactor**. In this entry, we illustrate some of the features of predict after using dfactor.

dfactor writes the specified model as a state-space model and estimates the parameters by maximum likelihood. The unobserved factors and the residuals are states in the state-space form of the model, and they are estimated by the Kalman filter or the Kalman smoother. The smethod() option controls how these states are estimated.

The Kalman filter or Kalman smoother is run over the specified sample. Changing the sample can alter the predicted value for a given observation, because the Kalman filter and Kalman smoother are recursive algorithms.

After estimating the parameters of a dynamic-factor model, there are many quantities of potential interest. Here we will discuss several of these statistics and illustrate how to use predict to compute them.

▷ Example 1

Let's begin by estimating the parameters of the dynamic-factor model considered in example 2 in [TS] **dfactor**.

```
. use http://www.stata-press.com/data/r12/dfex
(St. Louis Fed (FRED) macro data)
. dfactor (D.(ipman income hours unemp) = , noconstant ar(1)) (f = , ar(1/2))
  (output omitted )
```

While several of the six statistics computed by `predict` might be of interest, we will look only at a few of these statistics for `D.ipman`. We begin by obtaining one-step predictions in the estimation sample and a six-month dynamic forecast for `D.ipman`. The graph of the in-sample predictions indicates that our model accounts only for a small fraction of the variability in `D.ipman`.

```
. tsappend, add(6)
. predict Dipman_f, dynamic(tm(2008m12)) equation(D.ipman)
(option y assumed; fitted values)
. tsline D.ipman Dipman_f if month<=tm(2008m11), lcolor(gs13) xtitle("")
> legend(rows(2))
```

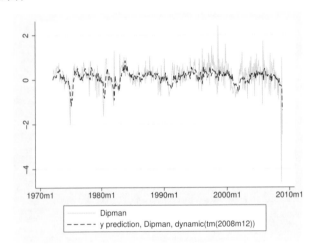

Graphing the last year of the sample and the six-month out-of-sample forecast yields

```
. tsline D.ipman Dipman_f if month>=tm(2008m1), xtitle("") legend(rows(2))
```

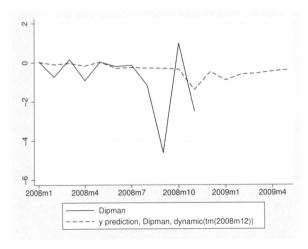

◁

▷ Example 2

Another common task is to estimate an unobserved factor. We can estimate the unobserved factor at each time period by using only previous information (the smethod(onestep) option), previous and contemporaneous information (the smethod(filter) option), or all the sample information (the smethod(smooth) option). We are interested in the one-step predictive power of the unobserved factor, so we use the default, smethod(onestep).

```
. predict fac if e(sample), factor
. tsline D.ipman fac, lcolor(gs10) xtitle("") legend(rows(2))
```

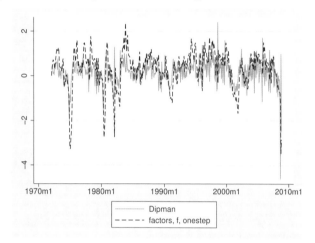

◁

Methods and formulas

All postestimation commands listed above are implemented as ado-files.

dfactor estimates the parameters by writing the model in state-space form and using sspace. Analogously, predict after dfactor uses the methods described in [TS] **sspace postestimation**. The unobserved factors and the residuals are states in the state-space form of the model.

See *Methods and formulas* of [TS] **sspace postestimation** for how predictions are made after estimating the parameters of a state-space model.

Also see

[TS] **dfactor** — Dynamic-factor models

[TS] **sspace** — State-space models

[TS] **sspace postestimation** — Postestimation tools for sspace

[U] **20 Estimation and postestimation commands**

Title

dfgls — DF-GLS unit-root test

Syntax

dfgls *varname* [*if*] [*in*] [*, options*]

options	Description
Main	
maxlag(*#*)	use # as the highest lag order for Dickey–Fuller GLS regressions
notrend	series is stationary around a mean instead of around a linear time trend
ers	present interpolated critical values from Elliott, Rothenberg, and Stock (1996)

You must tsset your data before using dfgls; see [TS] **tsset**.

varname may contain time-series operators; see [U] **11.4.4 Time-series varlists**.

Menu

Statistics > Time series > Tests > DF-GLS test for a unit root

Description

dfgls performs a modified Dickey–Fuller t test for a unit root in which the series has been transformed by a generalized least-squares regression.

Options

⌐ Main ⌐

maxlag(*#*) sets the value of k, the highest lag order for the first-differenced, detrended variable in the Dickey–Fuller regression. By default, dfgls sets k according to the method proposed by Schwert (1989); that is, dfgls sets $k_{max} = \text{int}[12\{(T+1)/100\}^{0.25}]$.

notrend specifies that the alternative hypothesis be that the series is stationary around a mean instead of around a linear time trend. By default, a trend is included.

ers specifies that dfgls should present interpolated critical values from tables presented by Elliott, Rothenberg, and Stock (1996), which they obtained from simulations. See *Critical values* under *Methods and formulas* for details.

Remarks

dfgls tests for a unit root in a time series. It performs the modified Dickey–Fuller t test (known as the DF-GLS test) proposed by Elliott, Rothenberg, and Stock (1996). Essentially, the test is an augmented Dickey–Fuller test, similar to the test performed by Stata's dfuller command, except that the time series is transformed via a generalized least squares (GLS) regression before performing the test. Elliott, Rothenberg, and Stock and later studies have shown that this test has significantly greater power than the previous versions of the augmented Dickey–Fuller test.

134

dfgls performs the DF-GLS test for the series of models that include 1 to k lags of the first-differenced, detrended variable, where k can be set by the user or by the method described in Schwert (1989). Stock and Watson (2011, 644–649) provide an excellent discussion of the approach.

As discussed in [TS] **dfuller**, the augmented Dickey–Fuller test involves fitting a regression of the form

$$\Delta y_t = \alpha + \beta y_{t-1} + \delta t + \zeta_1 \Delta y_{t-1} + \zeta_2 \Delta y_{t-2} + \cdots + \zeta_k \Delta y_{t-k} + \epsilon_t$$

and then testing the null hypothesis H_0: $\beta = 0$. The DG-GLS test is performed analogously but on GLS-detrended data. The null hypothesis of the test is that y_t is a random walk, possibly with drift. There are two possible alternative hypotheses: y_t is stationary about a linear time trend or y_t is stationary with a possibly nonzero mean but with no linear time trend. The default is to use the former. To specify the latter alternative, use the notrend option.

▷ Example 1

Here we use the German macroeconomic dataset and test whether the natural log of income exhibits a unit root. We use the default options with dfgls.

```
. use http://www.stata-press.com/data/r12/lutkepohl2
(Quarterly SA West German macro data, Bil DM, from Lutkepohl 1993 Table E.1)

. dfgls ln_inv
DF-GLS for ln_inv                                   Number of obs =     80
Maxlag = 11 chosen by Schwert criterion
               DF-GLS tau      1% Critical     5% Critical     10% Critical
    [lags]    Test Statistic      Value           Value           Value

      11         -2.925          -3.610          -2.763          -2.489
      10         -2.671          -3.610          -2.798          -2.523
       9         -2.766          -3.610          -2.832          -2.555
       8         -3.259          -3.610          -2.865          -2.587
       7         -3.536          -3.610          -2.898          -2.617
       6         -3.115          -3.610          -2.929          -2.646
       5         -3.054          -3.610          -2.958          -2.674
       4         -3.016          -3.610          -2.986          -2.699
       3         -2.071          -3.610          -3.012          -2.723
       2         -1.675          -3.610          -3.035          -2.744
       1         -1.752          -3.610          -3.055          -2.762
Opt Lag (Ng-Perron seq t) =  7 with RMSE   .0388771
Min SC   = -6.169137 at lag  4 with RMSE   .0398949
Min MAIC = -6.136371 at lag  1 with RMSE   .0440319
```

The null hypothesis of a unit root is not rejected for lags 1–3, it is rejected at the 10% level for lags 9–10, and it is rejected at the 5% level for lags 4–8 and 11. For comparison, we also test for a unit root in log income by using dfuller with two different lag specifications. We need to use the trend option with dfuller because it is not included by default.

```
. dfuller ln_inv, lag(4) trend
Augmented Dickey-Fuller test for unit root          Number of obs   =     87
                          ———————— Interpolated Dickey-Fuller ————————
                  Test        1% Critical     5% Critical     10% Critical
               Statistic         Value           Value           Value

    Z(t)         -3.133          -4.069          -3.463          -3.158

MacKinnon approximate p-value for Z(t) = 0.0987
```

```
. dfuller ln_inv, lag(7) trend
Augmented Dickey-Fuller test for unit root        Number of obs   =        84
                                    ———————— Interpolated Dickey-Fuller ————————
                    Test          1% Critical       5% Critical      10% Critical
                 Statistic           Value             Value             Value
```

	Test Statistic	1% Critical Value	5% Critical Value	10% Critical Value
Z(t)	-3.994	-4.075	-3.466	-3.160

```
MacKinnon approximate p-value for Z(t) = 0.0090
```

The critical values and the test statistic produced by `dfuller` with 4 lags do not support rejecting the null hypothesis, although the approximate p-value is less than 0.1. With 7 lags, the critical values and the test statistic reject the null hypothesis at the 5% level, and the approximate p-value is less than 0.01.

That the `dfuller` results are not as strong as those produced by `dfgls` is not surprising because the DF-GLS test with a trend has been shown to be more powerful than the standard augmented Dickey–Fuller test.

◁

Saved results

If `maxlag(0)` is specified, `dfgls` saves the following in `r()`:

Scalars
    r(rmse0)   RMSE
    r(dft0)   DF-GLS statistic

Otherwise, `dfgls` saves the following in `r()`:

Scalars
    r(maxlag)   highest lag order k
    r(N)   number of observations
    r(sclag)   lag chosen by Schwarz criterion
    r(maiclag)   lag chosen by modified AIC method
    r(optlag)   lag chosen by sequential-t method
Matrices
    r(results) k, MAIC, SIC, RMSE, and DF-GLS statistics

Methods and formulas

`dfgls` is implemented as an ado-file.

`dfgls` tests for a unit root. There are two possible alternative hypotheses: y_t is stationary around a linear trend or y_t is stationary with no linear time trend. Under the first alternative hypothesis, the DF-GLS test is performed by first estimating the intercept and trend via GLS. The GLS estimation is performed by generating the new variables, $\widetilde{y}_t$, x_t, and z_t, where

$$\widetilde{y}_1 = y_1$$
$$\widetilde{y}_t = y_t - \alpha^* y_{t-1}, \qquad t = 2, \ldots, T$$
$$x_1 = 1$$
$$x_t = 1 - \alpha^*, \qquad t = 2, \ldots, T$$
$$z_1 = 1$$
$$z_t = t - \alpha^*(t - 1)$$

and $\alpha^* = 1 - (13.5/T)$. An OLS regression is then estimated for the equation

$$\widetilde{y}_t = \delta_0 x_t + \delta_1 z_t + \epsilon_t$$

The OLS estimators $\widehat{\delta}_0$ and $\widehat{\delta}_1$ are then used to remove the trend from y_t; that is, we generate

$$y^* = y_t - (\widehat{\delta}_0 + \widehat{\delta}_1 t)$$

Finally, we perform an augmented Dickey–Fuller test on the transformed variable by fitting the OLS regression

$$\Delta y_t^* = \alpha + \beta y_{t-1}^* + \sum_{j=1}^{k} \zeta_j \Delta y_{t-j}^* + \epsilon_t$$

and then test the null hypothesis H_0: $\beta = 0$ by using tabulated critical values.

To perform the DF-GLS test under the second alternative hypothesis, we proceed as before but define $\alpha^* = 1 - (7/T)$, eliminate z from the GLS regression, compute $y^* = y_t - \delta_0$, fit the augmented Dickey–Fuller regression by using the newly transformed variable, and perform a test of the null hypothesis that $\beta = 0$ by using the tabulated critical values.

dfgls reports the DF-GLS statistic and its critical values obtained from the regression in (1) for $k \in \{1, 2, \ldots, k_{max}\}$. By default, dfgls sets $k_{max} = \text{int}[12\{(T + 1)/100\}^{0.25}]$ as proposed by Schwert (1989), although you can override this choice with another value. The sample size available with k_{max} lags is used in all the regressions. Because there are k_{max} lags of the first-differenced series, $k_{max} + 1$ observations are lost, leaving $T - k_{max}$ observations. dfgls requires that the sample of $T + 1$ observations on $y_t = (y_0, y_1, \ldots, y_T)$ have no gaps.

dfgls reports the results of three different methods for choosing which value of k to use. These are method 1 the Ng–Perron sequential t, method 2 the minimum Schwarz information criterion (SIC), and method 3 the Ng–Perron modified Akaike information criterion (MAIC). Although the SIC has a long history in time-series modeling, the Ng–Perron sequential t was developed by Ng and Perron (1995), and the MAIC was developed by Ng and Perron (2000).

The SIC can be calculated using either the log likelihood or the sum-of-squared errors from a regression; dfgls uses the latter definition. Specifically, for each k

$$\text{SIC} = \ln(\widehat{\text{rmse}}^2) + (k + 1)\frac{\ln(T - k_{max})}{(T - k_{max})}$$

where

$$\widehat{\text{rmse}} = \frac{1}{(T - k_{\max})} \sum_{t=k_{\max}+1}^{T} \widehat{e}_t^2$$

`dfgls` reports the value of the smallest SIC and the k that produced it.

Ng and Perron (1995) derived a sequential-t algorithm for choosing k:

i. Set $n = 0$ and run the regression in method 2 with all $k_{\max} - n$ lags. If the coefficient on $\beta_{k_{\max}}$ is significantly different from zero at level α, choose k to $k_{\max}$. Otherwise, continue to ii.

ii. If $n < k_{\max}$, set $n = n + 1$ and continue to iii. Otherwise, set $k = 0$ and stop.

iii. Run the regression in method 2 with $k_{\max} - n$ lags. If the coefficient on $\beta_{k_{\max}-n}$ is significantly different from zero at level α, choose k to $k_{\max} - n$. Otherwise, return to ii.

Per Ng and Perron (1995), `dfgls` uses $\alpha = 10\%$. `dfgls` reports the k selected by this sequential-t algorithm and the $\widehat{\text{rmse}}$ from the regression.

Method (3) is based on choosing k to minimize the MAIC. The MAIC is calculated as

$$\text{MAIC}(k) = \ln(\widehat{\text{rmse}}^2) + \frac{2\{\tau(k) + k\}}{T - k_{\max}}$$

where

$$\tau(k) = \frac{1}{\widehat{\text{rmse}}^2} \widehat{\beta}_0^2 \sum_{t=k_{\max}+1}^{T} \widetilde{y}_t^2$$

and $\widetilde{y}$ was defined previously.

Critical values

By default, `dfgls` uses the 5% and 10% critical values computed from the response surface analysis of Cheung and Lai (1995). Because Cheung and Lai (1995) did not present results for the 1% case, the 1% critical values are always interpolated from the critical values presented by ERS.

ERS presented critical values, obtained from simulations, for the DF-GLS test with a linear trend and showed that the critical values for the mean-only DF-GLS test were the same as those for the ADF test. If `dfgls` is run with the `ers` option, `dfgls` will present interpolated critical values from these tables. The method of interpolation is standard. For the trend case, below 50 observations and above 200 there is no interpolation; the values for 50 and ∞ are reported from the tables. For a value N that lies between two values in the table, say, N_1 and N_2, with corresponding critical values CV_1 and CV_2, the critical value

$$\text{cv} = \text{CV}_1 + \frac{N - N_1}{N_1}(\text{CV}_2 - \text{CV}_1)$$

is presented. The same method is used for the mean-only case, except that interpolation is possible for values between 50 and 500.

Acknowledgments

We thank Christopher F. Baum of Boston College and Richard Sperling for a previous version of `dfgls`.

References

Cheung, Y.-W., and K. S. Lai. 1995. Lag order and critical values of a modified Dickey–Fuller test. *Oxford Bulletin of Economics and Statistics* 57: 411–419.

Dickey, D. A., and W. A. Fuller. 1979. Distribution of the estimators for autoregressive time series with a unit root. *Journal of the American Statistical Association* 74: 427–431.

Elliott, G., T. J. Rothenberg, and J. H. Stock. 1996. Efficient tests for an autoregressive unit root. *Econometrica* 64: 813–836.

Ng, S., and P. Perron. 1995. Unit root tests in ARMA models with data-dependent methods for the selection of the truncation lag. *Journal of the American Statistical Association* 90: 268–281.

——. 2000. Lag length selection and the construction of unit root tests with good size and power. *Econometrica* 69: 1519–1554.

Schwert, G. W. 1989. Tests for unit roots: A Monte Carlo investigation. *Journal of Business and Economic Statistics* 2: 147–159.

Stock, J. H., and M. W. Watson. 2011. *Introduction to Econometrics.* 3rd ed. Boston: Addison–Wesley.

Also see

[TS] **dfuller** — Augmented Dickey–Fuller unit-root test

[TS] **pperron** — Phillips–Perron unit-root test

[TS] **tsset** — Declare data to be time-series data

[XT] **xtunitroot** — Panel-data unit-root tests

Title

> **dfuller** — Augmented Dickey–Fuller unit-root test

Syntax

dfuller *varname* $\begin{bmatrix} if \end{bmatrix}$ $\begin{bmatrix} in \end{bmatrix}$ $\begin{bmatrix} , options \end{bmatrix}$

options	Description
Main	
<u>nocon</u>stant	suppress constant term in regression
<u>trend</u>	include trend term in regression
<u>dri</u>ft	include drift term in regression
<u>regress</u>	display regression table
<u>lags</u>(#)	include # lagged differences

You must tsset your data before using dfuller; see [TS] **tsset**.
varname may contain time-series operators; see [U] **11.4.4 Time-series varlists**.

Menu

Statistics > Time series > Tests > Augmented Dickey-Fuller unit-root test

Description

dfuller performs the augmented Dickey–Fuller test that a variable follows a unit-root process. The null hypothesis is that the variable contains a unit root, and the alternative is that the variable was generated by a stationary process. You may optionally exclude the constant, include a trend term, and include lagged values of the difference of the variable in the regression.

Options

> Main

noconstant suppresses the constant term (intercept) in the model and indicates that the process under the null hypothesis is a random walk without drift. noconstant cannot be used with the trend or drift option.

trend specifies that a trend term be included in the associated regression and that the process under the null hypothesis is a random walk, perhaps with drift. This option may not be used with the noconstant or drift option.

drift indicates that the process under the null hypothesis is a random walk with nonzero drift. This option may not be used with the noconstant or trend option.

regress specifies that the associated regression table appear in the output. By default, the regression table is not produced.

lags(#) specifies the number of lagged difference terms to include in the covariate list.

Remarks

Dickey and Fuller (1979) developed a procedure for testing whether a variable has a unit root or, equivalently, that the variable follows a random walk. Hamilton (1994, 528–529) describes the four different cases to which the augmented Dickey–Fuller test can be applied. The null hypothesis is always that the variable has a unit root. They differ in whether the null hypothesis includes a drift term and whether the regression used to obtain the test statistic includes a constant term and time trend.

The true model is assumed to be

$$y_t = \alpha + y_{t-1} + u_t$$

where u_t is an independently and identically distributed zero-mean error term. In cases one and two, presumably $\alpha = 0$, which is a random walk without drift. In cases three and four, we allow for a drift term by letting α be unrestricted.

The Dickey–Fuller test involves fitting the model

$$y_t = \alpha + \rho y_{t-1} + \delta t + u_t$$

by ordinary least squares (OLS), perhaps setting $\alpha = 0$ or $\delta = 0$. However, such a regression is likely to be plagued by serial correlation. To control for that, the augmented Dickey–Fuller test instead fits a model of the form

$$\Delta y_t = \alpha + \beta y_{t-1} + \delta t + \zeta_1 \Delta y_{t-1} + \zeta_2 \Delta y_{t-2} + \cdots + \zeta_k \Delta y_{t-k} + \epsilon_t \tag{1}$$

where k is the number of lags specified in the `lags()` option. The `noconstant` option removes the constant term α from this regression, and the `trend` option includes the time trend δt, which by default is not included. Testing $\beta = 0$ is equivalent to testing $\rho = 1$, or, equivalently, that y_t follows a unit root process.

In the first case, the null hypothesis is that y_t follows a random walk without drift, and (1) is fit without the constant term α and the time trend δt. The second case has the same null hypothesis as the first, except that we include α in the regression. In both cases, the population value of α is zero under the null hypothesis. In the third case, we hypothesize that y_t follows a unit root with drift, so that the population value of α is nonzero; we do not include the time trend in the regression. Finally, in the fourth case, the null hypothesis is that y_t follows a unit root with or without drift so that α is unrestricted, and we include a time trend in the regression.

The following table summarizes the four cases.

Case	Process under null hypothesis	Regression restrictions	dfuller option
1	Random walk without drift	$\alpha = 0,\ \delta = 0$	`noconstant`
2	Random walk without drift	$\delta = 0$	(default)
3	Random walk with drift	$\delta = 0$	`drift`
4	Random walk with or without drift	(none)	`trend`

Except in the third case, the t-statistic used to test H_0: $\beta = 0$ does not have a standard distribution. Hamilton (1994, chap. 17) derives the limiting distributions, which are different for each of the three other cases. The critical values reported by `dfuller` are interpolated based on the tables in Fuller (1996). MacKinnon (1994) shows how to approximate the p-values on the basis of a regression surface, and `dfuller` also reports that p-value. In the third case, where the regression includes a constant term and under the null hypothesis the series has a nonzero drift parameter α, the t statistic has the usual t distribution; `dfuller` reports the one-sided critical values and p-value for the test of H_0 against the alternative H_a: $\beta < 0$, which is equivalent to $\rho < 1$.

Deciding which case to use involves a combination of theory and visual inspection of the data. If economic theory favors a particular null hypothesis, the appropriate case can be chosen based on that. If a graph of the data shows an upward trend over time, then case four may be preferred. If the data do not show a trend but do have a nonzero mean, then case two would be a valid alternative.

▷ Example 1

In this example, we examine the international airline passengers dataset from Box, Jenkins, and Reinsel (2008, Series G). This dataset has 144 observations on the monthly number of international airline passengers from 1949 through 1960. Because the data show a clear upward trend, we use the trend option with dfuller to include a constant and time trend in the augmented Dickey–Fuller regression.

```
. use http://www.stata-press.com/data/r12/air2
(TIMESLAB: Airline passengers)
. dfuller air, lags(3) trend regress
```

Augmented Dickey-Fuller test for unit root   Number of obs   =   140

	Test Statistic	1% Critical Value	Interpolated Dickey-Fuller 5% Critical Value	10% Critical Value
Z(t)	-6.936	-4.027	-3.445	-3.145

MacKinnon approximate p-value for Z(t) = 0.0000

D.air	Coef.	Std. Err.	t	P>|t|	[95% Conf. Interval]	
air						
L1.	-.5217089	.0752195	-6.94	0.000	-.67048	-.3729379
LD.	.5572871	.0799894	6.97	0.000	.399082	.7154923
L2D.	.095912	.0876692	1.09	0.276	-.0774825	.2693065
L3D.	.14511	.0879922	1.65	0.101	-.0289232	.3191433
_trend	1.407534	.2098378	6.71	0.000	.9925118	1.822557
_cons	44.49164	7.78335	5.72	0.000	29.09753	59.88575

Here we can overwhelmingly reject the null hypothesis of a unit root at all common significance levels. From the regression output, the estimated β of -0.522 implies that $\rho = (1 - 0.522) = 0.478$. Experiments with fewer or more lags in the augmented regression yield the same conclusion.

◁

▷ Example 2

In this example, we use the German macroeconomic dataset to determine whether the log of consumption follows a unit root. We will again use the trend option, because consumption grows over time.

```
. use http://www.stata-press.com/data/r12/lutkepohl2
(Quarterly SA West German macro data, Bil DM, from Lutkepohl 1993 Table E.1)

. tsset qtr
        time variable:  qtr, 1960q1 to 1982q4
            delta:  1 quarter

. dfuller ln_consump, lags(4) trend
Augmented Dickey-Fuller test for unit root     Number of obs   =        87
```

	Test Statistic	1% Critical Value	Interpolated Dickey-Fuller 5% Critical Value	10% Critical Value
Z(t)	-1.318	-4.069	-3.463	-3.158

```
MacKinnon approximate p-value for Z(t) = 0.8834
```

As we might expect from economic theory, here we cannot reject the null hypothesis that log consumption exhibits a unit root. Again using different numbers of lag terms yield the same conclusion.

◁

Saved results

dfuller saves the following in r():

Scalars

r(N)	number of observations	r(Zt)	Dickey–Fuller test statistic
r(lags)	number of lagged differences	r(p)	MacKinnon approximate p-value (if there is a constant or trend in associated regression)

Methods and formulas

dfuller is implemented as an ado-file.

In the OLS estimation of an AR(1) process with Gaussian errors,

$$y_t = \rho y_{t-1} + \epsilon_t$$

where ϵ_t are independently and identically distributed as $N(0, \sigma^2)$ and $y_0 = 0$, the OLS estimate (based on an n-observation time series) of the autocorrelation parameter ρ is given by

$$\widehat{\rho}_n = \frac{\sum_{t=1}^{n} y_{t-1} y_t}{\sum_{t=1}^{n} y_t^2}$$

If $|\rho| < 1$, then

$$\sqrt{n}(\widehat{\rho}_n - \rho) \to N(0, 1 - \rho^2)$$

If this result were valid when $\rho = 1$, the resulting distribution would have a variance of zero. When $\rho = 1$, the OLS estimate $\widehat{\rho}$ still converges in probability to one, though we need to find a suitable nondegenerate distribution so that we can perform hypothesis tests of H_0: $\rho = 1$. Hamilton (1994, chap. 17) provides a superb exposition of the requisite theory.

To compute the test statistics, we fit the augmented Dickey–Fuller regression

$$\Delta y_t = \alpha + \beta y_{t-1} + \delta t + \sum_{j=1}^{k} \zeta_j \Delta y_{t-j} + e_t$$

via OLS where, depending on the options specified, the constant term α or time trend δt is omitted and k is the number of lags specified in the lags() option. The test statistic for $H_0\colon \beta = 0$ is $Z_t = \widehat{\beta}/\widehat{\sigma}_\beta$, where $\widehat{\sigma}_\beta$ is the standard error of $\widehat{\beta}$.

The critical values included in the output are linearly interpolated from the table of values that appears in Fuller (1996), and the MacKinnon approximate p-values use the regression surface published in MacKinnon (1994).

David Alan Dickey (1945–) was born in Ohio and obtained degrees in mathematics at Miami University and a PhD in statistics at Iowa State University in 1976 as a student of Wayne Fuller. He works at North Carolina State University and specializes in time-series analysis.

Wayne Arthur Fuller (1931–) was born in Iowa, obtained three degrees at Iowa State University and then served on the faculty between 1959 and 2001. He has made many distinguished contributions to time series, measurement-error models, survey sampling, and econometrics.

References

Box, G. E. P., G. M. Jenkins, and G. C. Reinsel. 2008. *Time Series Analysis: Forecasting and Control*. 4th ed. Hoboken, NJ: Wiley.

Dickey, D. A., and W. A. Fuller. 1979. Distribution of the estimators for autoregressive time series with a unit root. *Journal of the American Statistical Association* 74: 427–431.

Fuller, W. A. 1996. *Introduction to Statistical Time Series*. 2nd ed. New York: Wiley.

Hamilton, J. D. 1994. *Time Series Analysis*. Princeton: Princeton University Press.

MacKinnon, J. G. 1994. Approximate asymptotic distribution functions for unit-root and cointegration tests. *Journal of Business and Economic Statistics* 12: 167–176.

Newton, H. J. 1988. *TIMESLAB: A Time Series Analysis Laboratory*. Belmont, CA: Wadsworth.

Also see

[TS] **tsset** — Declare data to be time-series data

[TS] **dfgls** — DF-GLS unit-root test

[TS] **pperron** — Phillips–Perron unit-root test

[XT] **xtunitroot** — Panel-data unit-root tests

Title

> **fcast compute** — Compute dynamic forecasts of dependent variables after var, svar, or vec

Syntax

After var *and* svar

> fcast compute *prefix* [, *options*₁]

After vec

> fcast compute *prefix* [, *options*₂]

prefix is the prefix appended to the names of the dependent variables to create the names of the variables holding the dynamic forecasts.

*options*₁	Description
Main	
step(#)	set # periods to forecast; default is step(1)
dynamic(*time_constant*)	begin dynamic forecasts at *time_constant*
estimates(*estname*)	use previously saved results *estname*; default is to use active results
replace	replace existing forecast variables that have the same prefix
Std. Errors	
nose	suppress asymptotic standard errors
bs	obtain standard errors from bootstrapped residuals
bsp	obtain standard errors from parametric bootstrap
bscentile	estimate bounds by using centiles of bootstrapped dataset
reps(#)	perform # bootstrap replications; default is reps(200)
nodots	suppress the usual dot after each bootstrap replication
saving(*filename*[, replace])	save bootstrap results as *filename*; use replace to overwrite existing *filename*
Reporting	
level(#)	set confidence level; default is level(95)

options$_2$	Description
Main	
step(*#*)	set *#* periods to forecast; default is step(1)
dynamic(*time_constant*)	begin dynamic forecasts at *time_constant*
estimates(*estname*)	use previously saved results *estname*; default is to use active results
replace	replace existing forecast variables that have the same prefix
differences	save dynamic predictions of the first-differenced variables
Std. Errors	
nose	suppress asymptotic standard errors
Reporting	
level(*#*)	set confidence level; default is level(95)

Default is to use asymptotic standard errors if no options are specified.

fcast compute can be used only after var, svar, and vec; see [TS] **var**, [TS] **var svar**, and [TS] **vec**. You must tsset your data before using fcast compute; see [TS] **tsset**.

Menu

Statistics > Multivariate time series > Dynamic forecasts > Compute forecasts (required for graph)

Description

fcast compute produces dynamic forecasts of the dependent variables in a model previously fit by var, svar, or vec. fcast compute creates new variables and, if necessary, extends the time frame of the dataset to contain the prediction horizon.

Options

⌐ Main ⌐

step(*#*) specifies the number of periods to be forecast. The default is step(1).

dynamic(*time_constant*) specifies the period to begin the dynamic forecasts. The default is the period after the last observation in the estimation sample. The dynamic() option accepts either a Stata date function that returns an integer or an integer that corresponds to a date using the current tsset format. dynamic() must specify a date in the range of two or more periods into the estimation sample to one period after the estimation sample.

estimates(*estname*) specifies that fcast compute use the estimation results stored as *estname*. By default, fcast compute uses the active estimation results. See [R] **estimates** for more information about saving and restoring previously obtained estimates.

replace causes fcast compute to replace the variables in memory with the specified predictions.

differences specifies that fcast compute also save dynamic predictions of the first-differenced variables. differences can be specified only with vec estimation results.

⌐ Std. Errors ⌐

nose specifies that the asymptotic standard errors of the forecasted levels and, thus the asymptotic confidence intervals for the levels, not be calculated. By default, the asymptotic standard errors and the asymptotic confidence intervals of the forecasted levels are calculated.

bs specifies that fcast compute use confidence bounds estimated by a simulation method based on bootstrapping the residuals.

bsp specifies that fcast compute use confidence bounds estimated via simulation in which the innovations are drawn from a multivariate normal distribution.

bscentile specifies that fcast compute use centiles of the bootstrapped dataset to estimate the bounds of the confidence intervals. By default, fcast compute uses the estimated standard errors and the quantiles of the standard normal distribution determined by level().

reps(#) gives the number of repetitions used in the simulations. The default is 200.

nodots specifies that no dots be displayed while obtaining the simulation-based standard errors. By default, for each replication, a dot is displayed.

saving(filename[, replace]) specifies the name of the file to hold the dataset that contains the bootstrap replications. The replace option overwrites any file with this name.

　replace specifies that filename be overwritten if it exists. This option is not shown in the dialog box.

⌐ Reporting ⌐

level(#) specifies the confidence level, as a percentage, for confidence intervals. The default is level(95) or as set by set level; see [U] **20.7 Specifying the width of confidence intervals**.

Remarks

Researchers often use VARs and VECMs to construct dynamic forecasts. fcast compute computes dynamic forecasts of the dependent variables in a VAR or VECM previously fit by var, svar, or vec. If you are interested in conditional, one-step-ahead predictions, use predict (see [TS] **var**, [TS] **var svar**, and [TS] **vec**).

To obtain and analyze dynamic forecasts, you fit a model, use fcast compute to compute the dynamic forecasts, and use fcast graph to graph the results.

▷ Example 1

Typing

```
. use http://www.stata-press.com/data/r12/lutkepohl2
. var dln_inc dln_consump dln_inv if qtr<tq(1979q1)
. fcast compute m2_, step(8)
. fcast graph m2_dln_inc m2_dln_inv m2_dln_consump, observed
```

fits a VAR with two lags, computes eight-step dynamic predictions for each endogenous variable, and produces the graph

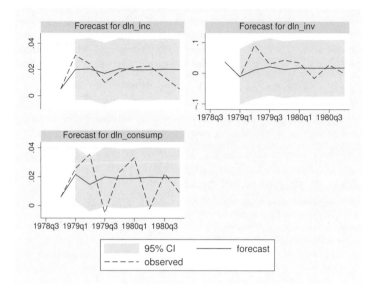

The graph shows that the model is better at predicting changes in income and investment than in consumption. The graph also shows how quickly the predictions from the two-lag model settle down to their mean values.

◁

fcast compute creates new variables in the dataset. If there are K dependent variables in the previously fitted model, fcast compute generates $4K$ new variables:

K new variables that hold the forecasted levels, named by appending the specified prefix to the name of the original variable

K estimated lower bounds for the forecast interval, named by appending the specified prefix and the suffix "_LB" to the name of the original variable

K estimated upper bounds for the forecast interval, named by appending the specified prefix and the suffix "_UB" to the name of the original variable

K estimated standard errors of the forecast, named by appending the specified prefix and the suffix "_SE" to the name of the original variable

If you specify options so that fcast compute does not calculate standard errors, the $3K$ variables that hold them and the bounds of the confidence intervals are not generated.

If the model previously fit is a VECM, specifying differences generates another K variables that hold the forecasts of the first differences of the dependent variables, named by appending the prefix "*prefix*D_" to the name of the original variable.

▷ Example 2

Plots of the forecasts from different models along with the observations from a holdout sample can provide insights to their relative forecasting performance. Continuing the previous example,

```
. var dln_inc dln_consump dln_inv if qtr<tq(1979q1), lags(1/6)
  (output omitted )

. fcast compute m6_, step(8)

. graph twoway line m6_dln_inv m2_dln_inv dln_inv qtr
> if m6_dln_inv < ., legend(cols(1))
```

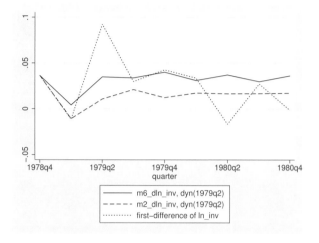

The model with six lags predicts changes in investment better than the two-lag model in some periods but markedly worse in other periods.

◁

Methods and formulas

`fcast compute` is implemented as an ado-file.

Predictions after var and svar

A VAR with endogenous variables $\mathbf{y}_t$ and exogenous variables $\mathbf{x}_t$ can be written as

$$\mathbf{y}_t = \mathbf{v} + \mathbf{A}_1 \mathbf{y}_{t-1} + \cdots + \mathbf{A}_p \mathbf{y}_{t-p} + \mathbf{B}\mathbf{x}_t + \mathbf{u}_t$$

where

$$t = 1, \ldots, T$$

$\mathbf{y}_t = (y_{1t}, \ldots, y_{Kt})'$ is a $K \times 1$ random vector,
the $\mathbf{A}_i$ are fixed $(K \times K)$ matrices of parameters,
$\mathbf{x}_t$ is an $(M \times 1)$ vector of exogenous variables,
$\mathbf{B}$ is a $(K \times M)$ matrix of coefficients,
$\mathbf{v}$ is a $(K \times 1)$ vector of fixed parameters, and
$\mathbf{u}_t$ is assumed to be white noise; that is,

$$E(\mathbf{u}_t) = \mathbf{0}_K$$
$$E(\mathbf{u}_t \mathbf{u}_t') = \mathbf{\Sigma}$$
$$E(\mathbf{u}_t \mathbf{u}_s') = \mathbf{0}_K \text{ for } t \neq s$$

`fcast compute` will dynamically predict the variables in the vector $\mathbf{y}_t$ conditional on p initial values of the endogenous variables and any exogenous $\mathbf{x}_t$. Adopting the notation from Lütkepohl (2005, 402) to fit the case at hand, the optimal h-step-ahead forecast of $\mathbf{y}_{t+h}$ conditional on $\mathbf{x}_t$ is

$$\mathbf{y}_t(h) = \widehat{\mathbf{v}} + \widehat{\mathbf{A}}_1 \mathbf{y}_t(h-1) + \cdots + \widehat{\mathbf{A}}_p \mathbf{y}_t(h-p) + \widehat{\mathbf{B}} \mathbf{x}_t \tag{1}$$

If there are no exogenous variables, (1) becomes

$$\mathbf{y}_t(h) = \widehat{\mathbf{v}} + \widehat{\mathbf{A}}_1 \mathbf{y}_t(h-1) + \cdots + \widehat{\mathbf{A}}_p \mathbf{y}_t(h-p)$$

When there are no exogenous variables, `fcast compute` can compute the asymptotic confidence bounds.

As shown by Lütkepohl (2005, 204–205), the asymptotic estimator of the covariance matrix of the prediction error is given by

$$\widehat{\mathbf{\Sigma}}_{\widehat{y}}(h) = \widehat{\mathbf{\Sigma}}_y(h) + \frac{1}{T}\widehat{\mathbf{\Omega}}(h) \tag{2}$$

where

$$\widehat{\mathbf{\Sigma}}_y(h) = \sum_{i=0}^{h-1} \widehat{\mathbf{\Phi}}_i \widehat{\mathbf{\Sigma}} \widehat{\mathbf{\Phi}}_i'$$

$$\widehat{\mathbf{\Omega}}(h) = \frac{1}{T}\sum_{t=0}^{T} \left\{ \sum_{i=0}^{h-1} \mathbf{Z}_t' \left(\widehat{\mathbf{B}}'\right)^{h-1-i} \otimes \widehat{\mathbf{\Phi}}_i \right\} \widehat{\mathbf{\Sigma}}_\beta \left\{ \sum_{i=0}^{h-1} \mathbf{Z}_t' \left(\widehat{\mathbf{B}}'\right)^{h-1-i} \otimes \widehat{\mathbf{\Phi}}_i \right\}' \tag{4}$$

$$\widehat{\mathbf{B}} = \begin{bmatrix} 1 & \mathbf{0} & \mathbf{0} & \cdots & \mathbf{0} & \mathbf{0} \\ \widehat{\mathbf{v}} & \widehat{\mathbf{A}}_1 & \widehat{\mathbf{A}}_2 & \cdots & \widehat{\mathbf{A}}_{p-1} & \widehat{\mathbf{A}}_p \\ \mathbf{0} & \mathbf{I}_K & \mathbf{0} & \cdots & \mathbf{0} & \mathbf{0} \\ \mathbf{0} & \mathbf{0} & \mathbf{I}_K & & \mathbf{0} & \mathbf{0} \\ \vdots & \vdots & & \ddots & & \vdots \\ \mathbf{0} & \mathbf{0} & \mathbf{0} & \cdots & \mathbf{I}_K & \mathbf{0} \end{bmatrix}$$

$$\mathbf{Z}_t = (1, \mathbf{y}_t', \ldots, \mathbf{y}_{t-p-1}')'$$

$$\widehat{\mathbf{\Phi}}_0 = \mathbf{I}_K$$

$$\widehat{\mathbf{\Phi}}_i = \sum_{j=1}^{i} \widehat{\mathbf{\Phi}}_{i-j}\widehat{\mathbf{A}}_j \qquad i = 1, 2, \ldots$$

$$\widehat{\mathbf{A}}_j = \mathbf{0} \quad \text{for } j > p$$

$\widehat{\mathbf{\Sigma}}$ is the estimate of the covariance matrix of the innovations, and $\widehat{\mathbf{\Sigma}}_\beta$ is the estimated VCE of the coefficients in the VAR. The formula in (3) is general enough to handle the case in which constraints are placed on the coefficients in the VAR(p).

Equation (2) is made up of two terms. $\widehat{\mathbf{\Sigma}}_y(h)$ is the estimated mean squared error (MSE) of the forecast. $\widehat{\mathbf{\Sigma}}_y(h)$ estimates the error in the forecast arising from the unseen innovations. $T^{-1}\widehat{\mathbf{\Omega}}(h)$ estimates the error in the forecast that is due to using estimated coefficients instead of the true coefficients. As the sample size grows, uncertainty with respect to the coefficient estimates decreases, and $T^{-1}\widehat{\mathbf{\Omega}}(h)$ goes to zero.

If $\mathbf{y}_t$ is normally distributed, the bounds for the asymptotic $(1 - \alpha)100\%$ interval around the forecast for the kth component of $\mathbf{y}_t$, h periods ahead, are

$$\widehat{\mathbf{y}}_{k,t}(h) \pm z_{\left(\frac{\alpha}{2}\right)}\widehat{\sigma}_k(h) \tag{3}$$

where $\widehat{\sigma}_k(h)$ is the kth diagonal element of $\widehat{\boldsymbol{\Sigma}}_{\widehat{y}}(h)$.

Specifying the bs option causes the standard errors to be computed via simulation, using bootstrapped residuals. Both var and svar contain estimators for the coefficients of a VAR that are conditional on the first p observations on the endogenous variables in the data. Similarly, these algorithms are conditional on the first p observations of the endogenous variables in the data. However, the simulation-based estimates of the standard errors are also conditional on the estimated coefficients. The asymptotic standard errors are not conditional on the coefficient estimates because the second term on the right-hand side of (2) accounts for the uncertainty arising from using estimated parameters.

For a simulation with R repetitions, this method uses the following algorithm:

1. Fit the model and save the estimated coefficients.

2. Use the estimated coefficients to calculate the residuals.

3. Repeat steps 3a–3c R times.

   3a. Draw a simple random sample with replacement of size $T + h$ from the residuals. When the tth observation is drawn, all K residuals are selected, preserving any contemporaneous correlation among the residuals.

   3b. Use the sampled residuals, p initial values of the endogenous variables, any exogenous variables, and the estimated coefficients to construct a new sample dataset.

   3c. Save the simulated endogenous variables for the h forecast periods in the bootstrapped dataset.

4. For each endogenous variable and each forecast period, the simulated standard error is the estimated standard error of the R simulated forecasts. By default, the upper and lower bounds of the $(1 - \alpha)100\%$ are estimated using the simulation-based estimates of the standard errors and the normality assumption, as in (3). If the bscentile option is specified, the sample centiles for the upper and lower bounds of the R simulated forecasts are used for the upper and lower bounds of the confidence intervals.

If the bsp option is specified, a parametric simulation algorithm is used. Specifically, everything is as above except that 3a is replaced by 3a(bsp) as follows:

   3a(bsp). Draw $T + h$ observations from a multivariate normal distribution with covariance matrix $\widehat{\boldsymbol{\Sigma}}$.

The algorithm above assumes that h forecast periods come after the original sample of T observations. If the h forecast periods lie within the original sample, smaller simulated datasets are sufficient.

Dynamic forecasts after vec

Methods and formulas of [TS] **vec** discusses how to obtain the one-step predicted differences and levels. fcast compute uses the previous dynamic predictions as inputs for later dynamic predictions.

Per Lütkepohl (2005, sec. 6.5), `fcast compute` uses

$$\widehat{\boldsymbol{\Sigma}}_{\widehat{y}}(h) = \left(\frac{T}{T-d}\right) \sum_{i=0}^{h-1} \widehat{\boldsymbol{\Phi}}_i \widehat{\boldsymbol{\Omega}} \widehat{\boldsymbol{\Phi}}_i$$

where the $\widehat{\boldsymbol{\Phi}}_i$ are the estimated matrices of impulse–response functions, T is the number of observations in the sample, d is the number of degrees of freedom, and $\widehat{\boldsymbol{\Omega}}$ is the estimated cross-equation variance matrix. The formulas for d and $\widehat{\boldsymbol{\Omega}}$ are given in *Methods and formulas* of [TS] **vec**.

The estimated standard errors at step h are the square roots of the diagonal elements of $\widehat{\boldsymbol{\Sigma}}_{\widehat{y}}(h)$.

Per Lütkepohl (2005), the estimated forecast-error variance does not consider parameter uncertainty. As the sample size gets infinitely large, the importance of parameter uncertainty diminishes to zero.

References

Hamilton, J. D. 1994. *Time Series Analysis*. Princeton: Princeton University Press.

Lütkepohl, H. 2005. *New Introduction to Multiple Time Series Analysis*. New York: Springer.

Also see

[TS] **fcast graph** — Graph forecasts of variables computed by fcast compute

[TS] **var intro** — Introduction to vector autoregressive models

[TS] **vec intro** — Introduction to vector error-correction models

Title

> **fcast graph** — Graph forecasts of variables computed by fcast compute

Syntax

fcast graph *varlist* [*if*] [*in*] [*, options*]

where *varlist* contains one or more forecasted variables generated by fcast compute.

options	Description
Main	
differences	graph forecasts of the first-differenced variables (vec only)
noci	suppress confidence bands
observed	include observed values of the predicted variables
Forecast plot	
cline_options	affect rendition of the forecast lines
CI plot	
ciopts(*area_options*)	affect rendition of the confidence bands
Observed plot	
obopts(*cline_options*)	affect rendition of the observed values
Y axis, Time axis, Titles, Legend, Overall	
twoway_options	any options other than by() documented in [G-3] ***twoway_options***
byopts(*by_option*)	affect appearance of the combined graph; see [G-3] ***by_option***

Menu

Statistics > Multivariate time series > Dynamic forecasts > Graph forecasts

Description

fcast graph graphs dynamic forecasts of the endogenous variables from a VAR(p) or VECM that has already been obtained from fcast compute; see [TS] **fcast compute**.

Options

Main

differences specifies that the forecasts of the first-differenced variables be graphed. This option is available only with forecasts computed by fcast compute after vec. The differences option implies noci.

noci specifies that the confidence intervals be suppressed. By default, the confidence intervals are included.

observed specifies that observed values of the predicted variables be included in the graph. By default, observed values are not graphed.

153

⌐ Forecast plot ⌐

cline_options affect the rendition of the plotted lines corresponding to the forecast; see [G-3] ***cline_options***.

⌐ CI plot ⌐

ciopts(*area_options*) affects the rendition of the confidence bands for the forecasts; see [G-3] ***area_options***.

⌐ Observed plot ⌐

obopts(*cline_options*) affects the rendition of the observed values of the predicted variables; see [G-3] ***cline_options***. This option implies the observed option.

⌐ Y axis, Time axis, Titles, Legend, Overall ⌐

twoway_options are any of the options documented in [G-3] ***twoway_options***, excluding by().

byopts(*by_option*) are documented in [G-3] ***by_option***. These options affect the appearance of the combined graph.

Remarks

fcast graph graphs dynamic forecasts created by fcast compute.

▷ Example 1

In this example, we use a cointegrating VECM to model the state-level unemployment rates in Missouri, Indiana, Kentucky, and Illinois, and we graph the forecasts against a 6-month holdout sample.

```
. use http://www.stata-press.com/data/r12/urates
. vec missouri indiana kentucky illinois if t < tm(2003m7), trend(rconstant)
> rank(2) lags(4)
  (output omitted )
. fcast compute m1_, step(6)
```

```
. fcast graph m1_missouri m1_indiana m1_kentucky m1_illinois, observed
```

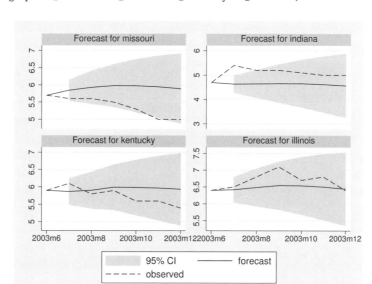

Because the 95% confidence bands for the predicted unemployment rates in Missouri and Indiana do not contain all their observed values, the model does not reliably predict these unemployment rates.

◁

Methods and formulas

fcast graph is implemented as an ado-file.

Also see

[TS] **fcast compute** — Compute dynamic forecasts of dependent variables after var, svar, or vec

[TS] **var intro** — Introduction to vector autoregressive models

[TS] **vec intro** — Introduction to vector error-correction models

Title

> **haver** — Load data from Haver Analytics database

Syntax

Describe contents of a Haver dataset

> haver <u>des</u>cribe *filename* [, <u>det</u>ail]

Describe specified variables in a Haver dataset

> haver <u>des</u>cribe *varlist* using *filename* [, <u>det</u>ail]

Load Haver dataset

> haver use *filename* [, *use_options*]

Load specified variables from a Haver dataset

> haver use *varlist* using *filename* [, *use_options*]

use_options	Description
<u>tin</u>([*constant*] , [*constant*])	load data within specified date range
<u>twithin</u>([*constant*] , [*constant*])	same as tin(), except exclude the endpoints of range
<u>tvar</u>(*varname*)	create time variable *varname*
<u>hm</u>issing(*misval*)	record missing values as *misval*
<u>fill</u>	include observations with missing data in resulting dataset and record missing values for them
clear	clear data in memory before loading the Haver dataset

Menu

File > Import > Haver Analytics database

Description

Haver Analytics (http://www.haver.com) provides economic and financial databases in the form of .dat files to which you can purchase access. The haver command allows you to use those datasets with Stata. The haver command is provided only with Stata for Windows.

haver describe describes the contents of a Haver file.

haver use loads the specified variables from a Haver file into Stata's memory.

If *filename* is specified without a suffix, .dat is assumed.

Option for use with haver describe

detail specifies that a detailed report of all information available on the variables be presented. By default, the Haver concepts data type, number of observations, aggregate type, difference type, magnitude, and date modified are not reported.

Options for use with haver use

tin([*constant*], [*constant*]) specifies the date range of the data to be loaded. *constant* refers to a date constant specified in the usual way, for example, tin(1jan1999, 31dec1999), which would mean from and including 1 January 1999 through 31 December 1999.

twithin([*constant*], [*constant*]) functions the same as tin(), except that the endpoints of the range will be excluded in the data loaded.

tvar(*varname*) specifies the name of the time variable Stata will create; the default is tvar(time). The tvar() variable is the name of the variable that you would use to tsset the data after loading, although doing is unnecessary because haver use automatically tssets the data for you.

hmissing(*misval*) specifies which of Stata's 27 missing values (., .a, ..., .z) is to be recorded when there are missing values in the Haver dataset.

Two kinds of missing values can occur when loading a Haver dataset. One is created when a variable's data does not span the entire time series, and these missing values are always recorded as . by Stata. The other corresponds to an actual missing value recorded in the Haver format. hmissing() sets what is recorded when this second kind of missing value is encountered. You could specify hmissing(.h) to cause Stata's .h code to be recorded here.

The default is to store . for both kinds of missing values. See [U] **12.2.1 Missing values**.

fill specifies that observations with no data be left in the resulting dataset formed in memory and that missing values be recorded for them. The default is to exclude such observations, which can result in the loaded time-series dataset having gaps.

Specifying fill has the same effect as issuing the tsfill command after loading; see [TS] **tsfill**.

clear clears the data in memory before loading the Haver dataset.

Remarks

Remarks are presented under the following headings:

> *Installation*
> *Determining the contents of a Haver dataset*
> *Loading a Haver dataset*
> *Combining variables*
> *Using subsets*

Installation

Haver Analytics (http://www.haver.com) provides more than 100 economic and financial databases in the form of .dat files to which you can purchase access. The haver command allows you to use those datasets with Stata.

haver is provided only with Stata for Windows, and it uses the DLXAPI32.DLL file, which Haver provides. That file is probably already installed. If it is not, Stata will complain that "DLXAPI32.DLL could not be found".

Stata accesses this file via the Windows operating system if it is installed where Windows expects. If Windows cannot find the file, the location can be specified via Stata's `set` command:

```
. set haverdll "C:\Windows\system32\DLXAPI32.DLL"
```

Determining the contents of a Haver dataset

`haver describe` displays the contents of a Haver dataset. If no varlist is specified, all variables are described:

```
. haver describe dailytst
Dataset:        dailytst (use)
```

Variable	Description	Time span	Format	Group	Source
FDB6	6-Month Eurod..	02jan1996-01aug2003	Daily	F01	FRB
FFED	Federal Funds..	01jan1996-01aug2003	Daily	F01	FRB
FFED2	Federal Funds..	01jan1988-29dec1995	Daily	F70	FRB
FFED3	Federal Funds..	01jan1980-31dec1987	Daily	F70	FRB
FFED4	Federal Funds..	03jan1972-31dec1979	Daily	F70	FRB
FFED5	Federal Funds..	01jan1964-31dec1971	Daily	F70	FRB
FFED6	Federal Funds..	02jan1962-31dec1963	Daily	F70	FRB

Above, we describe the Haver dataset `dailytst.dat`, which we already have on our computer and in our current directory.

By default, each line of the output corresponds to one Haver variable. Specifying `detail` would display more information about each variable, and specifying the optional varlist allows us to restrict the output to the variables that interest us:

```
. haver describe FDB6 using dailytst, detail
```

FDB6	6-Month Eurodollar Deposits (London Bid) (% p.a.)

Frequency: Daily	Time span: 02jan1996 - 01aug2003
Group: F01	Number of Observations: 1979
Source: FRB	Date Modified: 04aug2003
Agg Type: 1	Decimal Precision: 2
Difference Type: 1	Magnitude: 0
Data Type: %	

Loading a Haver dataset

`haver use` loads Haver datasets. If no varlist is specified, all variables are loaded:

```
. haver use dailytst
. list in 1/5
```

	FDB6	FFED	FFED2	FFED3	FFED4	FFED5	FFED6	time
1.	.	.	.	.	.	.	2.75	02jan1962
2.	.	.	.	.	.	.	2.5	03jan1962
3.	.	.	.	.	.	.	2.75	04jan1962
4.	.	.	.	.	.	.	2.5	05jan1962
5.	.	.	.	.	.	.	2	08jan1962

The entire dataset contains 10,849 observations. The variables FDB6 through FFED5 contain missing because there is no information about them for these first few dates; later in the dataset, nonmissing values for them will be found.

haver use requires that the variables loaded all be of the same time frequency (for example, annual, quarterly) Haver's weekly data are treated by Stata as daily data with a 7-day interval, so weekly data is loaded as daily data with gaps.

Haver datasets include the following information about each variable:

- Source — the Haver code associated with the source for the data
- Group — the Haver group to which the variable belongs
- DataType — whether the variable is a ratio, index, etc.
- AggType — the Haver code for the level of aggregation
- DifType — the Haver code for the type of difference
- DecPrecision — the number of decimals to which the variable is recorded
- Magnitude — the Haver magnitude code
- DateTimeMod — the date the series was last modified by Haver

When a variable is loaded, this information is stored in variable characteristics (see [P] **char**). Those characteristics can be viewed using char list:

```
. char list FDB6[]
FDB6[Source]:           FRB
FDB6[Group]:            F01
FDB6[DataType]:         %
FDB6[AggType]:          1
FDB6[DifType]:          1
FDB6[DecPrecision]:     2
FDB6[Magnitude]:        0
FDB6[DateTimeMod]:      04aug2003
```

Combining variables

In fact, the variables FFED through FFED6 all contain information about the same thing (the Federal Funds rate) but over different times:

```
. haver describe dailytst
Dataset:      dailytst (use)
```

Variable	Description	Time span	Format	Group	Source
FDB6	6-Month Eurod..	02jan1996-01aug2003	Daily	F01	FRB
FFED	Federal Funds..	01jan1996-01aug2003	Daily	F01	FRB
FFED2	Federal Funds..	01jan1988-29dec1995	Daily	F70	FRB
FFED3	Federal Funds..	01jan1980-31dec1987	Daily	F70	FRB
FFED4	Federal Funds..	03jan1972-31dec1979	Daily	F70	FRB
FFED5	Federal Funds..	01jan1964-31dec1971	Daily	F70	FRB
FFED6	Federal Funds..	02jan1962-31dec1963	Daily	F70	FRB

In Haver datasets, variables may contain up to 2,100 data points, and if more are needed, Haver Analytics breaks the variables into multiple variables. You must put them back together.

To determine when variables need to be combined, look at the description and time span. To combine variables, use egen's rowfirst() function and then drop the originals:

```
. egen fedfunds = rowfirst(FFED FFED2 FFED3 FFED4 FFED5 FFED6)
. drop FFED*
```

You can also combine variables by hand:

```
. gen fedfunds = FFED
(8869 missing values generated)
. replace fedfunds = FFED2 if fedfunds >= .
(2086 real changes made)
. replace fedfunds = FFED3 if fedfunds >= .
(2088 real changes made)
. replace fedfunds = FFED4 if fedfunds >= .
(2086 real changes made)
. replace fedfunds = FFED5 if fedfunds >= .
(2088 real changes made)
. replace fedfunds = FFED6 if fedfunds >= .
(521 real changes made)
. drop FFED*
```

Using subsets

You can use a subset of the Haver dataset by specifying the variables to be loaded or the observations to be included:

```
. haver use FFED FFED2 using dailytst, tin(1jan1990,31dec1999)
(data from dailytst.dat loaded)
```

Above, we load only the variables FFED and FFED2 and the observations from 1jan1990 to 31dec1999.

Also see

[TS] **tsset** — Declare data to be time-series data

[D] **import** — Overview of importing data into Stata

[D] **insheet** — Read text data created by a spreadsheet

[D] **odbc** — Load, write, or view data from ODBC sources

Title

> **irf** — Create and analyze IRFs, dynamic-multiplier functions, and FEVDs

Syntax

$$\texttt{irf } \textit{subcommand } \dots \ \big[\ , \ \dots \big]$$

subcommand	Description
create	create IRF file containing IRFs, dynamic-multiplier functions, and FEVDs
set	set the active IRF file
graph	graph results from active file
cgraph	combine graphs of IRFs, dynamic-multiplier functions, and FEVDs
ograph	graph overlaid IRFs, dynamic-multiplier functions, and FEVDs
table	create tables of IRFs, dynamic-multiplier functions, and FEVDs from active file
ctable	combine tables of IRFs, dynamic-multiplier functions, and FEVDs
describe	describe contents of active file
add	add results from an IRF file to the active IRF file
drop	drop IRF results from active file
rename	rename IRF results within a file

IRF stands for impulse–response function; FEVD stands for forecast-error variance decomposition.

irf can be used only after var, svar, or vec; see [TS] **var**, [TS] **var svar**, and [TS] **vec**.

See [TS] **irf create**, [TS] **irf set**, [TS] **irf graph**, [TS] **irf cgraph**, [TS] **irf ograph**, [TS] **irf table**, [TS] **irf ctable**, [TS] **irf describe**, [TS] **irf add**, [TS] **irf drop**, and [TS] **irf rename** for details about subcommands.

Description

irf creates and manipulates IRF files that contain estimates of the IRFs, dynamic-multiplier functions, and forecast-error variance decompositions (FEVDs) created after estimation by var, svar, or vec; see [TS] **var**, [TS] **var svar**, or [TS] **vec**.

IRFs and FEVDs are described below, and the process of analyzing them is outlined. After reading this entry, please see [TS] **irf create**.

Remarks

An IRF measures the effect of a shock to an endogenous variable on itself or on another endogenous variable; see Lütkepohl (2005, 51–63) and Hamilton (1994, 318–323) for formal definitions. Of the many types of IRFs, irf create estimates the five most important: simple IRFs, orthogonalized IRFs, cumulative IRFs, cumulative orthogonalized IRFs, and structural IRFs.

A dynamic-multiplier function, or transfer function, measures the impact of a unit increase in an exogenous variable on the endogenous variables over time; see Lütkepohl (2005, chap. 10) for formal definitions. irf create estimates simple and cumulative dynamic-multiplier functions after var.

The forecast-error variance decomposition (FEVD) measures the fraction of the forecast-error variance of an endogenous variable that can be attributed to orthogonalized shocks to itself or to another endogenous variable; see Lütkepohl (2005, 63–66) and Hamilton (1994, 323–324) for formal definitions. Of the many types of FEVDs, `irf create` estimates the two most important: Cholesky and structural.

To analyze IRFs and FEVDs in Stata, you first fit a model, then use `irf create` to estimate the IRFs and FEVDs and store them in a file, and finally use `irf graph` or any of the other `irf` analysis commands to examine results:

```
. use http://www.stata-press.com/data/r12/lutkepohl2
(Quarterly SA West German macro data, Bil DM, from Lutkepohl 1993 Table E.1)
. var dln_inv dln_inc dln_consump if qtr<=tq(1978q4), lags(1/2) dfk
  (output omitted )
. irf create order1, step(10) set(myirf1)
(file myirf1.irf created)
(file myirf1.irf now active)
(file myirf1.irf updated)
. irf graph oirf, impulse(dln_inc) response(dln_consump)
```

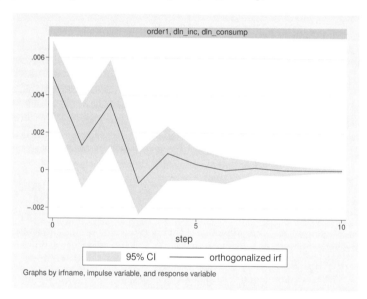

Graphs by irfname, impulse variable, and response variable

Multiple sets of IRFs and FEVDs can be placed in the same file, with each set of results in a file bearing a distinct name. The `irf create` command above created file `myirf1.irf` and put one set of results in it, named `order1`. The `order1` results include estimates of the simple IRFs, orthogonalized IRFs, cumulative IRFs, cumulative orthogonalized IRFs, and Cholesky FEVDs.

Below we use the same estimated `var` but use a different Cholesky ordering to create a second set of IRF results, which we will store as `order2` in the same file, and then we will graph both results:

```
. irf create order2, step(10) order(dln_inc dln_inv dln_consump)
(file myirf1.irf updated)
. irf graph oirf, irf(order1 order2) impulse(dln_inc) response(dln_consump)
```

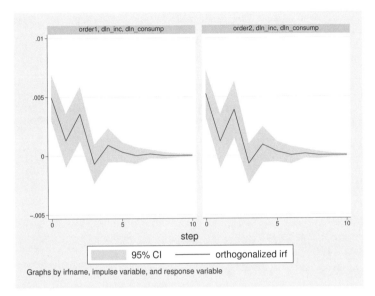

We have compared results for one model under two different identification schemes. We could just as well have compared results of two different models. We now use irf table to display the results tabularly:

```
. irf table oirf, irf(order1 order2) impulse(dln_inc) response(dln_consump)
                       Results from order1 order2
```

step	(1) oirf	(1) Lower	(1) Upper	(2) oirf	(2) Lower	(2) Upper
0	.004934	.003016	.006852	.005244	.003252	.007237
1	.001309	-.000931	.003549	.001235	-.001011	.003482
2	.003573	.001285	.005862	.00391	.001542	.006278
3	-.000692	-.002333	.00095	-.000677	-.002347	.000993
4	.000905	-.000541	.002351	.00094	-.000576	.002456
5	.000328	-.0005	.001156	.000341	-.000518	.001201
6	.000021	-.000675	.000717	.000042	-.000693	.000777
7	.000154	-.000206	.000515	.000161	-.000218	.00054
8	.000026	-.000248	.0003	.000027	-.000261	.000315
9	.000026	-.000121	.000174	.00003	-.000125	.000184
10	.000026	-.000061	.000113	.000027	-.000065	.00012

```
95% lower and upper bounds reported
(1) irfname = order1, impulse = dln_inc, and response = dln_consump
(2) irfname = order2, impulse = dln_inc, and response = dln_consump
```

Both the table and the graph show that the two orthogonalized IRFs are essentially the same. In both functions, an increase in the orthogonalized shock to dln_inc causes a short series of increases in dln_consump that dies out after four or five periods.

References

Hamilton, J. D. 1994. *Time Series Analysis*. Princeton: Princeton University Press.

Lütkepohl, H. 1993. *Introduction to Multiple Time Series Analysis*. 2nd ed. New York: Springer.

——. 2005. *New Introduction to Multiple Time Series Analysis*. New York: Springer.

Also see

[TS] **var** — Vector autoregressive models

[TS] **var svar** — Structural vector autoregressive models

[TS] **varbasic** — Fit a simple VAR and graph IRFs or FEVDs

[TS] **vec** — Vector error-correction models

[TS] **var intro** — Introduction to vector autoregressive models

[TS] **vec intro** — Introduction to vector error-correction models

Title

> **irf add** — Add results from an IRF file to the active IRF file

Syntax

> irf <u>a</u>dd { _all | [*newname*=]*oldname* ... }, using(*irf_filename*)

Menu

Statistics > Multivariate time series > Manage IRF results and files > Add IRF results

Description

irf add copies results from one IRF file to another—from the specified using() file to the active IRF file, set by irf set; see [TS] **irf set**.

Option

using(*irf_filename*) specifies the file from which results are to be obtained and is required. If *irf_filename* is specified without an extension, .irf is assumed.

Remarks

If you have not read [TS] **irf**, please do so.

▷ Example 1

After fitting a VAR model, we create two separate IRF files:

```
. use http://www.stata-press.com/data/r12/lutkepohl2
(Quarterly SA West German macro data, Bil DM, from Lutkepohl 1993 Table E.1)
. var dln_inv dln_inc dln_consump if qtr<=tq(1978q4), lags(1/2) dfk
 (output omitted )
. irf create original, set(irf1, replace)
(file irf1.irf created)
(file irf1.irf now active)
(file irf1.irf updated)
. irf create order2, order(dln_inc dln_inv dln_consump) set(irf2, replace)
(file irf2.irf created)
(file irf2.irf now active)
(file irf2.irf updated)
```

We copy IRF results original to the active file giving them the name order1.

```
. irf add order1 = original, using(irf1)
(file irf2.irf updated)
```

Here we create new IRF results and save them in the new file irf3.

```
. irf create order3, order(dln_inc dln_consump dln_inv) set(irf3, replace)
(file irf3.irf created)
(file irf3.irf now active)
(file irf3.irf updated)
```

Now we copy all the IRF results in file irf2 into the active file.

```
. irf add _all, using(irf2)
(file irf3.irf updated)
```

◁

Methods and formulas

irf add is implemented as an ado-file.

Also see

[TS] **irf** — Create and analyze IRFs, dynamic-multiplier functions, and FEVDs

[TS] **var intro** — Introduction to vector autoregressive models

[TS] **vec intro** — Introduction to vector error-correction models

Title

> **irf cgraph** — Combine graphs of IRFs, dynamic-multiplier functions, and FEVDs

Syntax

irf cgraph (*spec₁*) [(*spec₂*) ... [(*spec_N*)]] [, *options*]

where (*spec_k*) is

(*irfname impulsevar responsevar stat* [, *spec_options*])

irfname is the name of a set of IRF results in the active IRF file. *impulsevar* should be specified as an endogenous variable for all statistics except dm and cdm; for those, specify as an exogenous variable. *responsevar* is an endogenous variable name. *stat* is one or more statistics from the list below:

stat	Description
Main	
irf	impulse–response function
oirf	orthogonalized impulse–response function
dm	dynamic-multiplier function
cirf	cumulative impulse–response function
coirf	cumulative orthogonalized impulse–response function
cdm	cumulative dynamic-multiplier function
fevd	Cholesky forecast-error variance decomposition
sirf	structural impulse–response function
sfevd	structural forecast-error variance decomposition

Notes: 1. No statistic may appear more than once.

     2. If confidence intervals are included (the default), only two statistics may be included.

     3. If confidence intervals are suppressed (option noci), up to four statistics may be included.

options	Description
Main	
set (*filename*)	make *filename* active
Options	
combine_options	affect appearance of combined graph
Y axis, X axis, Titles, Legend, Overall	
twoway_options	any options other than by() documented in [G-3] ***twoway_options***
* *spec_options*	level, steps, and rendition of plots and their CIs
individual	graph each combination individually

* *spec_options* appear on multiple tabs in the dialog box.

individual does not appear in the dialog box.

167

spec_options	Description
Main	
noci	suppress confidence bands
Options	
level(#)	set confidence level; default is level(95)
lstep(#)	use # for first step
ustep(#)	use # for maximum step
Plots	
plot#opts(*line_options*)	affect rendition of the line plotting the # *stat*
CI plots	
ci#opts(*area_options*)	affect rendition of the confidence interval for the # *stat*

spec_options may be specified within a graph specification, globally, or in both. When specified in a graph specification, the *spec_options* affect only the specification in which they are used. When supplied globally, the *spec_options* affect all graph specifications. When supplied in both places, options in the graph specification take precedence.

Menu

Statistics > Multivariate time series > IRF and FEVD analysis > Combined graphs

Description

irf cgraph makes a graph or a combined graph of IRF results. Each block within a pair of matching parentheses—each (*spec_k*)—specifies the information for a specific graph. irf cgraph combines these graphs into one image, unless the individual option is also specified, in which case separate graphs for each block are created.

To become familiar with this command, we recommend that you type db irf cgraph.

Options

    ⌐ Main ⌐

noci suppresses graphing the confidence interval for each statistic. noci is assumed when the model was fit by vec because no confidence intervals were estimated.

set(*filename*) specifies the file to be made active; see [TS] **irf set**. If set() is not specified, the active file is used.

    ⌐ Options ⌐

level(#) specifies the default confidence level, as a percentage, for confidence intervals, when they are reported. The default is level(95) or as set by set level; see [U] **20.7 Specifying the width of confidence intervals**. The value set of an overall level() can be overridden by the level() inside a (*spec_k*).

lstep(#) specifies the first step, or period, to be included in the graph. lstep(0) is the default.

ustep(#), # ≥ 1, specifies the maximum step, or period, to be included in the graph.

combine_options affect the appearance of the combined graph; see [G-2] **graph combine**.

    ⌐ Plots ⌐

plot1opts(*cline_options*), ..., plot4opts(*cline_options*) affect the rendition of the plotted statistics. plot1opts() affects the rendition of the first statistic; plot2opts(), the second; and so on. *cline_options* are as described in [G-3] ***cline_options***.

    ⌐ CI plots ⌐

ci1opts1(*area_options*) and ci2opts2(*area_options*) affect the rendition of the confidence intervals for the first (ci1opts()) and second (ci2opts()) statistics. See [TS] **irf graph** for a description of this option and [G-3] ***area_options*** for the suboptions that change the look of the CI.

    ⌐ Y axis, X axis, Titles, Legend, Overall ⌐

twoway_options are any of the options documented in [G-3] ***twoway_options***, excluding by(). These include options for titling the graph (see [G-3] ***title_options***) and for saving the graph to disk (see [G-3] ***saving_option***).

The following option is available with irf cgraph but is not shown in the dialog box:

individual specifies that each graph be displayed individually. By default, irf cgraph combines the subgraphs into one image.

Remarks

If you have not read [TS] **irf**, please do so.

The relationship between irf cgraph and irf graph is syntactically and conceptually the same as that between irf ctable and irf table; see [TS] **irf ctable** for a description of the syntax.

irf cgraph is much the same as using irf graph to make individual graphs and then using graph combine to put them together. If you cannot use irf cgraph to do what you want, consider the other approach.

▷ Example 1

You have previously issued the commands

```
. use http://www.stata-press.com/data/r12/lutkepohl2
. mat a = (., 0, 0\0,.,0\.,.,.)
. mat b = I(3)
. svar dln_inv dln_inc dln_consump, aeq(a) beq(b)
. irf create modela, set(results3) step(8)
. svar dln_inc dln_inv dln_consump, aeq(a) beq(b)
. irf create modelb, step(8)
```

You now type

```
. irf cgraph (modela dln_inc dln_consump oirf sirf)
>            (modelb dln_inc dln_consump oirf sirf)
>            (modela dln_inc dln_consump fevd sfevd, lstep(1))
>            (modelb dln_inc dln_consump fevd sfevd, lstep(1)),
>            title("Results from modela and modelb")
```

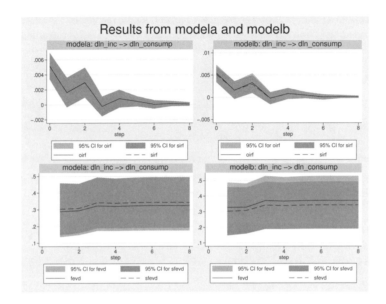

◁

Saved results

irf cgraph saves the following in r():

Scalars
    r(k)   number of specific graph commands

Macros
    r(individual)   individual, if specified
    r(save)   *filename*, replace from saving() option for combined graph
    r(name)   *name*, replace from name() option for combined graph
    r(title)   title of the combined graph
    r(save#)   *filename*, replace from saving() option for individual graphs
    r(name#)   *name*, replace from name() option for individual graphs
    r(title#)   title for the #th graph
    r(ci#)   level applied to the #th confidence interval or noci
    r(response#)   response specified in the #th command
    r(impulse#)   impulse specified in the #th command
    r(irfname#)   IRF name specified in the #th command
    r(stats#)   statistics specified in the #th command

Methods and formulas

irf cgraph is implemented as an ado-file.

Also see

[TS] **irf** — Create and analyze IRFs, dynamic-multiplier functions, and FEVDs

[TS] **var intro** — Introduction to vector autoregressive models

[TS] **vec intro** — Introduction to vector error-correction models

Title

> **irf create** — Obtain IRFs, dynamic-multiplier functions, and FEVDs

Syntax

After var

> irf c̲reate *irfname* [, *var_options*]

After svar

> irf c̲reate *irfname* [, *svar_options*]

After vec

> irf c̲reate *irfname* [, *vec_options*]

irfname is any valid name that does not exceed 15 characters.

var_options	Description
Main	
set(*filename*[, replace])	make *filename* active
replace	replace *irfname* if it already exists
s̲tep(*#*)	set forecast horizon to *#*; default is step(8)
o̲rder(*varlist*)	specify Cholesky ordering of endogenous variables
est̲imates(*estname*)	use previously saved results *estname*; default is to use active results
Std. errors	
nose	do not calculate standard errors
bs	obtain standard errors from bootstrapped residuals
bsp	obtain standard errors from parametric bootstrap
nod̲ots	do not display "." for each bootstrap replication
r̲eps(*#*)	use *#* bootstrap replications; default is reps(200)
bsa̲ving(*filename*[, replace])	save bootstrap results in *filename*

svar_options	Description
Main	
set(*filename*[, replace])	make *filename* active
replace	replace *irfname* if it already exists
s̲tep(*#*)	set forecast horizon to *#*; default is step(8)
est̲imates(*estname*)	use previously saved results *estname*; default is to use active results

Std. errors

nose	do not calculate standard errors
bs	obtain standard errors from bootstrapped residual
bsp	obtain standard errors from parametric bootstrap
nodots	do not display "." for each bootstrap replication
reps(#)	use # bootstrap replications; default is reps(200)
bsaving(filename[, replace])	save bootstrap results in filename

vec_options	Description
Main	
set(filename[, replace])	make filename active
replace	replace irfname if it already exists
step(#)	set forecast horizon to #; default is step(8)
estimates(estname)	use previously saved results estname; default is to use active results

The default is to use asymptotic standard errors if no options are specified.

irf create is for use after fitting a model with the var, svar, or vec commands; see [TS] **var**, [TS] **var svar**, and [TS] **vec**.

You must tsset your data before using var, svar, or vec and, hence, before using irf create; see [TS] **tsset**.

Menu

Statistics > Multivariate time series > IRF and FEVD analysis > Obtain IRFs, dynamic-multiplier functions, and FEVDs

Description

irf create estimates multiple sets of impulse–response functions (IRFs), dynamic-multiplier functions, and forecast-error variance decompositions (FEVDs) after estimation by var, svar, or vec; see [TS] **var**, [TS] **var svar**, or [TS] **vec**. These estimates and their standard errors are known collectively as IRF results and are stored in an IRF file under the specified *irfname*.

The following types of IRFs and dynamic-multiplier functions are stored:

simple IRFs	after svar, var, or vec
orthogonalized IRFs	after svar, var, or vec
dynamic multipliers	after var
cumulative IRFs	after svar, var, or vec
cumulative orthogonalized IRFs	after svar, var, or vec
cumulative dynamic multipliers	after var
structural IRFs	after svar only

The following types of FEVDs are stored:

Cholesky FEVDs	after svar, var, or vec
structural FEVDs	after svar only

Once you have created a set of IRF results, use the other irf commands to analyze them.

Options

set(*filename*[, replace]) specifies the IRF file to be used. If set() is not specified, the active IRF file is used; see [TS] **irf set**.

If set() is specified, the specified file becomes the active file, just as if you had issued an irf set command.

replace specifies that the results stored under *irfname* may be replaced, if they already exist. IRF results are saved in files, and one file may contain multiple IRF results.

step(*#*) specifies the step (forecast) horizon; the default is eight periods.

order(*varlist*) is allowed only after estimation by var; it specifies the Cholesky ordering of the endogenous variables to be used when estimating the orthogonalized IRFs. By default, the order in which the variables were originally specified on the var command is used.

estimates(*estname*) specifies that estimation results previously estimated by svar, var, or vec, and stored by estimates, be used. This option is rarely specified; see [R] **estimates**.

nose, bs, and bsp are alternatives that specify how (whether) standard errors are to be calculated. If none of these options is specified, asymptotic standard errors are calculated, except in two cases: after estimation by vec and after estimation by svar in which long-run constraints were applied. In those two cases, the default is as if nose were specified, although in the second case, you could specify bs or bsp. After estimation by vec, standard errors are simply not available.

nose specifies that no standard errors be calculated.

bs specifies that standard errors be calculated by bootstrapping the residuals. bs may not be specified if there are gaps in the data.

bsp specifies that standard errors be calculated via a multivariate-normal parametric bootstrap. bsp may not be specified if there are gaps in the data.

nodots, reps(*#*), and bsaving(*filename*[, replace]) are relevant only if bs or bsp is specified.

nodots specifies that dots not be displayed each time irf create performs a bootstrap replication.

reps(*#*), *# > 50*, specifies the number of bootstrap replications to be performed. reps(200) is the default.

bsaving(*filename*[, replace]) specifies that file *filename* be created and that the bootstrap replications be saved in it. New file *filename* is just a .dta dataset than can be loaded later using use; see [D] **use**. If *filename* is specified without an extension, .dta is assumed.

Remarks

If you have not read [TS] **irf**, please do so. An introductory example using IRFs is presented there.

Remarks are presented under the following headings:

> *Introductory examples*
> *Technical aspects of IRF files*
> *IRFs and FEVDs*
> *IRF results for VARs*
>> *An introduction to impulse–response functions for VARs*
>> *An introduction to dynamic-multiplier functions for VARs*
>> *An introduction to forecast-error variance decompositions for VARs*
> *IRF results for VECMs*
>> *An introduction to impulse–response functions for VECMs*
>> *An introduction to forecast-error variance decompositions for VECMs*

Introductory examples

▷ Example 1: After var

Below we compare bootstrap and asymptotic standard errors for a specific FEVD. We begin by fitting a VAR(2) model to the Lütkepohl data (we use the `var` command). We next use the `irf create` command twice, first to create results with asymptotic standard errors (saved under the name `asymp`) and then to recreate the same results, this time with bootstrap standard errors (saved under the name `bs`). Because bootstrapping is a random process, we set the random-number seed (`set seed 123456`) before using `irf create` the second time; this makes our results reproducible. Finally, we compare results by using the IRF analysis command `irf ctable`.

```
. use http://www.stata-press.com/data/r12/lutkepohl2
(Quarterly SA West German macro data, Bil DM, from Lutkepohl 1993 Table E.1)
. var dln_inv dln_inc dln_consump if qtr>=tq(1961q2) & qtr<=tq(1978q4), lags(1/2)
  (output omitted )
. irf create asymp, step(8) set(results1)
(file results1.irf created)
(file results1.irf now active)
(file results1.irf updated)
. set seed 123456
. irf create bs, step(8) bs reps(250) nodots
(file results1.irf updated)
. irf ctable (asymp dln_inc dln_consump fevd)
> (bs dln_inc dln_consump fevd), noci stderror
```

step	(1) fevd	(1) S.E.	(2) fevd	(2) S.E.
0	0	0	0	0
1	.282135	.087373	.282135	.104073
2	.278777	.083782	.278777	.096954
3	.33855	.090006	.33855	.100452
4	.339942	.089207	.339942	.099085
5	.342813	.090494	.342813	.099326
6	.343119	.090517	.343119	.09934
7	.343079	.090499	.343079	.099325
8	.34315	.090569	.34315	.099368

(1) irfname = asymp, impulse = dln_inc, and response = dln_consump
(2) irfname = bs, impulse = dln_inc, and response = dln_consump

Point estimates are, of course, the same. The bootstrap estimates of the standard errors, however, are larger than the asymptotic estimates, which suggests that the sample size of 71 is not large enough for the distribution of the estimator of the FEVD to be well approximated by the asymptotic distribution. Here we would expect the bootstrap confidence interval to be more reliable than the confidence interval that is based on the asymptotic standard error.

◁

❑ Technical note

The details of the bootstrap algorithms are given in *Methods and formulas*. These algorithms are conditional on the first p observations, where p is the order of the fitted VAR. (In an SVAR model, p is the order of the VAR that underlies the SVAR.) The bootstrapped estimates are conditional on the first p observations, just as the estimators of the coefficients in VAR models are conditional on the first p observations. With bootstrap standard errors (option bs), the p initial observations are used with resampling the residuals to produce the bootstrap samples used for estimation. With the more parametric bootstrap (option bsp), the p initial observations are used with draws from a multivariate-normal distribution with variance–covariance matrix $\widehat{\Sigma}$ to generate the bootstrap samples.

❑

❑ Technical note

For var and svar e() results, irf uses $\widehat{\Sigma}$, the estimated variance matrix of the disturbances, in computing the asymptotic standard errors of all the functions. The point estimates of the orthogonalized impulse–response functions, the structural impulse–response functions, and all the variance decompositions also depend on $\widehat{\Sigma}$. As discussed in [TS] **var**, var and svar use the ML estimator of this matrix by default, but they have option dfk, which will instead use an estimator that includes a small-sample correction. Specifying dfk when the model is fit—when the var or svar command is given—changes the estimate of $\widehat{\Sigma}$ and will change the IRF results that depend on it.

❑

▷ Example 2: After var with exogenous variables

After fitting a VAR, irf create computes estimates of the dynamic multipliers, which describe the impact of a unit change in an exogenous variable on each endogenous variable. For instance, below we estimate and report the cumulative dynamic multipliers from a model in which changes in investment are exogenous. The results indicate that both of the cumulative dynamic multipliers are significant.

```
. var dln_inc dln_consump if qtr>=tq(1961q2) & qtr<=tq(1978q4), lags(1/2)
> exog(L(0/2).dln_inv)
  (output omitted )
. irf create dm, step(8)
(file results1.irf updated)
```

```
. irf table cdm, impulse(dln_inv) irf(dm)
```

Results from dm

step	(1) cdm	(1) Lower	(1) Upper
0	.032164	-.027215	.091544
1	.096568	.003479	.189656
2	.140107	.022897	.257317
3	.150527	.032116	.268938
4	.148979	.031939	.26602
5	.151247	.033011	.269482
6	.150267	.033202	.267331
7	.150336	.032858	.267813
8	.150525	.033103	.267948

step	(2) cdm	(2) Lower	(2) Upper
0	.058681	.012529	.104832
1	.062723	-.005058	.130504
2	.126167	.032497	.219837
3	.136583	.038691	.234476
4	.146482	.04442	.248543
5	.146075	.045201	.24695
6	.145542	.044988	.246096
7	.146309	.045315	.247304
8	.145786	.045206	.246365

```
95% lower and upper bounds reported
(1) irfname = dm, impulse = dln_inv, and response = dln_inc
(2) irfname = dm, impulse = dln_inv, and response = dln_consump
```

◁

▷ Example 3: After vec

Although all IRFs and orthogonalized IRFs (OIRFs) from models with stationary variables will taper off to zero, some of the IRFs and OIRFs from models with first-difference stationary variables will not. This is the key difference between IRFs and OIRFs from systems of stationary variables fit by var or svar and those obtained from systems of first-difference stationary variables fit by vec. When the effect of the innovations dies out over time, the shocks are said to be transitory. In contrast, when the effect does not taper off, shocks are said to be permanent.

In this example, we look at the OIRF from one of the VECMs fit to the unemployment-rate data analyzed in example 2 of [TS] vec. We see that an orthogonalized shock to Indiana has a permanent effect on the unemployment rate in Missouri:

```
. use http://www.stata-press.com/data/r12/urates
. vec missouri indiana kentucky illinois, trend(rconstant) rank(2) lags(4)
  (output omitted)
. irf create vec1, set(vecirfs) step(50)
(file vecirfs.irf created)
(file vecirfs.irf now active)
(file vecirfs.irf updated)
```

Now we can use `irf graph` to graph the OIRF of interest:

```
. irf graph oirf, impulse(indiana) response(missouri)
```

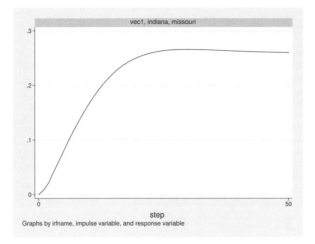

The graph shows that the estimated OIRF converges to a positive asymptote, which indicates that an orthogonalized innovation to the unemployment rate in Indiana has a permanent effect on the unemployment rate in Missouri.

◁

Technical aspects of IRF files

This section is included for programmers wishing to extend the `irf` system.

`irf create` estimates a series of impulse–response functions and their standard errors. Although these estimates are stored in an IRF file, most users will never need to look at the contents of this file. The IRF commands fill in, analyze, present, and manage IRF results.

IRF files are just Stata datasets that have names ending in `.irf` instead of `.dta`. The dataset in the file has a nested panel structure.

Variable `irfname` contains the *irfname* specified by the user. Variable `impulse` records the name of the endogenous variable whose innovations are the impulse. Variable `response` records the name of the endogenous variable that is responding to the innovations. In a model with K endogenous variables, there are K^2 combinations of `impulse` and `response`. Variable `step` records the periods for which these estimates were computed.

Below is a catalog of the statistics that `irf create` estimates and the variable names under which they are stored in the IRF file.

Statistic	Name
impulse–response functions	irf
orthogonalized impulse–response functions	oirf
dynamic-multiplier functions	dm
cumulative impulse–response functions	cirf
cumulative orthogonalized impulse–response functions	coirf
cumulative dynamic-multiplier functions	cdm
Cholesky forecast-error decomposition	fevd
structural impulse–response functions	sirf
structural forecast-error decomposition	sfevd
standard error of the impulse–response functions	stdirf
standard error of the orthogonalized impulse–response functions	stdoirf
standard error of the cumulative impulse–response functions	stdcirf
standard error of the cumulative orthogonalized impulse–response functions	stdcoirf
standard error of the Cholesky forecast-error decomposition	stdfevd
standard error of the structural impulse–response functions	stdsirf
standard error of the structural forecast-error decomposition	stdsfevd

In addition to the variables, information is stored in _dta characteristics. Much of the following information is also available in r() after irf describe, where it is often more convenient to obtain the information. Characteristic _dta[version] contains the version number of the IRF file, which is currently 1.1. Characteristic _dta[irfnames] contains a list of all the *irfnames* in the IRF file. For each *irfname*, there are a series of additional characteristics:

Name	Contents
_dta[*irfname*_model]	var, sr var, lr var, or vec
_dta[*irfname*_order]	Cholesky order used in IRF estimates
_dta[*irfname*_exog]	exogenous variables, and their lags, in VAR
_dta[*irfname*_exogvars]	exogenous variables in VAR
_dta[*irfname*_constant]	constant or noconstant, depending on whether noconstant was specified in var or svar
_dta[*irfname*_lags]	lags in model
_dta[*irfname*_exlags]	lags of exogenous variables in model
_dta[*irfname*_tmin]	minimum value of timevar in the estimation sample
_dta[*irfname*_tmax]	maximum value of timevar in the estimation sample
_dta[*irfname*_timevar]	name of tsset timevar
_dta[*irfname*_tsfmt]	format of timevar
_dta[*irfname*_varcns]	constrained or colon-separated list of constraints placed on VAR coefficients
_dta[*irfname*_svarcns]	constrained or colon-separated list of constraints placed on VAR coefficients
_dta[*irfname*_step]	maximum step in IRF estimates
_dta[*irfname*_stderror]	asymptotic, bs, bsp, or none, depending on the type of standard errors requested
_dta[*irfname*_reps]	number of bootstrap replications performed
_dta[*irfname*_version]	version of the IRF file that originally held *irfname* IRF results
_dta[*irfname*_rank]	number of cointegrating equations
_dta[*irfname*_trend]	trend() specified in vec
_dta[*irfname*_veccns]	constraints placed on VECM parameters
_dta[*irfname*_sind]	normalized seasonal indicators included in vec

IRFs and FEVDs

irf create can estimate several types of IRFs and FEVDs for VARs and VECMs. We first discuss IRF results for VAR and SVAR models, and then we discuss them in the context of VECMs. Because the cointegrating VECM is an extension of the stationary VAR framework, the section that discusses the IRF results for VECMs draws on the earlier VAR material.

IRF results for VARs

An introduction to impulse–response functions for VARs

A pth-order vector autoregressive model (VAR) with exogenous variables is given by

$$\mathbf{y}_t = \mathbf{v} + \mathbf{A}_1\mathbf{y}_{t-1} + \cdots + \mathbf{A}_p\mathbf{y}_{t-p} + \mathbf{B}\mathbf{x}_t + \mathbf{u}_t$$

where

$\mathbf{y}_t = (y_{1t}, \ldots, y_{Kt})'$ is a $K \times 1$ random vector,
the $\mathbf{A}_i$ are fixed $K \times K$ matrices of parameters,
$\mathbf{x}_t$ is an $R_0 \times 1$ vector of exogenous variables,
$\mathbf{B}$ is a $K \times R_0$ matrix of coefficients,
$\mathbf{v}$ is a $K \times 1$ vector of fixed parameters, and
$\mathbf{u}_t$ is assumed to be white noise; that is,
$$E(\mathbf{u}_t) = \mathbf{0}$$
$$E(\mathbf{u}_t\mathbf{u}_t') = \mathbf{\Sigma}$$
$$E(\mathbf{u}_t\mathbf{u}_s') = \mathbf{0} \text{ for } t \neq s$$

As discussed in [TS] **varstable**, a VAR can be rewritten in moving-average form only if it is stable. Any exogenous variables are assumed to be covariance stationary. Because the functions of interest in this section depend only on the exogenous variables through their effect on the estimated $\mathbf{A}_i$, we can simplify the notation by dropping them from the analysis. All the formulas given below still apply, although the $\mathbf{A}_i$ are estimated jointly with $\mathbf{B}$ on the exogenous variables.

Below we discuss conditions under which the IRFs and forecast-error variance decompositions have a causal interpretation. Although estimation requires only that the exogenous variables be predetermined, that is, that $E[\mathbf{x}_{jt}u_{it}] = 0$ for all i, j, and t, assigning a causal interpretation to IRFs and FEVDs requires that the exogenous variables be strictly exogenous, that is, that $E[\mathbf{x}_{js}u_{it}] = 0$ for all i, j, s, and t.

IRFs describe how the innovations to one variable affect another variable after a given number of periods. For an example of how IRFs are interpreted, see Stock and Watson (2001). They use IRFs to investigate the effect of surprise shocks to the Federal Funds rate on inflation and unemployment. In another example, Christiano, Eichenbaum, and Evans (1999) use IRFs to investigate how shocks to monetary policy affect other macroeconomic variables.

Consider a VAR without exogenous variables:

$$\mathbf{y}_t = \mathbf{v} + \mathbf{A}_1\mathbf{y}_{t-1} + \cdots + \mathbf{A}_p\mathbf{y}_{t-p} + \mathbf{u}_t \tag{1}$$

The VAR represents the variables in $\mathbf{y}_t$ as functions of its own lags and serially uncorrelated innovations $\mathbf{u}_t$. All the information about contemporaneous correlations among the K variables in $\mathbf{y}_t$ is contained in $\mathbf{\Sigma}$. In fact, as discussed in [TS] **var svar**, a VAR can be viewed as the reduced form of a dynamic simultaneous-equation model.

To see how the innovations affect the variables in $\mathbf{y}_t$ after, say, i periods, rewrite the model in its moving-average form

$$\mathbf{y}_t = \boldsymbol{\mu} + \sum_{i=0}^{\infty} \boldsymbol{\Phi}_i \mathbf{u}_{t-i} \tag{2}$$

where $\boldsymbol{\mu}$ is the $K \times 1$ time-invariant mean of $\mathbf{y}_t$, and

$$\boldsymbol{\Phi}_i = \begin{cases} \mathbf{I}_K & \text{if } i = 0 \\ \sum_{j=1}^{i} \boldsymbol{\Phi}_{i-j} \mathbf{A}_j & \text{if } i = 1, 2, \ldots \end{cases}$$

We can rewrite a VAR in the moving-average form only if it is stable. Essentially, a VAR is stable if the variables are covariance stationary and none of the autocorrelations are too high (the issue of stability is discussed in greater detail in [TS] **varstable**).

The $\boldsymbol{\Phi}_i$ are the simple IRFs. The j, k element of $\boldsymbol{\Phi}_i$ gives the effect of a 1–time unit increase in the kth element of $\mathbf{u}_t$ on the jth element of $\mathbf{y}_t$ after i periods, holding everything else constant. Unfortunately, these effects have no causal interpretation, which would require us to be able to answer the question, "How does an innovation to variable k, holding everything else constant, affect variable j after i periods?" Because the $\mathbf{u}_t$ are contemporaneously correlated, we cannot assume that everything else is held constant. Contemporaneous correlation among the $\mathbf{u}_t$ implies that a shock to one variable is likely to be accompanied by shocks to some of the other variables, so it does not make sense to shock one variable and hold everything else constant. For this reason, (2) cannot provide a causal interpretation.

This shortcoming may be overcome by rewriting (2) in terms of mutually uncorrelated innovations. Suppose that we had a matrix $\mathbf{P}$, such that $\boldsymbol{\Sigma} = \mathbf{PP}'$. If we had such a $\mathbf{P}$, then $\mathbf{P}^{-1}\boldsymbol{\Sigma}\mathbf{P}'^{-1} = \mathbf{I}_K$, and

$$E\{\mathbf{P}^{-1}\mathbf{u}_t(\mathbf{P}^{-1}\mathbf{u}_t)'\} = \mathbf{P}^{-1}E\{(\mathbf{u}_t\mathbf{u}_t')\}\mathbf{P}'^{-1} = \mathbf{P}^{-1}\boldsymbol{\Sigma}\mathbf{P}'^{-1} = \mathbf{I}_K$$

We can thus use $\mathbf{P}^{-1}$ to orthogonalize the $\mathbf{u}_t$ and rewrite (2) as

$$\mathbf{y}_t = \boldsymbol{\mu} + \sum_{i=0}^{\infty} \boldsymbol{\Phi}_i \mathbf{PP}^{-1} \mathbf{u}_{t-i}$$

$$= \boldsymbol{\mu} + \sum_{i=0}^{\infty} \boldsymbol{\Theta}_i \mathbf{P}^{-1} \mathbf{u}_{t-i}$$

$$= \boldsymbol{\mu} + \sum_{i=0}^{\infty} \boldsymbol{\Theta}_i \mathbf{w}_{t-i}$$

where $\boldsymbol{\Theta}_i = \boldsymbol{\Phi}_i \mathbf{P}$ and $\mathbf{w}_t = \mathbf{P}^{-1}\mathbf{u}_t$. If we had such a $\mathbf{P}$, the $\mathbf{w}_k$ would be mutually orthogonal, and no information would be lost in the holding-everything-else-constant assumption, implying that the $\boldsymbol{\Theta}_i$ would have the causal interpretation that we seek.

Choosing a $\mathbf{P}$ is similar to placing identification restrictions on a system of dynamic simultaneous equations. The simple IRFs do not identify the causal relationships that we wish to analyze. Thus we seek at least as many identification restrictions as necessary to identify the causal IRFs.

So, where do we get such a $\mathbf{P}$? Sims (1980) popularized the method of choosing $\mathbf{P}$ to be the Cholesky decomposition of $\widehat{\boldsymbol{\Sigma}}$. The IRFs based on this choice of $\mathbf{P}$ are known as the *orthogonalized IRFs*. Choosing $\mathbf{P}$ to be the Cholesky decomposition of $\widehat{\boldsymbol{\Sigma}}$ is equivalent to imposing a recursive structure for the corresponding dynamic structural equation model. The ordering of the recursive structure is the same as the ordering imposed in the Cholesky decomposition. Because this choice is arbitrary, some researchers will look at the OIRFs with different orderings assumed in the Cholesky decomposition. The order() option available with `irf create` facilitates this type of analysis.

The SVAR approach integrates the need to identify the causal IRFs into the model specification and estimation process. Sufficient identification restrictions can be obtained by placing either short-run or long-run restrictions on the model. The VAR in (1) can be rewritten as

$$\mathbf{y}_t - \mathbf{v} - \mathbf{A}_1 \mathbf{y}_{t-1} - \cdots - \mathbf{A}_p \mathbf{y}_{t-p} = \mathbf{u}_t \tag{3}$$

Similarly, a short-run SVAR model can be written as

$$\mathbf{A}(\mathbf{y}_t - \mathbf{v} - \mathbf{A}_1 \mathbf{y}_{t-1} - \cdots - \mathbf{A}_p \mathbf{y}_{t-p}) = \mathbf{A}\mathbf{u}_t = \mathbf{B}\mathbf{e}_t \tag{4}$$

where $\mathbf{A}$ and $\mathbf{B}$ are $K \times K$ nonsingular matrices of parameters to be estimated, $\mathbf{e}_t$ is a $K \times 1$ vector of disturbances with $\mathbf{e}_t \sim N(\mathbf{0}, \mathbf{I}_K)$, and $E[\mathbf{e}_t \mathbf{e}_s'] = \mathbf{0}_K$ for all $s \neq t$. Sufficient constraints must be placed on $\mathbf{A}$ and $\mathbf{B}$ so that $\mathbf{P}$ is identified. One way to see the connection is to draw out the implications of the latter equality in (4). From (4) it can be shown that

$$\boldsymbol{\Sigma} = \mathbf{A}^{-1}\mathbf{B}(\mathbf{A}^{-1}\mathbf{B})'$$

As discussed in [TS] **var svar**, the estimates $\widehat{\mathbf{A}}$ and $\widehat{\mathbf{B}}$ are obtained by maximizing the concentrated log-likelihood function on the basis of the $\widehat{\boldsymbol{\Sigma}}$ obtained from the underlying VAR. The short-run SVAR approach chooses $\mathbf{P} = \widehat{\mathbf{A}}^{-1}\widehat{\mathbf{B}}$ to identify the causal IRFs. The long-run SVAR approach works similarly, with $\mathbf{P} = \widehat{\mathbf{C}} = \widehat{\overline{\mathbf{A}}}^{-1}\widehat{\mathbf{B}}$, where $\widehat{\overline{\mathbf{A}}}^{-1}$ is the matrix of estimated long-run or accumulated effects of the reduced-form VAR shocks.

There is one important difference between long-run and short-run SVAR models. As discussed by Amisano and Giannini (1997, chap. 6), in the short-run model the constraints are applied directly to the parameters in $\mathbf{A}$ and $\mathbf{B}$. Then $\mathbf{A}$ and $\mathbf{B}$ interact with the estimated parameters of the underlying VAR. In contrast, in a long-run model, the constraints are placed on functions of the estimated VAR parameters. Although estimation and inference of the parameters in $\mathbf{C}$ is straightforward, obtaining the asymptotic standard errors of the structural IRFs requires untenable assumptions. For this reason, `irf create` does not estimate the asymptotic standard errors of the structural IRFs generated by long-run SVAR models. However, bootstrap standard errors are still available.

An introduction to dynamic-multiplier functions for VARs

A dynamic-multiplier function measures the effect of a unit change in an exogenous variable on the endogenous variables over time. Per Lütkepohl (2005, chap. 10), if the VAR with exogenous variables is stable, it can be rewritten as

$$\mathbf{y}_t = \sum_{i=0}^{\infty} \mathbf{D}_i x_{t-i} + \sum_{i=0}^{\infty} \boldsymbol{\Phi}_i u_{t-i}$$

where the $\mathbf{D}_i$ are the dynamic-multiplier functions. (See *Methods and formulas* for details.) Some authors refer to the dynamic-multiplier functions as transfer functions because they specify how a unit change in an exogenous variable is "transferred" to the endogenous variables.

❏ Technical note

irf create computes dynamic-multiplier functions only after var. After short-run SVAR models, the dynamic multipliers from the VAR are the same as those from the SVAR. The dynamic multipliers for long-run SVARs have not yet been worked out.

❏

An introduction to forecast-error variance decompositions for VARs

Another measure of the effect of the innovations in variable k on variable j is the FEVD. This method, which is also known as *innovation accounting*, measures the fraction of the error in forecasting variable j after h periods that is attributable to the orthogonalized innovations in variable k. Because deriving the FEVD requires orthogonalizing the $\mathbf{u}_t$ innovations, the FEVD is always predicated upon a choice of $\mathbf{P}$.

Lütkepohl (2005, sec. 2.2.2) shows that the h-step forecast error can be written as

$$\mathbf{y}_{t+h} - \widehat{\mathbf{y}}_t(h) = \sum_{i=0}^{h-1} \mathbf{\Phi}_i \mathbf{u}_{t+h-i} \tag{6}$$

where y_{t+h} is the value observed at time $t + h$ and $\widehat{y}_t(h)$ is the h-step-ahead predicted value for y_{t+h} that was made at time t.

Because the $\mathbf{u}_t$ are contemporaneously correlated, their distinct contributions to the forecast error cannot be ascertained. However, if we choose a $\mathbf{P}$ such that $\mathbf{\Sigma} = \mathbf{PP}'$, as above, we can orthogonalize the $\mathbf{u}_t$ into $\mathbf{w}_t = \mathbf{P}^{-1}\mathbf{u}_t$. We can then ascertain the relative contribution of the distinct elements of $\mathbf{w}_t$. Thus we can rewrite (6) as

$$\mathbf{y}_{t+h} - \widehat{\mathbf{y}}_t(h) = \sum_{i=0}^{h-1} \mathbf{\Phi}_i \mathbf{PP}^{-1} \mathbf{u}_{t+h-i}$$

$$= \sum_{i=0}^{h-1} \mathbf{\Theta}_i \mathbf{w}_{t+h-i} \tag{7}$$

Because the forecast errors can be written in terms of the orthogonalized errors, the forecast-error variance can be written in terms of the orthogonalized error variances. Forecast-error variance decompositions measure the fraction of the total forecast-error variance that is attributable to each orthogonalized shock.

❏ Technical note

The details in this note are not critical to the discussion that follows. A forecast-error variance decomposition is derived for a given $\mathbf{P}$. Per Lütkepohl (2005, sec. 2.3.3), letting $\theta_{mn,i}$ be the m, nth element of $\mathbf{\Theta}_i$, we can express the h-step forecast error of the jth component of $\mathbf{y}_t$ as

$$\mathbf{y}_{j,t+h} - \widehat{\mathbf{y}}_j(h) = \sum_{i=0}^{h-1} \theta_{j1,1} \mathbf{w}_{1,t+h-i} + \cdots + \theta_{jK,i} \mathbf{w}_{K,t+h-i}$$

$$= \sum_{k=1}^{K} \theta_{jk,0} \mathbf{w}_{k,t+h} + \cdots + \theta_{jk,h-1} \mathbf{w}_{k,t+1}$$

The $\mathbf{w}_t$, which were constructed using $\mathbf{P}$, are mutually orthogonal with unit variance. This allows us to compute easily the mean squared error (MSE) of the forecast of variable j at horizon h in terms of the contributions of the components of $\mathbf{w}_t$. Specifically,

$$E[\{y_{j,t+h} - y_{j,t}(h)\}^2] = \sum_{k=1}^{K} (\theta_{jk,0}^2 + \cdots + \theta_{jk,h-1}^2)$$

The kth term in the sum above is interpreted as the contribution of the orthogonalized innovations in variable k to the h-step forecast error of variable j. Note that the kth element in the sum above can be rewritten as

$$(\theta_{jk,0}^2 + \cdots + \theta_{jk,h-1}^2) = \sum_{i=0}^{h-1} \left(\mathbf{e}_j' \mathbf{\Theta}_k \mathbf{e}_k \right)^2$$

where $\mathbf{e}_i$ is the ith column of $\mathbf{I}_K$. Normalizing by the forecast error for variable j at horizon h yields

$$\omega_{jk,h} = \frac{\sum_{i=0}^{h-1} \left(\mathbf{e}_j' \mathbf{\Theta}_k \mathbf{e}_k \right)^2}{\mathrm{MSE}[y_{j,t}(h)]}$$

where $\mathrm{MSE}[y_{j,t}(h)] = \sum_{i=0}^{h-1} \sum_{k=1}^{K} \theta_{jk,i}^2$.

❏

Because the FEVD depends on the choice of $\mathbf{P}$, there are different forecast-error variance decompositions associated with each distinct $\mathbf{P}$. irf create can estimate the FEVD for a VAR or an SVAR. For a VAR, $\mathbf{P}$ is the Cholesky decomposition of $\widehat{\mathbf{\Sigma}}$. For an SVAR, $\mathbf{P}$ is the estimated structural decomposition, $\mathbf{P} = \widehat{\mathbf{A}}^{-1}\widehat{\mathbf{B}}$ for short-run models and $\mathbf{P} = \widehat{\mathbf{C}}$ for long-run SVAR models. Due to the same complications that arose with the structural impulse–response functions, the asymptotic standard errors of the structural FEVD are not available after long-run SVAR models, but bootstrap standard errors are still available.

IRF results for VECMs

An introduction to impulse–response functions for VECMs

As discussed in [TS] **vec intro**, the VECM is a reparameterization of the VAR that is especially useful for fitting VARs with cointegrating variables. This implies that the estimated parameters for the corresponding VAR model can be backed out from the estimated parameters of the VECM model. This relationship means we can use the VAR form of the cointegrating VECM to discuss the IRFs for VECMs.

Consider a cointegrating VAR with one lag with no constant or trend,

$$\mathbf{y}_t = \mathbf{A}\mathbf{y}_{t-1} + \mathbf{u}_t \tag{8}$$

where $\mathbf{y}_t$ is a $K \times 1$ vector of endogenous, first-difference stationary variables among which there are $1 \leq r < K$ cointegration equations; $\mathbf{A}$ is $K \times K$ matrix of parameters; and $\mathbf{u}_t$ is a $K \times 1$ vector of i.i.d. disturbances.

We developed intuition for the IRFs from a stationary VAR by rewriting the VAR as an infinite-order vector moving-average (VMA) process. While the Granger representation theorem establishes the existence of a VMA formulation of this model, because the cointegrating VAR is not stable, the inversion is not nearly so intuitive. (See Johansen [1995, chapters 3 and 4] for more details.) For this reason, we use (8) to develop intuition for the IRFs from a cointegrating VAR.

Suppose that K is 3, that $\mathbf{u}_1 = (1, 0, 0)$, and that we want to analyze the time paths of the variables in $\mathbf{y}$ conditional on the initial values $\mathbf{y}_0 = \mathbf{0}$, $\mathbf{A}$, and the condition that there are no more shocks to the system, that is, $\mathbf{0} = \mathbf{u}_2 = \mathbf{u}_3 = \cdots$. These assumptions and (8) imply that

$$\mathbf{y}_1 = \mathbf{u}_1$$

$$\mathbf{y}_2 = \mathbf{A}\mathbf{y}_1 = \mathbf{A}\mathbf{u}_1$$

$$\mathbf{y}_3 = \mathbf{A}\mathbf{y}_2 = \mathbf{A}^2\mathbf{u}_1$$

and so on. The ith-row element of the first column of $\mathbf{A}^s$ contains the effect of the unit shock to the first variable after s periods. The first column of $\mathbf{A}^s$ contains the IRF of a unit impulse to the first variable after s periods. We could deduce the IRFs of a unit impulse to any of the other variables by administering the unit shock to one of them instead of to the first variable. Thus we can see that the (i, j)th element of $\mathbf{A}^s$ contains the unit IRF from variable j to variable i after s periods. By starting with orthogonalized shocks of the form $\mathbf{P}^{-1}\mathbf{u}_t$, we can use the same logic to derive the OIRFs to be $\mathbf{A}^s\mathbf{P}$.

For the stationary VAR, stability implies that all the eigenvalues of $\mathbf{A}$ have moduli strictly less than one, which in turn implies that all the elements of $\mathbf{A}^s \rightarrow \mathbf{0}$ as $s \rightarrow \infty$. This implies that all the IRFs from a stationary VAR taper off to zero as $s \rightarrow \infty$. In contrast, in a cointegrating VAR, some of the eigenvalues of $\mathbf{A}$ are 1, while the remaining eigenvalues have moduli strictly less than 1. This implies that in cointegrating VARs some of the elements of $\mathbf{A}^s$ are not going to zero as $s \rightarrow \infty$, which in turn implies that some of the IRFs and OIRFs are not going to zero as $s \rightarrow \infty$. The fact that the IRFs and OIRFs taper off to zero for stationary VARs but not for cointegrating VARs is one of the key differences between the two models.

When the IRF or OIRF from the innovation in one variable to another tapers off to zero as time goes on, the innovation to the first variable is said to have a transitory effect on the second variable. When the IRF or OIRF does not go to zero, the effect is said to be permanent.

Note that, because some of the IRFs and OIRFs do not taper off to zero, some of the cumulative IRFs and OIRFs diverge over time.

An introduction to forecast-error variance decompositions for VECMs

The results from *An introduction to impulse–response functions for VECMs* can be used to show that the interpretation of FEVDs for a finite number of steps in cointegrating VARs is essentially the same as in the stationary case. Because the MSE of the forecast is diverging, this interpretation is valid only for a finite number of steps. (See [TS] **vec intro** and [TS] **fcast compute** for more information on this point.)

Methods and formulas

irf create is implemented as an ado-file.

Methods and formulas are presented under the following headings:

Impulse–response function formulas for VARs
Dynamic-multiplier function formulas for VARs
Forecast-error variance decomposition formulas for VARs
Impulse–response function formulas for VECMs
Algorithms for bootstrapping the VAR IRF and FEVD standard errors

Impulse–response function formulas for VARs

The previous discussion implies that there are three different choices of $\mathbf{P}$ that can be used to obtain distinct $\boldsymbol{\Theta}_i$. $\mathbf{P}$ is the Cholesky decomposition of $\boldsymbol{\Sigma}$ for the OIRFs. For the structural IRFs, $\mathbf{P} = \mathbf{A}^{-1}\mathbf{B}$ for short-run models, and $\mathbf{P} = \mathbf{C}$ for long-run models. We will distinguish between the three by defining $\boldsymbol{\Theta}_i^o$ to be the OIRFs, $\boldsymbol{\Theta}_i^{\mathrm{sr}}$ to be the short-run structural IRFs, and $\boldsymbol{\Theta}_i^{lr}$ to be the long-run structural IRFs.

We also define $\widehat{\mathbf{P}}_c$ to be the Cholesky decomposition of $\widehat{\boldsymbol{\Sigma}}$, $\widehat{\mathbf{P}}_{\mathrm{sr}} = \widehat{\mathbf{A}}^{-1}\widehat{\mathbf{B}}$ to be the short-run structural decomposition, and $\widehat{\mathbf{P}}_{lr} = \widehat{\mathbf{C}}$ to be the long-run structural decomposition.

Given estimates of the $\widehat{\mathbf{A}}_i$ and $\widehat{\boldsymbol{\Sigma}}$ from `var` or `svar`, the estimates of the simple IRFs and the OIRFs are, respectively,

$$\widehat{\boldsymbol{\Phi}}_i = \sum_{j=1}^{i} \widehat{\boldsymbol{\Phi}}_{i-j}\widehat{\mathbf{A}}_j$$

and

$$\widehat{\boldsymbol{\Theta}}_i^o = \widehat{\boldsymbol{\Phi}}_i\widehat{\mathbf{P}}_c$$

where $\widehat{\mathbf{A}}_j = \mathbf{0}_K$ for $j > p$.

Given the estimates $\widehat{\mathbf{A}}$ and $\widehat{\mathbf{B}}$, or $\widehat{\mathbf{C}}$, from `svar`, the estimates of the structural IRFs are either

$$\widehat{\boldsymbol{\Theta}}_i^{\mathrm{sr}} = \widehat{\boldsymbol{\Phi}}_i\widehat{\mathbf{P}}_{\mathrm{sr}}$$

or

$$\widehat{\boldsymbol{\Theta}}_i^{lr} = \widehat{\boldsymbol{\Phi}}_i\widehat{\mathbf{P}}_{lr}$$

The estimated structural IRFs stored in an IRF file with the variable name `sirf` may be from either a short-run model or a long-run model, depending on the estimation results used to create the IRFs. As discussed in [TS] **irf describe**, you can easily determine whether the structural IRFs were generated from a short-run or a long-run SVAR model using `irf describe`.

Following Lütkepohl (2005, sec. 3.7), estimates of the cumulative IRFs and the cumulative orthogonalized impulse–response functions (COIRFs) at period n are, respectively,

$$\widehat{\boldsymbol{\Psi}}_n = \sum_{i=0}^{n} \widehat{\boldsymbol{\Phi}}_i$$

and

$$\widehat{\boldsymbol{\Xi}}_n = \sum_{i=0}^{n} \widehat{\boldsymbol{\Theta}}_i$$

The asymptotic standard errors of the different impulse–response functions are obtained by applications of the delta method. See Lütkepohl (2005, sec. 3.7) and Amisano and Giannini (1997, chap. 4) for the derivations. See Serfling (1980, sec. 3.3) for a discussion of the delta method. In presenting the variance–covariance matrix estimators, we make extensive use of the vec() operator, where vec($\mathbf{X}$) is the vector obtained by stacking the columns of $\mathbf{X}$.

Lütkepohl (2005, sec. 3.7) derives the asymptotic VCEs of vec($\mathbf{\Phi}_i$), vec($\mathbf{\Theta}_i^o$), vec($\widehat{\mathbf{\Psi}}_n$), and vec($\widehat{\mathbf{\Xi}}_n$). Because vec($\mathbf{\Phi}_i$) is $K^2 \times 1$, the asymptotic VCE of vec($\mathbf{\Phi}_i$) is $K^2 \times K^2$, and it is given by

$$\mathbf{G}_i \widehat{\mathbf{\Sigma}}_{\widehat{\alpha}} \mathbf{G}_i'$$

where

$$\mathbf{G}_i = \sum_{m=0}^{i-1} \mathbf{J}(\widehat{\mathbf{M}}')^{(i-1-m)} \otimes \widehat{\mathbf{\Phi}}_m \qquad \mathbf{G}_i \text{ is } K^2 \times K^2 p$$

$$\mathbf{J} = (\mathbf{I}_K, \mathbf{0}_K, \ldots, \mathbf{0}_K) \qquad \mathbf{J} \text{ is } K \times Kp$$

$$\widehat{\mathbf{M}} = \begin{bmatrix} \widehat{\mathbf{A}}_1 & \widehat{\mathbf{A}}_2 & \ldots & \widehat{\mathbf{A}}_{p-1} & \widehat{\mathbf{A}}_p \\ \mathbf{I}_K & \mathbf{0}_K & \ldots & \mathbf{0}_K & \mathbf{0}_K \\ \mathbf{0}_K & \mathbf{I}_K & & \mathbf{0}_K & \mathbf{0}_K \\ \vdots & & \ddots & \vdots & \vdots \\ \mathbf{0}_K & \mathbf{0}_K & \ldots & \mathbf{I}_K & \mathbf{0}_K \end{bmatrix} \qquad \widehat{\mathbf{M}} \text{ is } Kp \times Kp$$

The $\widehat{\mathbf{A}}_i$ are the estimates of the coefficients on the lagged variables in the VAR, and $\widehat{\mathbf{\Sigma}}_{\widehat{\alpha}}$ is the VCE matrix of $\widehat{\alpha} = \text{vec}(\widehat{\mathbf{A}}_1, \ldots, \widehat{\mathbf{A}}_p)$. $\widehat{\mathbf{\Sigma}}_{\widehat{\alpha}}$ is a $K^2 p \times K^2 p$ matrix whose elements come from the VCE of the VAR coefficient estimator. As such, this VCE is the VCE of the constrained estimator if there are any constraints placed on the VAR coefficients.

The $K^2 \times K^2$ asymptotic VCE matrix for vec($\widehat{\mathbf{\Psi}}_n$) after n periods is given by

$$\mathbf{F}_n \widehat{\mathbf{\Sigma}}_{\widehat{\alpha}} \mathbf{F}_n'$$

where

$$\mathbf{F}_n = \sum_{i=1}^{n} \mathbf{G}_i$$

The $K^2 \times K^2$ asymptotic VCE matrix of the vectorized, orthogonalized, IRFs at horizon i, vec($\mathbf{\Theta}_i^o$), is

$$\mathbf{C}_i \widehat{\mathbf{\Sigma}}_{\widehat{\alpha}} \mathbf{C}_i' + \overline{\mathbf{C}}_i \widehat{\mathbf{\Sigma}}_{\widehat{\sigma}} \overline{\mathbf{C}}_i'$$

where

$$\mathbf{C}_0 = \mathbf{0} \qquad\qquad \mathbf{C}_0 \text{ is } K^2 \times K^2 p$$

$$\mathbf{C}_i = (\widehat{\mathbf{P}}'_c \otimes \mathbf{I}_K)\mathbf{G}_i, \quad i = 1, 2, \ldots \qquad\qquad \mathbf{C}_i \text{ is } K^2 \times K^2 p$$

$$\overline{\mathbf{C}}_i = (\mathbf{I}_K \otimes \mathbf{\Phi}_i)\mathbf{H}, \quad i = 0, 1, \ldots \qquad\qquad \overline{\mathbf{C}}_i \text{ is } K^2 \times K^2$$

$$\mathbf{H} = \mathbf{L}'_K \left\{ \mathbf{L}_K \mathbf{N}_K (\widehat{\mathbf{P}}_c \otimes \mathbf{I}_K)\mathbf{L}'_K \right\}^{-1} \qquad\qquad \mathbf{H} \text{ is } K^2 \times K\frac{(K+1)}{2}$$

$$\mathbf{L}_K \text{ solves } \qquad \mathrm{vech}(\mathbf{F}) = \mathbf{L}_K \, \mathrm{vec}(\mathbf{F}) \qquad\qquad \mathbf{L}_K \text{ is } K\frac{(K+1)}{2} \times K^2$$

$$\text{for } \mathbf{F} \ K \times K \text{ and symmetric}$$

$$\mathbf{K}_K \text{ solves } \qquad \mathbf{K}_K \mathrm{vec}(\mathbf{G}) = \mathrm{vec}(\mathbf{G}') \text{ for any } K \times K \text{ matrix } \mathbf{G} \qquad\qquad \mathbf{K}_K \text{ is } K^2 \times K^2$$

$$\mathbf{N}_K = \tfrac{1}{2}\left(\mathbf{I}_{K^2} + \mathbf{K}_K\right) \qquad\qquad \mathbf{N}_K \text{ is } K^2 \times K^2$$

$$\widehat{\mathbf{\Sigma}}_{\widehat{\sigma}} = 2\mathbf{D}_K^+(\widehat{\mathbf{\Sigma}} \otimes \widehat{\mathbf{\Sigma}})\mathbf{D}_K^+ \qquad\qquad \widehat{\mathbf{\Sigma}}_{\widehat{\sigma}} \text{ is } K\frac{(K+1)}{2} \times K\frac{(K+1)}{2}$$

$$\mathbf{D}_K^+ = \left(\mathbf{D}'_K \mathbf{D}_K\right)^{-1}\mathbf{D}'_K \qquad\qquad \mathbf{D}_K^+ \text{ is } K\frac{(K+1)}{2} \times K^2$$

$$\mathbf{D}_K \text{ solves } \quad \mathbf{D}_K \mathrm{vech}(\mathbf{F}) = \mathrm{vec}(\mathbf{F}) \quad \text{ for } \mathbf{F} \ K \times K \text{ and symmetric} \qquad \mathbf{D}_K \text{ is } K^2 \times K\frac{(K+1)}{2}$$

$$\mathrm{vech}(\mathbf{X}) = \begin{bmatrix} x_{11} \\ x_{21} \\ \vdots \\ x_{K1} \\ x_{22} \\ \vdots \\ x_{K2} \\ \vdots \\ x_{KK} \end{bmatrix} \qquad \text{for } \mathbf{X} \ K \times K \qquad\qquad \mathrm{vech}(\mathbf{X}) \text{ is } K\frac{(K+1)}{2} \times 1$$

Note that $\widehat{\mathbf{\Sigma}}_{\widehat{\sigma}}$ is the VCE of $\mathrm{vech}(\widehat{\mathbf{\Sigma}})$. More details about $\mathbf{L}_K$, $\mathbf{K}_K$, $\mathbf{D}_K$ and vech() are available in Lütkepohl (2005, sec. A.12). Finally, as Lütkepohl (2005, 113–114) discusses, $\mathbf{D}_K^+$ is the Moore–Penrose inverse of $\mathbf{D}_K$.

As discussed in Amisano and Giannini (1997, chap. 6), the asymptotic standard errors of the structural IRFs are available for short-run SVAR models but not for long-run SVAR models. Following Amisano and Giannini (1997, chap. 5), the asymptotic $K^2 \times K^2$ VCE of the short-run structural IRFs after i periods, when a maximum of h periods are estimated, is the i, i block of

$$\widehat{\mathbf{\Sigma}}(h)_{ij} = \widetilde{\mathbf{G}}_i \widehat{\mathbf{\Sigma}}_{\widehat{\alpha}} \widetilde{\mathbf{G}}'_j + \left\{ \mathbf{I}_K \otimes (\mathbf{J}\widehat{\mathbf{M}}^i \mathbf{J}') \right\} \mathbf{\Sigma}(0) \left\{ \mathbf{I}_K \otimes (\mathbf{J}\widehat{\mathbf{M}}^j \mathbf{J}') \right\}'$$

where

$$\widetilde{\mathbf{G}}_0 = \mathbf{0}_K \qquad\qquad \mathbf{G}_0 \text{ is } K^2 \times K^2 p$$

$$\widetilde{\mathbf{G}}_i = \sum_{k=0}^{i-1}\left\{\widehat{\mathbf{P}}'_{\mathrm{sr}}\mathbf{J}(\widehat{\mathbf{M}}')^{i-1-k} \otimes \left(\mathbf{J}\widehat{\mathbf{M}}^k\mathbf{J}'\right)\right\} \qquad \mathbf{G}_i \text{ is } K^2 \times K^2 p$$

$$\widehat{\mathbf{\Sigma}}(0) = \mathbf{Q}_2\widehat{\mathbf{\Sigma}}_W\mathbf{Q}'_2 \qquad\qquad \widehat{\mathbf{\Sigma}}(0) \text{ is } K^2 \times K^2$$

$$\widehat{\mathbf{\Sigma}}_W = \mathbf{Q}_1\widehat{\mathbf{\Sigma}}_{AB}\mathbf{Q}'_1 \qquad\qquad \widehat{\mathbf{\Sigma}}_W \text{ is } K^2 \times K^2$$

$$\mathbf{Q}_2 = \widehat{\mathbf{P}}'_{\mathrm{sr}} \otimes \widehat{\mathbf{P}}_{\mathrm{sr}} \qquad\qquad \mathbf{Q}_2 \text{ is } K^2 \times K^2$$

$$\mathbf{Q}_1 = \left\{(\mathbf{I}_K \otimes \widehat{\mathbf{B}}^{-1}), (-\widehat{\mathbf{P}}'^{-1}_{\mathrm{sr}} \otimes \mathbf{B}^{-1})\right\} \qquad \mathbf{Q}_1 \text{ is } K^2 \times 2K^2$$

and $\widehat{\mathbf{\Sigma}}_{AB}$ is the $2K^2 \times 2K^2$ VCE of the estimator of $\mathrm{vec}(\mathbf{A}, \mathbf{B})$.

Dynamic-multiplier function formulas for VARs

This section provides the details of how irf create estimates the dynamic-multiplier functions and their asymptotic standard errors.

A pth order vector autoregressive model (VAR) with exogenous variables may be written as

$$\mathbf{y}_t = \mathbf{v} + \mathbf{A}_1\mathbf{y}_{t-1} + \cdots + \mathbf{A}_p\mathbf{y}_{t-p} + \mathbf{B}_0\mathbf{x}_t + \mathbf{B}_1\mathbf{x}_{t-1} + \cdots + \mathbf{B}_s\mathbf{x}_{t-s} + \mathbf{u}_t$$

where all the notation is the same as above except that the s $K \times R$ matrices $\mathbf{B}_1, \mathbf{B}_2, \ldots, \mathbf{B}_s$ are explicitly included and s is the number of lags of the R exogenous variables in the model.

Lütkepohl (2005) shows that the dynamic-multipliers $\mathbf{D}_i$ are consistently estimated by

$$\widehat{\mathbf{D}}_i = \mathbf{J}_x\widetilde{\mathbf{A}}^i_x\widehat{\mathbf{B}}_x \qquad i \in \{0, 1, \ldots\}$$

where

$$\mathbf{J}_x = (\mathbf{I}_K, \mathbf{0}_K, \ldots, \mathbf{0}_K) \qquad\qquad \mathbf{J} \text{ is } K \times (Kp+Rs)$$

$$\widetilde{\mathbf{A}}_x = \begin{bmatrix} \widehat{\mathbf{M}} & \widehat{\mathbf{B}} \\ \widetilde{\mathbf{0}} & \widetilde{\mathbf{I}} \end{bmatrix} \qquad\qquad \widetilde{\mathbf{A}}_x \text{ is } (Kp+Rs)\times(Kp+Rs)$$

$$\widehat{\mathbf{B}} = \begin{bmatrix} \widehat{\mathbf{B}}_1 & \widehat{\mathbf{B}}_2 & \ldots & \widehat{\mathbf{B}}_s \\ \ddot{\mathbf{0}} & \ddot{\mathbf{0}} & \ldots & \ddot{\mathbf{0}} \\ \vdots & \vdots & \ddots & \vdots \\ \ddot{\mathbf{0}} & \ddot{\mathbf{0}} & \ldots & \ddot{\mathbf{0}} \end{bmatrix} \qquad\qquad \widehat{\mathbf{B}} \text{ is } Kp \times Rs$$

$$\widetilde{\mathbf{I}} = \begin{bmatrix} \mathbf{0}_R & \mathbf{0}_R & \ldots & \mathbf{0}_R & \mathbf{0}_R \\ \mathbf{I}_R & \mathbf{0}_R & \ldots & \mathbf{0}_R & \mathbf{0}_R \\ \mathbf{0}_R & \mathbf{I}_R & & \mathbf{0}_R & \mathbf{0}_R \\ \vdots & & \ddots & \vdots & \vdots \\ \mathbf{0}_R & \mathbf{0}_R & \ldots & \mathbf{I}_R & \mathbf{0}_R \end{bmatrix} \qquad \widetilde{\mathbf{I}} \text{ is } Rs \times Rs$$

$$\widehat{\mathbf{B}}'_x = \begin{bmatrix} \widetilde{\mathbf{B}}' & \ddot{\mathbf{I}}' \end{bmatrix} \qquad\qquad \widehat{\mathbf{B}}'_x \text{ is } R \times (Kp+Rs)$$

$$\widetilde{\mathbf{B}}' = \begin{bmatrix} \widehat{\mathbf{B}}'_0 & \ddot{\mathbf{0}}' \cdots \ddot{\mathbf{0}}' \end{bmatrix} \qquad\qquad \widetilde{\mathbf{B}} \text{ is } R \times Kp$$

$$\ddot{\mathbf{I}}' = \begin{bmatrix} \mathbf{I}_R & \mathbf{0}_R \cdots \mathbf{0}_R \end{bmatrix} \qquad\qquad \ddot{\mathbf{i}} \text{ is } R \times Rs$$

and $\ddot{\mathbf{0}}$ is a $K \times R$ matrix of 0s and $\widetilde{\mathbf{0}}$ is a $Rs \times Kp$ matrix of 0s.

Consistent estimators of the cumulative dynamic-multiplier functions are given by

$$\overline{\mathbf{D}}_i = \sum_{j=0}^{i} \widehat{\mathbf{D}}_j$$

Letting $\boldsymbol{\beta}_x = \mathrm{vec}[\mathbf{A}_1\mathbf{A}_2\cdots\mathbf{A}_p\mathbf{B}_1\mathbf{B}_2\cdots\mathbf{B}_s\mathbf{B}_0]$ and letting $\boldsymbol{\Sigma}_{\widehat{\boldsymbol{\beta}}_x}$ be the asymptotic variance–covariance estimator (VCE) of $\widehat{\boldsymbol{\beta}}_x$, Lütkepohl shows that an asymptotic VCE of $\widehat{\mathbf{D}}_i$ is $\widetilde{\mathbf{G}}_i \boldsymbol{\Sigma}_{\widehat{\boldsymbol{\beta}}_x} \widetilde{\mathbf{G}}_i'$ where

$$\widetilde{\mathbf{G}}_i = \left[\sum_{j=0}^{i-1} \mathbf{B}_x' \widetilde{\mathbf{A}}_x^{i-1-j} \otimes \mathbf{J}_x \widetilde{\mathbf{A}}_x^{j} \mathbf{J}_x', \ \mathbf{I}_R \otimes \mathbf{J}_x \widetilde{\mathbf{A}}_x^{j} \mathbf{J}_x \right]$$

Similarly, an asymptotic VCE of $\overline{\mathbf{D}}_i$ is $\left(\sum_{j=0}^{i} \widetilde{\mathbf{G}}_j \right) \boldsymbol{\Sigma}_{\widehat{\boldsymbol{\beta}}_x} \left(\sum_{j=0}^{i} \widetilde{\mathbf{G}}_j' \right)$.

Forecast-error variance decomposition formulas for VARs

This section provides details of how `irf create` estimates the Cholesky FEVD, the structural FEVD, and their standard errors. Beginning with the Cholesky-based forecast-error decompositions, the fraction of the h-step-ahead forecast-error variance of variable j that is attributable to the Cholesky orthogonalized innovations in variable k can be estimated as

$$\widehat{\omega}_{jk,h} = \frac{\sum_{i=0}^{h-1}(\mathbf{e}_j'\widehat{\boldsymbol{\Theta}}_i\mathbf{e}_k)^2}{\widehat{\mathrm{MSE}}_j(h)}$$

where $\mathrm{MSE}_j(h)$ is the jth diagonal element of

$$\sum_{i=0}^{h-1} \widehat{\boldsymbol{\Phi}}_i \widehat{\boldsymbol{\Sigma}} \widehat{\boldsymbol{\Phi}}_i'$$

(See Lütkepohl [2005, 109] for a discussion of this result.) $\widehat{\omega}_{jk,h}$ and $\mathrm{MSE}_j(h)$ are scalars. The square of the standard error of $\widehat{\omega}_{jk,h}$ is

$$\mathbf{d}_{jk,h}\widehat{\boldsymbol{\Sigma}}_\alpha \mathbf{d}_{jk,h}' + \overline{\mathbf{d}}_{jk,h}\widehat{\boldsymbol{\Sigma}}_\sigma \overline{\mathbf{d}}_{jk,h}$$

where

$$\mathbf{d}_{jk,h} = \frac{2}{\mathrm{MSE}_j(h)^2} \sum_{i=0}^{h-1} \Bigg\{ \mathrm{MSE}_j(h)(\mathbf{e}_j'\widehat{\boldsymbol{\Phi}}_i\widehat{\mathbf{P}}_c\mathbf{e}_k)(\mathbf{e}_k'\widehat{\mathbf{P}}_c' \otimes \mathbf{e}_j')\mathbf{G}_i$$

$$-(\mathbf{e}_j'\boldsymbol{\Phi}_i\widehat{\mathbf{P}}_c\mathbf{e}_k)^2 \sum_{m=0}^{h-1}(\mathbf{e}_j'\widehat{\boldsymbol{\Phi}}_m\widehat{\boldsymbol{\Sigma}} \otimes \mathbf{e}_j')\mathbf{G}_m \Bigg\} \qquad \mathbf{d}_{jk,h} \text{ is } 1 \times K^2 p$$

$$\overline{\mathbf{d}}_{jk,h} = \sum_{i=0}^{h-1}\Bigg\{ \mathrm{MSE}_j(h)(\mathbf{e}_j'\widehat{\boldsymbol{\Phi}}_i\mathbf{P}_c\mathbf{e}_k)(\mathbf{e}_k' \otimes \mathbf{e}_j'\widehat{\boldsymbol{\Phi}}_i)\mathbf{H}$$

$$-(\mathbf{e}_j'\widehat{\boldsymbol{\Phi}}_i\widehat{\mathbf{P}}_c\mathbf{e}_k)^2 \sum_{m=0}^{h-1}(\mathbf{e}_j'\widehat{\boldsymbol{\Phi}}_m \otimes \mathbf{e}_j\widehat{\boldsymbol{\Phi}}_m)\mathbf{D}_K \Bigg\} \frac{1}{\mathrm{MSE}_j(h)^2} \qquad \overline{\mathbf{d}}_{jk,h} \text{ is } 1 \times K\frac{(K+1)}{2}$$

$$\mathbf{G}_0 = \mathbf{0} \qquad\qquad\qquad \mathbf{G}_0 \text{ is } K^2 \times K^2 p$$

and $\mathbf{D}_K$ is the $K^2 \times K\{(K+1)/2\}$ duplication matrix defined previously.

For the structural forecast-error decompositions, we follow Amisano and Giannini (1997, sec. 5.2). They define the matrix of structural forecast-error decompositions at horizon s, when a maximum of h periods are estimated, as

$$\widehat{\mathbf{W}}_s = \widehat{\mathbf{F}}_s^{-1} \widehat{\widetilde{\mathbf{M}}}_s \qquad \text{for } s = 1, \ldots, h+1$$

$$\widehat{\mathbf{F}}_s = \left(\sum_{i=0}^{s-1} \widehat{\boldsymbol{\Theta}}_i^{\mathrm{sr}} \widehat{\boldsymbol{\Theta}}_i^{\mathrm{sr}\prime} \right) \odot \mathbf{I}_K$$

$$\widehat{\widetilde{\mathbf{M}}}_s = \sum_{i=0}^{s-1} \widehat{\boldsymbol{\Theta}}_i^{\mathrm{sr}} \odot \widehat{\boldsymbol{\Theta}}_i^{\mathrm{sr}}$$

where $\odot$ is the Hadamard, or element-by-element, product.

The $K^2 \times K^2$ asymptotic VCE of $\mathrm{vec}(\widehat{\mathbf{W}}_s)$ is given by

$$\widetilde{\mathbf{Z}}_s \boldsymbol{\Sigma}(h) \widetilde{\mathbf{Z}}_s'$$

where $\widehat{\boldsymbol{\Sigma}}(h)$ is as derived previously, and

$$\widetilde{\mathbf{Z}}_s = \left\{ \frac{\partial \mathrm{vec}(\widehat{\mathbf{W}}_s)}{\partial \mathrm{vec}(\widehat{\boldsymbol{\Theta}}_0^{\mathrm{sr}})}, \frac{\partial \mathrm{vec}(\widehat{\mathbf{W}}_s)}{\partial \mathrm{vec}(\widehat{\boldsymbol{\Theta}}_1^{\mathrm{sr}})}, \ldots, \frac{\partial \mathrm{vec}(\widehat{\mathbf{W}}_s)}{\partial \mathrm{vec}(\widehat{\boldsymbol{\Theta}}_h^{\mathrm{sr}})} \right\}$$

$$\frac{\partial \mathrm{vec}(\widehat{\mathbf{W}}_s)}{\partial \mathrm{vec}(\widehat{\boldsymbol{\Theta}}_j^{\mathrm{sr}})} = 2 \left\{ (\mathbf{I}_K \otimes \widehat{\mathbf{F}}_s^{-1}) \widetilde{\mathbf{D}}(\widehat{\boldsymbol{\Theta}}_j^{\mathrm{sr}}) - (\widehat{\mathbf{W}}_s' \otimes \widehat{\mathbf{F}}_s^{-1}) \widetilde{\mathbf{D}}(\mathbf{I}_K) \mathbf{N}_K (\widehat{\boldsymbol{\Theta}}_j^{\mathrm{sr}} \otimes I_K) \right\}$$

If $\mathbf{X}$ is an $n \times n$ matrix, then $\widetilde{\mathbf{D}}(\mathbf{X})$ is the $n^2 \times n^2$ matrix with $\mathrm{vec}(\mathbf{X})$ on the diagonal and zeros in all the off-diagonal elements, and $\mathbf{N}_K$ is as defined previously.

Impulse–response function formulas for VECMs

We begin by providing the formulas for backing out the estimates of the $\mathbf{A}_i$ from the $\boldsymbol{\Gamma}_i$ estimated by vec. As discussed in [TS] **vec intro**, the VAR in (1) can be rewritten as a VECM:

$$\Delta y_t = v + \boldsymbol{\Pi} y_{t-1} + \Gamma_1 \Delta y_{t-1} + \Gamma_{p-1} \Delta y_{p-2} + \epsilon_t$$

vec estimates $\boldsymbol{\Pi}$ and the $\boldsymbol{\Gamma}_i$. Johansen (1995, 25) notes that

$$\boldsymbol{\Pi} = \sum_{i=1}^{p} \mathbf{A}_i - \mathbf{I}_K \tag{9}$$

where $\mathbf{I}_K$ is the K-dimensional identity matrix, and

$$\boldsymbol{\Gamma}_i = - \sum_{j=i+1}^{p} \mathbf{A}_j \tag{10}$$

Defining

$$\boldsymbol{\Gamma} = \mathbf{I}_K - \sum_{i=1}^{p-1} \boldsymbol{\Gamma}_i$$

and using (9) and (10) allow us to solve for the $\mathbf{A}_i$ as

$$\mathbf{A}_1 = \boldsymbol{\Pi} + \boldsymbol{\Gamma}_1 + \mathbf{I}_K$$

$$\mathbf{A}_i = \boldsymbol{\Gamma}_i - \boldsymbol{\Gamma}_{i-1} \quad \text{for } i = \{2, \ldots, p-1\}$$

and

$$\mathbf{A}_p = -\boldsymbol{\Gamma}_{p-1}$$

Using these formulas, we can back out estimates of $\mathbf{A}_i$ from the estimates of the $\boldsymbol{\Gamma}_i$ and $\boldsymbol{\Pi}$ produced by vec. Then we simply use the formulas for the IRFs and OIRFs presented in *Impulse–response function formulas for VARs*.

The running sums of the IRFs and OIRFs over the steps within each impulse–response pair are the cumulative IRFs and OIRFs.

Algorithms for bootstrapping the VAR IRF and FEVD standard errors

irf create offers two bootstrap algorithms for estimating the standard errors of the various IRFs and FEVDs. Both var and svar contain estimators for the coefficients in a VAR that are conditional on the first p observations. The two bootstrap algorithms are also conditional on the first p observations.

Specifying the bs option calculates the standard errors by bootstrapping the residuals. For a bootstrap with R repetitions, this method uses the following algorithm:

1. Fit the model and save the estimated parameters.

2. Use the estimated coefficients to calculate the residuals.

3. Repeat steps 3a to 3c R times.

   3a. Draw a simple random sample of size T with replacement from the residuals. The random samples are drawn over the $K \times 1$ vectors of residuals. When the tth vector is drawn, all K residuals are selected. This preserves the contemporaneous correlations among the residuals.

   3b. Use the p initial observations, the sampled residuals, and the estimated coefficients to construct a new sample dataset.

   3c. Fit the model and calculate the different IRFs and FEVDs.

   3d. Save these estimates as observation r in the bootstrapped dataset.

4. For each IRF and FEVD, the estimated standard deviation from the R bootstrapped estimates is the estimated standard error of that impulse–response function or forecast-error variance decomposition.

Specifying the bsp option estimates the standard errors by a multivariate normal parametric bootstrap. The algorithm for the multivariate normal parametric bootstrap is identical to the one above, with the exception that 3a is replaced by 3a(bsp):

   3a(bsp). Draw T pseudovariates from a multivariate normal distribution with covariance matrix $\widehat{\boldsymbol{\Sigma}}$.

References

Amisano, G., and C. Giannini. 1997. *Topics in Structural VAR Econometrics.* 2nd ed. Heidelberg: Springer.

Christiano, L. J., M. Eichenbaum, and C. L. Evans. 1999. Monetary policy shocks: What have we learned and to what end? In *Handbook of Macroeconomics: Volume 1A*, ed. J. B. Taylor and M. Woodford. New York: Elsevier.

Hamilton, J. D. 1994. *Time Series Analysis.* Princeton: Princeton University Press.

Johansen, S. 1995. *Likelihood-Based Inference in Cointegrated Vector Autoregressive Models.* Oxford: Oxford University Press.

Lütkepohl, H. 1993. *Introduction to Multiple Time Series Analysis.* 2nd ed. New York: Springer.

———. 2005. *New Introduction to Multiple Time Series Analysis.* New York: Springer.

Serfling, R. J. 1980. *Approximation Theorems of Mathematical Statistics.* New York: Wiley.

Sims, C. A. 1980. Macroeconomics and reality. *Econometrica* 48: 1–48.

Stock, J. H., and M. W. Watson. 2001. Vector autoregressions. *Journal of Economic Perspectives* 15: 101–115.

Also see

[TS] **irf** — Create and analyze IRFs, dynamic-multiplier functions, and FEVDs

[TS] **var intro** — Introduction to vector autoregressive models

[TS] **vec intro** — Introduction to vector error-correction models

Title

irf ctable — Combine tables of IRFs, dynamic-multiplier functions, and FEVDs

Syntax

irf ctable (*spec₁*) [(*spec₂*) ... [(*spec_N*)]] [, *options*]

where (*spec_k*) is

(*irfname impulsevar responsevar stat* [, *spec_options*])

irfname is the name of a set of IRF results in the active IRF file. *impulsevar* should be specified as an endogenous variable for all statistics except dm and cdm; for those, specify as an exogenous variable. *responsevar* is an endogenous variable name. *stat* is one or more statistics from the list below:

stat	Description
irf	impulse–response function
oirf	orthogonalized impulse–response function
dm	dynamic-multiplier function
cirf	cumulative impulse–response function
coirf	cumulative orthogonalized impulse–response function
cdm	cumulative dynamic-multiplier function
fevd	Cholesky forecast-error variance decomposition
sirf	structural impulse–response function
sfevd	structural forecast-error variance decomposition

options	Description
set(*filename*)	make *filename* active
noci	do not report confidence intervals
stderror	include standard errors for each statistic
individual	make an individual table for each combination
title("*text*")	use *text* as overall table title
step(#)	set common maximum step
level(#)	set confidence level; default is level(95)

spec_options	Description
noci	do not report confidence intervals
stderror	include standard errors for each statistic
level(#)	set confidence level; default is level(95)
ititle("*text*")	use *text* as individual subtitle for specific table

spec_options may be specified within a table specification, globally, or both. When specified in a table specification, the *spec_options* affect only the specification in which they are used. When supplied globally, the *spec_options* affect all table specifications. When specified in both places, options for the table specification take precedence. ititle() does not appear in the dialog box.

194

Menu

Statistics > Multivariate time series > IRF and FEVD analysis > Combined tables

Description

irf ctable makes a table or a combined table of IRF results. Each block within a pair of matching parentheses—each (*spec_k*)—specifies the information for a specific table. irf ctable combines these tables into one table, unless the individual option is specified, in which case separate tables for each block are created.

irf ctable operates on the active IRF file; see [TS] **irf set**.

Options

set(*filename*) specifies the file to be made active; see [TS] **irf set**. If set() is not specified, the active file is used.

noci suppresses reporting of the confidence intervals for each statistic. noci is assumed when the model was fit by vec because no confidence intervals were estimated.

stderror specifies that standard errors for each statistic also be included in the table.

individual places each block, or (*spec_k*), in its own table. By default, irf ctable combines all the blocks into one table.

title("*text*") specifies a title for the table or the set of tables.

step(*#*) specifies the maximum number of steps to use for all tables. By default, each table is constructed using all steps available.

level(*#*) specifies the default confidence level, as a percentage, for confidence intervals, when they are reported. The default is level(95) or as set by set level; see [U] **20.7 Specifying the width of confidence intervals**.

The following option is available with irf ctable but is not shown in the dialog box:

ititle("*text*") specifies an individual subtitle for a specific table. ititle() may be specified only when the individual option is also specified.

Remarks

If you have not read [TS] **irf**, please do so.

Also see [TS] **irf table** for a slightly easier to use, but less powerful, table command.

irf ctable creates a series of tables from IRF results. The information enclosed within each set of parentheses,

(*irfname impulsevar responsevar stat* [, *spec_options*])

forms a request for a specific table.

The first part—*irfname impulsevar responsevar*—identifies a set of IRF estimates or a set of variance decomposition estimates. The next part—*stat*—specifies which statistics are to be included in the table. The last part—*spec_options*—includes the noci, level(), and stderror options, and places (or suppresses) additional columns in the table.

Each specific table displays the requested statistics corresponding to the specified combination of *irfname*, *impulsevar*, and *responsevar* over the step horizon. By default, all the individual tables are combined into one table. Also by default, all the steps, or periods, available are included in the table. You can use the step() option to impose a common maximum for all tables.

▷ Example 1

In example 1 of [TS] **irf table**, we fit a model using var and we saved the IRFs for two different orderings. The commands we used were

```
. use http://www.stata-press.com/data/r12/lutkepohl2
. var dln_inv dln_inc dln_consump
. irf set results4
. irf create ordera, step(8)
. irf create orderb, order(dln_inc dln_inv dln_consump) step(8)
```

We then formed the desired table by typing

```
. irf table oirf fevd, impulse(dln_inc) response(dln_consump) noci std
> title("Ordera versus orderb")
```

Using irf ctable, we can form the equivalent table by typing

```
. irf ctable (ordera dln_inc dln_consump oirf fevd)
>            (orderb dln_inc dln_consump oirf fevd),
>            noci std title("Ordera versus orderb")
```

Ordera versus orderb

step	(1) oirf	(1) S.E.	(1) fevd	(1) S.E.
0	.005123	.000878	0	0
1	.001635	.000984	.288494	.077483
2	.002948	.000993	.294288	.073722
3	-.000221	.000662	.322454	.075562
4	.000811	.000586	.319227	.074063
5	.000462	.000333	.322579	.075019
6	.000044	.000275	.323552	.075371
7	.000151	.000162	.323383	.075314
8	.000091	.000114	.323499	.075386

step	(2) oirf	(2) S.E.	(2) fevd	(2) S.E.
0	.005461	.000925	0	0
1	.001578	.000988	.327807	.08159
2	.003307	.001042	.328795	.077519
3	-.00019	.000676	.370775	.080604
4	.000846	.000617	.366896	.079019
5	.000491	.000349	.370399	.079941
6	.000069	.000292	.371487	.080323
7	.000158	.000172	.371315	.080287
8	.000096	.000122	.371438	.080366

(1) irfname = ordera, impulse = dln_inc, and response = dln_consump
(2) irfname = orderb, impulse = dln_inc, and response = dln_consump

The output is displayed in one table. Because the table did not fit horizontally, it automatically wrapped. At the bottom of the table is a list of keys that appear at the top of each column. The results in the table above indicate that the orthogonalized IRFs do not change by much. Because the estimated forecast-error variances do change, we might want to produce two tables that contain the estimated forecast-error variance decompositions and their 95% confidence intervals:

```
. irf ctable (ordera dln_inc dln_consump fevd)
>            (orderb dln_inc dln_consump fevd), individual
Table 1
```

step	(1) fevd	(1) Lower	(1) Upper
0	0	0	0
1	.288494	.13663	.440357
2	.294288	.149797	.43878
3	.322454	.174356	.470552
4	.319227	.174066	.464389
5	.322579	.175544	.469613
6	.323552	.175826	.471277
7	.323383	.17577	.470995
8	.323499	.175744	.471253

```
95% lower and upper bounds reported
(1) irfname = ordera, impulse = dln_inc, and response = dln_consump
Table 2
```

step	(2) fevd	(2) Lower	(2) Upper
0	0	0	0
1	.327807	.167893	.487721
2	.328795	.17686	.48073
3	.370775	.212794	.528757
4	.366896	.212022	.52177
5	.370399	.213718	.52708
6	.371487	.214058	.528917
7	.371315	.213956	.528674
8	.371438	.213923	.528953

```
95% lower and upper bounds reported
(2) irfname = orderb, impulse = dln_inc, and response = dln_consump
```

Because we specified the individual option, the output contains two tables, one for each specific table command. At the bottom of each table is a list of the keys used in that table and a note indicating the level of the confidence intervals that we requested. The results from table 1 and table 2 indicate that each estimated function is well within the confidence interval of the other, so we conclude that the functions are not significantly different.

◁

Saved results

irf ctable saves the following in r():

Scalars
    r(ncols)   number of columns in all tables
    r(k_umax)   number of distinct keys
    r(k)   number of specific table commands
Macros
    r(key#)   #th key
    r(tnotes)   list of keys applied to each column

Methods and formulas

irf ctable is implemented as an ado-file.

Also see

[TS] **irf** — Create and analyze IRFs, dynamic-multiplier functions, and FEVDs

[TS] **var intro** — Introduction to vector autoregressive models

[TS] **vec intro** — Introduction to vector error-correction models

Title

> **irf describe** — Describe an IRF file

Syntax

irf <u>d</u>escribe [*irf_resultslist*] [, *options*]

options	Description
<u>set</u>(*filename*)	make *filename* active
<u>using</u>(*irf_filename*)	describe *irf_filename* without making active
<u>detail</u>	show additional details of IRF results
<u>variables</u>	show underlying structure of the IRF dataset

Menu

Statistics > Multivariate time series > Manage IRF results and files > Describe IRF file

Description

irf describe describes the IRF results stored in an IRF file.

If set() or using() is not specified, the IRF results of the active IRF file are described.

Options

set(*filename*) specifies the IRF file to be described and set; see [TS] **irf set**. If *filename* is specified without an extension, .irf is assumed.

using(*irf_filename*) specifies the IRF file to be described. The active IRF file, if any, remains unchanged. If *irf_filename* is specified without an extension, .irf is assumed.

detail specifies that irf describe display detailed information about each set of IRF results. detail is implied when *irf_resultslist* is specified.

variables is a programmer's option; additionally displays the output produced by the describe command.

Remarks

If you have not read [TS] **irf**, please do so.

> Example 1

```
. use http://www.stata-press.com/data/r12/lutkepohl2
(Quarterly SA West German macro data, Bil DM, from Lutkepohl 1993 Table E.1)
. var dln_inv dln_inc dln_consump if qtr<=tq(1978q4), lags(1/2) dfk
(output omitted)
```

We create three sets of IRF results:

```
. irf create order1, set(myirfs, replace)
(file myirfs.irf created)
(file myirfs.irf now active)
(file myirfs.irf updated)
. irf create order2, order(dln_inc dln_inv dln_consump)
(file myirfs.irf updated)
. irf create order3, order(dln_inc dln_consump dln_inv)
(file myirfs.irf updated)
. irf describe
Contains irf results from myirfs.irf (dated 4 Apr 2011 12:36)
```

irfname	model	endogenous variables and order (*)
order1	var	dln_inv dln_inc dln_consump
order2	var	dln_inc dln_inv dln_consump
order3	var	dln_inc dln_consump dln_inv

```
(*) order is relevant only when model is var
```

The output reveals the order in which we specified the variables.

```
. irf describe order1
```

irf results for order1

```
   Estimation specification
         model:  var
         endog:  dln_inv dln_inc dln_consump
        sample:  quarterly data from 1960q4 to 1978q4
          lags:  1 2
      constant:  constant
          exog:  none
      exogvars:  none
        exlags:  none
        varcns:  unconstrained
   IRF specification
          step:  8
         order:  dln_inv dln_inc dln_consump
     std error:  asymptotic
          reps:  none
```

Here we see a summary of the model we fit as well as the specification of the IRFs.

◁

Saved results

irf describe saves the following in r():

Scalars
r(N)	number of observations in the IRF file
r(k)	number of variables in the IRF file
r(width)	width of dataset in the IRF file
r(N_max)	maximum number of observations
r(k_max)	maximum number of variables
r(widthmax)	maximum width of the dataset
r(changed)	flag indicating that data have changed since last saved

Macros
r(_version)	version of IRF results file
r(irfnames)	names of IRF results in the IRF file
r(*irfname*_model)	var, sr var, lr var, or vec
r(*irfname*_order)	Cholesky order assumed in IRF estimates
r(*irfname*_exog)	exogenous variables, and their lags, in VAR or underlying VAR
r(*irfname*_exogvar)	exogenous variables in VAR or underlying VAR
r(*irfname*_constant)	constant or noconstant
r(*irfname*_lags)	lags in model
r(*irfname*_exlags)	lags of exogenous variables in model
r(*irfname*_tmin)	minimum value of timevar in the estimation sample
r(*irfname*_tmax)	maximum value of timevar in the estimation sample
r(*irfname*_timevar)	name of tsset timevar
r(*irfname*_tsfmt)	format of timevar in the estimation sample
r(*irfname*_varcns)	unconstrained or colon-separated list of constraints placed on VAR coefficients
r(*irfname*_svarcns)	"." or colon-separated list of constraints placed on SVAR coefficients
r(*irfname*_step)	maximum step in IRF estimates
r(*irfname*_stderror)	asymptotic, bs, bsp, or none, depending on type of standard errors specified to irf create
r(*irfname*_reps)	"." or number of bootstrap replications performed
r(*irfname*_version)	version of IRF file that originally held *irfname* IRF results
r(*irfname*_rank)	"." or number of cointegrating equations
r(*irfname*_trend)	"." or trend() specified in vec
r(*irfname*_veccns)	"." or constraints placed on VECM parameters
r(*irfname*_sind)	"." or normalized seasonal indicators included in vec

Methods and formulas

irf describe is implemented as an ado-file.

Also see

[TS] **irf** — Create and analyze IRFs, dynamic-multiplier functions, and FEVDs

[TS] **var intro** — Introduction to vector autoregressive models

[TS] **vec intro** — Introduction to vector error-correction models

Title

> **irf drop** — Drop IRF results from the active IRF file

Syntax

irf drop *irf_resultslist* [, set(*filename*)]

Menu

Statistics > Multivariate time series > Manage IRF results and files > Drop IRF results

Description

irf drop removes IRF results from the active IRF file.

Option

set(*filename*) specifies the file to be made active; see [TS] **irf set**. If set() is not specified, the active file is used.

Remarks

If you have not read [TS] **irf**, please do so.

▷ Example 1

```
. use http://www.stata-press.com/data/r12/lutkepohl2
(Quarterly SA West German macro data, Bil DM, from Lutkepohl 1993 Table E.1)
. var dln_inv dln_inc dln_consump if qtr<=tq(1978q4), lags(1/2) dfk
(output omitted)
```

We create three sets of IRF results:

```
. irf create order1, set(myirfs, replace)
(file myirfs.irf created)
(file myirfs.irf now active)
(file myirfs.irf updated)
. irf create order2, order(dln_inc dln_inv dln_consump)
(file myirfs.irf updated)
. irf create order3, order(dln_inc dln_consump dln_inv)
(file myirfs.irf updated)
. irf describe
Contains irf results from myirfs.irf (dated 4 Apr 2011 12:59)
```

irfname	model	endogenous variables and order (*)
order1	var	dln_inv dln_inc dln_consump
order2	var	dln_inc dln_inv dln_consump
order3	var	dln_inc dln_consump dln_inv

(*) order is relevant only when model is var

Now let's remove order1 and order2 from myirfs.irf.

```
. irf drop order1 order2
(order1 dropped)
(order2 dropped)
file myirfs.irf updated
. irf describe
Contains irf results from myirfs.irf (dated 4 Apr 2011 12:59)
```

irfname	model	endogenous variables and order (*)
order3	var	dln_inc dln_consump dln_inv

```
(*) order is relevant only when model is var
```

order1 and order2 have been dropped.

◁

Methods and formulas

irf drop is implemented as an ado-file.

Also see

[TS] **irf** — Create and analyze IRFs, dynamic-multiplier functions, and FEVDs

[TS] **var intro** — Introduction to vector autoregressive models

[TS] **vec intro** — Introduction to vector error-correction models

Title

irf graph — Graph IRFs, dynamic-multiplier functions, and FEVDs

Syntax

irf graph *stat* [, *options*]

stat	Description
irf	impulse–response function
oirf	orthogonalized impulse–response function
dm	dynamic-multiplier function
cirf	cumulative impulse–response function
coirf	cumulative orthogonalized impulse–response function
cdm	cumulative dynamic-multiplier function
fevd	Cholesky forecast-error variance decomposition
sirf	structural impulse–response function
sfevd	structural forecast-error variance decomposition

Notes: 1. No statistic may appear more than once.
2. If confidence intervals are included (the default), only two statistics may be included.
3. If confidence intervals are suppressed (option noci), up to four statistics may be included.

options	Description
Main	
set(*filename*)	make *filename* active
irf(*irfnames*)	use *irfnames* IRF result sets
impulse(*impulsevar*)	use *impulsevar* as impulse variables
response(*endogvars*)	use endogenous variables as response variables
noci	suppress confidence bands
level(#)	set confidence level; default is level(95)
lstep(#)	use # for first step
ustep(#)	use # for maximum step
Advanced	
individual	graph each combination individually
iname(*namestub* [, replace])	stub for naming the individual graphs
isaving(*filenamestub* [, replace])	stub for saving the individual graphs to files
Plots	
plot#opts(*cline_options*)	affect rendition of the line plotting the # *stat*
CI plots	
ci#opts(*area_options*)	affect rendition of the confidence interval for the # *stat*

Y axis, X axis, Titles, Legend, Overall

twoway_options	any options other than by() documented in [G-3] ***twoway_options***
<u>byopts</u>(*by_option*)	how subgraphs are combined, labeled, etc.

Menu

Statistics > Multivariate time series > IRF and FEVD analysis > Graphs by impulse or response

Description

irf graph graphs impulse–response functions (IRFs), dynamic-multiplier functions, and forecast-error variance decompositions (FEVDs) over time.

Options

   ⌐ Main ⌐

set(*filename*) specifies the file to be made active; see [TS] **irf set**. If set() is not specified, the active file is used.

irf(*irfnames*) specifies the IRF result sets to be used. If irf() is not specified, each of the results in the active IRF file is used. (Files often contain just one set of IRF results stored under one *irfname*; in that case, those results are used.)

impulse(*impulsevar*) and response(*endogvars*) specify the impulse and response variables. Usually one of each is specified, and one graph is drawn. If multiple variables are specified, a separate subgraph is drawn for each impulse–response combination. If impulse() and response() are not specified, subgraphs are drawn for all combination of impulse and response variables.

impulsevar should be specified as an endogenous variable for all statistics except dm or cdm; for those, specify as an exogenous variable.

noci suppresses graphing the confidence interval for each statistic. noci is assumed when the model was fit by vec because no confidence intervals were estimated.

level(#) specifies the default confidence level, as a percentage, for confidence intervals, when they are reported. The default is level(95) or as set by set level; see [U] **20.7 Specifying the width of confidence intervals**. Also see [TS] **irf cgraph** for a graph command that allows the confidence level to vary over the graphs.

lstep(#) specifies the first step, or period, to be included in the graphs. lstep(0) is the default.

ustep(#), # ≥ 1, specifies the maximum step, or period, to be included in the graphs.

   ⌐ Advanced ⌐

individual specifies that each graph be displayed individually. By default, irf graph combines the subgraphs into one image. When individual is specified, byopts() may not be specified, but the isaving() and iname() options may be specified.

iname(*namestub* [, replace]) specifies that the *i*th individual graph be saved in memory under the name *namestubi*, which must be a valid Stata name of 24 characters or fewer. iname() may be specified only with the individual option.

isaving(*filenamestub* [, replace]) specifies that the *i*th individual graph should be saved to disk in the current working directory under the name *filenamestubi*.gph. isaving() may be specified only when the individual option is also specified.

⌐ Plots ⌐

plot1opts(*cline_options*), ..., plot4opts(*cline_options*) affect the rendition of the plotted statistics (the *stat*). plot1opts() affects the rendition of the first statistic; plot2opts(), the second; and so on. *cline_options* are as described in [G-3] ***cline_options***.

⌐ CI plots ⌐

ci1opts(*area_options*) and ci2opts(*area_options*) affect the rendition of the confidence intervals for the first (ci1opts()) and second (ci2opts()) statistics in *stat*. *area_options* are as described in [G-3] ***area_options***.

⌐ Y axis, X axis, Titles, Legend, Overall ⌐

twoway_options are any of the options documented in [G-3] ***twoway_options***, excluding by(). These include options for titling the graph (see [G-3] ***title_options***) and for saving the graph to disk (see [G-3] ***saving_option***). Note that the saving() and name() options may not be combined with the individual option.

byopts(*by_option*) is as documented in [G-3] ***by_option*** and may not be specified when individual is specified. byopts() affects how the subgraphs are combined, labeled, etc.

Remarks

If you have not read [TS] **irf**, please do so.

Also see [TS] **irf cgraph**, which produces combined graphs; [TS] **irf ograph**, which produces overlaid graphs; and [TS] **irf table**, which displays results in tabular form.

irf graph produces one or more graphs and displays them arrayed into one image unless the individual option is specified, in which case the individual graphs are displayed separately. Each individual graph consists of all the specified *stat* and represents one impulse–response combination.

Because all the specified *stat* appear on the same graph, putting together statistics with very different scales is not recommended. For instance, sometimes sirf and oirf are on similar scales while irf is on a different scale. In such cases, combining sirf and oirf on the same graph looks fine, but combining either with irf produces an uninformative graph.

▷ Example 1

Suppose that we have results generated from two different SVAR models. We want to know whether the shapes of the structural IRFs and the structural FEVDs are similar in the two models. We are also interested in knowing whether the structural IRFs and the structural FEVDs differ significantly from their Cholesky counterparts.

Filling in the background, we have previously issued the commands

```
. use http://www.stata-press.com/data/r12/lutkepohl2
. mat a = (., 0, 0\0,.,0\.,.,.)
. mat b = I(3)
. svar dln_inv dln_inc dln_consump, aeq(a) beq(b)
. irf create modela, set(results3) step(8)
. svar dln_inc dln_inv dln_consump, aeq(a) beq(b)
. irf create modelb, step(8)
```

To see whether the shapes of the structural IRFs and the structural FEVDs are similar in the two models, we type

```
. irf graph oirf sirf, impulse(dln_inc) response(dln_consump)
```

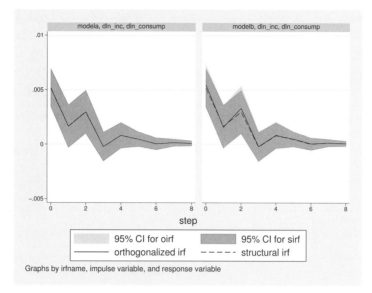

The graph reveals that the `oirf` and the `sirf` estimates are essentially the same for both models and that the shapes of the functions are very similar for the two models.

To see whether the structural IRFs and the structural FEVDs differ significantly from their Cholesky counterparts, we type

```
. irf graph fevd sfevd, impulse(dln_inc) response(dln_consump) lstep(1)
> legend(cols(1))
```

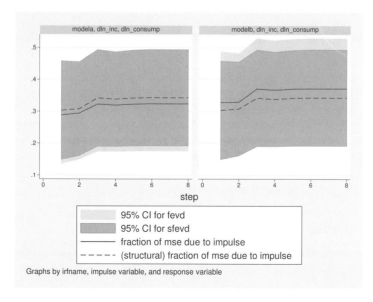

This combined graph reveals that the shapes of these functions are also similar for the two models. However, the graph illuminates one minor difference between them: In `modela`, the estimated structural

FEVD is slightly larger than the Cholesky-based estimates, whereas in modelb the Cholesky-based estimates are slightly larger than the structural estimates. For both models, however, the structural estimates are close to the center of the wide confidence intervals for the two estimates.

◁

▷ Example 2

Let's focus on the results from modela. Suppose that we were interested in examining how dln_consump responded to impulses in its own structural innovations, structural innovations to dln_inc, and structural innovations to dln_inv. We type

    . irf graph sirf, irf(modela) response(dln_consump)

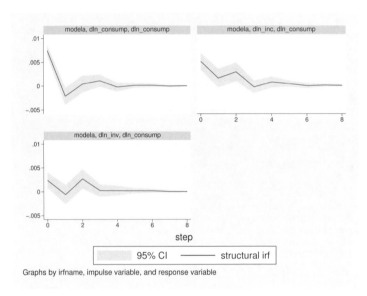

The upper-left graph shows the structural IRF of an innovation in dln_consump on dln_consump. It indicates that the identification restrictions used in modela imply that a positive shock to dln_consump causes an increase in dln_consump, followed by a decrease, followed by an increase, and so on, until the effect dies out after roughly 5 periods.

The upper-right graph shows the structural IRF of an innovation in dln_inc on dln_consump, indicating that a positive shock to dln_inc causes an increase in dln_consump, which dies out after 4 or 5 periods.

◁

❑ Technical note

[TS] **irf table** contains a technical note warning you to be careful in naming variables when you fit models. What is said there applies equally here.

❑

Saved results

irf graph saves the following in r():

Scalars
  r(k)   number of graphs

Macros

r(stats)	*statlist*	r(byopts)	contents of byopts()
r(irfname)	*resultslist*	r(saving)	supplied saving() option
r(impulse)	*impulselist*	r(name)	supplied name() option
r(response)	*responselist*	r(individual)	individual or blank
r(plot#)	contents of plot#opts()	r(isaving)	contents of saving()
r(ci)	level applied to confidence	r(iname)	contents of name()
	intervals or noci	r(subtitle#)	subtitle for individual graph #
r(ciopts#)	contents of ci#opts()		

Methods and formulas

irf graph is implemented as an ado-file.

Also see

[TS] **irf** — Create and analyze IRFs, dynamic-multiplier functions, and FEVDs

[TS] **var intro** — Introduction to vector autoregressive models

[TS] **vec intro** — Introduction to vector error-correction models

Title

> **irf ograph** — Graph overlaid IRFs, dynamic-multiplier functions, and FEVDs

Syntax

irf o̲graph (*spec₁*) [(*spec₂*) ...[(*spec₁₅*)]] [, *options*]

where (*specₖ*) is

(*irfname impulsevar responsevar stat* [, *spec_options*])

irfname is the name of a set of IRF results in the active IRF file or ".", which means the first named result in the active IRF file. *impulsevar* should be specified as an endogenous variable for all statistics except dm and cdm; for those, specify as an exogenous variable. *responsevar* is an endogenous variable name. *stat* is one or more statistics from the list below:

stat	Description
irf	impulse–response function
oirf	orthogonalized impulse–response function
dm	dynamic-multiplier function
cirf	cumulative impulse–response function
coirf	cumulative orthogonalized impulse–response function
cdm	cumulative dynamic-multiplier function
fevd	Cholesky forecast-error variance decomposition
sirf	structural impulse–response function
sfevd	structural forecast-error variance decomposition

options	Description
Plots	
plot_options	define the IRF plots
set(*filename*)	make *filename* active
Options	
common_options	level and steps
Y axis, X axis, Titles, Legend, Overall	
twoway_options	any options other than by() documented in [G-3] ***twoway_options***

plot_options	Description
Main	
set(*filename*)	make *filename* active
i̲rf(*irfnames*)	use *irfnames* IRF result sets
i̲mpulse(*impulsevar*)	use *impulsevar* as impulse variables
r̲esponse(*endogvars*)	use endogenous variables as response variables
ci	add confidence bands to the graph

spec_options	Description
Options	
common_options	level and steps
Plot	
cline_options	affect rendition of the plotted lines
CI plot	
<u>ciopts</u>(*area_options*)	affect rendition of the confidence intervals

common_options	Description
Options	
<u>level</u>(*#*)	set confidence level; default is level(95)
<u>lstep</u>(*#*)	use *#* for first step
<u>ustep</u>(*#*)	use *#* for maximum step

common_options may be specified within a plot specification, globally, or in both. When specified in a plot specification, the *common_options* affect only the specification in which they are used. When supplied globally, the *common_options* affect all plot specifications. When supplied in both places, options in the plot specification take precedence.

Menu

Statistics > Multivariate time series > IRF and FEVD analysis > Overlaid graph

Description

irf ograph displays plots of irf results on one graph (one pair of axes).

To become familiar with this command, type db irf ograph.

Options

⌐ Plots ⌐

plot_options defines the IRF plots and are found under the **Main**, **Plot**, and **CI plot** tabs.

set(*filename*) specifies the file to be made active; see [TS] **irf set**. If set() is not specified, the active file is used.

⌐ Main ⌐

set(*filename*) specifies the file to be made active; see [TS] **irf set**. If set() is not specified, the active file is used.

irf(*irfnames*) specifies the IRF result sets to be used. If irf() is not specified, each of the results in the active IRF file is used. (Files often contain just one set of IRF results stored under one *irfname*; in that case, those results are used.)

impulse(*varlist*) and response(*endogvars*) specify the impulse and response variables. Usually one of each is specified, and one graph is drawn. If multiple variables are specified, a separate subgraph is drawn for each impulse–response combination. If impulse() and response() are not specified, subgraphs are drawn for all combination of impulse and response variables.

ci adds confidence bands to the graph. The noci option may be used within a plot specification to suppress its confidence bands when the ci option is supplied globally.

⌐‾‾‾‾‾⌐ Plot ⌐

cline_options affect the rendition of the plotted lines; see [G-3] *cline_options*.

⌐‾‾‾‾‾⌐ CI plot ⌐

ciopts(*area_options*) affects the rendition of the confidence bands for the plotted statistic; see [G-3] *area_options*. ciopts() implies ci.

⌐‾‾‾‾‾⌐ Options ⌐

level(#) specifies the confidence level, as a percentage, for confidence bands; see [U] **20.7 Specifying the width of confidence intervals**.

lstep(#) specifies the first step, or period, to be included in the graph. lstep(0) is the default.

ustep(#), $\# \geq 1$, specifies the maximum step, or period, to be included.

⌐‾‾‾‾‾⌐ Y axis, X axis, Titles, Legend, Overall ⌐

twoway_options are any of the options documented in [G-3] *twoway_options*, excluding by(). These include options for titling the graph (see [G-3] *title_options*) and for saving the graph to disk (see [G-3] *saving_option*).

Remarks

If you have not read [TS] **irf**, please do so.

irf ograph overlays plots of IRFs and FEVDs on one graph.

▷ Example 1

We have previously issued the commands

```
. use http://www.stata-press.com/data/r12/lutkepohl2
. var dln_inv dln_inc dln_consump if qtr<=tq(1978q4), lags(1/2) dfk
. irf create order1, step(10) set(myirf1, new)
. irf create order2, step(10) order(dln_inc dln_inv dln_consump)
```

We now wish to compare the oirf for impulse dln_inc and response dln_consump for two different Cholesky orderings:

```
. irf ograph (order1 dln_inc dln_consump oirf)
>               (order2 dln_inc dln_consump oirf)
```

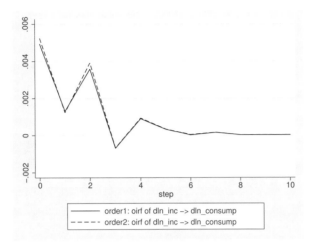

◁

❏ Technical note

Graph options allow you to change the appearance of each plot. The following graph contains the plots of the FEVDs (FEVDs) for impulse `dln_inc` and each response using the results from the first collection of results in the active IRF file (using the "." shortcut). In the second plot, we supply the `clpat(dash)` option (an abbreviation for `clpattern(dash)`) to give the line a dashed pattern. In the third plot, we supply the `m(o) clpat(dashdot) recast(connected)` options to get small circles connected by a line with a dash–dot pattern; the `cilines` option plots the confidence bands by using lines instead of areas. We use the `title()` option to add a descriptive title to the graph and supply the `ci` option globally to add confidence bands to all the plots.

```
. irf ograph (. dln_inc dln_inc fevd)
>            (. dln_inc dln_consump fevd, clpat(dash))
>            (. dln_inc dln_inv fevd, cilines m(o) clpat(dashdot)
>                                         recast(connected))
>         , ci title("Comparison of forecast-error variance decomposition")
```

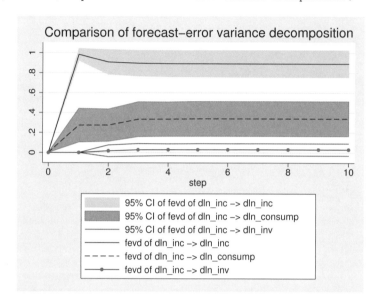

The clpattern() option is described in [G-3] **connect_options**, msymbol() is described in [G-3] **marker_options**, title() is described in [G-3] **title_options**, and recast() is described in [G-3] **advanced_options**.

❑

Saved results

irf ograph saves the following in r():

Scalars
r(plots)	number of plot specifications
r(ciplots)	number of plotted confidence bands

Macros
r(irfname#)	*irfname* from (*spec#*)
r(impulse#)	impulse from (*spec#*)
r(response#)	response from (*spec#*)
r(stat#)	statistics from (*spec#*)
r(ci#)	level from (*spec#*) or noci

Methods and formulas

irf ograph is implemented as an ado-file.

Also see

[TS] **irf** — Create and analyze IRFs, dynamic-multiplier functions, and FEVDs

[TS] **var intro** — Introduction to vector autoregressive models

[TS] **vec intro** — Introduction to vector error-correction models

Title

> **irf rename** — Rename an IRF result in an IRF file

Syntax

irf <u>rena</u>me *oldname* *newname* $\left[\,,\, \text{set}(\textit{filename})\,\right]$

Menu

Statistics > Multivariate time series > Manage IRF results and files > Rename IRF results

Description

irf rename changes the name of a set of IRF results stored in the active IRF file.

Option

set(*filename*) specifies the file to be made active; see [TS] **irf set**. If set() is not specified, the active file is used.

Remarks

If you have not read [TS] **irf**, please do so.

▷ Example 1

```
. use http://www.stata-press.com/data/r12/lutkepohl2
(Quarterly SA West German macro data, Bil DM, from Lutkepohl 1993 Table E.1)
. var dln_inv dln_inc dln_consump if qtr<=tq(1978q4), lags(1/2) dfk
(output omitted)
```

We create three sets of IRF results:

```
. irf create original, set(myirfs, replace)
(file myirfs.irf created)
(file myirfs.irf now active)
(file myirfs.irf updated)
. irf create order2, order(dln_inc dln_inv dln_consump)
(file myirfs.irf updated)
. irf create order3, order(dln_inc dln_consump dln_inv)
(file myirfs.irf updated)
. irf describe
Contains irf results from myirfs.irf (dated 4 Apr 2011 13:06)
```

irfname	model	endogenous variables and order (*)
original	var	dln_inv dln_inc dln_consump
order2	var	dln_inc dln_inv dln_consump
order3	var	dln_inc dln_consump dln_inv

(*) order is relevant only when model is var

Now let's rename IRF result original to order1.

```
. irf rename original order1
(81 real changes made)
original renamed to order1

. irf describe
Contains irf results from myirfs.irf (dated 4 Apr 2011 13:06)
        irfname | model    endogenous variables and order (*)
```

irfname	model	endogenous variables and order (*)
order1	var	dln_inv dln_inc dln_consump
order2	var	dln_inc dln_inv dln_consump
order3	var	dln_inc dln_consump dln_inv

(*) order is relevant only when model is var

original has been renamed to order1.

◁

Saved results

irf rename saves the following in r():

Macros
    r(irfnames)   *irfnames* after rename
    r(oldnew)   *oldname newname*

Methods and formulas

irf rename is implemented as an ado-file.

Also see

[TS] **irf** — Create and analyze IRFs, dynamic-multiplier functions, and FEVDs

[TS] **var intro** — Introduction to vector autoregressive models

[TS] **vec intro** — Introduction to vector error-correction models

Title

irf set — Set the active IRF file

Syntax

Report identity of active file

    irf set

Set, and if necessary create, active file

    irf set *irf_filename*

Create, and if necessary replace, active file

    irf set *irf_filename*, replace

Clear any active IRF file

    irf set, clear

Menu

Statistics > Multivariate time series > Manage IRF results and files > Set active IRF file

Description

In the first syntax, `irf set` reports the identity of the active file, if there is one. Also see [TS] **irf describe** for obtaining reports on the contents of an IRF file.

In the second syntax, `irf set` *irf_filename* specifies that the file be set as the active file and, if the file does not exist, that it be created as well.

In the third syntax, `irf set` *irf_filename*, `replace` specifies that even if file *irf_filename* exists, a new, empty file is to be created and set.

In the rarely used fourth syntax, `irf set, clear` specifies that, if any IRF file is set, it be unset and that there be no active IRF file.

IRF files are just files: they can be erased by `erase`, listed by `dir`, and copied by `copy`; see [D] **erase**, [D] **dir**, and [D] **copy**.

If *irf_filename* is specified without an extension, `.irf` is assumed.

Options

`replace` specifies that if *irf_filename* already exists, the file is to be erased and a new, empty IRF file is to be created in its place. If it does not already exist, a new, empty file is created.

`clear` unsets the active IRF file.

Remarks

If you have not read [TS] **irf**, please do so.

`irf set` reports the identity of the active IRF file:

```
. irf set
no irf file active
```

`irf set` *irf_filename* creates and sets an IRF file:

```
. irf set results1
(file results1.irf now active)
```

We specified the name `results1`, and `results1.irf` became the active file. The suffix `.irf` was added for us.

`irf set` *irf_filename* can also be used to create a new file:

```
. use http://www.stata-press.com/data/r12/lutkepohl2
(Quarterly SA West German macro data, Bil DM, from Lutkepohl 1993 Table E.1)
. var dln_inc dln_consump, exog(l.dln_inv)
  (output omitted )
. irf set results2
(file results2.irf created)
(file results2.irf now active)
. irf create order1
(file results2.irf updated)
```

Saved results

`irf set` saves the following in `r()`:

Macros
    r(Orville)   name of active IRF file, if there is an active IRF

Methods and formulas

`irf set` is implemented as an ado-file.

Also see

[TS] **irf** — Create and analyze IRFs, dynamic-multiplier functions, and FEVDs

[TS] **var intro** — Introduction to vector autoregressive models

[TS] **vec intro** — Introduction to vector error-correction models

Title

> **irf table** — Create tables of IRFs, dynamic-multiplier functions, and FEVDs

Syntax

irf table [*stat*] [, *options*]

stat	Description
Main	
irf	impulse–response function
oirf	orthogonalized impulse–response function
dm	dynamic-multiplier function
cirf	cumulative impulse–response function
coirf	cumulative orthogonalized impulse–response function
cdm	cumulative dynamic-multiplier function
fevd	Cholesky forecast-error variance decomposition
sirf	structural impulse–response function
sfevd	structural forecast-error variance decomposition

If *stat* is not specified, all statistics are included, unless option nostructural is also specified, in which case sirf and sfevd are excluded. You may specify more than one *stat*.

options	Description
Main	
set(*filename*)	make *filename* active
irf(*irfnames*)	use *irfnames* IRF result sets
impulse(*impulsevar*)	use *impulsevar* as impulse variables
response(*endogvars*)	use endogenous variables as response variables
individual	make an individual table for each result set
title("*text*")	use *text* for overall table title
Options	
level(#)	set confidence level; default is level(95)
noci	suppress confidence intervals
stderror	include standard errors in the tables
nostructural	suppress sirf and sfevd from the default list of statistics
step(#)	use common maximum step horizon # for all tables

Menu

Statistics > Multivariate time series > IRF and FEVD analysis > Tables by impulse or response

Description

irf table makes a table from the specified IRF results.

The rows of the tables are the time since impulse. Each column represents a combination of impulse() variable and response() variable for a *stat* from the irf() results.

Options

┌─ Main ┐

set(*filename*) specifies the file to be made active; see [TS] **irf set**. If set() is not specified, the active file is used.

All results are obtained from one IRF file. If you have results in different files that you want in one table, use irf add to copy results into one file; see [TS] **irf add**.

irf(*irfnames*) specifies the IRF result sets to be used. If irf() is not specified, all the results in the active IRF file are used. (Files often contain just one set of IRF results, stored under one *irfname*; in that case, those results are used. When there are multiple IRF results, you may also wish to specify the individual option.)

impulse(*impulsevar*) specifies the impulse variables for which the statistics are to be reported. If impulse() is not specified, each model variable, in turn, is used. *impulsevar* should be specified as an endogenous variable for all statistics except dm or cdm; for those, specify as an exogenous variable.

response(*endogvars*) specifies the response variables for which the statistics are to be reported. If response() is not specified, each endogenous variable, in turn, is used.

individual specifies that each set of IRF results be placed in its own table, with its own title and footer. By default, irf table places all the IRF results in one table with one title and one footer. individual may not be combined with title().

title("*text*") specifies a title for the overall table.

┌─ Options ┐

level(*#*) specifies the default confidence level, as a percentage, for confidence intervals, when they are reported. The default is level(95) or as set by set level; see [U] **20.7 Specifying the width of confidence intervals**.

noci suppresses reporting of the confidence intervals for each statistic. noci is assumed when the model was fit by vec because no confidence intervals were estimated.

stderror specifies that standard errors for each statistic also be included in the table.

nostructural specifies that *stat*, when not specified, exclude sirf and sfevd.

step(*#*) specifies the maximum step horizon for all tables. If step() is not specified, each table is constructed using all steps available.

Remarks

If you have not read [TS] **irf**, please do so.

Also see [TS] **irf graph**, which produces output in graphical form, and see [TS] **irf ctable**, which also produces tabular output. irf ctable is more difficult to use but provides more control over how tables are formed.

▷ Example 1

We have fit a model with `var`, and we saved the IRFs from two different orderings. The commands we previously used were

```
. use http://www.stata-press.com/data/r12/lutkepohl2
. var dln_inv dln_inc dln_consump
. irf set results4
. irf create ordera, step(8)
. irf create orderb, order(dln_inc dln_inv dln_consump) step(8)
```

We now wish to compare the two orderings:

```
. irf table oirf fevd, impulse(dln_inc) response(dln_consump) noci std
> title("Ordera versus orderb")
```

                        Ordera versus orderb

step	(1) oirf	(1) S.E.	(1) fevd	(1) S.E.
0	.005123	.000878	0	0
1	.001635	.000984	.288494	.077483
2	.002948	.000993	.294288	.073722
3	−.000221	.000662	.322454	.075562
4	.000811	.000586	.319227	.074063
5	.000462	.000333	.322579	.075019
6	.000044	.000275	.323552	.075371
7	.000151	.000162	.323383	.075314
8	.000091	.000114	.323499	.075386

step	(2) oirf	(2) S.E.	(2) fevd	(2) S.E.
0	.005461	.000925	0	0
1	.001578	.000988	.327807	.08159
2	.003307	.001042	.328795	.077519
3	−.00019	.000676	.370775	.080604
4	.000846	.000617	.366896	.079019
5	.000491	.000349	.370399	.079941
6	.000069	.000292	.371487	.080323
7	.000158	.000172	.371315	.080287
8	.000096	.000122	.371438	.080366

```
(1) irfname = ordera, impulse = dln_inc, and response = dln_consump
(2) irfname = orderb, impulse = dln_inc, and response = dln_consump
```

The output is displayed as a "single" table; because the table did not fit horizontally, it wrapped automatically. At the bottom of the table is a definition of the keys that appear at the top of each column. The results in the table above indicate that the orthogonalized IRFs do not change by much.

◁

▷ Example 2

Because the estimated FEVDs do change significantly, we might want to produce two tables that contain the estimated FEVDs and their 95% confidence intervals:

```
. irf table fevd, impulse(dln_inc) response(dln_consump) individual
        Results from ordera
```

	(1)	(1)	(1)
step	fevd	Lower	Upper
0	0	0	0
1	.288494	.13663	.440357
2	.294288	.149797	.43878
3	.322454	.174356	.470552
4	.319227	.174066	.464389
5	.322579	.175544	.469613
6	.323552	.175826	.471277
7	.323383	.17577	.470995
8	.323499	.175744	.471253

```
95% lower and upper bounds reported
(1) irfname = ordera, impulse = dln_inc, and response = dln_consump
        Results from orderb
```

	(1)	(1)	(1)
step	fevd	Lower	Upper
0	0	0	0
1	.327807	.167893	.487721
2	.328795	.17686	.48073
3	.370775	.212794	.528757
4	.366896	.212022	.52177
5	.370399	.213718	.52708
6	.371487	.214058	.528917
7	.371315	.213956	.528674
8	.371438	.213923	.528953

```
95% lower and upper bounds reported
(1) irfname = orderb, impulse = dln_inc, and response = dln_consump
```

Because we specified the individual option, the output contains two tables, one for each set of IRF results. Examining the results in the tables indicates that each of the estimated functions is well within the confidence interval of the other, so we conclude that the functions are not significantly different.

◁

❑ Technical note

Be careful in how you name variables when you fit models. Say that you fit one model with var and used time-series operators to form one of the endogenous variables

```
. var d.ln_inv ...
```

and in another model, you created a new variable:

```
. gen dln_inv = d.ln_inv
. var dln_inv ...
```

Say that you saved IRF results from both (perhaps they differ in the number of lags). Now you wish to use irf table to compare them. You would not be able to specify response(d.ln_inv) or response(dln_inv) because neither variable is in both models. Similarly, you could not specify impulse(d.ln_inv) or impulse(dln_inv) for the same reason.

All is not lost; if impulse() is not specified, all endogenous variables are used, and similarly if response() is not specified, so you could obtain the result you desired by simply not specifying the options, but you will also obtain a lot more, besides. If you want to specify the impulse() or response() options, be sure to name variables consistently.

Also, you may forget how the endogenous variables were named. If so, irf describe, detail can provide the answer. In irf describe's output, the endogenous variables are listed next to endog.

❑

Saved results

If the individual option is not specified, irf table saves the following in r():

Scalars
r(ncols)	number of columns in table
r(k_umax)	number of distinct keys
r(k)	number of specific table commands

Macros
r(key#)	#th key
r(tnotes)	list of keys applied to each column

If the individual option is specified, then for each *irfname*, irf table saves the following in r():

Scalars
r(*irfname*_ncols)	number of columns in table for *irfname*
r(*irfname*_k_umax)	number of distinct keys in table for *irfname*
r(*irfname*_k)	number of specific table commands used to create table for *irfname*

Macros
r(*irfname*_key#)	#th key for *irfname* table
r(*irfname*_tnotes)	list of keys applied to each column in table for *irfname*

Methods and formulas

irf table is implemented as an ado-file.

Also see

[TS] **irf** — Create and analyze IRFs, dynamic-multiplier functions, and FEVDs

[TS] **var intro** — Introduction to vector autoregressive models

[TS] **vec intro** — Introduction to vector error-correction models

Title

> **mgarch** — Multivariate GARCH models

Syntax

mgarch *model eq* [*eq* ... *eq*] [*if*] [*in*] [, ...]

Family	*model*
Vech	
Diagonal vech	dvech
Conditional correlation	
constant conditional correlation	ccc
dynamic conditional correlation	dcc
varying conditional correlation	vcc

See [TS] **mgarch dvech**, [TS] **mgarch ccc**, [TS] **mgarch dcc**, and [TS] **mgarch vcc** for details.

Description

mgarch estimates the parameters of multivariate generalized autoregressive conditional-heteroskedasticity (MGARCH) models. MGARCH models allow both the conditional mean and the conditional covariance to be dynamic.

The general MGARCH model is so flexible that not all the parameters can be estimated. For this reason, there are many MGARCH models that parameterize the problem more parsimoniously.

mgarch implements four commonly used parameterizations: the diagonal vech model, the constant conditional correlation model, the dynamic conditional correlation model, and the time-varying conditional correlation model.

Remarks

Remarks are presented under the following headings:

An introduction to MGARCH models
Diagonal vech MGARCH models
Conditional correlation MGARCH models
    Constant conditional correlation MGARCH model
    Dynamic conditional correlation MGARCH model
    Varying conditional correlation MGARCH model
Error distributions and quasimaximum likelihood
Treatment of missing data

An introduction to MGARCH models

Multivariate GARCH models allow the conditional covariance matrix of the dependent variables to follow a flexible dynamic structure and allow the conditional mean to follow a vector-autoregressive (VAR) structure.

The general MGARCH model is too flexible for most problems. There are many restricted MGARCH models in the literature because there is no parameterization that always provides an optimal trade-off between flexibility and parsimony.

`mgarch` implements four commonly used parameterizations: the diagonal vech (DVECH) model, the constant conditional correlation (CCC) model, the dynamic conditional correlation (DCC) model, and the time-varying conditional correlation (VCC) model.

Bollerslev, Engle, and Wooldridge (1988); Bollerslev, Engle, and Nelson (1994); Bauwens, Laurent, and Rombouts (2006); Silvennoinen and Teräsvirta (2009); and Engle (2009) provide general introductions to MGARCH models. We provide a quick introduction organized around the models implemented in `mgarch`.

We give a formal definition of the general MGARCH model to establish notation that facilitates comparisons of the models. The general MGARCH model is given by

$$\mathbf{y}_t = \mathbf{C}\mathbf{x}_t + \epsilon_t$$
$$\epsilon_t = \mathbf{H}_t^{1/2}\boldsymbol{\nu}_t$$

where

$\mathbf{y}_t$ is an $m \times 1$ vector of dependent variables;

$\mathbf{C}$ is an $m \times k$ matrix of parameters;

$\mathbf{x}_t$ is a $k \times 1$ vector of independent variables, which may contain lags of $\mathbf{y}_t$;

$\mathbf{H}_t^{1/2}$ is the Cholesky factor of the time-varying conditional covariance matrix $\mathbf{H}_t$; and

$\boldsymbol{\nu}_t$ is an $m \times 1$ vector of zero-mean, unit-variance, and independent and identically distributed innovations.

In the general MGARCH model, $\mathbf{H}_t$ is a matrix generalization of univariate GARCH models. For example, in a general MGARCH model with one autoregressive conditional heteroskedastic (ARCH) term and one GARCH term,

$$\text{vech}\left(\mathbf{H}_t\right) = \mathbf{s} + \mathbf{A}\text{vech}\left(\epsilon_{t-1}\epsilon_{t-1}'\right) + \mathbf{B}\text{vech}\left(\mathbf{H}_{t-1}\right) \tag{1}$$

where the vech() function stacks the unique elements that lie on or below the main diagonal in a symmetric matrix into a vector, $\mathbf{s}$ is a vector of parameters, and $\mathbf{A}$ and $\mathbf{B}$ are conformable matrices of parameters. Because this model uses the vech() function to extract and model the unique elements of $\mathbf{H}_t$, it is also known as the VECH model.

Because it is a conditional covariance matrix, $\mathbf{H}_t$ must be positive definite. Equation (1) can be used to show that the parameters in $\mathbf{s}$, $\mathbf{A}$, and $\mathbf{B}$ are not uniquely identified and that further restrictions must be placed on $\mathbf{s}$, $\mathbf{A}$, and $\mathbf{B}$ to ensure that $\mathbf{H}_t$ is positive definite for all t.

The various MGARCH models proposed in the literature differ in how they trade off flexibility and parsimony in their specifications for $\mathbf{H}_t$. Increased flexibility allows a model to capture more complex $\mathbf{H}_t$ processes. Increased parsimony makes parameter estimation feasible for more datasets. An important measure of the flexibility–parsimony trade-off is how fast the number of model parameters increases with the number of time series m, because many applied models use multiple time series.

Diagonal vech MGARCH models

Bollerslev, Engle, and Wooldridge (1988) derived the diagonal vech (DVECH) model by restricting $\mathbf{A}$ and $\mathbf{B}$ to be diagonal. Although the DVECH model is much more parsimonious than the general model, it can only handle a few series because the number of parameters grows quadratically with the number of series. For example, there are $3m(m+1)/2$ parameters in a DVECH(1,1) model for $\mathbf{H}_t$.

Despite the large number of parameters, the diagonal structure implies that each conditional variance and each conditional covariance depends on its own past but not on the past of the other conditional variances and covariances. Formally, in the DVECH(1,1) model each element of $\mathbf{H}_t$ is modeled by

$$h_{ij,t} = s_{ij} + a_{ij}\epsilon_{i,(t-1)}\epsilon_{j,(t-1)} + b_{ij}h_{ij,(t-1)}$$

Parameter estimation can be difficult because it requires that $\mathbf{H}_t$ be positive definite for each t. The requirement that $\mathbf{H}_t$ be positive definite for each t imposes complicated restrictions on the off-diagonal elements.

See [TS] **mgarch dvech** for more details about this model.

Conditional correlation MGARCH models

Conditional correlation (CC) models use nonlinear combinations of univariate GARCH models to represent the conditional covariances. In each of the conditional correlation models, the conditional covariance matrix is positive definite by construction and has a simple structure, which facilitates parameter estimation. CC models have a slower parameter growth rate than DVECH models as the number of time series increases.

In CC models, $\mathbf{H}_t$ is decomposed into a matrix of conditional correlations $\mathbf{R}_t$ and a diagonal matrix of conditional variances $\mathbf{D}_t$:

$$\mathbf{H}_t = \mathbf{D}_t^{1/2}\mathbf{R}_t\mathbf{D}_t^{1/2} \tag{2}$$

where each conditional variance follows a univariate GARCH process and the parameterizations of $\mathbf{R}_t$ vary across models.

Equation (2) implies that

$$h_{ij,t} = \rho_{ij,t}\sigma_{i,t}\sigma_{j,t} \tag{3}$$

where $\sigma_{i,t}^2$ is modeled by a univariate GARCH process. Equation (3) highlights that CC models use nonlinear combinations of univariate GARCH models to represent the conditional covariances and that the parameters in the model for $\rho_{ij,t}$ describe the extent to which the errors from equations i and j move together.

Comparing (1) and (2) shows that the number of parameters increases more slowly with the number of time series in a CC model than in a DVECH model.

The three CC models implemented in `mgarch` differ in how they parameterize $\mathbf{R}_t$.

Constant conditional correlation MGARCH model

Bollerslev (1990) proposed a CC MGARCH model in which the correlation matrix is time invariant. It is for this reason that the model is known as a constant conditional correlation (CCC) MGARCH model. Restricting $\mathbf{R}_t$ to a constant matrix reduces the number of parameters and simplifies the estimation but may be too strict in many empirical applications.

See [TS] **mgarch ccc** for more details about this model.

Dynamic conditional correlation MGARCH model

Engle (2002) introduced a dynamic conditional correlation (DCC) MGARCH model in which the conditional quasicorrelations $\mathbf{R}_t$ follow a GARCH(1,1)-like process. (As described by Engle [2009] and Aielli [2009], the parameters in $\mathbf{R}_t$ are not standardized to be correlations and are thus known as quasicorrelations.) To preserve parsimony, all the conditional quasicorrelations are restricted to follow the same dynamics. The DCC model is significantly more flexible than the CCC model without introducing an unestimable number of parameters for a reasonable number of series.

See [TS] **mgarch dcc** for more details about this model.

Varying conditional correlation MGARCH model

Tse and Tsui (2002) derived the varying conditional correlation (VCC) MGARCH model in which the conditional correlations at each period are a weighted sum of a time-invariant component, a measure of recent correlations among the residuals, and last period's conditional correlations. For parsimony, all the conditional correlations are restricted to follow the same dynamics.

See [TS] **mgarch vcc** for more details about this model.

Error distributions and quasimaximum likelihood

By default, `mgarch dvech`, `mgarch ccc`, `mgarch dcc`, and `mgarch vcc` estimate the parameters of MGARCH models by maximum likelihood (ML), assuming that the errors come from a multivariate normal distribution. Both the ML estimator and the quasi–maximum likelihood (QML) estimator, which drops the normality assumption, are assumed to be consistent and normally distributed in large samples; see Jeantheau (1998), Berkes and Horváth (2003), Comte and Lieberman (2003), Ling and McAleer (2003), and Fiorentini and Sentana (2007). Specify `vce(robust)` to estimate the parameters by QML. The QML parameter estimates are the same as the ML estimates, but the VCEs are different.

Based on low-level assumptions, Jeantheau (1998), Comte and Lieberman (2003), and Ling and McAleer (2003) prove that some of the ML and QML estimators implemented in `mgarch` are consistent and asymptotically normal. Based on higher-level assumptions, Fiorentini and Sentana (2007) prove that all the ML and QML estimators implemented in `mgarch` are consistent and asymptotically normal. The low-level assumption proofs specify the technical restrictions on the data-generating processes more precisely than the high-level proofs, but they do not cover as many models or cases as the high-level proofs.

It is generally accepted that there could be more low-level theoretical work done to substantiate the claims that the ML and QML estimators are consistent and asymptotically normally distributed. These widely applied estimators have been subjected to many Monte Carlo studies that show that the large-sample theory performs well in finite samples.

The distribution(t) option causes the mgarch commands to estimate the parameters of the corresponding model by ML assuming that the errors come from a multivariate Student t distribution.

The choice between the multivariate normal and the multivariate t distributions is one between robustness and efficiency. If the disturbances come from a multivariate Student t, then the ML estimates based on the multivariate Student t assumption will be consistent and efficient, while the QML estimates based on the multivariate normal assumption will be consistent but not efficient. In contrast, if the disturbances come from a well-behaved distribution that is neither multivariate Student t nor multivariate normal, then the ML estimates based on the multivariate Student t assumption will not be consistent, while the QML estimates based on the multivariate normal assumption will be consistent but not efficient.

Fiorentini and Sentana (2007) compare the ML and QML estimators implemented in mgarch and provide many useful technical results pertaining to the estimators.

Treatment of missing data

mgarch allows for gaps due to missing data. The unconditional expectations are substituted for the dynamic components that cannot be computed because of gaps. This method of handling gaps can only handle the case in which g/T goes to zero as T goes to infinity, where g is the number of observations lost to gaps in the data and T is the number of nonmissing observations.

References

Aielli, G. P. 2009. Dynamic Conditional Correlations: On Properties and Estimation. Working paper, Dipartimento di Statistica, University of Florence, Florence, Italy.

Bauwens, L., S. Laurent, and J. V. K. Rombouts. 2006. Multivariate GARCH models: A survey. *Journal of Applied Econometrics* 21: 79–109.

Berkes, I., and L. Horváth. 2003. The rate of consistency of the quasi-maximum likelihood estimator. *Statistics and Probability Letters* 61: 133–143.

Bollerslev, T. 1990. Modelling the coherence in short-run nominal exchange rates: A multivariate generalized ARCH model. *Review of Economics and Statistics* 72: 498–505.

Bollerslev, T., R. F. Engle, and D. B. Nelson. 1994. ARCH models. In *Handbook of Econometrics, Volume IV*, ed. R. F. Engle and D. L. McFadden. New York: Elsevier.

Bollerslev, T., R. F. Engle, and J. M. Wooldridge. 1988. A capital asset pricing model with time-varying covariances. *Journal of Political Economy* 96: 116–131.

Comte, F., and O. Lieberman. 2003. Asymptotic theory for multivariate GARCH processes. *Journal of Multivariate Analysis* 84: 61–84.

Engle, R. F. 2002. Dynamic conditional correlation: A simple class of multivariate generalized autoregressive conditional heteroskedasticity models. *Journal of Business & Economic Statistics* 20: 339–350.

——. 2009. *Anticipating Correlations: A New Paradigm for Risk Management*. Princeton, NJ: Princeton University Press.

Fiorentini, G., and E. Sentana. 2007. On the efficiency and consistency of likelihood estimation in multivariate conditionally heteroskedastic dynamic regression models. Working paper 0713, CEMFI, Madrid, Spain. ftp://ftp.cemfi.es/wp/07/0713.pdf.

Jeantheau, T. 1998. Strong consistency of estimators for multivariate ARCH models. *Economic Theory* 14: 70–86.

Ling, S., and M. McAleer. 2003. Asymptotic theory for a vector ARM–GARCH model. *Economic Theory* 19: 280–310.

Silvennoinen, A., and T. Teräsvirta. 2009. Multivariate GARCH models. In *Handbook of Financial Time Series*, ed. T. G. Andersen, R. A. Davis, J.-P. Kreiß, and T. Mikosch, 201–229. New York: Springer.

Tse, Y. K., and A. K. C. Tsui. 2002. A multivariate generalized autoregressive conditional heteroscedasticity model with time-varying correlations. *Journal of Business & Economic Statistics* 20: 351–362.

Also see

[TS] **arch** — Autoregressive conditional heteroskedasticity (ARCH) family of estimators

[TS] **var** — Vector autoregressive models

[U] **20 Estimation and postestimation commands**

Title

> **mgarch ccc** — Constant conditional correlation multivariate GARCH models

Syntax

mgarch ccc *eq* $\big[$ *eq* ... *eq* $\big]$ $\big[$ *if* $\big]$ $\big[$ *in* $\big]$ $\big[$, *options* $\big]$

where each *eq* has the form

(*depvars* = $\big[$ *indepvars* $\big]$ $\big[$, *eqoptions* $\big]$)

options	Description
Model	
<u>ar</u>ch(*numlist*)	ARCH terms for all equations
<u>g</u>arch(*numlist*)	GARCH terms for all equations
het(*varlist*)	include *varlist* in the specification of the conditional variance for all equations
<u>d</u>istribution(*dist* $\big[$ *#* $\big]$)	use *dist* distribution for errors [may be <u>g</u>aussian (synonym <u>norm</u>al) or t; default is gaussian]
<u>unc</u>oncentrated	perform optimization on unconcentrated log likelihood
<u>constr</u>aints(*numlist*)	apply linear constraints
SE/Robust	
vce(*vcetype*)	*vcetype* may be oim or <u>r</u>obust
Reporting	
<u>l</u>evel(*#*)	set confidence level; default is level(95)
<u>nocns</u>report	do not display constraints
display_options	control column formats, row spacing, line width, and display of omitted variables and base and empty cells
Maximization	
maximize_options	control the maximization process; seldom used
from(*matname*)	initial values for the coefficients; seldom used
<u>coefl</u>egend	display legend instead of statistics

eqoptions	Description
<u>nocons</u>tant	suppress constant term in the mean equation
<u>ar</u>ch(*numlist*)	ARCH terms
<u>g</u>arch(*numlist*)	GARCH terms
het(*varlist*)	include *varlist* in the specification of the conditional variance

You must tsset your data before using mgarch ccc; see [TS] **tsset**.

indepvars and *varlist* may contain factor variables; see [U] **11.4.3 Factor variables**.

depvars, *indepvars*, and *varlist* may contain time-series operators; see [U] **11.4.4 Time-series varlists**.

by, statsby, and rolling are allowed; see [U] **11.1.10 Prefix commands**.

coeflegend does not appear in the dialog box.

See [U] **20 Estimation and postestimation commands** for more capabilities of estimation commands.

Menu

Statistics > Multivariate time series > Multivariate GARCH

Description

mgarch ccc estimates the parameters of constant conditional correlation (CCC) multivariate generalized autoregressive conditionally heteroskedastic (MGARCH) models in which the conditional variances
are modeled as univariate generalized autoregressive conditionally heteroskedastic (GARCH) models
and the conditional covariances are modeled as nonlinear functions of the conditional variances. The
conditional correlation parameters that weight the nonlinear combinations of the conditional variance
are constant in the CCC MGARCH model.

The CCC MGARCH model is less flexible than the dynamic conditional correlation MGARCH model
(see [TS] **mgarch dcc**) and varying conditional correlation MGARCH model (see [TS] **mgarch vcc**),
which specify GARCH-like processes for the conditional correlations. The conditional correlation
MGARCH models are more parsimonious than the diagonal vech MGARCH model (see [TS] **mgarch
dvech**).

Options

⌐ Model ⌐

arch(*numlist*) specifies the ARCH terms for all equations in the model. By default, no ARCH terms
are specified.

garch(*numlist*) specifies the GARCH terms for all equations in the model. By default, no GARCH
terms are specified.

het(*varlist*) specifies that *varlist* be included in the model in the specification of the conditional
variance for all equations. This varlist enters the variance specification collectively as multiplicative
heteroskedasticity.

distribution(*dist* [*#*]) specifies the assumed distribution for the errors. *dist* may be gaussian,
normal, or t.

gaussian and normal are synonyms; each causes mgarch ccc to assume that the errors come
from a multivariate normal distribution. # cannot be specified with either of them.

t causes mgarch ccc to assume that the errors follow a multivariate Student t distribution, and
the degree-of-freedom parameter is estimated along with the other parameters of the model. If
distribution(t #) is specified, then mgarch ccc uses a multivariate Student t distribution
with # degrees of freedom. # must be greater than 2.

unconcentrated specifies that optimization be performed on the unconcentrated log likelihood. The
default is to start with the concentrated log likelihood.

constraints(*numlist*) specifies linear constraints to apply to the parameter estimates.

⌐ SE/Robust ⌐

vce(*vcetype*) specifies the estimator for the variance–covariance matrix of the estimator.

vce(oim), the default, specifies to use the observed information matrix (OIM) estimator.

vce(robust) specifies to use the Huber/White/sandwich estimator.

> ⌐ Reporting ⌐

`level(#)`; see [R] **estimation options**.

`nocnsreport`; see [R] **estimation options**.

display_options: <u>noomit</u>ted, vsquish, noemptycells, <u>base</u>levels, <u>allbase</u>levels,
     `cformat(%fmt)`, `pformat(%fmt)`, `sformat(%fmt)`, and `nolstretch`; see [R] **estimation options**.

> ⌐ Maximization ⌐

maximize_options: <u>diff</u>icult, <u>tech</u>nique(*algorithm_spec*), <u>iter</u>ate(#), [<u>no</u>]<u>log</u>, <u>tra</u>ce,
     gradient, showstep, <u>hess</u>ian, <u>showtol</u>erance, <u>tol</u>erance(#), <u>ltol</u>erance(#),
     <u>nrtol</u>erance(#), and <u>nonrtol</u>erance, and from(*matname*); see [R] **maximize** for all options
     except from(), and see below for information on from(). These options are seldom used.

from(*matname*) specifies initial values for the coefficients. from(b0) causes mgarch ccc to begin
     the optimization algorithm with the values in b0. b0 must be a row vector, and the number of
     columns must equal the number of parameters in the model.

The following option is available with mgarch ccc but is not shown in the dialog box:

`coeflegend`; see [R] **estimation options**.

Eqoptions

noconstant suppresses the constant term in the mean equation.

arch(*numlist*) specifies the ARCH terms in the equation. By default, no ARCH terms are specified.
     This option may not be specified with model-level arch().

garch(*numlist*) specifies the GARCH terms in the equation. By default, no GARCH terms are specified.
     This option may not be specified with model-level garch().

het(*varlist*) specifies that *varlist* be included in the specification of the conditional variance. This
     varlist enters the variance specification collectively as multiplicative heteroskedasticity. This option
     may not be specified with model-level het().

Remarks

We assume that you have already read [TS] **mgarch**, which provides an introduction to MGARCH
models and the methods implemented in mgarch ccc.

MGARCH models are dynamic multivariate regression models in which the conditional variances
and covariances of the errors follow an autoregressive-moving-average structure. The CCC MGARCH
model uses a nonlinear combination of univariate GARCH models in which the cross-equation weights
are time invariant to model the conditional covariance matrix of the disturbances.

As discussed in [TS] **mgarch**, MGARCH models differ in the parsimony and flexibility of their
specifications for a time-varying conditional covariance matrix of the disturbances, denoted by $\mathbf{H}_t$.
In the conditional correlation family of MGARCH models, the diagonal elements of $\mathbf{H}_t$ are modeled
as univariate GARCH models, whereas the off-diagonal elements are modeled as nonlinear functions
of the diagonal terms. In the CCC MGARCH model,

$$h_{ij,t} = \rho_{ij}\sqrt{h_{ii,t}h_{jj,t}}$$

where the diagonal elements $h_{ii,t}$ and $h_{jj,t}$ follow univariate GARCH processes and ρ_{ij} is a time-invariate weight interpreted as a conditional correlation.

In the dynamic conditional correlation (DCC) and varying conditional correlation (VCC) MGARCH models discussed in [TS] **mgarch dcc** and [TS] **mgarch vcc**, the ρ_{ij} are allowed to vary over time. Although the conditional-correlation structure provides a useful trade-off between parsimony and flexibility in the DCC MGARCH and VCC MGARCH models, the time-invariant parameterization used in the CCC MGARCH model is generally viewed as too restrictive for many applications; see Silvennoinen and Teräsvirta (2009). The baseline CCC MGARCH estimates are frequently compared with DCC MGARCH and VCC MGARCH estimates.

❑ Technical note

Formally, the CCC MGARCH model derived by Bollerslev (1990) can be written as

$$\mathbf{y}_t = \mathbf{C}\mathbf{x}_t + \boldsymbol{\epsilon}_t$$
$$\boldsymbol{\epsilon}_t = \mathbf{H}_t^{1/2}\boldsymbol{\nu}_t$$
$$\mathbf{H}_t = \mathbf{D}_t^{1/2}\mathbf{R}\mathbf{D}_t^{1/2}$$

where

$\mathbf{y}_t$ is an $m \times 1$ vector of dependent variables;

$\mathbf{C}$ is an $m \times k$ matrix of parameters;

$\mathbf{x}_t$ is a $k \times 1$ vector of independent variables, which may contain lags of $\mathbf{y}_t$;

$\mathbf{H}_t^{1/2}$ is the Cholesky factor of the time-varying conditional covariance matrix $\mathbf{H}_t$;

$\boldsymbol{\nu}_t$ is an $m \times 1$ vector of normal, independent, and identically distributed innovations;

$\mathbf{D}_t$ is a diagonal matrix of conditional variances,

$$\mathbf{D}_t = \begin{pmatrix} \sigma_{1,t}^2 & 0 & \cdots & 0 \\ 0 & \sigma_{2,t}^2 & \cdots & 0 \\ \vdots & \vdots & \ddots & \vdots \\ 0 & 0 & \cdots & \sigma_{m,t}^2 \end{pmatrix}$$

in which each $\sigma_{i,t}^2$ evolves according to a univariate GARCH model of the form

$$\sigma_{i,t}^2 = s_i + \sum_{j=1}^{p_i} \alpha_j \epsilon_{i,t-j}^2 + \sum_{j=1}^{q_i} \beta_j \sigma_{i,t-j}^2$$

by default, or

$$\sigma_{i,t}^2 = \exp(\boldsymbol{\gamma}_i \mathbf{z}_{i,t}) + \sum_{j=1}^{p_i} \alpha_j \epsilon_{i,t-j}^2 + \sum_{j=1}^{q_i} \beta_j \sigma_{i,t-j}^2$$

when the het() option is specified, where $\boldsymbol{\gamma}_t$ is a $1 \times p$ vector of parameters, $\mathbf{z}_i$ is a $p \times 1$ vector of independent variables including a constant term, the α_j's are ARCH parameters, and the β_j's are GARCH parameters; and

$\mathbf{R}$ is a matrix of time-invariant unconditional correlations of the standardized residuals $\mathbf{D}_t^{-1/2}\boldsymbol{\epsilon}_t$,

$$\mathbf{R} = \begin{pmatrix} 1 & \rho_{12} & \cdots & \rho_{1m} \\ \rho_{12} & 1 & \cdots & \rho_{2m} \\ \vdots & \vdots & \ddots & \vdots \\ \rho_{1m} & \rho_{2m} & \cdots & 1 \end{pmatrix}$$

This model is known as the constant conditional correlation MGARCH model because $\mathbf{R}$ is time invariant.

❑

Some examples

▷ Example 1

We have daily data on the stock returns of three car manufacturers—Toyota, Nissan, and Honda, from January 2, 2003, to December 31, 2010—in the variables `toyota`, `nissan`, and `honda`. We model the conditional means of the returns as a first-order vector autoregressive process and the conditional covariances as a CCC MGARCH process in which the variance of each disturbance term follows a GARCH(1,1) process. We specify the `noconstant` option, because the returns have mean zero. The estimated constants in the variance equations are near zero in this example because of how the data are scaled.

```
. use http://www.stata-press.com/data/r12/stocks
(Data from Yahoo! Finance)

. mgarch ccc (toyota nissan honda = L.toyota L.nissan L.honda, noconstant),
> arch(1) garch(1)

Calculating starting values....

Optimizing concentrated log likelihood

(setting technique to bhhh)
Iteration 0:   log likelihood =  16898.994
Iteration 1:   log likelihood =  17008.914
Iteration 2:   log likelihood =  17156.946
Iteration 3:   log likelihood =  17249.527
Iteration 4:   log likelihood =  17287.251
Iteration 5:   log likelihood =    17313.5
Iteration 6:   log likelihood =  17335.087
Iteration 7:   log likelihood =  17356.534
Iteration 8:   log likelihood =  17376.051
Iteration 9:   log likelihood =  17400.035
(switching technique to nr)
Iteration 10:  log likelihood =  17423.634
Iteration 11:  log likelihood =  17440.774
Iteration 12:  log likelihood =  17446.838
Iteration 13:  log likelihood =  17447.635
Iteration 14:  log likelihood =  17447.645
Iteration 15:  log likelihood =  17447.645

Optimizing unconcentrated log likelihood

Iteration 0:   log likelihood =  17447.645
Iteration 1:   log likelihood =  17447.651
Iteration 2:   log likelihood =  17447.651

Constant conditional correlation MGARCH model

Sample: 1 - 2015                            Number of obs    =       2014
Distribution: Gaussian                      Wald chi2(9)     =      17.46
Log likelihood =  17447.65                  Prob > chi2      =     0.0420
```

	Coef.	Std. Err.	z	P>|z|	[95% Conf. Interval]	
toyota						
toyota						
L1.	-.0537817	.0353211	-1.52	0.128	-.1230098	.0154463
nissan						
L1.	.026686	.024841	1.07	0.283	-.0220015	.0753734
honda						
L1.	-.0043073	.0302761	-0.14	0.887	-.0636473	.0550327
ARCH_toyota						
arch						
L1.	.0615321	.0087313	7.05	0.000	.0444191	.0786452
garch						
L1.	.9213798	.0110412	83.45	0.000	.8997395	.9430201
_cons	4.42e-06	1.12e-06	3.93	0.000	2.21e-06	6.62e-06
nissan						
toyota						
L1.	-.0232321	.0400563	-0.58	0.562	-.1017411	.0552769
nissan						
L1.	-.0299552	.0309362	-0.97	0.333	-.0905891	.0306787
honda						
L1.	.0369229	.0360532	1.02	0.306	-.0337402	.1075859
ARCH_nissan						
arch						
L1.	.0740294	.0119353	6.20	0.000	.0506366	.0974222
garch						
L1.	.9102548	.0142328	63.95	0.000	.8823589	.9381506
_cons	6.36e-06	1.76e-06	3.61	0.000	2.91e-06	9.81e-06
honda						
toyota						
L1.	-.0378616	.036792	-1.03	0.303	-.1099727	.0342495
nissan						
L1.	.0551649	.0272559	2.02	0.043	.0017444	.1085855
honda						
L1.	-.0431919	.0331268	-1.30	0.192	-.1081193	.0217354
ARCH_honda						
arch						
L1.	.0433036	.0070224	6.17	0.000	.0295399	.0570673
garch						
L1.	.939117	.010131	92.70	0.000	.9192605	.9589735
_cons	5.02e-06	1.31e-06	3.83	0.000	2.45e-06	7.59e-06

Correlation						
toyota						
nissan	.6532264	.0128035	51.02	0.000	.628132	.6783208
honda	.7185412	.0108132	66.45	0.000	.6973477	.7397346
nissan						
honda	.6298972	.0135336	46.54	0.000	.6033717	.6564226

The iteration log has three parts: the dots from the search for initial values, the iteration log from optimizing the concentrated log likelihood, and the iteration log from maximizing the unconcentrated log likelihood. A detailed discussion of the optimization methods can be found in *Methods and formulas*.

The header describes the estimation sample and reports a Wald test against the null hypothesis that all the coefficients on the independent variables in the mean equations are zero. Here the null hypothesis is rejected at the 5% level.

The output table first presents results for the mean or variance parameters used to model each dependent variable. Subsequently, the output table presents results for the conditional correlation parameters. For example, the conditional correlation between the standardized residuals for Toyota and Nissan is estimated to be 0.65.

The output above indicates that we may not need all the vector autoregressive parameters, but that each of the univariate ARCH, univariate GARCH, and conditional correlation parameters are statistically significant. That the estimated conditional correlation parameters are positive and significant indicates that the returns on these stocks rise or fall together.

That the conditional correlations are time invariant is a restrictive assumption. The DCC MGARCH model and the VCC MGARCH model nest the CCC MGARCH model. When we test the time-invariance assumption with Wald tests on the parameters of these more general models in [TS] **mgarch dcc** and [TS] **mgarch vcc**, we reject the null hypothesis that these conditional correlations are time invariant.

◁

▷ Example 2

We improve the previous example by removing the insignificant parameters from the model. To remove these parameters, we specify the honda equation separately from the toyota and nissan equations:

```
. mgarch ccc (toyota nissan = , noconstant) (honda = L.nissan, noconstant),
> arch(1) garch(1)

Calculating starting values....

Optimizing concentrated log likelihood

(setting technique to bhhh)
Iteration 0:   log likelihood =   16886.88
Iteration 1:   log likelihood =  16974.779
Iteration 2:   log likelihood =  17147.893
Iteration 3:   log likelihood =  17247.473
Iteration 4:   log likelihood =  17285.549
Iteration 5:   log likelihood =  17311.153
Iteration 6:   log likelihood =  17333.588
Iteration 7:   log likelihood =  17353.717
Iteration 8:   log likelihood =  17374.895
Iteration 9:   log likelihood =  17400.669
(switching technique to nr)
```

```
Iteration 10:  log likelihood =  17425.661
Iteration 11:  log likelihood =  17436.768
Iteration 12:  log likelihood =  17439.738
Iteration 13:  log likelihood =  17439.865
Iteration 14:  log likelihood =  17439.866

Optimizing unconcentrated log likelihood

Iteration 0:   log likelihood =  17439.866
Iteration 1:   log likelihood =  17439.872
Iteration 2:   log likelihood =  17439.872

Constant conditional correlation MGARCH model
```

Sample: 1 - 2015   Number of obs   =   2014
Distribution: Gaussian   Wald chi2(1)   =   1.81
Log likelihood = 17439.87   Prob > chi2   =   0.1781

	Coef.	Std. Err.	z	P>|z|	[95% Conf.	Interval]
ARCH_toyota						
arch						
L1.	.0619604	.0087942	7.05	0.000	.044724	.0791968
garch						
L1.	.9208961	.0110996	82.97	0.000	.8991414	.9426508
_cons	4.43e-06	1.13e-06	3.94	0.000	2.23e-06	6.64e-06
ARCH_nissan						
arch						
L1.	.0773095	.012328	6.27	0.000	.0531471	.1014719
garch						
L1.	.906088	.0147303	61.51	0.000	.8772171	.9349589
_cons	6.77e-06	1.85e-06	3.66	0.000	3.14e-06	.0000104
honda						
nissan						
L1.	.0186628	.0138575	1.35	0.178	-.0084975	.0458231
ARCH_honda						
arch						
L1.	.0433741	.006996	6.20	0.000	.0296622	.0570861
garch						
L1.	.9391094	.0100707	93.25	0.000	.9193712	.9588477
_cons	5.02e-06	1.31e-06	3.83	0.000	2.45e-06	7.60e-06
Correlation						
toyota						
nissan	.652299	.0128271	50.85	0.000	.6271583	.6774396
honda	.7189531	.0108005	66.57	0.000	.6977845	.7401218
nissan						
honda	.628435	.0135653	46.33	0.000	.6018475	.6550225

It turns out that the coefficient on L1.nissan in the honda equation is now statistically insignificant. We could further improve the model by removing L1.nissan from the model.

As expected, removing the insignificant parameters from conditional mean equations had almost no effect on the estimated conditional variance parameters.

There is no mean equation for Toyota or Nissan. In [TS] **mgarch ccc postestimation**, we discuss prediction from models without covariates.

◁

▷ Example 3

Here we fit a bivariate CCC MGARCH model for the Toyota and Nissan shares. We believe that the shares of these car manufacturers follow the same process, so we impose the constraints that the ARCH and the GARCH coefficients are the same for the two companies.

```
. constraint 1 _b[ARCH_toyota:L.arch] = _b[ARCH_nissan:L.arch]

. constraint 2 _b[ARCH_toyota:L.garch] = _b[ARCH_nissan:L.garch]

. mgarch ccc (toyota nissan = , noconstant), arch(1) garch(1) constraints(1 2)
Calculating starting values....

Optimizing concentrated log likelihood

(setting technique to bhhh)
Iteration 0:   log likelihood =   10317.225
Iteration 1:   log likelihood =   10630.464
Iteration 2:   log likelihood =   10865.964
Iteration 3:   log likelihood =   11063.329
 (output omitted )
Iteration 8:   log likelihood =   11273.962
Iteration 9:   log likelihood =   11274.409
(switching technique to nr)
Iteration 10:  log likelihood =   11274.494
Iteration 11:  log likelihood =   11274.499
Iteration 12:  log likelihood =   11274.499

Optimizing unconcentrated log likelihood

Iteration 0:   log likelihood =   11274.499
Iteration 1:   log likelihood =   11274.501
Iteration 2:   log likelihood =   11274.501

Constant conditional correlation MGARCH model
```

Sample: 1 - 2015	Number of obs =	2015
Distribution: Gaussian	Wald chi2(.) =	.
Log likelihood = 11274.5	Prob > chi2 =	.

```
 ( 1)   [ARCH_toyota]L.arch - [ARCH_nissan]L.arch = 0
 ( 2)   [ARCH_toyota]L.garch - [ARCH_nissan]L.garch = 0
```

	Coef.	Std. Err.	z	P>|z|	[95% Conf. Interval]	
ARCH_toyota						
arch						
L1.	.0742678	.0095464	7.78	0.000	.0555572	.0929785
garch						
L1.	.9131674	.0111558	81.86	0.000	.8913024	.9350323
_cons	3.77e-06	1.02e-06	3.71	0.000	1.78e-06	5.77e-06
ARCH_nissan						
arch						
L1.	.0742678	.0095464	7.78	0.000	.0555572	.0929785
garch						
L1.	.9131674	.0111558	81.86	0.000	.8913024	.9350323
_cons	5.30e-06	1.36e-06	3.89	0.000	2.63e-06	7.97e-06

	Coef.	Std. Err.	z	P>|z|	[95% Conf. Interval]
Correlation						
toyota						
nissan	.651389	.0128482	50.70	0.000	.6262071	.6765709

We could test our constraints by fitting the unconstrained model and performing a likelihood-ratio test. The results indicate that the restricted model is preferable.

◁

⊳ Example 4

In this example, we have data on fictional stock returns for the Acme and Anvil corporations and we believe that the movement of the two stocks is governed by different processes. We specify one ARCH and one GARCH term for the conditional variance equation for Acme and two ARCH terms for the conditional variance equation for Anvil. In addition, we include the lagged value of the stock return for Apex, the main subsidiary of Anvil corporation, in the variance equation of Anvil. For Acme, we have data on the changes in an index of futures prices of products related to those produced by Acme in afrelated. For Anvil, we have data on the changes in an index of futures prices of inputs used by Anvil in afinputs.

```
. use http://www.stata-press.com/data/r12/acmeh
. mgarch ccc (acme = afrelated, noconstant arch(1) garch(1))
> (anvil = afinputs, arch(1/2) het(L.apex))
Calculating starting values....
Optimizing concentrated log likelihood
(setting technique to bhhh)
Iteration 0:   log likelihood = -12996.245
Iteration 1:   log likelihood = -12609.982
Iteration 2:   log likelihood = -12563.103
Iteration 3:   log likelihood =  -12554.73
Iteration 4:   log likelihood = -12554.542
Iteration 5:   log likelihood = -12554.534
Iteration 6:   log likelihood = -12554.534
Iteration 7:   log likelihood = -12554.534
Optimizing unconcentrated log likelihood
Iteration 0:   log likelihood = -12554.534
Iteration 1:   log likelihood = -12554.533
Constant conditional correlation MGARCH model
Sample: 1 - 2500                           Number of obs    =       2499
Distribution: Gaussian                     Wald chi2(2)     =    2212.30
Log likelihood = -12554.53                 Prob > chi2      =     0.0000
```

	Coef.	Std. Err.	z	P>|z|	[95% Conf. Interval]
acme						
afrelated	.9175148	.0651088	14.09	0.000	.7899039	1.045126
ARCH_acme						
arch						
L1.	.0798719	.0169526	4.71	0.000	.0466455	.1130983
garch						
L1.	.7336823	.0601569	12.20	0.000	.6157768	.8515877
_cons	2.880836	.760206	3.79	0.000	1.390859	4.370812

anvil						
afinputs	-1.015561	.0226437	-44.85	0.000	-1.059942	-.97118
_cons	.0703606	.0211689	3.32	0.001	.0288703	.1118508
ARCH_anvil						
arch						
L1.	.4893288	.0286012	17.11	0.000	.4332714	.5453862
L2.	.2782296	.0208172	13.37	0.000	.2374287	.3190305
apex						
L1.	1.894972	.0616293	30.75	0.000	1.774181	2.015763
_cons	.1034111	.0735512	1.41	0.160	-.0407466	.2475688
Correlation						
acme						
anvil	-.5354047	.0143275	-37.37	0.000	-.563486	-.5073234

The results indicate that increases in the futures prices for related products lead to higher returns on the Acme stock, and increased input prices lead to lower returns on the Anvil stock. In the conditional variance equation for Anvil, the coefficient on L1.apex is positive and significant, which indicates that an increase in the return on the Apex stock leads to more variability in the return on the Anvil stock. That the estimated conditional correlation between the two returns is -0.54 indicates that these returns tend to move in opposite directions; in other words, an increase in the return for the Acme stock tends to be associated with a decrease in the return for the Anvil stock, and vice versa.

◁

Saved results

mgarch ccc saves the following in e():

Scalars

e(N)	number of observations
e(k)	number of parameters
e(k_aux)	number of auxiliary parameters
e(k_extra)	number of extra estimates added to _b
e(k_eq)	number of equations in e(b)
e(k_dv)	number of dependent variables
e(df_m)	model degrees of freedom
e(ll)	log likelihood
e(chi2)	χ^2
e(p)	significance
e(estdf)	1 if distribution parameter was estimated, 0 otherwise
e(usr)	user-provided distribution parameter
e(tmin)	minimum time in sample
e(tmax)	maximum time in sample
e(N_gaps)	number of gaps
e(rank)	rank of e(V)
e(ic)	number of iterations
e(rc)	return code
e(converged)	1 if converged, 0 otherwise

Macros
    e(cmd)   mgarch
    e(model)   ccc
    e(cmdline)   command as typed
    e(depvar)   names of dependent variables
    e(covariates)   list of covariates
    e(dv_eqs)   dependent variables with mean equations
    e(indeps)   independent variables in each equation
    e(tvar)   time variable
    e(title)   title in estimation output
    e(chi2type)   Wald; type of model χ^2 test
    e(vce)   *vcetype* specified in vce()
    e(vcetype)   title used to label Std. Err.
    e(tmins)   formatted minimum time
    e(tmaxs)   formatted maximum time
    e(dist)   distribution for error term: gaussian or t
    e(arch)   specified ARCH terms
    e(garch)   specified GARCH terms
    e(technique)   maximization technique
    e(properties)   b V
    e(estat_cmd)   program used to implement estat
    e(predict)   program used to implement predict
    e(marginsok)   predictions allowed by margins
    e(marginsnotok)   predictions disallowed by margins

Matrices
    e(b)   coefficient vector
    e(Cns)   constraints matrix
    e(ilog)   iteration log (up to 20 iterations)
    e(gradient)   gradient vector
    e(hessian)   Hessian matrix
    e(V)   variance–covariance matrix of the estimators
    e(pinfo)   parameter information, used by predict
Functions
    e(sample)   marks estimation sample

Methods and formulas

mgarch ccc is implemented as an ado-file.

mgarch ccc estimates the parameters of the CCC MGARCH model by maximum likelihood. The unconcentrated log-likelihood function based on the multivariate normal distribution for observation t is

$$l_t = -0.5m \log(2\pi) - 0.5\log\left\{\det\left(\mathbf{R}\right)\right\} - \log\left\{\det\left(\mathbf{D}_t^{1/2}\right)\right\} - 0.5\widetilde{\boldsymbol{\epsilon}}_t \mathbf{R}^{-1}\widetilde{\boldsymbol{\epsilon}}_t' \tag{1}$$

where $\widetilde{\boldsymbol{\epsilon}}_t = \mathbf{D}_t^{-1/2}\boldsymbol{\epsilon}_t$ is an $m \times 1$ vector of standardized residuals, $\boldsymbol{\epsilon}_t = \mathbf{y}_t - \mathbf{C}\mathbf{x}_t$. The log-likelihood function is $\sum_{t=1}^{T} l_t$.

If we assume that $\boldsymbol{\nu}_t$ follow a multivariate t distribution with degrees of freedom (df) greater than 2, then the unconcentrated log-likelihood function for observation t is

$$l_t = \log \Gamma\left(\frac{\text{df} + m}{2}\right) - \log \Gamma\left(\frac{\text{df}}{2}\right) - \frac{m}{2}\log\left\{(\text{df} - 2)\pi\right\}$$

$$- 0.5\log\left\{\det\left(\mathbf{R}\right)\right\} - \log\left\{\det\left(\mathbf{D}_t^{1/2}\right)\right\} - \frac{\text{df} + m}{2}\log\left(1 + \frac{\widetilde{\boldsymbol{\epsilon}}_t \mathbf{R}^{-1}\widetilde{\boldsymbol{\epsilon}}_t'}{\text{df} - 2}\right) \tag{2}$$

The correlation matrix $\mathbf{R}$ can be concentrated out of (1) and (2) by defining the (i, j)th element of $\mathbf{R}$ as

$$\widehat{\rho}_{ij} = \left(\sum_{t=1}^{T} \widetilde{\epsilon}_{it} \widetilde{\epsilon}_{jt} \right) \left(\sum_{t=1}^{T} \widetilde{\epsilon}_{it}^2 \right)^{-\frac{1}{2}} \left(\sum_{t=1}^{T} \widetilde{\epsilon}_{jt}^2 \right)^{-\frac{1}{2}}$$

`mgarch ccc` starts the optimization process with the concentrated log-likelihood function.

The starting values for the parameters in the mean equations and the initial residuals $\widehat{\epsilon}_t$ are obtained by least-squares regression. The starting values for the parameters in the variance equations are obtained by a procedure proposed by Gourieroux and Monfort (1997, sec. 6.2.2). If the optimization is started with the unconcentrated log likelihood, then the initial values for the parameters in $\mathbf{R}$ are calculated from the standardized residuals $\widetilde{\epsilon}_t$.

GARCH estimators require initial values that can be plugged in for $\epsilon_{t-i}\epsilon_{t-i}'$ and $\mathbf{H}_{t-j}$ when $t - i < 1$ and $t - j < 1$. `mgarch ccc` substitutes an estimator of the unconditional covariance of the disturbances

$$\widehat{\boldsymbol{\Sigma}} = T^{-1} \sum_{t=1}^{T} \widehat{\widehat{\epsilon}}_t \widehat{\widehat{\epsilon}}_t' \tag{3}$$

for $\epsilon_{t-i}\epsilon_{t-i}'$ when $t - i < 1$ and for $\mathbf{H}_{t-j}$ when $t - j < 1$, where $\widehat{\widehat{\epsilon}}_t$ is the vector of residuals calculated using the estimated parameters.

`mgarch ccc` requires a sample size that at the minimum is equal to the number of parameters in the model plus twice the number of equations.

`mgarch ccc` uses numerical derivatives in maximizing the log-likelihood function.

References

Bollerslev, T. 1990. Modelling the coherence in short-run nominal exchange rates: A multivariate generalized ARCH model. *Review of Economics and Statistics* 72: 498–505.

Gourieroux, C., and A. Monfort. 1997. *Time Series and Dynamic Models*. Trans. ed. G. M. Gallo. Cambridge: Cambridge University Press.

Silvennoinen, A., and T. Teräsvirta. 2009. Multivariate GARCH models. In *Handbook of Financial Time Series*, ed. T. G. Andersen, R. A. Davis, J.-P. Kreiß, and T. Mikosch, 201–229. Berlin: Springer.

Also see

[TS] **mgarch ccc postestimation** — Postestimation tools for mgarch ccc

[TS] **mgarch** — Multivariate GARCH models

[TS] **tsset** — Declare data to be time-series data

[TS] **arch** — Autoregressive conditional heteroskedasticity (ARCH) family of estimators

[TS] **var** — Vector autoregressive models

[U] **20 Estimation and postestimation commands**

Title

> **mgarch ccc postestimation** — Postestimation tools for mgarch ccc

Description

The following standard postestimation commands are available after `mgarch ccc`:

Command	Description
contrast	contrasts and ANOVA-style joint tests of estimates
estat	AIC, BIC, VCE, and estimation sample summary
estimates	cataloging estimation results
lincom	point estimates, standard errors, testing, and inference for linear combinations of coefficients
lrtest	likelihood-ratio test
margins	marginal means, predictive margins, marginal effects, and average marginal effects
marginsplot	graph the results from margins (profile plots, interaction plots, etc.)
nlcom	point estimates, standard errors, testing, and inference for nonlinear combinations of coefficients
predict	predictions, residuals, influence statistics, and other diagnostic measures
predictnl	point estimates, standard errors, testing, and inference for generalized predictions
pwcompare	pairwise comparisons of estimates
test	Wald tests of simple and composite linear hypotheses
testnl	Wald tests of nonlinear hypotheses

See the corresponding entries in the *Base Reference Manual* for details.

Syntax for predict

predict [*type*] { *stub** | *newvarlist* } [*if*] [*in*] [, *statistic options*]

statistic	Description
Main	
xb	linear prediction; the default
residuals	residuals
variance	conditional variances and covariances

These statistics are available both in and out of sample; type `predict ... if e(sample) ...` if wanted only for the estimation sample.

options	Description
Options	
equation(*eqnames*)	names of equations for which predictions are made
dynamic(*time_constant*)	begin dynamic forecast at specified time

Menu

Statistics > Postestimation > Predictions, residuals, etc.

Options for predict

⌐ Main ⌐

xb, the default, calculates the linear predictions of the dependent variables.

residuals calculates the residuals.

variance predicts the conditional variances and conditional covariances.

⌐ Options ⌐

equation(*eqnames*) specifies the equation for which the predictions are calculated. Use this option to predict a statistic for a particular equation. Equation names, such as equation(income), are used to identify equations.

One equation name may be specified when predicting the dependent variable, the residuals, or the conditional variance. For example, specifying equation(income) causes predict to predict income, and specifying variance equation(income) causes predict to predict the conditional variance of income.

Two equations may be specified when predicting a conditional variance or covariance. For example, specifying equation(income, consumption) variance causes predict to predict the conditional covariance of income and consumption.

dynamic(*time_constant*) specifies when predict starts producing dynamic forecasts. The specified *time_constant* must be in the scale of the time variable specified in tsset, and the *time_constant* must be inside a sample for which observations on the dependent variables are available. For example, dynamic(tq(2008q4)) causes dynamic predictions to begin in the fourth quarter of 2008, assuming that your time variable is quarterly; see [D] **datetime**. If the model contains exogenous variables, they must be present for the whole predicted sample. dynamic() may not be specified with residuals.

Remarks

We assume that you have already read [TS] **mgarch ccc**. In this entry, we use predict after mgarch ccc to make in-sample and out-of-sample forecasts.

▷ Example 1

In this example, we obtain dynamic forecasts for the Toyota, Nissan, and Honda stock returns modeled in example 2 of [TS] **mgarch ccc**. In the output below, we reestimate the parameters of the model, use tsappend (see [TS] **tsappend**) to extend the data, and use predict to obtain in-sample one-step-ahead forecasts and dynamic forecasts of the conditional variances of the returns. We graph the forecasts below.

```
. use http://www.stata-press.com/data/r12/stocks
(Data from Yahoo! Finance)
. quietly mgarch ccc (toyota nissan = , noconstant) (honda = L.nissan, noconstant),
> arch(1) garch(1)
. tsappend, add(50)
. predict H*, variance dynamic(2016)
```

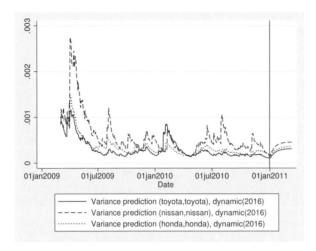

Recent in-sample one-step-ahead forecasts are plotted to the left of the vertical line in the above graph, and the dynamic out-of-sample forecasts appear to the right of the vertical line. The graph shows the tail end of the huge increase in return volatility that took place in 2008 and 2009. It also shows that the dynamic forecasts quickly converge.

◁

Methods and formulas

All postestimation commands listed above are implemented as ado-files.

All one-step predictions are obtained by substituting the parameter estimates into the model. The estimated unconditional variance matrix of the disturbances, $\widehat{\Sigma}$, is the initial value for the ARCH and GARCH terms. The postestimation routines recompute $\widehat{\Sigma}$ using the prediction sample, the parameter estimates stored in e(b), and (3) in *Methods and formulas* of [TS] **mgarch ccc**.

For observations in which the residuals are missing, the estimated unconditional variance matrix of the disturbances is used in place of the outer product of the residuals.

Dynamic predictions of the dependent variables use previously predicted values beginning in the period specified by dynamic().

Dynamic variance predictions are implemented by substituting $\widehat{\Sigma}$ for the outer product of the residuals beginning in the period specified in dynamic().

Also see

[TS] **mgarch ccc** — Constant conditional correlation multivariate GARCH models

[U] **20 Estimation and postestimation commands**

Title

> **mgarch dcc** — Dynamic conditional correlation multivariate GARCH models

Syntax

mgarch dcc *eq* [*eq* ... *eq*] [*if*] [*in*] [, *options*]

where each *eq* has the form

(*depvars* = [*indepvars*] [, *eqoptions*])

options	Description
Model	
<u>ar</u>ch(*numlist*)	ARCH terms for all equations
<u>garch</u>(*numlist*)	GARCH terms for all equations
het(*varlist*)	include *varlist* in the specification of the conditional variance for all equations
<u>d</u>istribution(*dist* [#])	use *dist* distribution for errors [may be <u>gaussian</u> (synonym <u>norm</u>al) or t; default is <u>gaussian</u>]
<u>cons</u>traints(*numlist*)	apply linear constraints
SE/Robust	
vce(*vcetype*)	*vcetype* may be oim or <u>r</u>obust
Reporting	
<u>l</u>evel(#)	set confidence level; default is level(95)
<u>nocnsr</u>eport	do not display constraints
display_options	control column formats, row spacing, line width, and display of omitted variables and base and empty cells
Maximization	
maximize_options	control the maximization process; seldom used
from(*matname*)	initial values for the coefficients; seldom used
<u>coefl</u>egend	display legend instead of statistics

eqoptions	Description
<u>nocons</u>tant	suppress constant term in the mean equation
<u>ar</u>ch(*numlist*)	ARCH terms
<u>garch</u>(*numlist*)	GARCH terms
het(*varlist*)	include *varlist* in the specification of the conditional variance

You must tsset your data before using mgarch dcc; see [TS] **tsset**.

indepvars and *varlist* may contain factor variables; see [U] **11.4.3 Factor variables**.

depvars, *indepvars*, and *varlist* may contain time-series operators; see [U] **11.4.4 Time-series varlists**.

by, statsby, and rolling are allowed; see [U] **11.1.10 Prefix commands**.

coeflegend does not appear in the dialog box.

See [U] **20 Estimation and postestimation commands** for more capabilities of estimation commands.

247

Menu

Statistics > Multivariate time series > Multivariate GARCH

Description

mgarch dcc estimates the parameters of dynamic conditional correlation (DCC) multivariate generalized autoregressive conditionally heteroskedastic (MGARCH) models in which the conditional variances are modeled as univariate generalized autoregressive conditionally heteroskedastic (GARCH) models and the conditional covariances are modeled as nonlinear functions of the conditional variances. The conditional quasicorrelation parameters that weight the nonlinear combinations of the conditional variances follow the GARCH-like process specified in Engle (2002).

The DCC MGARCH model is about as flexible as the closely related varying conditional correlation MGARCH model (see [TS] **mgarch vcc**), more flexible than the conditional correlation MGARCH model (see [TS] **mgarch ccc**), and more parsimonious than the diagonal vech MGARCH model (see [TS] **mgarch dvech**).

Options

‎ ⌐ Model ⌐

arch(*numlist*) specifies the ARCH terms for all equations in the model. By default, no ARCH terms are specified.

garch(*numlist*) specifies the GARCH terms for all equations in the model. By default, no GARCH terms are specified.

het(*varlist*) specifies that *varlist* be included in the model in the specification of the conditional variance for all equations. This varlist enters the variance specification collectively as multiplicative heteroskedasticity.

distribution(*dist* $\left[\,\#\,\right]$) specifies the assumed distribution for the errors. *dist* may be gaussian, normal, or t.

  gaussian and normal are synonyms; each causes mgarch dcc to assume that the errors come from a multivariate normal distribution. # may not be specified with either of them.

  t causes mgarch dcc to assume that the errors follow a multivariate Student t distribution, and the degree-of-freedom parameter is estimated along with the other parameters of the model. If distribution(t #) is specified, then mgarch dcc uses a multivariate Student t distribution with # degrees of freedom. # must be greater than 2.

constraints(*numlist*) specifies linear constraints to apply to the parameter estimates.

‎ ⌐ SE/Robust ⌐

vce(*vcetype*) specifies the estimator for the variance–covariance matrix of the estimator.

  vce(oim), the default, specifies to use the observed information matrix (OIM) estimator.

  vce(robust) specifies to use the Huber/White/sandwich estimator.

‎ ⌐ Reporting ⌐

level(#); see [R] **estimation options**.

nocnsreport; see [R] **estimation options**.

display_options: <u>noomit</u>ted, vsquish, noemptycells, <u>base</u>levels, <u>allbase</u>levels, cformat(%*fmt*), pformat(%*fmt*), sformat(%*fmt*), and nolstretch; see [R] **estimation options**.

⌐____ Maximization ⌐_____

maximize_options: <u>dif</u>ficult, <u>tech</u>nique(*algorithm_spec*), <u>iter</u>ate(#), [<u>no</u>]<u>log</u>, <u>trac</u>e, gradient, showstep, <u>hess</u>ian, <u>showtol</u>erance, <u>tol</u>erance(#), <u>ltol</u>erance(#), <u>nrtol</u>erance(#), and <u>nonrtol</u>erance, and from(*matname*); see [R] **maximize** for all options except from(), and see below for information on from(). These options are seldom used.

from(*matname*) specifies initial values for the coefficients. from(b0) causes mgarch dcc to begin the optimization algorithm with the values in b0. b0 must be a row vector, and the number of columns must equal the number of parameters in the model.

The following option is available with mgarch dcc but is not shown in the dialog box:

coeflegend; see [R] **estimation options**.

Eqoptions

noconstant suppresses the constant term in the mean equation.

arch(*numlist*) specifies the ARCH terms in the equation. By default, no ARCH terms are specified. This option may not be specified with model-level arch().

garch(*numlist*) specifies the GARCH terms in the equation. By default, no GARCH terms are specified. This option may not be specified with model-level garch().

het(*varlist*) specifies that *varlist* be included in the specification of the conditional variance. This varlist enters the variance specification collectively as multiplicative heteroskedasticity. This option may not be specified with model-level het().

Remarks

We assume that you have already read [TS] **mgarch**, which provides an introduction to MGARCH models and the methods implemented in mgarch dcc.

MGARCH models are dynamic multivariate regression models in which the conditional variances and covariances of the errors follow an autoregressive-moving-average structure. The DCC MGARCH model uses a nonlinear combination of univariate GARCH models with time-varying cross-equation weights to model the conditional covariance matrix of the errors.

As discussed in [TS] **mgarch**, MGARCH models differ in the parsimony and flexibility of their specifications for a time-varying conditional covariance matrix of the disturbances, denoted by $\mathbf{H}_t$. In the conditional correlation family of MGARCH models, the diagonal elements of $\mathbf{H}_t$ are modeled as univariate GARCH models, whereas the off-diagonal elements are modeled as nonlinear functions of the diagonal terms. In the DCC MGARCH model,

$$h_{ij,t} = \rho_{ij,t}\sqrt{h_{ii,t}h_{jj,t}}$$

where the diagonal elements $h_{ii,t}$ and $h_{jj,t}$ follow univariate GARCH processes and $\rho_{ij,t}$ follows the dynamic process specified in Engle (2002) and discussed below.

Because the $\rho_{ij,t}$ varies with time, this model is known as the DCC GARCH model.

❑ Technical note

The DCC GARCH model proposed by Engle (2002) can be written as

$$
\begin{aligned}
\mathbf{y}_t &= \mathbf{C}\mathbf{x}_t + \boldsymbol{\epsilon}_t \\
\boldsymbol{\epsilon}_t &= \mathbf{H}_t^{1/2}\boldsymbol{\nu}_t \\
\mathbf{H}_t &= \mathbf{D}_t^{1/2}\mathbf{R}_t\mathbf{D}_t^{1/2} \\
\mathbf{R}_t &= \operatorname{diag}(\mathbf{Q}_t)^{-1/2}\mathbf{Q}_t\operatorname{diag}(\mathbf{Q}_t)^{-1/2} \\
\mathbf{Q}_t &= (1 - \lambda_1 - \lambda_2)\bar{\mathbf{R}} + \lambda_1 \widetilde{\boldsymbol{\epsilon}}_{t-1}\widetilde{\boldsymbol{\epsilon}}'_{t-1} + \lambda_2 \mathbf{Q}_{t-1}
\end{aligned}
\tag{1}
$$

where

$\mathbf{y}_t$ is an $m \times 1$ vector of dependent variables;

$\mathbf{C}$ is an $m \times k$ matrix of parameters;

$\mathbf{x}_t$ is a $k \times 1$ vector of independent variables, which may contain lags of $\mathbf{y}_t$;

$\mathbf{H}_t^{1/2}$ is the Cholesky factor of the time-varying conditional covariance matrix $\mathbf{H}_t$;

$\boldsymbol{\nu}_t$ is an $m \times 1$ vector of normal, independent, and identically distributed innovations;

$\mathbf{D}_t$ is a diagonal matrix of conditional variances,

$$
\mathbf{D}_t =
\begin{pmatrix}
\sigma_{1,t}^2 & 0 & \cdots & 0 \\
0 & \sigma_{2,t}^2 & \cdots & 0 \\
\vdots & \vdots & \ddots & \vdots \\
0 & 0 & \cdots & \sigma_{m,t}^2
\end{pmatrix}
$$

in which each $\sigma_{i,t}^2$ evolves according to a univariate GARCH model of the form

$$
\sigma_{i,t}^2 = s_i + \sum_{j=1}^{p_i} \alpha_j \epsilon_{i,t-j}^2 + \sum_{j=1}^{q_i} \beta_j \sigma_{i,t-j}^2
$$

by default, or

$$
\sigma_{i,t}^2 = \exp(\boldsymbol{\gamma}_i \mathbf{z}_{i,t}) + \sum_{j=1}^{p_i} \alpha_j \epsilon_{i,t-j}^2 + \sum_{j=1}^{q_i} \beta_j \sigma_{i,t-j}^2
$$

when the het() option is specified, where $\boldsymbol{\gamma}_t$ is a $1 \times p$ vector of parameters, $\mathbf{z}_i$ is a $p \times 1$ vector of independent variables including a constant term, the α_j's are ARCH parameters, and the β_j's are GARCH parameters;

$\mathbf{R}_t$ is a matrix of conditional quasicorrelations,

$$
\mathbf{R}_t =
\begin{pmatrix}
1 & \rho_{12,t} & \cdots & \rho_{1m,t} \\
\rho_{12,t} & 1 & \cdots & \rho_{2m,t} \\
\vdots & \vdots & \ddots & \vdots \\
\rho_{1m,t} & \rho_{2m,t} & \cdots & 1
\end{pmatrix}
$$

$\widetilde{\boldsymbol{\epsilon}}_t$ is an $m \times 1$ vector of standardized residuals, $\mathbf{D}_t^{-1/2}\boldsymbol{\epsilon}_t$; and

λ_1 and λ_2 are parameters that govern the dynamics of conditional quasicorrelations. λ_1 and λ_2 are nonnegative and satisfy $0 \le \lambda_1 + \lambda_2 < 1$.

When $\mathbf{Q}_t$ is stationary, the $\mathbf{R}$ matrix in (1) is a weighted average of the unconditional covariance matrix of the standardized residuals $\widetilde{\epsilon}_t$, denoted by $\overline{\mathbf{R}}$, and the unconditional mean of $\mathbf{Q}_t$, denoted by $\overline{\mathbf{Q}}$. Because $\overline{\mathbf{R}} \neq \overline{\mathbf{Q}}$, as shown by Aielli (2009), $\mathbf{R}$ is neither the unconditional correlation matrix nor the unconditional mean of $\mathbf{Q}_t$. For this reason, the parameters in $\mathbf{R}$ are known as quasicorrelations; see Aielli (2009) and Engle (2009) for discussions.

❑

Some examples

▷ Example 1

We have daily data on the stock returns of three car manufacturers—Toyota, Nissan, and Honda, from January 2, 2003, to December 31, 2010—in the variables toyota, nissan and honda. We model the conditional means of the returns as a first-order vector autoregressive process and the conditional covariances as a DCC MGARCH process in which the variance of each disturbance term follows a GARCH(1,1) process.

```
. use http://www.stata-press.com/data/r12/stocks
(Data from Yahoo! Finance)

. mgarch dcc (toyota nissan honda = L.toyota L.nissan L.honda, noconstant),
> arch(1) garch(1)

Calculating starting values....

Optimizing log likelihood

(setting technique to bhhh)
Iteration 0:    log likelihood =  16902.435
Iteration 1:    log likelihood =  17005.448
Iteration 2:    log likelihood =  17157.958
Iteration 3:    log likelihood =  17267.363
Iteration 4:    log likelihood =   17318.29
Iteration 5:    log likelihood =  17353.029
Iteration 6:    log likelihood =  17369.115
Iteration 7:    log likelihood =  17388.035
Iteration 8:    log likelihood =  17401.254
Iteration 9:    log likelihood =  17435.556
(switching technique to nr)
Iteration 10:   log likelihood =  17451.739
Iteration 11:   log likelihood =  17475.511
Iteration 12:   log likelihood =  17483.259
Iteration 13:   log likelihood =  17484.906
Iteration 14:   log likelihood =   17484.95
Iteration 15:   log likelihood =   17484.95

Refining estimates

Iteration 0:    log likelihood =   17484.95
Iteration 1:    log likelihood =   17484.95
```

Dynamic conditional correlation MGARCH model

Sample: 1 - 2015				Number of obs	=	2014
Distribution: Gaussian				Wald chi2(9)	=	19.54
Log likelihood = 17484.95				Prob > chi2	=	0.0210

	Coef.	Std. Err.	z	P>|z|	[95% Conf. Interval]	
toyota						
toyota						
L1.	-.0510871	.0339825	-1.50	0.133	-.1176915	.0155174
nissan						
L1.	.0297832	.0247455	1.20	0.229	-.0187171	.0782835
honda						
L1.	-.0162822	.0300323	-0.54	0.588	-.0751445	.0425801
ARCH_toyota						
arch						
L1.	.0608225	.0086687	7.02	0.000	.0438321	.0778128
garch						
L1.	.9222202	.0111055	83.04	0.000	.9004538	.9439865
_cons	4.47e-06	1.15e-06	3.90	0.000	2.22e-06	6.72e-06
nissan						
toyota						
L1.	-.0056724	.0389348	-0.15	0.884	-.0819832	.0706384
nissan						
L1.	-.0287095	.0309379	-0.93	0.353	-.0893467	.0319276
honda						
L1.	.0154978	.0358802	0.43	0.666	-.0548261	.0858217
ARCH_nissan						
arch						
L1.	.0844239	.0128192	6.59	0.000	.0592988	.109549
garch						
L1.	.8994209	.0151124	59.52	0.000	.8698011	.9290407
_cons	7.21e-06	1.93e-06	3.74	0.000	3.43e-06	.000011
honda						
toyota						
L1.	-.0272421	.0361819	-0.75	0.451	-.0981573	.0436731
nissan						
L1.	.0617495	.0271378	2.28	0.023	.0085603	.1149387
honda						
L1.	-.063507	.0332919	-1.91	0.056	-.1287578	.0017438

ARCH_honda						
arch						
L1.	.0490142	.0073697	6.65	0.000	.0345698	.0634586
garch						
L1.	.9331114	.0103689	89.99	0.000	.9127888	.9534341
_cons	5.35e-06	1.35e-06	3.95	0.000	2.69e-06	8.00e-06
Correlation						
toyota						
nissan	.6689551	.016802	39.81	0.000	.6360239	.7018864
honda	.7259628	.0140155	51.80	0.000	.6984929	.7534327
nissan						
honda	.6335662	.018041	35.12	0.000	.5982064	.6689259
Adjustment						
lambda1	.0315267	.008838	3.57	0.000	.0142045	.0488489
lambda2	.8704173	.0613324	14.19	0.000	.750208	.9906266

The iteration log has three parts: the dots from the search for initial values, the iteration log from optimizing the log likelihood, and the iteration log from the refining step. A detailed discussion of the optimization methods is in *Methods and formulas*.

The header describes the estimation sample and reports a Wald test against the null hypothesis that all the coefficients on the independent variables in the mean equations are zero. Here the null hypothesis is rejected at the 5% level.

The output table first presents results for the mean or variance parameters used to model each dependent variable. Subsequently, the output table presents results for the conditional quasicorrelations. For example, the conditional quasicorrelation between the standardized residuals for Toyota and Nissan is estimated to be 0.67. Finally, the output table presents results for the adjustment parameters λ_1 and λ_2. In the example at hand, the estimates for both λ_1 and λ_2 are statistically significant.

The DCC MGARCH model reduces to the CCC MGARCH model when $\lambda_1 = \lambda_2 = 0$. The output below shows that a Wald test rejects the null hypothesis that $\lambda_1 = \lambda_2 = 0$ at all conventional levels.

```
. test _b[Adjustment:lambda1] = _b[Adjustment:lambda2] = 0

 ( 1)  [Adjustment]lambda1 - [Adjustment]lambda2 = 0
 ( 2)  [Adjustment]lambda1 = 0

          chi2(  2) = 1102.33
        Prob > chi2 =    0.0000
```

These results indicate that the assumption of time-invariant conditional correlations maintained in the CCC MGARCH model is too restrictive for these data.

◁

▷ Example 2

We improve the previous example by removing the insignificant parameters from the model. To remove these parameters, we specify the honda equation separately from the toyota and nissan equations:

```
. mgarch dcc (toyota nissan = , noconstant) (honda = L.nissan, noconstant),
> arch(1) garch(1)

Calculating starting values....

Optimizing log likelihood

(setting technique to bhhh)
Iteration 0:    log likelihood =  16884.502
Iteration 1:    log likelihood =  16970.755
Iteration 2:    log likelihood =  17140.318
Iteration 3:    log likelihood =  17237.807
Iteration 4:    log likelihood =   17306.12
Iteration 5:    log likelihood =  17342.533
Iteration 6:    log likelihood =  17363.511
Iteration 7:    log likelihood =  17392.501
Iteration 8:    log likelihood =  17407.242
Iteration 9:    log likelihood =  17448.702
(switching technique to nr)
Iteration 10:   log likelihood =  17472.199
Iteration 11:   log likelihood =  17475.839
Iteration 12:   log likelihood =  17476.345
Iteration 13:   log likelihood =   17476.35
Iteration 14:   log likelihood =   17476.35

Refining estimates

Iteration 0:    log likelihood =   17476.35
Iteration 1:    log likelihood =   17476.35
```

Dynamic conditional correlation MGARCH model

Sample: 1 - 2015	Number of obs =	2014
Distribution: Gaussian	Wald chi2(1) =	2.21
Log likelihood = 17476.35	Prob > chi2 =	0.1374

	Coef.	Std. Err.	z	P>|z|	[95% Conf. Interval]	
ARCH_toyota						
arch						
L1.	.0608188	.0086675	7.02	0.000	.0438309	.0778067
garch						
L1.	.9219957	.0111066	83.01	0.000	.9002271	.9437642
_cons	4.49e-06	1.14e-06	3.95	0.000	2.27e-06	6.72e-06
ARCH_nissan						
arch						
L1.	.0876162	.01302	6.73	0.000	.0620974	.1131349
garch						
L1.	.8950964	.0152908	58.54	0.000	.865127	.9250657
_cons	7.69e-06	1.99e-06	3.86	0.000	3.79e-06	.0000116
honda						
nissan						
L1.	.019978	.0134488	1.49	0.137	-.0063811	.0463371

ARCH_honda						
arch						
L1.	.0488799	.0073767	6.63	0.000	.0344218	.063338
garch						
L1.	.9330047	.0103944	89.76	0.000	.912632	.9533774
_cons	5.42e-06	1.36e-06	3.98	0.000	2.75e-06	8.08e-06
Correlation						
toyota						
nissan	.6668433	.0163209	40.86	0.000	.6348548	.6988317
honda	.7258101	.0137072	52.95	0.000	.6989446	.7526757
nissan						
honda	.6313515	.0175454	35.98	0.000	.5969631	.6657399
Adjustment						
lambda1	.0324493	.0074013	4.38	0.000	.0179429	.0469556
lambda2	.8574681	.0476273	18.00	0.000	.7641202	.950816

It turns out that the coefficient on L1.nissan in the honda equation is now statistically insignificant. We could further improve the model by removing L1.nissan from the model.

There is no mean equation for Toyota or Nissan. In [TS] **mgarch dcc postestimation**, we discuss prediction from models without covariates.

◁

> ## Example 3

Here we fit a bivariate DCC MGARCH model for the Toyota and Nissan shares. We believe that the shares of these car manufacturers follow the same process, so we impose the constraints that the ARCH coefficients are the same for the two companies and that the GARCH coefficients are also the same.

```
. constraint 1 _b[ARCH_toyota:L.arch] = _b[ARCH_nissan:L.arch]
. constraint 2 _b[ARCH_toyota:L.garch] = _b[ARCH_nissan:L.garch]
. mgarch dcc (toyota nissan = , noconstant), arch(1) garch(1) constraints(1 2)
Calculating starting values....
Optimizing log likelihood

(setting technique to bhhh)
Iteration 0:   log likelihood =  10307.609
Iteration 1:   log likelihood =  10656.153
Iteration 2:   log likelihood =  10862.137
Iteration 3:   log likelihood =  10987.457
Iteration 4:   log likelihood =  11062.347
Iteration 5:   log likelihood =  11135.207
Iteration 6:   log likelihood =  11245.619
Iteration 7:   log likelihood =   11253.56
Iteration 8:   log likelihood =      11294
Iteration 9:   log likelihood =  11296.364
(switching technique to nr)
Iteration 10:  log likelihood =   11296.76
Iteration 11:  log likelihood =  11297.087
Iteration 12:  log likelihood =  11297.091
Iteration 13:  log likelihood =  11297.091
```

```
Refining estimates
Iteration 0:   log likelihood =  11297.091
Iteration 1:   log likelihood =  11297.091
Dynamic conditional correlation MGARCH model
```

Sample: 1 - 2015			Number of obs	=		2015
Distribution: Gaussian			Wald chi2(.)	=		.
Log likelihood = 11297.09			Prob > chi2	=		.

```
 ( 1)  [ARCH_toyota]L.arch - [ARCH_nissan]L.arch = 0
 ( 2)  [ARCH_toyota]L.garch - [ARCH_nissan]L.garch = 0
```

	Coef.	Std. Err.	z	P>|z|	[95% Conf. Interval]	
ARCH_toyota						
arch						
L1.	.080889	.0103227	7.84	0.000	.060657	.1011211
garch						
L1.	.9060711	.0119107	76.07	0.000	.8827267	.9294156
_cons	4.21e-06	1.10e-06	3.83	0.000	2.05e-06	6.36e-06
ARCH_nissan						
arch						
L1.	.080889	.0103227	7.84	0.000	.060657	.1011211
garch						
L1.	.9060711	.0119107	76.07	0.000	.8827267	.9294156
_cons	5.92e-06	1.47e-06	4.03	0.000	3.04e-06	8.80e-06
Correlation						
toyota						
nissan	.6646283	.0187793	35.39	0.000	.6278215	.7014351
Adjustment						
lambda1	.0446559	.0123017	3.63	0.000	.020545	.0687668
lambda2	.8686054	.0510884	17.00	0.000	.7684739	.968737

We could test our constraints by fitting the unconstrained model and performing a likelihood-ratio test. The results indicate that the restricted model is preferable.

◁

▷ Example 4

In this example, we have data on fictional stock returns for the Acme and Anvil corporations, and we believe that the movement of the two stocks is governed by different processes. We specify one ARCH and one GARCH term for the conditional variance equation for Acme and two ARCH terms for the conditional variance equation for Anvil. In addition, we include the lagged value of the stock return for Apex, the main subsidiary of Anvil corporation, in the variance equation of Anvil. For Acme, we have data on the changes in an index of futures prices of products related to those produced by Acme in afrelated. For Anvil, we have data on the changes in an index of futures prices of inputs used by Anvil in afinputs.

```
. use http://www.stata-press.com/data/r12/acmeh

. mgarch dcc (acme = afrelated, noconstant arch(1) garch(1))
> (anvil = afinputs, arch(1/2) het(L.apex))

Calculating starting values....

Optimizing log likelihood

(setting technique to bhhh)
Iteration 0:   log likelihood = -13260.522
  (output omitted )
Iteration 9:   log likelihood = -12362.876
(switching technique to nr)
Iteration 10:  log likelihood = -12362.876

Refining estimates

Iteration 0:   log likelihood = -12362.876
Iteration 1:   log likelihood = -12362.876

Dynamic conditional correlation MGARCH model
```

Sample: 1 - 2500   Number of obs   =   2499
Distribution: Gaussian   Wald chi2(2)   =   2596.18
Log likelihood = -12362.88   Prob > chi2   =   0.0000

	Coef.	Std. Err.	z	P>|z|	[95% Conf. Interval]	
acme						
afrelated	.950805	.0557082	17.07	0.000	.841619	1.059991
ARCH_acme						
arch						
L1.	.1063295	.015716	6.77	0.000	.0755266	.1371324
garch						
L1.	.7556294	.0391568	19.30	0.000	.6788836	.8323752
_cons	2.197566	.458343	4.79	0.000	1.29923	3.095901
anvil						
afinputs	-1.015657	.0209959	-48.37	0.000	-1.056808	-.9745054
_cons	.0808653	.019445	4.16	0.000	.0427538	.1189767
ARCH_anvil						
arch						
L1.	.5261675	.0281586	18.69	0.000	.4709777	.5813572
L2.	.2866454	.0196504	14.59	0.000	.2481314	.3251595
apex						
L1.	1.953173	.0594862	32.83	0.000	1.836582	2.069763
_cons	-.0062964	.0710842	-0.09	0.929	-.1456188	.1330261
Correlation						
acme						
anvil	-.5600358	.0326358	-17.16	0.000	-.6240008	-.4960708
Adjustment						
lambda1	.1904321	.0154449	12.33	0.000	.1601607	.2207035
lambda2	.7147267	.0226204	31.60	0.000	.6703916	.7590618

The results indicate that increases in the futures prices for related products lead to higher returns on the Acme stock, and increased input prices lead to lower returns on the Anvil stock. In the conditional variance equation for Anvil, the coefficient on L1.apex is positive and significant, which indicates that an increase in the return on the Apex stock leads to more variability in the return on the Anvil stock.   ◁

Saved results

mgarch dcc saves the following in e():

Scalars
e(N)	number of observations
e(k)	number of parameters
e(k_aux)	number of auxiliary parameters
e(k_extra)	number of extra estimates added to _b
e(k_eq)	number of equations in e(b)
e(k_dv)	number of dependent variables
e(df_m)	model degrees of freedom
e(ll)	log likelihood
e(chi2)	χ^2
e(p)	significance
e(estdf)	1 if distribution parameter was estimated, 0 otherwise
e(usr)	user-provided distribution parameter
e(tmin)	minimum time in sample
e(tmax)	maximum time in sample
e(N_gaps)	number of gaps
e(rank)	rank of e(V)
e(ic)	number of iterations
e(rc)	return code
e(converged)	1 if converged, 0 otherwise

Macros
e(cmd)	mgarch
e(model)	dcc
e(cmdline)	command as typed
e(depvar)	names of dependent variables
e(covariates)	list of covariates
e(dv_eqs)	dependent variables with mean equations
e(indeps)	independent variables in each equation
e(tvar)	time variable
e(title)	title in estimation output
e(chi2type)	Wald; type of model χ^2 test
e(vce)	*vcetype* specified in vce()
e(vcetype)	title used to label Std. Err.
e(tmins)	formatted minimum time
e(tmaxs)	formatted maximum time
e(dist)	distribution for error term: gaussian or t
e(arch)	specified ARCH terms
e(garch)	specified GARCH terms
e(technique)	maximization technique
e(properties)	b V
e(estat_cmd)	program used to implement estat
e(predict)	program used to implement predict
e(marginsok)	predictions allowed by margins
e(marginsnotok)	predictions disallowed by margins

Matrices
e(b)	coefficient vector
e(Cns)	constraints matrix
e(ilog)	iteration log (up to 20 iterations)
e(gradient)	gradient vector
e(hessian)	Hessian matrix
e(V)	variance–covariance matrix of the estimators
e(pinfo)	parameter information, used by predict

Functions
e(sample)	marks estimation sample

Methods and formulas

`mgarch dcc` is implemented as an ado-file.

`mgarch dcc` estimates the parameters of the DCC MGARCH model by maximum likelihood. The log-likelihood function based on the multivariate normal distribution for observation t is

$$l_t = -0.5m \log(2\pi) - 0.5\log\left\{\det\left(\mathbf{R}_t\right)\right\} - \log\left\{\det\left(\mathbf{D}_t^{1/2}\right)\right\} - 0.5\widetilde{\boldsymbol{\epsilon}}_t\mathbf{R}_t^{-1}\widetilde{\boldsymbol{\epsilon}}_t'$$

where $\widetilde{\boldsymbol{\epsilon}}_t = \mathbf{D}_t^{-1/2}\boldsymbol{\epsilon}_t$ is an $m \times 1$ vector of standardized residuals, $\boldsymbol{\epsilon}_t = \mathbf{y}_t - \mathbf{C}\mathbf{x}_t$. The log-likelihood function is $\sum_{t=1}^{T} l_t$.

If we assume that $\boldsymbol{\nu}_t$ follow a multivariate t distribution with degrees of freedom (df) greater than 2, then the log-likelihood function for observation t is

$$l_t = \log \Gamma\left(\frac{\mathrm{df} + m}{2}\right) - \log \Gamma\left(\frac{\mathrm{df}}{2}\right) - \frac{m}{2}\log\left\{(\mathrm{df} - 2)\pi\right\}$$

$$- 0.5\log\left\{\det\left(\mathbf{R}_t\right)\right\} - \log\left\{\det\left(\mathbf{D}_t^{1/2}\right)\right\} - \frac{\mathrm{df} + m}{2}\log\left(1 + \frac{\widetilde{\boldsymbol{\epsilon}}_t\mathbf{R}_t^{-1}\widetilde{\boldsymbol{\epsilon}}_t'}{\mathrm{df} - 2}\right)$$

The starting values for the parameters in the mean equations and the initial residuals $\widehat{\boldsymbol{\epsilon}}_t$ are obtained by least-squares regression. The starting values for the parameters in the variance equations are obtained by a procedure proposed by Gourieroux and Monfort (1997, sec. 6.2.2). The starting values for the quasicorrelation parameters are calculated from the standardized residuals $\widetilde{\boldsymbol{\epsilon}}_t$. Given the starting values for the mean and variance equations, the starting values for the parameters λ_1 and λ_2 are obtained from a grid search performed on the log likelihood.

The initial optimization step is performed in the unconstrained space. Once the maximum is found, we impose the constraints $\lambda_1 \geq 0$, $\lambda_2 \geq 0$, and $0 \leq \lambda_1 + \lambda_2 < 1$, and maximize the log likelihood in the constrained space. This step is reported in the iteration log as the refining step.

GARCH estimators require initial values that can be plugged in for $\boldsymbol{\epsilon}_{t-i}\boldsymbol{\epsilon}_{t-i}'$ and $\mathbf{H}_{t-j}$ when $t - i < 1$ and $t - j < 1$. `mgarch dcc` substitutes an estimator of the unconditional covariance of the disturbances

$$\widehat{\boldsymbol{\Sigma}} = T^{-1}\sum_{t=1}^{T}\widehat{\boldsymbol{\epsilon}}_t\widehat{\widetilde{\boldsymbol{\epsilon}}}_t' \tag{2}$$

for $\boldsymbol{\epsilon}_{t-i}\boldsymbol{\epsilon}_{t-i}'$ when $t - i < 1$ and for $\mathbf{H}_{t-j}$ when $t - j < 1$, where $\widehat{\boldsymbol{\epsilon}}_t$ is the vector of residuals calculated using the estimated parameters.

`mgarch dcc` uses numerical derivatives in maximizing the log-likelihood function.

References

Aielli, G. P. 2009. Dynamic Conditional Correlations: On Properties and Estimation. Working paper, Dipartimento di Statistica, University of Florence, Florence, Italy.

Engle, R. F. 2002. Dynamic conditional correlation: A simple class of multivariate generalized autoregressive conditional heteroskedasticity models. *Journal of Business & Economic Statistics* 20: 339–350.

———. 2009. *Anticipating Correlations: A New Paradigm for Risk Management*. Princeton, NJ: Princeton University Press.

Gourieroux, C., and A. Monfort. 1997. *Time Series and Dynamic Models*. Trans. ed. G. M. Gallo. Cambridge: Cambridge University Press.

Also see

[TS] **mgarch dcc postestimation** — Postestimation tools for mgarch dcc

[TS] **mgarch** — Multivariate GARCH models

[TS] **tsset** — Declare data to be time-series data

[TS] **arch** — Autoregressive conditional heteroskedasticity (ARCH) family of estimators

[TS] **var** — Vector autoregressive models

[U] **20 Estimation and postestimation commands**

Title

---
mgarch dcc postestimation — Postestimation tools for mgarch dcc

---

Description

The following standard postestimation commands are available after `mgarch dcc`:

Command	Description
contrast	contrasts and ANOVA-style joint tests of estimates
estat	AIC, BIC, VCE, and estimation sample summary
estimates	cataloging estimation results
lincom	point estimates, standard errors, testing, and inference for linear combinations of coefficients
lrtest	likelihood-ratio test
margins	marginal means, predictive margins, marginal effects, and average marginal effects
marginsplot	graph the results from margins (profile plots, interaction plots, etc.)
nlcom	point estimates, standard errors, testing, and inference for nonlinear combinations of coefficients
predict	predictions, residuals, influence statistics, and other diagnostic measures
predictnl	point estimates, standard errors, testing, and inference for generalized predictions
pwcompare	pairwise comparisons of estimates
test	Wald tests of simple and composite linear hypotheses
testnl	Wald tests of nonlinear hypotheses

See the corresponding entries in the *Base Reference Manual* for details.

Syntax for predict

$$\texttt{predict} \ \left[\textit{type}\right] \ \left\{\textit{stub*} \,|\, \textit{newvarlist}\right\} \ \left[\textit{if}\right] \ \left[\textit{in}\right] \ \left[\texttt{,} \ \textit{statistic} \ \textit{options}\right]$$

statistic	Description
Main	
xb	linear prediction; the default
residuals	residuals
variance	conditional variances and covariances

These statistics are available both in and out of sample; type `predict ... if e(sample) ...` if wanted only for the estimation sample.

options	Description
Options	
equation(*eqnames*)	names of equations for which predictions are made
dynamic(*time_constant*)	begin dynamic forecast at specified time

Menu

Statistics > Postestimation > Predictions, residuals, etc.

Options for predict

Main

xb, the default, calculates the linear predictions of the dependent variables.

residuals calculates the residuals.

variance predicts the conditional variances and conditional covariances.

Options

equation(*eqnames*) specifies the equation for which the predictions are calculated. Use this option to predict a statistic for a particular equation. Equation names, such as equation(income), are used to identify equations.

One equation name may be specified when predicting the dependent variable, the residuals, or the conditional variance. For example, specifying equation(income) causes predict to predict income, and specifying variance equation(income) causes predict to predict the conditional variance of income.

Two equations may be specified when predicting a conditional variance or covariance. For example, specifying equation(income, consumption) variance causes predict to predict the conditional covariance of income and consumption.

dynamic(*time_constant*) specifies when predict starts producing dynamic forecasts. The specified *time_constant* must be in the scale of the time variable specified in tsset, and the *time_constant* must be inside a sample for which observations on the dependent variables are available. For example, dynamic(tq(2008q4)) causes dynamic predictions to begin in the fourth quarter of 2008, assuming that your time variable is quarterly; see [D] **datetime**. If the model contains exogenous variables, they must be present for the whole predicted sample. dynamic() may not be specified with residuals.

Remarks

We assume that you have already read [TS] **mgarch dcc**. In this entry, we use predict after mgarch dcc to make in-sample and out-of-sample forecasts.

▷ Example 1

In this example, we obtain dynamic forecasts for the Toyota, Nissan, and Honda stock returns modeled in example 2 of [TS] **mgarch dcc**. In the output below, we reestimate the parameters of the model, use tsappend (see [TS] **tsappend**) to extend the data, and use predict to obtain in-sample one-step-ahead forecasts and dynamic forecasts of the conditional variances of the returns. We graph the forecasts below.

```
. use http://www.stata-press.com/data/r12/stocks
(Data from Yahoo! Finance)
. quietly mgarch dcc (toyota nissan = , noconstant) (honda = L.nissan, noconstant),
> arch(1) garch(1)
. tsappend, add(50)
. predict H*, variance dynamic(2016)
```

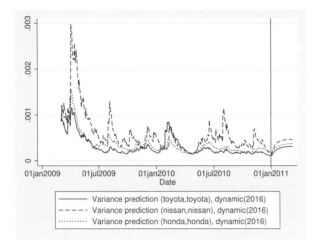

Recent in-sample one-step-ahead forecasts are plotted to the left of the vertical line in the above graph, and the dynamic out-of-sample forecasts appear to the right of the vertical line. The graph shows the tail end of the huge increase in return volatility that took place in 2008 and 2009. It also shows that the dynamic forecasts quickly converge.

◁

Methods and formulas

All postestimation commands listed above are implemented as ado-files.

All one-step predictions are obtained by substituting the parameter estimates into the model. The estimated unconditional variance matrix of the disturbances, $\widehat{\Sigma}$, is the initial value for the ARCH and GARCH terms. The postestimation routines recompute $\widehat{\Sigma}$ using the prediction sample, the parameter estimates stored in e(b), and (2) in *Methods and formulas* of [TS] **mgarch dcc**.

For observations in which the residuals are missing, the estimated unconditional variance matrix of the disturbances is used in place of the outer product of the residuals.

Dynamic predictions of the dependent variables use previously predicted values beginning in the period specified by dynamic().

Dynamic variance predictions are implemented by substituting $\widehat{\Sigma}$ for the outer product of the residuals beginning in the period specified in dynamic().

Also see

[TS] **mgarch dcc** — Dynamic conditional correlation multivariate GARCH models

[U] **20 Estimation and postestimation commands**

Title

mgarch dvech — Diagonal vech multivariate GARCH models

Syntax

mgarch dvech *eq* [*eq* ... *eq*] [*if*] [*in*] [, *options*]

where each *eq* has the form

(*depvars* = [*indepvars*] [, <u>noconst</u>ant])

options	Description
Model	
<u>arc</u>h(*numlist*)	ARCH terms
<u>garc</u>h(*numlist*)	GARCH terms
<u>d</u>istribution(*dist* [*#*])	use *dist* distribution for errors (may be <u>gaus</u>sian, <u>norm</u>al, or t; default is gaussian)
<u>constr</u>aints(*numlist*)	apply linear constraints
SE/Robust	
<u>vce</u>(*vcetype*)	*vcetype* may be oim or <u>r</u>obust
Reporting	
<u>l</u>evel(*#*)	set confidence level; default is level(95)
nocnsreport	do not display constraints
display_options	control column formats, row spacing, line width, and display of omitted variables and base and empty cells
Maximization	
maximize_options	control the maximization process; seldom used
from(*matname*)	initial values for the coefficients; seldom used
<u>svtech</u>nique(*algorithm_spec*)	starting-value maximization algorithm
<u>sviter</u>ate(*#*)	number of starting-value iterations; default is sviterate(25)
<u>coefl</u>egend	display legend instead of statistics

You must tsset your data before using mgarch dvech; see [TS] **tsset**.
indepvars may contain factor variables; see [U] **11.4.3 Factor variables**.
depvars and *indepvars* may contain time-series operators; see [U] **11.4.4 Time-series varlists**.
by, statsby, and rolling are allowed; see [U] **11.1.10 Prefix commands**.
coeflegend does not appear in the dialog box.
See [U] **20 Estimation and postestimation commands** for more capabilities of estimation commands.

Menu

Statistics > Multivariate time series > Multivariate GARCH

Description

mgarch dvech estimates the parameters of diagonal vech (DVECH) multivariate generalized autoregressive conditionally heteroskedastic (MGARCH) models in which each element of the conditional correlation matrix is parameterized as a linear function of its own past and past shocks.

DVECH MGARCH models are less parsimonious than the conditional correlation models discussed in [TS] **mgarch ccc**, [TS] **mgarch dcc**, and [TS] **mgarch vcc** because the number of parameters in DVECH MGARCH models increases more rapidly with the number of series modeled.

Options

_____⌐ Model ⌐_____

noconstant suppresses the constant term(s).

arch(*numlist*) specifies the ARCH terms in the model. By default, no ARCH terms are specified.

garch(*numlist*) specifies the GARCH terms in the model. By default, no GARCH terms are specified.

distribution(*dist* [*#*]) specifies the assumed distribution for the errors. *dist* may be gaussian, normal, or t.

   gaussian and normal are synonyms; each causes mgarch dvech to assume that the errors come from a multivariate normal distribution. # cannot be specified with either of them.

   t causes mgarch dvech to assume that the errors follow a multivariate Student t distribution, and the degree-of-freedom parameter is estimated along with the other parameters of the model. If distribution(t #) is specified, then mgarch dvech uses a multivariate Student t distribution with # degrees of freedom. # must be greater than 2.

constraints(*numlist*) specifies linear constraints to apply to the parameter estimates.

_____⌐ SE/Robust ⌐_____

vce(*vcetype*) specifies the estimator for the variance–covariance matrix of the estimator.

   vce(oim), the default, specifies to use the observed information matrix (OIM) estimator.

   vce(robust) specifies to use the Huber/White/sandwich estimator.

_____⌐ Reporting ⌐_____

level(*#*); see [R] **estimation options**.

nocnsreport; see [R] **estimation options**.

display_options: noomitted, vsquish, noemptycells, baselevels, allbaselevels, cformat(%*fmt*), pformat(%*fmt*), sformat(%*fmt*), and nolstretch; see [R] **estimation options**.

_____⌐ Maximization ⌐_____

maximize_options: difficult, technique(*algorithm_spec*), iterate(*#*), [no] log, trace, gradient, showstep, hessian, showtolerance, tolerance(*#*), ltolerance(*#*), nrtolerance(*#*), and nonrtolerance, and from(*matname*); see [R] **maximize** for all options except from(), and see below for information on from(). These options are seldom used.

from(*matname*) specifies initial values for the coefficients. from(b0) causes mgarch dvech to begin the optimization algorithm with the values in b0. b0 must be a row vector, and the number of columns must equal the number of parameters in the model.

svtechnique(*algorithm_spec*) and sviterate(*#*) specify options for the starting-value search process.

> svtechnique(*algorithm_spec*) specifies the algorithm used to search for initial values. The syntax for *algorithm_spec* is the same as for the technique() option; see [R] **maximize**. svtechnique(bhhh 5 nr 16000) is the default. This option may not be specified with from().

> sviterate(*#*) specifies the maximum number of iterations that the search algorithm may perform. The default is sviterate(25). This option may not be specified with from().

The following option is available with mgarch dvech but is not shown in the dialog box:

coeflegend; see [R] **estimation options**.

Remarks

We assume that you have already read [TS] **mgarch**, which provides an introduction to MGARCH models and the methods implemented in mgarch dvech.

MGARCH models are dynamic multivariate regression models in which the conditional variances and covariances of the errors follow an autoregressive-moving-average structure. The DVECH MGARCH model parameterizes each element of the current conditional covariance matrix as a linear function of its own past and past shocks.

As discussed in [TS] **mgarch**, MGARCH models differ in the parsimony and flexibility of their specifications for a time-varying conditional covariance matrix of the disturbances, denoted by $\mathbf{H}_t$. In a DVECH MGARCH model with one ARCH term and one GARCH term, the (i, j)th element of conditional covariance matrix is modeled by

$$h_{ij,t} = s_{ij} + a_{ij}\epsilon_{i,t-1}\epsilon_{j,t-1} + b_{ij}h_{ij,t-1}$$

where s_{ij}, a_{ij}, and b_{ij} are parameters and ϵ_{t-1} is the vector of errors from the previous period. This expression shows the linear form in which each element of the current conditional covariance matrix is a function of its own past and past shocks.

❑ Technical note

The general vech MGARCH model developed by Bollerslev, Engle, and Wooldridge (1988) can be written as

$$\mathbf{y}_t = \mathbf{C}\mathbf{x}_t + \epsilon_t \tag{1}$$

$$\epsilon_t = \mathbf{H}_t^{1/2}\boldsymbol{\nu}_t \tag{2}$$

$$\mathbf{h}_t = \mathbf{s} + \sum_{i=1}^{p}\mathbf{A}_i\text{vech}(\epsilon_{t-i}\epsilon_{t-i}') + \sum_{j=1}^{q}\mathbf{B}_j\mathbf{h}_{t-j} \tag{3}$$

where

> $\mathbf{y}_t$ is an $m \times 1$ vector of dependent variables;

> $\mathbf{C}$ is an $m \times k$ matrix of parameters;

> $\mathbf{x}_t$ is a $k \times 1$ vector of independent variables, which may contain lags of $\mathbf{y}_t$;

$\mathbf{H}_t^{1/2}$ is the Cholesky factor of the time-varying conditional covariance matrix $\mathbf{H}_t$;

$\boldsymbol{\nu}_t$ is an $m \times 1$ vector of independent and identically distributed innovations;

$\mathbf{h}_t = \text{vech}(\mathbf{H}_t)$;

the vech() function stacks the lower diagonal elements of a symmetric matrix into a column vector, for example,

$$\text{vech} \begin{pmatrix} 1 & 2 \\ 2 & 3 \end{pmatrix} = (1, \ 2, \ 3)'$$

s is an $m(m+1)/2 \times 1$ vector of parameters;

each $\mathbf{A}_i$ is an $\{m(m+1)/2\} \times \{m(m+1)/2\}$ matrix of parameters; and

each $\mathbf{B}_j$ is an $\{m(m+1)/2\} \times \{m(m+1)/2\}$ matrix of parameters.

Bollerslev, Engle, and Wooldridge (1988) argued that the general-vech MGARCH model in (1)–(3) was too flexible to fit to data, so they proposed restricting the matrices $\mathbf{A}_i$ and $\mathbf{B}_j$ to be diagonal matrices. It is for this restriction that the model is known as a diagonal vech MGARCH model. The diagonal vech MGARCH model can also be expressed by replacing (3) with

$$\mathbf{H}_t = \mathbf{S} + \sum_{i=1}^{p} \mathbf{A}_i \odot \boldsymbol{\epsilon}_{t-i} \boldsymbol{\epsilon}_{t-i}' + \sum_{j=1}^{q} \mathbf{B}_j \odot \mathbf{H}_{t-j} \tag{3'}$$

where $\mathbf{S}$ is an $m \times m$ symmetric parameter matrix; each $\mathbf{A}_i$ is an $m \times m$ symmetric parameter matrix; $\odot$ is the elementwise or Hadamard product; and each $\mathbf{B}_j$ is an $m \times m$ symmetric parameter matrix. In (3'), $\mathbf{A}$ and $\mathbf{B}$ are symmetric but not diagonal matrices because we used the Hadamard product. The matrices are diagonal in the vech representation of (3) but not in the Hadamard-product representation of (3').

The Hadamard-product representation in (3') clarifies that each element in $\mathbf{H}_t$ depends on its past values and the past values of the corresponding ARCH terms. Although this representation does not allow cross-covariance effects, it is still quite flexible. The rapid rate at which the number of parameters grows with m, p, or q is one aspect of the model's flexibility.

❑

Some examples

▷ Example 1

We have data on a secondary market rate of a six-month U.S. Treasury bill, tbill, and on Moody's seasoned AAA corporate bond yield, bond. We model the first-differences of tbill and the first-differences of bond as a VAR(1) with an ARCH(1) term.

```
. use http://www.stata-press.com/data/r12/irates4
(St. Louis Fed (FRED) financial data)

. mgarch dvech (D.bond D.tbill = LD.bond LD.tbill), arch(1)

Getting starting values
(setting technique to bhhh)
Iteration 0:   log likelihood =  3569.2723
Iteration 1:   log likelihood =  3708.4561
  (output omitted )
Iteration 6:   log likelihood =  4183.8853
Iteration 7:   log likelihood =  4184.2424
(switching technique to nr)
Iteration 8:   log likelihood =  4184.4141
Iteration 9:   log likelihood =  4184.5973
Iteration 10:  log likelihood =  4184.5975

Estimating parameters
(setting technique to bhhh)
Iteration 0:   log likelihood =  4184.5975
Iteration 1:   log likelihood =  4200.6303
Iteration 2:   log likelihood =  4208.5342
Iteration 3:   log likelihood =   4212.426
Iteration 4:   log likelihood =  4215.2373
(switching technique to nr)
Iteration 5:   log likelihood =  4217.0676
Iteration 6:   log likelihood =  4221.5706
Iteration 7:   log likelihood =  4221.6576
Iteration 8:   log likelihood =  4221.6577
```

Diagonal vech MGARCH model

Sample: 3 - 2456   Number of obs   =   2454
Distribution: Gaussian   Wald chi2(4)   =   1183.52
Log likelihood = 4221.658   Prob > chi2   =   0.0000

	Coef.	Std. Err.	z	P>|z|	[95% Conf. Interval]	
D.bond						
bond						
LD.	.2967674	.0247149	12.01	0.000	.2483271	.3452077
tbill						
LD.	.0947949	.0098683	9.61	0.000	.0754533	.1141364
_cons	.0003991	.00143	0.28	0.780	-.0024036	.0032019
D.tbill						
bond						
LD.	.0108373	.0301501	0.36	0.719	-.0482558	.0699304
tbill						
LD.	.4344747	.0176497	24.62	0.000	.3998819	.4690675
_cons	.0011611	.0021033	0.55	0.581	-.0029612	.0052835
Sigma0						
1_1	.004894	.0002006	24.40	0.000	.0045008	.0052871
2_1	.0040986	.0002396	17.10	0.000	.0036289	.0045683
2_2	.0115149	.0005227	22.03	0.000	.0104904	.0125395
L.ARCH						
1_1	.4514942	.0456835	9.88	0.000	.3619562	.5410323
2_1	.2518879	.036736	6.86	0.000	.1798866	.3238893
2_2	.843368	.0608055	13.87	0.000	.7241914	.9625446

The output has three parts: an iteration log, a header, and an output table. The iteration log has two parts: the first part reports the iterations from the process of searching for starting values, and the second part reports the iterations from maximizing the log-likelihood function.

The header describes the estimation sample and reports a Wald test against the null hypothesis that all the coefficients on the independent variables in each equation are zero. Here the null hypothesis is rejected at all conventional levels.

The output table reports point estimates, standard errors, tests against zero, and confidence intervals for the estimated coefficients, the estimated elements of S, and any estimated elements of A or B. Here the output indicates that in the equation for D.tbill, neither the coefficient on LD.bond nor the constant are statistically significant. The elements of S are reported in the Sigma0 equation. The estimate of $S[1, 1]$ is 0.005, and the estimate of $S[2, 1]$ is 0.004. The ARCH term results are reported in the L.ARCH equation. In the L.ARCH equation, 1_1 is the coefficient on the ARCH term for the conditional variance of the first dependent variable, 2_1 is the coefficient on the ARCH term for the conditional covariance between the first and second dependent variables, and 2_2 is the coefficient on the ARCH term for the conditional variance of the second dependent variable.

◁

▷ Example 2

We improve the previous example by removing the insignificant parameters from the model:

```
. mgarch dvech (D.bond = LD.bond LD.tbill, noconstant)
> (D.tbill = LD.tbill, noconstant), arch(1)

Getting starting values
(setting technique to bhhh)
Iteration 0:   log likelihood =  3566.8824
Iteration 1:   log likelihood =  3701.6181
Iteration 2:   log likelihood =  3952.8048
Iteration 3:   log likelihood =  4076.5164
Iteration 4:   log likelihood =  4166.6842
Iteration 5:   log likelihood =  4180.2998
Iteration 6:   log likelihood =  4182.4545
Iteration 7:   log likelihood =  4182.9563
(switching technique to nr)
Iteration 8:   log likelihood =  4183.0293
Iteration 9:   log likelihood =  4183.1112
Iteration 10:  log likelihood =  4183.1113

Estimating parameters
(setting technique to bhhh)
Iteration 0:   log likelihood =  4183.1113
Iteration 1:   log likelihood =  4202.0304
Iteration 2:   log likelihood =  4210.2929
Iteration 3:   log likelihood =  4215.7798
Iteration 4:   log likelihood =  4217.7755
(switching technique to nr)
Iteration 5:   log likelihood =  4219.0078
Iteration 6:   log likelihood =  4221.4197
Iteration 7:   log likelihood =   4221.433
Iteration 8:   log likelihood =   4221.433
```

```
Diagonal vech MGARCH model
Sample: 3 - 2456                              Number of obs    =       2454
Distribution: Gaussian                        Wald chi2(3)     =    1197.76
Log likelihood =  4221.433                    Prob > chi2      =     0.0000
```

	Coef.	Std. Err.	z	P>|z|	[95% Conf.	Interval]
D.bond						
bond						
LD.	.2941649	.0234734	12.53	0.000	.2481579	.3401718
tbill						
LD.	.0953158	.0098077	9.72	0.000	.076093	.1145386
D.tbill						
tbill						
LD.	.4385945	.0136672	32.09	0.000	.4118072	.4653817
Sigma0						
1_1	.0048922	.0002005	24.40	0.000	.0044993	.0052851
2_1	.0040949	.0002394	17.10	0.000	.0036256	.0045641
2_2	.0115043	.0005184	22.19	0.000	.0104883	.0125203
L.ARCH						
1_1	.4519233	.045671	9.90	0.000	.3624099	.5414368
2_1	.2515474	.0366701	6.86	0.000	.1796752	.3234195
2_2	.8437212	.0600839	14.04	0.000	.7259589	.9614836

We specified each equation separately to remove the insignificant parameters. All the parameter estimates are statistically significant.

◁

▷ Example 3

Here we analyze some fictional weekly data on the percentages of bad widgets found in the factories of Acme Inc. and Anvil Inc. We model the levels as a first-order autoregressive process. We believe that the adaptive management style in these companies causes the variances to follow a diagonal vech MGARCH process with one ARCH term and one GARCH term. Furthermore, these close competitors follow essentially the same process, so we impose the constraints that the ARCH coefficients are the same for the two companies and that the GARCH coefficients are also the same.

Imposing these constraints yields

```
. use http://www.stata-press.com/data/r12/acme, clear

. constraint 1 [L.ARCH]1_1 = [L.ARCH]2_2

. constraint 2 [L.GARCH]1_1 = [L.GARCH]2_2

. mgarch dvech (acme = L.acme) (anvil = L.anvil), arch(1) garch(1) constraints(1 2)

Getting starting values
(setting technique to bhhh)
Iteration 0:    log likelihood = -6087.0665   (not concave)
Iteration 1:    log likelihood = -6022.2046
Iteration 2:    log likelihood = -5986.6152
Iteration 3:    log likelihood = -5976.5739
Iteration 4:    log likelihood = -5974.4342
Iteration 5:    log likelihood = -5974.4046
Iteration 6:    log likelihood = -5974.4036
Iteration 7:    log likelihood = -5974.4035

Estimating parameters
(setting technique to bhhh)
Iteration 0:    log likelihood = -5974.4035
Iteration 1:    log likelihood =  -5973.812
Iteration 2:    log likelihood = -5973.8004
Iteration 3:    log likelihood = -5973.7999
Iteration 4:    log likelihood = -5973.7999

Diagonal vech MGARCH model

Sample: 1969w35 - 1998w25                      Number of obs    =      1499
Distribution: Gaussian                         Wald chi2(2)     =    272.47
Log likelihood =   -5973.8                      Prob > chi2      =    0.0000

 ( 1)   [L.ARCH]1_1 - [L.ARCH]2_2 = 0
 ( 2)   [L.GARCH]1_1 - [L.GARCH]2_2 = 0
```

		Coef.	Std. Err.	z	P>|z|	[95% Conf. Interval]	
acme							
acme	L1.	.3365278	.0255134	13.19	0.000	.2865225	.3865331
	_cons	1.124611	.060085	18.72	0.000	1.006847	1.242376
anvil							
anvil	L1.	.3151955	.0263287	11.97	0.000	.2635922	.3667988
	_cons	1.215786	.0642052	18.94	0.000	1.089947	1.341626
Sigma0							
	1_1	1.889237	.2168733	8.71	0.000	1.464173	2.314301
	2_1	.4599576	.1139843	4.04	0.000	.2365525	.6833626
	2_2	2.063113	.2454633	8.40	0.000	1.582014	2.544213
L.ARCH							
	1_1	.2813443	.0299124	9.41	0.000	.222717	.3399716
	2_1	.181877	.0335393	5.42	0.000	.1161412	.2476128
	2_2	.2813443	.0299124	9.41	0.000	.222717	.3399716
L.GARCH							
	1_1	.1487581	.0697531	2.13	0.033	.0120445	.2854716
	2_1	.085404	.1446524	0.59	0.555	-.1981094	.3689175
	2_2	.1487581	.0697531	2.13	0.033	.0120445	.2854716

We could test our constraints by fitting the unconstrained model and performing either a Wald or a likelihood-ratio test. The results indicate that we might further restrict the time-invariant components of the conditional variances to be the same across companies.

◁

▷ Example 4

Some models of financial data include no covariates or constant terms. For example, in modeling fictional data on the stock returns of Acme Inc. and Anvil Inc., we found it best not to include any covariates or constant terms. We include two ARCH terms and one GARCH term to model the conditional variances.

```
. use http://www.stata-press.com/data/r12/aacmer

. mgarch dvech (acme anvil = , noconstant), arch(1/2) garch(1)

Getting starting values
(setting technique to bhhh)
Iteration 0:    log likelihood = -18417.243  (not concave)
Iteration 1:    log likelihood = -18215.005
Iteration 2:    log likelihood = -18199.691
Iteration 3:    log likelihood = -18136.699
Iteration 4:    log likelihood = -18084.256
Iteration 5:    log likelihood = -17993.662
Iteration 6:    log likelihood =   -17731.1
Iteration 7:    log likelihood = -17629.505
(switching technique to nr)
Iteration 8:    log likelihood = -17548.172
Iteration 9:    log likelihood = -17544.987
Iteration 10:   log likelihood = -17544.937
Iteration 11:   log likelihood = -17544.937

Estimating parameters
(setting technique to bhhh)
Iteration 0:    log likelihood = -17544.937
Iteration 1:    log likelihood = -17544.937

Diagonal vech MGARCH model
Sample: 1 - 5000                          Number of obs    =       5000
Distribution: Gaussian                    Wald chi2(.)     =          .
Log likelihood = -17544.94                Prob > chi2      =          .
```

		Coef.	Std. Err.	z	P>|z|	[95% Conf. Interval]	
Sigma0							
	1_1	1.026283	.0823348	12.46	0.000	.8649096	1.187656
	2_1	.4300997	.0590294	7.29	0.000	.3144042	.5457952
	2_2	1.019753	.0837146	12.18	0.000	.8556751	1.18383
L.ARCH							
	1_1	.2878739	.02157	13.35	0.000	.2455975	.3301504
	2_1	.1036685	.0161446	6.42	0.000	.0720256	.1353114
	2_2	.2034196	.019855	10.25	0.000	.1645044	.2423347
L2.ARCH							
	1_1	.1837825	.0274555	6.69	0.000	.1299706	.2375943
	2_1	.0884425	.02208	4.01	0.000	.0451665	.1317185
	2_2	.2025718	.0272639	7.43	0.000	.1491355	.256008
L.GARCH							
	1_1	.0782467	.053944	1.45	0.147	-.0274816	.183975
	2_1	.2888104	.0818303	3.53	0.000	.1284261	.4491948
	2_2	.201618	.0470584	4.28	0.000	.1093853	.2938508

The model test is omitted from the output, because there are no covariates in the model. The univariate tests indicate that the included parameters fit the data well. In [TS] **mgarch dvech postestimation**, we discuss prediction from models without covariates.

◁

Saved results

mgarch dvech saves the following in e():

Scalars
e(N)	number of observations
e(k)	number of parameters
e(k_extra)	number of extra estimates added to _b
e(k_eq)	number of equations in e(b)
e(k_dv)	number of dependent variables
e(df_m)	model degrees of freedom
e(ll)	log likelihood
e(chi2)	χ^2
e(p)	significance
e(estdf)	1 if distribution parameter was estimated, 0 otherwise
e(usr)	user-provided distribution parameter
e(tmin)	minimum time in sample
e(tmax)	maximum time in sample
e(N_gaps)	number of gaps
e(rank)	rank of e(V)
e(ic)	number of iterations
e(rc)	return code
e(converged)	1 if converged, 0 otherwise

Macros
e(cmd)	mgarch
e(model)	dvech
e(cmdline)	command as typed
e(depvar)	names of dependent variables
e(covariates)	list of covariates
e(dv_eqs)	dependent variables with mean equations
e(indeps)	independent variables in each equation
e(tvar)	time variable
e(title)	title in estimation output
e(chi2type)	Wald; type of model χ^2 test
e(vce)	*vcetype* specified in vce()
e(vcetype)	title used to label Std. Err.
e(tmins)	formatted minimum time
e(tmaxs)	formatted maximum time
e(dist)	distribution for error term: gaussian or t
e(arch)	specified ARCH terms
e(garch)	specified GARCH terms
e(svtechnique)	maximization technique(s) for starting values
e(technique)	maximization technique
e(properties)	b V
e(estat_cmd)	program used to implement estat
e(predict)	program used to implement predict
e(marginsok)	predictions allowed by margins
e(marginsnotok)	predictions disallowed by margins

Matrices
  e(b)   coefficient vector
  e(Cns)   constraints matrix
  e(ilog)   iteration log (up to 20 iterations)
  e(gradient)   gradient vector
  e(hessian)   Hessian matrix
  e(A)   estimates of A matrices
  e(B)   estimates of B matrices
  e(S)   estimates of Sigma0 matrix
  e(Sigma)   Sigma hat
  e(V)   variance–covariance matrix of the estimators
  e(pinfo)   parameter information, used by predict

Functions
  e(sample)   marks estimation sample

Methods and formulas

mgarch dvech is implemented as an ado-file.

Recall that the diagonal vech MGARCH model can be written as

$$\mathbf{y}_t = \mathbf{C}\mathbf{x}_t + \epsilon_t$$

$$\epsilon_t = \mathbf{H}_t^{1/2}\boldsymbol{\nu}_t$$

$$\mathbf{H}_t = \mathbf{S} + \sum_{i=1}^{p}\mathbf{A}_i \odot \epsilon_{t-i}\epsilon_{t-i}' + \sum_{j=1}^{q}\mathbf{B}_j \odot \mathbf{H}_{t-j}$$

where

  $\mathbf{y}_t$ is an $m \times 1$ vector of dependent variables;

  $\mathbf{C}$ is an $m \times k$ matrix of parameters;

  $\mathbf{x}_t$ is a $k \times 1$ vector of independent variables, which may contain lags of $\mathbf{y}_t$;

  $\mathbf{H}_t^{1/2}$ is the Cholesky factor of the time-varying conditional covariance matrix $\mathbf{H}_t$;

  $\boldsymbol{\nu}_t$ is an $m \times 1$ vector of normal, independent, and identically distributed innovations;

  $\mathbf{S}$ is an $m \times m$ symmetric matrix of parameters;

  each $\mathbf{A}_i$ is an $m \times m$ symmetric matrix of parameters;

  $\odot$ is the elementwise or Hadamard product; and

  each $\mathbf{B}_j$ is an $m \times m$ symmetric matrix of parameters.

mgarch dvech estimates the parameters by maximum likelihood. The log-likelihood function based on the multivariate normal distribution for observation t is

$$l_t = -0.5m\log(2\pi) - 0.5\log\left\{\det\left(\mathbf{H}_t\right)\right\} - 0.5\epsilon_t\mathbf{H}_t^{-1}\epsilon_t'$$

where $\epsilon_t = \mathbf{y}_t - \mathbf{C}\mathbf{x}_t$. The log-likelihood function is $\sum_{t=1}^{T} l_t$.

If we assume that ν_t follow a multivariate t distribution with degrees of freedom (df) greater than 2, then the log-likelihood function for observation t is

$$l_t = \log \Gamma \left(\frac{\mathrm{df} + m}{2} \right) - \log \Gamma \left(\frac{\mathrm{df}}{2} \right) - \frac{m}{2} \log \left\{ (\mathrm{df} - 2)\pi \right\}$$
$$- 0.5 \log \left\{ \det \left(\mathbf{H}_t \right) \right\} - \frac{\mathrm{df} + m}{2} \log \left(1 + \frac{\epsilon_t \mathbf{H}_t^{-1} \epsilon_t'}{\mathrm{df} - 2} \right)$$

mgarch dvech ensures that $\mathbf{H}_t$ is positive definite for each t.

By default, mgarch dvech performs an iterative search for starting values. mgarch dvech estimates starting values for $\mathbf{C}$ by seemingly unrelated regression, uses these estimates to compute residuals $\widehat{\epsilon}_t$, plugs $\widehat{\epsilon}_t$ into the above log-likelihood function, and optimizes this log-likelihood function over the parameters in $\mathbf{H}_t$. This starting-value method plugs in consistent estimates of the parameters for the conditional means of the dependent variables and then iteratively searches for the variance parameters that maximize the log-likelihood function. Lütkepohl (2005, chap. 16) discusses this method as an estimator for the variance parameters.

GARCH estimators require initial values that can be plugged in for $\epsilon_{t-i}\epsilon_{t-i}'$ and $\mathbf{H}_{t-j}$ when $t - i < 1$ and $t - j < 1$. mgarch dvech substitutes an estimator of the unconditional covariance of the disturbances,

$$\widehat{\boldsymbol{\Sigma}} = T^{-1} \sum_{t=1}^{T} \widehat{\widehat{\epsilon}}_t \widehat{\widehat{\epsilon}}_t' \tag{4}$$

for $\epsilon_{t-i}\epsilon_{t-i}'$ when $t - i < 1$ and for $\mathbf{H}_{t-j}$ when $t - j < 1$, where $\widehat{\widehat{\epsilon}}_t$ is the vector of residuals calculated using the estimated parameters.

mgarch dvech uses analytic first and second derivatives in maximizing the log-likelihood function based on the multivariate normal distribution. mgarch dvech uses numerical derivatives in maximizing the log-likelihood function based on the multivariate t distribution.

References

Bollerslev, T., R. F. Engle, and J. M. Wooldridge. 1988. A capital asset pricing model with time-varying covariances. *Journal of Political Economy* 96: 116–131.

Lütkepohl, H. 2005. *New Introduction to Multiple Time Series Analysis*. New York: Springer.

Also see

[TS] **mgarch dvech postestimation** — Postestimation tools for mgarch dvech

[TS] **mgarch** — Multivariate GARCH models

[TS] **tsset** — Declare data to be time-series data

[TS] **arch** — Autoregressive conditional heteroskedasticity (ARCH) family of estimators

[TS] **var** — Vector autoregressive models

[U] **20 Estimation and postestimation commands**

Title

mgarch dvech postestimation — Postestimation tools for mgarch dvech

Description

The following standard postestimation commands are available after `mgarch dvech`:

Command	Description
contrast	contrasts and ANOVA-style joint tests of estimates
estat	AIC, BIC, VCE, and estimation sample summary
estimates	cataloging estimation results
lincom	point estimates, standard errors, testing, and inference for linear combinations of coefficients
lrtest	likelihood-ratio test
margins	marginal means, predictive margins, marginal effects, and average marginal effects
marginsplot	graph the results from margins (profile plots, interaction plots, etc.)
nlcom	point estimates, standard errors, testing, and inference for nonlinear combinations of coefficients
predict	predictions, residuals, influence statistics, and other diagnostic measures
predictnl	point estimates, standard errors, testing, and inference for generalized predictions
pwcompare	pairwise comparisons of estimates
test	Wald tests of simple and composite linear hypotheses
testnl	Wald tests of nonlinear hypotheses

See the corresponding entries in the *Base Reference Manual* for details.

Syntax for predict

predict [*type*] { *stub** | *newvarlist* } [*if*] [*in*] [, *statistic options*]

statistic	Description
Main	
xb	linear prediction; the default
<u>res</u>iduals	residuals
<u>var</u>iance	conditional variances and covariances

These statistics are available both in and out of sample; type `predict ... if e(sample) ...` if wanted only for the estimation sample.

options	Description
Options	
<u>eq</u>uation(*eqnames*)	names of equations for which predictions are made
<u>dyn</u>amic(*time_constant*)	begin dynamic forecast at specified time

Menu

Statistics > Postestimation > Predictions, residuals, etc.

Options for predict

Main

xb, the default, calculates the linear predictions of the dependent variables.

residuals calculates the residuals.

variance predicts the conditional variances and conditional covariances.

Options

equation(*eqnames*) specifies the equation for which the predictions are calculated. Use this option to predict a statistic for a particular equation. Equation names, such as equation(income), are used to identify equations.

One equation name may be specified when predicting the dependent variable, the residuals, or the conditional variance. For example, specifying equation(income) causes predict to predict income, and specifying variance equation(income) causes predict to predict the conditional variance of income.

Two equations may be specified when predicting a conditional variance or covariance. For example, specifying equation(income, consumption) variance causes predict to predict the conditional covariance of income and consumption.

dynamic(*time_constant*) specifies when predict starts producing dynamic forecasts. The specified *time_constant* must be in the scale of the time variable specified in tsset, and the *time_constant* must be inside a sample for which observations on the dependent variables are available. For example, dynamic(tq(2008q4)) causes dynamic predictions to begin in the fourth quarter of 2008, assuming that your time variable is quarterly; see [D] **datetime**. If the model contains exogenous variables, they must be present for the whole predicted sample. dynamic() may not be specified with residuals.

Remarks

We assume that you have already read [TS] **mgarch dvech**. In this entry, we illustrate some of the features of predict after using mgarch dvech to estimate the parameters of diagonal vech MGARCH models.

> Example 1

In this example, we obtain dynamic predictions for the Acme Inc. and Anvil Inc. fictional widget data modeled in example 3 of [TS] **mgarch dvech**. We begin by reestimating the parameters of the model.

```
. use http://www.stata-press.com/data/r12/acme
. constraint 1 [L.ARCH]1_1 = [L.ARCH]2_2
. constraint 2 [L.GARCH]1_1 = [L.GARCH]2_2
. mgarch dvech (acme = L.acme) (anvil = L.anvil), arch(1) garch(1) constraints(1 2)
Getting starting values
(setting technique to bhhh)
Iteration 0:   log likelihood = -6087.0665  (not concave)
Iteration 1:   log likelihood = -6022.2046
Iteration 2:   log likelihood = -5986.6152
Iteration 3:   log likelihood = -5976.5739
Iteration 4:   log likelihood = -5974.4342
Iteration 5:   log likelihood = -5974.4046
Iteration 6:   log likelihood = -5974.4036
Iteration 7:   log likelihood = -5974.4035

Estimating parameters
(setting technique to bhhh)
Iteration 0:   log likelihood = -5974.4035
Iteration 1:   log likelihood =  -5973.812
Iteration 2:   log likelihood = -5973.8004
Iteration 3:   log likelihood = -5973.7999
Iteration 4:   log likelihood = -5973.7999
Diagonal vech MGARCH model
```

Sample: 1969w35 - 1998w25	Number of obs	=	1499
Distribution: Gaussian	Wald chi2(2)	=	272.47
Log likelihood = -5973.8	Prob > chi2	=	0.0000

```
 ( 1)  [L.ARCH]1_1 - [L.ARCH]2_2 = 0
 ( 2)  [L.GARCH]1_1 - [L.GARCH]2_2 = 0
```

		Coef.	Std. Err.	z	P>|z|	[95% Conf. Interval]	
acme							
acme							
L1.		.3365278	.0255134	13.19	0.000	.2865225	.3865331
_cons		1.124611	.060085	18.72	0.000	1.006847	1.242376
anvil							
anvil							
L1.		.3151955	.0263287	11.97	0.000	.2635922	.3667988
_cons		1.215786	.0642052	18.94	0.000	1.089947	1.341626
Sigma0							
	1_1	1.889237	.2168733	8.71	0.000	1.464173	2.314301
	2_1	.4599576	.1139843	4.04	0.000	.2365525	.6833626
	2_2	2.063113	.2454633	8.40	0.000	1.582014	2.544213
L.ARCH							
	1_1	.2813443	.0299124	9.41	0.000	.222717	.3399716
	2_1	.181877	.0335393	5.42	0.000	.1161412	.2476128
	2_2	.2813443	.0299124	9.41	0.000	.222717	.3399716
L.GARCH							
	1_1	.1487581	.0697531	2.13	0.033	.0120445	.2854716
	2_1	.085404	.1446524	0.59	0.555	-.1981094	.3689175
	2_2	.1487581	.0697531	2.13	0.033	.0120445	.2854716

Now we use tsappend (see [TS] **tsappend**) to extend the data, use predict to obtain the dynamic predictions, and graph the predictions.

```
. tsappend, add(12)

. predict H*, variance dynamic(tw(1998w26))

. tsline H_acme_acme H_anvil_anvil if t>=tw(1995w25), legend(rows(2))
```

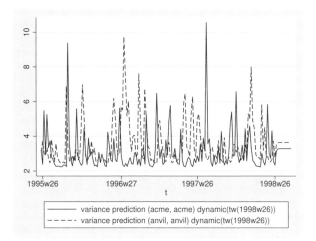

The graph shows that the in-sample predictions are similar for the conditional variances of Acme Inc. and Anvil Inc. and that the dynamic forecasts converge to similar levels. It also shows that the ARCH and GARCH parameters cause substantial time-varying volatility. The predicted conditional variance of acme ranges from lows of just over 2 to highs above 10.

◁

▷ Example 2

In this example, we obtain the in-sample predicted conditional variances of the returns for the fictional Acme Inc., which we modeled in example 4 of [TS] **mgarch dvech**. First, we reestimate the parameters of the model.

```
. use http://www.stata-press.com/data/r12/aacmer, clear

. mgarch dvech (acme anvil = , noconstant), arch(1/2) garch(1)
Getting starting values
(setting technique to bhhh)
Iteration 0:   log likelihood = -18417.243  (not concave)
Iteration 1:   log likelihood = -18215.005
Iteration 2:   log likelihood = -18199.691
Iteration 3:   log likelihood = -18136.699
Iteration 4:   log likelihood = -18084.256
Iteration 5:   log likelihood = -17993.662
Iteration 6:   log likelihood =   -17731.1
Iteration 7:   log likelihood = -17629.505
(switching technique to nr)
Iteration 8:   log likelihood = -17548.172
Iteration 9:   log likelihood = -17544.987
Iteration 10:  log likelihood = -17544.937
Iteration 11:  log likelihood = -17544.937

Estimating parameters
(setting technique to bhhh)
Iteration 0:   log likelihood = -17544.937
Iteration 1:   log likelihood = -17544.937
```

Diagonal vech MGARCH model

Sample: 1 - 5000   Number of obs   =   5000
Distribution: Gaussian   Wald chi2(.)   =   .
Log likelihood = -17544.94   Prob > chi2   =   .

		Coef.	Std. Err.	z	P>|z|	[95% Conf. Interval]	
Sigma0							
	1_1	1.026283	.0823348	12.46	0.000	.8649096	1.187656
	2_1	.4300997	.0590294	7.29	0.000	.3144042	.5457952
	2_2	1.019753	.0837146	12.18	0.000	.8556751	1.18383
L.ARCH							
	1_1	.2878739	.02157	13.35	0.000	.2455975	.3301504
	2_1	.1036685	.0161446	6.42	0.000	.0720256	.1353114
	2_2	.2034196	.019855	10.25	0.000	.1645044	.2423347
L2.ARCH							
	1_1	.1837825	.0274555	6.69	0.000	.1299706	.2375943
	2_1	.0884425	.02208	4.01	0.000	.0451665	.1317185
	2_2	.2025718	.0272639	7.43	0.000	.1491355	.256008
L.GARCH							
	1_1	.0782467	.053944	1.45	0.147	-.0274816	.183975
	2_1	.2888104	.0818303	3.53	0.000	.1284261	.4491948
	2_2	.201618	.0470584	4.28	0.000	.1093853	.2938508

Now we use `predict` to obtain the in-sample conditional variances of `acme` and use `tsline` (see [TS] **tsline**) to graph the results.

```
. predict h_acme, variance eq(acme, acme)

. tsline h_acme
```

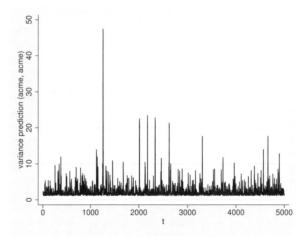

The graph shows that the predicted conditional variances vary substantially over time, as the parameter estimates indicated.

Because there are no covariates in the model for `acme`, specifying `xb` puts a prediction of 0 in each observation, and specifying `residuals` puts the value of the dependent variable into the prediction. ◁

Methods and formulas

All postestimation commands listed above are implemented as ado-files.

All one-step predictions are obtained by substituting the parameter estimates into the model. The estimated unconditional variance matrix of the disturbances, $\widehat{\Sigma}$, is the initial value for the ARCH and GARCH terms. The postestimation routines recompute $\widehat{\Sigma}$ using the prediction sample, the parameter estimates stored in e(b), and (4) in *Methods and formulas* of [TS] **mgarch dvech**.

For observations in which the residuals are missing, the estimated unconditional variance matrix of the disturbances is used in place of the outer product of the residuals.

Dynamic predictions of the dependent variables use previously predicted values beginning in the period specified by dynamic().

Dynamic variance predictions are implemented by substituting $\widehat{\Sigma}$ for the outer product of the residuals beginning in the period specified by dynamic().

Also see

[TS] **mgarch dvech** — Diagonal vech multivariate GARCH models

[U] **20 Estimation and postestimation commands**

Title

> **mgarch vcc** — Varying conditional correlation multivariate GARCH models

Syntax

mgarch vcc *eq* [*eq* ... *eq*] [*if*] [*in*] [, *options*]

where each *eq* has the form

(*depvars* = [*indepvars*] [, *eqoptions*])

options	Description
Model	
<u>ar</u>ch(*numlist*)	ARCH terms for all equations
<u>g</u>arch(*numlist*)	GARCH terms for all equations
het(*varlist*)	include *varlist* in the specification of the conditional variance for all equations
<u>distr</u>ibution(*dist* [#])	use *dist* distribution for errors [may be <u>gaussian</u> (synonym <u>normal</u>) or t; default is gaussian]
<u>constr</u>aints(*numlist*)	apply linear constraints
SE/Robust	
vce(*vcetype*)	*vcetype* may be oim or <u>r</u>obust
Reporting	
<u>level</u>(#)	set confidence level; default is level(95)
<u>nocnsr</u>eport	do not display constraints
display_options	control column formats, row spacing, line width, and display of omitted variables and base and empty cells
Maximization	
maximize_options	control the maximization process; seldom used
from(*matname*)	initial values for the coefficients; seldom used
<u>coefl</u>egend	display legend instead of statistics

eqoptions	Description
<u>nocons</u>tant	suppress constant term in the mean equation
<u>ar</u>ch(*numlist*)	ARCH terms
<u>g</u>arch(*numlist*)	GARCH terms
het(*varlist*)	include *varlist* in the specification of the conditional variance

You must tsset your data before using mgarch vcc; see [TS] tsset.

indepvars and *varlist* may contain factor variables; see [U] **11.4.3 Factor variables**.

depvars, *indepvars*, and *varlist* may contain time-series operators; see [U] **11.4.4 Time-series varlists**.

by, statsby, and rolling are allowed; see [U] **11.1.10 Prefix commands**.

coeflegend does not appear in the dialog box.

See [U] **20 Estimation and postestimation commands** for more capabilities of estimation commands.

Menu

Statistics > Multivariate time series > Multivariate GARCH

Description

mgarch vcc estimates the parameters of varying conditional correlation (VCC) multivariate generalized autoregressive conditionally heteroskedastic (MGARCH) models in which the conditional variances are modeled as univariate generalized autoregressive conditionally heteroskedastic (GARCH) models and the conditional covariances are modeled as nonlinear functions of the conditional variances. The conditional correlation parameters that weight the nonlinear combinations of the conditional variance follow the GARCH-like process specified in Tse and Tsui (2002).

The VCC MGARCH model is about as flexible as the closely related dynamic conditional correlation MGARCH model (see [TS] **mgarch dcc**), more flexible than the conditional correlation MGARCH model (see [TS] **mgarch ccc**), and more parsimonious than the diagonal vech model (see [TS] **mgarch dvech**).

Options

   ⌐ Model ⌐

arch(*numlist*) specifies the ARCH terms for all equations in the model. By default, no ARCH terms are specified.

garch(*numlist*) specifies the GARCH terms for all equations in the model. By default, no GARCH terms are specified.

het(*varlist*) specifies that *varlist* be included in the model in the specification of the conditional variance for all equations. This varlist enters the variance specification collectively as multiplicative heteroskedasticity.

distribution(*dist* [#]) specifies the assumed distribution for the errors. *dist* may be gaussian, normal, or t.

    gaussian and normal are synonyms; each causes mgarch vcc to assume that the errors come from a multivariate normal distribution. # may not be specified with either of them.

    t causes mgarch vcc to assume that the errors follow a multivariate Student t distribution, and the degree-of-freedom parameter is estimated along with the other parameters of the model. If distribution(t #) is specified, then mgarch vcc uses a multivariate Student t distribution with # degrees of freedom. # must be greater than 2.

constraints(*numlist*) specifies linear constraints to apply to the parameter estimates.

   ⌐ SE/Robust ⌐

vce(*vcetype*) specifies the estimator for the variance–covariance matrix of the estimator.

    vce(oim), the default, specifies to use the observed information matrix (OIM) estimator.

    vce(robust) specifies to use the Huber/White/sandwich estimator.

   ⌐ Reporting ⌐

level(#); see [R] **estimation options**.

nocnsreport; see [R] **estimation options**.

display_options: noomitted, vsquish, noemptycells, baselevels, allbaselevels, cformat(% *fmt*), pformat(% *fmt*), sformat(% *fmt*), and nolstretch; see [R] **estimation options**.

_____⌐ Maximization ⌐_____

maximize_options: difficult, technique(*algorithm_spec*), iterate(*#*), [no]log, trace, gradient, showstep, hessian, showtolerance, tolerance(*#*), ltolerance(*#*), nrtolerance(*#*), and nonrtolerance, and from(*matname*); see [R] **maximize** for all options except from(), and see below for information on from(). These options are seldom used.

from(*matname*) specifies initial values for the coefficients. from(b0) causes mgarch vcc to begin the optimization algorithm with the values in b0. b0 must be a row vector, and the number of columns must equal the number of parameters in the model.

The following option is available with mgarch vcc but is not shown in the dialog box:

coeflegend; see [R] **estimation options**.

Eqoptions

noconstant suppresses the constant term in the mean equation.

arch(*numlist*) specifies the ARCH terms in the equation. By default, no ARCH terms are specified. This option may not be specified with model-level arch().

garch(*numlist*) specifies the GARCH terms in the equation. By default, no GARCH terms are specified. This option may not be specified with model-level garch().

het(*varlist*) specifies that *varlist* be included in the specification of the conditional variance. This varlist enters the variance specification collectively as multiplicative heteroskedasticity. This option may not be specified with model-level het().

Remarks

We assume that you have already read [TS] **mgarch**, which provides an introduction to MGARCH models and the methods implemented in mgarch vcc.

MGARCH models are dynamic multivariate regression models in which the conditional variances and covariances of the errors follow an autoregressive-moving-average structure. The VCC MGARCH model uses a nonlinear combination of univariate GARCH models with time-varying cross-equation weights to model the conditional covariance matrix of the errors.

As discussed in [TS] **mgarch**, MGARCH models differ in the parsimony and flexibility of their specifications for a time-varying conditional covariance matrix of the disturbances, denoted by $\mathbf{H}_t$. In the conditional correlation family of MGARCH models, the diagonal elements of $\mathbf{H}_t$ are modeled as univariate GARCH models, whereas the off-diagonal elements are modeled as nonlinear functions of the diagonal terms. In the VCC MGARCH model,

$$h_{ij,t} = \rho_{ij,t} \sqrt{h_{ii,t} h_{jj,t}}$$

where the diagonal elements $h_{ii,t}$ and $h_{jj,t}$ follow univariate GARCH processes and $\rho_{ij,t}$ follows the dynamic process specified in Tse and Tsui (2002) and discussed below.

Because the $\rho_{ij,t}$ varies with time, this model is known as the VCC GARCH model.

❏ Technical note

The VCC GARCH model proposed by Tse and Tsui (2002) can be written as

$$
\begin{aligned}
\mathbf{y}_t &= \mathbf{C}\mathbf{x}_t + \boldsymbol{\epsilon}_t \\
\boldsymbol{\epsilon}_t &= \mathbf{H}_t^{1/2}\boldsymbol{\nu}_t \\
\mathbf{H}_t &= \mathbf{D}_t^{1/2}\mathbf{R}_t\mathbf{D}_t^{1/2} \\
\mathbf{R}_t &= (1 - \lambda_1 - \lambda_2)\mathbf{R} + \lambda_1\boldsymbol{\Psi}_{t-1} + \lambda_2\mathbf{R}_{t-1}
\end{aligned}
\tag{1}
$$

where

$\mathbf{y}_t$ is an $m \times 1$ vector of dependent variables;

$\mathbf{C}$ is an $m \times k$ matrix of parameters;

$\mathbf{x}_t$ is a $k \times 1$ vector of independent variables, which may contain lags of $\mathbf{y}_t$;

$\mathbf{H}_t^{1/2}$ is the Cholesky factor of the time-varying conditional covariance matrix $\mathbf{H}_t$;

$\boldsymbol{\nu}_t$ is an $m \times 1$ vector of independent and identically distributed innovations;

$\mathbf{D}_t$ is a diagonal matrix of conditional variances,

$$
\mathbf{D}_t = \begin{pmatrix}
\sigma_{1,t}^2 & 0 & \cdots & 0 \\
0 & \sigma_{2,t}^2 & \cdots & 0 \\
\vdots & \vdots & \ddots & \vdots \\
0 & 0 & \cdots & \sigma_{m,t}^2
\end{pmatrix}
$$

in which each $\sigma_{i,t}^2$ evolves according to a univariate GARCH model of the form

$$
\sigma_{i,t}^2 = s_i + \sum_{j=1}^{p_i} \alpha_j \epsilon_{i,t-j}^2 + \sum_{j=1}^{q_i} \beta_j \sigma_{i,t-j}^2
$$

by default, or

$$
\sigma_{i,t}^2 = \exp(\boldsymbol{\gamma}_i \mathbf{z}_{i,t}) + \sum_{j=1}^{p_i} \alpha_j \epsilon_{i,t-j}^2 + \sum_{j=1}^{q_i} \beta_j \sigma_{i,t-j}^2
$$

when the het() option is specified, where $\boldsymbol{\gamma}_t$ is a $1 \times p$ vector of parameters, $\mathbf{z}_i$ is a $p \times 1$ vector of independent variables including a constant term, the α_j's are ARCH parameters, and the β_j's are GARCH parameters;

$\mathbf{R}_t$ is a matrix of conditional correlations,

$$
\mathbf{R}_t = \begin{pmatrix}
1 & \rho_{12,t} & \cdots & \rho_{1m,t} \\
\rho_{12,t} & 1 & \cdots & \rho_{2m,t} \\
\vdots & \vdots & \ddots & \vdots \\
\rho_{1m,t} & \rho_{2m,t} & \cdots & 1
\end{pmatrix}
$$

$\mathbf{R}$ is the matrix of means to which the dynamic process in (1) reverts;

$\boldsymbol{\Psi}_t$ is the rolling estimator of the correlation matrix of $\widetilde{\boldsymbol{\epsilon}}_t$, which uses the previous $m + 1$ observations; and

λ_1 and λ_2 are parameters that govern the dynamics of conditional correlations. λ_1 and λ_2 are nonnegative and satisfy $0 \leq \lambda_1 + \lambda_2 < 1$.

To differentiate this model from Engle (2002), Tse and Tsui (2002) call their model a VCC MGARCH model.

❑

Some examples

▷ Example 1

We have daily data on the stock returns of three car manufacturers—Toyota, Nissan, and Honda, from January 2, 2003, to December 31, 2010—in the variables `toyota`, `nissan`, and `honda`. We model the conditional means of the returns as a first-order vector autoregressive process and the conditional covariances as a VCC MGARCH process in which the variance of each disturbance term follows a GARCH(1,1) process.

```
. use http://www.stata-press.com/data/r12/stocks
(Data from Yahoo! Finance)
. mgarch vcc (toyota nissan honda = L.toyota L.nissan L.honda, noconstant),
> arch(1) garch(1)

Calculating starting values....

Optimizing log likelihood

(setting technique to bhhh)
Iteration 0:   log likelihood =    16901.2
Iteration 1:   log likelihood =   17028.644
Iteration 2:   log likelihood =   17145.905
Iteration 3:   log likelihood =   17251.485
Iteration 4:   log likelihood =   17306.115
Iteration 5:   log likelihood =    17332.59
Iteration 6:   log likelihood =   17353.617
Iteration 7:   log likelihood =    17374.86
Iteration 8:   log likelihood =   17398.526
Iteration 9:   log likelihood =   17418.748
(switching technique to nr)
Iteration 10:  log likelihood =   17442.552
Iteration 11:  log likelihood =    17455.66
Iteration 12:  log likelihood =   17463.597
Iteration 13:  log likelihood =   17463.922
Iteration 14:  log likelihood =   17463.925
Iteration 15:  log likelihood =   17463.925

Refining estimates

Iteration 0:   log likelihood =   17463.925
Iteration 1:   log likelihood =   17463.925

Varying conditional correlation MGARCH model
Sample: 1 - 2015                          Number of obs    =       2014
Distribution: Gaussian                    Wald chi2(9)     =      17.67
Log likelihood = 17463.92                 Prob > chi2      =     0.0392
```

	Coef.	Std. Err.	z	P>|z|	[95% Conf. Interval]
toyota					
toyota L1.	-.0565645	.0335696	-1.68	0.092	-.1223597 .0092307
nissan L1.	.0248101	.0252701	0.98	0.326	-.0247184 .0743385
honda L1.	.0035836	.0298895	0.12	0.905	-.0549986 .0621659

ARCH_toyota						
arch						
L1.	.0602807	.0086799	6.94	0.000	.0432684	.077293
garch						
L1.	.9224689	.0110316	83.62	0.000	.9008473	.9440906
_cons	4.38e-06	1.12e-06	3.91	0.000	2.18e-06	6.58e-06
nissan						
toyota						
L1.	-.0196399	.0387112	-0.51	0.612	-.0955124	.0562326
nissan						
L1.	-.0306663	.031051	-0.99	0.323	-.091525	.0301925
honda						
L1.	.038315	.0354691	1.08	0.280	-.0312031	.1078331
ARCH_nissan						
arch						
L1.	.0774228	.0119642	6.47	0.000	.0539733	.1008723
garch						
L1.	.9076856	.0139339	65.14	0.000	.8803756	.9349956
_cons	6.20e-06	1.70e-06	3.65	0.000	2.87e-06	9.53e-06
honda						
toyota						
L1.	-.0358292	.0340492	-1.05	0.293	-.1025645	.030906
nissan						
L1.	.0544071	.0276156	1.97	0.049	.0002814	.1085327
honda						
L1.	-.0424383	.0326249	-1.30	0.193	-.106382	.0215054
ARCH_honda						
arch						
L1.	.0458673	.0072714	6.31	0.000	.0316157	.0601189
garch						
L1.	.9369253	.0101755	92.08	0.000	.9169816	.9568689
_cons	4.99e-06	1.29e-06	3.85	0.000	2.45e-06	7.52e-06
Correlation						
toyota						
nissan	.6643028	.0151086	43.97	0.000	.6346905	.6939151
honda	.7302093	.0126361	57.79	0.000	.705443	.7549755
nissan						
honda	.6347321	.0159738	39.74	0.000	.603424	.6660401
Adjustment						
lambda1	.0277374	.0086942	3.19	0.001	.010697	.0447778
lambda2	.8255525	.0755882	10.92	0.000	.6774024	.9737026

The output has three parts: an iteration log, a header, and an output table.

The iteration log has three parts: the dots from the search for initial values, the iteration log from optimizing the log likelihood, and the iteration log from the refining step. A detailed discussion of the optimization methods is in *Methods and formulas*.

The header describes the estimation sample and reports a Wald test against the null hypothesis that all the coefficients on the independent variables in the mean equations are zero. Here the null hypothesis is rejected at the 5% level.

The output table first presents results for the mean or variance parameters used to model each dependent variable. Subsequently, the output table presents results for the parameters in $\mathbf{R}$. For example, the estimate of the mean of the process that associates Toyota and Nissan is 0.66. Finally, the output table presents results for the adjustment parameters λ_1 and λ_2. In the example at hand, the estimates for both λ_1 and λ_2 are statistically significant.

The VCC MGARCH model reduces to the CCC MGARCH model when $\lambda_1 = \lambda_2 = 0$. The output below shows that a Wald test rejects the null hypothesis that $\lambda_1 = \lambda_2 = 0$ at all conventional levels.

```
. test _b[Adjustment:lambda1] = _b[Adjustment:lambda2] = 0

 ( 1)  [Adjustment]lambda1 - [Adjustment]lambda2 = 0
 ( 2)  [Adjustment]lambda1 = 0

           chi2(  2) =    482.80
         Prob > chi2 =     0.0000
```

These results indicate that the assumption of time-invariant conditional correlations maintained in the CCC MGARCH model is too restrictive for these data.

◁

▷ Example 2

We improve the previous example by removing the insignificant parameters from the model. To accomplish that, we specify the honda equation separately from the toyota and nissan equations:

```
. mgarch vcc (toyota nissan = , noconstant) (honda = L.nissan, noconstant),
> arch(1) garch(1)

Calculating starting values....

Optimizing log likelihood

(setting technique to bhhh)
Iteration 0:   log likelihood =   16889.43
Iteration 1:   log likelihood =  17002.567
Iteration 2:   log likelihood =  17134.525
Iteration 3:   log likelihood =  17233.192
Iteration 4:   log likelihood =  17295.342
Iteration 5:   log likelihood =  17326.347
Iteration 6:   log likelihood =  17348.063
Iteration 7:   log likelihood =  17363.988
Iteration 8:   log likelihood =  17387.216
Iteration 9:   log likelihood =  17404.734
(switching technique to nr)
Iteration 10:  log likelihood =  17438.432  (not concave)
Iteration 11:  log likelihood =  17450.001
Iteration 12:  log likelihood =  17455.443
Iteration 13:  log likelihood =  17455.971
Iteration 14:  log likelihood =   17455.98
Iteration 15:  log likelihood =   17455.98

Refining estimates

Iteration 0:   log likelihood =   17455.98
Iteration 1:   log likelihood =   17455.98  (backed up)
```

Varying conditional correlation MGARCH model

Sample: 1 - 2015   Number of obs   =   2014
Distribution: Gaussian   Wald chi2(1)   =   1.62
Log likelihood = 17455.98   Prob > chi2   =   0.2032

	Coef.	Std. Err.	z	P>|z|	[95% Conf.	Interval]
ARCH_toyota						
arch						
L1.	.0609064	.0087784	6.94	0.000	.0437011	.0781117
garch						
L1.	.921703	.0111493	82.67	0.000	.8998509	.9435552
_cons	4.42e-06	1.13e-06	3.91	0.000	2.20e-06	6.64e-06
ARCH_nissan						
arch						
L1.	.0806598	.0123529	6.53	0.000	.0564486	.104871
garch						
L1.	.9035239	.014421	62.65	0.000	.8752592	.9317886
_cons	6.61e-06	1.79e-06	3.70	0.000	3.11e-06	.0000101
honda						
nissan						
L1.	.0175566	.0137982	1.27	0.203	-.0094874	.0446005
ARCH_honda						
arch						
L1.	.0461398	.0073048	6.32	0.000	.0318226	.060457
garch						
L1.	.9366096	.0102021	91.81	0.000	.9166139	.9566052
_cons	5.03e-06	1.31e-06	3.85	0.000	2.47e-06	7.59e-06
Correlation						
toyota						
nissan	.6635251	.0150293	44.15	0.000	.6340682	.692982
honda	.7299703	.0124828	58.48	0.000	.7055045	.754436
nissan						
honda	.6338207	.0158681	39.94	0.000	.6027198	.6649217
Adjustment						
lambda1	.0285319	.0092448	3.09	0.002	.0104124	.0466514
lambda2	.8113924	.0854955	9.49	0.000	.6438243	.9789604

It turns out that the coefficient on L1.nissan in the honda equation is now statistically insignificant. We could further improve the model by removing L1.nissan from the model.

There is no mean equation for Toyota or Nissan. In [TS] **mgarch vcc postestimation**, we discuss prediction from models without covariates.

◁

▷ Example 3

Here we fit a bivariate VCC MGARCH model for the Toyota and Nissan shares. We believe that the shares of these car manufacturers follow the same process, so we impose the constraints that the ARCH coefficients are the same for the two companies and that the GARCH coefficients are also the same.

```
. constraint 1 _b[ARCH_toyota:L.arch] = _b[ARCH_nissan:L.arch]
. constraint 2 _b[ARCH_toyota:L.garch] = _b[ARCH_nissan:L.garch]
. mgarch vcc (toyota nissan = , noconstant), arch(1) garch(1) constraints(1 2)
Calculating starting values....

Optimizing log likelihood

(setting technique to bhhh)
Iteration 0:   log likelihood =  10326.298
Iteration 1:   log likelihood =   10680.73
Iteration 2:   log likelihood =  10881.388
Iteration 3:   log likelihood =  11043.345
Iteration 4:   log likelihood =  11122.459
Iteration 5:   log likelihood =  11202.411
Iteration 6:   log likelihood =  11253.657
Iteration 7:   log likelihood =  11276.325
Iteration 8:   log likelihood =  11279.823
Iteration 9:   log likelihood =  11281.704
(switching technique to nr)
Iteration 10:  log likelihood =  11282.313
Iteration 11:  log likelihood =   11282.46
Iteration 12:  log likelihood =  11282.461

Refining estimates

Iteration 0:   log likelihood =  11282.461
Iteration 1:   log likelihood =  11282.461  (backed up)

Varying conditional correlation MGARCH model

Sample: 1 - 2015                               Number of obs     =       2015
Distribution: Gaussian                         Wald chi2(.)      =          .
Log likelihood = 11282.46                      Prob > chi2       =          .

 ( 1)  [ARCH_toyota]L.arch - [ARCH_nissan]L.arch = 0
 ( 2)  [ARCH_toyota]L.garch - [ARCH_nissan]L.garch = 0
```

	Coef.	Std. Err.	z	P>|z|	[95% Conf. Interval]	
ARCH_toyota						
arch						
L1.	.0797459	.0101634	7.85	0.000	.059826	.0996659
garch						
L1.	.9063808	.0118211	76.67	0.000	.883212	.9295497
_cons	4.24e-06	1.10e-06	3.85	0.000	2.08e-06	6.40e-06
ARCH_nissan						
arch						
L1.	.0797459	.0101634	7.85	0.000	.059826	.0996659
garch						
L1.	.9063808	.0118211	76.67	0.000	.883212	.9295497
_cons	5.91e-06	1.47e-06	4.03	0.000	3.03e-06	8.79e-06

Correlation						
toyota						
nissan	.6720056	.0162585	41.33	0.000	.6401394	.7038718
Adjustment						
lambda1	.0343012	.0128097	2.68	0.007	.0091945	.0594078
lambda2	.7945548	.101067	7.86	0.000	.5964671	.9926425

We could test our constraints by fitting the unconstrained model and performing a likelihood-ratio test. The results indicate that the restricted model is preferable.

◁

▷ Example 4

In this example, we have data on fictional stock returns for the Acme and Anvil corporations, and we believe that the movement of the two stocks is governed by different processes. We specify one ARCH and one GARCH term for the conditional variance equation for Acme and two ARCH terms for the conditional variance equation for Anvil. In addition, we include the lagged value of the stock return for Apex, the main subsidiary of Anvil corporation, in the variance equation of Anvil. For Acme, we have data on the changes in an index of futures prices of products related to those produced by Acme in `afrelated`. For Anvil, we have data on the changes in an index of futures prices of inputs used by Anvil in `afinputs`.

```
. use http://www.stata-press.com/data/r12/acmeh
. mgarch vcc (acme = afrelated, noconstant arch(1) garch(1))
> (anvil = afinputs, arch(1/2) het(L.apex))
Calculating starting values....
Optimizing log likelihood
(setting technique to bhhh)
Iteration 0:   log likelihood = -13252.793
Iteration 1:   log likelihood = -12859.124
Iteration 2:   log likelihood =  -12522.14
Iteration 3:   log likelihood = -12406.487
Iteration 4:   log likelihood = -12304.275
Iteration 5:   log likelihood = -12273.103
Iteration 6:   log likelihood = -12256.104
Iteration 7:   log likelihood =  -12254.55
Iteration 8:   log likelihood = -12254.482
Iteration 9:   log likelihood = -12254.478
(switching technique to nr)
Iteration 10:  log likelihood = -12254.478
Iteration 11:  log likelihood = -12254.478
Refining estimates
Iteration 0:   log likelihood = -12254.478
Iteration 1:   log likelihood = -12254.478
```

```
Varying conditional correlation MGARCH model
Sample: 1 - 2500                            Number of obs   =        2499
Distribution: Gaussian                      Wald chi2(2)    =     5226.19
Log likelihood = -12254.48                  Prob > chi2     =      0.0000
```

	Coef.	Std. Err.	z	P>|z|	[95% Conf. Interval]	
acme						
afrelated	.9672465	.0510066	18.96	0.000	.8672753	1.067218
ARCH_acme						
arch						
L1.	.0949142	.0147302	6.44	0.000	.0660435	.1237849
garch						
L1.	.7689442	.038885	19.77	0.000	.6927309	.8451574
_cons	2.129468	.464916	4.58	0.000	1.218249	3.040687
anvil						
afinputs	-1.018629	.0145027	-70.24	0.000	-1.047053	-.9902037
_cons	.1015986	.0177952	5.71	0.000	.0667205	.1364766
ARCH_anvil						
arch						
L1.	.4990272	.0243531	20.49	0.000	.4512959	.5467584
L2.	.2839812	.0181966	15.61	0.000	.2483165	.3196459
apex						
L1.	1.897144	.0558791	33.95	0.000	1.787623	2.006665
_cons	.0682724	.0662257	1.03	0.303	-.0615276	.1980724
Correlation						
acme						
anvil	-.6574256	.0294259	-22.34	0.000	-.7150994	-.5997518
Adjustment						
lambda1	.2375029	.0179114	13.26	0.000	.2023971	.2726086
lambda2	.6492072	.0254493	25.51	0.000	.5993274	.6990869

The results indicate that increases in the futures prices for related products lead to higher returns on the Acme stock, and increased input prices lead to lower returns on the Anvil stock. In the conditional variance equation for Anvil, the coefficient on L1.apex is positive and significant, which indicates that an increase in the return on the Apex stock leads to more variability in the return on the Anvil stock.

◁

Saved results

mgarch vcc saves the following in e():

Scalars
e(N)	number of observations
e(k)	number of parameters
e(k_aux)	number of auxiliary parameters
e(k_extra)	number of extra estimates added to _b
e(k_eq)	number of equations in e(b)
e(k_dv)	number of dependent variables
e(df_m)	model degrees of freedom
e(ll)	log likelihood
e(chi2)	χ^2
e(p)	significance
e(estdf)	1 if distribution parameter was estimated, 0 otherwise
e(usr)	user-provided distribution parameter
e(tmin)	minimum time in sample
e(tmax)	maximum time in sample
e(N_gaps)	number of gaps
e(rank)	rank of e(V)
e(ic)	number of iterations
e(rc)	return code
e(converged)	1 if converged, 0 otherwise

Macros
e(cmd)	mgarch
e(model)	vcc
e(cmdline)	command as typed
e(depvar)	names of dependent variables
e(covariates)	list of covariates
e(dv_eqs)	dependent variables with mean equations
e(indeps)	independent variables in each equation
e(tvar)	time variable
e(title)	title in estimation output
e(chi2type)	Wald; type of model χ^2 test
e(vce)	*vcetype* specified in vce()
e(vcetype)	title used to label Std. Err.
e(tmins)	formatted minimum time
e(tmaxs)	formatted maximum time
e(dist)	distribution for error term: gaussian or t
e(arch)	specified ARCH terms
e(garch)	specified GARCH terms
e(technique)	maximization technique
e(properties)	b V
e(estat_cmd)	program used to implement estat
e(predict)	program used to implement predict
e(marginsok)	predictions allowed by margins
e(marginsnotok)	predictions disallowed by margins

Matrices
e(b)	coefficient vector
e(Cns)	constraints matrix
e(ilog)	iteration log (up to 20 iterations)
e(gradient)	gradient vector
e(hessian)	Hessian matrix
e(V)	variance–covariance matrix of the estimators
e(pinfo)	parameter information, used by predict

Functions
e(sample)	marks estimation sample

Methods and formulas

mgarch vcc is implemented as an ado-file.

mgarch vcc estimates the parameters of the varying conditional correlation MGARCH model by maximum likelihood. The log-likelihood function based on the multivariate normal distribution for observation t is

$$l_t = -0.5m \log(2\pi) - 0.5 \log \left\{ \det\left(\mathbf{R}_t\right) \right\} - \log \left\{ \det\left(\mathbf{D}_t^{1/2}\right) \right\} - 0.5 \widetilde{\epsilon}_t \mathbf{R}_t^{-1} \widetilde{\epsilon}_t'$$

where $\widetilde{\epsilon}_t = \mathbf{D}_t^{-1/2} \epsilon_t$ is an $m \times 1$ vector of standardized residuals, $\epsilon_t = \mathbf{y}_t - \mathbf{C}\mathbf{x}_t$. The log-likelihood function is $\sum_{t=1}^{T} l_t$.

If we assume that ν_t follow a multivariate t distribution with degrees of freedom (df) greater than 2, then the log-likelihood function for observation t is

$$l_t = \log \Gamma \left(\frac{\mathrm{df} + m}{2} \right) - \log \Gamma \left(\frac{\mathrm{df}}{2} \right) - \frac{m}{2} \log \left\{ (\mathrm{df} - 2)\pi \right\}$$

$$- 0.5 \log \left\{ \det\left(\mathbf{R}_t\right) \right\} - \log \left\{ \det\left(\mathbf{D}_t^{1/2}\right) \right\} - \frac{\mathrm{df} + m}{2} \log \left(1 + \frac{\widetilde{\epsilon}_t \mathbf{R}_t^{-1} \widetilde{\epsilon}_t'}{\mathrm{df} - 2} \right)$$

The starting values for the parameters in the mean equations and the initial residuals $\widehat{\epsilon}_t$ are obtained by least-squares regression. The starting values for the parameters in the variance equations are obtained by a procedure proposed by Gourieroux and Monfort (1997, sec. 6.2.2). The starting values for the parameters in $\mathbf{R}$ are calculated from the standardized residuals $\widetilde{\epsilon}_t$. Given the starting values for the mean and variance equations, the starting values for the parameters λ_1 and λ_2 are obtained from a grid search performed on the log likelihood.

The initial optimization step is performed in the unconstrained space. Once the maximum is found, we impose the constraints $\lambda_1 \geq 0$, $\lambda_2 \geq 0$, and $0 \leq \lambda_1 + \lambda_2 < 1$, and maximize the log likelihood in the constrained space. This step is reported in the iteration log as the refining step.

GARCH estimators require initial values that can be plugged in for $\epsilon_{t-i}\epsilon_{t-i}'$ and $\mathbf{H}_{t-j}$ when $t - i < 1$ and $t - j < 1$. mgarch vcc substitutes an estimator of the unconditional covariance of the disturbances

$$\widehat{\boldsymbol{\Sigma}} = T^{-1} \sum_{t=1}^{T} \widehat{\widetilde{\epsilon}}_t \widehat{\widetilde{\epsilon}}_t' \tag{2}$$

for $\epsilon_{t-i}\epsilon_{t-i}'$ when $t - i < 1$ and for $\mathbf{H}_{t-j}$ when $t - j < 1$, where $\widehat{\widetilde{\epsilon}}_t$ is the vector of residuals calculated using the estimated parameters.

mgarch vcc uses numerical derivatives in maximizing the log-likelihood function.

References

Engle, R. F. 2002. Dynamic conditional correlation: A simple class of multivariate generalized autoregressive conditional heteroskedasticity models. *Journal of Business & Economic Statistics* 20: 339–350.

Gourieroux, C., and A. Monfort. 1997. *Time Series and Dynamic Models*. Trans. ed. G. M. Gallo. Cambridge: Cambridge University Press.

Tse, Y. K., and A. K. C. Tsui. 2002. A multivariate generalized autoregressive conditional heteroscedasticity model with time-varying correlations. *Journal of Business & Economic Statistics* 20: 351–362.

Also see

[TS] **mgarch vcc postestimation** — Postestimation tools for mgarch vcc

[TS] **mgarch** — Multivariate GARCH models

[TS] **tsset** — Declare data to be time-series data

[TS] **arch** — Autoregressive conditional heteroskedasticity (ARCH) family of estimators

[TS] **var** — Vector autoregressive models

[U] **20 Estimation and postestimation commands**

Title

> **mgarch vcc postestimation** — Postestimation tools for mgarch vcc

Description

The following standard postestimation commands are available after `mgarch vcc`:

Command	Description
contrast	contrasts and ANOVA-style joint tests of estimates
estat	AIC, BIC, VCE, and estimation sample summary
estimates	cataloging estimation results
lincom	point estimates, standard errors, testing, and inference for linear combinations of coefficients
lrtest	likelihood-ratio test
margins	marginal means, predictive margins, marginal effects, and average marginal effects
marginsplot	graph the results from margins (profile plots, interaction plots, etc.)
nlcom	point estimates, standard errors, testing, and inference for nonlinear combinations of coefficients
predict	predictions and residuals
predictnl	point estimates, standard errors, testing, and inference for generalized predictions
pwcompare	pairwise comparisons of estimates
test	Wald tests of simple and composite linear hypotheses
testnl	Wald tests of nonlinear hypotheses

See the corresponding entries in the *Base Reference Manual* for details.

Syntax for predict

> predict [*type*] { *stub** | *newvarlist* } [*if*] [*in*] [, *statistic options*]

statistic	Description
Main	
xb	linear prediction; the default
<u>res</u>iduals	residuals
<u>var</u>iance	conditional variances and covariances

These statistics are available both in and out of sample; type `predict ... if e(sample) ...` if wanted only for the estimation sample.

options	Description
Options	
<u>eq</u>uation(*eqnames*)	names of equations for which predictions are made
<u>dynamic</u>(*time_constant*)	begin dynamic forecast at specified time

Menu

Statistics > Postestimation > Predictions, residuals, etc.

Options for predict

⌐ Main ⌐

xb, the default, calculates the linear predictions of the dependent variables.

residuals calculates the residuals.

variance predicts the conditional variances and conditional covariances.

⌐ Options ⌐

equation(*eqnames*) specifies the equation for which the predictions are calculated. Use this option to predict a statistic for a particular equation. Equation names, such as equation(income), are used to identify equations.

One equation name may be specified when predicting the dependent variable, the residuals, or the conditional variance. For example, specifying equation(income) causes predict to predict income, and specifying variance equation(income) causes predict to predict the conditional variance of income.

Two equations may be specified when predicting a conditional variance or covariance. For example, specifying equation(income, consumption) variance causes predict to predict the conditional covariance of income and consumption.

dynamic(*time_constant*) specifies when predict starts producing dynamic forecasts. The specified *time_constant* must be in the scale of the time variable specified in tsset, and the *time_constant* must be inside a sample for which observations on the dependent variables are available. For example, dynamic(tq(2008q4)) causes dynamic predictions to begin in the fourth quarter of 2008, assuming that your time variable is quarterly; see [D] **datetime**. If the model contains exogenous variables, they must be present for the whole predicted sample. dynamic() may not be specified with residuals.

Remarks

We assume that you have already read [TS] **mgarch vcc**. In this entry, we use predict after mgarch vcc to make in-sample and out-of-sample forecasts.

> Example 1

In this example, we obtain dynamic forecasts for the Toyota, Nissan, and Honda stock returns modeled in example 2 of [TS] **mgarch vcc**. In the output below, we reestimate the parameters of the model, use tsappend (see [TS] **tsappend**) to extend the data, and use predict to obtain in-sample one-step-ahead forecasts and dynamic forecasts of the conditional variances of the returns. We graph the forecasts below.

```
. use http://www.stata-press.com/data/r12/stocks
(Data from Yahoo! Finance)
. quietly mgarch vcc (toyota nissan = , noconstant) (honda = L.nissan, noconstant),
> arch(1) garch(1)
. tsappend, add(50)
. predict H*, variance dynamic(2016)
```

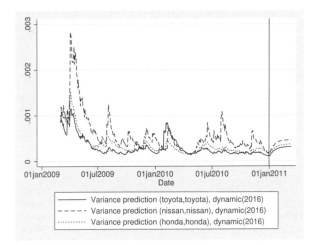

Recent in-sample one-step-ahead forecasts are plotted to the left of the vertical line in the above graph, and the dynamic out-of-sample forecasts appear to the right of the vertical line. The graph shows the tail end of the huge increase in return volatility that took place in 2008 and 2009. It also shows that the dynamic forecasts quickly converge.

◁

Methods and formulas

All postestimation commands listed above are implemented as ado-files.

All one-step predictions are obtained by substituting the parameter estimates into the model. The estimated unconditional variance matrix of the disturbances, $\widehat{\Sigma}$, is the initial value for the ARCH and GARCH terms. The postestimation routines recompute $\widehat{\Sigma}$ using the prediction sample, the parameter estimates stored in e(b), and (2) in *Methods and formulas* of [TS] **mgarch vcc**.

For observations in which the residuals are missing, the estimated unconditional variance matrix of the disturbances is used in place of the outer product of the residuals.

Dynamic predictions of the dependent variables use previously predicted values beginning in the period specified by dynamic().

Dynamic variance predictions are implemented by substituting $\widehat{\Sigma}$ for the outer product of the residuals beginning in the period specified in dynamic().

Also see

[TS] **mgarch vcc** — Varying conditional correlation multivariate GARCH models

[U] **20 Estimation and postestimation commands**

Title

> **newey** — Regression with Newey–West standard errors

Syntax

newey *depvar* [*indepvars*] [*if*] [*in*] [*weight*] , lag(#) [*options*]

options	Description
Model	
*lag(#)	set maximum lag order of autocorrelation
<u>nocon</u>stant	suppress constant term
Reporting	
<u>level</u>(#)	set confidence level; default is level(95)
display_options	control column formats, row spacing, line width, and display of omitted variables and base and empty cells
<u>coefl</u>egend	display legend instead of statistics

*lag(#) is required.

You must tsset your data before using newey; see [TS] **tsset**.

indepvars may contain factor variables; see [U] **11.4.3 Factor variables**.

depvar and *indepvars* may contain time-series operators; see [U] **11.4.4 Time-series varlists**.

by, rolling, and statsby are allowed; see [U] **11.1.10 Prefix commands**.

aweights are allowed; see [U] **11.1.6 weight**.

coeflegend does not appear in the dialog box.

See [U] **20 Estimation and postestimation commands** for more capabilities of estimation commands.

Menu

Statistics > Time series > Regression with Newey-West std. errors

Description

newey produces Newey–West standard errors for coefficients estimated by OLS regression. The error structure is assumed to be heteroskedastic and possibly autocorrelated up to some lag.

Options

> ⌐ Model ⌐

lag(#) specifies the maximum lag to be considered in the autocorrelation structure. If you specify lag(0), the output is the same as regress, vce(robust). lag() is required.

noconstant; see [R] **estimation options**.

```
     ┌ Reporting ┐
```

level(#); see [R] **estimation options**.

display_options: <u>noomit</u>ted, vsquish, noemptycells, <u>baselevels</u>, <u>allbaselevels</u>,
cformat(%*fmt*), pformat(%*fmt*), sformat(%*fmt*), and nolstretch; see [R] **estimation options**.

The following option is available with newey but is not shown in the dialog box:

coeflegend; see [R] **estimation options**.

Remarks

The Huber/White/sandwich robust variance estimator (see White [1980]) produces consistent standard errors for OLS regression coefficient estimates in the presence of heteroskedasticity. The Newey–West (1987) variance estimator is an extension that produces consistent estimates when there is autocorrelation in addition to possible heteroskedasticity.

The Newey–West variance estimator handles autocorrelation up to and including a lag of m, where m is specified by stipulating the lag() option. Thus, it assumes that any autocorrelation at lags greater than m can be ignored.

If lag(0) is specified, the variance estimates produced by newey are simply the Huber/White/sandwich robust variances estimates calculated by regress, vce(robust); see [R] **regress**.

▷ Example 1

newey, lag(0) is equivalent to regress, vce(robust):

```
. use http://www.stata-press.com/data/r12/auto
(1978 Automobile Data)
. regress price weight displ, vce(robust)
Linear regression                               Number of obs =      74
                                                F(  2,    71) =   14.44
                                                Prob > F      =  0.0000
                                                R-squared     =  0.2909
                                                Root MSE      =  2518.4
```

price	Coef.	Robust Std. Err.	t	P>|t|	[95% Conf. Interval]	
weight	1.823366	.7808755	2.34	0.022	.2663445	3.380387
displacement	2.087054	7.436967	0.28	0.780	-12.74184	16.91595
_cons	247.907	1129.602	0.22	0.827	-2004.455	2500.269

```
. generate t = _n
. tsset t
        time variable:  t, 1 to 74
                delta:  1 unit
```

```
. newey price weight displ, lag(0)
```

Regression with Newey-West standard errors   Number of obs =   74
maximum lag: 0   F(2,   71) =   14.44
                                    Prob > F   =   0.0000

price	Coef.	Newey-West Std. Err.	t	P>|t|	[95% Conf. Interval]	
weight	1.823366	.7808755	2.34	0.022	.2663445	3.380387
displacement	2.087054	7.436967	0.28	0.780	-12.74184	16.91595
_cons	247.907	1129.602	0.22	0.827	-2004.455	2500.269

Because `newey` requires the dataset to be `tsset`, we generated a dummy time variable `t`, which in this example played no role in the estimation.

◁

▷ Example 2

Say that we have time-series measurements on variables `usr` and `idle` and now wish to fit an OLS model but obtain Newey–West standard errors allowing for a lag of up to 3:

```
. use http://www.stata-press.com/data/r12/idle2, clear
. tsset time
        time variable:  time, 1 to 30
                delta:  1 unit
. newey usr idle, lag(3)
```

Regression with Newey-West standard errors   Number of obs =   30
maximum lag: 3   F(1,   28) =   10.90
                                      Prob > F   =   0.0026

usr	Coef.	Newey-West Std. Err.	t	P>|t|	[95% Conf. Interval]	
idle	-.2281501	.0690927	-3.30	0.003	-.3696801	-.08662
_cons	23.13483	6.327031	3.66	0.001	10.17449	36.09516

◁

Saved results

newey saves the following in e():

Scalars
e(N)	number of observations
e(df_m)	model degrees of freedom
e(df_r)	residual degrees of freedom
e(F)	F statistic
e(lag)	maximum lag
e(rank)	rank of e(V)

Macros
e(cmd)	newey
e(cmdline)	command as typed
e(depvar)	name of dependent variable
e(wtype)	weight type
e(wexp)	weight expression
e(title)	title in estimation output
e(vcetype)	title used to label Std. Err.
e(properties)	b V
e(estat_cmd)	program used to implement estat
e(predict)	program used to implement predict
e(asbalanced)	factor variables fvset as asbalanced
e(asobserved)	factor variables fvset as asobserved

Matrices
e(b)	coefficient vector
e(Cns)	constraints matrix
e(V)	variance–covariance matrix of the estimators

Functions
e(sample)	marks estimation sample

Methods and formulas

newey is implemented as an ado-file.

newey calculates the estimates

$$\widehat{\beta}_{\text{OLS}} = (\mathbf{X}'\mathbf{X})^{-1}\mathbf{X}'\mathbf{y}$$
$$\widehat{\text{Var}}(\widehat{\beta}_{\text{OLS}}) = (\mathbf{X}'\mathbf{X})^{-1}\mathbf{X}'\widehat{\mathbf{\Omega}}\mathbf{X}(\mathbf{X}'\mathbf{X})^{-1}$$

That is, the coefficient estimates are simply those of OLS linear regression.

For lag(0) (no autocorrelation), the variance estimates are calculated using the White formulation:

$$\mathbf{X}'\widehat{\mathbf{\Omega}}\mathbf{X} = \mathbf{X}'\widehat{\mathbf{\Omega}}_0\mathbf{X} = \frac{n}{n-k}\sum_i \widehat{e}_i^2 \mathbf{x}_i'\mathbf{x}_i$$

Here $\widehat{e}_i = y_i - \mathbf{x}_i\widehat{\beta}_{\text{OLS}}$, where $\mathbf{x}_i$ is the ith row of the $\mathbf{X}$ matrix, n is the number of observations, and k is the number of predictors in the model, including the constant if there is one. The above formula is the same as that used by regress, vce(robust) with the regression-like formula (the default) for the multiplier q_c; see *Methods and formulas* of [R] **regress**.

For $\text{lag}(m)$, $m > 0$, the variance estimates are calculated using the Newey–West (1987) formulation

$$\mathbf{X}'\widehat{\mathbf{\Omega}}\mathbf{X} = \mathbf{X}'\widehat{\mathbf{\Omega}}_0\mathbf{X} + \frac{n}{n-k} \sum_{l=1}^{m} \left(1 - \frac{l}{m+1}\right) \sum_{t=l+1}^{n} \widehat{e}_t\widehat{e}_{t-l}(\mathbf{x}_t'\mathbf{x}_{t-l} + \mathbf{x}_{t-l}'\mathbf{x}_t)$$

where $\mathbf{x}_t$ is the row of the $\mathbf{X}$ matrix observed at time t.

Whitney K. Newey (1954–) earned degrees in economics at Brigham Young University and MIT. After a period at Princeton, he returned to MIT as a professor in 1990. His interests in theoretical and applied econometrics include bootstrapping, nonparametric estimation of models, semiparametric models, and choosing the number of instrumental variables.

Kenneth D. West (1953–) earned a bachelor's degree in economics and mathematics at Wesleyan University and then a PhD in economics at MIT. After a period at Princeton, he joined the University of Wisconsin in 1988. His interests include empirical macroeconomics and time-series econometrics.

References

Hardin, J. W. 1997. sg72: Newey–West standard errors for probit, logit, and poisson models. *Stata Technical Bulletin* 39: 32–35. Reprinted in *Stata Technical Bulletin Reprints*, vol. 7, pp. 182–186. College Station, TX: Stata Press.

Newey, W. K., and K. D. West. 1987. A simple, positive semi-definite, heteroskedasticity and autocorrelation consistent covariance matrix. *Econometrica* 55: 703–708.

White, H. 1980. A heteroskedasticity-consistent covariance matrix estimator and a direct test for heteroskedasticity. *Econometrica* 48: 817–838.

Also see

[TS] **newey postestimation** — Postestimation tools for newey

[TS] **tsset** — Declare data to be time-series data

[TS] **arima** — ARIMA, ARMAX, and other dynamic regression models

[R] **regress** — Linear regression

[U] **20 Estimation and postestimation commands**

Title

newey postestimation — Postestimation tools for newey

Description

The following postestimation commands are available after `newey`:

Command	Description
contrast	contrasts and ANOVA-style joint tests of estimates
estat	VCE and estimation sample summary
estimates	cataloging estimation results
lincom	point estimates, standard errors, testing, and inference for linear combinations of coefficients
linktest	link test for model specification
margins	marginal means, predictive margins, marginal effects, and average marginal effects
marginsplot	graph the results from margins (profile plots, interaction plots, etc.)
nlcom	point estimates, standard errors, testing, and inference for nonlinear combinations of coefficients
predict	predictions, residuals, influence statistics, and other diagnostic measures
predictnl	point estimates, standard errors, testing, and inference for generalized predictions
pwcompare	pairwise comparisons of estimates
test	Wald tests of simple and composite linear hypotheses
testnl	Wald tests of nonlinear hypotheses

See the corresponding entries in the *Base Reference Manual* for details.

Syntax for predict

predict [*type*] *newvar* [*if*] [*in*] [, *statistic*]

statistic	Description
Main	
xb	linear prediction; the default
stdp	standard error of the linear prediction
<u>res</u>iduals	residuals

These statistics are available both in and out of sample; type predict ... if e(sample) ... if wanted only for the estimation sample.

Menu

Statistics > Postestimation > Predictions, residuals, etc.

304

Options for predict

‸⌐ Main ⌐

xb, the default, calculates the linear prediction.

stdp calculates the standard error of the linear prediction.

residuals calculates the residuals.

Methods and formulas

All postestimation commands listed above are implemented as ado-files.

Also see

[TS] **newey** — Regression with Newey–West standard errors

[U] **20 Estimation and postestimation commands**

Title

> **pergram** — Periodogram

Syntax

> pergram *varname* [*if*] [*in*] [, *options*]

options	Description
Main	
<u>g</u>enerate(*newvar*)	generate *newvar* to contain the raw periodogram values
Plot	
cline_options	affect rendition of the plotted points connected by lines
marker_options	change look of markers (color, size, etc.)
marker_label_options	add marker labels; change look or position
Add plots	
addplot(*plot*)	add other plots to the generated graph
Y axis, X axis, Titles, Legend, Overall	
twoway_options	any options other than by() documented in [G-3] ***twoway_options***
nograph	suppress the graph

> You must tsset your data before using pergram; see [TS] **tsset**. Also, the time series must be dense (nonmissing with no gaps in the time variable) in the specified sample.
>
> *varname* may contain time-series operators; see [U] **11.4.4 Time-series varlists**.
>
> nograph does not appear in the dialog box.

Menu

Statistics > Time series > Graphs > Periodogram

Description

pergram plots the log-standardized periodogram for a dense time series.

Options

> ⌐ Main ⌐

generate(*newvar*) specifies a new variable to contain the raw periodogram values. The generated graph log-transforms and scales the values by the sample variance and then truncates them to the $[-6, 6]$ interval before graphing them.

306

_____ ⌐ Plot ⌐_____

cline_options affect the rendition of the plotted points connected by lines; see [G-3] ***cline_options***.

marker_options specify the look of markers. This look includes the marker symbol, the marker size, and its color and outline; see [G-3] ***marker_options***.

marker_label_options specify if and how the markers are to be labeled; see [G-3] ***marker_label_options***.

_____ ⌐ Add plots ⌐_____

addplot(*plot*) adds specified plots to the generated graph; see [G-3] ***addplot_option***.

_____ ⌐ Y axis, X axis, Titles, Legend, Overall ⌐_____

twoway_options are any of the options documented in [G-3] ***twoway_options***, excluding by(). These include options for titling the graph (see [G-3] ***title_options***) and for saving the graph to disk (see [G-3] ***saving_option***).

The following option is available with pergram but is not shown in the dialog box:

nograph prevents pergram from constructing a graph.

Remarks

A good discussion of the periodogram is provided in Chatfield (2004), Hamilton (1994), and Newton (1988). Chatfield is also a good introductory reference for time-series analysis. Another classic reference is Box, Jenkins, and Reinsel (2008). pergram produces a scatterplot in which the points of the scatterplot are connected. The points themselves represent the log-standardized periodogram, and the connections between points represent the (continuous) log-standardized sample spectral density.

In the following examples, we present the periodograms with an interpretation of the main features of the plots.

▷ Example 1

We have time-series data consisting of 144 observations on the monthly number of international airline passengers (in thousands) between 1949 and 1960 (Box, Jenkins, and Reinsel 2008, Series G). We can graph the raw series and the log periodogram for these data by typing

```
. use http://www.stata-press.com/data/r12/air2
(TIMESLAB: Airline passengers)

. scatter air time, m(o) c(l)
```

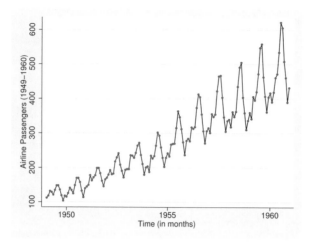

```
. pergram air
```

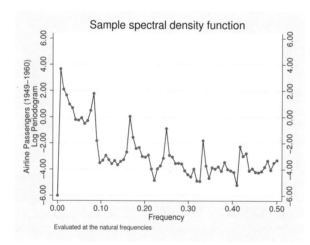

The periodogram highlights the annual cycle together with the harmonics. Notice the peak at a frequency of about 0.08 cycles per month (cpm). The period is the reciprocal of frequency, and the reciprocal of 0.08 cpm is approximately 12 months per cycle. The similarity in shape of each group of 12 observations reveals the annual cycle. The magnitude of the cycle is increasing, resulting in the peaks in the periodogram at the harmonics of the principal annual cycle.

◁

▷ Example 2

This example uses 215 observations on the annual number of sunspots from 1749 to 1963 (Box and Jenkins 1976, Series E). The graph of the raw series and the log periodogram for these data are given as

```
. use http://www.stata-press.com/data/r12/sunspot
(TIMESLAB: Wolfer sunspot data)

. scatter spot time, m(o) c(l)
```

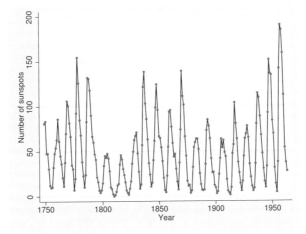

```
. pergram spot
```

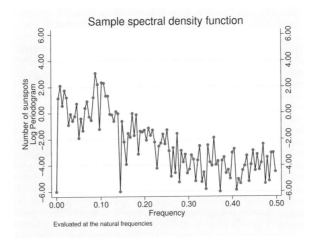

The periodogram peaks at a frequency of slightly less than 0.10 cycles per year, indicating a 10- to 12-year cycle in sunspot activity.

◁

▷ Example 3

Here we examine the number of trapped Canadian lynx from 1821 through 1934 (Newton 1988, 587). The raw series and the log periodogram are given as

```
. use http://www.stata-press.com/data/r12/lynx2
(TIMESLAB: Canadian lynx)

. scatter lynx time, m(o) c(l)
```

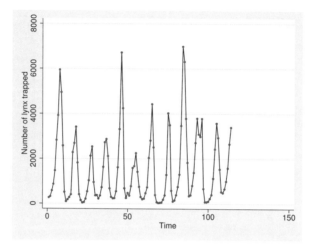

```
. pergram lynx
```

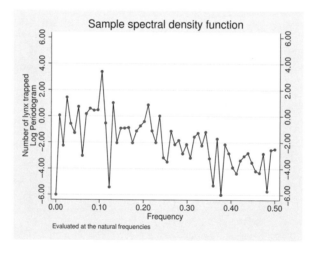

The periodogram indicates that there is a cycle with a duration of about 10 years for these data but that it is otherwise random.

◁

▷ Example 4

To more clearly highlight what the periodogram depicts, we present the result of analyzing a time series of the sum of four sinusoids (of different periods). The periodogram should be able to decompose the time series into four different sinusoids whose periods may be determined from the plot.

```
. use http://www.stata-press.com/data/r12/cos4
(TIMESLAB: Sum of 4 Cosines)

. scatter sumfc time, m(o) c(l)
```

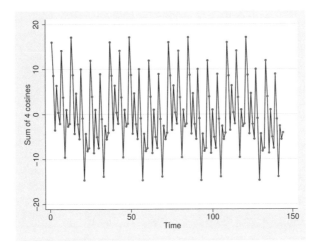

```
. pergram sumfc, gen(ordinate)
```

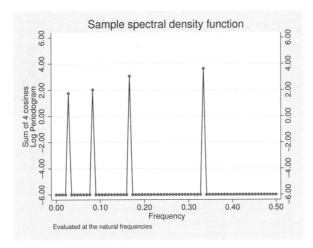

The periodogram clearly shows the four contributions to the original time series. From the plot, we can see that the periods of the summands were 3, 6, 12, and 36, although you can confirm this by using

```
. generate double omega = (_n-1)/144

. generate double period = 1/omega
(1 missing value generated)

. list period omega if ordinate> 1e-5 & omega <=.5
```

	period	omega
5.	36	.02777778
13.	12	.08333333
25.	6	.16666667
49.	3	.33333333

◁

Methods and formulas

pergram is implemented as an ado-file.

We use the notation of Newton (1988) in the following discussion.

A time series of interest is decomposed into a unique set of sinusoids of various frequencies and amplitudes.

A plot of the sinusoidal amplitudes (ordinates) versus the frequencies for the sinusoidal decomposition of a time series gives us the spectral density of the time series. If we calculate the sinusoidal amplitudes for a discrete set of "natural" frequencies $(1/n, 2/n, \ldots, q/n)$, we obtain the periodogram.

Let $x(1), \ldots, x(n)$ be a time series, and let $\omega_k = (k-1)/n$ denote the natural frequencies for $k = 1, \ldots, (n/2) + 1$. Define

$$C_k^2 = \frac{1}{n^2} \left| \sum_{t=1}^{n} x(t) e^{2\pi i (t-1)\omega_k} \right|^2$$

A plot of nC_k^2 versus ω_k is then called the periodogram.

The sample spectral density is defined for a continuous frequency ω as

$$\widehat{f}(\omega) = \begin{cases} \frac{1}{n} \left| \sum_{t=1}^{n} x(t) e^{2\pi i (t-1)\omega} \right|^2 & \text{if } \omega \in [0, .5] \\ \widehat{f}(1 - \omega) & \text{if } \omega \in [.5, 1] \end{cases}$$

The periodogram (and sample spectral density) is symmetric about $\omega = 0.5$. Further standardize the periodogram such that

$$\frac{1}{n} \sum_{k=2}^{n} \frac{nC_k^2}{\widehat{\sigma}^2} = 1$$

where $\widehat{\sigma}^2$ is the sample variance of the time series so that the average value of the ordinate is one.

Once the amplitudes are standardized, we may then take the natural log of the values and produce the log periodogram. In doing so, we truncate the graph at ± 6. We drop the word "log" and simply refer to the "log periodogram" as the "periodogram" in text.

References

Box, G. E. P., and G. M. Jenkins. 1976. *Time Series Analysis: Forecasting and Control.* Oakland, CA: Holden–Day.

Box, G. E. P., G. M. Jenkins, and G. C. Reinsel. 2008. *Time Series Analysis: Forecasting and Control.* 4th ed. Hoboken, NJ: Wiley.

Chatfield, C. 2004. *The Analysis of Time Series: An Introduction.* 6th ed. Boca Raton, FL: Chapman & Hall/CRC.

Hamilton, J. D. 1994. *Time Series Analysis.* Princeton: Princeton University Press.

Newton, H. J. 1988. *TIMESLAB: A Time Series Analysis Laboratory.* Belmont, CA: Wadsworth.

Also see

[TS] **tsset** — Declare data to be time-series data

[TS] **corrgram** — Tabulate and graph autocorrelations

[TS] **cumsp** — Cumulative spectral distribution

[TS] **wntestb** — Bartlett's periodogram-based test for white noise

Title

pperron — Phillips–Perron unit-root test

Syntax

pperron *varname* $[\textit{if}]$ $[\textit{in}]$ $[\,, \textit{options}]$

options	Description
Main	
<u>no</u>constant	suppress constant term
<u>trend</u>	include trend term in regression
<u>regress</u>	display regression table
<u>lags</u>(#)	use # Newey–West lags

You must tsset your data before using pperron; see [TS] **tsset**.

varname may contain time-series operators; see [U] **11.4.4 Time-series varlists**.

Menu

Statistics > Time series > Tests > Phillips-Perron unit-root test

Description

pperron performs the Phillips–Perron (1988) test that a variable has a unit root. The null hypothesis is that the variable contains a unit root, and the alternative is that the variable was generated by a stationary process. pperron uses Newey–West (1987) standard errors to account for serial correlation, whereas the augmented Dickey–Fuller test implemented in dfuller (see [TS] **dfuller**) uses additional lags of the first-differenced variable.

Options

⌐ Main ⌐

noconstant suppresses the constant term (intercept) in the model.

trend specifies that a trend term be included in the associated regression. This option may not be specified if noconstant is specified.

regress specifies that the associated regression table appear in the output. By default, the regression table is not produced.

lags(#) specifies the number of Newey–West lags to use in calculating the standard error. The default is to use $\text{int}\left\{4(T/100)^{2/9}\right\}$ lags.

Remarks

As noted in [TS] **dfuller**, the Dickey–Fuller test involves fitting the regression model

$$y_t = \alpha + \rho y_{t-1} + \delta t + u_t \tag{1}$$

by ordinary least squares (OLS), but serial correlation will present a problem. To account for this, the augmented Dickey–Fuller test's regression includes lags of the first differences of y_t.

The Phillips–Perron test involves fitting (1), and the results are used to calculate the test statistics. Phillips and Perron (1988) proposed two alternative statistics, which pperron presents. Phillips and Perron's test statistics can be viewed as Dickey–Fuller statistics that have been made robust to serial correlation by using the Newey–West (1987) heteroskedasticity- and autocorrelation-consistent covariance matrix estimator.

Hamilton (1994, chap. 17) and [TS] **dfuller** discuss four different cases into which unit-root tests can be classified. The Phillips–Perron test applies to cases one, two, and four but not to case three. Cases one and two assume that the variable has a unit root without drift under the null hypothesis, the only difference being whether the constant term α is included in regression (1). Case four assumes that the variable has a random walk, with or without drift, under the null hypothesis. Case three, which assumes that the variable has a random walk with drift under the null hypothesis, is just a special case of case four, so the fact that the Phillips–Perron test does not apply is not restrictive. The table below summarizes the relevant cases:

Case	Process under null hypothesis	Regression restrictions	dfuller option
1	Random walk without drift	$\alpha = 0,\ \delta = 0$	noconstant
2	Random walk without drift	$\delta = 0$	(default)
4	Random walk with or without drift	(none)	trend

The critical values for the Phillips–Perron test are the same as those for the augmented Dickey–Fuller test. See Hamilton (1994, chap. 17) for more information.

▷ Example 1

Here we use the international airline passengers dataset (Box, Jenkins, and Reinsel 2008, Series G). This dataset has 144 observations on the monthly number of international airline passengers from 1949 through 1960. Because the data exhibit a clear upward trend over time, we will use the trend option.

```
. use http://www.stata-press.com/data/r12/air2
(TIMESLAB: Airline passengers)

. pperron air, lags(4) trend regress
```

| Phillips-Perron test for unit root | | Number of obs = | 143 |
| | | Newey-West lags = | 4 |

	Test Statistic	1% Critical Value	5% Critical Value	10% Critical Value
		── Interpolated Dickey-Fuller ──		
Z(rho)	-46.405	-27.687	-20.872	-17.643
Z(t)	-5.049	-4.026	-3.444	-3.144

MacKinnon approximate p-value for Z(t) = 0.0002

air	Coef.	Std. Err.	t	P>|t|	[95% Conf. Interval]	
air						
L1.	.7318116	.0578092	12.66	0.000	.6175196	.8461035
_trend	.7107559	.1670563	4.25	0.000	.3804767	1.041035
_cons	25.95168	7.325951	3.54	0.001	11.46788	40.43547

Just as in the example in [TS] **dfuller**, we reject the null hypothesis of a unit root at all common significance levels. The interpolated critical values for Z_t differ slightly from those shown in the example in [TS] **dfuller** because the sample sizes are different: with the augmented Dickey–Fuller regression we lose observations because of the inclusion of lagged difference terms as regressors.

◁

Saved results

pperron saves the following in r():

Scalars

r(N)	number of observations	r(Zt)	Phillips–Perron τ test statistic
r(lags)	number of lagged differences used	r(Zrho)	Phillips–Perron ρ test statistic (not included if noconstant specified)
r(pval)	MacKinnon approximate p-value		

Methods and formulas

pperron is implemented as an ado-file.

In the OLS estimation of an AR(1) process with Gaussian errors,

$$y_i = \rho y_{i-1} + \epsilon_i$$

where ϵ_i are independently and identically distributed as $N(0, \sigma^2)$ and $y_0 = 0$, the OLS estimate (based on an n-observation time series) of the autocorrelation parameter ρ is given by

$$\widehat{\rho}_n = \frac{\sum_{i=1}^{n} y_{i-1} y_i}{\sum_{i=1}^{n} y_i^2}$$

If $|\rho| < 1$, then $\sqrt{n}(\widehat{\rho}_n - \rho) \to N(0, 1 - \rho^2)$. If this result were valid for when $\rho = 1$, then the resulting distribution would have a variance of zero. When $\rho = 1$, the OLS estimate $\widehat{\rho}$ still converges to one, though we need to find a nondegenerate distribution so that we can test H_0: $\rho = 1$. See Hamilton (1994, chap. 17).

The Phillips–Perron test involves fitting the regression

$$y_i = \alpha + \rho y_{i-1} + \epsilon_i$$

where we may exclude the constant or include a trend term. There are two statistics, Z_ρ and Z_τ, calculated as

$$Z_\rho = n(\widehat{\rho}_n - 1) - \frac{1}{2} \frac{n^2 \widehat{\sigma}^2}{s_n^2} \left(\widehat{\lambda}_n^2 - \widehat{\gamma}_{0,n} \right)$$

$$Z_\tau = \sqrt{\frac{\widehat{\gamma}_{0,n}}{\widehat{\lambda}_n^2}} \frac{\widehat{\rho}_n - 1}{\widehat{\sigma}} - \frac{1}{2} \left(\widehat{\lambda}_n^2 - \widehat{\gamma}_{0,n} \right) \frac{1}{\widehat{\lambda}_n} \frac{n\widehat{\sigma}}{s_n}$$

$$\widehat{\gamma}_{j,n} = \frac{1}{n} \sum_{i=j+1}^{n} \widehat{u}_i \widehat{u}_{i-j}$$

$$\widehat{\lambda}_n^2 = \widehat{\gamma}_{0,n} + 2 \sum_{j=1}^{q} \left(1 - \frac{j}{q+1} \right) \widehat{\gamma}_{j,n}$$

$$s_n^2 = \frac{1}{n-k} \sum_{i=1}^{n} \widehat{u}_i^2$$

where u_i is the OLS residual, k is the number of covariates in the regression, q is the number of Newey–West lags to use in calculating $\widehat{\lambda}_n^2$, and $\widehat{\sigma}$ is the OLS standard error of $\widehat{\rho}$.

The critical values, which have the same distribution as the Dickey–Fuller statistic (see Dickey and Fuller 1979) included in the output, are linearly interpolated from the table of values that appear in Fuller (1996), and the MacKinnon approximate p-values use the regression surface published in MacKinnon (1994).

Peter Charles Bonest Phillips (1948–) was born in Weymouth, England, and earned degrees in economics at the University of Auckland in New Zealand, and the London School of Economics. After periods at the Universities of Essex and Birmingham, Phillips moved to Yale in 1979. He also holds appointments at the University of Auckland and the University of York. His main research interests are in econometric theory, financial econometrics, time-series and panel-data econometrics, and applied macroeconomics.

Pierre Perron (1959–) was born in Québec, Canada, and earned degrees at McGill, Queen's, and Yale in economics. After posts at Princeton and the Université de Montréal, he joined Boston University in 1997. His research interests include time-series analysis, econometrics, and applied macroeconomics.

References

Box, G. E. P., G. M. Jenkins, and G. C. Reinsel. 2008. *Time Series Analysis: Forecasting and Control*. 4th ed. Hoboken, NJ: Wiley.

Dickey, D. A., and W. A. Fuller. 1979. Distribution of the estimators for autoregressive time series with a unit root. *Journal of the American Statistical Association* 74: 427–431.

Fuller, W. A. 1996. *Introduction to Statistical Time Series.* 2nd ed. New York: Wiley.

Hamilton, J. D. 1994. *Time Series Analysis.* Princeton: Princeton University Press.

MacKinnon, J. G. 1994. Approximate asymptotic distribution functions for unit-root and cointegration tests. *Journal of Business and Economic Statistics* 12: 167–176.

Newey, W. K., and K. D. West. 1987. A simple, positive semi-definite, heteroskedasticity and autocorrelation consistent covariance matrix. *Econometrica* 55: 703–708.

Phillips, P. C. B., and P. Perron. 1988. Testing for a unit root in time series regression. *Biometrika* 75: 335–346.

Also see

[TS] **tsset** — Declare data to be time-series data

[TS] **dfgls** — DF-GLS unit-root test

[TS] **dfuller** — Augmented Dickey–Fuller unit-root test

[XT] **xtunitroot** — Panel-data unit-root tests

Title

> **prais** — Prais–Winsten and Cochrane–Orcutt regression

Syntax

prais *depvar* [*indepvars*] [*if*] [*in*] [, *options*]

options	Description
Model	
<u>rho</u>type(<u>regress</u>)	base ρ on single-lag OLS of residuals; the default
<u>rho</u>type(freg)	base ρ on single-lead OLS of residuals
<u>rho</u>type(<u>ts</u>corr)	base ρ on autocorrelation of residuals
<u>rho</u>type(dw)	base ρ on autocorrelation based on Durbin–Watson
<u>rho</u>type(<u>the</u>il)	base ρ on adjusted autocorrelation
<u>rho</u>type(<u>nagar</u>)	base ρ on adjusted Durbin–Watson
<u>cor</u>c	use Cochrane–Orcutt transformation
<u>sse</u>search	search for ρ that minimizes SSE
<u>two</u>step	stop after the first iteration
<u>nocon</u>stant	suppress constant term
<u>has</u>cons	has user-defined constant
<u>save</u>space	conserve memory during estimation
SE/Robust	
vce(*vcetype*)	*vcetype* may be ols, <u>r</u>obust, <u>cl</u>uster *clustvar*, hc2, or hc3
Reporting	
<u>l</u>evel(*#*)	set confidence level; default is level(95)
nodw	do not report the Durbin–Watson statistic
display_options	control column formats, row spacing, line width, and display of omitted variables and base and empty cells
Optimization	
optimize_options	control the optimization process; seldom used
<u>coefl</u>egend	display legend instead of statistics

You must tsset your data before using prais; see [TS] **tsset**.

indepvars may contain factor variables; see [U] **11.4.3 Factor variables**.

depvar and *indepvars* may contain time-series operators; see [U] **11.4.4 Time-series varlists**.

by, rolling, and statsby are allowed; see [U] **11.1.10 Prefix commands**.

coeflegend does not appear in the dialog box.

See [U] **20 Estimation and postestimation commands** for more capabilities of estimation commands.

Menu

Statistics > Time series > Prais-Winsten regression

Description

    prais uses the generalized least-squares method to estimate the parameters in a linear regression model in which the errors are serially correlated. Specifically, the errors are assumed to follow a first-order autoregressive process.

Options

    ⌐—————⌐ Model ⌐——

rhotype(*rhomethod*) selects a specific computation for the autocorrelation parameter ρ, where *rhomethod* can be

regress	$\rho_{\text{reg}} = \beta$ from the residual regression $\epsilon_t = \beta\epsilon_{t-1}$
freg	$\rho_{\text{freg}} = \beta$ from the residual regression $\epsilon_t = \beta\epsilon_{t+1}$
tscorr	$\rho_{\text{tscorr}} = \epsilon'\epsilon_{t-1}/\epsilon'\epsilon$, where ϵ is the vector of residuals
dw	$\rho_{\text{dw}} = 1 - \text{dw}/2$, where dw is the Durbin–Watson d statistic
theil	$\rho_{\text{theil}} = \rho_{\text{tscorr}}(N - k)/N$
nagar	$\rho_{\text{nagar}} = (\rho_{\text{dw}} * N^2 + k^2)/(N^2 - k^2)$

    The prais estimator can use any consistent estimate of ρ to transform the equation, and each of these estimates meets that requirement. The default is regress, which produces the minimum sum-of-squares solution (ssesearch option) for the Cochrane–Orcutt transformation—none of these computations will produce the minimum sum-of-squares solution for the full Prais–Winsten transformation. See Judge et al. (1985) for a discussion of each estimate of ρ.

corc specifies that the Cochrane–Orcutt transformation be used to estimate the equation. With this option, the Prais–Winsten transformation of the first observation is not performed, and the first observation is dropped when estimating the transformed equation; see *Methods and formulas* below.

ssesearch specifies that a search be performed for the value of ρ that minimizes the sum-of-squared errors of the transformed equation (Cochrane–Orcutt or Prais–Winsten transformation). The search method is a combination of quadratic and modified bisection searches using golden sections.

twostep specifies that prais stop on the first iteration after the equation is transformed by ρ—the two-step efficient estimator. Although iterating these estimators to convergence is customary, they are efficient at each step.

noconstant; see [R] **estimation options**.

hascons indicates that a user-defined constant, or a set of variables that in linear combination forms a constant, has been included in the regression. For some computational concerns, see the discussion in [R] **regress**.

savespace specifies that prais attempt to save as much space as possible by retaining only those variables required for estimation. The original data are restored after estimation. This option is rarely used and should be used only if there is insufficient space to fit a model without the option.

    ⌐—————⌐ SE/Robust ⌐————————————————————————————————————

vce(*vcetype*) specifies the type of standard error reported, which includes types that are derived from asymptotic theory, that are robust to some kinds of misspecification, and that allow for intragroup correlation; see [R] *vce_option*.

    vce(ols), the default, uses the standard variance estimator for ordinary least squares regression.

prais also allows the following:

vce(hc2) and vce(hc3) specify an alternative bias correction for the vce(robust) variance calculation; for more information, see [R] **regress**. You may specify only one of vce(hc2), vce(hc3), or vce(robust).

All estimates from prais are conditional on the estimated value of ρ. Robust variance estimates here are robust only to heteroskedasticity and are not generally robust to misspecification of the functional form or omitted variables. The estimation of the functional form is intertwined with the estimation of ρ, and all estimates are conditional on ρ. Thus estimates cannot be robust to misspecification of functional form. For these reasons, it is probably best to interpret vce(robust) in the spirit of White's (1980) original paper on estimation of heteroskedastic-consistent covariance matrices.

Reporting

level(#); see [R] **estimation options**.

nodw suppresses reporting of the Durbin–Watson statistic.

display_options: noomitted, vsquish, noemptycells, baselevels, allbaselevels, cformat(%*fmt*), pformat(%*fmt*), sformat(%*fmt*), and nolstretch; see [R] **estimation options**.

Optimization

optimize_options: iterate(#), [no]log, tolerance(#). iterate() specifies the maximum number of iterations. log/nolog specifies whether to show the iteration log. tolerance() specifies the tolerance for the coefficient vector; tolerance(1e-6) is the default. These options are seldom used.

The following option is available with prais but is not shown in the dialog box:

coeflegend; see [R] **estimation options**.

Remarks

prais fits a linear regression of *depvar* on *indepvars* that is corrected for first-order serially correlated residuals by using the Prais–Winsten (1954) transformed regression estimator, the Cochrane–Orcutt (1949) transformed regression estimator, or a version of the search method suggested by Hildreth and Lu (1960). Davidson and MacKinnon (1993) provide theoretical details on the three methods (see pages 333–335 for the latter two and pages 343–351 for Prais–Winsten).

The most common autocorrelated error process is the first-order autoregressive process. Under this assumption, the linear regression model can be written as

$$y_t = \mathbf{x}_t \beta + u_t$$

where the errors satisfy

$$u_t = \rho\, u_{t-1} + e_t$$

and the e_t are independently and identically distributed as $N(0, \sigma^2)$. The covariance matrix $\boldsymbol{\Psi}$ of the error term e can then be written as

$$
\boldsymbol{\Psi} = \frac{1}{1-\rho^2}
\begin{bmatrix}
1 & \rho & \rho^2 & \cdots & \rho^{T-1} \\
\rho & 1 & \rho & \cdots & \rho^{T-2} \\
\rho^2 & \rho & 1 & \cdots & \rho^{T-3} \\
\vdots & \vdots & \vdots & \ddots & \vdots \\
\rho^{T-1} & \rho^{T-2} & \rho^{T-3} & \cdots & 1
\end{bmatrix}
$$

The Prais–Winsten estimator is a generalized least squares (GLS) estimator. The Prais–Winsten method (as described in Judge et al. 1985) is derived from the AR(1) model for the error term described above. Whereas the Cochrane–Orcutt method uses a lag definition and loses the first observation in the iterative method, the Prais–Winsten method preserves that first observation. In small samples, this can be a significant advantage.

❑ Technical note

To fit a model with autocorrelated errors, you must specify your data as time series and have (or create) a variable denoting the time at which an observation was collected. The data for the regression should be equally spaced in time.

❑

▷ Example 1

Say that we wish to fit a time-series model of usr on idle but are concerned that the residuals may be serially correlated. We will declare the variable t to represent time by typing

```
. use http://www.stata-press.com/data/r12/idle
. tsset t
        time variable:  t, 1 to 30
                delta:  1 unit
```

We can obtain Cochrane–Orcutt estimates by specifying the corc option:

```
. prais usr idle, corc
Iteration 0:  rho = 0.0000
Iteration 1:  rho = 0.3518
  (output omitted )
Iteration 13:  rho = 0.5708
```

Cochrane-Orcutt AR(1) regression -- iterated estimates

Source	SS	df	MS		
Model	40.1309584	1	40.1309584	Number of obs =	29
Residual	166.898474	27	6.18142498	F(1, 27) =	6.49
				Prob > F =	0.0168
				R-squared =	0.1938
				Adj R-squared =	0.1640
Total	207.029433	28	7.39390831	Root MSE =	2.4862

usr	Coef.	Std. Err.	t	P>|t|	[95% Conf. Interval]	
idle	-.1254511	.0492356	-2.55	0.017	-.2264742	-.024428
_cons	14.54641	4.272299	3.40	0.002	5.78038	23.31245
rho	.5707918					

```
Durbin-Watson statistic (original)     1.295766
Durbin-Watson statistic (transformed)  1.466222
```

The fitted model is

$$\text{usr}_t = -0.1255\,\text{idle}_t + 14.55 + u_t \quad \text{and} \quad u_t = 0.5708\,u_{t-1} + e_t$$

We can also fit the model with the Prais–Winsten method,

```
. prais usr idle

Iteration 0:  rho = 0.0000
Iteration 1:  rho = 0.3518
 (output omitted )
Iteration 14:  rho = 0.5535

Prais-Winsten AR(1) regression -- iterated estimates
```

Source	SS	df	MS
Model	43.0076941	1	43.0076941
Residual	169.165739	28	6.04163354
Total	212.173433	29	7.31632528

```
Number of obs =      30
F( 1,    28) =     7.12
Prob > F      =  0.0125
R-squared     =  0.2027
Adj R-squared =  0.1742
Root MSE      =   2.458
```

usr	Coef.	Std. Err.	t	P>|t|	[95% Conf.	Interval]
idle	-.1356522	.0472195	-2.87	0.008	-.2323769	-.0389275
_cons	15.20415	4.160391	3.65	0.001	6.681978	23.72633
rho	.5535476					

```
Durbin-Watson statistic (original)    1.295766
Durbin-Watson statistic (transformed) 1.476004
```

where the Prais–Winsten fitted model is

$$\text{usr}_t = -.1357\,\text{idle}_t + 15.20 + u_t \quad \text{and} \quad u_t = .5535\,u_{t-1} + e_t$$

As the results indicate, for these data there is little difference between the Cochrane–Orcutt and Prais–Winsten estimators, whereas the OLS estimate of the slope parameter is substantially different.

◁

▷ Example 2

We have data on quarterly sales, in millions of dollars, for 5 years, and we would like to use this information to model sales for company X. First, we fit a linear model by OLS and obtain the Durbin–Watson statistic by using estat dwatson; see [R] **regress postestimation time series**.

```
. use http://www.stata-press.com/data/r12/qsales
. regress csales isales
```

Source	SS	df	MS
Model	110.256901	1	110.256901
Residual	.133302302	18	.007405683
Total	110.390204	19	5.81001072

```
Number of obs =      20
F( 1,    18) =14888.15
Prob > F      =  0.0000
R-squared     =  0.9988
Adj R-squared =  0.9987
Root MSE      =  .08606
```

csales	Coef.	Std. Err.	t	P>|t|	[95% Conf.	Interval]
isales	.1762828	.0014447	122.02	0.000	.1732475	.1793181
_cons	-1.454753	.2141461	-6.79	0.000	-1.904657	-1.004849

```
. estat dwatson
Durbin-Watson d-statistic( 2,   20) =  .7347276
```

Because the Durbin–Watson statistic is far from 2 (the expected value under the null hypothesis of no serial correlation) and well below the 5% lower limit of 1.2, we conclude that the disturbances are serially correlated. (Upper and lower bounds for the d statistic can be found in most econometrics texts; for example, Harvey [1990]. The bounds have been derived for only a limited combination of regressors and observations.) To reinforce this conclusion, we use two other tests to test for serial correlation in the error distribution.

```
. estat bgodfrey, lags(1)

Breusch-Godfrey LM test for autocorrelation
```

lags(p)	chi2	df	Prob > chi2
1	7.998	1	0.0047

HO: no serial correlation

```
. estat durbinalt

Durbin's alternative test for autocorrelation
```

lags(p)	chi2	df	Prob > chi2
1	11.329	1	0.0008

HO: no serial correlation

`estat bgodfrey` reports the Breusch–Godfrey Lagrange-multiplier test statistic, and `estat durbinalt` reports the Durbin's alternative test statistic. Both tests give a small p-value and thus reject the null hypothesis of no serial correlation. These two tests are asymptotically equivalent when testing for AR(1) process. See [R] **regress postestimation time series** if you are not familiar with these two tests.

We correct for autocorrelation with the `ssesearch` option of `prais` to search for the value of ρ that minimizes the sum-of-squared residuals of the Cochrane–Orcutt transformed equation. Normally, the default Prais–Winsten transformations is used with such a small dataset, but the less-efficient Cochrane–Orcutt transformation allows us to demonstrate an aspect of the estimator's convergence.

```
. prais csales isales, corc ssesearch

Iteration 1:  rho = 0.8944, criterion =  -.07298558
Iteration 2:  rho = 0.8944, criterion =  -.07298558
  (output omitted)
Iteration 15:  rho = 0.9588, criterion =  -.07167037

Cochrane-Orcutt AR(1) regression -- SSE search estimates
```

Source	SS	df	MS		
Model	2.33199178	1	2.33199178	Number of obs = 19
Residual	.071670369	17	.004215904	F(1, 17) = 553.14
				Prob > F = 0.0000
				R-squared = 0.9702
				Adj R-squared = 0.9684
Total	2.40366215	18	.133536786	Root MSE = .06493

csales	Coef.	Std. Err.	t	P>|t|	[95% Conf. Interval]
isales	.1605233	.0068253	23.52	0.000	.1461233 .1749234
_cons	1.738946	1.432674	1.21	0.241	-1.283732 4.761624
rho	.9588209				

```
Durbin-Watson statistic (original)     0.734728
Durbin-Watson statistic (transformed)  1.724419
```

We noted in *Options* that, with the default computation of ρ, the Cochrane–Orcutt method produces an estimate of ρ that minimizes the sum-of-squared residuals—the same criterion as the ssesearch option. Given that the two methods produce the same results, why would the search method ever be preferred? It turns out that the back-and-forth iterations used by Cochrane–Orcutt may have difficulty converging if the value of ρ is large. Using the same data, the Cochrane–Orcutt iterative procedure requires more than 350 iterations to converge, and a higher tolerance must be specified to prevent premature convergence:

```
. prais csales isales, corc tol(1e-9) iterate(500)

Iteration 0:   rho = 0.0000
Iteration 1:   rho = 0.6312
Iteration 2:   rho = 0.6866
 (output omitted )
Iteration 377:  rho = 0.9588
Iteration 378:  rho = 0.9588
Iteration 379:  rho = 0.9588

Cochrane-Orcutt AR(1) regression -- iterated estimates
```

Source	SS	df	MS
Model	2.33199171	1	2.33199171
Residual	.071670369	17	.004215904
Total	2.40366208	18	.133536782

```
Number of obs =      19
F(  1,    17) =  553.14
Prob > F      =  0.0000
R-squared     =  0.9702
Adj R-squared =  0.9684
Root MSE      =  .06493
```

csales	Coef.	Std. Err.	t	P>|t|	[95% Conf. Interval]
isales	.1605233	.0068253	23.52	0.000	.1461233 .1749234
_cons	1.738946	1.432674	1.21	0.241	-1.283732 4.761625
rho	.9588209				

```
Durbin-Watson statistic (original)      0.734728
Durbin-Watson statistic (transformed)  1.724419
```

Once convergence is achieved, the two methods produce identical results.

◁

Saved results

prais saves the following in e():

Scalars

e(N)	number of observations
e(N_gaps)	number of gaps
e(mss)	model sum of squares
e(df_m)	model degrees of freedom
e(rss)	residual sum of squares
e(df_r)	residual degrees of freedom
e(r2)	R^2
e(r2_a)	adjusted R^2
e(F)	F statistic
e(rmse)	root mean squared error
e(ll)	log likelihood
e(N_clust)	number of clusters
e(rho)	autocorrelation parameter ρ
e(dw)	Durbin–Watson d statistic for transformed regression
e(dw_0)	Durbin–Watson d statistic of untransformed regression
e(tol)	target tolerance
e(max_ic)	maximum number of iterations
e(ic)	number of iterations

Macros

e(cmd)	prais
e(cmdline)	command as typed
e(depvar)	name of dependent variable
e(title)	title in estimation output
e(clustvar)	name of cluster variable
e(cons)	noconstant or not reported
e(method)	twostep, iterated, or SSE search
e(tranmeth)	corc or prais
e(rhotype)	method specified in rhotype() option
e(vce)	*vcetype* specified in vce()
e(vcetype)	title used to label Std. Err.
e(properties)	b V
e(predict)	program used to implement predict
e(marginsok)	predictions allowed by margins
e(asbalanced)	factor variables fvset as asbalanced
e(asobserved)	factor variables fvset as asobserved

Matrices

e(b)	coefficient vector
e(V)	variance–covariance matrix of the estimators
e(V_modelbased)	model-based variance

Functions

e(sample)	estimation sample

Methods and formulas

prais is implemented as an ado-file.

Consider the command 'prais y x z'. The 0th iteration is obtained by estimating a, b, and c from the standard linear regression:

$$y_t = ax_t + bz_t + c + u_t$$

An estimate of the correlation in the residuals is then obtained. By default, prais uses the auxiliary regression:

$$u_t = \rho u_{t-1} + e_t$$

This can be changed to any computation noted in the rhotype() option.

Next we apply a Cochrane–Orcutt transformation (1) for observations $t = 2, \ldots, n$

$$y_t - \rho y_{t-1} = a(x_t - \rho x_{t-1}) + b(z_t - \rho z_{t-1}) + c(1 - \rho) + v_t \tag{1}$$

and the transformation $(1')$ for $t = 1$

$$\sqrt{1 - \rho^2}\, y_1 = a(\sqrt{1 - \rho^2}\, x_1) + b(\sqrt{1 - \rho^2}\, z_1) + c\sqrt{1 - \rho^2} + \sqrt{1 - \rho^2}\, v_1 \tag{1'}$$

Thus the differences between the Cochrane–Orcutt and the Prais–Winsten methods are that the latter uses $(1')$ in addition to (1), whereas the former uses only (1), necessarily decreasing the sample size by one.

Equations (1) and $(1')$ are used to transform the data and obtain new estimates of a, b, and c.

When the `twostep` option is specified, the estimation process stops at this point and reports these estimates. Under the default behavior of iterating to convergence, this process is repeated until the change in the estimate of ρ is within a specified tolerance.

The new estimates are used to produce fitted values

$$\widehat{y}_t = \widehat{a} x_t + \widehat{b} z_t + \widehat{c}$$

and then ρ is reestimated using, by default, the regression defined by

$$y_t - \widehat{y}_t = \rho(y_{t-1} - \widehat{y}_{t-1}) + u_t \tag{2}$$

We then reestimate (1) by using the new estimate of ρ and continue to iterate between (1) and (2) until the estimate of ρ converges.

Convergence is declared after `iterate()` iterations or when the absolute difference in the estimated correlation between two iterations is less than `tol()`; see [R] **maximize**. Sargan (1964) has shown that this process will always converge.

Under the `ssesearch` option, a combined quadratic and bisection search using golden sections searches for the value of ρ that minimizes the sum-of-squared residuals from the transformed equation. The transformation may be either the Cochrane–Orcutt (1 only) or the Prais–Winsten (1 and $1'$).

All reported statistics are based on the ρ-transformed variables, and ρ is assumed to be estimated without error. See Judge et al. (1985) for details.

The Durbin–Watson d statistic reported by `prais` and `estat dwatson` is

$$d = \frac{\sum_{j=1}^{n-1} (u_{j+1} - u_j)^2}{\sum_{j=1}^{n} u_j^2}$$

where u_j represents the residual of the jth observation.

This command supports the Huber/White/sandwich estimator of the variance and its clustered version using `vce(robust)` and `vce(cluster clustvar)`, respectively. See [P] _robust, particularly *Introduction* and *Methods and formulas*.

All estimates from `prais` are conditional on the estimated value of ρ. Robust variance estimates here are robust only to heteroskedasticity and are not generally robust to misspecification of the functional form or omitted variables. The estimation of the functional form is intertwined with the estimation of ρ, and all estimates are conditional on ρ. Thus estimates cannot be robust to misspecification of functional form. For these reasons, it is probably best to interpret `vce(robust)` in the spirit of White's original paper on estimation of heteroskedastic-consistent covariance matrices.

Acknowledgment

We thank Richard Dickens of the Centre for Economic Performance at the London School of Economics and Political Science for testing and assistance with an early version of this command.

Sigbert Jon Prais (1928–) was born in Frankfurt and moved to Britain in 1934 as a refugee. After earning degrees at the universities of Birmingham and Cambridge and serving in various posts in research and industry, he settled at the National Institute of Economic and Social Research. Prais's interests extend widely across economics, including studies of the influence of education on economic progress.

Christopher Blake Winsten (1923–2005) was born in Welwyn Garden City, England; the son of the writer Stephen Winsten and the painter and sculptress Clare Blake. He was educated at the University of Cambridge and worked with the Cowles Commission at the University of Chicago and at the universities of Oxford, London (Imperial College) and Essex, making many contributions to economics and statistics, including the Prais–Winsten transformation and joint authorship of a celebrated monograph on transportation economics.

Donald Cochrane (1917–1983) was an Australian economist and econometrician. He was born in Melbourne and earned degrees at Melbourne and Cambridge. After wartime service in the Royal Australian Air Force, he held chairs at Melbourne and Monash, being active also in work for various international organizations and national committees.

Guy Henderson Orcutt (1917–) was born in Michigan and earned degrees in physics and economics at the University of Michigan. He worked at Harvard, the University of Wisconsin, and Yale. He has contributed to econometrics and economics in several fields, most distinctively in developing microanalytical models of economic behavior.

References

Chatterjee, S., and A. S. Hadi. 2006. *Regression Analysis by Example*. 4th ed. New York: Wiley.

Cochrane, D., and G. H. Orcutt. 1949. Application of least squares regression to relationships containing auto-correlated error terms. *Journal of the American Statistical Association* 44: 32–61.

Davidson, R., and J. G. MacKinnon. 1993. *Estimation and Inference in Econometrics*. New York: Oxford University Press.

Durbin, J., and G. S. Watson. 1950. Testing for serial correlation in least squares regression. I. *Biometrika* 37: 409–428.

———. 1951. Testing for serial correlation in least squares regression. II. *Biometrika* 38: 159–177.

Hardin, J. W. 1995. sts10: Prais–Winsten regression. *Stata Technical Bulletin* 25: 26–29. Reprinted in *Stata Technical Bulletin Reprints*, vol. 5, pp. 234–237. College Station, TX: Stata Press.

Harvey, A. C. 1990. *The Econometric Analysis of Time Series*. 2nd ed. Cambridge, MA: MIT Press.

Hildreth, C., and J. Y. Lu. 1960. Demand relations with autocorrelated disturbances. Reprinted in *Agricultural Experiment Station Technical Bulletin*, No. 276. East Lansing, MI: Michigan State University Press.

Judge, G. G., W. E. Griffiths, R. C. Hill, H. Lütkepohl, and T.-C. Lee. 1985. *The Theory and Practice of Econometrics*. 2nd ed. New York: Wiley.

King, M. L., and D. E. A. Giles, ed. 1987. *Specification Analysis in the Linear Model: Essays in Honor of Donald Cochrane*. London: Routledge & Kegan Paul.

Kmenta, J. 1997. *Elements of Econometrics*. 2nd ed. Ann Arbor: University of Michigan Press.

Prais, S. J., and C. B. Winsten. 1954. Trend estimators and serial correlation. Working paper 383, Cowles Commission. http://cowles.econ.yale.edu/P/ccdp/st/s-0383.pdf.

Sargan, J. D. 1964. Wages and prices in the United Kingdom: A study in econometric methodology. In *Econometric Analysis for National Economic Planning*, ed. P. E. Hart, G. Mills, and J. K. Whitaker, 25–64. London: Butterworths.

Theil, H. 1971. *Principles of Econometrics*. New York: Wiley.

White, H. 1980. A heteroskedasticity-consistent covariance matrix estimator and a direct test for heteroskedasticity. *Econometrica* 48: 817–838.

Wooldridge, J. M. 2009. *Introductory Econometrics: A Modern Approach*. 4th ed. Cincinnati, OH: South-Western.

Zellner, A. 1990. Guy H. Orcutt: Contributions to economic statistics. *Journal of Economic Behavior and Organization* 14: 43–51.

Also see

Title

prais postestimation — Postestimation tools for prais

Description

The following standard postestimation commands are available after prais:

Command	Description
contrast	contrasts and ANOVA-style joint tests of estimates
estat	AIC, BIC, VCE, and estimation sample summary
estimates	cataloging estimation results
lincom	point estimates, standard errors, testing, and inference for linear combinations of coefficients
linktest	link test for model specification
margins	marginal means, predictive margins, marginal effects, and average marginal effects
marginsplot	graph the results from margins (profile plots, interaction plots, etc.)
nlcom	point estimates, standard errors, testing, and inference for nonlinear combinations of coefficients
predict	predictions, residuals, influence statistics, and other diagnostic measures
predictnl	point estimates, standard errors, testing, and inference for generalized predictions
pwcompare	pairwise comparisons of estimates
test	Wald tests of simple and composite linear hypotheses
testnl	Wald tests of nonlinear hypotheses

See the corresponding entries in the *Base Reference Manual* for details.

Syntax for predict

predict [*type*] *newvar* [*if*] [*in*] [, *statistic*]

statistic	Description
Main	
xb	linear prediction; the default
stdp	standard error of the linear prediction
<u>r</u>esiduals	residuals

These statistics are available both in and out of sample; type predict ... if e(sample) ... if wanted only for the estimation sample.

Menu

Statistics > Postestimation > Predictions, residuals, etc.

Options for predict

⌐ Main ⌐

xb, the default, calculates the fitted values—the prediction of $x_j b$ for the specified equation. This is the linear predictor from the fitted regression model; it does not apply the estimate of ρ to prior residuals.

stdp calculates the standard error of the prediction for the specified equation, that is, the standard error of the predicted expected value or mean for the observation's covariate pattern. The standard error of the prediction is also referred to as the standard error of the fitted value.

As computed for prais, this is strictly the standard error from the variance in the estimates of the parameters of the linear model and assumes that ρ is estimated without error.

residuals calculates the residuals from the linear prediction.

Methods and formulas

All postestimation commands listed above are implemented as ado-files.

Also see

[TS] **prais** — Prais–Winsten and Cochrane–Orcutt regression

[U] **20 Estimation and postestimation commands**

Title

> **psdensity** — Parametric spectral density estimation after arima, arfima, and ucm

Syntax

> psdensity [*type*] *newvar*$_{sd}$ *newvar*$_f$ [*if*] [*in*] [, *options*]

where *newvar*$_{sd}$ is the name of the new variable that will contain the estimated spectral density and *newvar*$_f$ is the name of the new variable that will contain the frequencies at which the spectral density estimate is computed.

options	Description
Model	
pspectrum	estimate the power spectrum rather than the spectral density
range(*a b*)	limit the frequency range to $[a, b)$
cycle(#)	estimate the spectral density from the specified stochastic cycle; only allowed after ucm
smemory	estimate the spectral density of the short-memory component of the ARFIMA process; only allowed after arfima

Menu

Statistics > Time series > Postestimation > Parametric spectral density

Description

psdensity estimates the spectral density of a stationary process using the parameters of a previously estimated parametric model.

psdensity works after arfima and ucm, and after arima when there are no multiplicative autoregressive moving-average (ARMA) terms; see [TS] **arfima**, [TS] **ucm**, and [TS] **arima**.

Options

───┐ Model ┌──

pspectrum causes psdensity to estimate the power spectrum rather than the spectral density. The power spectrum is equal to the spectral density times the variance of the process.

range(*a b*) limits the frequency range. By default, the spectral density is computed over $[0, \pi)$. Specifying range(*a b*) causes the spectral density to be computed over $[a, b)$. We require that $0 \le a < b < \pi$.

cycle(#) causes psdensity to estimate the spectral density from the specified stochastic cycle after ucm. By default, the spectral density from the first stochastic cycle is estimated. cycle(#) must specify an integer that corresponds to a cycle in the model fit by ucm.

smemory causes psdensity to ignore the ARFIMA fractional integration parameter. The spectral density computed is for the short-memory ARMA component of the model.

Remarks

Remarks are presented under the following headings:

         The frequency-domain approach to time series
         Some ARMA examples

The frequency-domain approach to time series

A stationary process can be decomposed into random components that occur at the frequencies $\omega \in [0, \pi]$. The spectral density of a stationary process describes the relative importance of these random components. psdensity uses the estimated parameters of a parametric model to estimate the spectral density of a stationary process.

We need some concepts from the frequency-domain approach to time-series analysis to interpret estimated spectral densities. Here we provide a simple, intuitive explanation. More technical presentations can be found in Priestley (1981), Harvey (1989, 1993), Hamilton (1994), Fuller (1996), and Wei (2006).

In the time domain, the dependent variable evolves over time because of random shocks. The autocovariances γ_j, $j \in \{0, 1, \ldots, \infty\}$, of a covariance-stationary process y_t specify its variance and dependence structure, and the autocorrelations ρ_j, $j \in \{1, 2, \ldots, \infty\}$, provide a scale-free measure of its dependence structure. The autocorrelation at lag j specifies whether realizations at time t and realizations at time $t - j$ are positively related, unrelated, or negatively related.

In the frequency domain, the dependent variable is generated by an infinite number of random components that occur at the frequencies $\omega \in [0, \pi]$. The spectral density specifies the relative importance of these random components. The area under the spectral density in the interval $(\omega, \omega + d\omega)$ is the fraction of the variance of the process than can be attributed to the random components that occur at the frequencies in the interval $(\omega, \omega + d\omega)$.

The spectral density and the autocorrelations provide the same information about the dependence structure, albeit in different domains. The spectral density can be written as a weighted average of the autocorrelations of y_t, and it can be inverted to retrieve the autocorrelations as a function of the spectral density.

Like autocorrelations, the spectral density is normalized by γ_0, the variance of y_t. Multiplying the spectral density by γ_0 yields the power spectrum of y_t, which changes with the units of y_t.

A peak in the spectral density around frequency ω implies that the random components around ω make an important contribution to the variance of y_t.

A random variable primarily generated by low-frequency components will tend to have more runs above or below its mean than an independent and identically distributed (i.i.d.) random variable, and its plot may look smoother than the plot of the i.i.d. variable. A random variable primarily generated by high-frequency components will tend to have fewer runs above or below its mean than an i.i.d. random variable, and its plot may look more jagged than the plot of the i.i.d. variable.

❑ Technical note

A more formal specification of the spectral density allows us to be more specific about how the spectral density specifies the relative importance of the random components.

If y_t is a covariance-stationary process with absolutely summable autocovariances, its spectrum is given by

$$g_y(\omega) = \frac{1}{2\pi}\gamma_0 + \frac{1}{\pi}\sum_{k=1}^{\infty}\gamma_k\cos(\omega k) \tag{1}$$

where $g_y(\omega)$ is the spectrum of y_t at frequency ω and γ_k is the kth autocovariance of y_t. Taking the inverse Fourier transform of each side of (1) yields

$$\gamma_k = \int_{-\pi}^{\pi} g_y(\omega)e^{i\omega k}d\omega \tag{2}$$

where i is the imaginary number $i = \sqrt{-1}$.

Evaluating (2) at $k = 0$ yields

$$\gamma_0 = \int_{-\pi}^{\pi} g_y(\omega)d\omega$$

which means that the variance of y_t can be decomposed in terms of the spectrum $g_y(\omega)$. In particular, $g_y(\omega)d\omega$ is the contribution to the variance of y_t attributable to the random components in the interval $(\omega, \omega + d\omega)$.

The spectrum depends on the units in which y_t is measured, because it depends on the γ_0. Dividing both sides of (1) by γ_0 gives us the scale-free spectral density of y_t:

$$f_y(\omega) = \frac{1}{2\pi} + \frac{1}{\pi}\sum_{k=1}^{\infty}\rho_k\cos(\omega k)$$

By construction,

$$\int_{-\pi}^{\pi} f_y(\omega)d\omega = 1$$

so $f_y(\omega)d\omega$ is the fraction of the variance of y_t attributable to the random components in the interval $(\omega, \omega + d\omega)$.

❑

Some ARMA examples

In this section, we estimate and interpret the spectral densities implied by the estimated ARMA parameters. The examples illustrate some of the essential relationships between covariance-stationary processes, the parameters of ARMA models, and the spectral densities implied by the ARMA-model parameters.

See [TS] **ucm** for a discussion of unobserved-components models and the stochastic-cycle model derived by Harvey (1989) for stationary processes. The stochastic-cycle model has a different parameterization of the spectral density, and it tends to produce spectral densities that look more like probability densities than ARMA models. See *Remarks* in [TS] **ucm** for an introduction to these models, some examples, and some comparisons between the stochastic-cycle model and ARMA models.

▷ Example 1

Let's consider the changes in the number of manufacturing employees in the United States, which we plot below.

```
. use http://www.stata-press.com/data/r12/manemp2
(FRED data: Number of manufacturing employees in U.S.)
. tsline D.manemp, yline(-0.206)
```

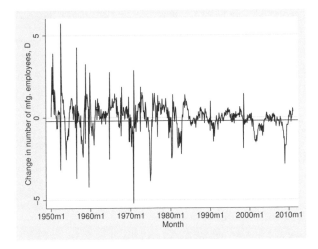

We added a horizontal line at the sample mean of -0.0206 to highlight that there appear to be more runs above or below the mean than we would expect in data generated by an i.i.d. process.

As a first pass at modeling this dependence, we use `arima` to estimate the parameters of a first-order autoregressive (AR(1)) model. Formally, the AR(1) model is given by

$$y_t = \alpha y_{t-1} + \epsilon_t$$

where y_t is the dependent variable, α is the autoregressive coefficient, and ϵ_t is an i.i.d. error term. See [TS] **arima** for an introduction to ARMA modeling and the `arima` command.

```
. arima D.manemp, ar(1) noconstant
(setting optimization to BHHH)
Iteration 0:   log likelihood = -870.64844
Iteration 1:   log likelihood = -870.64794
Iteration 2:   log likelihood = -870.64789
Iteration 3:   log likelihood = -870.64787
Iteration 4:   log likelihood = -870.64786
(switching optimization to BFGS)
Iteration 5:   log likelihood = -870.64786
Iteration 6:   log likelihood = -870.64786
ARIMA regression
Sample:  1950m2 - 2011m2                   Number of obs    =        733
                                           Wald chi2(1)     =     730.51
Log likelihood = -870.6479                 Prob > chi2      =     0.0000
```

D.manemp	Coef.	OPG Std. Err.	z	P>|z|	[95% Conf. Interval]	
ARMA						
ar L1.	.5179561	.0191638	27.03	0.000	.4803959	.5555164
/sigma	.7934554	.0080636	98.40	0.000	.777651	.8092598

Note: The test of the variance against zero is one sided, and the two-sided
    confidence interval is truncated at zero.

The statistically significant estimate of 0.518 for the autoregressive coefficient indicates that there is an important amount of positive autocorrelation in this series.

The spectral density of a covariance-stationary process is symmetric around 0. Following convention, psdensity estimates the spectral density over the interval $[0, \pi)$ at the points given in *Methods and formulas*.

Now we use psdensity to estimate the spectral density of the process implied by the estimated ARMA parameters. We specify the names of two new variables in the call to psdensity. The first new variable will contain the estimated spectral density. The second new variable will contain the frequencies at which the spectral density is estimated.

```
. psdensity psden1 omega

. line psden1 omega
```

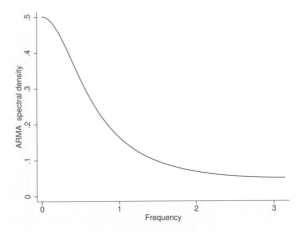

The above graph is typical of a spectral density of an AR(1) process with a positive coefficient. The curve is highest at frequency 0, and it tapers off toward zero or a positive asymptote. The estimated spectral density is telling us that the low-frequency random components are the most important random components of an AR(1) process with a positive autoregressive coefficient.

The closer the α is to 1, the more important are the low-frequency components relative to the high-frequency components. To illustrate this point, we plot the spectral densities implied by AR(1) models with $\alpha = 0.1$ and $\alpha = 0.9$.

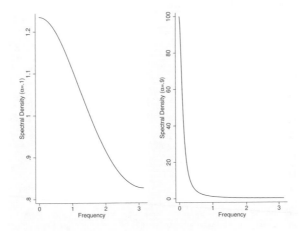

As α gets closer to 1, the plot of the spectral density gets closer to being a spike at frequency 0, implying that only the lowest-frequency components are important.

▷ Example 2

Now let's consider a dataset for which the estimated coefficient from an AR(1) model is negative. Below we plot the changes in initial claims for unemployment insurance in the United States.

```
. use http://www.stata-press.com/data/r12/icsa1, clear
. tsline D.icsa, yline(0.08)
```

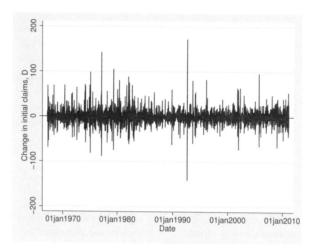

The plot looks a little more jagged than we would expect from an i.i.d. process, but it is hard to tell. Below we estimate the AR(1) coefficient.

```
. arima D.icsa, ar(1) noconstant
(setting optimization to BHHH)
Iteration 0:    log likelihood = -9934.0659
Iteration 1:    log likelihood = -9934.0657
Iteration 2:    log likelihood = -9934.0657
ARIMA regression
Sample:  14jan1967 - 19feb2011           Number of obs     =       2302
                                         Wald chi2(1)      =     666.06
Log likelihood = -9934.066               Prob > chi2       =     0.0000
```

D.icsa	Coef.	OPG Std. Err.	z	P>|z|	[95% Conf. Interval]
ARMA						
ar L1.	-.2756024	.0106789	-25.81	0.000	-.2965326 -.2546722
/sigma	18.10988	.1176556	153.92	0.000	17.87928 18.34048

```
Note: The test of the variance against zero is one sided, and the two-sided
      confidence interval is truncated at zero.
```

The estimated coefficient is negative and statistically significant.

The spectral density implied by the estimated parameters is

```
. psdensity psden2 omega2
. line psden2 omega2
```

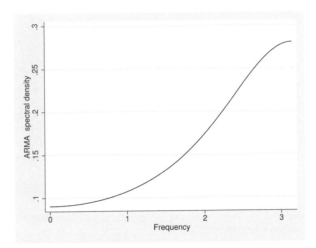

The above graph is typical of a spectral density of an AR(1) process with a negative coefficient. The curve is lowest at frequency 0, and it monotonically increases to its highest point, which occurs when the frequency is π.

When the coefficient of an AR(1) model is negative, the high-frequency random components are the most important random components of the process. The closer the α is to -1, the more important are the high-frequency components relative to the low-frequency components. To illustrate this point, we plot the spectral densities implied by AR(1) models with $\alpha = -0.1$, and $\alpha = -0.9$.

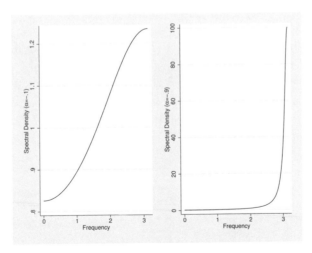

As α gets closer to -1, the plot of the spectral density shifts toward becoming a spike at frequency π, implying that only the highest-frequency components are important.

◁

For examples of psdensity after arfima and ucm, see [TS] **arfima** and [TS] **ucm**.

Methods and formulas

`psdensity` is implemented as an ado-file.

Methods and formulas are presented under the following headings:

> *Introduction*
> *Spectral density after arima or arfima*
> *Spectral density after ucm*

Introduction

The spectral density $f(\omega)$ is estimated at the values $\omega \in \{\omega_1, \omega_2, \ldots, \omega_N\}$ using one of the formulas given below. Given a sample of size N, after accounting for any if or in restrictions, the N values of ω are given by $\omega_i = \pi(i-1)/(N-1)$ for $i \in \{1, 2, \ldots, N\}$.

In the rare case in which the dataset in memory has insufficient observations for the desired resolution of the estimated spectral density, you may use tsappend or set obs (see [TS] **tsappend** or [D] **obs**) to increase the number of observations in the current dataset.

You may use an if restriction or an in restriction to restrict the observations to handle panel data or to compute the estimates for a subset of the observations.

Spectral density after arima or arfima

Let ϕ_k and θ_k denote the p autoregressive and q moving-average parameters of an ARMA model, respectively. Box, Jenkins, and Reinsel (2008) show that the spectral density implied by the ARMA parameters is

$$f_{\text{ARMA}}(\omega; \boldsymbol{\phi}, \boldsymbol{\theta}, \sigma_\epsilon^2, \gamma_0) = \frac{\sigma_\epsilon^2}{2\pi\gamma_0} \frac{|1 + \theta_1 e^{-i\omega} + \theta_2 e^{-i2\omega} + \cdots + \theta_q e^{-iq\omega}|^2}{|1 - \phi_1 e^{-i\omega} - \phi_2 e^{-i2\omega} - \cdots - \phi_p e^{-ip\omega}|^2}$$

where $\omega \in [0, \pi]$ and σ_ϵ^2 is the variance of the idiosyncratic error and γ_0 is the variance of the dependent variable. We estimate γ_0 using the arima parameter estimates.

The spectral density for the ARFIMA model is

$$f_{\text{ARFIMA}}(\omega; \boldsymbol{\phi}, \boldsymbol{\theta}, d, \sigma_\epsilon^2, \gamma_0) = |1 - e^{i\omega}|^{-2d} f_{\text{ARMA}}(\omega; \boldsymbol{\phi}, \boldsymbol{\theta}, \sigma_\epsilon^2)$$

where d, $-1/2 < d < 1/2$, is the fractional integration parameter. The spectral density goes to infinity as the frequency approaches 0 for $0 < d < 1/2$, and it is zero at frequency 0 for $-1/2 < d < 0$.

The smemory option causes psdensity to perform the estimation with $d = 0$, which is equivalent to estimating the spectral density of the fractionally differenced series.

The power spectrum omits scaling by γ_0.

Spectral density after ucm

The spectral density of an order-k stochastic cycle with frequency λ and damping ρ is (Trimbur 2006)

$$f(\omega; \rho, \lambda, \sigma_\kappa^2) = \left\{ \frac{(1-\rho^2)^{2k-1}}{\sigma_\kappa^2 \sum_{i=0}^{k-1} \binom{k-1}{i}^2 \rho^{2i}} \right\} \times$$

$$\frac{\sum_{j=0}^{k} \sum_{i=0}^{k} (-1)^{j+i} \binom{k}{j}\binom{k}{i} \rho^{j+i} \cos\lambda(j-i) \cos\omega(j-i)}{2\pi \left\{1 + 4\rho^2\cos^2\lambda + \rho^4 - 4\rho(1+\rho^2)\cos\lambda\cos\omega + 2\rho^2\cos 2\omega\right\}^k}$$

where σ_κ^2 is the variance of the cycle error term.

The variance of the cycle is

$$\sigma_\omega^2 = \sigma_\kappa^2 \frac{\sum_{i=0}^{k-1} \binom{k-1}{i}^2 \rho^{2i}}{(1-\rho^2)^{2k-1}}$$

and the power spectrum omits scaling by σ_ω^2.

References

Box, G. E. P., G. M. Jenkins, and G. C. Reinsel. 2008. *Time Series Analysis: Forecasting and Control.* 4th ed. Hoboken, NJ: Wiley.

Fuller, W. A. 1996. *Introduction to Statistical Time Series.* 2nd ed. New York: Wiley.

Hamilton, J. D. 1994. *Time Series Analysis.* Princeton: Princeton University Press.

Harvey, A. C. 1989. *Forecasting, Structural Time Series Models and the Kalman Filter.* Cambridge: Cambridge University Press.

——. 1993. *Time Series Models.* 2nd ed. Cambridge, MA: MIT Press.

Priestley, M. B. 1981. *Spectral Analysis and Time Series.* London: Academic Press.

Trimbur, T. M. 2006. Properties of higher order stochastic cycles. *Journal of Time Series Analysis* 27: 1–17.

Wei, W. W. S. 2006. *Time Series Analysis: Univariate and Multivariate Methods.* 2nd ed. Boston: Pearson.

Also see

[TS] **arfima** — Autoregressive fractionally integrated moving-average models

[TS] **arima** — ARIMA, ARMAX, and other dynamic regression models

[TS] **ucm** — Unobserved-components model

Title

rolling — Rolling-window and recursive estimation

Syntax

rolling [*exp_list*] [*if*] [*in*] [, *options*] : *command*

options	Description
Main	
* window(#)	number of consecutive data points in each sample
recursive	use recursive samples
rrecursive	use reverse recursive samples
Options	
clear	replace data in memory with results
saving(*filename*, ...)	save results to *filename*; save statistics in double precision; save results to *filename* every # replications
stepsize(#)	number of periods to advance window
start(*time_constant*)	period at which rolling is to start
end(*time_constant*)	period at which rolling is to end
keep(*varname*[, start])	save *varname* along with results; optionally, use value at left edge of window
Reporting	
nodots	suppress replication dots
noisily	display any output from *command*
trace	trace *command*'s execution
Advanced	
reject(*exp*)	identify invalid results

* window(#) is required.

You must tsset your data before using rolling; see [TS] tsset.

aweights are allowed in *command* if *command* accepts aweights; see [U] **11.1.6 weight**.

exp_list contains	(*name*: *elist*)
	elist
	eexp
elist contains	*newvar* = (*exp*)
	(*exp*)
eexp is	*specname*
	[*eqno*]*specname*
specname is	_b
	_b[]
	_se
	_se[]

342

eqno is ## ##

name

exp is a standard Stata expression; see [U] **13 Functions and expressions**.

Distinguish between [ ], which are to be typed, and [ ], which indicate optional arguments.

Menu

Statistics > Time series > Rolling-window and recursive estimation

Description

rolling is a moving sampler that collects statistics from *command* after executing *command* on subsets of the data in memory. Typing

. rolling *exp_list*, window(50) clear: *command*

executes *command* on sample windows of span 50. That is, rolling will first execute *command* by using periods 1–50 of the dataset, and then using periods 2–51, 3–52, and so on. rolling can also perform recursive and reverse recursive analyses, in which the starting or ending period is held fixed and the window size grows.

command defines the statistical command to be executed. Most Stata commands and user-written programs can be used with rolling, as long as they follow standard Stata syntax and allow the if qualifier; see [U] **11 Language syntax**. The by prefix cannot be part of *command*.

exp_list specifies the statistics to be collected from the execution of *command*. If no expressions are given, *exp_list* assumes a default of _b if *command* stores results in e() and of all the scalars if *command* stores results in r() and not in e(). Otherwise, not specifying an expression in *exp_list* is an error.

Options

___ Main ___

window(#) defines the window size used each time *command* is executed. The window size refers to calendar periods, not the number of observations. If there are missing data (for example, because of weekends), the actual number of observations used by *command* may be less than window(#). window(#) is required.

recursive specifies that a recursive analysis be done. The starting period is held fixed, the ending period advances, and the window size grows.

rrecursive specifies that a reverse recursive analysis be done. Here the ending period is held fixed, the starting period advances, and the window size shrinks.

___ Options ___

clear specifies that Stata replace the data in memory with the collected statistics even though the current data in memory have not been saved to disk.

saving(*filename* [, *suboptions*]) creates a Stata data file (.dta file) consisting of (for each statistic in *exp_list*) a variable containing the window replicates.

double specifies that the results for each replication be stored as doubles, meaning 8-byte reals. By default, they are stored as floats, meaning 4-byte reals.

every(#) specifies that results be written to disk every #th replication. every() should be specified in conjunction only with saving() when *command* takes a long time for each replication. This will allow recovery of partial results should your computer crash. See [P] **postfile**.

stepsize(#) specifies the number of periods the window is to be advanced each time *command* is executed.

start(*time_constant*) specifies the date on which rolling is to start. start() may be specified as an integer or as a date literal.

end(*time_constant*) specifies the date on which rolling is to end. end() may be specified as an integer or as a date literal.

keep(*varname* [, start]) specifies a variable to be posted along with the results. The value posted is the value that corresponds to the right edge of the window. Specifying the start() option requests that the value corresponding to the left edge of the window be posted instead. This option is often used to record calendar dates.

⌐Reporting⌐

nodots suppresses display of the replication dot for each window on which *command* is executed. By default, one dot character is printed for each window. A red 'x' is printed if *command* returns with an error or if any value in *exp_list* is missing.

noisily causes the output of *command* to be displayed for each window on which *command* is executed. This option implies the nodots option.

trace causes a trace of the execution of *command* to be displayed. This option implies the noisily and nodots options.

⌐Advanced⌐

reject(*exp*) identifies an expression that indicates when results should be rejected. When *exp* is true, the saved statistics are set to missing values.

Remarks

rolling executes a command on each of a series of windows of observations and stores the results. rolling can perform what are commonly called rolling regressions, recursive regressions, and reverse recursive regressions. However, rolling is not limited to just linear regression analysis: any command that saves results in e() or r() can be used with rolling.

Suppose that you have data collected at 100 consecutive points in time, numbered 1–100, and you wish to perform a rolling regression with a window size of 20 periods. Typing

```
. rolling _b, window(20) clear: regress depvar indepvar
```

causes Stata to regress *depvar* on *indepvar* using periods 1–20, store the regression coefficients (_b), run the regression using periods 2–21, and so on, finishing with a regression using periods 81–100 (the last 20 periods).

The stepsize() option specifies how far ahead the window is moved each time. For example, if you specify step(2), then *command* is executed on periods 1–20, and then 3–22, 5–24, etc. By default, rolling replaces the dataset in memory with the computed statistics unless the saving() option is specified, in which case the computed statistics are stored in the filename specified. If the dataset in memory has been changed since it was last saved and you do not specify saving(), you must use clear.

rolling can also perform recursive and reverse recursive analyses. In a recursive analysis, the starting date is held fixed, and the window size grows as the ending date is advanced. In a reverse recursive analysis, the ending date is held fixed, and the window size shrinks as the starting date is advanced.

> Example 1

We have data on the daily returns to IBM stock (ibm), the S&P 500 (spx), and short-term interest rates (irx), and we want to create a series containing the beta of IBM by using the previous 200 trading days at each date. We will also record the standard errors, so that we can obtain 95% confidence intervals for the betas. See, for example, Stock and Watson (2011, 118) for more information on estimating betas. We type

```
. use http://www.stata-press.com/data/r12/ibm
(Source: Yahoo! Finance)
. tsset t
        time variable:  t, 1 to 494
                delta:  1 unit
. generate ibmadj = ibm - irx
(1 missing value generated)
. generate spxadj = spx - irx
(1 missing value generated)
. rolling _b _se, window(200) saving(betas, replace) keep(date): regress ibmadj
> spxadj
(running regress on estimation sample)
(note: file betas.dta not found)
Rolling replications (295)
──┼── 1 ──┼── 2 ──┼── 3 ──┼── 4 ──┼── 5
................................................     50
................................................    100
................................................    150
................................................    200
................................................    250
..........................................
file betas.dta saved
```

Our dataset has both a time variable t that runs consecutively and a date variable date that measures the calendar date and therefore has gaps at weekends and holidays. Had we used the date variable as our time variable, rolling would have used windows consisting of 200 calendar days instead of 200 trading days, and each window would not have exactly 200 observations. We used the keep(date) option so that we could refer to the date variable when working with the results dataset.

We can list a portion of the dataset created by rolling to see what it contains:

```
. use betas, clear
(rolling: regress)
. sort date
. list in 1/3, abbrev(10)
```

	start	end	date	_b_spxadj	_b_cons	_se_spxadj	_se_cons
1.	1	200	16oct2003	1.043422	-.0181504	.0658531	.0748295
2.	2	201	17oct2003	1.039024	-.0126876	.0656893	.074609
3.	3	202	20oct2003	1.038371	-.0235616	.0654591	.0743851

The variables `start` and `end` indicate the first and last observations used each time that `rolling` called `regress`, and the `date` variable contains the calendar date corresponding the period represented by `end`. The remaining variables are the estimated coefficients and standard errors from the regression. In our example , `_b_spxadj` contains the estimated betas, and `_b_cons` contains the estimated alphas. The variables `_se_spxadj` and `_se_cons` have the corresponding standard errors.

Finally, we compute the confidence intervals for the betas and examine how they have changed over time:

```
. generate lower = _b_spxadj - 1.96*_se_spxadj
. generate upper = _b_spxadj + 1.96*_se_spxadj
. twoway (line _b_spxadj date) (rline lower upper date) if date>=td(1oct2003),
> ytitle("Beta")
```

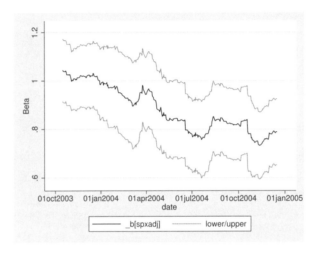

As 2004 progressed, IBM's stock returns were less influenced by returns in the broader market. Beginning in June of 2004, IBM's beta became significantly different from unity at the 95% confidence level, as indicated by the fact that the confidence interval does not contain one from then onward.

◁

In addition to rolling-window analyses, `rolling` can also perform recursive ones. Suppose again that you have data collected at 100 consecutive points in time, and now you type

```
. rolling _b, window(20) recursive clear: regress depvar indepvar
```

Stata will first `regress` *depvar* on *indepvar* by using observations 1–20, store the coefficients, run the regression using observations 1–21, observations 1–22, and so on, finishing with a regression using all 100 observations. Unlike a rolling regression, in which case the number of observations is held constant and the starting and ending points are shifted, a recursive regression holds the starting point fixed and increases the number of observations. Recursive analyses are often used in forecasting situations. As time goes by, more information becomes available that can be used in making forecasts. See Kmenta (1997, 423–424).

> ## Example 2

Using the same dataset, we type

```
. use http://www.stata-press.com/data/r12/ibm, clear
(Source: Yahoo! Finance)

. tsset t
        time variable:  t, 1 to 494
                delta:  1 unit

. generate ibmadj = ibm - irx
(1 missing value generated)

. generate spxadj = spx - irx
(1 missing value generated)

. rolling _b _se, recursive window(200) clear: regress ibmadj spxadj
  (output omitted )

. list in 1/3, abbrev(10)
```

	start	end	_b_spxadj	_b_cons	_se_spxadj	_se_cons
1.	1	200	1.043422	-.0181504	.0658531	.0748295
2.	1	201	1.039024	-.0126876	.0656893	.074609
3.	1	202	1.037687	-.016475	.0655896	.0743481

Here the starting period remains fixed and the window grows larger.

◁

In a reverse recursive analysis, the ending date is held fixed, and the window size becomes smaller as the starting date is advanced. For example, with a dataset that has observations numbered 1–100, typing

```
. rolling _b, window(20) reverse recursive clear: regress depvar indepvar
```

creates a dataset in which the first observation has the results based on periods 1–100, the second observation has the results based on 2–100, the third having 3–100, and so on, up to the last observation having results based on periods 81–100 (the last 20 observations).

> ## Example 3

Using the data on stock returns, we want to build a model in which we predict today's IBM stock return on the basis of yesterday's returns on IBM and the S&P 500. That is, letting i_t and s_t denote the returns to IBM and the S&P 500 on date t, we want to fit the regression model

$$i_t = \beta_0 + \beta_1 i_{t-1} + \beta_2 s_{t-1} + \epsilon_t$$

where ϵ_t is a regression error term, and then compute

$$\widehat{i_{t+1}} = \widehat{\beta_0} + \widehat{\beta_1} i_t + \widehat{\beta_2} s_t$$

We will use recursive regression because we suspect that the more data we have to fit the regression model, the better the model will predict returns. We will use at least 20 periods in fitting the regression.

```
. use http://www.stata-press.com/data/r12/ibm, clear
(Source: Yahoo! Finance)

. tsset t
        time variable:  t, 1 to 494
        delta:  1 unit
```

One alternative would be to use `rolling` with the `recursive` option to fit the regressions, collect the coefficients, and then compute the predicted values afterward. However, we will instead write a short program that computes the forecasts automatically and then use `rolling, recursive` on that program. The program must accept an `if` expression so that `rolling` can indicate to the program which observations are to be used. Our program is

```
program myforecast, rclass
        syntax [if]
        regress ibm L.ibm L.spx `if'
        // Find last time period of estimation sample and
        // make forecast for period just after that
        summ t if e(sample)
        local last = r(max)
        local fcast = _b[_cons] + _b[L.ibm]*ibm[`last'] + ///
                      _b[L.spx]*spx[`last']
        return scalar forecast = `fcast'
        // Next period's actual return
        // Will return missing value for final period
        return scalar actual = ibm[`last'+1]
end
```

Now we call `rolling`:

```
. rolling actual=r(actual) forecast=r(forecast), recursive window(20): myforecast
  (output omitted )
. corr actual forecast
(obs=474)
```

	actual	forecast
actual	1.0000	
forecast	-0.0957	1.0000

Our model does not work too well—the correlation between actual returns and our forecasts is negative.

◁

Saved results

`rolling` sets no r- or e-class macros. The results from the command used with `rolling`, depending on the last window of data used, are available after `rolling` has finished.

Methods and formulas

`rolling` is implemented as an ado-file.

Acknowledgment

We thank Christopher F. Baum for an earlier rolling regression command.

References

Kmenta, J. 1997. *Elements of Econometrics*. 2nd ed. Ann Arbor: University of Michigan Press.

Stock, J. H., and M. W. Watson. 2011. *Introduction to Econometrics*. 3rd ed. Boston: Addison–Wesley.

Also see

[D] **statsby** — Collect statistics for a command across a by list

[R] **saved results** — Saved results

Title

sspace — State-space models

Syntax

Covariance-form syntax

sspace *state_ceq* [*state_ceq* ... *state_ceq*] *obs_ceq* [*obs_ceq* ... *obs_ceq*]

[*if*] [*in*] [, *options*]

where each *state_ceq* is of the form

(*statevar* [*lagged_statevars*] [*indepvars*], state [noerror noconstant])

and each *obs_ceq* is of the form

(*depvar* [*statevars*] [*indepvars*] [, noerror noconstant])

Error-form syntax

sspace *state_efeq* [*state_efeq* ... *state_efeq*] *obs_efeq* [*obs_efeq* ... *obs_efeq*]

[*if*] [*in*] [, *options*]

where each *state_efeq* is of the form

(*statevar* [*lagged_statevars*] [*indepvars*] [*state_errors*], state [noconstant])

and each *obs_efeq* is of the form

(*depvar* [*statevars*] [*indepvars*] [*obs_errors*] [, noconstant])

statevar is the name of an unobserved state, not a variable. If there happens to be a variable of the same name, the variable is ignored and plays no role in the estimation.

lagged_statevars is a list of lagged *statevars*. Only first lags are allowed.

state_errors is a list of state-equation errors that enter a state equation. Each state error has the form e.*statevar*, where *statevar* is the name of a state in the model.

obs_errors is a list of observation-equation errors that enter an equation for an observed variable. Each error has the form e.*depvar*, where *depvar* is an observed dependent variable in the model.

equation-level options	Description
Model	
state	specifies that the equation is a state equation
noerror	specifies that there is no error term in the equation
noconstant	suppresses the constant term from the equation

350

options	Description
Model	
<u>covst</u>ate(*covform*)	specifies the covariance structure for the errors in the state variables
<u>covob</u>served(*covform*)	specifies the covariance structure for the errors in the observed dependent variables
<u>constr</u>aints(*constraints*)	apply specified linear constraints
SE/Robust	
vce(*vcetype*)	*vcetype* may be oim or <u>r</u>obust
Reporting	
<u>level</u>(#)	set confidence level; default is level(95)
<u>nocnsr</u>eport	do not display constraints
display_options	control column formats, row spacing, and display of omitted variables and base and empty cells
Maximization	
maximize_options	control the maximization process; seldom used
Advanced	
<u>meth</u>od(*method*)	specify the method for calculating the log likelihood; seldom used
<u>coefl</u>egend	display legend instead of statistics

covform	Description
<u>id</u>entity	identity matrix; the default for error-form syntax
<u>ds</u>calar	diagonal scalar matrix
<u>di</u>agonal	diagonal matrix; the default for covariance-form syntax
<u>un</u>structured	symmetric, positive-definite matrix; not allowed with error-form syntax

method	Description
<u>hy</u>brid	use the stationary Kalman filter and the De Jong diffuse Kalman filter; the default
<u>dej</u>ong	use the stationary De Jong Kalman filter and the De Jong diffuse Kalman filter
<u>kd</u>iffuse	use the stationary Kalman filter and the nonstationary large-κ diffuse Kalman filter; seldom used

You must tsset your data before using sspace; see [TS] **tsset**.

indepvars may contain factor variables; see [U] **11.4.3 Factor variables**.

indepvars and *depvar* may contain time-series operators; see [U] **11.4.4 Time-series varlists**.

by, statsby, and rolling are allowed; see [U] **11.1.10 Prefix commands**.

coeflegend does not appear in the dialog box.

See [U] **20 Estimation and postestimation commands** for more capabilities of estimation commands.

Menu

Statistics > State-space models

Description

sspace estimates the parameters of linear state-space models by maximum likelihood. Linear state-space models are very flexible and many linear time-series models can be written as linear state-space models.

sspace uses two forms of the Kalman filter to recursively obtain conditional means and variances of both the unobserved states and the measured dependent variables that are used to compute the likelihood.

The covariance-form syntax and the error-form syntax of sspace reflect the two different forms in which researchers specify state-space models. Choose the syntax that is easier for you; the two forms are isomorphic.

Options

Equation-level options

⌐ Model ⌐

state specifies that the equation is a state equation.

noerror specifies that there is no error term in the equation. noerror may not be specified in the error-form syntax.

noconstant suppresses the constant term from the equation.

Options

⌐ Model ⌐

covstate(*covform*) specifies the covariance structure for the state errors.

covstate(identity) specifies a covariance matrix equal to an identity matrix, and it is the default for the error-form syntax.

covstate(dscalar) specifies a covariance matrix equal to σ_{state}^2 times an identity matrix.

covstate(diagonal) specifies a diagonal covariance matrix, and it is the default for the covariance-form syntax.

covstate(unstructured) specifies a symmetric, positive-definite covariance matrix with parameters for all variances and covariances. covstate(unstructured) may not be specified with the error-form syntax.

covobserved(*covform*) specifies the covariance structure for the observation errors.

covobserved(identity) specifies a covariance matrix equal to an identity matrix, and it is the default for the error-form syntax.

covobserved(dscalar) specifies a covariance matrix equal to $\sigma_{\text{observed}}^2$ times an identity matrix.

covobserved(diagonal) specifies a diagonal covariance matrix, and it is the default for the covariance-form syntax.

covobserved(unstructured) specifies a symmetric, positive-definite covariance matrix with parameters for all variances and covariances. covobserved(unstructured) may not be specified with the error-form syntax.

constraints(*constraints*); see [R] **estimation options**.

───────┤ SE/Robust ├──

vce(*vcetype*) specifies the estimator for the variance–covariance matrix of the estimator.

vce(oim), the default, causes sspace to use the observed information matrix estimator.

vce(robust) causes sspace to use the Huber/White/sandwich estimator.

───────┤ Reporting ├──

level(*#*), nocnsreport; see [R] **estimation options**.

display_options: noomitted, vsquish, noemptycells, baselevels, allbaselevels, cformat(%*fmt*), pformat(%*fmt*), and sformat(%*fmt*); see [R] **estimation options**.

───────┤ Maximization ├──

maximize_options: difficult, technique(*algorithm_spec*), iterate(*#*), [no]log, trace, gradient, showstep, hessian, showtolerance, tolerance(*#*), ltolerance(*#*), nrtolerance(*#*), and from(*matname*); see [R] **maximize** for all options except from(), and see below for information on from(). These options are seldom used.

from(*matname*) specifies initial values for the maximization process. from(b0) causes sspace to begin the maximization algorithm with the values in b0. b0 must be a row vector; the number of columns must equal the number of parameters in the model; and the values in b0 must be in the same order as the parameters in e(b).

───────┤ Advanced ├──

method(*method*) specifies how to compute the log likelihood. This option is seldom used.

method(hybrid), the default, uses the Kalman filter with model-based initial values for the states when the model is stationary and uses the De Jong (1988, 1991) diffuse Kalman filter when the model is nonstationary.

method(dejong) uses the Kalman filter with the De Jong (1988) method for estimating the initial values for the states when the model is stationary and uses the De Jong (1988, 1991) diffuse Kalman filter when the model is nonstationary.

method(kdiffuse) is a seldom used method that uses the Kalman filter with model-based initial values for the states when the model is stationary and uses the large-κ diffuse Kalman filter when the model is nonstationary.

The following option is available with sspace but is not shown in the dialog box:

coeflegend; see [R] **estimation options**.

Remarks

Remarks are presented under the following headings:

> *An introduction to state-space models*
> *Some stationary state-space models*
> *Some nonstationary state-space models*

An introduction to state-space models

Many linear time-series models can be written as linear state-space models, including vector autoregressive moving-average (VARMA) models, dynamic-factor (DF) models, and structural time-series (STS) models. The solutions to some stochastic dynamic-programming problems can also be written in the form of linear state-space models. We can estimate the parameters of a linear state-space model by maximum likelihood (ML). The Kalman filter or a diffuse Kalman filter is used to write the likelihood function in prediction-error form, assuming normally distributed errors. The quasi–maximum likelihood (QML) estimator, which drops the normality assumption, is consistent and asymptotically normal when the model is stationary. Chang, Miller, and Park (2009) establish consistency and asymptotic normality of the QML estimator for a class of nonstationary state-space models. The QML estimator differs from the ML estimator only in the VCE; specify the `vce(robust)` option to obtain the QML estimator.

Hamilton (1994a, 1994b), Harvey (1989), and Brockwell and Davis (1991) provide good introductions to state-space models. Anderson and Moore's (1979) text is a classic reference; they produced many results used subsequently. Caines (1988) and Hannan and Deistler (1988) provide excellent, more advanced, treatments.

`sspace` estimates linear state-space models with time-invariant coefficient matrices, which cover the models listed above and many others. `sspace` can estimate parameters from state-space models of the form

$$\mathbf{z}_t = \mathbf{A}\mathbf{z}_{t-1} + \mathbf{B}\mathbf{x}_t + \mathbf{C}\epsilon_t$$
$$\mathbf{y}_t = \mathbf{D}\mathbf{z}_t + \mathbf{F}\mathbf{w}_t + \mathbf{G}\nu_t$$

where

$\mathbf{z}_t$ is an $m \times 1$ vector of unobserved state variables;

$\mathbf{x}_t$ is a $k_x \times 1$ vector of exogenous variables;

ϵ_t is a $q \times 1$ vector of state-error terms, $(q \leq m)$;

$\mathbf{y}_t$ is an $n \times 1$ vector of observed endogenous variables;

$\mathbf{w}_t$ is a $k_w \times 1$ vector of exogenous variables;

ν_t is an $r \times 1$ vector of observation-error terms, $(r \leq n)$; and

$\mathbf{A}, \mathbf{B}, \mathbf{C}, \mathbf{D}, \mathbf{F}$, and $\mathbf{G}$ are parameter matrices.

The equations for $\mathbf{z}_t$ are known as the state equations, and the equations for $\mathbf{y}_t$ are known as the observation equations.

The error terms are assumed to be zero mean, normally distributed, serially uncorrelated, and uncorrelated with each other;

$$\epsilon_t \sim N(0, \mathbf{Q})$$
$$\nu_t \sim N(0, \mathbf{R})$$
$$E[\epsilon_t \epsilon_s'] = \mathbf{0} \text{ for all } s \neq t$$
$$E[\epsilon_t \nu_s'] = \mathbf{0} \text{ for all } s \text{ and } t$$

The state-space form is used to derive the log likelihood of the observed endogenous variables conditional on their own past and any exogenous variables. When the model is stationary, a method for recursively predicting the current values of the states and the endogenous variables, known as the Kalman filter, is used to obtain the prediction error form of the log-likelihood function. When the model is nonstationary, a diffuse Kalman filter is used. How the Kalman filter and the diffuse

Kalman filter initialize their recursive computations depends on the `method()` option; see *Methods and formulas*.

The linear state-space models with time-invariant coefficient matrices defined above can be specified in the covariance-form syntax and the error-form syntax. The covariance-form syntax requires that $\mathbf{C}$ and $\mathbf{G}$ be selection matrices, but places no restrictions on $\mathbf{Q}$ or $\mathbf{R}$. In contrast, the error-form syntax places no restrictions $\mathbf{C}$ or $\mathbf{G}$, but requires that $\mathbf{Q}$ and $\mathbf{R}$ be either diagonal, diagonal-scalar, or identity matrices. Some models are more easily specified in the covariance-form syntax, while others are more easily specified in the error-form syntax. Choose the syntax that is easiest for your application.

Some stationary state-space models

▷ Example 1

Following Hamilton (1994b, 373–374), we can write the first-order autoregressive (AR(1)) model

$$y_t - \mu = \alpha(y_{t-1} - \mu) + \epsilon_t$$

as a state-space model with the observation equation

$$y_t = \mu + u_t$$

and the state equation

$$u_t = \alpha u_{t-1} + \epsilon_t$$

where the unobserved state is $u_t = y_t - \mu$.

Here we fit this model to data on the capacity utilization rate. The variable `lncaputil` contains data on the natural log of the capacity utilization rate for the manufacturing sector of the U.S. economy. We treat the series as first-difference stationary and fit its first-difference to an AR(1) process. Here we estimate the parameters of the above state-space form of the AR(1) model:

```
. use http://www.stata-press.com/data/r12/manufac
(St. Louis Fed (FRED) manufacturing data)
. constraint 1 [D.lncaputil]u = 1
. sspace (u L.u, state noconstant) (D.lncaputil u,  noerror), constraints(1)
searching for initial values .........
(setting technique to bhhh)
Iteration 0:   log likelihood =    1505.36
Iteration 1:   log likelihood =  1512.0581
  (output omitted )
Refining estimates:
Iteration 0:   log likelihood =    1516.44
Iteration 1:   log likelihood =    1516.44
```

State-space model

Sample: 1972m2 - 2008m12

	Number of obs	=	443
	Wald chi2(1)	=	61.73
	Prob > chi2	=	0.0000

Log likelihood =   1516.44
(1) [D.lncaputil]u = 1

lncaputil	Coef.	OIM Std. Err.	z	P>|z|	[95% Conf. Interval]	
u						
u						
L1.	.3523983	.0448539	7.86	0.000	.2644862	.4403104
D.lncaputil						
u	1	(constrained)				
_cons	-.0003558	.0005781	-0.62	0.538	-.001489	.0007773
Variance						
u	.0000622	4.18e-06	14.88	0.000	.000054	.0000704

Note: Tests of variances against zero are one sided, and the two-sided
   confidence intervals are truncated at zero.

The iteration log has three parts: the dots from the search for initial values, the log from finding the maximum, and the log from a refining step. Here is a description of the logic behind each part:

1. The quality of the initial values affect the speed and robustness of the optimization algorithm. sspace takes a few iterations in a nonlinear least-squares (NLS) algorithm to find good initial values and reports a dot for each (NLS) iteration.

2. This iteration log is the standard method by which Stata reports the search for the maximum likelihood estimates of the parameters in a nonlinear model.

3. Some of the parameters are transformed in the maximization process that sspace reports in part 2. After a maximum candidate is found in part 2, sspace looks for a maximum in the unconstrained space, checks that the Hessian of the log-likelihood function is of full rank, and reports these iterations as the refining step.

The header in the output describes the estimation sample, reports the log-likelihood function at the maximum, and gives the results of a Wald test against the null hypothesis that the coefficients on all the independent variables, state variables, and lagged state variables are zero. In this example, the null hypothesis that the coefficient on L1.u is zero is rejected at all conventional levels.

The estimation table reports results for the state equations, the observation equations, and the variance–covariance parameters. The estimated autoregressive coefficient of 0.3524 indicates that there is persistence in the first-differences of the log of the manufacturing rate. The estimated mean of the differenced series is -0.0004, which is smaller in magnitude than its standard error, indicating that there is no deterministic linear trend in the series.

Typing

```
. arima D.lncaputil, ar(1) technique(nr)
(output omitted)
```

produces nearly identical parameter estimates and standard errors for the mean and the autoregressive parameter. Because `sspace` estimates the variance of the state error while `arima` estimates the standard deviation, calculations are required to obtain the same results. The different parameterization of the variance parameter can cause small numerical differences.

◁

❏ Technical note

In some situations, the second part of the iteration log terminates but the refining step never converges. Only when the refining step converges does the maximization algorithm find interpretable estimates. If the refining step iterates without convergence, the parameters of the specified model are not identified by the data. (See Rothenberg [1971], Drukker and Wiggins [2004], and Davidson and MacKinnon [1993, sec. 5.2] for discussions of identification.)

❏

▷ Example 2

Following Harvey (1993, 95–96), we can write a zero-mean, first-order, autoregressive moving-average (ARMA(1,1)) model

$$y_t = \alpha y_{t-1} + \theta \epsilon_{t-1} + \epsilon_t \tag{1}$$

as a state-space model with state equations

$$\begin{pmatrix} y_t \\ \theta \epsilon_t \end{pmatrix} = \begin{pmatrix} \alpha & 1 \\ 0 & 0 \end{pmatrix} \begin{pmatrix} y_{t-1} \\ \theta \epsilon_{t-1} \end{pmatrix} + \begin{pmatrix} 1 \\ \theta \end{pmatrix} \epsilon_t \tag{2}$$

and observation equation

$$y_t = \begin{pmatrix} 1 & 0 \end{pmatrix} \begin{pmatrix} y_t \\ \theta \epsilon_t \end{pmatrix} \tag{3}$$

The unobserved states in this model are $u_{1t} = y_t$ and $u_{2t} = \theta \epsilon_t$. We set the process mean to zero because economic theory and the previous example suggest that we should do so. Below we estimate the parameters in the state-space model by using the error-form syntax:

```
. constraint 2 [u1]L.u2 = 1

. constraint 3 [u1]e.u1 = 1

. constraint 4 [D.lncaputil]u1 = 1

. sspace (u1 L.u1 L.u2 e.u1, state noconstant) (u2 e.u1, state noconstant)
> (D.lncaputil u1, noconstant), constraints(2/4) covstate(diagonal)
searching for initial values ...........
(setting technique to bhhh)
Iteration 0:   log likelihood =  1506.0947
Iteration 1:   log likelihood =   1514.014
  (output omitted )
Refining estimates:
Iteration 0:   log likelihood =   1531.255
Iteration 1:   log likelihood =   1531.255
```

State-space model

Sample: 1972m2 – 2008m12

		Number of obs	=	443
		Wald chi2(2)	=	333.84
		Prob > chi2	=	0.0000

```
Log likelihood =   1531.255
 ( 1)   [u1]L.u2 = 1
 ( 2)   [u1]e.u1 = 1
 ( 3)   [D.lncaputil]u1 = 1
```

lncaputil	Coef.	OIM Std. Err.	z	P>|z|	[95% Conf. Interval]	
u1						
u1 L1.	.8056815	.0522661	15.41	0.000	.7032418	.9081212
u2 L1.	1	(constrained)				
e.u1	1	(constrained)				
u2						
e.u1	−.5188453	.0701985	−7.39	0.000	−.6564317	−.3812588
D.lncaputil u1	1	(constrained)				
Variance u1	.0000582	3.91e−06	14.88	0.000	.0000505	.0000659

Note: Tests of variances against zero are one sided, and the two-sided
    confidence intervals are truncated at zero.

The command in the above output specifies two state equations, one observation equation, and two options. The first state equation defines u_{1t} and the second defines u_{2t} according to (2) above. The observation equation defines the process for D.lncaputil according to the one specified in (3) above. Several coefficients in (2) and (3) are set to 1, and constraints 2–4 place these restrictions on the model.

The estimated coefficient on L.u1 in equation u1, 0.806, is the estimate of α in (2), which is the autoregressive coefficient in the ARMA model in (1). The estimated coefficient on e.u1 in equation u2, −0.519, is the estimate of θ, which is the moving-average term in the ARMA model in (1).

This example highlights a difference between the error-form syntax and the covariance-form syntax. The error-form syntax used in this example includes only explicitly included errors. In contrast, the covariance-form syntax includes an error term in each equation, unless the noerror option is specified.

The default for covstate() also differs between the error-form syntax and the covariance-form syntax. Because the coefficients on the errors in the error-form syntax are frequently used to estimate the standard deviation of the errors, covstate(identity) is the default for the error-form syntax. In contrast, unit variances are less common in the covariance-form syntax, for which covstate(diagonal) is the default. In this example, we specified covstate(diagonal) to estimate a nonunitary variance for the state.

Typing

```
. arima D.lncaputil, noconstant ar(1) ma(1) technique(nr)
(output omitted)
```

produces nearly identical results. As in the AR(1) example above, arima estimates the standard deviation of the error term, while sspace estimates the variance. Although they are theoretically equivalent, the different parameterizations give rise to small numerical differences in the other parameters.

◁

▷ Example 3

The variable lnhours contains data on the log of manufacturing hours, which we treat as first-difference stationary. We have a theory in which the process driving the changes in the log utilization rate affects the changes in the log of hours, but changes in the log hours do not affect changes in the log utilization rate. In line with this theory, we estimate the parameters of a lower triangular, first-order vector autoregressive (VAR(1)) process

$$\begin{pmatrix} \Delta \texttt{lncaputil}_t \\ \Delta \texttt{lnhours}_t \end{pmatrix} = \begin{pmatrix} \alpha_1 & 0 \\ \alpha_2 & \alpha_3 \end{pmatrix} \begin{pmatrix} \Delta \texttt{lncaputil}_{t-1} \\ \Delta \texttt{lnhours}_{t-1} \end{pmatrix} + \begin{pmatrix} \epsilon_{1t} \\ \epsilon_{2t} \end{pmatrix} \tag{4}$$

where $\Delta y_t = y_t - y_{t-1}$, $\epsilon_t = (\epsilon_{1t}, \epsilon_{2t})'$ and $\mathrm{Var}(\epsilon) = \Sigma$. We can write this VAR(1) process as a state-space model with state equations

$$\begin{pmatrix} u_{1t} \\ u_{2t} \end{pmatrix} = \begin{pmatrix} \alpha_1 & 0 \\ \alpha_2 & \alpha_3 \end{pmatrix} \begin{pmatrix} u_{1(t-1)} \\ u_{2(t-1)} \end{pmatrix} + \begin{pmatrix} \epsilon_{1t} \\ \epsilon_{2t} \end{pmatrix} \tag{5}$$

with $\mathrm{Var}(\epsilon) = \Sigma$ and observation equations

$$\begin{pmatrix} \Delta \texttt{lncaputil} \\ \Delta \texttt{lnhours} \end{pmatrix} = \begin{pmatrix} u_{1t} \\ u_{2t} \end{pmatrix}$$

Below we estimate the parameters of the state-space model:

```
. constraint 5 [D.lncaputil]u1 = 1

. constraint 6 [D.lnhours]u2   = 1

. sspace (u1 L.u1, state noconstant)
>        (u2 L.u1 L.u2, state noconstant)
>        (D.lncaputil u1, noconstant noerror)
>        (D.lnhours u2, noconstant noerror),
>        constraints(5/6) covstate(unstructured)
searching for initial values ...........
(setting technique to bhhh)
Iteration 0:   log likelihood =  2993.6647
Iteration 1:   log likelihood =  3088.7416
  (output omitted )
Refining estimates:
Iteration 0:   log likelihood =  3211.7532
Iteration 1:   log likelihood =  3211.7532

State-space model

Sample: 1972m2 - 2008m12                    Number of obs   =          443
                                            Wald chi2(3)    =       166.87
Log likelihood =  3211.7532                 Prob > chi2     =       0.0000
  ( 1)  [D.lncaputil]u1 = 1
  ( 2)  [D.lnhours]u2 = 1
```

		Coef.	OIM Std. Err.	z	P>|z|	[95% Conf. Interval]	
u1							
u1							
L1.		.353257	.0448456	7.88	0.000	.2653612	.4411528
u2							
u1							
L1.		.1286218	.0394742	3.26	0.001	.0512537	.2059899
u2							
L1.		-.3707083	.0434255	-8.54	0.000	-.4558208	-.2855959
D.lncaputil							
u1		1	(constrained)				
D.lnhours							
u2		1	(constrained)				
Variance							
u1		.0000623	4.19e-06	14.88	0.000	.0000541	.0000705
Covariance							
u1							
u2		.000026	2.67e-06	9.75	0.000	.0000208	.0000312
Variance							
u2		.0000386	2.61e-06	14.76	0.000	.0000335	.0000437

Note: Tests of variances against zero are one sided, and the two-sided
      confidence intervals are truncated at zero.

Specifying covstate(unstructured) caused sspace to estimate the off-diagonal element of Σ. The output indicates that this parameter, cov(u2,u1):_cons, is small but statistically significant.

The estimated coefficient on L.u1 in equation u1, 0.353, is the estimate of α_1 in (5). The estimated coefficient on L.u1 in equation u2, 0.129, is the estimate of α_2 in (5). The estimated coefficient on L.u1 in equation u2, -0.371, is the estimate of α_3 in (5).

For the VAR(1) model in (4), the estimated autoregressive coefficient for D.lncaputil is similar to the corresponding estimate in the univariate results in example 1. The estimated effect of LD.lncaputil on D.lnhours is 0.129, the estimated autoregressive coefficient of D.lnhours is -0.371, and both are statistically significant.

These estimates can be compared with those produced by typing

```
. constraint 101 [D_lncaputil]LD.lnhours = 0
. var D.lncaputil D.lnhours, lags(1) noconstant constraints(101)
(output omitted)
. matrix list e(Sigma)
(output omitted)
```

The var estimates are not the same as the sspace estimates because the generalized least-squares estimator implemented in var is only asymptotically equivalent to the ML estimator implemented in sspace, but the point estimates are similar. The comparison is useful for pedagogical purposes because the var estimator is relatively simple.

Some problems require constraining a covariance term to zero. If we wanted to constrain cov(u2,u1):_cons to zero, we could type

```
. constraint 7 [cov(u2,u1)]_cons = 0
. sspace (u1 L.u1, state noconstant)
>        (u2 L.u1 L.u2, state noconstant)
>        (D.lncaputil u1, noconstant noerror)
>        (D.lnhours u2, noconstant noerror),
>        constraints(5/7) covstate(unstructured)
(output omitted)
```

◁

▷ Example 4

We now extend the previous example by modeling D.lncaputil and D.lnhours as a first-order vector autoregressive moving-average (VARMA(1,1)) process. Building on the previous examples, we allow the lag of D.lncaputil to affect D.lnhours but we do not allow the lag of D.lnhours to affect the lag of D.lncaputil. Previous univariate analysis revealed that D.lnhours is better modeled as an autoregressive process than as an ARMA(1,1) process. As a result, we estimate the parameters of

$$\begin{pmatrix} \Delta\text{lncaputil}_t \\ \Delta\text{lnhours}_t \end{pmatrix} = \begin{pmatrix} \alpha_1 & 0 \\ \alpha_2 & \alpha_3 \end{pmatrix} \begin{pmatrix} \Delta\text{lncaputil}_{t-1} \\ \Delta\text{lnhours}_{t-1} \end{pmatrix} + \begin{pmatrix} \theta_1 & 0 \\ 0 & 0 \end{pmatrix} \begin{pmatrix} \epsilon_{1(t-1)} \\ \epsilon_{2(t-1)} \end{pmatrix} + \begin{pmatrix} \epsilon_{1t} \\ \epsilon_{2t} \end{pmatrix}$$

We can write this VARMA(1,1) process as a state-space model with state equations

$$\begin{pmatrix} s_{1t} \\ s_{2t} \\ s_{3t} \end{pmatrix} = \begin{pmatrix} \alpha_1 & 1 & 0 \\ 0 & 0 & 0 \\ \alpha_2 & 0 & \alpha_3 \end{pmatrix} \begin{pmatrix} s_{1(t-1)} \\ s_{2(t-1)} \\ s_{3(t-1)} \end{pmatrix} + \begin{pmatrix} 1 & 0 \\ \theta_1 & 0 \\ 0 & 1 \end{pmatrix} \begin{pmatrix} \epsilon_{1t} \\ \epsilon_{2t} \end{pmatrix}$$

where the states are

$$\begin{pmatrix} s_{1t} \\ s_{2t} \\ s_{3t} \end{pmatrix} = \begin{pmatrix} \Delta\text{lncaputil}_t \\ \theta_1\epsilon_{1t} \\ \Delta\text{lnhours}_t \end{pmatrix}$$

and we simplify the problem by assuming that

$$\text{Var}\begin{pmatrix} \epsilon_{1t} \\ \epsilon_{2t} \end{pmatrix} = \begin{pmatrix} \sigma_1^2 & 0 \\ 0 & \sigma_2^2 \end{pmatrix}$$

Below we estimate the parameters of this model by using sspace:

```
. constraint 7  [u1]L.u2    = 1
. constraint 8  [u1]e.u1    = 1
. constraint 9  [u3]e.u3    = 1
. constraint 10  [D.lncaputil]u1 = 1
. constraint 11 [D.lnhours]u3 = 1
. sspace (u1 L.u1 L.u2 e.u1, state noconstant)
>        (u2 e.u1, state noconstant)
>        (u3 L.u1 L.u3 e.u3, state noconstant)
>        (D.lncaputil u1, noconstant)
>        (D.lnhours u3, noconstant),
>        constraints(7/11) technique(nr) covstate(diagonal)
searching for initial values .........
 (output omitted )
Refining estimates:
Iteration 0:   log likelihood =  3156.0564
Iteration 1:   log likelihood =  3156.0564

State-space model
```

Sample: 1972m2 - 2008m12

Number of obs	= 443
Wald chi2(4)	= 427.55
Prob > chi2	= 0.0000

```
Log likelihood =  3156.0564
 ( 1)  [u1]L.u2 = 1
 ( 2)  [u1]e.u1 = 1
 ( 3)  [u3]e.u3 = 1
 ( 4)  [D.lncaputil]u1 = 1
 ( 5)  [D.lnhours]u3 = 1
```

		OIM					
	Coef.	Std. Err.	z	P>|z|	[95% Conf. Interval]	
u1							
u1							
L1.	.8058031	.0522493	15.42	0.000	.7033964	.9082098
u2							
L1.	1	(constrained)					
e.u1	1	(constrained)					
u2							
e.u1	-.518907	.0701848	-7.39	0.000	-.6564667	-.3813474
u3							
u1							
L1.	.1734868	.0405156	4.28	0.000	.0940776	.252896
u3							
L1.	-.4809376	.0498574	-9.65	0.000	-.5786563	-.3832188
e.u3	1	(constrained)					
D.lncaputil							
u1	1	(constrained)					
D.lnhours							
u3	1	(constrained)					
Variance							
u1	.0000582	3.91e-06	14.88	0.000	.0000505	.0000659
u3	.0000382	2.56e-06	14.88	0.000	.0000331	.0000432

Note: Tests of variances against zero are one sided, and the two-sided
      confidence intervals are truncated at zero.

The estimates of the parameters in the model for D.lncaputil are similar to those in the univariate model fit in example 2. The estimates of the parameters in the model for D.lnhours indicate that the lag of D.lncaputil has a positive effect on D.lnhours.

◁

❑ Technical note

The technique(nr) option facilitates convergence in example 4. Fitting state-space models is notoriously difficult. Convergence problems are common. Four methods for overcoming convergence problems are 1) selecting an alternate optimization algorithm by using the technique() option, 2) using alternative starting values by specifying the from() option, 3) using starting values obtained by estimating the parameters of a restricted version of the model of interest, or 4) putting the variables on the same scale.

❑

▷ Example 5

Stock and Watson (1989, 1991) wrote a simple macroeconomic model as a dynamic-factor model, estimated the parameters by ML, and extracted an economic indicator. In this example, we estimate the parameters of a dynamic-factor model. In [TS] **sspace postestimation**, we extend this example and extract an economic indicator for the differenced series.

We have data on an industrial-production index, ipman; an aggregate weekly hours index, hours; and aggregate unemployment, unemp. income is real disposable income divided by 100. We rescaled real disposable income to avoid convergence problems.

We postulate a latent factor that follows an AR(2) process. Each measured variable is then related to the current value of that latent variable by a parameter. The state-space form of our model is

$$\begin{pmatrix} f_t \\ f_{t-1} \end{pmatrix} = \begin{pmatrix} \theta_1 & \theta_2 \\ 1 & 0 \end{pmatrix} \begin{pmatrix} f_{t-1} \\ f_{t-2} \end{pmatrix} + \begin{pmatrix} \nu_t \\ 0 \end{pmatrix}$$

$$\begin{pmatrix} \Delta\text{ipman}_t \\ \Delta\text{income}_t \\ \Delta\text{hours}_t \\ \Delta\text{unemp}_t \end{pmatrix} = \begin{pmatrix} \gamma_1 \\ \gamma_2 \\ \gamma_3 \\ \gamma_4 \end{pmatrix} f_t + \begin{pmatrix} \epsilon_{1t} \\ \epsilon_{2t} \\ \epsilon_{3t} \\ \epsilon_{4t} \end{pmatrix}$$

where

$$\text{Var}\begin{pmatrix} \epsilon_{1t} \\ \epsilon_{2t} \\ \epsilon_{3t} \\ \epsilon_{4t} \end{pmatrix} = \begin{pmatrix} \sigma_1^2 & 0 & 0 & 0 \\ 0 & \sigma_2^2 & 0 & 0 \\ 0 & 0 & \sigma_3^2 & 0 \\ 0 & 0 & 0 & \sigma_4^2 \end{pmatrix}$$

The parameter estimates are

```
. use http://www.stata-press.com/data/r12/dfex
(St. Louis Fed (FRED) macro data)

. constraint 12 [lf]L.f = 1

. sspace (f L.f L.lf, state noconstant)
>       (lf L.f, state noconstant noerror)
>       (D.ipman f, noconstant)
>       (D.income f, noconstant)
>       (D.hours f, noconstant)
>       (D.unemp f, noconstant),
>       covstate(identity) constraints(12)
searching for initial values ................
(setting technique to bhhh)
Iteration 0:   log likelihood = -676.30903
Iteration 1:   log likelihood = -665.61104
  (output omitted )
Refining estimates:
Iteration 0:   log likelihood = -662.09507
Iteration 1:   log likelihood = -662.09507

State-space model

Sample: 1972m2 - 2008m11                    Number of obs    =         442
                                            Wald chi2(6)     =      751.95
Log likelihood = -662.09507                 Prob > chi2      =      0.0000
 ( 1)  [lf]L.f = 1
```

	Coef.	OIM Std. Err.	z	P>|z|	[95% Conf. Interval]	
f						
f L1.	.2651932	.0568663	4.66	0.000	.1537372	.3766491
lf L1.	.4820398	.0624635	7.72	0.000	.3596136	.604466
lf						
f L1.	1	(constrained)				
D.ipman						
f	.3502249	.0287389	12.19	0.000	.2938976	.4065522
D.income						
f	.0746338	.0217319	3.43	0.001	.0320401	.1172276
D.hours						
f	.2177469	.0186769	11.66	0.000	.1811407	.254353
D.unemp						
f	-.0676016	.0071022	-9.52	0.000	-.0815217	-.0536816
Variance						
D.ipman	.1383158	.0167086	8.28	0.000	.1055675	.1710641
D.income	.2773808	.0188302	14.73	0.000	.2404743	.3142873
D.hours	.0911446	.0080847	11.27	0.000	.0752988	.1069903
D.unemp	.0237232	.0017932	13.23	0.000	.0202086	.0272378

Note: Tests of variances against zero are one sided, and the two-sided
      confidence intervals are truncated at zero.

The output indicates that the unobserved factor is quite persistent and that it is a significant predictor for each of the observed variables.

These models are frequently used to forecast the dependent variables and to estimate the unobserved factors. We present some illustrative examples in [TS] **sspace postestimation**. The dfactor command estimates the parameters of dynamic-factor models; see [TS] **dfactor**.

◁

Some nonstationary state-space models

▷ Example 6

Harvey (1989) advocates the use of STS models. These models parameterize the trends and seasonal components of a set of time series. The simplest STS model is the local-level model, which is given by

$$y_t = \mu_t + \epsilon_t$$

where

$$\mu_t = \mu_{t-1} + \nu_t$$

The model is called a local-level model because the level of the series is modeled as a random walk plus an idiosyncratic noise term. (The model is also known as the random-walk-plus-noise model.) The local-level model is nonstationary because of the random-walk component. When the variance of the idiosyncratic-disturbance ϵ_t is zero and the variance of the level-disturbance ν_t is not zero, the local-level model reduces to a random walk. When the variance of the level-disturbance ν_t is zero and the variance of the idiosyncratic-disturbance ϵ_t is not zero,

$$\mu_t = \mu_{t-1} = \mu$$

and the local-level model reduces to

$$y_t = \mu + \epsilon_t$$

which is a simple regression with a time-invariant mean. The parameter μ is not estimated in the state-space formulation below.

In this example, we fit weekly levels of the Standard and Poor's 500 Index to a local-level model. Because this model is already in state-space form, we fit close by typing

```
. use http://www.stata-press.com/data/r12/sp500w, clear

. constraint 13 [z]L.z  = 1

. constraint 14 [close]z = 1

. sspace (z L.z, state noconstant) (close z, noconstant), constraints(13 14)
searching for initial values ..........
(setting technique to bhhh)
Iteration 0:   log likelihood = -12581.763
Iteration 1:   log likelihood = -12577.727
  (output omitted )
Refining estimates:
Iteration 0:   log likelihood =  -12576.99
Iteration 1:   log likelihood =  -12576.99

State-space model

Sample: 1 - 3093                          Number of obs    =       3093
Log likelihood =  -12576.99
 ( 1)   [z]L.z = 1
 ( 2)   [close]z = 1
```

close	Coef.	OIM Std. Err.	z	P>|z|	[95% Conf. Interval]
z					
z L1.	1	(constrained)			
close					
z	1	(constrained)			
Variance					
z	170.3456	7.584909	22.46	0.000	155.4794 185.2117
close	15.24858	3.392457	4.49	0.000	8.599486 21.89767

```
Note: Model is not stationary.
Note: Tests of variances against zero are one sided, and the two-sided
      confidence intervals are truncated at zero.
```

The results indicate that both components have nonzero variances. The output footer informs us that the model is nonstationary at the estimated parameter values.

◁

❑ Technical note

In the previous example, we estimated the parameters of a nonstationary state-space model. The model is nonstationary because one of the eigenvalues of the $\mathbf{A}$ matrix has unit modulus. That all the coefficients in the $\mathbf{A}$ matrix are fixed is also important. See Lütkepohl (2005, 636–637) for why the ML estimator for the parameters of a nonstationary state-model that is nonstationary because of eigenvalues with unit moduli from a fixed $\mathbf{A}$ matrix is still consistent and asymptotically normal.

❑

▷ Example 7

In another basic STS model, known as the local linear-trend model, both the level and the slope of a linear time trend are random walks. Here are the state equations and the observation equation for a local linear-trend model for the level of industrial production contained in variable ipman:

$$\begin{pmatrix} \mu_t \\ \beta_t \end{pmatrix} = \begin{pmatrix} 1 & 1 \\ 0 & 1 \end{pmatrix} \begin{pmatrix} \mu_{t-1} \\ \beta_{t-1} \end{pmatrix} + \begin{pmatrix} \nu_{1t} \\ \nu_{2t} \end{pmatrix}$$

$$\text{ipman}_t = \mu_t + \epsilon_t$$

The estimated parameters are

```
. use http://www.stata-press.com/data/r12/dfex, clear
(St. Louis Fed (FRED) macro data)

. constraint 15 [f1]L.f1 = 1

. constraint 16 [f1]L.f2 = 1

. constraint 17 [f2]L.f2 = 1

. constraint 18 [ipman]f1 = 1

. sspace  (f1 L.f1 L.f2, state noconstant)
>         (f2 L.f2, state noconstant)
>         (ipman f1, noconstant), constraints(15/18)
searching for initial values ..........
(setting technique to bhhh)
Iteration 0:   log likelihood = -362.93861
Iteration 1:   log likelihood = -362.12048
  (output omitted )
Refining estimates:
Iteration 0:   log likelihood =  -359.1266
Iteration 1:   log likelihood =  -359.1266

State-space model

Sample: 1972m1 - 2008m11                      Number of obs    =        443
Log likelihood =  -359.1266
 ( 1)   [f1]L.f1 = 1
 ( 2)   [f1]L.f2 = 1
 ( 3)   [f2]L.f2 = 1
 ( 4)   [ipman]f1 = 1
```

ipman	Coef.	OIM Std. Err.	z	P>|z|	[95% Conf. Interval]
f1					
f1 L1.	1	(constrained)			
f2 L1.	1	(constrained)			
f2					
f2 L1.	1	(constrained)			
ipman					
f1	1	(constrained)			
Variance					
f1	.1473071	.0407156	3.62	0.000	.067506	.2271082
f2	.0178752	.0065743	2.72	0.003	.0049898	.0307606
ipman	.0354429	.0148186	2.39	0.008	.0063989	.0644868

Note: Model is not stationary.
Note: Tests of variances against zero are one sided, and the two-sided
      confidence intervals are truncated at zero.

There is little evidence that either of the variance parameters are zero. The fit obtained indicates that we could now proceed with specification testing and checks to see how well this model forecasts these data.

◁

Saved results

sspace saves the following in e():

Scalars

e(N)	number of observations
e(k)	number of parameters
e(k_aux)	number of auxiliary parameters
e(k_eq)	number of equations in e(b)
e(k_dv)	number of dependent variables
e(k_obser)	number of observation equations
e(k_state)	number of state equations
e(k_obser_err)	number of observation-error terms
e(k_state_err)	number of state-error terms
e(df_m)	model degrees of freedom
e(ll)	log likelihood
e(chi2)	χ^2
e(p)	significance
e(tmin)	minimum time in sample
e(tmax)	maximum time in sample
e(stationary)	1 if the estimated parameters indicate a stationary model, 0 otherwise
e(rank)	rank of VCE
e(ic)	number of iterations
e(rc)	return code
e(converged)	1 if converged, 0 otherwise

Macros

e(cmd)	sspace
e(cmdline)	command as typed
e(depvar)	unoperated names of dependent variables in observation equations
e(obser_deps)	names of dependent variables in observation equations
e(state_deps)	names of dependent variables in state equations
e(covariates)	list of covariates
e(indeps)	independent variables
e(tvar)	variable denoting time within groups
e(eqnames)	names of equations
e(title)	title in estimation output
e(tmins)	formatted minimum time
e(tmaxs)	formatted maximum time
e(R_structure)	structure of observed-variable-error covariance matrix
e(Q_structure)	structure of state-error covariance matrix
e(chi2type)	Wald; type of model χ^2 test
e(vce)	vcetype specified in vce()
e(vcetype)	title used to label Std. Err.
e(opt)	type of optimization
e(method)	likelihood method
e(initial_values)	type of initial values
e(technique)	maximization technique
e(tech_steps)	iterations taken in maximization technique
e(datasignature)	the checksum
e(datasignaturevars)	variables used in calculation of checksum
e(properties)	b V
e(estat_cmd)	program used to implement estat
e(predict)	program used to implement predict
e(marginsok)	predictions allowed by margins
e(marginsnotok)	predictions disallowed by margins

Matrices
    e(b)   parameter vector

e(b)	parameter vector
e(Cns)	constraints matrix
e(ilog)	iteration log (up to 20 iterations)
e(gradient)	gradient vector
e(gamma)	mapping from parameter vector to state-space matrices
e(A)	estimated A matrix
e(B)	estimated B matrix
e(C)	estimated C matrix
e(D)	estimated D matrix
e(F)	estimated F matrix
e(G)	estimated G matrix
e(chol_R)	Cholesky factor of estimated R matrix
e(chol_Q)	Cholesky factor of estimated Q matrix
e(chol_Sz0)	Cholesky factor of initial state covariance matrix
e(z0)	initial state vector augmented with a matrix identifying nonstationary components
e(d)	additional term in diffuse initial state vector, if nonstationary model
e(T)	inner part of quadratic form for initial state covariance in a partially nonstationary model
e(M)	outer part of quadratic form for initial state covariance in a partially nonstationary model
e(V)	variance–covariance matrix of the estimators
e(V_modelbased)	model-based variance

Functions

e(sample)	marks estimation sample

Methods and formulas

sspace is implemented as an ado-file.

Recall that our notation for linear state-space models with time-invariant coefficient matrices is

$$\mathbf{z}_t = \mathbf{A}\mathbf{z}_{t-1} + \mathbf{B}\mathbf{x}_t + \mathbf{C}\boldsymbol{\epsilon}_t$$
$$\mathbf{y}_t = \mathbf{D}\mathbf{z}_t + \mathbf{F}\mathbf{w}_t + \mathbf{G}\boldsymbol{\nu}_t$$

where

$\mathbf{z}_t$ is an $m \times 1$ vector of unobserved state variables;

$\mathbf{x}_t$ is a $k_x \times 1$ vector of exogenous variables;

$\boldsymbol{\epsilon}_t$ is a $q \times 1$ vector of state-error terms, $(q \leq m)$;

$\mathbf{y}_t$ is an $n \times 1$ vector of observed endogenous variables;

$\mathbf{w}_t$ is a $k_w \times 1$ vector of exogenous variables;

$\boldsymbol{\nu}_t$ is an $r \times 1$ vector of observation-error terms, $(r \leq n)$; and

$\mathbf{A}$, $\mathbf{B}$, $\mathbf{C}$, $\mathbf{D}$, $\mathbf{F}$, and $\mathbf{G}$ are parameter matrices.

The equations for $\mathbf{z}_t$ are known as the state equations, and the equations for $\mathbf{y}_t$ are known as the observation equations.

The error terms are assumed to be zero mean, normally distributed, serially uncorrelated, and uncorrelated with each other;

$$\epsilon_t \sim N(0, \mathbf{Q})$$
$$\nu_t \sim N(0, \mathbf{R})$$
$$E[\epsilon_t \epsilon_s'] = \mathbf{0} \text{ for all } s \neq t$$
$$E[\epsilon_t \nu_s'] = \mathbf{0} \text{ for all } s \text{ and } t$$

sspace estimates the parameters of linear state-space models by maximum likelihood. The Kalman filter is a method for recursively obtaining linear, least-squares forecasts of $\mathbf{y}_t$ conditional on past information. These forecasts are used to construct the log likelihood, assuming normality and stationarity. When the model is nonstationary, a diffuse Kalman filter is used.

Hamilton (1994a; 1994b, 389) shows that the QML estimator, obtained when the normality assumption is dropped, is consistent and asymptotically normal, although the variance–covariance matrix of the estimator (VCE) must be estimated by the Huber/White/sandwich estimator. Hamilton's discussion applies to stationary models, and specifying vce(robust) produces a consistent estimator of the VCE when the errors are not normal.

Methods for computing the log likelihood differ in how they calculate initial values for the Kalman filter when the model is stationary, how they compute a diffuse Kalman filter when the model is nonstationary, and whether terms for initial states are included. sspace offers the method(hybrid), method(dejong), and method(kdiffuse) options for computing the log likelihood. All three methods handle both stationary and nonstationary models.

method(hybrid), the default, uses the initial values for the states implied by stationarity to initialize the Kalman filter when the model is stationary. Hamilton (1994b, 378) discusses this method of computing initial values for the states and derives a log-likelihood function that does not include terms for the initial states. When the model is nonstationary, method(hybrid) uses the De Jong (1988, 1991) diffuse Kalman filter and log-likelihood function, which includes terms for the initial states.

method(dejong) uses the stationary De Jong (1988) method when the model is stationary and the De Jong (1988, 1991) diffuse Kalman filter when the model is nonstationary. The stationary De Jong (1988) method estimates initial values for the Kalman filter as part of the log-likelihood computation, as in De Jong (1988).

method(kdiffuse) implements the seldom-used large-κ diffuse approximation to the diffuse Kalman filter when the model is nonstationary and uses initial values for the states implied by stationarity when the model is stationary. The log likelihood does not include terms for the initial states in either case. We recommend that you do not use method(kdiffuse) except to replicate older results computed using this method.

De Jong (1988, 1991) and De Jong and Chu-Chun-Lin (1994) derive the log likelihood and a diffuse Kalman filter for handling nonstationary data. De Jong (1988) replaces the stationarity assumption with a time-immemorial assumption, which he uses to derive the log-likelihood function, an initial state vector, and a covariance of the initial state vector when the model is nonstationary. By default, and when method(hybrid) or method(dejong) is specified, sspace uses the diffuse Kalman filter given in definition 5 of De Jong and Chu-Chun-Lin (1994). This method uses theorem 3 of De Jong and Chu-Chun-Lin (1994) to compute the covariance of the initial states. When using this method, sspace saves the matrices from their theorem 3 in e(), although the names are changed. e(Z) is their $\mathbf{U}_1$, e(T) is their $\mathbf{U}_2$, e(A) is their $\mathbf{T}$, and e(M) is their $\mathbf{M}$.

See De Jong (1988, 1991) and De Jong and Chu-Chun-Lin (1994) for the details of the De Jong diffuse Kalman filter.

Practical estimation and inference require that the maximum likelihood estimator be consistent and normally distributed in large samples. These statistical properties of the maximum likelihood estimator are well established when the model is stationary; see Caines (1988, chap. 5 and 7), Hamilton (1994b, 388–389), and Hannan and Deistler (1988, chap. 4). When the model is nonstationary, additional assumptions must hold for the maximum likelihood estimator to be consistent and asymptotically normal; see Harvey (1989, sec. 3.4), Lütkepohl (2005, 636–637), and Schneider (1988). Chang, Miller, and Park (2009) show that the ML and the QML estimators are consistent and asymptotically normal for a class of nonstationary state-space models.

We now give an intuitive version of the Kalman filter. sspace uses theoretically equivalent, but numerically more stable, methods. For each time t, the Kalman filter produces the conditional expected state vector $\mathbf{z}_{t|t}$ and the conditional covariance matrix $\mathbf{\Omega}_{t|t}$; both are conditional on information up to and including time t. Using the model and previous period results, for each t we begin with

$$\mathbf{z}_{t|t-1} = \mathbf{A}\mathbf{z}_{t-1|t-1} + \mathbf{B}\mathbf{x}_t$$
$$\mathbf{\Omega}_{t|t-1} = \mathbf{A}\mathbf{\Omega}_{t-1|t-1}\mathbf{A}' + \mathbf{C}\mathbf{Q}\mathbf{C}' \tag{6}$$
$$\mathbf{y}_{t|t-1} = \mathbf{D}\mathbf{z}_{t|t-1} + \mathbf{F}\mathbf{w}_t$$

The residuals and the mean squared error (MSE) matrix of the forecast error are

$$\widetilde{\boldsymbol{\nu}}_{t|t} = \mathbf{y}_t - \mathbf{y}_{t|t-1}$$
$$\mathbf{\Sigma}_{t|t} = \mathbf{D}\mathbf{\Omega}_{t|t-1}\mathbf{D}' + \mathbf{G}\mathbf{R}\mathbf{G}' \tag{7}$$

In the last steps, we update the conditional expected state vector and the conditional covariance with the time t information:

$$\mathbf{z}_{t|t} = \mathbf{z}_{t|t-1} + \mathbf{\Omega}_{t|t-1}\mathbf{D}\mathbf{\Sigma}_{t|t}^{-1}\widetilde{\boldsymbol{\nu}}_{t|t}$$
$$\mathbf{\Omega}_{t|t} = \mathbf{\Omega}_{t|t-1} - \mathbf{\Omega}_{t|t-1}\mathbf{D}\mathbf{\Sigma}_{t|t}^{-1}\mathbf{D}'\mathbf{\Omega}_{t|t-1} \tag{8}$$

Equations (6)–(8) are the Kalman filter. The equations denoted by (6) are the one-step predictions. The one-step predictions do not use contemporaneous values of $\mathbf{y}_t$; only past values of $\mathbf{y}_t$, past values of the exogenous $\mathbf{x}_t$, and contemporaneous values of $\mathbf{x}_t$ are used. Equations (7) and (8) form the update step of the Kalman filter; they incorporate the contemporaneous dependent variable information into the predicted states.

The Kalman filter requires initial values for the states and a covariance matrix for the initial states to start off the recursive process. Hamilton (1994b) discusses how to compute initial values for the Kalman filter assuming stationarity. This method is used by default when the model is stationary. De Jong (1988) discusses how to estimate initial values by maximum likelihood; this method is used when method(dejong) is specified.

Letting $\boldsymbol{\delta}$ be the vector of parameters in the model, Lütkepohl (2005) and Harvey (1989) show that the log-likelihood function for the parameters of a stationary model is given by

$$\ln L(\boldsymbol{\delta}) = -0.5 \left\{ nT \ln(2\pi) + \sum_{t=1}^{T} \ln(|\boldsymbol{\Sigma}_{t|t-1}|) + \sum_{t=1}^{T} \mathbf{e}_t' \boldsymbol{\Sigma}_{t|t-1}^{-1} \mathbf{e}_t \right\}$$

where $\mathbf{e}_t = (\mathbf{y}_t - \mathbf{y}_{t|t-1})$ depends on $\boldsymbol{\delta}$ and $\boldsymbol{\Sigma}$ also depends on $\boldsymbol{\delta}$.

The variance–covariance matrix of the estimator (VCE) is estimated by the observed information matrix (OIM) estimator by default. Specifying vce(robust) causes sspace to use the Huber/White/sandwich estimator. Both estimators of the VCE are standard and documented in Hamilton (1994b).

Hamilton (1994b), Hannan and Deistler (1988), and Caines (1988) show that the ML estimator is consistent and asymptotically normal when the model is stationary. Schneider (1988) establishes consistency and asymptotic normality when the model is nonstationary because $\mathbf{A}$ has some eigenvalues with modulus 1 and there are no unknown parameters in $\mathbf{A}$.

Not all state-space models are identified, as discussed in Hamilton (1994b) and Lütkepohl (2005). sspace checks for local identification at the optimum. sspace will not declare convergence unless the Hessian is full rank. This check for local identifiability is due to Rothenberg (1971).

Specifying method(dejong) causes sspace to maximize the log-likelihood function given in section 2 (vii) of De Jong (1988). This log-likelihood function includes the initial states as parameters to be estimated. We use some of the methods in Casals, Sotoca, and Jerez (1999) for computing the De Jong (1988) log-likelihood function.

References

Anderson, B. D. O., and J. B. Moore. 1979. *Optimal Filtering*. Englewood Cliffs, NJ: Prentice Hall.

Brockwell, P. J., and R. A. Davis. 1991. *Time Series: Theory and Methods*. 2nd ed. New York: Springer.

Caines, P. E. 1988. *Linear Stochastic Systems*. New York: Wiley.

Casals, J., S. Sotoca, and M. Jerez. 1999. A fast and stable method to compute the likelihood of time invariant state-space models. *Economics Letters* 65: 329–337.

Chang, Y., J. I. Miller, and J. Y. Park. 2009. Extracting a common stochastic trend: Theory with some applications. *Journal of Econometrics* 150: 231–247.

Davidson, R., and J. G. MacKinnon. 1993. *Estimation and Inference in Econometrics*. New York: Oxford University Press.

De Jong, P. 1988. The likelihood for a state space model. *Biometrika* 75: 165–169.

———. 1991. The diffuse Kalman filter. *Annals of Statistics* 19: 1073–1083.

De Jong, P., and S. Chu-Chun-Lin. 1994. Stationary and non-stationary state space models. *Journal of Time Series Analysis* 15: 151–166.

Drukker, D. M., and V. L. Wiggins. 2004. Verifying the solution from a nonlinear solver: A case study: Comment. *American Economic Review* 94: 397–399.

Hamilton, J. D. 1994a. State-space models. In Vol. 4 of *Handbook of Econometrics*, ed. R. F. Engle and D. L. McFadden, 3039–3080. New York: Elsevier.

———. 1994b. *Time Series Analysis*. Princeton: Princeton University Press.

Hannan, E. J., and M. Deistler. 1988. *The Statistical Theory of Linear Systems*. New York: Wiley.

Harvey, A. C. 1989. *Forecasting, Structural Time Series Models and the Kalman Filter*. Cambridge: Cambridge University Press.

———. 1993. *Time Series Models*. 2nd ed. Cambridge, MA: MIT Press.

Lütkepohl, H. 2005. *New Introduction to Multiple Time Series Analysis*. New York: Springer.

Rothenberg, T. J. 1971. Identification in parametric models. *Econometrica* 39: 577–591.

Schneider, W. 1988. Analytical uses of Kalman filtering in econometrics: A survey. *Statistical Papers* 29: 3–33.

Stock, J. H., and M. W. Watson. 1989. New indexes of coincident and leading economic indicators. In *NBER Macroeconomics Annual 1989*, ed. O. J. Blanchard and S. Fischer, vol. 4, 351–394. Cambridge, MA: MIT Press.

——. 1991. A probability model of the coincident economic indicators. In *Leading Economic Indicators: New Approaches and Forecasting Records*, ed. K. Lahiri and G. H. Moore, 63–89. Cambridge: Cambridge University Press.

Also see

[TS] **sspace postestimation** — Postestimation tools for sspace

[TS] **tsset** — Declare data to be time-series data

[TS] **dfactor** — Dynamic-factor models

[TS] **ucm** — Unobserved-components model

[TS] **var** — Vector autoregressive models

[TS] **arima** — ARIMA, ARMAX, and other dynamic regression models

[U] **20 Estimation and postestimation commands**

Title

> **sspace postestimation** — Postestimation tools for sspace

Description

The following standard postestimation commands are available after `sspace`:

Command	Description
estat	AIC, BIC, VCE, and estimation sample summary
estimates	cataloging estimation results
lincom	point estimates, standard errors, testing and inference for linear combinations of coefficients
lrtest	likelihood-ratio test
nlcom	point estimates, standard errors, testing and inference for nonlinear combinations of coefficients
predict	predictions, residuals, influence statistics, and other diagnostic measures
predictnl	point estimates, standard errors, testing, and inference for generalized predictions
test	Wald tests of simple and composite linear hypotheses
testnl	Wald tests of nonlinear hypotheses

See the corresponding entries in the *Base Reference Manual* for details.

Syntax for predict

predict [*type*] { *stub* * | *newvarlist* } [*if*] [*in*] [, *statistic options*]

statistic	Description
Main	
xb	observable variables
states	latent state variables
residuals	residuals
rstandard	standardized residuals

These statistics are available both in and out of sample; type `predict ... if e(sample) ...` if wanted only for the estimation sample.

options	Description
Options	
equation(*eqnames*)	name(s) of equation(s) for which predictions are to be made
rmse(*stub* | newvarlist*)	put estimated root mean squared errors of predicted statistics in new variables
dynamic(*time_constant*)	begin dynamic forecast at specified time
Advanced	
smethod(*method*)	method for predicting unobserved states

method	Description
onestep	predict using past information
smooth	predict using all sample information
filter	predict using past and contemporaneous information

Menu

Statistics > Postestimation > Predictions, residuals, etc.

Options for predict

   ⌐ Main ⌐

xb, states, residuals, and rstandard specify the statistic to be predicted.

   xb, the default, calculates the linear predictions of the observed variables.

   states calculates the linear predictions of the latent state variables.

   residuals calculates the residuals in the equations for observable variables. residuals may not be specified with dynamic().

   rstandard calculates the standardized residuals, which are the residuals normalized to be uncorrelated and to have unit variances. rstandard may not be specified with smethod(filter), smethod(smooth), or dynamic().

   ⌐ Options ⌐

equation(*eqnames*) specifies the equation(s) for which the predictions are to be calculated. If you do not specify equation() or *stub**, the results are the same as if you had specified the name of the first equation for the predicted statistic.

   You specify a list of equation names, such as equation(income consumption) or equation(factor1 factor2), to identify the equations. Specify names of state equations when predicting states and names of observable equations in all other cases.

   equation() may not be specified with *stub**.

rmse(*stub* \| newvarlist*) puts the root mean squared errors of the predicted statistics into the specified new variables. The root mean squared errors measure the variances due to the disturbances but do not account for estimation error.

dynamic(*time_constant*) specifies when `predict` starts producing dynamic forecasts. The specified *time_constant* must be in the scale of the time variable specified in `tsset`, and the *time_constant* must be inside a sample for which observations on the dependent variables are available. For example, `dynamic(tq(2008q4))` causes dynamic predictions to begin in the fourth quarter of 2008, assuming that your time variable is quarterly; see [D] **datetime**. If the model contains exogenous variables, they must be present for the whole predicted sample. `dynamic()` may not be specified with `rstandard`, `residuals`, or `smethod(smooth)`.

◀─────┐ Advanced └──

smethod(*method*) specifies the method for predicting the unobserved states; `smethod(onestep)`, `smethod(filter)`, and `smethod(smooth)` cause different amounts of information on the dependent variables to be used in predicting the states at each time period.

> smethod(onestep), the default, causes `predict` to estimate the states at each time period using previous information on the dependent variables. The Kalman filter is performed on previous periods, but only the one-step predictions are made for the current period.

> smethod(smooth) causes `predict` to estimate the states at each time period using all the sample data by the Kalman smoother. `smethod(smooth)` may not be specified with `rstandard`.

> smethod(filter) causes `predict` to estimate the states at each time period using previous and contemporaneous data by the Kalman filter. The Kalman filter is performed on previous periods and the current period. `smethod(filter)` may be specified only with `states`.

Remarks

We assume that you have already read [TS] **sspace**. In this entry, we illustrate some of the features of `predict` after using `sspace` to estimate the parameters of a state-space model.

All the predictions after `sspace` depend on the unobserved states, which are estimated recursively. Changing the sample can alter the state estimates, which can change all other predictions.

▷ Example 1

In example 5 of [TS] **sspace**, we estimated the parameters of the dynamic-factor model

$$
\begin{pmatrix} f_t \\ f_{t-1} \end{pmatrix} = \begin{pmatrix} \theta_1 & \theta_2 \\ 1 & 0 \end{pmatrix} \begin{pmatrix} f_{t-1} \\ f_{t-2} \end{pmatrix} + \begin{pmatrix} \nu_t \\ 0 \end{pmatrix}
$$

$$
\begin{pmatrix} \Delta \text{ipman}_t \\ \Delta \text{income}_t \\ \Delta \text{hours}_t \\ \Delta \text{unemp}_t \end{pmatrix} = \begin{pmatrix} \gamma_1 \\ \gamma_2 \\ \gamma_3 \\ \gamma_4 \end{pmatrix} f_t + \begin{pmatrix} \epsilon_{1t} \\ \epsilon_{2t} \\ \epsilon_{3t} \\ \epsilon_{4t} \end{pmatrix}
$$

where

$$
\text{Var} \begin{pmatrix} \epsilon_{1t} \\ \epsilon_{2t} \\ \epsilon_{3t} \\ \epsilon_{4t} \end{pmatrix} = \begin{pmatrix} \sigma_1^2 & 0 & 0 & 0 \\ 0 & \sigma_2^2 & 0 & 0 \\ 0 & 0 & \sigma_3^2 & 0 \\ 0 & 0 & 0 & \sigma_4^2 \end{pmatrix}
$$

by typing

```
. use http://www.stata-press.com/data/r12/dfex
(St. Louis Fed (FRED) macro data)
. constraint 1 [lf]L.f = 1
. sspace (f L.f L.lf, state noconstant)
>         (lf L.f, state noconstant noerror)
>         (D.ipman f, noconstant)
>         (D.income f, noconstant)
>         (D.hours f, noconstant)
>         (D.unemp f, noconstant),
>         covstate(identity) constraints(1)
  (output omitted )
```

Below we obtain the one-step predictions for each of the four dependent variables in the model, and then we graph the actual and predicted `ipman`:

```
. predict dep*
(option xb assumed; fitted values)
. tsline D.ipman dep1, lcolor(gs10) xtitle("") legend(rows(2))
```

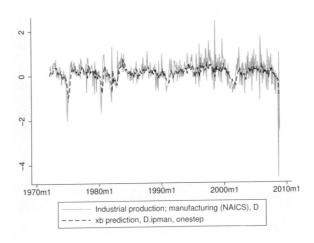

The graph shows that the one-step predictions account for only a small part of the swings in the realized `ipman`.

◁

▷ Example 2

We use the estimates from the previous example to make out-of-sample predictions. After using `tsappend` to extend the dataset by six periods, we use `predict` with the `dynamic()` option and graph the result.

```
. tsappend, add(6)
. predict Dipman_f, dynamic(tm(2008m12)) equation(D.ipman)
. tsline D.ipman Dipman_f if month>=tm(2008m1), xtitle("") legend(rows(2))
```

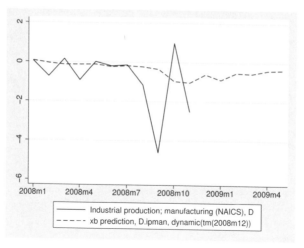

The model predicts that the changes in industrial production will remain negative for the forecast horizon, although they increase toward zero.

◁

▷ Example 3

In this example, we want to estimate the unobserved factor instead of predicting a dependent variable. Specifying smethod(smooth) causes predict to use all sample information in estimating the states by the Kalman smoother.

Below we estimate the unobserved factor by using the estimation sample, and we graph ipman and the estimated factor:

```
. predict fac if e(sample), states smethod(smooth) equation(f)
. tsline D.ipman fac, xtitle("") legend(rows(2))
```

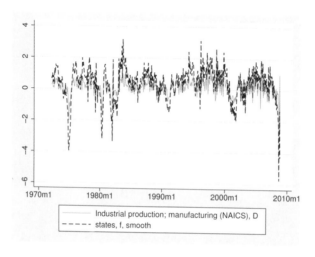

◁

▷ Example 4

The residuals and the standardized residuals are frequently used to review the specification of the model.

Below we calculate the standardized residuals for each of the series and display them in a combined graph:

```
. predict sres1-sres4 if e(sample), rstandard
. tsline sres1, xtitle("") name(sres1)
. tsline sres2, xtitle("") name(sres2)
. tsline sres3, xtitle("") name(sres3)
. tsline sres4, xtitle("") name(sres4)
. graph combine sres1 sres2 sres3 sres4, name(combined)
```

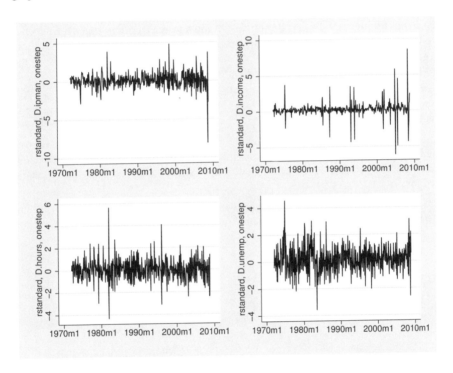

◁

Methods and formulas

All postestimation commands listed above are implemented as ado-files.

Estimating the unobserved states is key to predicting the dependent variables.

By default and with the `smethod(onestep)` option, `predict` estimates the states in each period by applying the Kalman filter to all previous periods and only making the one-step predictions to the current period. (See *Methods and formulas* of [TS] **sspace** for the Kalman filter equations.)

With the `smethod(filter)` option, `predict` estimates the states in each period by applying the Kalman filter on all previous periods and the current period. The computational difference between `smethod(onestep)` and `smethod(filter)` is that `smethod(filter)` performs the update step on the current period while `smethod(onestep)` does not. The statistical difference between `smethod(onestep)` and `smethod(filter)` is that `smethod(filter)` uses contemporaneous information on the dependent variables while `smethod(onestep)` does not.

As noted in [TS] **sspace**, `sspace` has both a stationary and a diffuse Kalman filter. `predict` uses the same Kalman filter used for estimation.

With the `smethod(smooth)` option, `predict` estimates the states in each period using all the sample information by applying the Kalman smoother. `predict` uses the Harvey (1989, sec. 3.6.2) fixed-interval smoother with model-based initial values to estimate the states when the estimated parameters imply a stationary model. De Jong (1989) provides a computationally efficient method. Hamilton (1994) discusses the model-based initial values for stationary state-space models. When the model is nonstationary, the De Jong (1989) diffuse Kalman smoother is used to predict the states. The smoothed estimates of the states are subsequently used to predict the dependent variables.

The dependent variables are predicted by plugging in the estimated states. The residuals are calculated as the differences between the predicted and the realized dependent variables. The root mean squared errors are the square roots of the diagonal elements of the mean squared error matrices that are computed by the Kalman filter. The standardized residuals are the residuals normalized by the Cholesky factor of their mean squared error produced by the Kalman filter.

`predict` uses the Harvey (1989, sec. 3.5) methods to compute the dynamic forecasts and the root mean squared errors. Let τ be the period at which the dynamic forecasts begin; τ must either be in the specified sample or be in the period immediately following the specified sample.

The dynamic forecasts depend on the predicted states in the period $\tau - 1$, which `predict` obtains by running the Kalman filter or the diffuse Kalman filter on the previous sample observations. The states in the periods prior to starting the dynamic predictions may be estimated using `smethod(onestep)` or `smethod(smooth)`.

Using an `if` or `in` qualifier to alter the prediction sample can change the estimate of the unobserved states in the period prior to beginning the dynamic predictions and hence alter the dynamic predictions. The initial states are estimated using `e(b)` and the prediction sample.

References

De Jong, P. 1988. The likelihood for a state space model. *Biometrika* 75: 165–169.

———. 1989. Smoothing and interpolation with the state-space model. *Journal of the American Statistical Association* 84: 1085–1088.

———. 1991. The diffuse Kalman filter. *Annals of Statistics* 19: 1073–1083.

Hamilton, J. D. 1994. *Time Series Analysis*. Princeton: Princeton University Press.

Harvey, A. C. 1989. *Forecasting, Structural Time Series Models and the Kalman Filter*. Cambridge: Cambridge University Press.

Lütkepohl, H. 2005. *New Introduction to Multiple Time Series Analysis*. New York: Springer.

Also see

Title

> **tsappend** — Add observations to a time-series dataset

Syntax

tsappend , { add(#) | last(*date* | *clock*) tsfmt(*string*) } [*options*]

options	Description
* add(#)	add # observations
* last(*date* | *clock*)	add observations at *date* or *clock*
* tsfmt(*string*)	use time-series function *string* with last(*date* | *clock*)
panel(*panel_id*)	add observations to panel *panel_id*

* Either add(#) is required, or last(*date* | *clock*) and tsfmt(*string*) are required.

You must tsset your data before using tsappend; see [TS] **tsset**.

Menu

Statistics > Time series > Setup and utilities > Add observations to time-series dataset

Description

tsappend appends observations to a time-series dataset or to a panel dataset. tsappend uses and updates the information set by tsset.

Options

add(#) specifies the number of observations to add.

last(*date* | *clock*) and tsfmt(*string*) must be specified together and are an alternative to add().

last(*date* | *clock*) specifies the date or the date and time of the last observation to add.

tsfmt(*string*) specifies the name of the Stata time-series function to use in converting the date specified in last() to an integer. The function names are tc (clock), tC (Clock), td (daily), tw (weekly), tm (monthly), tq (quarterly), th (half-yearly), and ty (yearly).

For clock times, the last time added (if any) will be earlier than the time requested in last(*date* | *clock*) if last() is not a multiple of delta units from the last time in the data.

For instance, you might specify last(17may2007) tsfmt(td), last(2001m1) tsfmt(tm), or last(17may2007 15:30:00) tsfmt(tc).

panel(*panel_id*) specifies that observations be added only to panels with the ID specified in panel().

Remarks

Remarks are presented under the following headings:

Introduction
Using tsappend with time-series data
Using tsappend with panel data

Introduction

tsappend adds observations to a time-series dataset or to a panel dataset. You must tsset your data before using tsappend. tsappend simultaneously removes any gaps from the dataset.

There are two ways to use tsappend: you can specify the add(#) option to request that # observations be added, or you can specify the last(*date* | *clock*) option to request that observations be appended until the date specified is reached. If you specify last(), you must also specify tsfmt(). tsfmt() specifies the Stata time-series date function that converts the date held in last() to an integer.

tsappend works with time series of panel data. With panel data, tsappend adds the requested observations to all the panels, unless the panel() option is also specified.

Using tsappend with time-series data

tsappend can be useful for appending observations when dynamically predicting a time series. Consider an example in which tsappend adds the extra observations before dynamically predicting from an AR(1) regression:

```
. use http://www.stata-press.com/data/r12/tsappend1
. regress y l.y
```

Source	SS	df	MS
Model	115.349555	1	115.349555
Residual	461.241577	477	.966963473
Total	576.591132	478	1.2062576

Number of obs =   479
F(1, 477) = 119.29
Prob > F   = 0.0000
R-squared   = 0.2001
Adj R-squared = 0.1984
Root MSE   = .98334

y	Coef.	Std. Err.	t	P>|t|	[95% Conf. Interval]
y					
L1.	.4493507	.0411417	10.92	0.000	.3685093 .5301921
_cons	11.11877	.8314581	13.37	0.000	9.484993 12.75254

```
. mat b = e(b)
. mat colnames b = L.xb one
. tsset
        time variable:  t2, 1960m2 to 2000m1
                delta:  1 month
. tsappend, add(12)
. tsset
        time variable:  t2, 1960m2 to 2001m1
                delta:  1 month
. predict xb if t2<=tm(2000m2)
(option xb assumed; fitted values)
(12 missing values generated)
. gen one=1
. mat score xb=b if t2>=tm(2000m2), replace
```

The calls to tsset before and after tsappend were unnecessary. Their output reveals that tsappend added another year of observations. We then used predict and matrix score to obtain the dynamic predictions, which allows us to produce the following graph:

```
. line y xb t2 if t2>=tm(1995m1), ytitle("") xtitle("time")
```

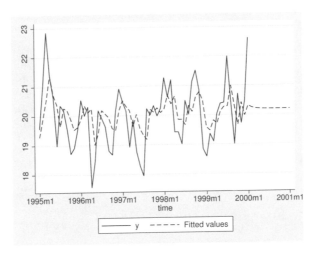

In the call to `tsappend`, instead of saying that we wanted to add 12 observations, we could have specified that we wanted to fill in observations through the first month of 2001:

```
. use http://www.stata-press.com/data/r12/tsappend1, clear
. tsset
        time variable:  t2, 1960m2 to 2000m1
                delta:  1 month
. tsappend, last(2001m1) tsfmt(tm)
. tsset
        time variable:  t2, 1960m2 to 2001m1
                delta:  1 month
```

We specified the `tm()` function in the `tsfmt()` option. [D] **functions** contains a list of time-series functions for translating date literals to integers. Because we have monthly data, and since [D] **functions** tells us that we want to use the `tm()` function, we specified the `tsfmt(tm)` option. The following table shows the most common types of time-series data, their formats, the appropriate translation functions, and the corresponding options for `tsappend`:

Description	Format	Function	Option
time	%tc	tc()	tsfmt(tc)
time	%tC	tC()	tsfmt(tC)
daily	%td	td()	tsfmt(td)
weekly	%tw	tw()	tsfmt(tw)
monthly	%tm	tm()	tsfmt(tm)
quarterly	%tq	tq()	tsfmt(tq)
half-yearly	%th	th()	tsfmt(th)
yearly	%ty	ty()	tsfmt(ty)

Using tsappend with panel data

tsappend's actions on panel data are similar to its action on time-series data, except that tsappend performs those actions on each time series within the panels.

If the end dates vary over panels, last() and add() will produce different results. add(#) always adds # observations to each panel. If the data end at different periods before tsappend, add() is used, the data will still end at different periods after tsappend, add(). In contrast, tsappend, last() tsfmt() will cause all the panels to end on the specified last date. If the beginning dates differ across panels, using tsappend, last() tsfmt() to provide a uniform ending date will not create balanced panels because the number of observations per panel will still differ.

Consider the panel data summarized in the output below:

```
. use http://www.stata-press.com/data/r12/tsappend3, clear
. xtdescribe
    id:  1, 2, ..., 3                            n =        3
    t2:  1998m1, 1998m2, ..., 2000m1             T =       25
         Delta(t2) = 1 month
         Span(t2)  = 25 periods
         (id*t2 uniquely identifies each observation)
  Distribution of T_i:   min      5%     25%      50%     75%     95%     max
                          13      13      13       20      24      24      24
      Freq.  Percent    Cum.  |  Pattern
  ---------------------------+----------------------------
         1    33.33    33.33  |  ............1111111111111
         1    33.33    66.67  |  1111.111111111111111111111
         1    33.33   100.00  |  1111111111111111111111.....
  ---------------------------+----------------------------
         3   100.00           |  XXXXXXXXXXXXXXXXXXXXXXXXXX
. by id: sum t2

-> id = 1
    Variable |     Obs       Mean    Std. Dev.      Min      Max
  -----------+-------------------------------------------------
          t2 |      13        474      3.89444      468      480

-> id = 2
    Variable |     Obs       Mean    Std. Dev.      Min      Max
  -----------+-------------------------------------------------
          t2 |      20      465.5      5.91608      456      475

-> id = 3
    Variable |     Obs       Mean    Std. Dev.      Min      Max
  -----------+-------------------------------------------------
          t2 |      24   468.3333     7.322786      456      480
```

The output from xtdescribe and summarize on these data tells us that one panel starts later than the other, that another panel ends before the other two, and that the remaining panel has a gap in the time variable but otherwise spans the entire time frame.

Now consider the data after a call to tsappend, add(6):

```
. tsappend, add(6)
. xtdescribe
        id:  1, 2, ..., 3                                   n =          3
        t2:  1998m1, 1998m2, ..., 2000m7                    T =         31
             Delta(t2) = 1 month
             Span(t2)  = 31 periods
             (id*t2 uniquely identifies each observation)
    Distribution of T_i:   min     5%     25%    50%    75%    95%    max
                            19     19      19     26     31     31     31

        Freq.  Percent   Cum.  |  Pattern
    ------------------------------+---------------------------------
            1    33.33    33.33  |  ............1111111111111111111
            1    33.33    66.67  |  1111111111111111111111111......
            1    33.33   100.00  |  1111111111111111111111111111111
    ------------------------------+---------------------------------
            3   100.00          |  XXXXXXXXXXXXXXXXXXXXXXXXXXXXXXX
. by id: sum t2
```

```
-> id = 1
    Variable |      Obs        Mean    Std. Dev.      Min        Max
    ---------+----------------------------------------------------------
          t2 |       19         477    5.627314       468        486
```

```
-> id = 2
    Variable |      Obs        Mean    Std. Dev.      Min        Max
    ---------+----------------------------------------------------------
          t2 |       26       468.5    7.648529       456        481
```

```
-> id = 3
    Variable |      Obs        Mean    Std. Dev.      Min        Max
    ---------+----------------------------------------------------------
          t2 |       31         471    9.092121       456        486
```

This output from xtdescribe and summarize after the call to tsappend shows that the call to tsappend, add(6) added 6 observations to each panel and filled in the gap in the time variable in the second panel. tsappend, add() did not cause a uniform end date over the panels.

The following output illustrates the contrast between tsappend, add() and tsappend, last() tsfmt() with panel data that end at different dates. The output from xtdescribe and summarize shows that the call to tsappend, last() tsfmt() filled in the gap in t2 and caused all the panels to end at the specified end date. The output also shows that the panels remain unbalanced because one panel has a later entry date than the other two.

```
. use http://www.stata-press.com/data/r12/tsappend2, clear
. tsappend, last(2000m7) tsfmt(tm)
```

```
. xtdescribe
     id:  1, 2, ..., 3                                      n =         3
     t2:  1998m1, 1998m2, ..., 2000m7                       T =        31
          Delta(t2) = 1 month
          Span(t2)  = 31 periods
          (id*t2 uniquely identifies each observation)
Distribution of T_i:   min     5%    25%    50%    75%    95%    max
                        19     19     19     31     31     31     31
     Freq.  Percent   Cum.  | Pattern
     ───────────────────────┼───────────────────────────────────────
        2    66.67   66.67  | 1111111111111111111111111111111
        1    33.33  100.00  | ............1111111111111111111
     ───────────────────────┼───────────────────────────────────────
        3   100.00         | XXXXXXXXXXXXXXXXXXXXXXXXXXXXXXX
. by id: sum t2
```

```
-> id = 1
     Variable |       Obs       Mean   Std. Dev.       Min        Max
     ─────────┼──────────────────────────────────────────────────────
           t2 |        19        477   5.627314        468        486
```

```
-> id = 2
     Variable |       Obs       Mean   Std. Dev.       Min        Max
     ─────────┼──────────────────────────────────────────────────────
           t2 |        31        471   9.092121        456        486
```

```
-> id = 3
     Variable |       Obs       Mean   Std. Dev.       Min        Max
     ─────────┼──────────────────────────────────────────────────────
           t2 |        31        471   9.092121        456        486
```

Saved results

tsappend saves the following in $r()$:

Scalars
    r(add)   number of observations added

Methods and formulas

tsappend is implemented as an ado-file.

Also see

[TS] **tsset** — Declare data to be time-series data

[TS] **tsfill** — Fill in gaps in time variable

Title

> **tsfill** — Fill in gaps in time variable

Syntax

tsfill [, _f_ull]

You must tsset your data before using tsfill; see [TS] **tsset**.

Menu

Statistics > Time series > Setup and utilities > Fill in gaps in time variable

Description

tsfill is used after tsset to fill in gaps in time-series data and gaps in panel data with new observations, which contain missing values. For instance, perhaps observations for *timevar* = $1, 3, 5, 6, \ldots, 22$ exist. tsfill would create observations for *timevar* = 2 and *timevar* = 4 containing all missing values. There is seldom reason to do this because Stata's time-series operators consider *timevar*, not the observation number. Referring to L.gnp to obtain lagged gnp values would correctly produce a missing value for *timevar* = 3, even if the data were not filled in. Referring to L2.gnp would correctly return the value of gnp in the first observation for *timevar* = 3, even if the data were not filled in.

Option

full is for use with panel data only. With panel data, tsfill by default fills in observations for each panel according to the minimum and maximum values of *timevar* for the panel. Thus if the first panel spanned the times 5–20 and the second panel the times 1–15, after tsfill they would still span the same periods; observations would be created to fill in any missing times from 5–20 in the first panel and from 1–15 in the second.

If full is specified, observations are created so that both panels span the time 1–20, the overall minimum and maximum of *timevar* across panels.

Remarks

Remarks are presented under the following headings:

> Using tsfill with time-series data
> Using tsfill with panel data

Using tsfill with time-series data

You have monthly data, with gaps:

```
. use http://www.stata-press.com/data/r12/tsfillxmpl
. tsset
        time variable:  mdate, 1995m7 to 1996m3, but with gaps
                delta:  1 month
. list mdate income
```

	mdate	income
1.	1995m7	1153
2.	1995m8	1181
3.	1995m11	1236
4.	1995m12	1297
5.	1996m1	1265
6.	1996m3	1282

You can fill in the gaps by interpolation easily with `tsfill` and `ipolate`. `tsfill` creates the missing observations:

```
. tsfill
. list mdate income
```

	mdate	income	
1.	1995m7	1153	
2.	1995m8	1181	
3.	1995m9	.	← new
4.	1995m10	.	← new
5.	1995m11	1236	
6.	1995m12	1297	
7.	1996m1	1265	
8.	1996m2	.	← new
9.	1996m3	1282	

We can now use `ipolate` (see [D] **ipolate**) to fill them in:

```
. ipolate income mdate, gen(ipinc)
. list mdate income ipinc
```

	mdate	income	ipinc
1.	1995m7	1153	1153
2.	1995m8	1181	1181
3.	1995m9	.	1199.3333
4.	1995m10	.	1217.6667
5.	1995m11	1236	1236
6.	1995m12	1297	1297
7.	1996m1	1265	1265
8.	1996m2	.	1273.5
9.	1996m3	1282	1282

Using tsfill with panel data

You have the following panel dataset:

```
. use http://www.stata-press.com/data/r12/tsfillxmpl2, clear

. tsset
      panel variable:  edlevel (unbalanced)
       time variable:  year, 1988 to 1992, but with a gap
               delta:  1 unit

. list edlevel year income
```

	edlevel	year	income
1.	1	1988	14500
2.	1	1989	14750
3.	1	1990	14950
4.	1	1991	15100
5.	2	1989	22100
6.	2	1990	22200
7.	2	1992	22800

Just as with nonpanel time-series datasets, you can use `tsfill` to fill in the gaps:

```
. tsfill

. list edlevel year income
```

	edlevel	year	income	
1.	1	1988	14500	
2.	1	1989	14750	
3.	1	1990	14950	
4.	1	1991	15100	
5.	2	1989	22100	
6.	2	1990	22200	
7.	2	1991	.	← new
8.	2	1992	22800	

You could instead use `tsfill` to produce fully balanced panels with the `full` option:

```
. tsfill, full

. list edlevel year income, sep(0)
```

	edlevel	year	income	
1.	1	1988	14500	
2.	1	1989	14750	
3.	1	1990	14950	
4.	1	1991	15100	
5.	1	1992	.	← new
6.	2	1988	.	← new
7.	2	1989	22100	
8.	2	1990	22200	
9.	2	1991	.	← new
10.	2	1992	22800	

Methods and formulas

`tsfill` is implemented as an ado-file.

Also see

[TS] **tsset** — Declare data to be time-series data

[TS] **tsappend** — Add observations to a time-series dataset

Title

---

tsfilter — Filter a time-series, keeping only selected periodicities

---

Syntax

Filter one variable

    tsfilter *filter* [*type*] *newvar* = *varname* [*if*] [*in*] [*, options*]

Filter multiple variables, unique names

    tsfilter *filter* [*type*] *newvarlist* = *varlist* [*if*] [*in*] [*, options*]

Filter multiple variables, common name stub

    tsfilter *filter* [*type*] *stub** = *varlist* [*if*] [*in*] [*, options*]

filter	Name	See
bk	Baxter–King	[TS] **tsfilter bk**
bw	Butterworth	[TS] **tsfilter bw**
cf	Christiano–Fitzgerald	[TS] **tsfilter cf**
hp	Hodrick–Prescott	[TS] **tsfilter hp**

You must tsset or xtset your data before using tsfilter; see [TS] **tsset** and [XT] **xtset**.
varname and *varlist* may contain time-series operators; see [U] **11.4.4 Time-series varlists**.
options differ across the filters and are documented in each *filter*'s manual entry.

Description

tsfilter separates a time series into trend and cyclical components. The trend component may contain a deterministic or a stochastic trend. The stationary cyclical component is driven by stochastic cycles at the specified periods.

Remarks

The time-series filters implemented in tsfilter separate a time-series y_t into trend and cyclical components:

$$y_t = \tau_t + c_t$$

where τ_t is the trend component and c_t is the cyclical component. τ_t may be nonstationary; it may contain a deterministic or a stochastic trend, as discussed below.

The primary objective of the methods implemented in tsfilter is to estimate c_t, a stationary cyclical component that is driven by stochastic cycles within a specified range of periods. The trend component τ_t is calculated by the difference $\tau_t = y_t - c_t$.

Although the filters implemented in tsfilter have been widely applied by macroeconomists, they are general time-series methods and may be of interest to other researchers.

Remarks are presented under the following headings:

> *An example dataset*
> *A baseline method: Symmetric moving-average (SMA) filters*
> *An overview of filtering in the frequency domain*
> *SMA revisited: The Baxter–King filter*
> *Filtering a random walk: The Christiano–Fitzgerald filter*
> *A one-parameter high-pass filter: The Hodrick–Prescott filter*
> *A two-parameter high-pass filter: The Butterworth filter*

An example dataset

Time series are frequently filtered to remove unwanted characteristics, such as trends and seasonal components, or to estimate components driven by stochastic cycles from a specific range of periods. Although the filters implemented in `tsfilter` can be used for both purposes, their primary purpose is the latter, and we restrict our discussion to that use.

We explain the methods implemented in `tsfilter` by estimating the business-cycle component of a macroeconomic variable, because they are frequently used for this purpose. We estimate the business-cycle component of the natural log of an index of the industrial production of the United States, which is plotted below.

▷ Example 1

```
. use http://www.stata-press.com/data/r12/ipq
(Federal Reserve Economic Data, St. Louis Fed)

. tsline ip_ln
```

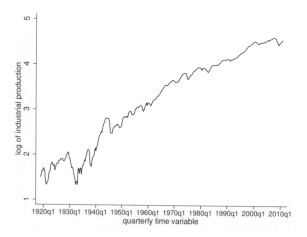

◁

The above graph shows that `ip_ln` contains a trend component. Time series may contain deterministic trends or stochastic trends. A polynomial function of time is the most common deterministic time trend. An integrated process is the most common stochastic trend. An integrated process is a random variable that must be differenced one or more times to be stationary; see Hamilton (1994) for a discussion. The different filters implemented in `tsfilter` allow for different orders of deterministic time trends or integrated processes.

We now illustrate the four methods implemented in `tsfilter`, each of which will remove the trend and estimate the business-cycle component. Burns and Mitchell (1946) defined oscillations in business data with recurring periods between 1.5 and 8 years to be business-cycle fluctuations; we use their commonly accepted definition.

A baseline method: Symmetric moving-average (SMA) filters

Symmetric moving-average (SMA) filters form a baseline method for estimating a cyclical component because of their properties and simplicity. An SMA filter of a time series y_t, $t \in \{1, \ldots, T\}$, is the data transform defined by

$$y_t^* = \sum_{j=-q}^{q} \alpha_j y_{t-j}$$

for each $t \in \{q+1, \ldots, T-q\}$, where $\alpha_{-j} = \alpha_j$ for $j \in \{-q, \ldots, q\}$. Although the original series has T observations, the filtered series has only $T - 2q$, where q is known as the order of the SMA filter.

SMA filters with weights that sum to zero remove deterministic and stochastic trends of order 2 or less, as shown by Fuller (1996) and Baxter and King (1999).

▷ Example 2

This trend-removal property of SMA filters with coefficients that sum to zero may surprise some readers. For illustration purposes, we filter `ip_ln` by the filter

$$-0.2\mathrm{ip\_ln}_{t-2} - 0.2\mathrm{ip\_ln}_{t-1} + 0.8\mathrm{ip\_ln}_t - 0.2\mathrm{ip\_ln}_{t+1} - 0.2\mathrm{ip\_ln}_{t+2}$$

and plot the filtered series. We do not even need `tsfilter` to implement this second-order SMA filter; we can use `generate`.

```
. generate ip_sma = -.2*L2.ip_ln-.2*L.ip_ln+.8*ip_ln-.2*F.ip_ln-.2*F2.ip_ln
(4 missing values generated)

. tsline ip_sma
```

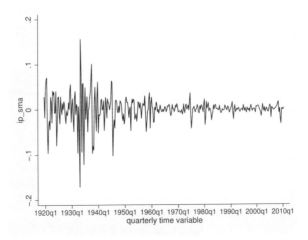

The filter has removed the trend.

◁

There is no good reason why we chose that particular SMA filter. Baxter and King (1999) derived a class of SMA filters with coefficients that sum to zero and get as close as possible to keeping only the specified cyclical component.

An overview of filtering in the frequency domain

We need some concepts from the frequency-domain approach to time-series analysis to motivate how Baxter and King (1999) defined "as close as possible". These concepts also motivate the other filters in `tsfilter`. The intuitive explanation presented here glosses over many technical details discussed by Priestley (1981), Hamilton (1994), Fuller (1996), and Wei (2006).

As with much time-series analysis, the basic results are for covariance-stationary processes with additional results handling some nonstationary cases. We present some useful results for covariance-stationary processes and discuss how to handle nonstationary series below.

The autocovariances γ_j, $j \in \{0, 1, \ldots, \infty\}$, of a covariance-stationary process y_t specify its variance and dependence structure. In the frequency-domain approach to time-series analysis, y_t and the autocovariances are specified in terms of independent stochastic cycles that occur at frequencies $\omega \in [-\pi, \pi]$. The spectral density function $f_y(\omega)$ specifies the contribution of stochastic cycles at each frequency ω relative to the variance of y_t, which is denoted by σ_y^2. The variance and the autocovariances can be expressed as an integral of the spectral density function. Formally,

$$\gamma_j = \int_{-\pi}^{\pi} e^{i\omega j} f_y(\omega) d\omega \tag{1}$$

where i is the imaginary number $i = \sqrt{-1}$.

Equation (1) can be manipulated to show what fraction of the variance of y_t is attributable to stochastic cycles in a specified range of frequencies. Hamilton (1994, 156) discusses this point in more detail.

Equation (1) implies that if $f_y(\omega) = 0$ for $\omega \in [\omega_1, \omega_2]$, then stochastic cycles at these frequencies contribute zero to the variance and autocovariances of y_t.

The goal of time-series filters is to transform the original series into a new series y_t^* for which the spectral density function of the filtered series $f_{y^*}(\omega)$ is zero for unwanted frequencies and equal to $f_y(\omega)$ for desired frequencies.

A linear filter of y_t can be written as

$$y_t^* = \sum_{j=-\infty}^{\infty} \alpha_j y_{t-j} = \alpha(L) y_t$$

where we let y_t be an infinitely long series as required by some of the results below. To see the impact of the filter on the components of y_t at each frequency ω, we need an expression for $f_{y^*}(\omega)$ in terms of $f_y(\omega)$ and the filter weights α_j. Wei (2006, 282) shows that for each ω,

$$f_{y^*}(\omega) = |\alpha(e^{i\omega})|^2 f_y(\omega) \tag{2}$$

where $|\alpha(e^{i\omega})|$ is known as the gain of the filter. Equation (2) makes explicit that the squared gain function $|a(e^{i\omega})|^2$ converts the spectral density of the original series, $f_y(\omega)$, into the spectral density of the filtered series, $f_{y^*}(\omega)$. In particular, (2) says that for each frequency ω, the spectral density of the filtered series is the product of the square of the gain of the filter and the spectral density of the original series.

As we will see in the examples below, the gain function provides a crucial interpretation of what a filter is doing. We want a filter for which $f_{y^*}(\omega) = 0$ for unwanted frequencies and for which $f_{y^*}(\omega) = f_y(\omega)$ for desired frequencies. So we seek a filter for which the gain is 0 for unwanted frequencies and for which the gain is 1 for desired frequencies.

In practice, we cannot find such an ideal filter exactly, because the constraints an ideal filter places on filter coefficients cannot be satisfied for time series with only a finite number of observations. The expansive literature on filters is a result of the trade-offs involved in designing implementable filters that approximate the ideal filter.

Ideally, filters pass or block the stochastic cycles at specified frequencies by having a gain of 1 or 0. Band-pass filters, such as the Baxter–King (BK) and the Christiano–Fitzgerald (CF) filters, pass through stochastic cycles in the specified range of frequencies and block all the other stochastic cycles. High-pass filters, such as the Hodrick–Prescott (HP) and Butterworth filters, only allow the stochastic cycles at or above a specified frequency to pass through and block the lower-frequency stochastic cycles. For band-pass filters, let $[\omega_0, \omega_1]$ be the set of desired frequencies with all other frequencies being undesired. For high-pass filters, let ω_0 be the cutoff frequency with only those frequencies $\omega \geq \omega_0$ being desired.

SMA revisited: The Baxter–King filter

We now return to the class of SMA filters with coefficients that sum to zero and get as close as possible to keeping only the specified cyclical component as derived by Baxter and King (1999).

For an infinitely long series, there is an ideal band-pass filter for which the gain function is 1 for $\omega \in [\omega_0, \omega_1]$ and 0 for all other frequencies. It just so happens that this ideal band-pass filter is an SMA filter with coefficients that sum to zero. Baxter and King (1999) derive the coefficients of this ideal band-pass filter and then define the BK filter to be the SMA filter with $2q + 1$ terms that are as close as possible to those of the ideal filter. There is a trade-off in choosing q: larger values of q cause the gain of the BK filter to be closer to the gain of the ideal filter, but larger values also increase the number of missing observations in the filtered series.

Although the mathematics of the frequency-domain approach to time-series analysis is in terms of stochastic cycles at frequencies $\omega \in [-\pi, \pi]$, applied work is generally in terms of periods p, where $p = 2\pi/\omega$. So the options for the `tsfilter` subcommands are in terms of periods.

▷ Example 3

Below we use `tsfilter bk`, which implements the BK filter, to estimate the business-cycle component composed of stochastic cycles between 6 and 32 periods, and then we graph the estimated component.

```
. tsfilter bk ip_bk = ip_ln, minperiod(6) maxperiod(32)
. tsline ip_bk
```

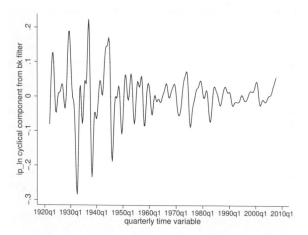

The above graph tells us what the estimated business-cycle component looks like, but it presents no evidence as to how well we have estimated the component. A periodogram is better for this purpose. A periodogram is an estimator of a transform of the spectral density function; see [TS] **pergram** for details. Below we plot the periodogram for the BK estimate of the business-cycle component. pergram displays the results in natural frequencies, which are the standard frequencies divided by 2π. We use the xline() option to draw vertical lines at the lower natural-frequency cutoff ($1/32 = 0.03125$) and the upper natural-frequency cutoff ($1/6 \approx 0.16667$).

```
. pergram ip_bk, xline(0.03125 0.16667)
```

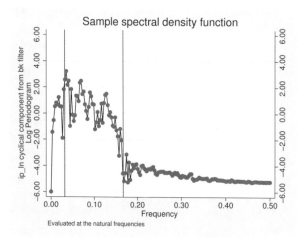

If the filter completely removed the stochastic cycles corresponding to the unwanted frequencies, the periodogram would be a flat line at the minimum value of -6 outside the range identified by the vertical lines. That the periodogram takes on values greater than -6 outside the specified range indicates the inability of the BK filter to pass through only stochastic cycles at frequencies inside the specified band.

We can also evaluate the BK filter by plotting its gain function against the gain function of an ideal filter. In the output below, we reestimate the business-cycle component to store the gain of the BK filter for the specified parameters. (The coefficients and the gain of the BK filter are completely determined by the specified minimum period, the maximum period, and the order of the SMA filter.) We label the variable bkgain for the graph below.

```
. drop ip_bk
. tsfilter bk ip_bk = ip_ln, minperiod(6) maxperiod(32) gain(bkgain abk)
. label variable bkgain "BK filter"
```

Below we generate ideal, the gain function of the ideal band-pass filter at the frequencies f. Then we plot the gain of the ideal filter and the gain of the BK filter.

```
. generate f = _pi*(_n-1)/_N
. generate ideal = cond(f<_pi/16, 0, cond(f<_pi/3, 1,0))
. label variable ideal "Ideal filter"
. twoway line ideal f || line bkgain abk
```

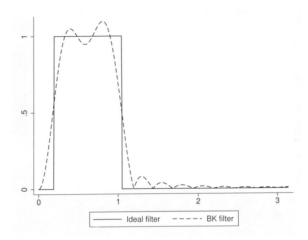

The graph reveals that the gain of the BK filter deviates markedly from the square-wave gain of the ideal filter. Increasing the symmetric moving average via the smaorder() option will cause the gain of the BK filter to more closely approximate the gain of the ideal filter at the cost of lost observations in the filtered series.

◁

Filtering a random walk: The Christiano–Fitzgerald filter

Although Baxter and King (1999) minimized the error between the coefficients in their filter and the ideal band-pass filter, Christiano and Fitzgerald (2003) minimized the mean squared error between the estimated component and the true component, assuming that the raw series is a random-walk process. Christiano and Fitzgerald (2003) give three important reasons for using their filter:

1. The true dependence structure of the data affects which filter is optimal.

2. Many economic time series are well approximated by random-walk processes.

3. Their filter does a good job passing through stochastic cycles of desired frequencies and blocking stochastic cycles from unwanted frequencies on a range of processes that are close to being a random-walk process.

The CF filter obtains its optimality properties at the cost of an additional parameter that must be estimated and a loss of robustness. The CF filter is optimal for a random-walk process. If the true process is a random walk with drift, then the drift term must be estimated and removed; see [TS] **tsfilter cf** for details. The CF filter is not symmetric, so it will not remove second-order deterministic or second-order integrated processes. `tsfilter cf` also implements another filter that Christiano and Fitzgerald (2003) derived that is an SMA filter with coefficients that sum to zero. This filter is designed to be as close as possible to the random-walk optimal filter under the constraint that it be an SMA filter with constraints that sum to zero; see [TS] **tsfilter cf** for details.

❏ Technical note

A random-walk process is a first-order integrated process; it must be differenced once to produce a stationary process. Formally, a random-walk process is given by $y_t = y_{t-1} + \epsilon_t$, where ϵ_t is a zero-mean stationary random variable. A random-walk-plus-drift process is given by $\widetilde{y}_t = \mu + \widetilde{y}_{t-1} + \epsilon_t$, where ϵ_t is a zero-mean stationary random variable.

❏

▷ Example 4

In this example, we use the CF filter to estimate the business-cycle component, and we plot the periodogram of the CF estimates. We specify the `drift` option because `ip_ln` is well approximated by a random-walk-plus-drift process.

```
. tsfilter cf ip_cf = ip_ln, minperiod(6) maxperiod(32) drift
. pergram ip_cf, xline(0.03125 0.16667)
```

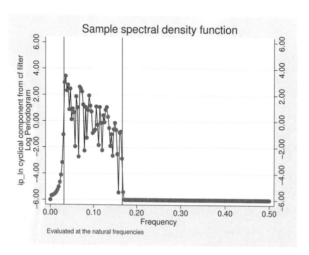

The periodogram of the CF estimates of the business-cycle component indicates that the CF filter did a better job than the BK filter of passing through only the desired stochastic cycles. Given that `ip_ln` is well approximated by a random-walk-plus-drift process, the relative performance of the CF filter is not surprising.

As with the BK filter, plotting the gain of the CF filter and the gain of the ideal filter gives an impression of how well the filter isolates the specified components. In the output below, we reestimate the business-cycle component, using the gain() option to store the gain of the CF filter, and we plot the gain functions.

```
. drop ip_cf
. tsfilter cf ip_cf = ip_ln, minperiod(6) maxperiod(32) drift gain(cfgain acf)
. label variable cfgain "CF filter"
. twoway line ideal f || line cfgain acf
```

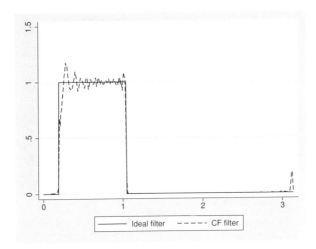

Comparing this graph with the graph of the BK gain function reveals that the CF filter is closer to the gain of the ideal filter than is the BK filter. The graph also reveals that the gain of the CF filter oscillates above and below 1 for desired frequencies.

◁

The choice between the BK or the CF filter is one between robustness or efficiency. The BK filter handles a broader class of stochastic processes, but the CF filter produces a better estimate of c_t if y_t is close to a random-walk process or a random-walk-plus-drift process.

A one-parameter high-pass filter: The Hodrick–Prescott filter

Hodrick and Prescott (1997) motivated the Hodrick–Prescott (HP) filter as a trend-removal technique that could be applied to data that came from a wide class of data-generating processes. In their view, the technique specified a trend in the data, and the data were filtered by removing the trend. The smoothness of the trend depends on a parameter λ. The trend becomes smoother as $\lambda \to \infty$. Hodrick and Prescott (1997) recommended setting λ to 1,600 for quarterly data.

King and Rebelo (1993) showed that removing a trend estimated by the HP filter is equivalent to a high-pass filter. They derived the gain function of this high-pass filter and showed that the filter would make integrated processes of order 4 or less stationary, making the HP filter comparable with the band-pass filters discussed above.

▷ Example 5

We begin by applying the HP high-pass filter to `ip_ln` and plotting the periodogram of the estimated business-cycle component. We specify the `gain()` option because will use the gain of the filter in the next example.

```
. tsfilter hp ip_hp = ip_ln, gain(hpg1600 ahp1600)
. label variable hpg1600 "HP(1600) filter"
. pergram ip_hp, xline(0.03125)
```

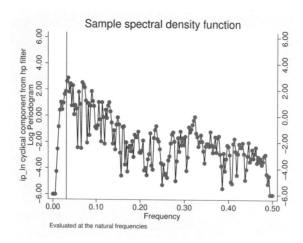

Because the HP filter is a high-pass filter, the high-frequency stochastic cycles corresponding to those periods below 6 remain in the estimated component. Of more concern is the presence of the low-frequency stochastic cycles that the filter should remove. We address this issue in the example below.

◁

▷ Example 6

Hodrick and Prescott (1997) argued that the smoothing parameter λ should be 1,600 on the basis of a heuristic argument that specified values for the variance of the cyclical component and the variance of the second difference of the trend component, both recorded at quarterly frequencies. In this example, we choose the smoothing parameter to be 677.13, which sets the gain of the filter to 0.5 at the frequency corresponding to 32 periods, as explained in the technical note below. We then plot the periodogram of the filtered series.

```
. tsfilter hp ip_hp2 = ip_ln, smooth(677.13)  gain(hpg677 ahp677)

. label variable hpg677 "HP(677) filter"

. pergram ip_hp, xline(0.03125)
```

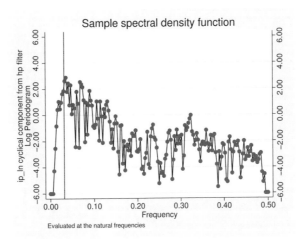

Although the periodogram looks better than the periodogram with the default smoothing, the HP filter still did not zero out the low-frequency stochastic cycles as well as the CF filter did. We take another look at this issue by plotting the gain functions for these filters along with the gain function from the ideal band-pass filter.

```
. twoway line ideal f || line hpg677 ahp677
```

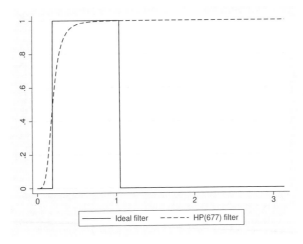

Comparing the gain graphs reveals that the gain of the CF filter is closest to the gain of the ideal filter. Both the BK and the HP filters allow some low-frequency stochastic cycles to pass through. The plot also illustrates that the HP filter is a high-pass filter because its gain is 1 for those stochastic cycles at frequencies above 6 periods, whereas the other gain functions go to zero.

◁

❑ Technical note

Conventionally, economists have used $\lambda = 1600$, which Hodrick and Prescott (1997) recommended for quarterly data. Ravn and Uhlig (2002) derived values for λ at monthly and annual frequencies that are rescalings of the conventional $\lambda = 1600$ for quarterly data. These heuristic values are the default values; see [TS] **tsfilter hp** for details. In the filter literature, filter parameters are set as functions of the cutoff frequency; see Pollock (2000, 324), for instance. This method finds the filter parameter that sets the gain of the filter equal to $1/2$ at the cutoff frequency. Applying this method to selecting λ at the cutoff frequency of 32 periods requires solving

$$1/2 = \frac{4\lambda \left\{1 - \cos(2\pi/32)\right\}^2}{1 + 4\lambda \left\{1 - \cos(2\pi/32)\right\}^2}$$

for λ, which yields $\lambda \approx 677.13$, which was used in the previous example.

The gain function of the HP filter is a function of the parameter λ, and λ sets both the location of the cutoff frequency and the slope of the gain function. The graph below illustrates this dependence by plotting the gain function of the HP filter for λ set to 10, 677.13, and 1,600 along with the gain function for the ideal band-pass filter with cutoff periods of 32 periods and 6 periods.

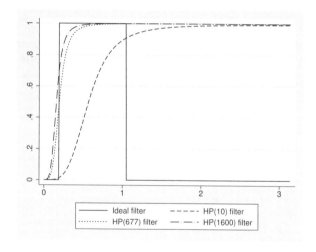

❑

A two-parameter high-pass filter: The Butterworth filter

Engineers have used Butterworth filters for a long time because they are "maximally flat". The gain functions of these filters are as close as possible to being a flat line at 0 for the unwanted periods and a flat line at 1 for the desired periods; see Butterworth (1930) and Bianchi and Sorrentino (2007, 17–20).

Pollock (2000) showed that Butterworth filters can be derived from some axioms that specify properties we would like a filter to have. Although the Butterworth and BK filters share the properties of symmetry and phase neutrality, the coefficients of Butterworth filters do not need to sum to zero. (Phase-neutral filters do not shift the signal forward or backward in time; see Pollock [1999].) Although the BK filter relies on the detrending properties of SMA filters with coefficients that sum to zero, Pollock (2000) shows that Butterworth filters have detrending properties that depend on the filters' parameters.

tsfilter bw implements the high-pass Butterworth filter using the computational method that Pollock (2000) derived. This filter has two parameters: the cutoff period and the order of the filter denoted by m. The cutoff period sets the location where the gain function starts to filter out the high-period (low-frequency) stochastic cycles, and m sets the slope of the gain function for a given cutoff period. For a given cutoff period, the slope of the gain function at the cutoff period increases with m. For a given m, the slope of the gain function at the cutoff period increases with the cutoff period.

We cannot obtain a vertical slope at the cutoff frequency, which is the ideal, because the computation becomes unstable; see Pollock (2000). The m for which the computation becomes unstable depends on the cutoff period.

Pollock (2000) and Gómez (1999) argue that the additional flexibility produced by the additional parameter makes the high-pass Butterworth filter a better filter than the HP filter for estimating the cyclical components.

Pollock (2000) shows that the high-pass Butterworth filter can estimate the desired components of the dth difference of a dth-order integrated process as long as $m \geq d$.

> ## Example 7

Below we use tsfilter bw to estimate the components driven by stochastic cycles greater than 32 periods using Butterworth filters of order 2 and order 6. We also compute, label, and plot the gain functions for each filter.

```
. tsfilter bw ip_bw1 = ip_ln, gain(bwgain1 abw1) maxperiod(32) order(2)
. label variable bwgain1 "BW 2"
. tsfilter bw ip_bw6 = ip_ln, gain(bwgain6 abw6) maxperiod(32) order(6)
. label variable bwgain6 "BW 6"
. twoway line ideal f || line bwgain1 abw1 || line bwgain6 abw6
```

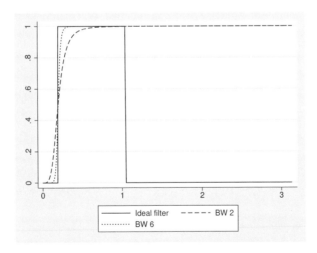

The graph illustrates that the slope of the gain function increases with the order of the filter.

The graph below provides another perspective by plotting the gain function from the ideal band-pass filter on a graph with plots of the gain functions from the Butterworth filter of order 6, the CF filter, and the HP(677) filter.

```
. twoway line ideal f || line bwgain6 abw6 || line cfgain acf
>        || line hpg677 ahp677
```

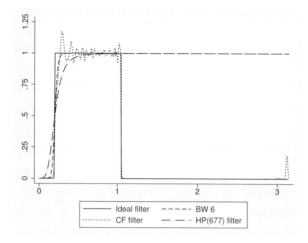

Although the slope of the gain function from the CF filter is closer to being vertical at the cutoff frequency, the gain function of the Butterworth filter does not oscillate above and below 1 after it first reaches the value of 1. The flatness of the Butterworth filter below and above the cutoff frequency is not an accident, it is one of the filter's properties.

◁

▷ Example 8

In the previous example, we used the Butterworth filter of order 6 to remove low-frequency stochastic cycles, and we stored the results in ip_bw6. The Butterworth filter did not address the high-frequency stochastic cycles below 6 periods because it is a high-pass filter. We remove those high-frequency stochastic cycles in this example by keeping the trend produced by refiltering the previously filtered series.

This example uses a common trick: keeping the trend produced by a high-pass filter turns that high-pass filter into a low-pass filter. Because we want to remove the high-frequency stochastic cycles still in the previously filtered series ip_bw6, we need a low-pass filter. So we keep the trend produced by refiltering the previously filtered series.

In the output below, we apply a Butterworth filter of order 20 to the previously filtered series ip_bw6. We explain why we used order 20 in the next example. We specify the trend() option to keep the low-frequency components from these filters. Then we compute and graph the periodogram for the trend variable.

```
. tsfilter bw ip_bwu20 = ip_bw6, gain(bwg20 fbw20) maxperiod(6) order(20)
> trend(ip_bwb)
. label variable bwg20 "BW upper filter 20"
. pergram ip_bwb, xline(0.03125 0.16667)
```

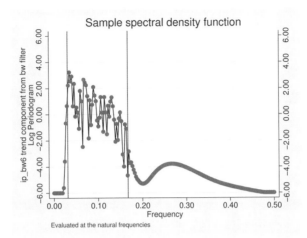

The periodogram reveals that the two-pass process has passed the original series ip_ln through a band-pass filter. It also reveals that the two-pass process did a reasonable job of filtering out the stochastic cycles corresponding to the unwanted frequencies.

◁

▷ Example 9

In the previous example, when the cutoff period was 6, we set the order of the Butterworth filter to 20. In contrast, in example 7, when the cutoff period was 32, we set the order of the Butterworth filter to 6. We had to increase filter order because the slope of the gain function of the Butterworth filter is increasing with the cutoff period. We needed a larger filter order to get an acceptable slope at the lower cutoff period.

We illustrate this point in the output below. We apply Butterworth filters of orders 1 and 6 to the previously filtered series ip_bw6, we compute the gain functions, we label the gain variables, and then we plot the gain functions from the ideal filter and the Butterworth filters.

```
. tsfilter bw ip_bwu1  = ip_bw6, gain(bwg1 fbw1)   maxperiod(6) order(2)
. label variable bwg1 "BW upper filter 2"
. tsfilter bw ip_bwu6  = ip_bw6, gain(bwg6 fbw6)   maxperiod(6) order(6)
. label variable bwg6 "BW upper filter 6"
. twoway line ideal f || line bwg1 fbw1 || line bwg6 fbw6 || line bwg20 fbw20
```

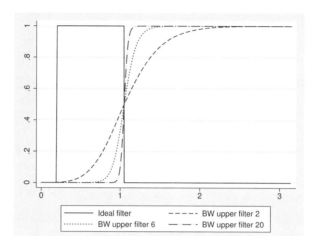

Because the cutoff period is 6, the gain functions for $m = 2$ and $m = 6$ are much flatter than the gain functions for $m = 2$ and $m = 6$ in example 7 when the cutoff period was 32. The gain function for $m = 20$ is reasonably close to vertical, so we used it in example 8. We mentioned above that for any given cutoff period, the computation eventually becomes unstable for larger values of m. For instance, when the cutoff period is 32, $m = 20$ is not numerically feasible.

◁

▷ Example 10

As a conclusion, we plot the business-cycle components estimated by the CF filter and by the two passes of Butterworth filters. The shaded areas identify recessions. The two estimates are close but the differences could be important. Which estimate is better depends on whether the oscillations around 1 in the graph of the CF gain function (the second graph of example 7) cause more problems than the nonvertical slopes at the cutoff periods that occur in the BW6 gain function of that same graph and the BW upper filter 20 gain function graphed above.

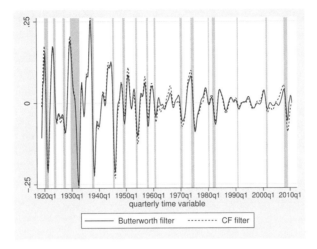

$\lhd$

There is a long tradition in economics of using models to estimate components. Instead of comparing filters by their gain functions, some authors compare filters by finding underlying models for which the filter parameters are the model parameters. For instance, Harvey and Jaeger (1993), Gómez (1999, 2001), Pollock (2000, 2006), and Harvey and Trimbur (2003) derive models that correspond to the HP or the Butterworth filter. Some of these references also compare components estimated by filters with components estimated by making predictions from estimated models. In effect, these references point out that `arima`, `dfactor`, `sspace`, and `ucm` (see [TS] **arima**, [TS] **dfactor**, [TS] **sspace**, and [TS] **ucm**) implement alternative methods to component estimation.

Methods and formulas

All filters work with both time-series data and panel data when there are many observations on each panel. When used with panel data, the calculations are performed separately within each panel.

For these filters, the default minimum and maximum periods of oscillation correspond to the boundaries used by economists (Burns and Mitchell 1946) for business cycles. Burns and Mitchell defined business cycles as oscillations in business data with recurring periods between 1.5 and 8 years. Their definition continues to be cited by economists investigating correlations between business cycles.

If y_t is a time series, then the cyclical component is

$$c_t = B(L)y_t = \sum_{j=-\infty}^{\infty} b_j y_{t-j}$$

where b_j are the coefficients of the impulse–response sequence of some ideal filter. The impulse–response sequence is the inverse Fourier transform of either a square wave or step function depending upon whether the filter is a band-pass or high-pass filter, respectively.

In finite sequences, it is necessary to approximate this calculation with a finite impulse–response sequence $\widehat{b}_j$:

$$\widehat{c}_t = \widehat{B}_t(L)y_t = \sum_{j=-n_1}^{n_2} \widehat{b}_j y_{t-j}$$

The infinite-order impulse–response sequence for the filters implemented in `tsfilter` are symmetric and time-invariant.

In the frequency domain, the relationships between the true cyclical component and its finite estimates respectively are

$$c(\omega) = B(\omega)y(\omega)$$

and

$$\widehat{c}(\omega) = \widehat{B}(\omega)y(\omega)$$

where $B(\omega)$ and $\widehat{B}(\omega)$ are the frequency transfer functions of the filters B and $\widehat{B}$.

The frequency transfer function for $B(\omega)$ can be expressed in polar form as

$$B(\omega) = |B(\omega)|\exp\{i\theta(\omega)\}$$

where $|B(\omega)|$ is the filter's gain function and $\theta(\omega)$ is the filter's phase function. The gain function determines whether the amplitude of the stochastic cycle is increased or decreased at a particular frequency. The phase function determines how a cycle at a particular frequency is shifted forward or backward in time.

In this form, it can be shown that the spectrum of the cyclical component, $f_c(\omega)$, is related to the spectrum of y_t series by the squared gain:

$$f_c(\omega) = |B(\omega)|^2 f_y(\omega)$$

Each of the four filters in `tsfilter` has an option for returning an estimate of the gain function together with its associated scaled frequency $a = \omega/\pi$, where $0 \leq \omega \leq \pi$. These are consistent estimates of $|B(\omega)|$, the gain from the ideal linear filter.

The band-pass filters implemented in `tsfilter`, the BK and CF filters, use a square wave as the ideal transfer function:

$$B(\omega) = \begin{cases} 1 & \text{if } |\omega| \in [\omega_l, \omega_h] \\ 0 & \text{if } |\omega| \notin [\omega_l, \omega_h] \end{cases}$$

The high-pass filters, the Hodrick–Prescott and Butterworth filters, use a step function as the ideal transfer function:

$$B(\omega) = \begin{cases} 1 & \text{if } |\omega| \geq \omega_h \\ 0 & \text{if } |\omega| < \omega_h \end{cases}$$

Acknowledgments

We thank Christopher F. Baum, Boston College, for his previous implementations of these filters: Baxter–King (`bking`), Christiano–Fitzgerald (`cfitzrw`), Hodrick–Prescott (`hprescott`), and Butterworth (`butterworth`).

We also thank D. S. G. Pollock for his helpful responses to our questions about Butterworth filters and the methods that he has developed.

References

Baxter, M., and R. G. King. 1999. Measuring business cycles: Approximate band-pass filters for economic time series. *Review of Economics and Statistics* 81: 575–593.

Bianchi, G., and R. Sorrentino. 2007. *Electronic Filter Simulation and Design*. New York: McGraw–Hill.

Burns, A. F., and W. C. Mitchell. 1946. *Measuring Business Cycles*. New York: National Bureau of Economic Research.

Butterworth, S. 1930. On the theory of filter amplifiers. *Experimental Wireless and the Wireless Engineer* 7: 536–541.

Christiano, L. J., and T. J. Fitzgerald. 2003. The band pass filter. *International Economic Review* 44: 435–465.

Fuller, W. A. 1996. *Introduction to Statistical Time Series*. 2nd ed. New York: Wiley.

Gómez, V. 1999. Three equivalent methods for filtering finite nonstationary time series. *Journal of Business and Economic Statistics* 17: 109–116.

——. 2001. The use of Butterworth filters for trend and cycle estimation in economic time series. *Journal of Business and Economic Statistics* 19: 365–373.

Hamilton, J. D. 1994. *Time Series Analysis*. Princeton: Princeton University Press.

Harvey, A. C., and A. Jaeger. 1993. Detrending, stylized facts and the business cycle. *Journal of Applied Econometrics* 8: 231–247.

Harvey, A. C., and T. M. Trimbur. 2003. General model-based filters for extracting cycles and trends in economic time series. *The Review of Economics and Statistics* 85: 244–255.

Hodrick, R. J., and E. C. Prescott. 1997. Postwar U.S. business cycles: An empirical investigation. *Journal of Money, Credit, and Banking* 29: 1–16.

King, R. G., and S. T. Rebelo. 1993. Low frequency filtering and real business cycles. *Journal of Economic Dynamics and Control* 17: 207–231.

Leser, C. E. V. 1961. A simple method of trend construction. *Journal of the Royal Statistical Society, Series B* 23: 91–107.

Pollock, D. S. G. 1999. *A Handbook of Time-Series Analysis, Signal Processing and Dynamics*. London: Academic Press.

——. 2000. Trend estimation and de-trending via rational square-wave filters. *Journal of Econometrics* 99: 317–334.

——. 2006. Econometric methods of signal extraction. *Computational Statistics & Data Analysis* 50: 2268–2292.

Priestley, M. B. 1981. *Spectral Analysis and Time Series*. London: Academic Press.

Ravn, M. O., and H. Uhlig. 2002. On adjusting the Hodrick–Prescott filter for the frequency of observations. *Review of Economics and Statistics* 84: 371–376.

Schmidt, T. J. 1994. sts5: Detrending with the Hodrick–Prescott filter. *Stata Technical Bulletin* 17: 22–24. Reprinted in *Stata Technical Bulletin Reprints*, vol. 3, pp. 216–219. College Station, TX: Stata Press.

Wei, W. W. S. 2006. *Time Series Analysis: Univariate and Multivariate Methods*. 2nd ed. Boston: Pearson.

Also see

[TS] **tsset** — Declare data to be time-series data

[XT] **xtset** — Declare data to be panel data

[TS] **tssmooth** — Smooth and forecast univariate time-series data

Title

> **tsfilter bk** — Baxter–King time-series filter

Syntax

Filter one variable

> tsfilter bk [*type*] *newvar* = *varname* [*if*] [*in*] [, *options*]

Filter multiple variables, unique names

> tsfilter bk [*type*] *newvarlist* = *varlist* [*if*] [*in*] [, *options*]

Filter multiple variables, common name stub

> tsfilter bk [*type*] *stub** = *varlist* [*if*] [*in*] [, *options*]

options	Description
Main	
<u>min</u>period(*#*)	filter out stochastic cycles at periods smaller than #
<u>max</u>period(*#*)	filter out stochastic cycles at periods larger than #
<u>sma</u>order(*#*)	number of observations in each direction that contribute to each filtered value
<u>stati</u>onary	use calculations for a stationary time series
Trend	
<u>tre</u>nd(*newvar* | *newvarlist* | *stub**)	save the trend component(s) in new variable(s)
Gain	
gain(*gainvar anglevar*)	save the gain and angular frequency

You must tsset or xtset your data before using tsfilter; see [TS] **tsset** and [XT] **xtset**.

varname and *varlist* may contain time-series operators; see [U] **11.4.4 Time-series varlists**.

Menu

Statistics > Time series > Filters for cyclical components > Baxter-King

Description

tsfilter bk uses the Baxter and King (1999) band-pass filter to separate a time series into trend and cyclical components. The trend component may contain a deterministic or a stochastic trend. The stationary cyclical component is driven by stochastic cycles at the specified periods.

See [TS] **tsfilter** for an introduction to the methods implemented in tsfilter bk.

Options

minperiod(#) filters out stochastic cycles at periods smaller than #, where # must be at least 2 and less than maxperiod(). By default, if the units of the time variable are set to daily, weekly, monthly, quarterly, or half-yearly, then # is set to the number of periods equivalent to 1.5 years; yearly data use minperiod(2); otherwise, the default value is minperiod(6).

maxperiod(#) filters out stochastic cycles at periods larger than #, where # must be greater than minperiod(). By default, if the units of the time variable are set to daily, weekly, monthly, quarterly, half-yearly, or yearly, then # is set to the number of periods equivalent to 8 years; otherwise, the default value is maxperiod(32).

smaorder(#) sets the order of the symmetric moving average, denoted by q. The order is an integer that specifies the number of observations in each direction used in calculating the symmetric moving average estimate of the cyclical component. This number must be an integer greater than zero and less than $(T-1)/2$. The estimate for the cyclical component for the tth observation, y_t, is based upon the $2q + 1$ values y_{t-q}, y_{t-q+1}, ..., y_t, y_{t+1}, ..., y_{t+q}. By default, if the units of the time variable are set to daily, weekly, monthly, quarterly, half-yearly, or yearly, then # is set to the equivalent of 3 years; otherwise, the default value is smaorder(12).

stationary modifies the filter calculations to those appropriate for a stationary series. By default, the series is assumed nonstationary.

trend(*newvar* | *newvarlist* | *stub**) saves the trend component(s) in the new variable(s) specified by *newvar*, *newvarlist*, or *stub**.

gain(*gainvar anglevar*) saves the gain in *gainvar* and its associated angular frequency in *anglevar*. Gains are calculated at the N angular frequencies that uniformly partition the interval $(0, \pi]$, where N is the sample size.

Remarks

We assume that you have already read [TS] **tsfilter**, which provides an introduction to filtering and the methods implemented in **tsfilter bk**, more examples using **tsfilter bk**, and a comparison of the four filters implemented by **tsfilter**. In particular, an understanding of gain functions as presented in [TS] **tsfilter** is required to understand these remarks.

tsfilter bk uses the Baxter–King (BK) band-pass filter to separate a time-series y_t into trend and cyclical components:

$$y_t = \tau_t + c_t$$

where τ_t is the trend component and c_t is the cyclical component. τ_t may be nonstationary; it may contain a deterministic or a stochastic trend, as discussed below.

The primary objective is to estimate c_t, a stationary cyclical component that is driven by stochastic cycles within a specified range of periods. The trend component τ_t is calculated by the difference $\tau_t = y_t - c_t$.

Although the BK band-pass filter implemented in **tsfilter bk** has been widely applied by macroeconomists, it is a general time-series method and may be of interest to other researchers.

Symmetric moving-average (SMA) filters with coefficients that sum to zero remove stochastic and deterministic trends of first and second order; see Fuller (1996), Baxter and King (1995), and Baxter and King (1999).

For an infinitely long series, there is an ideal band-pass filter for which the gain function is 1 for $\omega \in [\omega_0, \omega_1]$ and 0 for all other frequencies; see [TS] **tsfilter** for an introduction to gain functions. It just so happens that this ideal band-pass filter is an SMA filter with coefficients that sum to zero. Baxter and King (1999) derive the coefficients of this ideal band-pass filter and then define the BK filter to be the SMA filter with $2q + 1$ terms that are as close as possible to those of the ideal filter. There is a trade-off in choosing q: larger values of q cause the gain of the BK filter to be closer to the gain of the ideal filter, but they also increase the number of missing observations in the filtered series.

The smaorder() option specifies q. The default value of smaorder() is the number of periods equivalent to 3 years, following the Baxter and King (1999) recommendation.

Although the mathematics of the frequency-domain approach to time-series analysis is in terms of stochastic cycles at frequencies $\omega \in [-\pi, \pi]$, applied work is generally in terms of periods p, where $p = 2\pi/\omega$. So tsfilter bk has the minperiod() and maxperiod() options to specify the desired range of stochastic cycles.

Among economists, the BK filter is commonly used for investigating business cycles. Burns and Mitchell (1946) defined business cycles as stochastic cycles in business data corresponding to periods between 1.5 and 8 years. The default values for minperiod() and maxperiod() are the Burns–Mitchell values of 1.5 and 8 years, scaled to the frequency of the dataset. The calculations of the default values assume that the time variable is formatted as daily, weekly, monthly, quarterly, half-yearly, or yearly; see [D] **format**.

For each variable, the band-pass BK filter estimate of c_t is put in the corresponding new variable, and when the trend() option is specified, the estimate of τ_t is put in the corresponding new variable.

tsfilter bk automatically detects panel data from the information provided when the dataset was tsset or xtset. All calculations are done separately on each panel. Missing values at the beginning and end of the sample are excluded from the sample. The sample may not contain gaps.

Baxter and King (1999) derived their method for nonstationary time series, but they noted that a small modification makes it applicable to stationary time series. Imposing the condition that the filter coefficients sum to zero is what makes their method applicable to nonstationary time series; dropping this condition yields a filter for stationary time series. Specifying the stationary option causes tsfilter bk to use coefficients calculated without the constraint that they sum to zero.

▷ Example 1

In this and the subsequent examples, we use tsfilter bk to estimate the business-cycle component of the natural log of real gross domestic product (GDP) of the United States. Our sample of quarterly data goes from 1952q1 to 2010q4. Below we read in and plot the data.

```
. use http://www.stata-press.com/data/r12/gdp2
(Federal Reserve Economic Data, St. Louis Fed)

. tsline gdp_ln
```

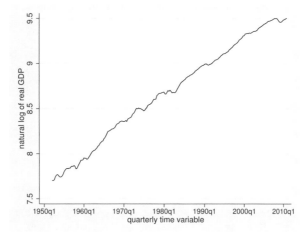

The series is nonstationary and is thus a candidate for the BK filter.

Below we use `tsfilter bk` to filter `gdp_ln`, and we use `pergram` (see [TS] **pergram**) to compute and to plot the periodogram of the estimated cyclical component.

```
. tsfilter bk gdp_bk = gdp_ln

. pergram gdp_bk, xline(.03125 .16667)
```

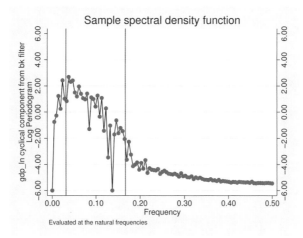

Because our sample is of quarterly data, `tsfilter bk` used the default values of `minperiod(6)`, `maxperiod(32)`, and `smaorder(12)`. The minimum and maximum periods are the Burns and Mitchell (1946) business-cycle periods for quarterly data. The default of `smaorder(12)` was recommend by Baxter and King (1999) for quarterly data.

In the periodogram, we added vertical lines at the natural frequencies corresponding to the conventional Burns and Mitchell (1946) values for business-cycle components. `pergram` displays the results in natural frequencies, which are the standard frequencies divided by 2π. We use the xline() option to draw vertical lines at the lower natural-frequency cutoff ($1/32 = 0.03125$) and the upper natural-frequency cutoff ($1/6 \approx 0.16667$).

If the filter completely removed the stochastic cycles at the unwanted frequencies, the periodogram would be a flat line at the minimum value of -6 outside the range identified by the vertical lines.

The periodogram reveals that the default value of smaorder(12) did not do a good job of filtering out the high-periodicity stochastic cycles, because there are too many points above -6.00 to the left of the left-hand vertical line. It also reveals that the filter did not remove enough low-periodicity stochastic cycles, because there are too many points above -6.00 to the right of the right-hand vertical line.

We address these problems in the next example.

◁

▷ Example 2

In this example, we change the symmetric moving average of the filter via the smaorder() option so that it will remove more of the unwanted stochastic cycles. As mentioned, larger values of q cause the gain of the BK filter to be closer to the gain of the ideal filter, but larger values also increase the number of missing observations in the filtered series.

In the output below, we estimate the business-cycle component and compute the gain functions when the SMA-order of the filter is 12 and when it is 20. We also generate ideal, the gain function of the ideal band-pass filter at the frequencies f. Then we plot the gain functions from all three filters.

```
. tsfilter bk gdp_bk12 = gdp_ln, gain(g12 a12)
. label variable g12 "BK SMA-order 12"
. tsfilter bk gdp_bk20 = gdp_ln, gain(g20 a20) smaorder(20)
. label variable g20 "BK SMA-order 20"
. generate f = _pi*(_n-1)/_N
. generate ideal = cond(f<_pi/16, 0, cond(f<_pi/3, 1,0))
. label variable ideal "Ideal filter"
. twoway line ideal f || line g12 a12 || line g20 a20
```

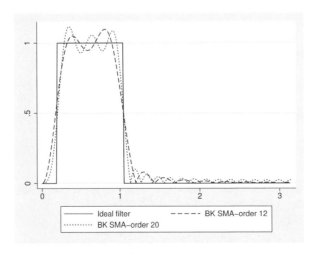

As discussed in [TS] **tsfilter**, the gain function of the ideal filter is a square wave with a value of 0 at the frequencies corresponding to unwanted frequencies and a value of 1 at the desired frequencies. The vertical lines in the gain function of the ideal filter occur at the frequencies $\pi/16$, corresponding to 32 periods, and at $\pi/3$, corresponding to 6 periods. (Given that $p = 2\pi/\omega$, where p is the period corresponding to the frequency ω, the frequency is given by $2\pi/p$.)

The differences between the gain function of the filter with SMA-order 12 and the gain function of the ideal band-pass filter is the root of the issues mentioned at the end of example 1. The filter with SMA-order 20 is closer to the gain function of the ideal band-pass filter at the cost of 16 more missing values in the filtered series.

Below we compute and graph the periodogram of the series filtered with SMA-order 20.

```
. pergram gdp_bk20, xline(.03125 .16667)
```

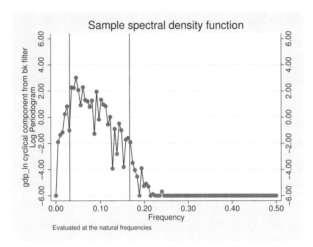

The above periodogram indicates that the filter of SMA-order 20 removed more of the stochastic cycles at the unwanted periodicities than did the filter of SMA-order 12. Whether removing the stochastic cycles at the unwanted periodicities is worth losing more observations in the filtered series is a judgment call.

Below we plot the estimated business-cycle component with recessions identified by the shaded areas.

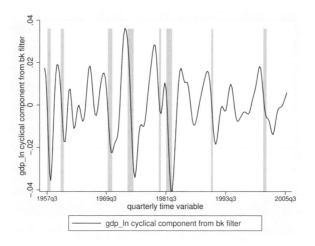

◁

Saved results

tsfilter bk saves the following in r():

Scalars
r(smaorder)	order of the symmetric moving average
r(minperiod)	minimum period of stochastic cycles
r(maxperiod)	maximum period of stochastic cycles

Macros
r(varlist)	original time-series variables
r(filterlist)	variables containing estimates of the cyclical components
r(trendlist)	variables containing estimates of the trend components, if trend() was specified
r(method)	Baxter-King
r(stationary)	yes or no, indicating whether the calculations assumed the series was or was not stationary
r(unit)	units of time variable set using tsset or xtset

Matrices
r(filter)	$(q+1) \times 1$ matrix of filter weights, where q is the order of the symmetric moving average

Methods and formulas

tsfilter bk is implemented as an ado-file.

Baxter and King (1999) showed that there is an infinite-order SMA filter with coefficients that sum to zero that can extract the specified components from a nonstationary time series. The components are specified in terms of the minimum and maximum periods of the stochastic cycles that drive these components in the frequency domain. This ideal filter is not feasible, because the constraints imposed on the filter can only be satisfied using an infinite number of coefficients, so Baxter and King (1999) derived a finite approximation to this ideal filter.

The infinite-order, ideal band-pass filter obtains the cyclical component with the calculation

$$c_t = \sum_{j=-\infty}^{\infty} b_j y_{t-j}$$

Letting p_l and p_h be the minimum and maximum periods of the stochastic cycles of interest, the weights b_j in this calculation are given by

$$
b_j = \begin{cases} \pi^{-1}(\omega_h - \omega_l) & \text{if } j = 0 \\[2ex] (j\pi)^{-1}\{\sin(j\omega_h) - \sin(j\omega_l)\} & \text{if } j \neq 0 \end{cases}
$$

where $\omega_l = 2\pi/p_l$ and $\omega_h = 2\pi/p_h$ are the lower and higher cutoff frequencies, respectively.

For the default case of nonstationary time series with finite length, the ideal band-pass filter cannot be used without modification. Baxter and King (1999) derived modified weights for a finite order SMA filter with coefficients that sum to zero.

As a result, Baxter and King (1999) estimate c_t by

$$
c_t = \sum_{j=-q}^{+q} \widehat{b}_j y_{t-j}
$$

The coefficients $\widehat{b}_j$ in this calculation are equal to $\widehat{b}_j = b_j - \overline{b}_q$, where $\widehat{b}_{-j} = \widehat{b}_j$ and $\overline{b}_q$ is the mean of the ideal coefficients truncated at $\pm q$:

$$
\overline{b}_q = (2q+1)^{-1} \sum_{j=-q}^{q} b_j
$$

Note that $\sum_{j=-q}^{+q} \widehat{b}_j = 0$ and that the first and last q values of the cyclical component cannot be estimated using this filter.

If the `stationary` option is set, the BK filter sets the coefficients to the ideal coefficients, that is, $\widehat{b}_j = b_j$. For these weights, $\widehat{b}_j = \widehat{b}_{-j}$, and although $\sum_{j=-\infty}^{\infty} \widehat{b}_j = 0$, for small q, $\sum_{-q}^{q} \widehat{b}_j \neq 0$.

References

Baxter, M., and R. G. King. 1995. Measuring business cycles approximate band-pass filters for economic time series. NBER Working Paper No. 5022, National Bureau of Economic Research. http://www.nber.org/papers/w5022.

———. 1999. Measuring business cycles: Approximate band-pass filters for economic time series. *Review of Economics and Statistics* 81: 575–593.

Burns, A. F., and W. C. Mitchell. 1946. *Measuring Business Cycles.* New York: National Bureau of Economic Research.

Fuller, W. A. 1996. *Introduction to Statistical Time Series.* 2nd ed. New York: Wiley.

Pollock, D. S. G. 1999. *A Handbook of Time-Series Analysis, Signal Processing and Dynamics.* London: Academic Press.

———. 2006. Econometric methods of signal extraction. *Computational Statistics & Data Analysis* 50: 2268–2292.

Also see

[TS] **tsset** — Declare data to be time-series data

[XT] **xtset** — Declare data to be panel data

[TS] **tsfilter** — Filter a time-series, keeping only selected periodicities

[D] **format** — Set variables' output format

[TS] **tssmooth** — Smooth and forecast univariate time-series data

Title

> **tsfilter bw** — Butterworth time-series filter

Syntax

Filter one variable

> tsfilter bw [*type*] *newvar* = *varname* [*if*] [*in*] [, *options*]

Filter multiple variables, unique names

> tsfilter bw [*type*] *newvarlist* = *varlist* [*if*] [*in*] [, *options*]

Filter multiple variables, common name stub

> tsfilter bw [*type*] *stub** = *varlist* [*if*] [*in*] [, *options*]

options	Description
Main	
<u>maxp</u>eriod(*#*)	filter out stochastic cycles at periods larger than *#*
<u>or</u>der(*#*)	set the order of the filter; default is order(2)
Trend	
<u>tr</u>end(*newvar* | *newvarlist* | *stub**)	save the trend component(s) in new variable(s)
Gain	
gain(*gainvar anglevar*)	save the gain and angular frequency

You must tsset or xtset your data before using tsfilter; see [TS] **tsset** and [XT] **xtset**.
varname and *varlist* may contain time-series operators; see [U] **11.4.4 Time-series varlists**.

Menu

Statistics > Time series > Filters for cyclical components > Butterworth

Description

tsfilter bw uses the Butterworth high-pass filter to separate a time series into trend and cyclical components. The trend component may contain a deterministic or a stochastic trend. The stationary cyclical component is driven by stochastic cycles at the specified periods.

See [TS] **tsfilter** for an introduction to the methods implemented in tsfilter bw.

418

Options

maxperiod(#) filters out stochastic cycles at periods larger than #, where # must be greater than 2. By default, if the units of the time variable are set to daily, weekly, monthly, quarterly, half-yearly, or yearly, then # is set to the number of periods equivalent to 8 years; otherwise, the default value is maxperiod(32).

order(#) sets the order of the Butterworth filter, which must be an integer. The default is order(2).

trend(*newvar* | *newvarlist* | *stub**) saves the trend component(s) in the new variable(s) specified by *newvar*, *newvarlist*, or *stub**.

gain(*gainvar anglevar*) saves the gain in *gainvar* and its associated angular frequency in *anglevar*. Gains are calculated at the N angular frequencies that uniformly partition the interval $(0, \pi]$, where N is the sample size.

Remarks

We assume that you have already read [TS] **tsfilter**, which provides an introduction to filtering and the methods implemented in tsfilter bw, more examples using tsfilter bw, and a comparison of the four filters implemented by tsfilter. In particular, an understanding of gain functions as presented in [TS] **tsfilter** is required to understand these remarks.

tsfilter bw uses the Butterworth high-pass filter to separate a time-series y_t into trend and cyclical components:

$$y_t = \tau_t + c_t$$

where τ_t is the trend component and c_t is the cyclical component. τ_t may be nonstationary; it may contain a deterministic or a stochastic trend, as discussed below.

The primary objective is to estimate c_t, a stationary cyclical component that is driven by stochastic cycles within a specified range of periods. The trend component τ_t is calculated by the difference $\tau_t = y_t - c_t$.

Although the Butterworth high-pass filter implemented in tsfilter bw has been widely applied by macroeconomists and engineers, it is a general time-series method and may be of interest to other researchers.

Engineers have used Butterworth filters for a long time because they are "maximally flat". The gain functions of these filters are as close as possible to being a flat line at 0 for the unwanted periods and a flat line at 1 for the desired periods; see Butterworth (1930) and Bianchi and Sorrentino (2007, 17–20). (See [TS] **tsfilter** for an introduction to gain functions.)

The high-pass Butterworth filter is a two-parameter filter. The maxperiod() option specifies the maximum period; the stochastic cycles of all higher periodicities are filtered out. The maxperiod() option sets the location of the cutoff period in the gain function. The order() option specifies the order of the filter, which determines the slope of the gain function at the cutoff frequency.

For a given cutoff period, the slope of the gain function at the cutoff period increases with filter order. For a given filter order, the slope of the gain function at the cutoff period increases with the cutoff period.

We cannot obtain a vertical slope at the cutoff frequency, which is the ideal, because the computation becomes unstable; see Pollock (2000). The filter order for which the computation becomes unstable depends on the cutoff period.

Among economists, the high-pass Butterworth filter is commonly used for investigating business cycles. Burns and Mitchell (1946) defined business cycles as stochastic cycles in business data corresponding to periods between 1.5 and 8 years. For this reason, the default value for maxperiod() is the number of periods in 8 years, if the time variable is formatted as daily, weekly, monthly, quarterly, half-yearly, or yearly; see [D] **format**. The default value for maxperiod() is 32 for all other time formats.

For each variable, the high-pass Butterworth filter estimate of c_t is put in the corresponding new variable, and when the trend() option is specified, the estimate of τ_t is put in the corresponding new variable.

tsfilter bw automatically detects panel data from the information provided when the dataset was tsset or xtset. All calculations are done separately on each panel. Missing values at the beginning and end of the sample are excluded from the sample. The sample may not contain gaps.

▷ Example 1

In this and the subsequent examples, we use tsfilter bw to estimate the business-cycle component of the natural log of the real gross domestic product (GDP) of the United States. Our sample of quarterly data goes from 1952q1 to 2010q4. Below we read in and plot the data.

```
. use http://www.stata-press.com/data/r12/gdp2
(Federal Reserve Economic Data, St. Louis Fed)

. tsline gdp_ln
```

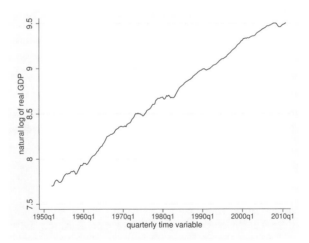

The series is nonstationary. Pollock (2000) shows that the high-pass Butterworth filter can estimate the components driven by the stochastic cycles at the specified frequencies when the original series is nonstationary.

Below we use tsfilter bw to filter gdp_ln and use pergram (see [TS] **pergram**) to compute and to plot the periodogram of the estimated cyclical component.

```
. tsfilter bw gdp_bw = gdp_ln

. pergram gdp_bw, xline(.03125 .16667)
```

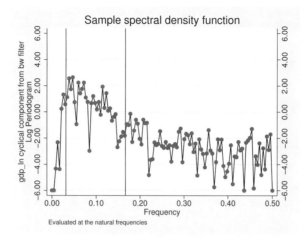

Sample spectral density function

Evaluated at the natural frequencies

tsfilter bw used the default value of maxperiod(32) because our sample is of quarterly data. In the periodogram, we added vertical lines at the natural frequencies corresponding to the conventional Burns and Mitchell (1946) values for business-cycle components. pergram displays the results in natural frequencies, which are the standard frequencies divided by 2π. We use option xline() to draw vertical lines at the lower natural-frequency cutoff ($1/32 = 0.03125$) and the upper natural-frequency cutoff ($1/6 \approx 0.16667$).

If the filter completely removed the stochastic cycles at the unwanted frequencies, the periodogram would be a flat line at the minimum value of -6 outside the range identified by the vertical lines.

The periodogram reveals two issues. First, it indicates that the default value of order(2) did not do a good job of filtering out the high-periodicity stochastic cycles, because there are too many points above -6.00 to the left of the left-hand vertical line. Second, it reveals the high-pass nature of the filter, because none of the low-period (high-frequency) stochastic cycles have been filtered out.

We cope with these two issues in the remaining examples.

◁

▷ Example 2

In this example, we change the order of the filter so that it will remove more of the unwanted low-frequency stochastic cycles. As previously mentioned, increasing the order of the filter increases the slope of the gain function at the cutoff period.

For orders 2 and 8, we compute the filtered series, compute the gain functions, and label the gain variables. We also generate ideal, the gain function of the ideal band-pass filter at the frequencies f. Then we plot the gain function of the ideal band-pass filter and the gain functions of the high-pass Butterworth filters of orders 2 and 8.

```
. tsfilter bw gdp_bw2 = gdp_ln, gain(g1 a1)
. label variable g1 "BW order 2"
. tsfilter bw gdp_bw8 = gdp_ln, gain(g8 a8) order(8)
. label variable g8 "BW order 8"
. generate f = _pi*(_n-1)/_N
. generate ideal = cond(f<_pi/16, 0, cond(f<_pi/3, 1,0))
. label variable ideal "Ideal filter"
```

```
. twoway line ideal f || line g1 a1 || line g8 a8
```

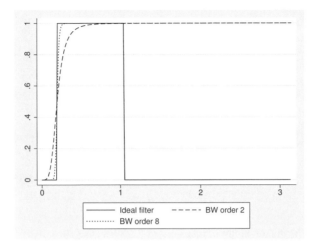

As discussed in [TS] **tsfilter**, the gain function of the ideal filter is a square wave with a value of 0 at the frequencies corresponding to unwanted frequencies and a value of 1 at the desired frequencies. The vertical lines in the gain function of the ideal filter occur at $\pi/16$, corresponding to 32 periods, and at $\pi/3$, corresponding to 6 periods. (Given that $p = 2\pi/\omega$, where p is the period corresponding to frequency ω, the frequency is given by $2\pi/p$.)

The distance between the gain function of the filter with order 2 and the gain function of the ideal band-pass filter at $\pi/16$ is the root of the first issue mentioned at the end of example 1. The filter with order 8 is much closer to the gain function of the ideal band-pass filter at $\pi/16$ than is the filter with order 2. That both gain functions are 1 to the right of the vertical line at $\pi/3$ reveals the high-pass nature of the filter.

◁

> Example 3

In this example, we use a common trick to resolve the second issue mentioned at the end of example 1. Keeping the trend produced by a high-pass filter turns that high-pass filter into a low-pass filter. Because we want to remove the high-frequency stochastic cycles still in the previously filtered series gdp_bw8, we need to run gdp_bw8 through a low-pass filter. So we keep the trend produced by refiltering the previously filtered series.

To determine an order for the filter, we run the filter with order(8), then with order(15), and then we plot the gain functions along with the gain function of the ideal filter.

```
. tsfilter bw gdp_bwn8  = gdp_bw8, gain(gc8 ac8) order(8)
> maxperiod(6) trend(gdp_bwc8)
. label variable gc8 "BW order 8"
. tsfilter bw gdp_bwn15 = gdp_bw8, gain(gc15 ac15) order(15)
> maxperiod(6) trend(gdp_bwc15)
. label variable gc15 "BW order 15"
. twoway line ideal f || line gc8 ac8 || line gc15 ac15
```

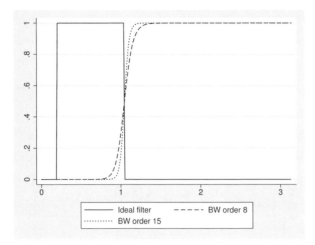

We specified much higher orders for the filter in this example because the cutoff period is 6 instead of 32. (As previously mentioned, holding the order of the filter constant, the slope of the gain function at the cutoff period decreases when the period decreases.) The above graph indicates that the filter with order(15) is reasonably close to the gain function of the ideal filter.

Now we compute and plot the periodogram of the estimated business-cycle component.

```
. pergram gdp_bwc15, xline(.03125 .16667)
```

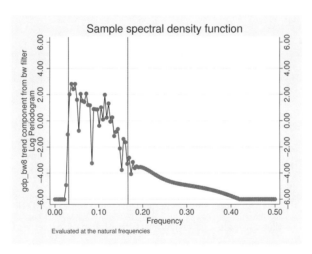

The graph indicates that the above applications of the Butterworth filter did a reasonable job of filtering out the high-periodicity stochastic cycles but that the low-periodicity stochastic cycles have not been completely removed.

Below we plot the estimated business-cycle component with recessions identified by the shaded areas.

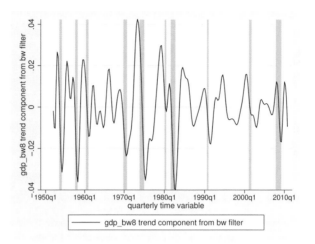

Saved results

tsfilter bw saves the following in r():

Scalars
    r(order)   order of the filter
    r(maxperiod) maximum period of stochastic cycles

Macros
    r(varlist)   original time-series variables
    r(filterlist) variables containing estimates of the cyclical components
    r(trendlist) variables containing estimates of the trend components, if trend() was specified
    r(method)   Butterworth
    r(unit)   units of time variable set using tsset or xtset

Methods and formulas

tsfilter bw is implemented as an ado-file.

tsfilter bw uses the computational methods described in Pollock (2000) to implement the filter.

Pollock (2000) shows that the gain of the Butterworth high-pass filter is given by

$$\psi(\omega) = \left[1 + \left\{ \frac{\tan(\omega_c/2)}{\tan(\omega/2)} \right\}^{2m} \right]^{-1}$$

where m is the order of the filter, $\omega_c = 2\pi/p_h$ is the cutoff frequency, and p_h is the maximum period.

Here is an outline of the computational procedure that Pollock (2000) derived.

Pollock (2000) showed that the Butterworth filter corresponds to a particular model. Actually, his model is more general than the Butterworth filter, but tsfilter bw restricts the computations to the case in which the model corresponds to the Butterworth filter.

The model represents the series to be filtered, y_t, in terms of zero mean, covariance stationary, and independent and identically distributed shocks ν_t and ε_t:

$$y_t = \frac{(1+L)^m}{(1-L)^m}\nu_t + \varepsilon_t$$

From this model, Pollock (2000) shows that the optimal estimate for the cyclical component is given by

$$\mathbf{c} = \lambda\mathbf{Q}(\mathbf{\Omega_L} + \lambda\mathbf{\Omega_H})^{-1}\mathbf{Q}'\mathbf{y}$$

where $\mathrm{Var}\{\mathbf{Q}'(\mathbf{y} - \mathbf{c})\} = \sigma_\nu^2\mathbf{\Omega_L}$ and $\mathrm{Var}\{\mathbf{Q}'\mathbf{c}\} = \sigma_\varepsilon^2\mathbf{\Omega_H}$. Here $\mathbf{\Omega_L}$ and $\mathbf{\Omega_H}$ are symmetric Toeplitz matrices with $2m+1$ nonzero diagonal bands and generating functions $(1+z)^m(1+z^{-1})^m$ and $(1-z)^m(1-z^{-1})^m$, respectively.

The parameter λ in this expression is a function of p_h (the maximum period of stochastic cycles filtered out) and the order of the filter:

$$\lambda = \{\tan(\pi/p_h)\}^{-2m}$$

The matrix $\mathbf{Q}'$ in this expression is a function of the coefficients in the polynomial $(1-L)^d = 1 + \delta_1 L + \cdots + \delta_d L^d$:

$$\mathbf{Q}' = \begin{pmatrix} \delta_d & \cdots & \delta_1 & 1 & \cdots & 0 & 0 & \cdots & 0 & 0 \\ \vdots & \ddots & \vdots & \vdots & \ddots & \vdots & \vdots & & \vdots & \vdots \\ 0 & \cdots & \delta_d & \delta_{d-1} & \cdots & 1 & 0 & \cdots & 0 & 0 \\ 0 & \cdots & 0 & \delta_d & \cdots & \delta_1 & 1 & \cdots & 0 & 0 \\ \vdots & & \vdots & \vdots & \ddots & & & \ddots & \vdots & \vdots \\ 0 & \cdots & 0 & 0 & \cdots & \delta_d & \delta_{d-1} & \cdots & 1 & 0 \\ 0 & \cdots & 0 & 0 & \cdots & 0 & \delta_d & \cdots & \delta_1 & 1 \end{pmatrix}^{(T-d)\times T}$$

It can be shown that $\Omega_H = \mathbf{Q}'\mathbf{Q}$ and $\mathbf{\Omega_L} = |\Omega_H|$, which simplifies the calculation of the cyclical component to

$$\mathbf{c} = \lambda\mathbf{Q}\{|\mathbf{Q}'\mathbf{Q}| + \lambda(\mathbf{Q}'\mathbf{Q})\}^{-1}\mathbf{Q}'\mathbf{y}$$

References

Bianchi, G., and R. Sorrentino. 2007. *Electronic Filter Simulation and Design.* New York: McGraw–Hill.

Burns, A. F., and W. C. Mitchell. 1946. *Measuring Business Cycles.* New York: National Bureau of Economic Research.

Butterworth, S. 1930. On the theory of filter amplifiers. *Experimental Wireless and the Wireless Engineer* 7: 536–541.

Pollock, D. S. G. 1999. *A Handbook of Time-Series Analysis, Signal Processing and Dynamics.* London: Academic Press.

——. 2000. Trend estimation and de-trending via rational square-wave filters. *Journal of Econometrics* 99: 317–334.

——. 2006. Econometric methods of signal extraction. *Computational Statistics & Data Analysis* 50: 2268–2292.

Also see

[TS] **tsset** — Declare data to be time-series data

[XT] **xtset** — Declare data to be panel data

[TS] **tsfilter** — Filter a time-series, keeping only selected periodicities

[D] **format** — Set variables' output format

[TS] **tssmooth** — Smooth and forecast univariate time-series data

Title

> **tsfilter cf** — Christiano–Fitzgerald time-series filter

Syntax

Filter one variable

> tsfilter cf [*type*] *newvar* = *varname* [*if*] [*in*] [, *options*]

Filter multiple variables, unique names

> tsfilter cf [*type*] *newvarlist* = *varlist* [*if*] [*in*] [, *options*]

Filter multiple variables, common name stub

> tsfilter cf [*type*] *stub** = *varlist* [*if*] [*in*] [, *options*]

options	Description
Main	
<u>min</u>period(#)	filter out stochastic cycles at periods smaller than #
<u>max</u>period(#)	filter out stochastic cycles at periods larger than #
<u>sma</u>order(#)	number of observations in each direction that contribute to each filtered value
stationary	use calculations for a stationary time series
drift	remove drift from the time series
Trend	
<u>trend</u>(*newvar* | *newvarlist* | *stub**)	save the trend component(s) in new variable(s)
Gain	
gain(*gainvar anglevar*)	save the gain and angular frequency

You must tsset or xtset your data before using tsfilter; see [TS] **tsset** and [XT] **xtset**.
varname and *varlist* may contain time-series operators; see [U] **11.4.4 Time-series varlists**.

Menu

Statistics > Time series > Filters for cyclical components > Christiano-Fitzgerald

Description

tsfilter cf uses the Christiano and Fitzgerald (2003) band-pass filter to separate a time series into trend and cyclical components. The trend component may contain a deterministic or a stochastic trend. The stationary cyclical component is driven by stochastic cycles at the specified periods.

See [TS] **tsfilter** for an introduction to the methods implemented in tsfilter cf.

427

Options

___ Main ___

minperiod(#) filters out stochastic cycles at periods smaller than #, where # must be at least 2 and less than maxperiod(). By default, if the units of the time variable are set to daily, weekly, monthly, quarterly, or half-yearly, then # is set to the number of periods equivalent to 1.5 years; yearly data use minperiod(2); otherwise, the default value is minperiod(6).

maxperiod(#) filters out stochastic cycles at periods larger than #, where # must be greater than minperiod(). By default, if the units of the time variable are set to daily, weekly, monthly, quarterly, half-yearly, or yearly, then # is set to the number of periods equivalent to 8 years; otherwise, the default value is maxperiod(32).

smaorder(#) sets the order of the symmetric moving average, denoted by q. By default, smaorder() is not set, which invokes the asymmetric calculations for the Christiano–Fitzgerald filter. The order is an integer that specifies the number of observations in each direction used in calculating the symmetric moving average estimate of the cyclical component. This number must be an integer greater than zero and less than $(T-1)/2$. The estimate of the cyclical component for the tth observation, y_t, is based upon the $2q+1$ values $y_{t-q}, y_{t-q+1}, \ldots, y_t, y_{t+1}, \ldots, y_{t+q}$.

stationary modifies the filter calculations to those appropriate for a stationary series. By default, the series is assumed nonstationary.

drift removes drift using the approach described in Christiano and Fitzgerald (2003). By default, drift is not removed.

___ Trend ___

trend(*newvar* | *newvarlist* | *stub**) saves the trend component(s) in the new variable(s) specified by *newvar*, *newvarlist*, or *stub**.

___ Gain ___

gain(*gainvar anglevar*) saves the gain in *gainvar* and its associated angular frequency in *anglevar*. Gains are calculated at the N angular frequencies that uniformly partition the interval $(0, \pi]$, where N is the sample size.

Remarks

We assume that you have already read [TS] **tsfilter**, which provides an introduction to filtering and the methods implemented in tsfilter cf, more examples using tsfilter cf, and a comparison of the four filters implemented by tsfilter. In particular, an understanding of gain functions as presented in [TS] **tsfilter** is required to understand these remarks.

tsfilter cf uses the Christiano–Fitzgerald (CF) band-pass filter to separate a time-series y_t into trend and cyclical components

$$y_t = \tau_t + c_t$$

where τ_t is the trend component and c_t is the cyclical component. τ_t may be nonstationary; it may contain a deterministic or a stochastic trend, as discussed below.

The primary objective is to estimate c_t, a stationary cyclical component that is driven by stochastic cycles at a specified range of periods. The trend component τ_t is calculated by the difference $\tau_t = y_t - c_t$.

Although the CF band-pass filter implemented in tsfilter cf has been widely applied by macroeconomists, it is a general time-series method and may be of interest to other researchers.

As discussed by Christiano and Fitzgerald (2003) and in [TS] **tsfilter**, if one had an infinitely long series, one could apply an ideal band-pass filter that perfectly separates out cyclical components driven by stochastic cycles at the specified periodicities. In finite samples, it is not possible to exactly satisfy the conditions that a filter must fulfill to perfectly separate out the specified stochastic cycles; the expansive filter literature reflects the trade-offs involved in choosing a finite-length filter to separate out the specified stochastic cycles.

Christiano and Fitzgerald (2003) derive a finite-length CF band-pass filter that minimizes the mean squared error between the filtered series and the series filtered by an ideal band-pass filter that perfectly separates out components driven by stochastic cycles at the specified periodicities. Christiano and Fitzgerald (2003) place two important restrictions on the mean squared error problem that their filter solves. First, the CF filter is restricted to be a linear filter. Second, y_t is assumed to be a random-walk process; in other words, $y_t = y_{t-1} + \epsilon_t$, where ϵ_t is independently and identically distributed with mean zero and finite variance. The CF filter is the best linear predictor of the series filtered by the ideal band-pass filter when y_t is a random walk.

Christiano and Fitzgerald (2003) make four points in support of the random-walk assumption. First, the mean squared error problem solved by their filter requires that the process for y_t be specified. Second, they provide a method for removing drift so that their filter handles cases in which y_t is a random walk with drift. Third, many economic time series are well approximated by a random-walk-plus-drift process. (We add that many time series encountered in applied statistics are well approximated by a random-walk-plus-drift process.) Fourth, they provide simulation evidence that their filter performs well when the process generating y_t is not a random-walk-plus-drift process but is close to being a random-walk-plus-drift process.

Comparing the CF filter with the Baxter–King (BK) filter provides some intuition and explains the `smaorder()` option in `tsfilter cf`. As discussed in [TS] **tsfilter** and Baxter and King (1999), symmetric moving-average (SMA) filters with coefficients that sum to zero can extract the components driven by stochastic cycles at specified periodicities when the series to be filtered has a deterministic or stochastic trend of order 1 or 2.

The coefficients of the finite-length BK filter are as close as possible to the coefficients of an ideal SMA band-pass filter under the constraints that the BK coefficients are symmetric and sum to zero. The coefficients of the CF filter are not symmetric nor do they sum to zero, but the CF filter was designed to filter out the specified periodicities when y_t has a first-order stochastic trend.

To be robust to second-order trends, Christiano and Fitzgerald (2003) derive a constrained version of the CF filter. The coefficients of the constrained filter are constrained to be symmetric and to sum to zero. Subject to these constraints, the coefficients of the constrained CF filter minimize the mean squared error between the filtered series and the series filtered by an ideal band-pass filter that perfectly separates out the components. Christiano and Fitzgerald (2003) note that the higher-order detrending properties of this constrained filter come at the cost of lost efficiency. If the constraints are binding, the constrained filter cannot predict the series filtered by the ideal filter as well as the unconstrained filter can.

Specifying the `smaorder()` option causes `tsfilter cf` to compute the SMA-constrained CF filter.

The choice between the BK and the CF filters is one between robustness and efficiency. The BK filter handles a broader class of stochastic processes than does the CF filter, but the CF filter produces a better estimate of c_t if y_t is close to a random-walk process or a random-walk-plus-drift process.

Among economists, the CF filter is commonly used for investigating business cycles. Burns and Mitchell (1946) defined business cycles as stochastic cycles in business data corresponding to periods between 1.5 and 8 years. The default values for `minperiod()` and `maxperiod()` are the Burns–Mitchell values of 1.5 and 8 years scaled to the frequency of the dataset. The calculations of the default values assume that the time variable is formatted as daily, weekly, monthly, quarterly, half-yearly, or yearly; see [D] **format**.

When y_t is assumed to be a random-walk-plus-drift process instead of a random-walk process, specify the `drift` option, which removes the linear drift in the series before applying the filter. Drift is removed by transforming the original series to a new series by using the calculation

$$z_t = y_t - \frac{(t-1)(y_T - y_1)}{T - 1}$$

The cyclical component c_t is calculated from drift-adjusted series z_t. The trend component τ_t is calculated by $\tau_t = y_t - c_t$.

By default, the CF filter assumes the series is nonstationary. If the series is stationary, the `stationary` option is used to change the calculations to those appropriate for a stationary series.

For each variable, the CF filter estimate of c_t is put in the corresponding new variable, and when the `trend()` option is specified, the estimate of τ_t is put in the corresponding new variable.

`tsfilter cf` automatically detects panel data from the information provided when the dataset was `tsset` or `xtset`. All calculations are done separately on each panel. Missing values at the beginning and end of the sample are excluded from the sample. The sample may not contain gaps.

▷ Example 1

In this and the subsequent examples, we use `tsfilter cf` to estimate the business-cycle component of the natural log of real gross domestic product (GDP) of the United States. Our sample of quarterly data goes from 1952q1 to 2010q4. Below we read in and plot the data.

```
. use http://www.stata-press.com/data/r12/gdp2
(Federal Reserve Economic Data, St. Louis Fed)
. tsline gdp_ln
```

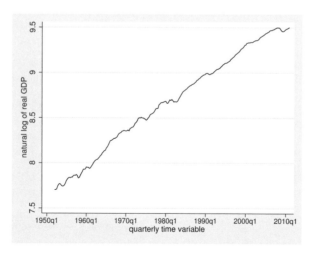

The series looks like it might be generated by a random-walk-plus-drift process and is thus a candidate for the CF filter.

Below we use `tsfilter cf` to filter `gdp_ln`, and we use `pergram` (see [TS] **pergram**) to compute and to plot the periodogram of the estimated cyclical component.

```
. tsfilter cf gdp_cf = gdp_ln
. pergram gdp_cf, xline(.03125 .16667)
```

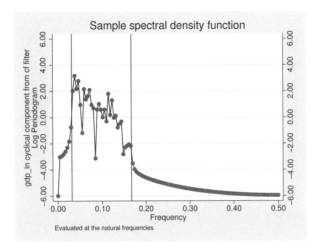

Because our sample is of quarterly data, `tsfilter cf` used the default values of `minperiod(6)` and `maxperiod(32)`. The minimum and maximum periods are the Burns and Mitchell (1946) business-cycle periods for quarterly data.

In the periodogram, we added vertical lines at the natural frequencies corresponding to the conventional Burns and Mitchell (1946) values for business-cycle components. `pergram` displays the results in natural frequencies, which are the standard frequencies divided by 2π. We use the `xline()` option to draw vertical lines at the lower natural-frequency cutoff ($1/32 = 0.03125$) and the upper natural-frequency cutoff ($1/6 \approx 0.16667$).

If the filter completely removed the stochastic cycles at the unwanted frequencies, the periodogram would be a flat line at the minimum value of -6 outside the range identified by the vertical lines.

The periodogram reveals that the CF did a reasonable job of filtering out the unwanted stochastic cycles.

Below we plot the estimated business-cycle component with recessions identified by the shaded areas.

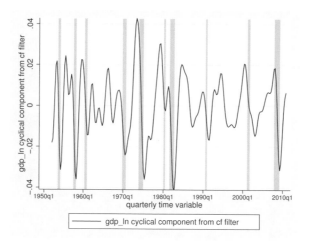

$\triangleleft$

Saved results

tsfilter cf saves the following in r():

Scalars
r(smaorder)	order of the symmetric moving average, if specified
r(minperiod)	minimum period of stochastic cycles
r(maxperiod)	maximum period of stochastic cycles

Macros
r(varlist)	original time-series variables
r(filterlist)	variables containing estimates of the cyclical components
r(trendlist)	variables containing estimates of the trend components, if trend() was specified
r(method)	Christiano-Fitzgerald
r(symmetric)	yes or no, indicating whether the symmetric version of the filter was or was not used
r(drift)	yes or no, indicating whether drift was or was not removed before filtering
r(stationary)	yes or no, indicating whether the calculations assumed the series was or was not stationary
r(unit)	units of time variable set using tsset or xtset

Matrices
r(filter)	$(q+1) \times 1$ matrix of weights $(\widehat{b_0}, \widehat{b_1}, \ldots, \widehat{b_q})'$, where q is the order of the symmetric moving average, and the weights are the Christiano–Fitzgerald coefficients; only returned when smaorder() is used to set q

Methods and formulas

tsfilter cf is implemented as an ado-file.

For an infinitely long series, there is an ideal band-pass filter that extracts the cyclical component by using the calculation

$$c_t = \sum_{j=-\infty}^{\infty} b_j y_{t-j}$$

If p_l and p_h are the minimum and maximum periods of the stochastic cycles of interest, the weights b_j in the ideal band-pass filter are given by

$$
b_j = \begin{cases} \pi^{-1}(\omega_h - \omega_l) & \text{if } j = 0 \\[2mm] (j\pi)^{-1}\{\sin(j\omega_h) - \sin(j\omega_l)\} & \text{if } j \neq 0 \end{cases}
$$

where $\omega_l = 2\pi/p_l$ and $\omega_h = 2\pi/p_h$ are the lower and higher cutoff frequencies, respectively.

Because our time series has finite length, the ideal band-pass filter cannot be computed exactly. Christiano and Fitzgerald (2003) derive the finite-length CF band-pass filter that minimizes the mean squared error between the filtered series and the series filtered by an ideal band-pass filter that perfectly separates out the components. This filter is not symmetric nor do the coefficients sum to zero. The formula for calculating the value of cyclical component c_t for $t = 2, 3, \ldots, T - 1$ using the asymmetric version of the CF filter can be expressed as

$$
c_t = b_0 y_t + \sum_{j=1}^{T-t-1} b_j y_{t+j} + \tilde{b}_{T-t} y_T + \sum_{j=1}^{t-2} b_j y_{t-j} + \tilde{b}_{t-1} y_1
$$

where $b_0, b_1, \ldots$ are the weights used by the ideal band-pass filter. $\tilde{b}_{T-t}$ and $\tilde{b}_{t-1}$ are linear functions of the ideal weights used in this calculation. The CF filter uses two different calculations for $\tilde{b}_t$ depending upon whether the series is assumed to be stationary or nonstationary.

For the default nonstationary case with $1 < t < T$, Christiano and Fitzgerald (2003) set $\tilde{b}_{T-t}$ and $\tilde{b}_{t-1}$ to

$$
\tilde{b}_{T-t} = -\frac{1}{2}b_0 - \sum_{j=1}^{T-t-1} b_j \quad \text{and} \quad \tilde{b}_{t-1} = -\frac{1}{2}b_0 - \sum_{j=1}^{t-2} b_j
$$

which forces the weights to sum to zero.

For the nonstationary case, when $t = 1$ or $t = T$, the two endpoints (c_1 and c_T) use only one modified weight, $\tilde{b}_{T-1}$:

$$
c_1 = \frac{1}{2}b_0 y_1 + \sum_{j=1}^{T-2} b_j y_{j+1} + \tilde{b}_{T-1} y_T \quad \text{and} \quad c_T = \frac{1}{2}b_0 y_T + \sum_{j=1}^{T-2} b_j y_{T-j} + \tilde{b}_{T-1} y_1
$$

When the `stationary` option is used to invoke the stationary calculations, all weights are set to the ideal filter weight, that is, $\tilde{b}_j = b_j$.

If the `smaorder()` option is set, the symmetric version of the CF filter is used. This option specifies the length of the symmetric moving average denoted by q. The symmetric calculations for c_t are similar to those used by the BK filter:

$$
c_t = \hat{b}_q \{L^{-q}(y_t) + L^q(y_t)\} + \sum_{j=-q+1}^{q-1} b_j L^j(y_t)
$$

where, for the default nonstationary calculations, $\hat{b}_q = -(1/2)b_0 - \sum_{j=1}^{q-1} b_j$. If the `smaorder()` and `stationary` options are set, then $\hat{b}_q$ is set equal to the ideal weight b_q.

References

Baxter, M., and R. G. King. 1999. Measuring business cycles: Approximate band-pass filters for economic time series. *Review of Economics and Statistics* 81: 575–593.

Burns, A. F., and W. C. Mitchell. 1946. *Measuring Business Cycles.* New York: National Bureau of Economic Research.

Christiano, L. J., and T. J. Fitzgerald. 2003. The band pass filter. *International Economic Review* 44: 435–465.

Pollock, D. S. G. 1999. *A Handbook of Time-Series Analysis, Signal Processing and Dynamics.* London: Academic Press.

——. 2006. Econometric methods of signal extraction. *Computational Statistics & Data Analysis* 50: 2268–2292.

Also see

[TS] **tsset** — Declare data to be time-series data

[XT] **xtset** — Declare data to be panel data

[TS] **tsfilter** — Filter a time-series, keeping only selected periodicities

[D] **format** — Set variables' output format

[TS] **tssmooth** — Smooth and forecast univariate time-series data

Title

tsfilter hp — Hodrick–Prescott time-series filter

Syntax

Filter one variable

> tsfilter hp [*type*] *newvar* = *varname* [*if*] [*in*] [, *options*]

Filter multiple variables, unique names

> tsfilter hp [*type*] *newvarlist* = *varlist* [*if*] [*in*] [, *options*]

Filter multiple variables, common name stub

> tsfilter hp [*type*] *stub** = *varlist* [*if*] [*in*] [, *options*]

options	Description
Main	
<u>smooth</u>(#)	smoothing parameter for the Hodrick–Prescott filter
Trend	
<u>trend</u>(*newvar* | *newvarlist* | *stub**)	save the trend component(s) in new variable(s)
Gain	
<u>gain</u>(*gainvar anglevar*)	save the gain and angular frequency

You must tsset or xtset your data before using tsfilter; see [TS] **tsset** and [XT] **xtset**.
varname and *varlist* may contain time-series operators; see [U] **11.4.4 Time-series varlists**.

Menu

Statistics > Time series > Filters for cyclical components > Hodrick-Prescott

Description

tsfilter hp uses the Hodrick–Prescott high-pass filter to separate a time series into trend and cyclical components. The trend component may contain a deterministic or a stochastic trend. The smoothing parameter determines the periods of the stochastic cycles that drive the stationary cyclical component.

See [TS] **tsfilter** for an introduction to the methods implemented in tsfilter hp.

Options

⌐ Main ⌐

smooth(*#*) sets the smoothing parameter for the Hodrick–Prescott filter. By default if the units of the time variable are set to daily, weekly, monthly, quarterly, half-yearly, or yearly, then the Ravn–Uhlig rule is used to set the smoothing parameter; otherwise, the default value is smooth(1600). The Ravn–Uhlig rule sets *#* to $1600p_q^4$, where p_q is the number of periods per quarter. The smoothing parameter must be greater than 0.

⌐ Trend ⌐

trend(*newvar* | *newvarlist* | *stub**) saves the trend component(s) in the new variable(s) specified by *newvar*, *newvarlist*, or *stub**.

⌐ Gain ⌐

gain(*gainvar anglevar*) saves the gain in *gainvar* and its associated angular frequency in *anglevar*. Gains are calculated at the N angular frequencies that uniformly partition the interval $(0, \pi]$, where N is the sample size.

Remarks

We assume that you have already read [TS] **tsfilter**, which provides an introduction to filtering and the methods implemented in tsfilter hp, more examples using tsfilter hp, and a comparison of the four filters implemented by tsfilter. In particular, an understanding of gain functions as presented in [TS] **tsfilter** is required to understand these remarks.

tsfilter hp uses the Hodrick–Prescott (HP) high-pass filter to separate a time-series y_t into trend and cyclical components

$$y_t = \tau_t + c_t$$

where τ_t is the trend component and c_t is the cyclical component. τ_t may be nonstationary; it may contain a deterministic or a stochastic trend, as discussed below.

The primary objective is to estimate c_t, a stationary cyclical component that is driven by stochastic cycles at a range of periods. The trend component τ_t is calculated by the difference $\tau_t = y_t - c_t$.

Although the HP high-pass filter implemented in tsfilter hp has been widely applied by macroeconomists, it is a general time-series method and may be of interest to other researchers.

Hodrick and Prescott (1997) motivated the HP filter as a trend-removal technique that could be applied to data that came from a wide class of data-generating processes. In their view, the technique specified a trend in the data and the data was filtered by removing the trend. The smoothness of the trend depends on a parameter λ. The trend becomes smoother as $\lambda \to \infty$, and Hodrick and Prescott (1997) recommended setting λ to 1,600 for quarterly data.

King and Rebelo (1993) showed that removing a trend estimated by the HP filter is equivalent to a high-pass filter. They derived the gain function of this high-pass filter and showed that the filter would make integrated processes of order 4 or less stationary, making the HP filter comparable to the other filters implemented in tsfilter.

▷ Example 1

In this and the subsequent examples, we use `tsfilter hp` to estimate the business-cycle component of the natural log of real gross domestic product (GDP) of the United States. Our sample of quarterly data goes from 1952q1 to 2010q4. Below we read in and plot the data.

```
. use http://www.stata-press.com/data/r12/gdp2
(Federal Reserve Economic Data, St. Louis Fed)
. tsline gdp_ln
```

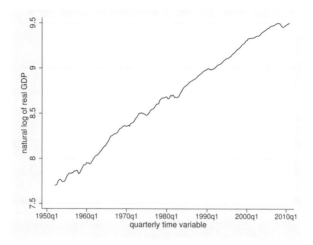

The series is nonstationary and is thus a candidate for the HP filter.

Below we use `tsfilter hp` to filter gdp_ln, and we use `pergram` (see [TS] **pergram**) to compute and to plot the periodogram of the estimated cyclical component.

```
. tsfilter hp gdp_hp = gdp_ln
. pergram gdp_hp, xline(.03125 .16667)
```

Because our sample is of quarterly data, `tsfilter hp` used the default value for the smoothing parameter of 1,600.

In the periodogram, we added vertical lines at the natural frequencies corresponding to the conventional Burns and Mitchell (1946) values for business-cycle components of 32 periods and 6 periods. `pergram` displays the results in natural frequencies, which are the standard frequencies divided by 2π. We use the `xline()` option to draw vertical lines at the lower natural-frequency cutoff ($1/32 = 0.03125$) and the upper natural-frequency cutoff ($1/6 \approx 0.16667$).

If the filter completely removed the stochastic cycles at the unwanted frequencies, the periodogram would be a flat line at the minimum value of -6 outside the range identified by the vertical lines.

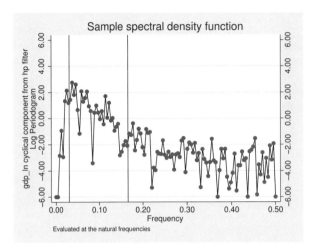

The periodogram reveals a high-periodicity issue and a low-periodicity issue. The points above -6.00 to the left of the left-hand vertical line in the periodogram reveal that the filter did not do a good job of filtering out the high-periodicity stochastic cycles with the default value smoothing parameter of 1,600. That there is no tendency of the points to the right of the right-hand vertical line to be smoothed toward -6.00 reveals that the HP filter did not remove any of the low-periodicity stochastic cycles. This result is not surprising, because the HP filter is a high-pass filter.

In the next example, we address the high-periodicity issue. See [TS] **tsfilter** and [TS] **tsfilter bw** for how to turn a high-pass filter into a band-pass filter.

◁

▷ Example 2

In the filter literature, filter parameters are set as functions of the cutoff frequency; see Pollock (2000, 324), for instance. This method finds the filter parameter that sets the gain of the filter equal to $1/2$ at the cutoff frequency. In a technical note in [TS] **tsfilter**, we showed that applying this method to selecting λ at the cutoff frequency of 32 periods suggests setting $\lambda \approx 677.13$. In the output below, we estimate the business-cycle component using this value for the smoothing parameter, and we compute and plot the periodogram of the estimated business-cycle component.

```
. tsfilter hp gdp_hp677 = gdp_ln, smooth(677.13)

. pergram gdp_hp677, xline(.03125 .16667)
```

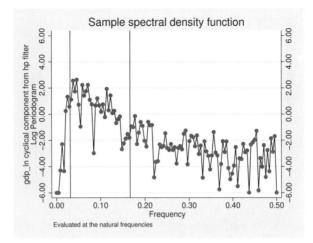

A comparison of the two periodograms reveals that setting the smoothing parameter to 677.13 removes more of the high-periodicity stochastic cycles than does the default 1,600. In [TS] **tsfilter**, we found that the HP filter was not as good at removing the high-periodicity stochastic cycles as was the Christiano–Fitzgerald filter implemented in tsfilter cf or as was the Butterworth filter implemented in tsfilter bw.

Below we plot the estimated business-cycle component with recessions identified by the shaded areas.

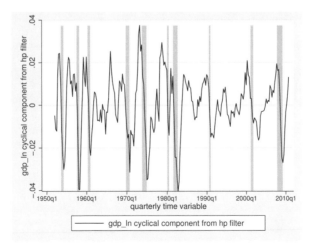

tsfilter hp automatically detects panel data from the information provided when the dataset was tsset or xtset. All calculations are done separately on each panel. Missing values at the beginning and end of the sample are excluded from the sample. The sample may not contain gaps.

Saved results

tsfilter hp saves the following in r():

Scalars
    r(smooth)   smoothing parameter λ

Macros
    r(varlist)   original time-series variables
    r(filterlist)   variables containing estimates of the cyclical components
    r(trendlist)   variables containing estimates of the trend components, if trend() was specified
    r(method)   Hodrick-Prescott
    r(unit)   units of time variable set using tsset or xtset

Methods and formulas

tsfilter hp is implemented as an ado-file.

Formally, the filter is defined as the solution to the following optimization problem for τ_t

$$\min_{\tau_t} \left[\sum_{t=1}^{T}(y_t - \tau_t)^2 + \lambda \sum_{t=2}^{T-1} \{(\tau_{t+1} - \tau_t) - (\tau_t - \tau_{t-1})\}^2 \right]$$

where the smoothing parameter λ is set fixed to a value.

If $\lambda = 0$, the solution degenerates to $\tau_t = y_t$, in which case the filter excludes all frequencies, that is, $c_t = 0$. On the other extreme, as $\lambda \to \infty$, the solution approaches the least-squares fit to the line $\tau_t = \beta_0 + \beta_1 t$; see Hodrick and Prescott (1997) for a discussion.

For a fixed λ, it can be shown that the cyclical component $\mathbf{c}' = (c_1, c_2, \ldots, c_T)$ is calculated by

$$\mathbf{c} = (\mathbf{I_T} - \mathbf{M}^{-1})\mathbf{y}$$

where $\mathbf{y}$ is the column vector $\mathbf{y}' = (y_1, y_2, \ldots, y_T)$, $\mathbf{I_T}$ is the $T \times T$ identity matrix, and $\mathbf{M}$ is the $T \times T$ matrix:

$$\mathbf{M} = \begin{pmatrix}
(1+\lambda) & -2\lambda & \lambda & 0 & 0 & 0 & \cdots & 0 \\
-2\lambda & (1+5\lambda) & -4\lambda & \lambda & 0 & 0 & \cdots & 0 \\
\lambda & -4\lambda & (1+6\lambda) & -4\lambda & \lambda & 0 & \cdots & 0 \\
0 & \lambda & -4\lambda & (1+6\lambda) & -4\lambda & \lambda & \cdots & 0 \\
\vdots & & \ddots & \ddots & \ddots & \ddots & \cdots & \vdots \\
\vdots & 0 & \ddots & \ddots & \ddots & \ddots & \ddots & 0 \\
0 & \cdots & \lambda & -4\lambda & (1+6\lambda) & -4\lambda & \lambda & 0 \\
0 & \cdots & 0 & \lambda & -4\lambda & (1+6\lambda) & -4\lambda & \lambda \\
0 & \cdots & 0 & 0 & \lambda & -4\lambda & (1+5\lambda) & -2\lambda \\
0 & \cdots & 0 & 0 & 0 & \lambda & -2\lambda & (1+\lambda)
\end{pmatrix}$$

The gain of the HP filter is given by (see King and Rebelo [1993], Maravall and del Rio [2007], or Harvey and Trimbur [2008])

$$\psi(\omega) = \frac{4\lambda\{1 - \cos(\omega)\}^2}{1 + 4\lambda\{1 - \cos(\omega)\}^2}$$

As discussed in [TS] **tsfilter**, there are two approaches to selecting λ. One method, based on the heuristic argument of Hodrick and Prescott (1997), is used to compute the default values for λ. The method sets λ to 1,600 for quarterly data and to the rescaled values worked out by Ravn and Uhlig (2002). The rescaled default values for λ are 6.25 for yearly data, 100 for half-yearly data, 129,600 for monthly data, 1600×12^4 for weekly data, and $1600 \times (365/4)^4$ for daily data.

The second method for selecting λ uses the recommendations of Pollock (2000, 324), who uses the gain function of the filter to identify a value for λ.

Additional literature critiques the HP filter by pointing out that the HP filter corresponds to a specific model. Harvey and Trimbur (2008) show that the cyclical component estimated by the HP filter is equivalent to one estimated by a particular unobserved-components model. Harvey and Jaeger (1993), Gómez (1999), Pollock (2000), and Gómez (2001) also show this result and provide interesting comparisons of estimating c_t by filtering and model-based methods.

References

Burns, A. F., and W. C. Mitchell. 1946. *Measuring Business Cycles.* New York: National Bureau of Economic Research.

Gómez, V. 1999. Three equivalent methods for filtering finite nonstationary time series. *Journal of Business and Economic Statistics* 17: 109–116.

——. 2001. The use of Butterworth filters for trend and cycle estimation in economic time series. *Journal of Business and Economic Statistics* 19: 365–373.

Harvey, A. C., and A. Jaeger. 1993. Detrending, stylized facts and the business cycle. *Journal of Applied Econometrics* 8: 231–247.

Harvey, A. C., and T. M. Trimbur. 2008. Trend estimation and the Hodrick–Prescott filter. *Journal of the Japanese Statistical Society* 38: 41–49.

Hodrick, R. J., and E. C. Prescott. 1997. Postwar U.S. business cycles: An empirical investigation. *Journal of Money, Credit, and Banking* 29: 1–16.

King, R. G., and S. T. Rebelo. 1993. Low frequency filtering and real business cycles. *Journal of Economic Dynamics and Control* 17: 207–231.

Leser, C. E. V. 1961. A simple method of trend construction. *Journal of the Royal Statistical Society, Series B* 23: 91–107.

Maravall, A., and A. del Rio. 2007. Temporal aggregation, systematic sampling, and the Hodrick–Prescott filter. Working Paper No. 0728, Banco de España. http://www.bde.es/webbde/Secciones/Publicaciones/PublicacionesSeriadas/DocumentosTrabajo/07/Fic/dt0728e.pdf.

Pollock, D. S. G. 1999. *A Handbook of Time-Series Analysis, Signal Processing and Dynamics.* London: Academic Press.

——. 2000. Trend estimation and de-trending via rational square-wave filters. *Journal of Econometrics* 99: 317–334.

——. 2006. Econometric methods of signal extraction. *Computational Statistics & Data Analysis* 50: 2268–2292.

Ravn, M. O., and H. Uhlig. 2002. On adjusting the Hodrick–Prescott filter for the frequency of observations. *Review of Economics and Statistics* 84: 371–376.

Also see

[TS] **tsset** — Declare data to be time-series data

[XT] **xtset** — Declare data to be panel data

[TS] **tsfilter** — Filter a time-series, keeping only selected periodicities

[D] **format** — Set variables' output format

[TS] **tssmooth** — Smooth and forecast univariate time-series data

Title

tsline — Plot time-series data

Syntax

Time-series line plot

$[$ <u>tw</u>oway $]$ tsline *varlist* $[$ *if* $]$ $[$ *in* $]$ $[$, *tsline_options* $]$

Time-series range plot with lines

$[$ <u>tw</u>oway $]$ tsrline y_1 y_2 $[$ *if* $]$ $[$ *in* $]$ $[$, *tsrline_options* $]$

where the time variable is assumed set by tsset (see [TS] **tsset**), *varlist* has the interpretation $y_1 [y_2 \ldots y_k]$.

tsline_options	Description
Plots	
scatter_options	any of the options documented in [G-2] **graph twoway scatter** with the exception of *marker_options*, *marker_placement_options*, and *marker_label_options*, which will be ignored if specified
Y axis, Time axis, Titles, Legend, Overall, By	
twoway_options	any options documented in [G-3] *twoway_options*

tsrline_options	Description
Plots	
rline_options	any of the options documented in [G-2] **graph twoway rline**
Y axis, Time axis, Titles, Legend, Overall, By	
twoway_options	any options documented in [G-3] *twoway_options*

Menu

Statistics > Time series > Graphs > Line plots

Description

tsline draws line plots for time-series data.

tsrline draws a range plot with lines for time-series data.

tsline and tsrline are both commands and *plottype*s as defined in [G-2] **graph twoway**. Thus the syntax for tsline is

```
. graph twoway tsline ...
```

442

```
. twoway tsline ...

. tsline ...
```

and similarly for `tsrline`. Being plot types, these commands may be combined with other plot types in the `twoway` family, as in,

```
. twoway (tsrline ...) (tsline ...) (lfit ...) ...
```

which can equivalently be written

```
. tsrline ... || tsline ... || lfit ... || ...
```

Options

─────┤ Plots ├──

scatter_options are any of the options allowed by the `graph twoway scatter` command except that *marker_options*, *marker_placement_option*, and *marker_label_options* will be ignored if specified; see [G-2] **graph twoway scatter**.

rline_options are any of the options allowed by the `graph twoway rline` command; see [G-2] **graph twoway rline**.

─────┤ Y axis, Time axis, Titles, Legend, Overall, By ├──────────────────

twoway_options are any of the options documented in [G-3] ***twoway_options***. These include options for titling the graph (see [G-3] ***title_options***), for saving the graph to disk (see [G-3] ***saving_option***), and the `by()` option, which will allow you to simultaneously plot different subsets of the data (see [G-3] ***by_option***).

Also see the `recast()` option discussed in [G-3] ***advanced_options*** for information on how to plot spikes, bars, etc., instead of lines.

Remarks

▷ Example 1

We simulated two separate time series (each of 200 observations) and placed them in a Stata dataset, `tsline1.dta`. The first series simulates an AR(2) process with $\phi_1 = 0.8$ and $\phi_2 = 0.2$; the second series simulates an MA(2) process with $\theta_1 = 0.8$ and $\theta_2 = 0.2$. We use `tsline` to graph these two series.

```
. use http://www.stata-press.com/data/r12/tsline1
. tsset lags
        time variable:  lags, 0 to 199
                delta:  1 unit
. tsline ar ma
```

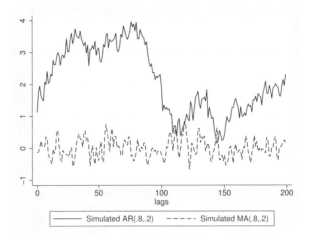

◁

> ## Example 2

Suppose that we kept a calorie log for an entire calendar year. At the end of the year, we would have a dataset (for example, `tsline2.dta`) that contains the number of calories consumed for 365 days. We could then use `tsset` to identify the date variable and `tsline` to plot calories versus time. Knowing that we tend to eat a little more food on Thanksgiving and Christmas day, we use the `ttick()` and `ttext()` options to point these days out on the time axis.

```
. use http://www.stata-press.com/data/r12/tsline2
. tsset day
        time variable:  day, 01jan2002 to 31dec2002
                delta:  1 day
. tsline calories, ttick(28nov2002 25dec2002, tpos(in))
> ttext(3470 28nov2002 "thanks" 3470 25dec2002 "x-mas", orient(vert))
```

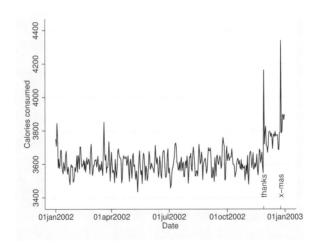

We were uncertain of the exact values we logged, so we also gave a range for each day. Here is a plot of the summer months.

```
. tsrline lcalories ucalories if tin(1may2002,31aug2002) || tsline cal ||
> if tin(1may2002,31aug2002), ytitle(Calories)
```

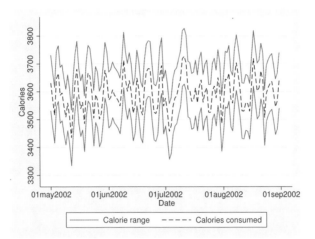

Options associated with the time axis allow dates (and times) to be specified in place of numeric date (and time) values. For instance, we used

```
ttick(28nov2002 25dec2002, tpos(in))
```

to place tick marks at the specified dates. This works similarly for `tlabel`, `tmlabel`, and `tmtick`.

Suppose that we wanted to place vertical lines for the previously mentioned holidays. We could specify the dates in the `tline()` option as follows:

```
. tsline calories, tline(28nov2002 25dec2002)
```

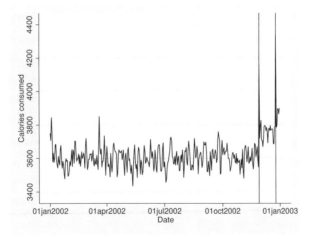

We could also modify the format of the time axis so that only the day in the year is displayed in the labeled ticks:

```
. tsline calories, tlabel(, format(%tdmd)) ttitle("Date (2002)")
```

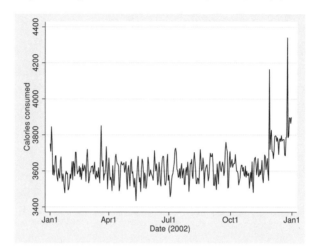

◁

Methods and formulas

tsline and tsrline are implemented as ado-files.

Reference

Cox, N. J. 2009. Stata tip 76: Separating seasonal time series. *Stata Journal* 9: 321–326.

Also see

[TS] **tsset** — Declare data to be time-series data

[G-2] **graph twoway** — Twoway graphs

[XT] **xtline** — Panel-data line plots

Title

> **tsreport** — Report time-series aspects of a dataset or estimation sample

Syntax

tsreport $\left[\,if\,\right]$ $\left[\,in\,\right]$ $\left[\,,\ options\,\right]$

options	Description
Main	
report	display number of gaps in the time series
report0	display count of gaps even if no gaps
list	display gaps data in tabular list
panel	ignore panel changes

tsreport typed without options produces no output but provides its standard saved results.

Menu

Statistics > Time series > Setup and utilities > Report time-series aspects of dataset

Description

tsreport reports time gaps in a sample of observations. The report option displays a one-line statement showing the count of the gaps, and list displays a list of records that follow gaps. A return value r(N_gaps) indicates the number of gaps in the sample.

Options

⌐ Main ⌐

report specifies that the number of gaps in the time series be reported, if any gaps exist.

report0 specifies that the count of gaps be reported, even if there are no gaps.

list specifies that a tabular list of gaps be displayed.

panel specifies that panel changes not be counted as gaps. Whether panel changes are counted as gaps usually depends on how the calling command handles panels.

Remarks

Time-series commands sometimes require that observations be on a fixed time interval with no gaps, or the command may not function properly. tsreport provides a tool for reporting the gaps in a sample.

▷ Example 1

The following monthly panel data have two panels and a missing month (March) in the second panel:

```
. use http://www.stata-press.com/data/r12/tsrptxmpl
. list edlevel month income in 1/6, sep(0)
```

	edlevel	month	income
1.	1	1998m1	687
2.	1	1998m2	783
3.	1	1998m3	790
4.	2	1998m1	1435
5.	2	1998m2	1522
6.	2	1998m4	1532

Invoking `tsreport` without the `panel` option gives us the following report:

```
. tsreport, report
Number of gaps in sample:  2   (gap count includes panel changes)
```

We could get a list of gaps and better see what has been counted as a gap by using the `list` option:

```
. tsreport, report list
Number of gaps in sample:  2   (gap count includes panel changes)
Observations with preceding time gaps
(gaps include panel changes)
```

Record	edlevel	month
4	2	1998m1
6	2	1998m4

We now see why `tsreport` is reporting two gaps. It is counting the known gap in March of the second panel and counting the change from the first to the second panel. (If we are programmers writing a procedure that does not account for panels, a change from one panel to the next represents a break in the time series just as a gap in the data does.)

We may prefer that the changes in panels not be counted as gaps. We can obtain a count without the panel change by using the `panel` option:

```
. tsreport, report panel
Number of gaps in sample:  1
```

To obtain a fuller report, we type

```
. tsreport, report list panel
Number of gaps in sample:  1
Observations with preceding time gaps
```

Record	edlevel	month
6	2	1998m4

◁

Saved results

tsreport saves the following in r():

Scalars
     r(N_gaps)   number of gaps in sample

Methods and formulas

tsreport is implemented as an ado-file.

Also see

[TS] **tsset** — Declare data to be time-series data

Title

> **tsrevar** — Time-series operator programming command

Syntax

> tsrevar [*varlist*] [*if*] [*in*] [, <u>substitute</u> <u>list</u>]

You must tsset your data before using tsrevar; see [TS] **tsset**.

Description

tsrevar, substitute takes a *varlist* that might contain *op.varname* combinations and substitutes equivalent temporary variables for the combinations.

tsrevar, list creates no new variables. It returns in r(varlist) the list of base variables corresponding to *varlist*.

Options

substitute specifies that tsrevar resolve *op.varname* combinations by creating temporary variables as described above. substitute is the default action taken by tsrevar; you do not need to specify the option.

list specifies that tsrevar return a list of base variable names.

Remarks

tsrevar substitutes temporary variables for any *op.varname* combinations in a variable list. For instance, the original *varlist* might be "gnp L.gnp r", and tsrevar, substitute would create *newvar* = L.gnp and create the equivalent varlist "gnp *newvar* r". This new varlist could then be used with commands that do not otherwise support time-series operators, or it could be used in a program to make execution faster at the expense of using more memory.

tsrevar, substitute might create no new variables, one new variable, or many new variables, depending on the number of *op.varname* combinations appearing in *varlist*. Any new variables created are temporary. The new, equivalent varlist is returned in r(varlist). The new varlist corresponds one to one with the original *varlist*.

tsrevar, list returns in r(varlist) the list of base variable names of *varlist* with the time-series operators removed. tsrevar, list creates no new variables. For instance, if the original *varlist* were "gnp l.gnp l2.gnp r l.cd", then r(varlist) would contain "gnp r cd". This is useful for programmers who might want to create programs to keep only the variables corresponding to *varlist*.

▷ Example 1

```
. use http://www.stata-press.com/data/r12/tsrevarex
. tsrevar l.gnp d.gnp r
```

creates two temporary variables containing the values for l.gnp and d.gnp. The variable r appears in the new variable list but does not require a temporary variable.

The resulting variable list is

```
. display "'r(varlist)'"
__00014P __00014Q r
```

(Your temporary variable names may be different, but that is of no consequence.)

We can see the results by listing the new variables alongside the original value of gnp.

```
. list gnp 'r(varlist)'  in 1/5
```

	gnp	__00014P	__00014Q	r
1.	128	.	.	3.2
2.	135	128	7	3.8
3.	132	135	-3	2.6
4.	138	132	6	3.9
5.	145	138	7	4.2

Temporary variables automatically vanish when the program concludes.

If we had needed only the base variable names, we could have specified

```
. tsrevar l.gnp d.gnp r, list
. display "'r(varlist)'"
gnp r
```

The order of the list will probably differ from that of the original list; base variables are listed only once and are listed in the order that they appear in the dataset.

◁

❑ Technical note

tsrevar, substitute avoids creating duplicate variables. Consider

```
. tsrevar gnp l.gnp r cd l.cd l.gnp
```

l.gnp appears twice in the varlist. tsrevar will create only one new variable for l.gnp and use that new variable twice in the resulting r(varlist). Moreover, tsrevar will even do this across multiple calls:

```
. tsrevar gnp l.gnp cd l.cd
. tsrevar cpi l.gnp
```

l.gnp appears in two separate calls. At the first call, tsrevar creates a temporary variable corresponding to l.gnp. At the second call, tsrevar remembers what it has done and uses that same temporary variable for l.gnp again.

❑

Saved results

tsrevar saves the following in r():

Macros
    r(varlist)   the modified variable list or list of base variable names

Also see

[P] **syntax** — Parse Stata syntax

[P] **unab** — Unabbreviate variable list

[U] **11 Language syntax**

[U] **11.4.4 Time-series varlists**

[U] **18 Programming Stata**

Title

tsset — Declare data to be time-series data

Syntax

Declare data to be time series

    tsset *timevar* [, *options*]

    tsset *panelvar timevar* [, *options*]

Display how data are currently tsset

    tsset

Clear time-series settings

    tsset, clear

In the declare syntax, *panelvar* identifies the panels and *timevar* identifies the times.

options	Description
Main	
unitoptions	specify units of *timevar*
Delta	
deltaoption	specify period of *timevar*
noquery	suppress summary calculations and output

noquery is not shown in the dialog box.

unitoptions	Description
(default)	*timevar*'s units to be obtained from *timevar*'s display format
clocktime	*timevar* is %tc: 0 = 1jan1960 00:00:00.000, 1 = 1jan1960 00:00:00.001, ...
daily	*timevar* is %td: 0 = 1jan1960, 1 = 2jan1960, ...
weekly	*timevar* is %tw: 0 = 1960w1, 1 = 1960w2, ...
monthly	*timevar* is %tm: 0 = 1960m1, 1 = 1960m2, ...
quarterly	*timevar* is %tq: 0 = 1960q1, 1 = 1960q2,...
halfyearly	*timevar* is %th: 0 = 1960h1, 1 = 1960h2,...
yearly	*timevar* is %ty: 1960 = 1960, 1961 = 1961, ...
generic	*timevar* is %tg: 0 = ?, 1 = ?, ...
format(% *fmt*)	specify *timevar*'s format and then apply default rule

In all cases, negative *timevar* values are allowed.

deltaoption specifies the period between observations in *timevar* units and may be specified as

deltaoption	Example
delta(#)	delta(1) or delta(2)
delta((*exp*))	delta((7*24))
delta(# *units*)	delta(7 days) or delta(15 minutes) or delta(7 days 15 minutes)
delta((*exp*) *units*)	delta((2+3) weeks)

Allowed units for %tc and %tC *timevars* are

seconds	second	secs	sec
minutes	minute	mins	min
hours	hour		
days	day		
weeks	week		

and for all other %t *timevars*, units specified must match the frequency of the data; for example, for %ty, units must be year or years.

Menu

Statistics > Time series > Setup and utilities > Declare dataset to be time-series data

Description

tsset declares the data in memory to be a time series. tssetting the data is what makes Stata's time-series operators such as L. and F. (lag and lead) work; the operators are discussed under *Remarks* below. Also, before using the other ts commands, you must tsset the data first. If you save the data after tsset, the data will be remembered to be time series and you will not have to tsset again.

There are two syntaxes for setting the data:

> tsset *timevar*
> tsset *panelvar timevar*

In the first syntax—tsset *timevar*—the data are set to be a straight time series.

In the second syntax—tsset *panelvar timevar*—the data are set to be a collection of time series, one for each value of *panelvar*, also known as panel data, cross-sectional time-series data, and xt data. Such datasets can be analyzed by xt commands as well as ts commands. If you tsset *panelvar timevar*, you do not need to xtset *panelvar timevar* to use the xt commands.

tsset without arguments—tsset—displays how the data are currently tsset and sorts the data on *timevar* or *panelvar timevar* if they are sorted differently from that.

tsset, clear is a rarely used programmer's command to declare that the data are no longer a time series.

Options

unitoptions clocktime, daily, weekly, monthly, quarterly, halfyearly, yearly, generic, and format(%*fmt*) specify the units in which *timevar* is recorded.

timevar will usually be a %t variable; see [D] **datetime**. If *timevar* already has a %t display format assigned to it, you do not need to specify a *unitoption*; tsset will obtain the units from the format. If you have not yet bothered to assign the appropriate %t format, however, you can use the *unitoptions* to tell tsset the units. Then tsset will set *timevar*'s display format for you. Thus, the *unitoptions* are convenience options; they allow you to skip formatting the time variable. The following all have the same net result:

Alternative 1	Alternative 2	Alternative 3
format t %td	*(t not formatted)*	*(t not formatted)*
tsset t	tsset t, daily	tsset t, format(%td)

timevar is not required to be a %t variable; it can be any variable of your own concocting so long as it takes on only integer values. In such cases, it is called generic and considered to be %tg. Specifying the *unitoption* generic or attaching a special format to *timevar*, however, is not necessary because tsset will assume that the variable is generic if it has any numerical format other than a %t format (or if it has a %tg format).

clear—used in tsset, clear—makes Stata forget that the data ever were tsset. This is a rarely used programmer's option.

delta() specifies the period of *timevar* and is commonly used when *timevar* is %tc. delta() is only sometimes used with the other %t formats or with generic time variables.

If delta() is not specified, delta(1) is assumed. This means that at *timevar* = 5, the previous time is *timevar* = 5 − 1 = 4 and the next time would be *timevar* = 5 + 1 = 6. Lag and lead operators, for instance, would work this way. This would be assumed regardless of the units of *timevar*.

If you specified delta(2), then at *timevar* = 5, the previous time would be *timevar* = 5 − 2 = 3 and the next time would be *timevar* = 5 + 2 = 7. Lag and lead operators would work this way. In the observation with *timevar* = 5, L.price would be the value of price in the observation for which *timevar* = 3 and F.price would be the value of price in the observation for which *timevar* = 7. If you then add an observation with *timevar* = 4, the operators will still work appropriately; that is, at *timevar* = 5, L.price will still have the value of price at *timevar* = 3.

There are two aspects of *timevar*: its units and its periodicity. The *unitoptions* set the units. delta() sets the periodicity.

We mentioned that delta() is commonly used with %tc *timevars* because Stata's %tc variables have units of milliseconds. If delta() is not specified and in some model you refer to L.price, you will be referring to the value of price 1 ms ago. Few people have data with periodicity of a millisecond. Perhaps your data are hourly. You could specify delta(3600000). Or you could specify delta((60*60*1000)), because delta() will allow expressions if you include an extra pair of parentheses. Or you could specify delta(1 hour). They all mean the same thing: *timevar* has periodicity of 3,600,000 ms. In an observation for which *timevar* = 1,489,572,000,000 (corresponding to 15mar2007 10:00:00), L.price would be the observation for which *timevar* = 1,489,572,000,000 − 3,600,000 = 1,489,568,400,000 (corresponding to 15mar2007 9:00:00).

When you `tsset` the data and specify `delta()`, `tsset` verifies that all the observations follow the specified periodicity. For instance, if you specified `delta(2)`, then *timevar* could contain any subset of $\{\ldots, -4, -2, 0, 2, 4, \ldots\}$ or it could contain any subset of $\{\ldots, -3, -1, 1, 3, \ldots\}$. If *timevar* contained a mix of values, `tsset` would issue an error message. If you also specify a *panelvar*—you type `tsset` *panelvar timevar*, `delta(2)`—the check is made on each panel independently. One panel might contain *timevar* values from one set and the next, another, and that would be fine.

The following option is available with `tsset` but is not shown in the dialog box:

`noquery` prevents `tsset` from performing most of its summary calculations and suppresses output. With this option, only the following results are posted:

r(tdelta)	r(tsfmt)
r(panelvar)	r(unit)
r(timevar)	r(unit1)

Remarks

`tsset` sets *timevar* so that Stata's time-series operators are understood in varlists and expressions. The time-series operators are

Operator	Meaning
L.	lag x_{t-1}
L2.	2-period lag x_{t-2}
...	
F.	lead x_{t+1}
F2.	2-period lead x_{t+2}
...	
D.	difference $x_t - x_{t-1}$
D2.	difference of difference $x_t - x_{t-1} - (x_{t-1} - x_{t-2}) = x_t - 2x_{t-1} + x_{t-2}$
...	
S.	"seasonal" difference $x_t - x_{t-1}$
S2.	lag-2 (seasonal) difference $x_t - x_{t-2}$
...	

Time-series operators may be repeated and combined. `L3.gnp` refers to the third lag of variable gnp, as do `LLL.gnp`, `LL2.gnp`, and `L2L.gnp`. `LF.gnp` is the same as gnp. `DS12.gnp` refers to the one-period difference of the 12-period difference. `LDS12.gnp` refers to the same concept, lagged once.

`D1.` = `S1.`, but `D2.` $\neq$ `S2.`, `D3.` $\neq$ `S3.`, and so on. `D2.` refers to the difference of the difference. `S2.` refers to the two-period difference. If you wanted the difference of the difference of the 12-period difference of gnp, you would write `D2S12.gnp`.

Operators may be typed in uppercase or lowercase. Most users would type `d2s12.gnp` instead of `D2S12.gnp`.

You may type operators however you wish; Stata internally converts operators to their canonical form. If you typed `1d2ls12d.gnp`, Stata would present the operated variable as `L2D3S12.gnp`.

Stata also understands *operator* (*numlist*) . to mean a set of operated variables. For instance, typing `L(1/3).gnp` in a varlist is the same as typing `L.gnp L2.gnp L3.gnp`. The operators can also be applied to a list of variables by enclosing the variables in parentheses; for example,

```
. list year L(1/3).(gnp cpi)
```

	year	L.gnp	L2.gnp	L3.gnp	L.cpi	L2.cpi	L3.cpi
1.	1989	.	.	.	.	.	.
2.	1990	5452.8	.	.	100	.	.
3.	1991	5764.9	5452.8	.	105	100	.
4.	1992	5932.4	5764.9	5452.8	108	105	100
			(output omitted)				
8.	1996	7330.1	6892.2	6519.1	122	119	112

In *operator#.*, making # zero returns the variable itself. L0.gnp is gnp. Thus, you can type list year l(0/3).gnp to mean list year gnp L.gnp L2.gnp L3.gnp.

The parenthetical notation may be used with any operator. Typing D(1/3).gnp would return the first through third differences.

The parenthetical notation may be used in operator lists with multiple operators, such as L(0/3)D2S12.gnp.

Operator lists may include up to one set of parentheses, and the parentheses may enclose a *numlist*; see [U] **11.1.8 numlist**.

Before you can use these time-series operators, however, the dataset must satisfy two requirements:

1. the dataset must be tsset and

2. the dataset must be sorted by *timevar* or, if it is a cross-sectional time-series dataset, by *panelvar timevar*.

tsset handles both requirements. As you use Stata, however, you may later use a command that re-sorts that data, and if you do, the time-series operators will not work:

```
. tsset time
(output omitted)
. regress y x l.x
(output omitted)
.  (you continue to use Stata and, sometime later:)
. regress y x l.x
not sorted
r(5);
```

Then typing tsset without arguments will reestablish the sort order:

```
. tsset
(output omitted)
. regress y x l.x
(output omitted)
```

Here typing tsset is the same as typing sort time. Had we previously tsset country time, however, typing tsset would be the same as typing sort country time. You can type the sort command or type tsset without arguments; it makes no difference.

There are two syntaxes for setting your data:

> tsset *timevar*
> tsset *panelvar timevar*

In both, *timevar* must contain integer values. If *panelvar* is specified, it too must contain integer values, and the dataset is declared to be a cross-section of time series, such as a collection of time series for different countries.

▷ Example 1: Numeric time variable

You have monthly data on personal income. Variable t records the time of an observation, but there is nothing special about the name of the variable. There is nothing special about the values of the variable, either. t is not required to be %tm variable—perhaps you do not even know what that means. t is just a numeric variable containing integer values that represent the month, and we will imagine that t takes on the values 1, 2, ..., 9, although it could just as well be $-3, -2 ..., 5$, or 1,023, 1,024, ..., 1,031. What is important is that the values are dense: adjacent months have a time value that differs by 1.

```
. use http://www.stata-press.com/data/r12/tssetxmpl
. list t income
```

```
          t    income

 1.       1      1153
 2.       2      1181
       (output omitted )
 9.       9      1282
```

```
. tsset t
        time variable:  t, 1 to 9
                delta:  1 unit
. regress income l.income
  (output omitted )
```

◁

▷ Example 2: Adjusting the starting date

In the example above, that t started at 1 was not important. As we said, the t variable could just as well be recorded $-3, -2 ..., 5$, or 1,023, 1,024, ..., 1,031. What is important is that the difference in t between observations be delta() when there are no gaps.

Although how time is measured makes no difference, Stata has formats to display time nicely if it is recorded in certain ways; you can learn about the formats by seeing [D] **datetime**. Stata likes time variables in which 1jan1960 is recorded as 0. In our previous example, if $t = 1$ corresponds to July 1995, then we could make a variable that fits Stata's preference by typing

```
. generate newt = tm(1995m7) + t - 1
```

tm() is the function that returns a month equivalent; tm(1995m7) evaluates to the constant 426, meaning 426 months after January 1960. We now have variable newt containing

```
. list t newt income
```

```
          t    newt    income

 1.       1     426      1153
 2.       2     427      1181
 3.       3     428      1208
           (output omitted )
 9.       9     434      1282
```

If we put a %tm format on newt, it will display more cleanly:

```
. format newt %tm

. list t newt income
```

	t	newt	income
1.	1	1995m7	1153
2.	2	1995m8	1181
3.	3	1995m9	1208
		(output omitted)	
9.	9	1996m3	1282

We could now `tsset newt` rather than `t`:

```
. tsset newt
        time variable:  newt, 1995m7 to 1996m3
                delta:  1 month
```
◁

❑ Technical note

In addition to monthly, Stata understands clock times (to the millisecond level) as well as daily, weekly, quarterly, half-yearly, and yearly data. See [D] **datetime** for a description of these capabilities.

Let's reconsider the previous example, but rather than monthly, let's assume the data are daily, weekly, etc. The only thing to know is that, corresponding to function `tm()`, there are functions `td()`, `tw()`, `tq()`, `th()`, and `ty()` and that, corresponding to format `%tm`, there are formats `%td`, `%tw`, `%tq`, `%th`, and `%ty`. Here is what we would have typed had our data been on a different time scale:

Daily: if your t variable had t=1 corresponding to 15mar1993
```
        . gen newt = td(15mar1993) + t - 1
        . tsset newt, daily
```

Weekly: if your t variable had t=1 corresponding to 1994w1:
```
        . gen newt = tw(1994w1) + t - 1
        . tsset newt, weekly
```

Monthly: if your t variable had t=1 corresponding to 2004m7:
```
        . gen newt = tm(2004m7) + t - 1
        . tsset newt, monthly
```

Quarterly: if your t variable had t=1 corresponding to 1994q1:
```
        . gen newt = tq(1994q1) + t - 1
        . tsset newt, quarterly
```

Half-yearly: if your t variable had t=1 corresponding to 1921h2:
```
        . gen newt = th(1921h2) + t - 1
        . tsset newt, halfyearly
```

Yearly: if your t variable had t=1 corresponding to 1842:
```
        . gen newt = 1842 + t - 1
        . tsset newt, yearly
```

In each example above, we subtracted one from our time variable in constructing the new time variable `newt` because we assumed that our starting time value was 1. For the quarterly example, if our starting time value were 5 and that corresponded to 1994q1, we would type

```
        . generate newt = tq(1994q1) + t - 5
```

Had our initial time value been $t = 742$ and that corresponded to 1994q1, we would have typed

```
        . generate newt = tq(1994q1) + t - 742
```
❑

> ▷ Example 3: Time-series data but no time variable

Perhaps we have the same time-series data but no time variable:

```
. use http://www.stata-press.com/data/r12/tssetxmpl2, clear
. list income
```

	income
1.	1153
2.	1181
3.	1208
4.	1272
5.	1236
6.	1297
7.	1265
8.	1230
9.	1282

Say that we know that the first observation corresponds to July 1995 and continues without gaps. We can create a monthly time variable and format it by typing

```
. generate t = tm(1995m7) + _n - 1
. format t %tm
```

We can now `tsset` our dataset and `list` it:

```
. tsset t
        time variable:  t, 1995m7 to 1996m3
                delta:  1 month
. list t income
```

	t	income
1.	1995m7	1153
2.	1995m8	1181
3.	1995m9	1208
	(output omitted)	
9.	1996m3	1282

◁

> ▷ Example 4: Time variable as a string

Your data might include a time variable that is encoded into a string. In the example below each monthly observation is identified by string variable `yrmo` containing the month and year of the observation, sometimes with punctuation between:

```
. use http://www.stata-press.com/data/r12/tssetxmpl, clear

. list yrmo income
```

	yrmo	income
1.	7/1995	1153
2.	8/1995	1181
3.	9-1995	1208
4.	10,1995	1272
5.	11 1995	1236
6.	12 1995	1297
7.	1/1996	1265
8.	2.1996	1230
9.	3- 1996	1282

The first step is to convert the string to a numeric representation. Doing so is easy using the monthly()
function; see [D] **datetime**.

```
. gen mdate = monthly(yrmo, "MY")

. list yrmo mdate income
```

	yrmo	mdate	income
1.	7/1995	426	1153
2.	8/1995	427	1181
3.	9-1995	428	1208
	(output omitted)		
9.	3- 1996	434	1282

Our new variable, mdate, contains the number of months from January 1960. Now that we have
numeric variable mdate, we can tsset the data:

```
. format mdate %tm

. tsset mdate
        time variable:  mdate, 1995m7 to 1996m3
                delta:  1 month
```

In fact, we can combine the two and type

```
. tsset mdate, format(%tm)
        time variable:  mdate, 1995m7 to 1996m3
                delta:  1 month
```

or type

```
. tsset mdate, monthly
        time variable:  mdate, 1995m7 to 1996m3
                delta:  1 month
```

In all cases, we obtain

```
. list yrmo mdate income
```

	yrmo	mdate	income
1.	7/1995	1995m7	1153
2.	8/1995	1995m8	1181
3.	9-1995	1995m9	1208
4.	10,1995	1995m10	1272
5.	11 1995	1995m11	1236
6.	12 1995	1995m12	1297
7.	1/1996	1996m1	1265
8.	2.1996	1996m2	1230
9.	3- 1996	1996m3	1282

Stata can translate many different date formats, including strings like 12jan2009; January 12, 2009; 12-01-2009; 01/12/2009; 01/12/09; 12jan2009 8:14; 12-01-2009 13:12; 01/12/09 1:12 pm; Wed Jan 31 13:03:25 CST 2009; 1998q1; and more. See [D] **datetime**.

◁

▷ Example 5: Time-series data with gaps

Gaps in the time series cause no difficulties:

```
. use http://www.stata-press.com/data/r12/tssetxmpl3, clear
. list yrmo income
```

	yrmo	income
1.	7/1995	1153
2.	8/1995	1181
3.	11 1995	1236
4.	12 1995	1297
5.	1/1996	1265
6.	3- 1996	1282

```
. gen mdate = monthly(yrmo, "MY")
. tsset mdate, monthly
        time variable:  mdate, 1995m7 to 1996m3, but with gaps
                delta:  1 month
```

Once the dataset has been tsset, we can use the time-series operators. The D operator specifies first differences:

```
. list mdate income d.income
```

	mdate	income	D.income
1.	1995m7	1153	.
2.	1995m8	1181	28
3.	1995m11	1236	.
4.	1995m12	1297	61
5.	1996m1	1265	-32
6.	1996m3	1282	.

We can use the operators in an expression or varlist context; we do not have to create a new variable to hold D.income. We can use D.income with the list command, with regress or any other Stata command that allows time-series varlists.

◁

▷ Example 6: Clock times

We have data from a large hotel in Las Vegas that changes the reservation prices for its rooms hourly. A piece of the data looks like

```
. use http://www.stata-press.com/data/r12/tssetxmpl4, clear
. list in 1/5
```

	time	price
1.	02.13.2007 08:00	140
2.	02.13.2007 09:00	155
3.	02.13.2007 10:00	160
4.	02.13.2007 11:00	155
5.	02.13.2007 12:00	160

Variable time is a string variable. The first step in making this dataset a time-series dataset is to translate the string to a numeric variable:

```
. generate double t = clock(time, "MDY hm")
. list in 1/5
```

	time	price	t
1.	02.13.2007 08:00	140	1.487e+12
2.	02.13.2007 09:00	155	1.487e+12
3.	02.13.2007 10:00	160	1.487e+12
4.	02.13.2007 11:00	155	1.487e+12
5.	02.13.2007 12:00	160	1.487e+12

See [D] **datetime** for an explanation of what is going on here. clock() is the function that converts strings to datetime (%tc) values. We typed clock(time, "MDY hm") to convert string variable time, and we told clock() that the values in time were in the order month, day, year, hour, and minute. We stored new variable t as a double because time values are large, and doing so is required to prevent rounding. Even so, the resulting values 1.487e+12 look rounded, but that is only because of the default display format for new variables. We can see the values better if we change the format:

```
. format t %20.0gc
. list in 1/5
```

	time	price	t
1.	02.13.2007 08:00	140	1,486,972,800,000
2.	02.13.2007 09:00	155	1,486,976,400,000
3.	02.13.2007 10:00	160	1,486,980,000,000
4.	02.13.2007 11:00	155	1,486,983,600,000
5.	02.13.2007 12:00	160	1,486,987,200,000

Even better would be to change the format to %tc—Stata's clock-time format:

```
. format t %tc
. list in 1/5
```

	time	price	t
1.	02.13.2007 08:00	140	13feb2007 08:00:00
2.	02.13.2007 09:00	155	13feb2007 09:00:00
3.	02.13.2007 10:00	160	13feb2007 10:00:00
4.	02.13.2007 11:00	155	13feb2007 11:00:00
5.	02.13.2007 12:00	160	13feb2007 12:00:00

We could drop variable `time`. New variable `t` contains the same information as `time` and `t` is better because it is a Stata time variable, the most important property of which being that it is numeric rather than string. We can `tsset` it. Here, however, we also need to specify the period with `tsset`'s `delta()` option. Stata's time variables are numeric, but they record milliseconds since 01jan1960 00:00:00. By default, `tsset` uses `delta(1)`, and that means the time-series operators would not work as we want them to work. For instance, `L.price` would look back only 1 ms (and find nothing). We want `L.price` to look back 1 hour (3,600,000 ms):

```
. tsset t, delta(1 hour)
        time variable:  t,
                        13feb2007 08:00:00.000 to 13feb2007 14:00:00.000
                delta:  1 hour
. list t price l.price in 1/5
```

	t	price	L.price
1.	13feb2007 08:00:00	140	.
2.	13feb2007 09:00:00	155	140
3.	13feb2007 10:00:00	160	155
4.	13feb2007 11:00:00	155	160
5.	13feb2007 12:00:00	160	155

◁

▷ Example 7: Clock times must be double

In the previous example, it was of vital importance that when we generated the %tc variable t,

```
. generate double t = clock(time, "MDY hm")
```

we generated it as a double. Let's see what would have happened had we forgotten and just typed `generate t = clock(time, "MDY hm")`. Let's go back and start with the same original data:

```
. use http://www.stata-press.com/data/r12/tssetxmpl4, clear
. list in 1/5
```

	time	price
1.	02.13.2007 08:00	140
2.	02.13.2007 09:00	155
3.	02.13.2007 10:00	160
4.	02.13.2007 11:00	155
5.	02.13.2007 12:00	160

Remember, variable time is a string variable, and we need to translate it to numeric. So we translate, but this time we forget to make the new variable a double:

```
. generate t = clock(time, "MDY hm")
. list in 1/5
```

	time	price	t
1.	02.13.2007 08:00	140	1.49e+12
2.	02.13.2007 09:00	155	1.49e+12
3.	02.13.2007 10:00	160	1.49e+12
4.	02.13.2007 11:00	155	1.49e+12
5.	02.13.2007 12:00	160	1.49e+12

We see the first difference—t now lists as 1.49e+12 rather than 1.487e+12 as it did previously—but this is nothing that would catch our attention. We would not even know that the value is different. Let's continue.

We next put a %20.0gc format on t to better see the numerical values. In fact, that is not something we would usually do in an analysis. We did that in the example to emphasize to you that the t values were really big numbers. We will repeat the exercise just to be complete, but in real analysis, we would not bother.

```
. format t %20.0gc
. list in 1/5
```

	time	price	t
1.	02.13.2007 08:00	140	1,486,972,780,544
2.	02.13.2007 09:00	155	1,486,976,450,560
3.	02.13.2007 10:00	160	1,486,979,989,504
4.	02.13.2007 11:00	155	1,486,983,659,520
5.	02.13.2007 12:00	160	1,486,987,198,464

Okay, we see big numbers in t. Let's continue.

Next we put a %tc format on t, and that is something we would usually do, and you should always do. You should also list a bit of the data, as we did:

```
. format t %tc
. list in 1/5
```

	time	price	t
1.	02.13.2007 08:00	140	13feb2007 07:59:40
2.	02.13.2007 09:00	155	13feb2007 09:00:50
3.	02.13.2007 10:00	160	13feb2007 09:59:49
4.	02.13.2007 11:00	155	13feb2007 11:00:59
5.	02.13.2007 12:00	160	13feb2007 11:59:58

By now, you should see a problem: the translated datetime values are off by a second or two. That was caused by rounding. Dates and times should be the same, not approximately the same, and when you see a difference like this, you should say to yourself, "The translation is off a little. Why is that?" and then you should think, "Of course, rounding. I bet that I did not create t as a double."

Let us assume, however, that you do not do this. You instead plow ahead:

```
. tsset t, delta(1 hour)
time values with period less than delta() found
r(451);
```

And that is what will happen when you forget to create t as a double. The rounding will cause uneven period, and tsset will complain.

By the way, it is only important that clock times (%tc and %tC variables) be stored as doubles. The other date values %td, %tw, %tm, %tq, %th, and %ty are small enough that they can safely be stored as floats, although forgetting and storing them as doubles does no harm.

◁

❑ Technical note

Stata provides two clock-time formats, %tc and %tC. %tC provides a clock with leap seconds. Leap seconds are occasionally inserted to account for randomness of the earth's rotation, which gradually slows. Unlike the extra day inserted in leap years, the timing of when leap seconds will be inserted cannot be foretold. The authorities in charge of such matters announce a leap second approximately 6 months before insertion. Leap seconds are inserted at the end of the day, and the leap second is called 23:59:60 (that is, 11:59:60 pm), which is then followed by the usual 00:00:00 (12:00:00 am). Most nonastronomers find these leap seconds vexing. The added seconds cause problems because of their lack of predictability—knowing how many seconds there will be between 01jan2012 and 01jan2013 is not possible—and because there are not necessarily 24 hours in a day. If you use a leap second adjusted–clock, most days have 24 hours, but a few have 24 hours and 1 second. You must look at a table to find out.

From a time-series analysis point of view, the nonconstant day causes the most problems. Let's say that you have data on blood pressure, taken hourly at 1:00, 2:00, ..., and that you have tsset your data with delta(1 hour). On most days, L24.bp would be blood pressure at the same time yesterday. If the previous day had a leap second, however, and your data were recorded using a leap second adjusted–clock, there would be no observation L24.bp because 86,400 seconds before the current reading does not correspond to an on-the-hour time; 86,401 seconds before the current reading corresponds to yesterday's time. Thus, whenever possible, using Stata's %tc encoding rather than %tC is better.

When times are recorded by computers using leap second–adjusted clocks, however, avoiding %tC is not possible. For performing most time-series analysis, the recommended procedure is to map the %tC values to %tc and then tsset those. You must ask yourself whether the process you are studying is based on the clock—the nurse does something at 2 o'clock every day—or the true passage of time—the emitter spits out an electron every 86,400,000 ms.

When dealing with computer-recorded times, first find out whether the computer (and its time-recording software) use a leap second–adjusted clock. If it does, translate that to a %tC value. Then use function cofC() to convert to a %tc value and tsset that. If variable T contains the %tC value,

```
. gen double t = cofC(T)
. format t %tc
. tsset t, delta(...)
```

Function cofC() moves leap seconds forward: 23:59:60 becomes 00:00:00 of the next day.

❑

Panel data

▷ Example 8: Time-series data for multiple groups

Assume that we have a time series on average annual income and that we have the series for two groups: individuals who have not completed high school (edlevel = 1) and individuals who have (edlevel = 2).

```
. use http://www.stata-press.com/data/r12/tssetxmpl5, clear
. list edlevel year income, sep(0)
```

	edlevel	year	income
1.	1	1988	14500
2.	1	1989	14750
3.	1	1990	14950
4.	1	1991	15100
5.	2	1989	22100
6.	2	1990	22200
7.	2	1992	22800

We declare the data to be a panel by typing

```
. tsset edlevel year, yearly
        panel variable:  edlevel, (unbalanced)
         time variable:  year, 1988 to 1992, but with a gap
                 delta:  1 year
```

Having tsset the data, we can now use time-series operators. The difference operator, for example, can be used to list annual changes in income:

```
. list edlevel year income d.income, sep(0)
```

	edlevel	year	income	D.income
1.	1	1988	14500	.
2.	1	1989	14750	250
3.	1	1990	14950	200
4.	1	1991	15100	150
5.	2	1989	22100	.
6.	2	1990	22200	100
7.	2	1992	22800	.

We see that in addition to producing missing values due to missing times, the difference operator correctly produced a missing value at the start of each panel. Once we have tsset our panel data, we can use time-series operators and be assured that they will handle missing time periods and panel changes correctly.

◁

Saved results

tsset saves the following in r():

Scalars
r(imin)	minimum panel ID
r(imax)	maximum panel ID
r(tmin)	minimum time
r(tmax)	maximum time
r(tdelta)	delta

Macros
r(panelvar)	name of panel variable
r(timevar)	name of time variable
r(tdeltas)	formatted delta
r(tmins)	formatted minimum time
r(tmaxs)	formatted maximum time
r(tsfmt)	%*fmt* of time variable
r(unit)	units of time variable: Clock, clock, daily, weekly, monthly, quarterly, halfyearly, yearly, or generic
r(unit1)	units of time variable: C, c, d, w, m, q, h, y, or ""
r(balanced)	unbalanced, weakly balanced, or strongly balanced; a set of panels are strongly balanced if they all have the same time values, otherwise balanced if same number of time values, otherwise unbalanced

Methods and formulas

tsset is implemented as an ado-file.

Reference

Baum, C. F. 2000. sts17: Compacting time series data. *Stata Technical Bulletin* 57: 44–45. Reprinted in *Stata Technical Bulletin Reprints*, vol. 10, pp. 369–370. College Station, TX: Stata Press.

Also see

[TS] **tsfill** — Fill in gaps in time variable

Title

tssmooth — Smooth and forecast univariate time-series data

Syntax

`tssmooth` *smoother* [*type*] *newvar* = *exp* [*if*] [*in*] [, ...]

Smoother category	*smoother*
Moving average	
with uniform weights	`ma`
with specified weights	`ma`
Recursive	
exponential	`exponential`
double exponential	`dexponential`
nonseasonal Holt–Winters	`hwinters`
seasonal Holt–Winters	`shwinters`
Nonlinear filter	`nl`

See [TS] **tssmooth ma**, [TS] **tssmooth exponential**, [TS] **tssmooth dexponential**, [TS] **tssmooth hwinters**, [TS] **tssmooth shwinters**, and [TS] **tssmooth nl**.

Description

`tssmooth` creates new variable *newvar* and fills it in by passing the specified expression (usually a variable name) through the requested smoother.

Remarks

The recursive smoothers may also be used for forecasting univariate time series; indeed, the Holt–Winters methods are used almost exclusively for this. All can perform dynamic out-of-sample forecasts, and the smoothing parameters may be chosen to minimize the in-sample sum-of-squared prediction errors.

The moving-average and nonlinear smoothers are generally used to extract the trend—or signal—from a time series while omitting the high-frequency or noise components.

All smoothers work both with time-series data and panel data. When used with panel data, the calculation is performed separately within panel.

Several texts provide good introductions to the methods available in `tssmooth`. Chatfield (2004) discusses how these methods fit into time-series analysis in general. Abraham and Ledolter (1983); Montgomery, Johnson, and Gardiner (1990); Bowerman, O'Connell, and Koehler (2005); and Chatfield (2001) discuss using these methods for modern time-series forecasting.

References

Abraham, B., and J. Ledolter. 1983. *Statistical Methods for Forecasting*. New York: Wiley.

Bowerman, B. L., R. T. O'Connell, and A. B. Koehler. 2005. *Forecasting, Time Series, and Regression: An Applied Approach*. 4th ed. Pacific Grove, CA: Brooks/Cole.

Chatfield, C. 2001. *Time-Series Forecasting*. London: Chapman & Hall/CRC.

———. 2004. *The Analysis of Time Series: An Introduction*. 6th ed. Boca Raton, FL: Chapman & Hall/CRC.

Chatfield, C., and M. Yar. 1988. Holt-Winters forecasting: Some practical issues. *Statistician* 37: 129–140.

Holt, C. C. 2004. Forecasting seasonals and trends by exponentially weighted moving averages. *International Journal of Forecasting* 20: 5–10.

Montgomery, D. C., L. A. Johnson, and J. S. Gardiner. 1990. *Forecasting and Time Series Analysis*. 2nd ed. New York: McGraw–Hill.

Winters, P. R. 1960. Forecasting sales by exponentially weighted moving averages. *Management Science* 6: 324–342.

Also see

[TS] **tsset** — Declare data to be time-series data

[TS] **arima** — ARIMA, ARMAX, and other dynamic regression models

[TS] **sspace** — State-space models

[TS] **tsfilter** — Filter a time-series, keeping only selected periodicities

[R] **smooth** — Robust nonlinear smoother

Title

> **tssmooth dexponential** — Double-exponential smoothing

Syntax

tssmooth dexponential [*type*] *newvar* = *exp* [*if*] [*in*] [, *options*]

options	Description
Main	
replace	replace *newvar* if it already exists
parms(#$_\alpha$)	use #$_\alpha$ as smoothing parameter
samp0(#)	use # observations to obtain initial values for recursions
s0(#$_1$ #$_2$)	use #$_1$ and #$_2$ as initial values for recursions
forecast(#)	use # periods for the out-of-sample forecast

You must tsset your data before using tssmooth dexponential; see [TS] **tsset**.

exp may contain time-series operators; see [U] **11.4.4 Time-series varlists**.

Menu

Statistics > Time series > Smoothers/univariate forecasters > Double-exponential smoothing

Description

tssmooth dexponential models the trend of a variable whose difference between changes from the previous values is serially correlated. More precisely, it models a variable whose second difference follows a low-order, moving-average process.

Options

◌ Main ◌

replace replaces *newvar* if it already exists.

parms(#$_\alpha$) specifies the parameter α for the double-exponential smoothers; $0 < \#_\alpha < 1$. If parms(#$_\alpha$) is not specified, the smoothing parameter is chosen to minimize the in-sample sum-of-squared forecast errors.

samp0(#) and s0(#$_1$ #$_2$) are mutually exclusive ways of specifying the initial values for the recursion.

By default, initial values are obtained by fitting a linear regression with a time trend, using the first half of the observations in the dataset; see *Remarks*.

samp0(#) specifies that the first # be used in that regression.

s0(#$_1$ #$_2$) specifies that #$_1$ #$_2$ be used as initial values.

forecast(#) specifies the number of periods for the out-of-sample prediction; $0 \leq \# \leq 500$. The default is forecast(0), which is equivalent to not performing an out-of-sample forecast.

471

Remarks

The double-exponential smoothing procedure is designed for series that can be locally approximated as

$$\widehat{x}_t = m_t + b_t t$$

where $\widehat{x}_t$ is the smoothed or predicted value of the series x, and the terms m_t and b_t change over time. Abraham and Ledolter (1983), Bowerman, O'Connell, and Koehler (2005), and Montgomery, Johnson, and Gardiner (1990) all provide good introductions to double-exponential smoothing. Chatfield (2001, 2004) provides helpful discussions of how double-exponential smoothing relates to modern time-series methods.

The double-exponential method has been used both as a smoother and as a prediction method. Recall from [TS] **tssmooth exponential** that the single-exponential smoothed series is given by

$$S_t = \alpha x_t + (1 - \alpha)S_{t-1}$$

where α is the smoothing constant and x_t is the original series. The double-exponential smoother is obtained by smoothing the smoothed series,

$$S_t^{[2]} = \alpha S_t + (1 - \alpha)S_{t-1}^{[2]}$$

Values of S_0 and $S_0^{[2]}$ are necessary to begin the process. Per Montgomery, Johnson, and Gardiner (1990), the default method is to obtain S_0 and $S_0^{[2]}$ from a regression of the first N_{pre} values of x_t on $\widetilde{t} = (1, \ldots, N_{\text{pre}} - t_0)'$. By default, N_{pre} is equal to one-half the number of observations in the sample. N_{pre} can be specified using the samp0() option.

The values of S_0 and $S_0^{[2]}$ can also be specified using the option s0().

▷ Example 1

Suppose that we had some data on the monthly sales of a book and that we wanted to smooth this series. The graph below illustrates that this series is locally trending over time, so we would not want to use single-exponential smoothing.

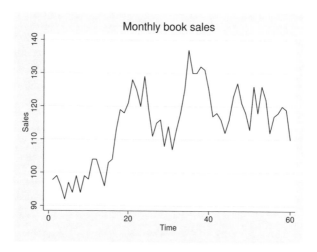

The following example illustrates that double-exponential smoothing is simply smoothing the smoothed series. Because the starting values are treated as time-zero values, we actually lose 2 observations when smoothing the smoothed series.

```
. use http://www.stata-press.com/data/r12/sales2

. tssmooth exponential double sm1=sales, p(.7) s0(1031)
exponential coefficient   =       0.7000
sum-of-squared residuals  =        13923
root mean squared error   =       13.192

. tssmooth exponential double sm2=sm1, p(.7) s0(1031)
exponential coefficient   =       0.7000
sum-of-squared residuals  =       7698.6
root mean squared error   =       9.8098

. tssmooth dexponential double sm2b=sales, p(.7) s0(1031 1031)
double-exponential coefficient  =       0.7000
sum-of-squared residuals        =       3724.4
root mean squared error         =       6.8231

. generate double sm2c = f2.sm2
(2 missing values generated)

. list sm2b sm2c in 1/10
```

	sm2b	sm2c
1.	1031	1031
2.	1028.3834	1028.3834
3.	1030.6306	1030.6306
4.	1017.8182	1017.8182
5.	1022.938	1022.938
6.	1026.0752	1026.0752
7.	1041.8587	1041.8587
8.	1042.8341	1042.8341
9.	1035.9571	1035.9571
10.	1030.6651	1030.6651

◁

The double-exponential method can also be viewed as a forecasting mechanism. The exponential forecast method is a constrained version of the Holt–Winters method implemented in [TS] **tssmooth hwinters** (as discussed by Gardner [1985] and Chatfield [2001]). Chatfield (2001) also notes that the double-exponential method arises when the underlying model is an ARIMA(0,2,2) with equal roots.

This method produces predictions $\widehat{x}_t$ for $t = t_1, \ldots, T + \text{forecast}()$. These predictions are obtained as a function of the smoothed series and the smoothed-smoothed series. For $t \in [t_0, T]$,

$$\widehat{x}_t = \left(2 + \frac{\alpha}{1 - \alpha}\right) S_t - \left(1 + \frac{\alpha}{1 - \alpha}\right) S_t^{[2]}$$

where S_t and $S_t^{[2]}$ are as given above.

The out-of-sample predictions are obtained as a function of the constant term, the linear term of the smoothed series at the last observation in the sample, and time. The constant term is $a_T = 2S_T - S_T^{[2]}$, and the linear term is $b_T = \frac{\alpha}{1-\alpha}(S_T - S_T^{[2]})$. The τth-step-ahead out-of-sample prediction is given by

$$\widehat{x}_t = a_t + \tau b_T$$

▷ Example 2

Specifying the `forecast` option puts the double-exponential forecast into the new variable instead of the double-exponential smoothed series. The code given below uses the smoothed series `sm1` and `sm2` that were generated above to illustrate how the double-exponential forecasts are computed.

```
. tssmooth dexponential double f1=sales, p(.7) s0(1031 1031) forecast(4)
double-exponential coefficient  =       0.7000
sum-of-squared residuals        =        20737
root mean squared error         =         16.1
. generate double xhat = (2 + .7/.3) * sm1 - (1 + .7/.3) * f.sm2
(5 missing values generated)
. list xhat f1 in 1/10
```

	xhat	f1
1.	1031	1031
2.	1031	1031
3.	1023.524	1023.524
4.	1034.8039	1034.8039
5.	994.0237	994.0237
6.	1032.4463	1032.4463
7.	1031.9015	1031.9015
8.	1071.1709	1071.1709
9.	1044.6454	1044.6454
10.	1023.1855	1023.1855

◁

▷ Example 3

Generally, when you are forecasting, you do not know the smoothing parameter. `tssmooth dexponential` computes the double-exponential forecasts of a series and obtains the optimal smoothing parameter by finding the smoothing parameter that minimizes the in-sample sum-of-squared forecast errors.

```
. tssmooth dexponential f2=sales, forecast(4)
computing optimal double-exponential coefficient (0,1)
optimal double-exponential coefficient =       0.3631
sum-of-squared residuals        =      16075.805
root mean squared error         =      14.175598
```

The following graph describes the fit that we obtained by applying the double-exponential forecast method to our sales data. The out-of-sample dynamic predictions are not constant, as in the single-exponential case.

```
. line f2 sales t, title("Double exponential forecast with optimal alpha")
> ytitle(Sales) xtitle(time)
```

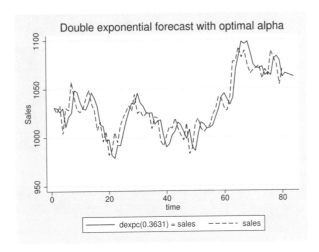

$\triangleleft$

tssmooth dexponential automatically detects panel data from the information provided when the dataset was tsset. The starting values are chosen separately for each series. If the smoothing parameter is chosen to minimize the sum-of-squared prediction errors, the optimization is performed separately on each panel. The saved results contain the results from the last panel. Missing values at the beginning of the sample are excluded from the sample. After at least one value has been found, missing values are filled in using the one-step-ahead predictions from the previous period.

Saved results

tssmooth dexponential saves the following in r():

Scalars
r(N)	number of observations
r(alpha)	α smoothing parameter
r(rss)	sum-of-squared errors
r(rmse)	root mean squared error
r(N_pre)	number of observations used in calculating starting values, if starting values calculated
r(s2_0)	initial value for linear term, i.e., $S_0^{[2]}$
r(s1_0)	initial value for constant term, i.e., S_0
r(linear)	final value of linear term
r(constant)	final value of constant term
r(period)	period, if filter is seasonal

Macros
r(method)	smoothing method
r(exp)	expression specified
r(timevar)	time variable specified in tsset
r(panelvar)	panel variable specified in tsset

Methods and formulas

tssmooth dexponential is implemented as an ado-file.

A truncated description of the specified double-exponential filter is used to label the new variable. See [D] **label** for more information on labels.

An untruncated description of the specified double-exponential filter is saved in the characteristic tssmooth for the new variable. See [P] **char** for more information on characteristics.

The updating equations for the smoothing and forecasting versions are as given previously.

The starting values for both the smoothing and forecasting versions of double-exponential are obtained using the same method, which begins with the model

$$x_t = \beta_0 + \beta_1 t$$

where x_t is the series to be smoothed and t is a time variable that has been normalized to equal 1 in the first period included in the sample. The regression coefficient estimates $\widehat{\beta}_0$ and $\widehat{\beta}_1$ are obtained via OLS. The sample is determined by the option samp0(). By default, samp0() includes the first half of the observations. Given the estimates $\widehat{\beta}_0$ and $\widehat{\beta}_1$, the starting values are

$$S_0 = \widehat{\beta}_0 - \{(1 - \alpha)/\alpha\}\widehat{\beta}_1$$
$$S_0^{[2]} = \widehat{\beta}_0 - 2\{(1 - \alpha)/\alpha\}\widehat{\beta}_1$$

References

Abraham, B., and J. Ledolter. 1983. *Statistical Methods for Forecasting*. New York: Wiley.

Bowerman, B. L., R. T. O'Connell, and A. B. Koehler. 2005. *Forecasting, Time Series, and Regression: An Applied Approach*. 4th ed. Pacific Grove, CA: Brooks/Cole.

Chatfield, C. 2001. *Time-Series Forecasting*. London: Chapman & Hall/CRC.

———. 2004. *The Analysis of Time Series: An Introduction*. 6th ed. Boca Raton, FL: Chapman & Hall/CRC.

Chatfield, C., and M. Yar. 1988. Holt-Winters forecasting: Some practical issues. *Statistician* 37: 129–140.

Gardner, E. S., Jr. 1985. Exponential smoothing: The state of the art. *Journal of Forecasting* 4: 1–28.

Holt, C. C. 2004. Forecasting seasonals and trends by exponentially weighted moving averages. *International Journal of Forecasting* 20: 5–10.

Montgomery, D. C., L. A. Johnson, and J. S. Gardiner. 1990. *Forecasting and Time Series Analysis*. 2nd ed. New York: McGraw–Hill.

Winters, P. R. 1960. Forecasting sales by exponentially weighted moving averages. *Management Science* 6: 324–342.

Also see

Title

> **tssmooth exponential** — Single-exponential smoothing

Syntax

tssmooth exponential [*type*] *newvar* = *exp* [*if*] [*in*] [, *options*]

options	Description
Main	
replace	replace *newvar* if it already exists
parms($\#_\alpha$)	use $\#_\alpha$ as smoothing parameter
samp0(#)	use # observations to obtain initial value for recursion
s0(#)	use # as initial value for recursion
forecast(#)	use # periods for the out-of-sample forecast

You must tsset your data before using tssmooth exponential; see [TS] **tsset**.

exp may contain time-series operators; see [U] **11.4.4 Time-series varlists**.

Menu

Statistics > Time series > Smoothers/univariate forecasters > Single-exponential smoothing

Description

tssmooth exponential models the trend of a variable whose change from the previous value is serially correlated. More precisely, it models a variable whose first difference follows a low-order, moving-average process.

Options

┌─ Main ┐

replace replaces *newvar* if it already exists.

parms($\#_\alpha$) specifies the parameter α for the exponential smoother; $0 < \#_\alpha < 1$. If parms($\#_\alpha$) is not specified, the smoothing parameter is chosen to minimize the in-sample sum-of-squared forecast errors.

samp0(#) and s0(#) are mutually exclusive ways of specifying the initial value for the recursion.

samp0(#) specifies that the initial value be obtained by calculating the mean over the first # observations of the sample.

s0(#) specifies the initial value to be used.

If neither option is specified, the default is to use the mean calculated over the first half of the sample.

forecast(#) gives the number of observations for the out-of-sample prediction; $0 \le \# \le 500$. The default value is forecast(0) and is equivalent to not forecasting out of sample.

477

Remarks

Introduction
Examples
Treatment of missing values

Introduction

Exponential smoothing can be viewed either as an adaptive-forecasting algorithm or, equivalently, as a geometrically weighted moving-average filter. Exponential smoothing is most appropriate when used with time-series data that exhibit no linear or higher-order trends but that do exhibit low-velocity, aperiodic variation in the mean. Abraham and Ledolter (1983), Bowerman, O'Connell, and Koehler (2005), and Montgomery, Johnson, and Gardiner (1990) all provide good introductions to single-exponential smoothing. Chatfield (2001, 2004) discusses how single-exponential smoothing relates to modern time-series methods. For example, simple exponential smoothing produces optimal forecasts for several underlying models, including ARIMA(0,1,1) and the random-walk-plus-noise state-space model. (See Chatfield [2001, sec. 4.3.1].)

The exponential filter with smoothing parameter α creates the series S_t, where

$$S_t = \alpha X_t + (1 - \alpha)S_{t-1} \qquad \text{for } t = 1, \ldots, T$$

and S_0 is the initial value. This is the adaptive forecast-updating form of the exponential smoother. This implies that

$$S_t = \alpha \sum_{k=0}^{T-1} (1 - \alpha)^K X_{T-k} + (1 - \alpha)^T S_0$$

which is the weighted moving-average representation, with geometrically declining weights. The choice of the smoothing constant α determines how quickly the smoothed series or forecast will adjust to changes in the mean of the unfiltered series. For small values of α, the response will be slow because more weight is placed on the previous estimate of the mean of the unfiltered series, whereas larger values of α will put more emphasis on the most recently observed value of the unfiltered series.

Examples

▷ Example 1

Let's consider some examples using sales data. Here we forecast sales for three periods with a smoothing parameter of 0.4:

```
. use http://www.stata-press.com/data/r12/sales1
. tssmooth exponential sm1=sales, parms(.4) forecast(3)
exponential coefficient   =     0.4000
sum-of-squared residuals  =       8345
root mean squared error   =     12.919
```

To compare our forecast with the actual data, we graph the series and the forecasted series over time.

```
. line sm1 sales t, title("Single exponential forecast") ytitle(Sales) xtitle(Time)
```

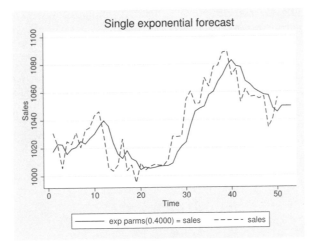

The graph indicates that our forecasted series may not be adjusting rapidly enough to the changes in the actual series. The smoothing parameter α controls the rate at which the forecast adjusts. Smaller values of α adjust the forecasts more slowly. Thus we suspect that our chosen value of 0.4 is too small. One way to investigate this suspicion is to ask `tssmooth exponential` to choose the smoothing parameter that minimizes the sum-of-squared forecast errors.

```
. tssmooth exponential sm2=sales, forecast(3)

computing optimal exponential   coefficient (0,1)

optimal exponential coefficient  =        0.7815
sum-of-squared residuals         =     6727.7056
root mean squared error          =     11.599746
```

The output suggests that the value of $\alpha = 0.4$ is too small. The graph below indicates that the new forecast tracks the series much more closely than the previous forecast.

```
. line sm2 sales t, title("Single exponential forecast with optimal alpha")
> ytitle(sales) xtitle(Time)
```

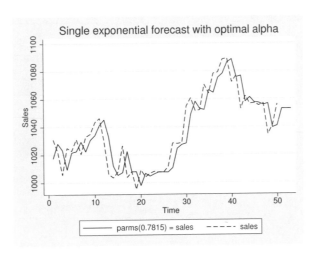

◁

We noted above that simple exponential forecasts are optimal for an ARIMA (0,1,1) model. (See [TS] **arima** for fitting ARIMA models in Stata.) Chatfield (2001, 90) gives the following useful derivation that relates the MA coefficient in an ARIMA (0,1,1) model to the smoothing parameter in single-exponential smoothing. An ARIMA (0,1,1) is given by

$$x_t - x_{t-1} = \epsilon_t + \theta\epsilon_{t-1}$$

where ϵ_t is an identically and independently distributed white-noise error term. Thus given $\widehat{\theta}$, an estimate of θ, an optimal one-step prediction of $\widehat{x}_{t+1}$ is $\widehat{x}_{t+1} = x_t + \widehat{\theta}\epsilon_t$. Because ϵ_t is not observable, it can be replaced by

$$\widehat{\epsilon}_t = x_t - \widehat{x}_{t-1}$$

yielding

$$\widehat{x}_{t+1} = x_t + \widehat{\theta}(x_t - \widehat{x}_{t-1})$$

Letting $\widehat{\alpha} = 1 + \widehat{\theta}$ and doing more rearranging implies that

$$\widehat{x}_{t+1} = (1 + \widehat{\theta})x_t - \widehat{\theta}\widehat{x}_{t-1}$$
$$\widehat{x}_{t+1} = \widehat{\alpha}x_t - (1 - \widehat{\alpha})\widehat{x}_{t-1}$$

▷ Example 2

Let's compare the estimate of the optimal smoothing parameter of 0.7815 with the one we could obtain using [TS] **arima**. Below we fit an ARIMA(0,1,1) to the sales data and then remove the estimate of α. The two estimates of α are quite close, given the large estimated standard error of $\widehat{\theta}$.

```
. arima sales, arima(0,1,1)

(setting optimization to BHHH)
Iteration 0:    log likelihood = -189.91037
Iteration 1:    log likelihood = -189.62405
Iteration 2:    log likelihood = -189.60468
Iteration 3:    log likelihood = -189.60352
Iteration 4:    log likelihood = -189.60343
(switching optimization to BFGS)
Iteration 5:    log likelihood = -189.60342

ARIMA regression
Sample:  2 - 50                          Number of obs    =         49
                                         Wald chi2(1)     =       1.41
Log likelihood = -189.6034               Prob > chi2      =     0.2347
```

		OPG				
D.sales	Coef.	Std. Err.	z	P>|z|	[95% Conf.	Interval]
sales						
_cons	.5025469	1.382727	0.36	0.716	-2.207548	3.212641
ARMA						
ma						
L1.	-.1986561	.1671699	-1.19	0.235	-.5263031	.1289908
/sigma	11.58992	1.240607	9.34	0.000	9.158378	14.02147

```
Note: The test of the variance against zero is one sided, and the two-sided
      confidence interval is truncated at zero.
. di 1 + _b[ARMA:L.ma]
.80134387
```

◁

▷ Example 3

tssmooth exponential automatically detects panel data. Suppose that we had sales figures for five companies in long form. Running tssmooth exponential on the variable that contains all five series puts the smoothed series and the predictions in one variable in long form. When the smoothing parameter is chosen to minimize the squared prediction error, an optimal value for the smoothing parameter is chosen separately for each panel.

```
. use http://www.stata-press.com/data/r12/sales_cert, clear
. tsset
       panel variable:  id (strongly balanced)
        time variable:  t, 1 to 100
               delta:  1 unit
. tssmooth exponential sm5=sales, forecast(3)

-> id = 1
computing optimal exponential  coefficient (0,1)

optimal exponential coefficient =      0.8702
sum-of-squared residuals         =   16070.567
root mean squared error          =   12.676974

-> id = 2
computing optimal exponential  coefficient (0,1)

optimal exponential coefficient =      0.7003
sum-of-squared residuals         =   20792.393
root mean squared error          =   14.419568

-> id = 3
computing optimal exponential  coefficient (0,1)

optimal exponential coefficient =      0.6927
sum-of-squared residuals         =      21629
root mean squared error          =   14.706801

-> id = 4
computing optimal exponential  coefficient (0,1)

optimal exponential coefficient =      0.3866
sum-of-squared residuals         =   22321.334
root mean squared error          =   14.940326

-> id = 5
computing optimal exponential  coefficient (0,1)

optimal exponential coefficient =      0.4540
sum-of-squared residuals         =   20714.095
root mean squared error          =   14.392392
```

tssmooth exponential computed starting values and chose an optimal α for each panel individually.

◁

Treatment of missing values

Missing values in the middle of the data are filled in with the one-step-ahead prediction using the previous values. Missing values at the beginning or end of the data are treated as if the observations were not there.

`tssmooth exponential` treats observations excluded from the sample by `if` and `in` just as if they were missing.

▷ Example 4

Here the 28th observation is missing. The prediction for the 29th observation is repeated in the new series.

```
. use http://www.stata-press.com/data/r12/sales1, clear
. tssmooth exponential sm1=sales, parms(.7) forecast(3)
  (output omitted)
. generate sales2=sales if t!=28
(4 missing values generated)
. tssmooth exponential sm3=sales2, parms(.7) forecast(3)
exponential coefficient    =        0.7000
sum-of-squared residuals   =        6842.4
root mean squared error    =        11.817
. list t sales2 sm3 if t>25 & t<31
```

	t	sales2	sm3
26.	26	1011.5	1007.5
27.	27	1028.3	1010.3
28.	28	.	1022.9
29.	29	1028.4	1022.9
30.	30	1054.8	1026.75

Because the data for $t = 28$ are missing, the prediction for period 28 has been used in its place. This implies that the updating equation for period 29 is

$$S_{29} = \alpha S_{28} + (1 - \alpha)S_{28} = S_{28}$$

which explains why the prediction for $t = 28$ is repeated.

Because this is a single-exponential procedure, the loss of that one observation will not be noticed several periods later.

```
. generate diff = sm3-sm1 if t>28
(28 missing values generated)
. list t diff if t>28 & t<39
```

	t	diff
29.	29	-3.5
30.	30	-1.050049
31.	31	-.3150635
32.	32	-.0946045
33.	33	-.0283203
34.	34	-.0085449
35.	35	-.0025635
36.	36	-.0008545
37.	37	-.0003662
38.	38	-.0001221

◁

▷ Example 5

Now consider an example in which there are data missing at the beginning and end of the sample.

```
. generate sales3=sales if t>2 & t<49
(7 missing values generated)
. tssmooth exponential sm4=sales3, parms(.7) forecast(3)
```

```
exponential coefficient   =        0.7000
sum-of-squared residuals  =        6215.3
root mean squared error   =        11.624
```

```
. list t sales sales3 sm4 if t<5 | t>45
```

	t	sales	sales3	sm4
1.	1	1031	.	.
2.	2	1022.1	.	.
3.	3	1005.6	1005.6	1016.787
4.	4	1025	1025	1008.956
46.	46	1055.2	1055.2	1057.2
47.	47	1056.8	1056.8	1055.8
48.	48	1034.5	1034.5	1056.5
49.	49	1041.1	.	1041.1
50.	50	1056.1	.	1041.1
51.	51	.	.	1041.1
52.	52	.	.	1041.1
53.	53	.	.	1041.1

The output above illustrates that missing values at the beginning or end of the sample cause the sample to be truncated. The new series begins with nonmissing data and begins predicting immediately after it stops.

One period after the actual data concludes, the exponential forecast becomes a constant. After the actual end of the data, the forecast at period t is substituted for the missing data. This also illustrates why the forecasted series is a constant.

◁

Saved results

tssmooth exponential saves the following in r():

Scalars
r(N)	number of observations
r(alpha)	α smoothing parameter
r(rss)	sum-of-squared prediction errors
r(rmse)	root mean squared error
r(N_pre)	number of observations used in calculating starting values
r(s1_0)	initial value for S_t

Macros
r(method)	smoothing method
r(exp)	expression specified
r(timevar)	time variable specified in tsset
r(panelvar)	panel variable specified in tsset

Methods and formulas

tssmooth exponential is implemented as an ado-file.

The formulas for deriving smoothed series are as given in the text. When the value of α is not specified, an optimal value is found that minimizes the mean squared forecast error. A method of bisection is used to find the solution to this optimization problem.

A truncated description of the specified exponential filter is used to label the new variable. See [D] **label** for more information about labels.

An untruncated description of the specified exponential filter is saved in the characteristic tssmooth for the new variable. See [P] **char** for more information about characteristics.

References

Abraham, B., and J. Ledolter. 1983. *Statistical Methods for Forecasting*. New York: Wiley.

Bowerman, B. L., R. T. O'Connell, and A. B. Koehler. 2005. *Forecasting, Time Series, and Regression: An Applied Approach*. 4th ed. Pacific Grove, CA: Brooks/Cole.

Chatfield, C. 2001. *Time-Series Forecasting*. London: Chapman & Hall/CRC.

———. 2004. *The Analysis of Time Series: An Introduction*. 6th ed. Boca Raton, FL: Chapman & Hall/CRC.

Chatfield, C., and M. Yar. 1988. Holt-Winters forecasting: Some practical issues. *Statistician* 37: 129–140.

Holt, C. C. 2004. Forecasting seasonals and trends by exponentially weighted moving averages. *International Journal of Forecasting* 20: 5–10.

Montgomery, D. C., L. A. Johnson, and J. S. Gardiner. 1990. *Forecasting and Time Series Analysis*. 2nd ed. New York: McGraw–Hill.

Winters, P. R. 1960. Forecasting sales by exponentially weighted moving averages. *Management Science* 6: 324–342.

Also see

[TS] **tsset** — Declare data to be time-series data

[TS] **tssmooth** — Smooth and forecast univariate time-series data

Title

tssmooth hwinters — Holt–Winters nonseasonal smoothing

Syntax

tssmooth hwinters [*type*] *newvar* = *exp* [*if*] [*in*] [, *options*]

options	Description
Main	
replace	replace *newvar* if it already exists
parms(#$_\alpha$ #$_\beta$)	use #$_\alpha$ and #$_\beta$ as smoothing parameters
samp0(#)	use # observations to obtain initial values for recursion
s0(#$_{cons}$ #$_{lt}$)	use #$_{cons}$ and #$_{lt}$ as initial values for recursion
forecast(#)	use # periods for the out-of-sample forecast
Options	
diff	alternative initial-value specification; see *Options*
Maximization	
maximize_options	control the maximization process; seldom used
from(#$_\alpha$ #$_\beta$)	use #$_\alpha$ and #$_\beta$ as starting values for the parameters

You must tsset your data before using tssmooth hwinters; see [TS] tsset.

exp may contain time-series operators; see [U] **11.4.4 Time-series varlists**.

Menu

Statistics > Time series > Smoothers/univariate forecasters > Holt-Winters nonseasonal smoothing

Description

tssmooth hwinters is used in smoothing or forecasting a series that can be modeled as a linear trend in which the intercept and the coefficient on time vary over time.

Options

┌─ Main ───

replace replaces *newvar* if it already exists.

parms(#$_\alpha$ #$_\beta$), $0 \leq$ #$_\alpha \leq 1$ and $0 \leq$ #$_\beta \leq 1$, specifies the parameters. If parms() is not specified, the values are chosen by an iterative process to minimize the in-sample sum-of-squared prediction errors.

If you experience difficulty converging (many iterations and "not concave" messages), try using from() to provide better starting values.

samp0(#) and s0(#$_{cons}$ #$_{lt}$) specify how the initial values #$_{cons}$ and #$_{lt}$ for the recursion are obtained.

485

By default, initial values are obtained by fitting a linear regression with a time trend using the first half of the observations in the dataset.

samp0(#) specifies that the first # observations be used in that regression.

s0($\#_{\text{cons}}$ $\#_{\text{lt}}$) specifies that $\#_{\text{cons}}$ and $\#_{\text{lt}}$ be used as initial values.

forecast(#) specifies the number of periods for the out-of-sample prediction; $0 \leq \# \leq 500$. The default is forecast(0), which is equivalent to not performing an out-of-sample forecast.

___

⌐ Options ⌐

diff specifies that the linear term is obtained by averaging the first difference of exp_t and the intercept is obtained as the difference of exp in the first observation and the mean of D.exp_t.

If the diff option is not specified, a linear regression of exp_t on a constant and t is fit.

___

⌐ Maximization ⌐

maximize_options controls the process for solving for the optimal α and β when parms() is not specified.

maximize_options: nodifficult, technique(*algorithm_spec*), iterate(#), [no]log, trace, gradient, showstep, hessian, showtolerance, tolerance(#), ltolerance(#), nrtolerance(#), and nonrtolerance; see [R] **maximize**. These options are seldom used.

from($\#_{\alpha}$ $\#_{\beta}$), $0 < \#_{\alpha} < 1$ and $0 < \#_{\beta} < 1$, specifies starting values from which the optimal values of α and β will be obtained. If from() is not specified, from(.5 .5) is used.

Remarks

The Holt–Winters method forecasts series of the form

$$\widehat{x}_{t+1} = a_t + b_t t$$

where $\widehat{x}_t$ is the forecast of the original series x_t, a_t is a mean that drifts over time, and b_t is a coefficient on time that also drifts. In fact, as Gardner (1985) has noted, the Holt–Winters method produces optimal forecasts for an ARIMA(0,2,2) model and some local linear models. See [TS] **arima** and the references in that entry for ARIMA models, and see Harvey (1989) for a discussion of the local linear model and its relationship to the Holt–Winters method. Abraham and Ledolter (1983), Bowerman, O'Connell, and Koehler (2005), and Montgomery, Johnson, and Gardiner (1990) all provide good introductions to the Holt–Winters method. Chatfield (2001, 2004) provides helpful discussions of how this method relates to modern time-series analysis.

The Holt–Winters method can be viewed as an extension of double-exponential smoothing with two parameters, which may be explicitly set or chosen to minimize the in-sample sum-of-squared forecast errors. In the latter case, as discussed in *Methods and formulas*, the smoothing parameters are chosen to minimize the in-sample sum-of-squared forecast errors plus a penalty term that helps to achieve convergence when one of the parameters is too close to the boundary.

Given the series x_t, the smoothing parameters α and β, and the starting values a_0 and b_0, the updating equations are

$$a_t = \alpha x_t + (1 - \alpha)(a_{t-1} + b_{t-1})$$
$$b_t = \beta(a_t - a_{t-1}) + (1 - \beta)b_{t-1}$$

After computing the series of constant and linear terms, a_t and b_t, respectively, the τ-step-ahead prediction of x_t is given by

$$\widehat{x}_{t+\tau} = a_t + b_t \tau$$

▷ Example 1

Below we show how to use `tssmooth hwinters` with specified smoothing parameters. This example also shows that the Holt–Winters method can closely follow a series in which both the mean and the time coefficient drift over time.

Suppose that we have data on the monthly sales of a book and that we want to forecast this series with the Holt–Winters method.

```
. use http://www.stata-press.com/data/r12/bsales
. tssmooth hwinters hw1=sales, parms(.7 .3) forecast(3)
Specified weights:
                    alpha = 0.7000
                     beta = 0.3000
sum-of-squared residuals = 2301.046
 root mean squared error = 6.192799
. line sales hw1 t, title("Holt-Winters Forecast with alpha=.7  and beta=.3")
> ytitle(Sales) xtitle(Time)
```

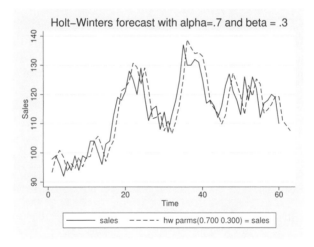

The graph indicates that the forecasts are for linearly decreasing sales. Given a_T and b_T, the out-of-sample predictions are linear functions of time. In this example, the slope appears to be too steep, probably because our choice of α and β.

◁

▷ Example 2

The graph in the previous example illustrates that the starting values for the linear and constant series can affect the in-sample fit of the predicted series for the first few observations. The previous example used the default method for obtaining the initial values for the recursion. The output below illustrates that, for some problems, the differenced-based initial values provide a better in-sample fit for the first few observations. However, the differenced-based initial values do not always outperform the regression-based initial values. Furthermore, as shown in the output below, for series of reasonable length, the predictions produced are nearly identical.

```
. tssmooth hwinters hw2=sales, parms(.7 .3) forecast(3) diff
Specified weights:
                          alpha = 0.7000
                           beta = 0.3000
sum-of-squared residuals = 2261.173
 root mean squared error = 6.13891

. list hw1 hw2 if _n<6 | _n>57
```

	hw1	hw2
1.	93.31973	97.80807
2.	98.40002	98.11447
3.	100.8845	99.2267
4.	98.50404	96.78276
5.	93.62408	92.2452
58.	116.5771	116.5771
59.	119.2146	119.2146
60.	119.2608	119.2608
61.	111.0299	111.0299
62.	109.2815	109.2815
63.	107.5331	107.5331

When the smoothing parameters are chosen to minimize the in-sample sum-of-squared forecast errors, changing the initial values can affect the choice of the optimal α and β. When changing the initial values results in different optimal values for α and β, the predictions will also differ.

◁

When the Holt–Winters model fits the data well, finding the optimal smoothing parameters generally proceeds well. When the model fits poorly, finding the α and β that minimize the in-sample sum-of-squared forecast errors can be difficult.

▷ Example 3

In this example, we forecast the book sales data using the α and β that minimize the in-sample squared forecast errors.

```
. tssmooth hwinters hw3=sales, forecast(3)
computing optimal weights
Iteration 0:   penalized RSS = -2632.2073  (not concave)
Iteration 1:   penalized RSS = -1982.8431
Iteration 2:   penalized RSS = -1976.4236
Iteration 3:   penalized RSS = -1975.9175
Iteration 4:   penalized RSS = -1975.9036
Iteration 5:   penalized RSS = -1975.9036
Optimal weights:
                                   alpha = 0.8209
                                    beta = 0.0067
penalized sum of squared residuals = 1975.904
        sum of squared residuals = 1975.904
        root mean-squared error = 5.738617
```

The following graph contains the data and the forecast using the optimal α and β. Comparing this graph with the one above illustrates how different choices of α and β can lead to very different forecasts. Instead of linearly decreasing sales, the new forecast is for linearly increasing sales.

```
. line sales hw3 t, title("Holt-Winters Forecast with optimal alpha and beta")
> ytitle(Sales) xtitle(Time)
```

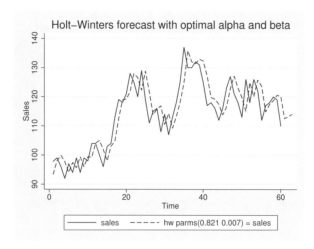

◁

Saved results

tssmooth hwinters saves the following in r():

Scalars

r(N)	number of observations	r(N_pre)	number of observations used
r(alpha)	α smoothing parameter		in calculating starting values
r(beta)	β smoothing parameter	r(s2_0)	initial value for linear term
r(rss)	sum-of-squared errors	r(s1_0)	initial value for constant term
r(prss)	penalized sum-of-squared errors,	r(linear)	final value of linear term
	if parms() not specified	r(constant)	final value of constant term
r(rmse)	root mean squared error		

Macros

r(method)	smoothing method	r(timevar)	time variables specified in tsset
r(exp)	expression specified	r(panelvar)	panel variables specified in tsset

Methods and formulas

tssmooth hwinters is implemented as an ado-file.

A truncated description of the specified Holt–Winters filter is used to label the new variable. See [D] **label** for more information on labels.

An untruncated description of the specified Holt–Winters filter is saved in the characteristic named tssmooth for the new variable. See [P] **char** for more information on characteristics.

Given the series, x_t; the smoothing parameters, α and β; and the starting values, a_0 and b_0, the updating equations are

$$a_t = \alpha x_t + (1 - \alpha)(a_{t-1} + b_{t-1})$$

$$b_t = \beta(a_t - a_{t-1}) + (1 - \beta) b_{t-1}$$

By default, the initial values are found by fitting a linear regression with a time trend. The time variable in this regression is normalized to equal one in the first period included in the sample. By default, one-half of the data is used in this regression, but this sample can be changed using `samp0()`. a_0 is then set to the estimate of the constant, and b_0 is set to the estimate of the coefficient on the time trend. Specifying the `diff` option sets b_0 to the mean of D.x and a_0 to $x_1 - b_0$. `s0()` can also be used to specify the initial values directly.

Sometimes, one or both of the optimal parameters may lie on the boundary of $[0, 1]$. To keep the estimates inside $[0, 1]$, `tssmooth hwinters` parameterizes the objective function in terms of their inverse logits, that is, in terms of $\exp(\alpha) / \{1 + \exp(\alpha)\}$ and $\exp(\beta) / \{1 + \exp(\beta)\}$. When one of these parameters is actually on the boundary, this can complicate the optimization. For this reason, `tssmooth hwinters` optimizes a penalized sum-of-squared forecast errors. Let $\widehat{x}_t(\widetilde{\alpha}, \widetilde{\beta})$ be the forecast for the series x_t, given the choices of $\widetilde{\alpha}$ and $\widetilde{\beta}$. Then the in-sample penalized sum-of-squared prediction errors is

$$P = \sum_{t=1}^{T} \left[\{x_t - \widehat{x}_t(\widetilde{\alpha}, \widetilde{\beta})\}^2 + I_{|f(\widetilde{\alpha})|>12}(|f(\widetilde{\alpha})| - 12)^2 + I_{|f(\widetilde{\beta})|>12}(|f(\widetilde{\beta})| - 12)^2 \right]$$

where $f(x) = \ln\{x(1 - x)\}$. The penalty term is zero unless one of the parameters is close to the boundary. When one of the parameters is close to the boundary, the penalty term will help to obtain convergence.

Acknowledgment

We thank Nicholas J. Cox of Durham University for his helpful comments.

References

Abraham, B., and J. Ledolter. 1983. *Statistical Methods for Forecasting*. New York: Wiley.

Bowerman, B. L., R. T. O'Connell, and A. B. Koehler. 2005. *Forecasting, Time Series, and Regression: An Applied Approach*. 4th ed. Pacific Grove, CA: Brooks/Cole.

Chatfield, C. 2001. *Time-Series Forecasting*. London: Chapman & Hall/CRC.

———. 2004. *The Analysis of Time Series: An Introduction*. 6th ed. Boca Raton, FL: Chapman & Hall/CRC.

Chatfield, C., and M. Yar. 1988. Holt-Winters forecasting: Some practical issues. *Statistician* 37: 129–140.

Gardner, E. S., Jr. 1985. Exponential smoothing: The state of the art. *Journal of Forecasting* 4: 1–28.

Harvey, A. C. 1989. *Forecasting, Structural Time Series Models and the Kalman Filter*. Cambridge: Cambridge University Press.

Holt, C. C. 2004. Forecasting seasonals and trends by exponentially weighted moving averages. *International Journal of Forecasting* 20: 5–10.

Montgomery, D. C., L. A. Johnson, and J. S. Gardiner. 1990. *Forecasting and Time Series Analysis*. 2nd ed. New York: McGraw–Hill.

Winters, P. R. 1960. Forecasting sales by exponentially weighted moving averages. *Management Science* 6: 324–342.

Also see

Title

tssmooth ma — Moving-average filter

Syntax

Moving average with uniform weights

> tssmooth ma [*type*] *newvar* = *exp* [*if*] [*in*] , <u>w</u>indow(#$_l$[#$_c$[#$_f$]]) [replace]

Moving average with specified weights

> tssmooth ma [*type*] *newvar* = *exp* [*if*] [*in*] , <u>we</u>ights([*numlist*$_l$] <#$_c$> [*numlist*$_f$])
>
> [replace]

You must tsset your data before using tssmooth ma; see [TS] **tsset**.

exp may contain time-series operators; see [U] **11.4.4 Time-series varlists**.

Menu

Statistics > Time series > Smoothers/univariate forecasters > Moving-average filter

Description

tssmooth ma creates a new series in which each observation is an average of nearby observations in the original series.

In the first syntax, window() is required and specifies the span of the filter. tssmooth ma constructs a uniformly weighted moving average of the expression.

In the second syntax, weights() is required and specifies the weights to be used. tssmooth ma then applies the specified weights to construct a weighted moving average of the expression.

Options

window(#$_l$ [#$_c$ [#$_f$]]) describes the span of the uniformly weighted moving average.

#$_l$ specifies the number of lagged terms to be included, $0 \leq$ #$_l \leq$ one-half the number of observations in the sample.

#$_c$ is optional and specifies whether to include the current observation in the filter. A 0 indicates exclusion and 1, inclusion. The current observation is excluded by default.

#$_f$ is optional and specifies the number of forward terms to be included, $0 \leq$ #$_f \leq$ one-half the number of observations in the sample.

weights([*numlist*$_l$] <#$_c$> [*numlist*$_f$]) is required for the weighted moving average and describes the span of the moving average, as well as the weights to be applied to each term in the average. The middle term literally is surrounded by < and >, so you might type weights(1/2 <3> 2/1).

numlist$_l$ is optional and specifies the weights to be applied to the lagged terms when computing the moving average.

$\#_c$ is required and specifies the weight to be applied to the current term.

numlist$_f$ is optional and specifies the weights to be applied to the forward terms when computing the moving average.

The number of elements in each *numlist* is limited to one-half the number of observations in the sample.

`replace` replaces *newvar* if it already exists.

Remarks

Moving averages are simple linear filters of the form

$$\widehat{x}_t = \frac{\sum_{i=-l}^{f} w_i x_{t+i}}{\sum_{i=-l}^{f} w_i}$$

where

$\widehat{x}_t$	is the moving average
x_t	is the variable or expression to be smoothed
w_i	are the weights being applied to the terms in the filter
l	is the longest lag in the span of the filter
f	is the longest lead in the span of the filter

Moving averages are used primarily to reduce noise in time-series data. Using moving averages to isolate signals is problematic, however, because the moving averages themselves are serially correlated, even when the underlying data series is not. Still, Chatfield (2004) discusses moving-average filters and provides several specific moving-average filters for extracting certain trends.

▷ Example 1

Suppose that we have a time series of sales data, and we want to separate the data into two components: signal and noise. To eliminate the noise, we apply a moving-average filter. In this example, we use a symmetric moving average with a span of 5. This means that we will average the first two lagged values, the current value, and the first two forward terms of the series, with each term in the average receiving a weight of 1.

```
. use http://www.stata-press.com/data/r12/sales1
. tsset
        time variable:  t, 1 to 50
                delta:  1 unit
. tssmooth ma sm1 = sales, window(2 1 2)
The smoother applied was
    (1/5)*[x(t-2) + x(t-1) + 1*x(t) + x(t+1) + x(t+2)]; x(t)= sales
```

We would like to smooth our series so that there is no autocorrelation in the noise. Below we compute the noise as the difference between the smoothed series and the series itself. Then we use ac (see [TS] **corrgram**) to check for autocorrelation in the noise.

```
. generate noise = sales-sm1

. ac noise
```

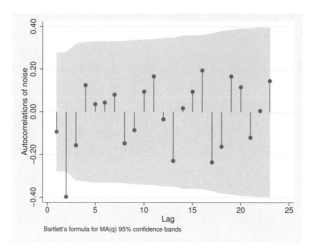

Bartlett's formula for MA(q) 95% confidence bands

◁

▷ Example 2

In the previous example, there is some evidence of negative second-order autocorrelation, possibly due to the uniform weighting or the length of the filter. We are going to specify a shorter filter in which the weights decline as the observations get farther away from the current observation.

The weighted moving-average filter requires that we supply the weights to apply to each element with the weights() option. In specifying the weights, we implicitly specify the span of the filter.

Below we use the filter

$$\widehat{x}_t = (1/9)(1x_{t-2} + 2x_{t-1} + 3x_t + 2x_{t+1} + 1x_{t+2})$$

In what follows, 1/2 does not mean one-half, it means the numlist 1 2:

```
. tssmooth ma sm2 = sales, weights( 1/2 <3> 2/1)
The smoother applied was
    (1/9)*[1*x(t-2) + 2*x(t-1) + 3*x(t) + 2*x(t+1) + 1*x(t+2)]; x(t)= sales
. generate noise2 = sales-sm2
```

We compute the noise and use ac to check for autocorrelation.

. ac noise2

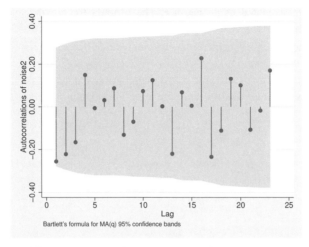

The graph shows no significant evidence of autocorrelation in the noise from the second filter.

◁

□ Technical note

tssmooth ma gives any missing observations a coefficient of zero in both the uniformly weighted and weighted moving-average filters. This simply means that missing values or missing periods are excluded from the moving average.

Sample restrictions, via if and in, cause the expression smoothed by tssmooth ma to be missing for the excluded observations. Thus sample restrictions have the same effect as missing values in a variable that is filtered in the expression. Also, gaps in the data that are longer than the span of the filter will generate missing values in the filtered series.

Because the first l observations and the last f observations will be outside the span of the filter, those observations will be set to missing in the moving-average series.

□

Saved results

tssmooth ma saves the following in r():

Scalars
    r(N)   number of observations
    r(w0)   weight on the current observation
    r(wlead#)   weight on lead #, if leads are specified
    r(wlag#)   weight on lag #, if lags are specified

Macros
    r(method)   smoothing method
    r(exp)   expression specified
    r(timevar)   time variable specified in tsset
    r(panelvar)   panel variable specified in tsset

Methods and formulas

`tssmooth ma` is implemented as an ado-file. The formula for moving averages is the same as previously given.

A truncated description of the specified moving-average filter labels the new variable. See [D] **label** for more information on labels.

An untruncated description of the specified moving-average filter is saved in the characteristic `tssmooth` for the new variable. See [P] **char** for more information on characteristics.

Reference

Chatfield, C. 2004. *The Analysis of Time Series: An Introduction.* 6th ed. Boca Raton, FL: Chapman & Hall/CRC.

Also see

[TS] **tsset** — Declare data to be time-series data

[TS] **tssmooth** — Smooth and forecast univariate time-series data

Title

tssmooth nl — Nonlinear filter

Syntax

tssmooth nl $[$ *type* $]$ *newvar* = *exp* $[$ *if* $]$ $[$ *in* $]$, <u>sm</u>oother(*smoother* $[$, <u>t</u>wice $]$)

$[$ replace $]$

where *smoother* is specified as $Sm \left[Sm \left[\dots \right] \right]$ and Sm is one of

$$\{ 1 \mid 2 \mid 3 \mid 4 \mid 5 \mid 6 \mid 7 \mid 8 \mid 9 \} \left[R \right]$$
$$3 \left[R \right] S \left[S \mid R \right] \left[S \mid R \right] \dots$$
$$E$$
$$H$$

The numbers specified in *smoother* represent the span of a running median smoother. For example, a number 3 specifies that each value be replaced by the median of the point and the two adjacent data values. The letter H indicates that a Hanning linear smoother, which is a span-3 smoother with binomial weights, be applied.

The letters E, S, and R are three refinements that can be combined with the running median and Hanning smoothers. First, the end points of a smooth can be given special treatment. This is specified by the E operator. Second, smoothing by 3, the span-3 running median, tends to produce flat-topped hills and valleys. The splitting operator, S, "splits" these repeated values, applies the end-point operator to them, and then "rejoins" the series. Third, it is sometimes useful to repeat an odd-span median smoother or the splitting operator until the smooth no longer changes. Following a digit or an S with an R specifies this type of repetition.

Finally, the `twice` operator specifies that after smoothing, the smoother be reapplied to the resulting rough, and any recovered signal be added back to the original smooth.

Letters may be specified in lowercase, if preferred. Examples of *smoother* $[$, `twice` $]$ include

3RSSH	3RSSH,twice	4253H	4253H,twice	43RSR2H,twice
3rssh	3rssh,twice	4253h	4253h,twice	43rsr2h,twice

You must `tsset` your data before using `tssmooth nl`; see [TS] **tsset**.

exp may contain time-series operators; see [U] **11.4.4 Time-series varlists**.

Menu

Statistics > Time series > Smoothers/univariate forecasters > Nonlinear filter

Description

`tssmooth nl` uses nonlinear smoothers to identify the underlying trend in a series.

Options

┌─ Main ┐

`smoother(`*smoother*`[ , twice ])` specifies the nonlinear smoother to be used.

`replace` replaces *newvar* if it already exists.

Remarks

`tssmooth nl` works as a front end to `smooth`. See [R] **smooth** for details.

Saved results

`tssmooth nl` saves the following in `r()`:

Scalars
    `r(N)`   number of observations

Macros
    `r(method)`   `nl`
    `r(smoother)`   specified smoother
    `r(timevar)`   time variable specified in `tsset`
    `r(panelvar)`   panel variable specified in `tsset`

Methods and formulas

`tssmooth nl` is implemented as an ado-file. The methods are documented in [R] **smooth**.

A truncated description of the specified nonlinear filter labels the new variable. See [D] **label** for more information on labels.

An untruncated description of the specified nonlinear filter is saved in the characteristic `tssmooth` for the new variable. See [P] **char** for more information on characteristics.

Also see

[TS] **tsset** — Declare data to be time-series data

[TS] **tssmooth** — Smooth and forecast univariate time-series data

Title

> **tssmooth shwinters** — Holt–Winters seasonal smoothing

Syntax

tssmooth shwinters $\begin{bmatrix} type \end{bmatrix}$ *newvar* = *exp* $\begin{bmatrix} if \end{bmatrix}$ $\begin{bmatrix} in \end{bmatrix}$ $\begin{bmatrix} , options \end{bmatrix}$

options	Description
Main	
replace	replace *newvar* if it already exists
parms($\#_\alpha$ $\#_\beta$ $\#_\gamma$)	use $\#_\alpha$, $\#_\beta$, and $\#_\gamma$ as smoothing parameters
samp0(#)	use # observations to obtain initial values for recursion
s0($\#_{cons}$ $\#_{lt}$)	use $\#_{cons}$ and $\#_{lt}$ as initial values for recursion
forecast(#)	use # periods for the out-of-sample forecast
period(#)	use # for period of the seasonality
additive	use additive seasonal Holt–Winters method
Options	
sn0_0(*varname*)	use initial seasonal values in *varname*
sn0_v(*newvar*)	store estimated initial values for seasonal terms in *newvar*
snt_v(*newvar*)	store final year's estimated seasonal terms in *newvar*
normalize	normalize seasonal values
altstarts	use alternative method for computing the starting values
Maximization	
maximize_options	control the maximization process; seldom used
from($\#_\alpha$ $\#_\beta$ $\#_\gamma$)	use $\#_\alpha$, $\#_\beta$, and $\#_\gamma$ as starting values for the parameters

You must tsset your data before using tssmooth shwinters; see [TS] **tsset**.

exp may contain time-series operators; see [U] **11.4.4 Time-series varlists**.

Menu

Statistics > Time series > Smoothers/univariate forecasters > Holt-Winters seasonal smoothing

Description

tssmooth shwinters performs the seasonal Holt–Winters method on a user-specified expression, which is usually just a variable name, and generates a new variable containing the forecasted series.

Options

> **Main**

replace replaces *newvar* if it already exists.

499

parms($\#_\alpha$ $\#_\beta$ $\#_\gamma$), $0 \leq \#_\alpha \leq 1$, $0 \leq \#_\beta \leq 1$, and $0 \leq \#_\gamma \leq 1$, specifies the parameters. If parms() is not specified, the values are chosen by an iterative process to minimize the in-sample sum-of-squared prediction errors.

If you experience difficulty converging (many iterations and "not concave" messages), try using from() to provide better starting values.

samp0(#) and s0($\#_{\mathrm{cons}}$ $\#_{\mathrm{lt}}$) have to do with how the initial values $\#_{\mathrm{cons}}$ and $\#_{\mathrm{lt}}$ for the recursion are obtained.

s0($\#_{\mathrm{cons}}$ $\#_{\mathrm{lt}}$) specifies the initial values to be used.

samp0(#) specifies that the initial values be obtained using the first # observations of the sample. This calculation is described under *Methods and formulas* and depends on whether the altstart and additive options are also specified.

If neither option is specified, the first half of the sample is used to obtain initial values.

forecast(#) specifies the number of periods for the out-of-sample prediction; $0 \leq \# \leq 500$. The default is forecast(0), which is equivalent to not performing an out-of-sample forecast.

period(#) specifies the period of the seasonality. If period() is not specified, the seasonality is obtained from the tsset options daily, weekly, ..., yearly; see [TS] **tsset**. If you did not specify one of those options when you tsset the data, you must specify the period() option. For instance, if your data are quarterly and you did not specify tsset's quarterly option, you must now specify period(4).

By default, seasonal values are calculated, but you may specify the initial seasonal values to be used via the sn0_0(*varname*) option. The first period() observations of *varname* are to contain the initial seasonal values.

additive uses the additive seasonal Holt–Winters method instead of the default multiplicative seasonal Holt–Winters method.

___
⌐ Options ⌐

sn0_0(*varname*) specifies the initial seasonal values to use. *varname* must contain a complete year's worth of seasonal values, beginning with the first observation in the estimation sample. For example, if you have monthly data, the first 12 observations of *varname* must contain nonmissing data. sn0_0() cannot be used with sn0_v().

sn0_v(*newvar*) stores in *newvar* the initial seasonal values after they have been estimated. sn0_v() cannot be used with sn0_0().

snt_v(*newvar*) stores in *newvar* the seasonal values for the final year's worth of data.

normalize specifies that the seasonal values be normalized. In the multiplicative model, they are normalized to sum to one. In the additive model, the seasonal values are normalized to sum to zero.

altstarts uses an alternative method to compute the starting values for the constant, the linear, and the seasonal terms. The default and the alternative methods are described in *Methods and formulas*. altstarts may not be specified with s0().

___
⌐ Maximization ⌐

maximize_options controls the process for solving for the optimal α, β, and γ when the parms() option is not specified.

maximize_options: <u>nodiff</u>icult, <u>tech</u>nique(*algorithm_spec*), <u>iter</u>ate(#), [<u>no</u>]<u>log</u>, <u>tr</u>ace, gradient, showstep, hessian, <u>showtol</u>erance, <u>tol</u>erance(#), <u>ltol</u>erance(#), <u>nrtol</u>erance(#), and <u>nonrtol</u>erance; see [R] **maximize**. These options are seldom used.

from(#$_\alpha$ #$_\beta$ #$_\gamma$), $0 < \#_\alpha < 1$, $0 < \#_\beta < 1$, and $0 < \#_\gamma < 1$, specifies starting values from which the optimal values of α, β, and γ will be obtained. If from() is not specified, from(.5 .5 .5) is used.

Remarks

Remarks are presented under the following headings:

> *Introduction*
> *Holt–Winters seasonal multiplicative method*
> *Holt–Winters seasonal additive method*

Introduction

The seasonal Holt–Winters methods forecast univariate series that have a seasonal component. If the amplitude of the seasonal component grows with the series, the Holt–Winters multiplicative method should be used. If the amplitude of the seasonal component is not growing with the series, the Holt–Winters additive method should be used. Abraham and Ledolter (1983), Bowerman, O'Connell, and Koehler (2005), and Montgomery, Johnson, and Gardiner (1990) provide good introductions to the Holt–Winters methods in recursive univariate forecasting methods. Chatfield (2001, 2004) provides introductions in the broader context of modern time-series analysis.

Like the other recursive methods in tssmooth, tssmooth shwinters uses the information stored by tsset to detect panel data. When applied to panel data, each series is smoothed separately, and the starting values are computed separately for each panel. If the smoothing parameters are chosen to minimize the in-sample sum-of-squared forecast errors, the optimization is performed separately on each panel.

When there are missing values at the beginning of the series, the sample begins with the first nonmissing observation. Missing values after the first nonmissing observation are filled in with forecasted values.

Holt–Winters seasonal multiplicative method

This method forecasts seasonal time series in which the amplitude of the seasonal component grows with the series. Chatfield (2001) notes that there are some nonlinear state-space models whose optimal prediction equations correspond to the multiplicative Holt–Winters method. This procedure is best applied to data that could be described by

$$x_{t+j} = (\mu_t + \beta j)S_{t+j} + \epsilon_{t+j}$$

where x_t is the series, μ_t is the time-varying mean at time t, β is a parameter, S_t is the seasonal component at time t, and ϵ_t is an idiosyncratic error. See *Methods and formulas* for the updating equations.

▷ Example 1

We have quarterly data on turkey sales by a new producer in the 1990s. The data have a strong seasonal component and an upward trend. We use the multiplicative Holt–Winters method to forecast sales for the year 2000. Because we have already `tsset` our data to the quarterly format, we do not need to specify the `period()` option.

```
. use http://www.stata-press.com/data/r12/turksales
. tssmooth shwinters shw1 = sales, forecast(4)
computing optimal weights

Iteration 0:    penalized RSS = -189.34609   (not concave)
Iteration 1:    penalized RSS = -108.68038   (not concave)
Iteration 2:    penalized RSS = -106.24548
Iteration 3:    penalized RSS =   -106.141
Iteration 4:    penalized RSS = -106.14093
Iteration 5:    penalized RSS = -106.14093

Optimal weights:
                        alpha = 0.1310
                         beta = 0.1428
                        gamma = 0.2999
penalized sum-of-squared residuals = 106.1409
        sum-of-squared residuals = 106.1409
        root mean squared error  = 1.628964
```

The graph below describes the fit and the forecast that was obtained.

```
. line sales shw1 t, title("Multiplicative Holt-Winters forecast")
> xtitle(Time) ytitle(Sales)
```

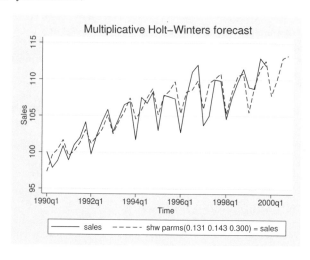

◁

Holt–Winters seasonal additive method

This method is similar to the previous one, but the seasonal effect is assumed to be additive rather than multiplicative. This method forecasts series that can be described by the equation

$$x_{t+j} = (\mu_t + \beta j) + S_{t+j} + \epsilon_{t+j}$$

See *Methods and formulas* for the updating equations.

▷ Example 2

In this example, we fit the data from the previous example to the additive model to forecast sales in the coming year. We use the snt_v() option to save the last year's seasonal terms in the new variable seas.

```
. tssmooth shwinters shwa = sales, forecast(4) snt_v(seas) normalize additive
computing optimal weights
Iteration 0:   penalized RSS = -190.90242  (not concave)
Iteration 1:   penalized RSS =  -108.8357
Iteration 2:   penalized RSS = -108.25359
Iteration 3:   penalized RSS = -107.68187
Iteration 4:   penalized RSS = -107.66444
Iteration 5:   penalized RSS = -107.66442
Iteration 6:   penalized RSS = -107.66442
Optimal weights:
                        alpha = 0.1219
                         beta = 0.1580
                        gamma = 0.3340
penalized sum-of-squared residuals = 107.6644
        sum-of-squared residuals = 107.6644
        root mean squared error = 1.640613
```

The output reveals that the multiplicative model has a better in-sample fit, and the graph below shows that the forecast from the multiplicative model is higher than that of the additive model.

```
. line shw1 shwa t if t>=tq(2000q1), title("Multiplicative and additive"
> "Holt-Winters forecasts") xtitle("Time") ytitle("Sales") legend(cols(1))
```

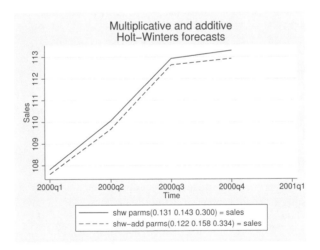

To check whether the estimated seasonal components are intuitively sound, we list the last year's seasonal components.

```
. list t seas if seas < .
```

	t	seas
37.	1999q1	-2.7533393
38.	1999q2	-.91752567
39.	1999q3	1.8082417
40.	1999q4	1.8626233

The output indicates that the signs of the estimated seasonal components agree with our intuition.

◁

Saved results

tssmooth shwinters saves the following in r():

Scalars

r(N)	number of observations	r(N_pre)	number of seasons used
r(alpha)	α smoothing parameter		in calculating starting values
r(beta)	β smoothing parameter	r(s2_0)	initial value for linear term
r(gamma)	γ smoothing parameter	r(s1_0)	initial value for constant term
r(prss)	penalized sum-of-squared errors	r(linear)	final value of linear term
r(rss)	sum-of-squared errors	r(constant)	final value of constant term
r(rmse)	root mean squared error	r(period)	period, if filter is seasonal

Macros

r(method)	shwinters, additive or	r(exp)	expression specified
	shwinters, multiplicative	r(timevar)	time variable specified in tsset
r(normalize)	normalize, if specified	r(panelvar)	panel variable specified in tsset

Methods and formulas

tssmooth shwinters is implemented as an ado-file.

A truncated description of the specified seasonal Holt–Winters filter labels the new variable. See [D] **label** for more information on labels.

An untruncated description of the specified seasonal Holt–Winters filter is saved in the characteristic named tssmooth for the new variable. See [P] **char** for more information on characteristics.

When the parms() option is not specified, the smoothing parameters are chosen to minimize the in-sample sum of penalized squared-forecast errors. Sometimes, one or more of the three optimal parameters lies on the boundary $[0, 1]$. To keep the estimates inside $[0, 1]$, tssmooth shwinters parameterizes the objective function in terms of their inverse logits, that is, $\exp(\alpha)/\{1 + \exp(\alpha)\}$, $\exp(\beta)/\{1 + \exp(\beta)\}$, and $\exp(\gamma)/\{1 + \exp(\gamma)\}$. When one of these parameters is actually on the boundary, this can complicate the optimization. For this reason, tssmooth shwinters optimizes a penalized sum-of-squared forecast errors. Let $\widehat{x}_t(\widetilde{\alpha}, \widetilde{\beta}, \widetilde{\gamma})$ be the forecast for the series x_t given the choices of $\widetilde{\alpha}$, $\widetilde{\beta}$, and $\widetilde{\gamma}$. Then the in-sample penalized sum-of-squared prediction errors is

$$P = \sum_{t=1}^{T} \left[\{x_t - \widehat{x}_t(\widetilde{\alpha}, \widetilde{\beta}, \widetilde{\gamma})\}^2 + I_{|f(\widetilde{\alpha})|>12)}(|f(\widetilde{\alpha})| - 12)^2 + I_{|f(\widetilde{\beta})|>12)}(|f(\widetilde{\beta})| - 12)^2 \right.$$

$$\left. + I_{|f(\widetilde{\gamma})|>12)}(|f(\widetilde{\gamma})| - 12)^2 \right]$$

where $f(x) = \ln\left(\frac{x}{1-x}\right)$. The penalty term is zero unless one of the parameters is close to the boundary. When one of the parameters is close to the boundary, the penalty term will help to obtain convergence.

Holt–Winters seasonal multiplicative procedure

As with the other recursive methods in tssmooth, there are three aspects to implementing the Holt–Winters seasonal multiplicative procedure: the forecasting equation, the initial values, and the updating equations. Unlike in the other methods, the data are now assumed to be seasonal with period L.

Given the estimates $a(t)$, $b(t)$, and $s(t + \tau - L)$, a τ step-ahead point forecast of x_t, denoted by $\widehat{y}_{t+\tau}$, is

$$\widehat{y}_{t+\tau} = \{a(t) + b(t)\tau\}\, s(t + \tau - L)$$

Given the smoothing parameters α, β, and γ, the updating equations are

$$a(t) = \alpha\frac{x_t}{s(t - L)} + (1 - \alpha)\{a(t - 1) + b(t - 1)\}$$

$$b(t) = \beta\{a(t) - a(t - 1)\} + (1 - \beta)\, b(t - 1)$$

and

$$s(t) = \gamma\left\{\frac{x_t}{a(t)}\right\} + (1 - \gamma)s(t - L)$$

To restrict the seasonal terms to sum to 1 over each year, specify the normalize option.

The updating equations require the $L + 2$ initial values $a(0)$, $b(0)$, $s(1 - L)$, $s(2 - L)$, ..., $s(0)$. Two methods calculate the initial values with the first m years, each of which contains L seasons. By default, m is set to the number of seasons in half the sample.

The initial value of the trend component, $b(0)$, can be estimated by

$$b(0) = \frac{\overline{x}_m - \overline{x}_1}{(m - 1)L}$$

where $\overline{x}_m$ is the average level of x_t in year m and $\overline{x}_1$ is the average level of x_t in the first year.

The initial value for the linear term, $a(0)$, is then calculated as

$$a(0) = \overline{x}_1 - \frac{L}{2}b(0)$$

To calculate the initial values for the seasons $1, 2, \ldots, L$, we first calculate the deviation-adjusted values,

$$S(t) = \frac{x_t}{\overline{x}_i - \left\{\frac{(L+1)}{2} - j\right\}b(0)}$$

where i is the year that corresponds to time t, j is the season that corresponds to time t, and $\overline{x}_i$ is the average level of x_t in year i.

Next, for each season $l = 1, 2, \ldots, L$, we define $\bar{s}_l$ as the average S_t over the years. That is,

$$\bar{s}_l = \frac{1}{m} \sum_{k=0}^{m-1} S_{l+kL} \qquad \text{for } l = 1, 2, \ldots, L$$

Then the initial seasonal estimates are

$$s_{0l} = \bar{s}_l \left(\frac{L}{\sum_{l=1}^{L} \bar{s}_l} \right) \qquad \text{for } l = 1, 2, \ldots, L$$

and these values are used to fill in $s(1 - L), \ldots, s(0)$.

If the `altstarts` option is specified, the starting values are computed based on a regression with seasonal indicator variables. Specifically, the series x_t is regressed on a time variable normalized to equal one in the first period in the sample and on a constant. Then $b(0)$ is set to the estimated coefficient on the time variable, and $a(0)$ is set to the estimated constant term. To calculate the seasonal starting values, x_t is regressed on a set of L seasonal dummy variables. The lth seasonal starting value is set to $(\frac{1}{\mu})\hat{\beta}_l$, where μ is the mean of x_t and $\hat{\beta}_l$ is the estimated coefficient on the lth seasonal dummy variable. The sample used in both regressions and the mean computation is restricted to include the first `samp0()` years. By default, `samp0()` includes half the data.

❑ Technical note

If there are missing values in the first few years, a small value of m can cause the starting value methods for seasonal term to fail. Here you should either specify a larger value of m by using `samp0()` or directly specify the seasonal starting values by using the `snt0_0()` option.

❑

Holt–Winters seasonal additive procedure

This procedure is similar to the previous one, except that the data are assumed to be described by

$$x_t = (\beta_0 + \beta_1 t) + s_t + \epsilon_t$$

As in the multiplicative case, there are three smoothing parameters, α, β, and γ, which can either be set or chosen to minimize the in-sample sum-of-squared forecast errors.

The updating equations are

$$a(t) = \alpha \left\{ x_t - s(t - L) \right\} + (1 - \alpha) \left\{ a(t - 1) + b(t - 1) \right\}$$

$$b(t) = \beta \left\{ a(t) - a(t - 1) \right\} + (1 - \beta) b(t - 1)$$

and

$$s(t) = \gamma \left\{ x_t - a(t) \right\} + (1 - \gamma) s(t - L)$$

To restrict the seasonal terms to sum to 0 over each year, specify the `normalize` option.

A τ-step-ahead forecast, denoted by $\hat{y}_{t+\tau}$, is given by

$$\hat{x}_{t+\tau} = a(t) + b(t)\tau + s(t + \tau - L)$$

As in the multiplicative case, there are two methods for setting the initial values.

The default method is to obtain the initial values for $a(0), b(0), s(1 - L), \ldots, s(0)$ from the regression

$$x_t = a(0) + b(0)t + \beta_{s,1-L}D_1 + \beta_{s,2-L}D_2 + \cdots + \beta_{s,0}D_L + e_t$$

where the $D_1, \ldots, D_L$ are dummy variables with

$$D_i = \left\{ \begin{array}{ll} 1 & \text{if } t \text{ corresponds to season } i \\ 0 & \text{otherwise} \end{array} \right\}$$

When `altstarts` is specified, an alternative method is used that regresses the x_t series on a time variable that has been normalized to equal one in the first period in the sample and on a constant term. $b(0)$ is set to the estimated coefficient on the time variable, and $a(0)$ is set to the estimated constant term. Then the demeaned series $\widetilde{x}_t = x_t - \mu$ is created, where μ is the mean of the x_t. The $\widetilde{x}_t$ are regressed on L seasonal dummy variables. The lth seasonal starting value is then set to β_l, where β_l is the estimated coefficient on the lth seasonal dummy variable. The sample in both the regression and the mean calculation is restricted to include the first `samp0` years, where, by default, `samp0()` includes half the data.

Acknowledgment

We thank Nicholas J. Cox of Durham University for his helpful comments.

References

Abraham, B., and J. Ledolter. 1983. *Statistical Methods for Forecasting.* New York: Wiley.

Bowerman, B. L., R. T. O'Connell, and A. B. Koehler. 2005. *Forecasting, Time Series, and Regression: An Applied Approach.* 4th ed. Pacific Grove, CA: Brooks/Cole.

Chatfield, C. 2001. *Time-Series Forecasting.* London: Chapman & Hall/CRC.

——. 2004. *The Analysis of Time Series: An Introduction.* 6th ed. Boca Raton, FL: Chapman & Hall/CRC.

Chatfield, C., and M. Yar. 1988. Holt-Winters forecasting: Some practical issues. *Statistician* 37: 129–140.

Holt, C. C. 2004. Forecasting seasonals and trends by exponentially weighted moving averages. *International Journal of Forecasting* 20: 5–10.

Montgomery, D. C., L. A. Johnson, and J. S. Gardiner. 1990. *Forecasting and Time Series Analysis.* 2nd ed. New York: McGraw–Hill.

Winters, P. R. 1960. Forecasting sales by exponentially weighted moving averages. *Management Science* 6: 324–342.

Also see

Title

> **ucm** — Unobserved-components model

Syntax

ucm *depvar* [*indepvars*] [*if*] [*in*] [, *options*]

options	Description
Model	
<u>mod</u>el(*model*)	specify trend and idiosyncratic components
<u>seasonal</u>(#)	include a seasonal component with a period of # time units
<u>cycle</u>(# [, <u>f</u>requency(#$_f$)])	include a cycle of order # and optionally set initial frequency to #$_f$, $0 < \#_f < \pi$; cycle() may be specified up to three times
<u>constraints</u>(*constraints*)	apply specified linear constraints
<u>coll</u>inear	keep collinear variables
SE/Robust	
vce(*vcetype*)	*vcetype* may be oim or <u>r</u>obust
Reporting	
<u>level</u>(#)	set confidence level; default is level(95)
<u>nocnsreport</u>	do not display constraints
display_options	control column formats, row spacing, and display of omitted variables and base and empty cells
Maximization	
maximize_options	control the maximization process
<u>coeflegend</u>	display legend instead of statistics

model	Description
rwalk	random-walk model; the default
none	no trend or idiosyncratic component
ntrend	no trend component but include idiosyncratic component
dconstant	deterministic constant with idiosyncratic component
llevel	local-level model
dtrend	deterministic-trend model with idiosyncratic component
lldtrend	local-level model with deterministic trend
rwdrift	random-walk-with-drift model
lltrend	local-linear-trend model
strend	smooth-trend model
rtrend	random-trend model

You must tsset your data before using ucm; see [TS] **tsset**.

indepvars may contain factor variables; see [U] **11.4.3 Factor variables**.

indepvars and *depvar* may contain time-series operators; see [U] **11.4.4 Time-series varlists**.

by, statsby, and rolling are allowed; see [U] **11.1.10 Prefix commands**.

coeflegend does not appear in the dialog box.

See [U] **20 Estimation and postestimation commands** for more capabilities of estimation commands.

Menu

Statistics > Time series > Unobserved-components model

Description

Unobserved-components models (UCMs) decompose a time series into trend, seasonal, cyclical, and idiosyncratic components and allow for exogenous variables. ucm estimates the parameters of UCMs by maximum likelihood.

All the components are optional. The trend component may be first-order deterministic or it may be first-order or second-order stochastic. The seasonal component is stochastic; the seasonal effects at each time period sum to a zero-mean finite-variance random variable. The cyclical component is modeled by the stochastic-cycle model derived by Harvey (1989).

Options

_____ Model _____

model(*model*) specifies the trend and idiosyncratic components. The default is model(rwalk). The available *models* are listed in *Syntax* and discussed in detail in *Models for the trend and idiosyncratic components* under *Remarks* below.

seasonal(#) adds a stochastic-seasonal component to the model. # is the period of the season, that is, the number of time-series observations required for the period to complete.

cycle(#) adds a stochastic-cycle component of order # to the model. The order # must be 1, 2, or 3. Multiple cycles are added by repeating the cycle(#) option with up to three cycles allowed.

cycle(#, frequency(#$_f$)) specifies #$_f$ as the initial value for the central-frequency parameter in the stochastic-cycle component of order #. #$_f$ must be in the interval $(0, \pi)$.

constraints(*constraints*), collinear; see [R] **estimation options**.

_____ SE/Robust _____

vce(*vcetype*) specifies the estimator for the variance–covariance matrix of the estimator.

vce(oim), the default, causes ucm to use the observed information matrix estimator.

vce(robust) causes ucm to use the Huber/White/sandwich estimator.

_____ Reporting _____

level(#), nocnsreport; see [R] **estimation options**.

display_options: noomitted, vsquish, noemptycells, baselevels, allbaselevels, cformat(%*fmt*), pformat(%*fmt*), and sformat(%*fmt*); see [R] **estimation options**.

⌐ Maximization ⌐

maximize_options: <u>difficult</u>, <u>tech</u>nique(*algorithm_spec*), <u>iter</u>ate(*#*), [<u>no</u>]<u>log</u>, <u>trace</u>,
<u>gradient</u>, showstep, <u>hessian</u>, <u>showtol</u>erance, <u>tol</u>erance(*#*), <u>ltol</u>erance(*#*),
<u>nrtol</u>erance(*#*), and from(*matname*); see [R] **maximize** for all options except from(), and
see below for information on from().

from(*matname*) specifies initial values for the maximization process. from(b0) causes ucm to
begin the maximization algorithm with the values in b0. b0 must be a row vector; the number
of columns must equal the number of parameters in the model; and the values in b0 must be
in the same order as the parameters in e(b).

If you model fails to converge, try using the difficult option. Also see the technical note below
example 5.

The following option is available with ucm but is not shown in the dialog box:

coeflegend; see [R] **estimation options**.

Remarks

Remarks are presented under the following headings:

An introduction to UCMs
A random-walk-model example
Frequency-domain concepts used in the stochastic-cycle model
Another random-walk-model example
Comparing UCM and ARIMA
A local-level-model example
Comparing UCM and ARIMA, revisited
Models for the trend and idiosyncratic components
Seasonal component

An introduction to UCMs

UCMs decompose a time series into trend, seasonal, cyclical, and idiosyncratic components and
allow for exogenous variables. Formally, UCMs can be written as

$$y_t = \tau_t + \gamma_t + \psi_t + \beta \mathbf{x}_t + \epsilon_t \tag{1}$$

where y_t is the dependent variable, τ_t is the trend component, γ_t is the seasonal component, ψ_t is
the cyclical component, β is a vector of fixed parameters, $\mathbf{x}_t$ is a vector of exogenous variables, and
ϵ_t is the idiosyncratic component.

By placing restrictions on τ_t and ϵ_t, Harvey (1989) derived a series of models for the trend and the
idiosyncratic components. These models are briefly described in *Syntax* and are further discussed in
Models for the trend and idiosyncratic components. To these models, Harvey (1989) added models for
the seasonal and cyclical components, and he also allowed for the presence of exogenous variables.

It is rare that a UCM contains all the allowed components. For instance, the seasonal component
is rarely needed when modeling deseasonalized data.

Harvey (1989) and Durbin and Koopman (2001) show that UCMs can be written as state-space
models that allow the parameters of a UCM to be estimated by maximum likelihood. In fact, ucm
uses sspace (see [TS] **sspace**) to perform the estimation calculations; see *Methods and formulas* for
details.

After estimating the parameters, `predict` can produce in-sample predictions or out-of-sample forecasts; see [TS] **ucm postestimation**. After estimating the parameters of a UCM that contains a cyclical component, `estat period` converts the estimated central frequency to an estimated central period and `psdensity` estimates the spectral density implied by the model; see [TS] **ucm postestimation** and the examples below.

We illustrate the basic approach of analyzing data with UCMs, and then we discuss the details of the different trend models in *Models for the trend and idiosyncratic components*.

Although the methods implemented in `ucm` have been widely applied by economists, they are general time-series techniques and may be of interest to researchers from other disciplines. In example 8, we analyze monthly data on the reported cases of mumps in New York City.

A random-walk-model example

▷ Example 1

We begin by plotting monthly data on the U.S. civilian unemployment rate.

```
. use http://www.stata-press.com/data/r12/unrate
. tsline unrate, name(unrate)
```

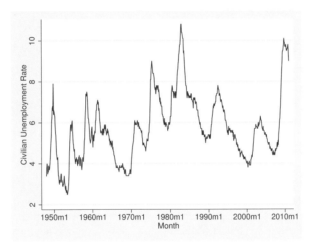

This series looks like it might be well approximated by a random-walk model. Formally, a random-walk model is given by

$$y_t = \mu_t$$
$$\mu_t = \mu_{t-1} + \eta_t$$

The random-walk is so frequently applied, at least as a starting model, that it is the default model for `ucm`. In the output below, we fit the random-walk model to the unemployment data.

```
. ucm unrate
searching for initial values ..........
(setting technique to bhhh)
Iteration 0:   log likelihood =  84.272992
Iteration 1:   log likelihood =  84.394942
Iteration 2:   log likelihood =  84.400923
Iteration 3:   log likelihood =  84.401282
Iteration 4:   log likelihood =  84.401305
(switching technique to nr)
Iteration 5:   log likelihood =  84.401306
Refining estimates:
Iteration 0:   log likelihood =  84.401306
Iteration 1:   log likelihood =  84.401307

Unobserved-components model
Components: random walk

Sample: 1948m1 - 2011m1                      Number of obs   =        757
Log likelihood =  84.401307
```

unrate	Coef.	OIM Std. Err.	z	P>|z|	[95% Conf. Interval]	
Variance level	.0467196	.002403	19.44	0.000	.0420098	.0514294

Note: Model is not stationary.
Note: Tests of variances against zero are one sided, and the two-sided
      confidence intervals are truncated at zero.

The output indicates that the model is nonstationary, as all random-walk models are.

We consider a richer model in the next example.

◁

▷ Example 2

We suspect that there should be a stationary cyclical component that produces serially correlated shocks around the random-walk trend. Harvey (1989) derived a stochastic-cycle model for these stationary cyclical components.

The stochastic-cycle model has three parameters: the frequency at which the random components are centered, a damping factor that parameterizes the dispersion of the random components around the central frequency, and the variance of the stochastic-cycle process that acts as a scale factor.

Fitting this model to unemployment data yields

```
. ucm unrate, cycle(1)
searching for initial values ...................
(setting technique to bhhh)
Iteration 0:   log likelihood =  84.273579
Iteration 1:   log likelihood =  87.852115
Iteration 2:   log likelihood =  88.253422
Iteration 3:   log likelihood =  89.191311
Iteration 4:   log likelihood =  94.675898
(switching technique to nr)
Iteration 5:   log likelihood =  98.394691  (not concave)
Iteration 6:   log likelihood =  98.983092
Iteration 7:   log likelihood =  99.983642
Iteration 8:   log likelihood =   104.8308
Iteration 9:   log likelihood =  114.27184
Iteration 10:  log likelihood =  116.47396
Iteration 11:  log likelihood =  118.45803
Iteration 12:  log likelihood =  118.88055
Iteration 13:  log likelihood =  118.88421
Iteration 14:  log likelihood =  118.88421
Refining estimates:
Iteration 0:   log likelihood =  118.88421
Iteration 1:   log likelihood =  118.88421

Unobserved-components model
Components: random walk, order 1 cycle

Sample: 1948m1 - 2011m1                    Number of obs   =        757
                                           Wald chi2(2)    =   26650.81
Log likelihood =  118.88421                Prob > chi2     =     0.0000
```

unrate	Coef.	OIM Std. Err.	z	P>|z|	[95% Conf. Interval]	
frequency	.0933466	.0103609	9.01	0.000	.0730397	.1136535
damping	.9820003	.0061121	160.66	0.000	.9700207	.9939798
Variance						
level	.0143786	.0051392	2.80	0.003	.004306	.0244511
cycle1	.0270339	.0054343	4.97	0.000	.0163829	.0376848

```
Note: Model is not stationary.
Note: Tests of variances against zero are one sided, and the two-sided
      confidence intervals are truncated at zero.
```

The estimated central frequency for the cyclical component is small, implying that the cyclical component is centered around low-frequency components. The high-damping factor indicates that all the components from this cyclical component are close to the estimated central frequency. The estimated variance of the stochastic-cycle process is small but significant.

We use estat period to convert the estimate of the central frequency to an estimated central period.

```
. estat period
```

cycle1	Coef.	Std. Err.	[95% Conf. Interval]	
period	67.31029	7.471004	52.66739	81.95319
frequency	.0933466	.0103609	.0730397	.1136535
damping	.9820003	.0061121	.9700207	.9939798

```
Note: Cycle time unit is monthly.
```

Because we have monthly data, the estimated central period of 67.31 implies that the cyclical component is composed of random components that occur around a central periodicity of around 5.61 years. This estimate falls within the conventional Burns and Mitchell (1946) definition of business-cycle shocks occurring between 1.5 and 8 years.

We can convert the estimated parameters of the cyclical component to an estimated spectral density of the cyclical component, as described by Harvey (1989). The spectral density of the cyclical component describes the relative importance of the random components at different frequencies; see *Frequency-domain concepts used in the stochastic-cycle model* for details. We use psdensity (see [TS] **psdensity**) to obtain the spectral density of the cyclical component implied by the estimated parameters, and we use twoway line (see [G-2] **graph twoway line**) to plot the estimated spectral density.

```
. psdensity sdensity omega
. line sdensity omega
```

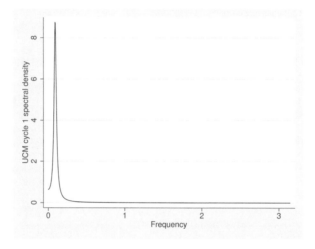

The estimated spectral density shows that the cyclical component is composed of random components that are tightly distributed around the low-frequency peak.

◁

Frequency-domain concepts used in the stochastic-cycle model

The parameters of the stochastic-cycle model are easiest to interpret in the frequency domain. We now provide a review of the useful concepts from the frequency domain. Crucial to understanding the stochastic-cycle model is the frequency-domain concept that a stationary process can be decomposed into random components that occur at the frequencies in the interval $[0, \pi]$.

We need some concepts from the frequency-domain approach to interpret the parameters in the stochastic-cycle model of the cyclical component. Here we provide a simple, intuitive explanation. More technical presentations can be found in Priestley (1981), Harvey (1989, 1993), Hamilton (1994), Fuller (1996), and Wei (2006).

As with much time-series analysis, the basic results are for covariance-stationary processes with additional results handling some nonstationary cases. We present some useful results for covariance-stationary processes. These results provide what we need to interpret the stochastic-cycle model for the stationary cyclical component.

The autocovariances γ_j, $j \in \{0, 1, \ldots, \infty\}$, of a covariance-stationary process y_t specify its variance and dependence structure. In the frequency-domain approach to time-series analysis, the spectral density describes the importance of the random components that occur at frequency ω relative to the components that occur at other frequencies.

The frequency-domain approach focuses on the relative contributions of random components that occur at the frequencies $[0, \pi]$.

The spectral density can be written as a weighted average of the autocorrelations of y_t. Like autocorrelations, the spectral density is normalized by γ_0, the variance of y_t. Multiplying the spectral density by γ_0 yields the power-spectrum of y_t.

In an independent and identically distributed (i.i.d.) process, the components at all frequencies are equally important, so the spectral density is a flat line.

In common parlance, we speak of high-frequency noise making a series look more jagged and of low-frequency components causing smoother plots. More formally, we say that a process composed primarily of high-frequency components will have fewer runs above or below the mean than an i.i.d. process and that a process composed primarily of low-frequency components will have more runs above or below the mean than an i.i.d. process.

To further formalize these ideas, consider the first-order autoregressive (AR(1)) process given by

$$y_t = \phi y_{t-1} + \epsilon_t$$

where ϵ_t is a zero-mean, covariance-stationary process with finite variance σ^2, and $|\phi| < 1$ so that y_t is covariance stationary. The first-order autocorrelation of this AR(1) process is ϕ.

Below are plots of simulated data when ϕ is set to 0, -0.8, and 0.8. When $\phi = 0$, the data are i.i.d. When $\phi = -0.8$, the value today is strongly negatively correlated with the value yesterday, so this case should be a prototypical high-frequency noise example. When $\phi = 0.8$, the value today is strongly positively correlated with the value yesterday, so this case should be a prototypical low-frequency shock example.

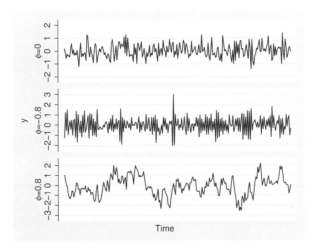

The plots above confirm our conjectures. The plot when $\phi = -0.8$ contains fewer runs above or below the mean, and it is more jagged than the i.i.d. plot. The plot when $\phi = 0.8$ contains more runs above or below the mean, and it is smoother than the i.i.d. plot.

Below we plot the spectral densities for the AR(1) model with $\phi = 0$, $\phi = -0.8$, and $\phi = 0.8$.

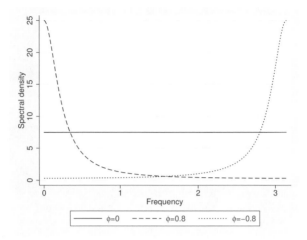

The high-frequency components are much more important to the AR(1) process with $\phi = -0.8$ than to the i.i.d. process with $\phi = 0$. The low-frequency components are much more important to the AR(1) process with $\phi = 0.8$ than to the i.i.d. process.

❑ Technical note

Autoregressive moving-average (ARMA) models parameterize the autocorrelation in a time series by allowing today's value to be a weighted average of past values and a weighted average of past i.i.d. shocks; see Hamilton (1994), Wei (2006), and [TS] **arima** for introductions and a Stata implementation. The intuitive ARMA parameterization has many nice features, including that one can easily rewrite the ARMA model as a weighted average of past i.i.d. shocks to trace how a shock feeds through the system.

Although it is easy to obtain the spectral density of an ARMA process, the parameters themselves provide limited information about the underlying spectral density.

In contrast, the parameters of the stochastic-cycle parameterization of autocorrelation in a time series directly provide information about the underlying spectral density. The parameter ω_0 is the central frequency around which the random components are clustered. If ω_0 is small, then the model is centered around low-frequency components. If ω_0 is close to π, then the model is centered around high-frequency components. The parameter ρ is the damping factor that indicates how tightly clustered the random components are around the central frequency ω_0. If ρ is close to 0, there is no clustering of the random components. If ρ is close to 1, the random components are tightly distributed around the central frequency ω_0.

In the graph below, we draw the spectral densities implied by stochastic-cycle models with four sets of parameters: $\omega_0 = \pi/4, \rho = 0.8$; $\omega_0 = \pi/4, \rho = 0.9$; $\omega_0 = 4\pi/5, \rho = 0.8$; and $\omega_0 = 4\pi/5, \rho = 0.9$. The graph below illustrates that ω_0 is the central frequency around which the other important random components are distributed. It also illustrates that the damping parameter ρ controls the dispersion of the important components around the central frequency.

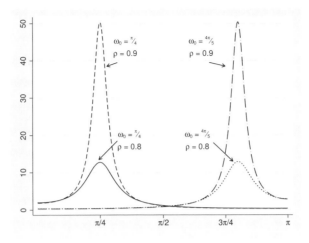

Another random-walk-model example

▷ Example 3

Now let's reconsider example 2. Although we might be happy with how our model has identified a stationary cyclical component that we could interpret in business-cycle terms, we suspect that there should also be a high-frequency cyclical component. It is difficult to estimate the parameters of a UCM with two or more stochastic-cycle models. Providing starting values for the central frequencies can be a crucial help to the optimization procedure. Below we estimate a UCM with two cyclical components. We use the frequency() suboption to provide starting values for the central frequencies; we specified the values below because we suspect one model will pick up the low-frequency components and the other will pick up the high-frequency components. We specified the low-frequency model to be order 2 to make it less peaked for any given damping factor. (Trimbur [2006] provides a nice introduction and some formal results for higher-order stochastic-cycle models.)

```
. ucm unrate, cycle(1, frequency(2.9)) cycle(2, frequency(.09))
searching for initial values ...................
(setting technique to bhhh)
Iteration 0:   log likelihood =   115.98563
Iteration 1:   log likelihood =   125.04043
Iteration 2:   log likelihood =   127.69387
Iteration 3:   log likelihood =   134.50864
Iteration 4:   log likelihood =   136.91353
(switching technique to nr)
Iteration 5:   log likelihood =    138.5091
Iteration 6:   log likelihood =   146.09273
Iteration 7:   log likelihood =   146.28132
Iteration 8:   log likelihood =   146.28326
Iteration 9:   log likelihood =   146.28326
Refining estimates:
Iteration 0:   log likelihood =   146.28326
Iteration 1:   log likelihood =   146.28326
```

```
Unobserved-components model
Components: random walk, 2 cycles of order 1 2
Sample: 1948m1 - 2011m1                     Number of obs    =        757
                                            Wald chi2(4)     =    7681.33
Log likelihood =   146.28326                Prob > chi2      =     0.0000
```

unrate	Coef.	OIM Std. Err.	z	P>|z|	[95% Conf. Interval]	
cycle1						
frequency	2.882382	.0668017	43.15	0.000	2.751453	3.013311
damping	.7004295	.1251571	5.60	0.000	.4551262	.9457328
cycle2						
frequency	.0667929	.0206849	3.23	0.001	.0262513	.1073345
damping	.9074708	.0142273	63.78	0.000	.8795858	.9353559
Variance						
level	.0207704	.0039669	5.24	0.000	.0129953	.0285454
cycle1	.0027886	.0014363	1.94	0.026	0	.0056037
cycle2	.002714	.001028	2.64	0.004	.0006991	.0047289

Note: Model is not stationary.
Note: Tests of variances against zero are one sided, and the two-sided
      confidence intervals are truncated at zero.

The output provides some support for the existence of a second, high-frequency cycle. The high-frequency components are centered around 2.88, whereas the low-frequency components are centered around 0.067. That the estimated damping factor is 0.70 for the high-frequency cycle whereas the estimated damping factor for the low-frequency cycle is 0.91 indicates that the high-frequency components are more diffusely distributed around 2.88 than the low-frequency components are around 0.067.

We obtain and plot the estimated spectral densities to get another look at these results.

```
. psdensity sdensity2a omega2a
. psdensity sdensity2b omega2b, cycle(2)
. line sdensity2a sdensity2b omega2a, legend(col(1))
```

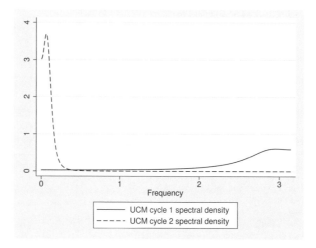

The estimated spectral densities indicate that we have found two distinct cyclical components.

It does not matter whether we specify omega2a or omega2b to be the x-axis variable, because they are equal to each other.

◁

❑ Technical note

That the estimated spectral densities in the previous example do not overlap is important for parameter identification. Although the parameters are identified in large-sample theory, we have found it difficult to estimate the parameters of two cyclical components when the spectral densities overlap. When the spectral densities of two cyclical components overlap, the parameters may not be well identified and the optimization procedure may not converge.

❑

Comparing UCM and ARIMA

▷ Example 4

This example provides some insight for readers familiar with autoregressive integrated moving-average (ARIMA) models but not with UCMs. If you are not familiar with ARIMA models, you may wish to skip this example. See [TS] **arima** for an introduction to ARIMA models in Stata.

UCMs provide an alternative to ARIMA models implemented in [TS] **arima**. Neither set of models is nested within the other, but there are some cases in which instructive comparisons can be made.

The random-walk model corresponds to an ARIMA model that is first-order integrated and has an i.i.d. error term. In other words, the random-walk UCM and the ARIMA(0,1,0) are asymptotically equivalent. Thus

```
ucm unrate
```

and

```
arima unrate, arima(0,1,0) noconstant
```

produce asymptotically equivalent results.

The stochastic-cycle model for the stationary cyclical component is an alternative functional form for stationary processes to stationary autoregressive moving-average (ARMA) models. Which model is preferred depends on the application and which parameters a researchers wants to interpret. Both the functional forms and the parameter interpretations differ between the stochastic-cycle model and the ARMA model. See Trimbur (2006, eq. 25) for some formal comparisons of the two models.

That both models can be used to estimate the stationary cyclical components for the random-walk model implies that we can compare the results in this case by comparing their estimated spectral densities. Below we estimate the parameters of an ARIMA(2,1,1) model and plot the estimated spectral density of the stationary component.

```
. arima unrate, noconstant arima(2,1,1)

(setting optimization to BHHH)
Iteration 0:    log likelihood =    129.8801
Iteration 1:    log likelihood =   134.61953
Iteration 2:    log likelihood =   137.04909
Iteration 3:    log likelihood =   137.71386
Iteration 4:    log likelihood =   138.25255
(switching optimization to BFGS)
Iteration 5:    log likelihood =   138.51924
Iteration 6:    log likelihood =   138.81638
Iteration 7:    log likelihood =   138.83615
Iteration 8:    log likelihood =    138.8364
Iteration 9:    log likelihood =   138.83642
Iteration 10:   log likelihood =   138.83642

ARIMA regression

Sample:  1948m2 - 2011m1                       Number of obs    =        756
                                               Wald chi2(3)     =     683.34
Log likelihood =  138.8364                     Prob > chi2      =     0.0000
```

D.unrate	Coef.	OPG Std. Err.	z	P>|z|	[95% Conf. Interval]	
ARMA						
ar						
L1.	.5398016	.0586304	9.21	0.000	.4248882	.6547151
L2.	.2468148	.0359396	6.87	0.000	.1763744	.3172551
ma						
L1.	-.5146506	.0632838	-8.13	0.000	-.6386845	-.3906167
/sigma	.2013332	.0032644	61.68	0.000	.1949351	.2077313

```
Note: The test of the variance against zero is one sided, and the two-sided
      confidence interval is truncated at zero.

. psdensity sdensity_arma omega_arma

. line sdensity_arma omega_arma
```

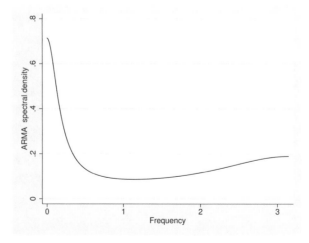

The estimated spectral density from the ARIMA(2,1,1) has a similar shape to the plot obtained by combining the two spectral densities estimated from the stochastic-cycle model in example 3. For this particular application, the estimated central frequencies of the two cyclical components from the stochastic-cycle model provide information about the business-cycle component and the high-frequency component that is not easily obtained from the ARIMA(2,1,1) model. On the other hand, it is easier to work out the impulse–response function for the ARMA model than for the stochastic-cycle model, implying that the ARMA model is easier to use when tracing the effect of a shock feeding through the system.

◁

A local-level-model example

We now consider the weekly series of initial claims for unemployment insurance in the United States, which is plotted below.

▷ Example 5

```
. use http://www.stata-press.com/data/r12/icsa1, clear
. tsline icsa
```

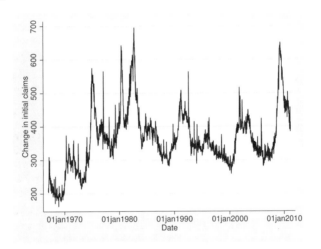

This series looks like it was generated by a random walk with extra noise, so we want to use a random-walk model that includes an additional random term. This structure causes the model to be occasionally known as the random-walk-plus-noise model, but it is more commonly known as the local-level model in the UCM literature.

The local-level model models the trend as a random walk and models the idiosyncratic components as independent and identically distributed components. Formally, the local-level model specifies the observed time-series y_t, for $t = 1, \ldots, T$, as

$$y_t = \mu_t + \epsilon_t$$
$$\mu_t = \mu_{t-1} + \eta_t$$

where $\epsilon_t \sim$ i.i.d. $N(0, \sigma_\epsilon^2)$ and $\eta_t \sim$ i.i.d. $N(0, \sigma_\eta^2)$ and are mutually independent.

We fit the local-level model in the output below:

```
. ucm icsa, model(llevel)
searching for initial values .........
(setting technique to bhhh)
Iteration 0:    log likelihood = -9982.7798
Iteration 1:    log likelihood = -9913.2745
Iteration 2:    log likelihood = -9894.9925
Iteration 3:    log likelihood = -9893.7191
Iteration 4:    log likelihood = -9893.2876
(switching technique to nr)
Iteration 5:    log likelihood = -9893.2614
Iteration 6:    log likelihood = -9893.2469
Iteration 7:    log likelihood = -9893.2469
Refining estimates:
Iteration 0:    log likelihood = -9893.2469
Iteration 1:    log likelihood = -9893.2469

Unobserved-components model
Components: local level

Sample: 07jan1967 - 19feb2011                  Number of obs    =        2303
Log likelihood = -9893.2469
```

icsa	Coef.	OIM Std. Err.	z	P>|z|	[95% Conf. Interval]	
Variance						
level	116.558	8.806587	13.24	0.000	99.29745	133.8186
icsa	124.2715	7.615506	16.32	0.000	109.3454	139.1976

Note: Model is not stationary.
Note: Tests of variances against zero are one sided, and the two-sided
      confidence intervals are truncated at zero.
Note: Time units are in 7 days.

The output indicates that both components are statistically significant.

◁

❑ Technical note

The estimation procedure will not always converge when estimating the parameters of the local-level model. If the series does not vary enough in the random level, modeled by the random walk, and in the stationary shocks around the random level, the estimation procedure will not converge because it will be unable to set the variance of one of the two components to 0.

Take another look at the graphs of unrate and icsa. The extra noise around the random level that can be seen in the graph of icsa allows us to estimate both variances.

A closely related point is that it is difficult to estimate the parameters of a local-level model with a stochastic-cycle component because the series must have enough variation to identify the variance of the random-walk component, the variance of the idiosyncratic term, and the parameters of the stochastic-cycle component. In some cases, series that look like candidates for the local-level model are best modeled as random-walk models with stochastic-cycle components.

In fact, convergence can be a problem for most of the models in ucm. Convergence problems occur most often when there is insufficient variation to estimate the variances of the components in the model. When there is insufficient variation to estimate the variances of the components in the model, the optimization routine will fail to converge as it attempts to set the variance equal to 0. This usually shows up in the iteration log when the log likelihood gets stuck at a particular value and the message (not concave) or (backed up) is displayed repeatedly. When this happens, use the iteration() option to limit the number of iterations, look to see which of the variances is being driven to 0, and drop that component from the model. (This technique is a method to obtain convergence to interpretable estimates, not a model-selection method.)

❏

▷ Example 6

We might suspect that there is some serial correlation in the idiosyncratic shock. Alternatively, we could include a cyclical component to model the stationary time-dependence in the series. In the example below, we add a stochastic-cycle model for the stationary cyclical process, but we drop the idiosyncratic term and use a random-walk model instead of the local-level model. We change the model because it is difficult to estimate the variance of the idiosyncratic term along with the parameters of a stationary cyclical component.

```
. ucm icsa, model(rwalk) cycle(1)
searching for initial values ...................
(setting technique to bhhh)
Iteration 0:   log likelihood = -10008.167
Iteration 1:   log likelihood = -10007.272
Iteration 2:   log likelihood = -10007.206  (backed up)
Iteration 3:   log likelihood =  -10007.17  (backed up)
Iteration 4:   log likelihood = -10007.148  (backed up)
(switching technique to nr)
Iteration 5:   log likelihood = -10007.137  (not concave)
Iteration 6:   log likelihood = -9885.1932  (not concave)
Iteration 7:   log likelihood = -9884.1636
Iteration 8:   log likelihood = -9881.6478
Iteration 9:   log likelihood = -9881.4496
Iteration 10:  log likelihood = -9881.4441
Iteration 11:  log likelihood = -9881.4441
Refining estimates:
Iteration 0:   log likelihood = -9881.4441
Iteration 1:   log likelihood = -9881.4441

Unobserved-components model
Components: random walk, order 1 cycle
Sample: 07jan1967 - 19feb2011                  Number of obs   =       2303
                                               Wald chi2(2)    =      23.04
Log likelihood = -9881.4441                    Prob > chi2     =     0.0000
```

| | | OIM | | | | |
icsa	Coef.	Std. Err.	z	P>|z|	[95% Conf.	Interval]
frequency	1.469633	.3855657	3.81	0.000	.7139385	2.225328
damping	.1644576	.0349537	4.71	0.000	.0959495	.2329656
Variance						
level	97.90982	8.320047	11.77	0.000	81.60282	114.2168
cycle1	149.7323	9.980798	15.00	0.000	130.1703	169.2943

Note: Model is not stationary.
Note: Tests of variances against zero are one sided, and the two-sided
      confidence intervals are truncated at zero.
Note: Time units are in 7 days.

Although the output indicates that the model fits well, the small estimate of the damping parameter indicates that the random components will be widely distributed around the central frequency. To get a better idea of the dispersion of the components, we look at the estimated spectral density of the stationary cyclical component.

```
. psdensity sdensity3 omega3
. line sdensity3 omega3
```

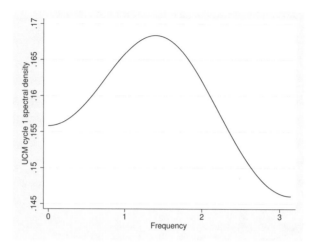

The graph shows that the random components that make up the cyclical component are diffusely distributed around a central frequency.

◁

Comparing UCM and ARIMA, revisited

▷ Example 7

Including lags of the dependent variable is an alternative method for modeling serially correlated errors. The estimated coefficients on the lags of the dependent variable estimate the coefficients in an autoregressive model for the stationary cyclical component; see Harvey (1989, 47–48) for a discussion. Including lags of the dependent variable should be viewed as an alternative to the stochastic-cycle model for the stationary cyclical component. In this example, we use the large-sample equivalence of the random-walk model with pth order autoregressive errors and an ARIMA$(p, 1, 0)$ to illustrate this point.

In the output below, we include 2 lags of the dependent variable in the random-walk UCM.

```
. ucm icsa L(1/2).icsa, model(rwalk)
searching for initial values ..........
(setting technique to bhhh)
Iteration 0:    log likelihood = -10044.209
Iteration 1:    log likelihood = -9975.8312
Iteration 2:    log likelihood = -9953.5727
Iteration 3:    log likelihood = -9936.7489
Iteration 4:    log likelihood = -9927.2306
(switching technique to nr)
Iteration 5:    log likelihood = -9918.9538
Iteration 6:    log likelihood = -9890.8306
Iteration 7:    log likelihood =  -9889.562
Iteration 8:    log likelihood = -9889.5608
Iteration 9:    log likelihood = -9889.5608
Refining estimates:
Iteration 0:    log likelihood = -9889.5608
Iteration 1:    log likelihood = -9889.5608

Unobserved-components model
Components: random walk

Sample: 21jan1967 - 19feb2011              Number of obs    =        2301
                                          Wald chi2(2)     =      271.88
Log likelihood = -9889.5608               Prob > chi2      =      0.0000
```

	Coef.	OIM Std. Err.	z	P>|z|	[95% Conf. Interval]	
icsa						
icsa						
L1.	-.3250633	.0205148	-15.85	0.000	-.3652715	-.2848551
L2.	-.1794686	.0205246	-8.74	0.000	-.2196961	-.1392411
Variance						
level	317.6474	9.36691	33.91	0.000	299.2886	336.0062

```
Note: Model is not stationary.
Note: Tests of variances against zero are one sided, and the two-sided
    confidence intervals are truncated at zero.
Note: Time units are in 7 days.
```

Now we use `arima` to estimate the parameters of an asymptotically equivalent ARIMA(2,1,0) model. (We specify the `technique(nr)` option so that `arima` will compute the observed information matrix standard errors that ucm computes.) We use `nlcom` to compute a point estimate and a standard error for the variance, which is directly comparable to the one produced by ucm.

```
. arima icsa, noconstant arima(2,1,0) technique(nr)

Iteration 0:    log likelihood = -9896.4584
Iteration 1:    log likelihood =  -9896.458

ARIMA regression

Sample:  14jan1967 - 19feb2011                  Number of obs    =      2302
                                                Wald chi2(2)     =    271.95
Log likelihood = -9896.458                      Prob > chi2      =    0.0000
```

D.icsa	Coef.	OIM Std. Err.	z	P>|z|	[95% Conf. Interval]	
ARMA						
ar						
L1.	-.3249383	.0205036	-15.85	0.000	-.3651246	-.284752
L2.	-.1793353	.0205088	-8.74	0.000	-.2195317	-.1391388
/sigma	17.81606	.2625695	67.85	0.000	17.30143	18.33068

```
Note: The test of the variance against zero is one sided, and the two-sided
      confidence interval is truncated at zero.

. nlcom _b[sigma:_cons]^2

     _nl_1:  _b[sigma:_cons]^2
```

D.icsa	Coef.	Std. Err.	z	P>|z|	[95% Conf. Interval]	
_nl_1	317.4119	9.355904	33.93	0.000	299.0746	335.7491

It is no accident that the parameter estimates and the standard errors from the two estimators are so close. As the sample size grows the differences in the parameter estimates and the estimated standard errors will go to 0, because the two estimators are equivalent in large samples.

◁

Models for the trend and idiosyncratic components

A general model that allows for fixed or stochastic trends in τ_t is given by

$$\tau_t = \tau_{t-1} + \beta_{t-1} + \eta_t \tag{2}$$
$$\beta_t = \beta_{t-1} + \xi_t \tag{3}$$

Following Harvey (1989), we define 11 flexible models for y_t that specify both τ_t and ϵ_t in (1). These models place restrictions on the general model specified in (2) and (3) and on ϵ_t in (1). In other words, these models jointly specify τ_t and ϵ_t.

To any of these models, a cyclical component, a seasonal component, or exogenous variables may be added.

Table 1. Models for the trend and idiosyncratic components

Model name	Syntax option	Model
No trend or idiosyncratic component	model(none)	
No trend	model(ntrend)	$y_t = \epsilon_t$
Deterministic constant	model(dconstant)	$y_t = \mu + \epsilon_t$ $\mu = \mu$
Local level	model(llevel)	$y_t = \mu_t + \epsilon_t$ $\mu_t = \mu_{t-1} + \eta_t$
Random walk	model(rwalk)	$y_t = \mu_t$ $\mu_t = \mu_{t-1} + \eta_t$
Deterministic trend	model(dtrend)	$y_t = \mu_t + \epsilon_t$ $\mu_t = \mu_{t-1} + \beta$ $\beta = \beta$
Local level with deterministic trend	model(lldtrend)	$y_t = \mu_t + \epsilon_t$ $\mu_t = \mu_{t-1} + \beta + \eta_t$ $\beta = \beta$
Random walk with drift	model(rwdrift)	$y_t = \mu_t$ $\mu_t = \mu_{t-1} + \beta + \eta_t$ $\beta = \beta$
Local linear trend	model(lltrend)	$y_t = \mu_t + \epsilon_t$ $\mu_t = \mu_{t-1} + \beta_{t-1} + \eta_t$ $\beta_t = \beta_{t-1} + \xi_t$
Smooth trend	model(strend)	$y_t = \mu_t + \epsilon_t$ $\mu_t = \mu_{t-1} + \beta_{t-1}$ $\beta_t = \beta_{t-1} + \xi_t$
Random trend	model(rtrend)	$y_t = \mu_t$ $\mu_t = \mu_{t-1} + \beta_{t-1}$ $\beta_t = \beta_{t-1} + \xi_t$

The majority of the models available in ucm are designed for nonstationary time series. The deterministic-trend model incorporates a first-order deterministic time-trend in the model. The local-level, random-walk, local-level-with-deterministic-trend, and random-walk-with-drift models are for modeling series with first-order stochastic trends. A series with a dth-order stochastic trend must be differenced d times to be stationary. The local-linear-trend, smooth-trend, and random-trend models are for modeling series with second-order stochastic trends.

The no-trend-or-idiosyncratic-component model is useful for using ucm to model stationary series with cyclical components or seasonal components and perhaps exogenous variables. The no-trend and the deterministic-constant models are useful for using ucm to model stationary series with seasonal components or exogenous variables.

Seasonal component

A seasonal component models cyclical behavior in a time series that occurs at known seasonal periodicities. A seasonal component is modeled in the time domain; the period of the cycle is specified as the number of time periods required for the cycle to complete.

▷ Example 8

Let's begin by considering a series that displays a seasonal effect. Below we plot a monthly series containing the number of new cases of mumps in New York City between January 1928 and December 1972. (See Hipel and McLeod [1994] for the source and further discussion of this dataset.)

```
. use http://www.stata-press.com/data/r12/mumps
(Hipel and Mcleod (1994), http://robjhyndman.com/tsdldata/epi/mumps.dat)

. tsline mumps
```

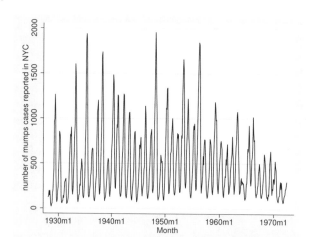

The graph reveals recurring spikes at regular intervals, which we suspect to be seasonal effects. The series may or may not be stationary; the graph evidence is not definitive.

Deterministic seasonal effects are a standard method of incorporating seasonality into a model. In a model with a constant term, the s deterministic seasonal effects are modeled as s parameters subject to the constraint that they sum to zero; formally, $\gamma_t + \gamma_{t-1} + \cdots + \gamma_{t-(s-1)} = 0$. A stochastic-seasonal model is a more flexible alternative that allows the seasonal effects at time t to sum to ζ_t, a zero-mean, finite-variance, i.i.d. random variable; formally, $\gamma_t + \gamma_{t-1} + \cdots + \gamma_{t-(s-1)} = \zeta_t$.

In the output below, we model the seasonal effects by a stochastic-seasonal model, we allow for the series to follow a random walk, and we include a stationary cyclical component.

```
. ucm mumps, seasonal(12) cycle(1)
searching for initial values ...................
(setting technique to bhhh)
Iteration 0:    log likelihood = -3268.1838
Iteration 1:    log likelihood = -3256.5253
Iteration 2:    log likelihood = -3254.6124
Iteration 3:    log likelihood = -3250.3465
Iteration 4:    log likelihood = -3249.3536
(switching technique to nr)
Iteration 5:    log likelihood = -3248.9212
Iteration 6:    log likelihood = -3248.7176
Iteration 7:    log likelihood = -3248.7138
Iteration 8:    log likelihood = -3248.7138
Refining estimates:
Iteration 0:    log likelihood = -3248.7138
Iteration 1:    log likelihood = -3248.7138
```

```
Unobserved-components model
Components: random walk, seasonal(12), order 1 cycle
```

Sample: 1928m1 - 1972m6	Number of obs	=	534
	Wald chi2(2)	=	2141.69
Log likelihood = -3248.7138	Prob > chi2	=	0.0000

mumps	Coef.	OIM Std. Err.	z	P>|z|	[95% Conf. Interval]	
frequency	.3863607	.0282037	13.70	0.000	.3310824	.4416389
damping	.8405622	.0197933	42.47	0.000	.8017681	.8793563
Variance						
level	221.2131	140.5179	1.57	0.058	0	496.6231
seasonal	4.151639	4.383442	0.95	0.172	0	12.74303
cycle1	12228.17	813.8394	15.03	0.000	10633.08	13823.27

```
Note: Model is not stationary.
Note: Tests of variances against zero are one sided, and the two-sided
      confidence intervals are truncated at zero.
```

The output indicates that the trend and seasonal variances may not be necessary. When the variance of the seasonal component is zero, the seasonal component becomes deterministic. Below we estimate the parameters of a model that includes deterministic seasonal effects and a stationary cyclical component.

```
. ucm mumps ibn.month, model(none) cycle(1)
searching for initial values .......
(setting technique to bhhh)
Iteration 0:    log likelihood = -4105.9711
Iteration 1:    log likelihood = -3676.1708
Iteration 2:    log likelihood = -3473.0335
Iteration 3:    log likelihood = -3412.5661
Iteration 4:    log likelihood = -3383.1948
(switching technique to nr)
Iteration 5:    log likelihood =  -3366.493
Iteration 6:    log likelihood = -3283.7964
Iteration 7:    log likelihood = -3283.0325
Iteration 8:    log likelihood = -3283.0284
Iteration 9:    log likelihood = -3283.0284
Refining estimates:
Iteration 0:    log likelihood = -3283.0284
Iteration 1:    log likelihood = -3283.0284
```

```
Unobserved-components model
Components: order 1 cycle
Sample: 1928m1 - 1972m6                   Number of obs    =        534
                                          Wald chi2(14)    =    3404.29
Log likelihood = -3283.0284               Prob > chi2      =     0.0000
```

mumps	Coef.	OIM Std. Err.	z	P>|z|	[95% Conf. Interval]	
cycle1						
frequency	.3272754	.0262922	12.45	0.000	.2757436	.3788071
damping	.844874	.0184994	45.67	0.000	.8086157	.8811322
mumps						
month						
1	480.5095	32.67128	14.71	0.000	416.475	544.544
2	561.9174	32.66999	17.20	0.000	497.8854	625.9494
3	832.8666	32.67696	25.49	0.000	768.8209	896.9122
4	894.0747	32.64568	27.39	0.000	830.0904	958.0591
5	869.6568	32.56282	26.71	0.000	805.8348	933.4787
6	770.1562	32.48587	23.71	0.000	706.4851	833.8274
7	433.839	32.50165	13.35	0.000	370.1369	497.541
8	218.2394	32.56712	6.70	0.000	154.409	282.0698
9	140.686	32.64138	4.31	0.000	76.7101	204.662
10	148.5876	32.69067	4.55	0.000	84.51508	212.6601
11	215.0958	32.70311	6.58	0.000	150.9989	279.1927
12	330.2232	32.68906	10.10	0.000	266.1538	394.2926
Variance						
cycle1	13031.53	798.2719	16.32	0.000	11466.95	14596.11

```
Note: Tests of variances against zero are one sided, and the two-sided
      confidence intervals are truncated at zero.
```

The output indicates that each of these components is statistically significant.

◁

❏ Technical note

In a stochastic model for the seasonal component, the seasonal effects sum to the random variable $\zeta_t \sim$ i.i.d. $N(0, \sigma_\zeta^2)$:

$$\gamma_t = -\sum_{j=1}^{s-1} \gamma_{t-j} + \zeta_t$$

❏

Saved results

Because ucm is estimated using sspace, most of the sspace saved results appear after ucm. Not all these results are relevant for ucm; programmers wishing to treat ucm results as sspace results should see *Saved results* of [TS] **sspace**. See *Methods and formulas* for the state-space representation of UCMs, and see [TS] **sspace** for more documentation that relates to all the saved results.

ucm saves the following in e():

Scalars
e(N)	number of observations
e(k)	number of parameters
e(k_aux)	number of auxiliary parameters
e(k_eq)	number of equations in e(b)
e(k_dv)	number of dependent variables
e(k_cycles)	number of stochastic cycles
e(df_m)	model degrees of freedom
e(ll)	log likelihood
e(chi2)	χ^2
e(p)	significance
e(tmin)	minimum time in sample
e(tmax)	maximum time in sample
e(stationary)	1 if the estimated parameters indicate a stationary model, 0 otherwise
e(rank)	rank of VCE
e(ic)	number of iterations
e(rc)	return code
e(converged)	1 if converged, 0 otherwise

Macros
e(cmd)	ucm
e(cmdline)	command as typed
e(depvar)	unoperated names of dependent variables in observation equations
e(covariates)	list of covariates
e(indeps)	independent variables
e(tvar)	variable denoting time within groups
e(eqnames)	names of equations
e(model)	type of model
e(title)	title in estimation output
e(tmins)	formatted minimum time
e(tmaxs)	formatted maximum time
e(chi2type)	Wald; type of model χ^2 test
e(vce)	*vcetype* specified in vce()
e(vcetype)	title used to label Std. Err.
e(opt)	type of optimization
e(initial_values)	type of initial values
e(technique)	maximization technique
e(tech_steps)	iterations taken in maximization technique
e(properties)	b V
e(estat_cmd)	program used to implement estat
e(predict)	program used to implement predict
e(marginsok)	predictions allowed by margins
e(marginsnotok)	predictions disallowed by margins

Matrices
e(b)	parameter vector
e(Cns)	constraints matrix
e(ilog)	iteration log (up to 20 iterations)
e(gradient)	gradient vector
e(V)	variance–covariance matrix of the estimators
e(V_modelbased)	model-based variance

Functions
e(sample)	marks estimation sample

Methods and formulas

ucm is implemented as an ado-file.

Methods and formulas are presented under the following headings:

> *Introduction*
> *State-space formulation*
> *Cyclical component extensions*

Introduction

The general form of UCMs can be expressed as

$$y_t = \tau_t + \gamma_t + \psi_t + \mathbf{x}_t\beta + \epsilon_t$$

where τ_t is the trend, γ_t is the seasonal component, ψ_t is the cycle, β is the regression coefficients for regressors $\mathbf{x}_t$, and ϵ_t is the idiosyncratic error with variance σ_ϵ^2.

We can decompose the trend as

$$\tau_t = \mu_t$$
$$\mu_t = \mu_{t-1} + \alpha_{t-1} + \eta_t$$
$$\alpha_t = \alpha_{t-1} + \xi_t$$

where μ_t is the local level, α_t is the local slope, and η_t and ξ_t are i.i.d. normal errors with mean 0 and variance σ_η^2 and σ_ξ^2, respectively.

Next consider the seasonal component, γ_t, with a period of s time units. Ignoring a seasonal disturbance term, the seasonal effects will sum to zero, $\sum_{j=0}^{s-1} \gamma_{t-j} = 0$. Adding a normal error term, ω_t, with mean 0 and variance σ_ω^2, we express the seasonal component as

$$\gamma_t = -\sum_{j=1}^{s-1} \gamma_{t-j} + \omega_t$$

Finally, the cyclical component, ψ_t, is a function of the frequency λ, in radians, and a unit-less scaling variable ρ, termed the damping effect, $0 < \rho < 1$. We require two equations to express the cycle:

$$\psi_t = \psi_{t-1}\rho\cos\lambda + \widetilde{\psi}_{t-1}\rho\sin\lambda + \kappa_t$$
$$\widetilde{\psi}_t = -\psi_{t-1}\rho\sin\lambda + \widetilde{\psi}_{t-1}\rho\cos\lambda + \widetilde{\kappa}_t$$

where the κ_t and $\widetilde{\kappa}_t$ disturbances are normally distributed with mean 0 and variance σ_κ^2.

The disturbance terms ϵ_t, η_t, ξ_t, ω_t, κ_t, and $\widetilde{\kappa}_t$ are independent.

State-space formulation

ucm is an easy-to-use implementation of the state-space command sspace, with special modifications, where the local linear trend components, seasonal components, and cyclical components are states of the state-space model. The state-space model can be expressed in matrix form as

$$\mathbf{y}_t = \mathbf{D}\mathbf{z}_t + \mathbf{F}\mathbf{x}_t + \epsilon_t$$

$$\mathbf{z}_t = \mathbf{A}\mathbf{z}_{t-1} + \mathbf{C}\zeta_t$$

where y_t, $t = 1, \ldots, T$, are the observations and $\mathbf{z}_t$ are the unobserved states. The number of states, m, depends on the model specified. The $k \times 1$ vector $\mathbf{x}_t$ contains the exogenous variables specified as *indepvars*, and the $1 \times k$ vector $\mathbf{F}$ contains the regression coefficients to be estimated. ϵ_t is the observation equation disturbance, and the $m_0 \times 1$ vector ζ_t contains the state equation disturbances, where $m_0 \leq m$. Finally, $\mathbf{C}$ is a $m \times m_0$ matrix of zeros and ones. These recursive equations are evaluated using the diffuse Kalman filter of De Jong (1991).

Below we give the state-space matrix structures for a local linear trend with a stochastic seasonal component, with a period of 4 time units, and an order-2 cycle. The state vector, $\mathbf{z}_t$, and its transition matrix, $\mathbf{A}$, have the structure

$$\mathbf{A} = \begin{pmatrix} 1 & 1 & 0 & 0 & 0 & 0 & 0 & 0 & 0 \\ 0 & 1 & 0 & 0 & 0 & 0 & 0 & 0 & 0 \\ 0 & 0 & -1 & -1 & -1 & 0 & 0 & 0 & 0 \\ 0 & 0 & 1 & 0 & 0 & 0 & 0 & 0 & 0 \\ 0 & 0 & 0 & 1 & 0 & 0 & 0 & 0 & 0 \\ 0 & 0 & 0 & 0 & 0 & \rho\cos\lambda & \rho\sin\lambda & 1 & 0 \\ 0 & 0 & 0 & 0 & 0 & -\rho\sin\lambda & \rho\cos\lambda & 0 & 1 \\ 0 & 0 & 0 & 0 & 0 & 0 & 0 & \rho\cos\lambda & \rho\sin\lambda \\ 0 & 0 & 0 & 0 & 0 & 0 & 0 & -\rho\sin\lambda & \rho\cos\lambda \end{pmatrix} \quad \mathbf{z}_t = \begin{pmatrix} \mu_t \\ \alpha_t \\ \gamma_t \\ \gamma_{t-1} \\ \gamma_{t-2} \\ \psi_{t,1} \\ \widetilde{\psi}_{t,1} \\ \psi_{t,2} \\ \widetilde{\psi}_{t,2} \end{pmatrix}$$

$$\mathbf{C} = \begin{pmatrix} 1 & 0 & 0 & 0 & 0 \\ 0 & 1 & 0 & 0 & 0 \\ 0 & 0 & 1 & 0 & 0 \\ 0 & 0 & 0 & 0 & 0 \\ 0 & 0 & 0 & 0 & 0 \\ 0 & 0 & 0 & 0 & 0 \\ 0 & 0 & 0 & 0 & 0 \\ 0 & 0 & 0 & 1 & 0 \\ 0 & 0 & 0 & 0 & 1 \end{pmatrix} \quad \zeta_t = \begin{pmatrix} \eta_t \\ \xi_t \\ \omega_t \\ \kappa_t \\ \widetilde{\kappa}_t \end{pmatrix}$$

$$\mathbf{D} = \begin{pmatrix} 1 & 0 & 1 & 0 & 0 & 1 & 0 & 0 & 0 \end{pmatrix}$$

Cyclical component extensions

Recall that the stochastic cyclical model is given by

$$\psi_t = \rho(\psi_{t-1} \cos \lambda_c + \psi_{t-1}^* \sin \lambda_c) + \kappa_{t,1}$$
$$\psi_t^* = \rho(-\psi_{t-1} \sin \lambda_c + \psi_{t-1}^* \cos \lambda_c) + \kappa_{t,2}$$

where $\kappa_{t,j} \sim$ i.i.d. $N(0, \sigma_\kappa^2)$ and $0 < \rho < 1$ is a damping effect. The cycle is variance-stationary when $\rho < 1$ because $\text{Var}(\psi_t) = \sigma_\kappa^2 / (1 - \rho)$. We will express a UCM with a cyclical component added to a trend as

$$y_t = \mu_t + \psi_t + \epsilon_t$$

where μ_t can be any of the trend parameterizations discussed earlier.

Higher-order cycles, $k = 2$ or $k = 3$, are defined as

$$\psi_{t,j} = \rho(\psi_{t-1,j} \cos \lambda_c + \psi_{t-1,j}^* \sin \lambda_c) + \psi_{t-1,j+1}$$
$$\psi_{t,j}^* = \rho(-\psi_{t-1,j} \sin \lambda_c + \psi_{t-1,j}^* \cos \lambda_c) + \psi_{t-1,j+1}^*$$

for $j < k$, and

$$\psi_{t,k} = \rho(\psi_{t-1,k} \cos \lambda_c + \psi_{t-1,k}^* \sin \lambda_c) + \kappa_{t,1}$$
$$\psi_{t,k}^* = \rho(-\psi_{t-1,k} \sin \lambda_c + \psi_{t-1,k}^* \cos \lambda_c) + \kappa_{t,2}$$

Harvey and Trimbur (2003) discuss the properties of this model and its state-space formulation.

References

Burns, A. F., and W. C. Mitchell. 1946. *Measuring Business Cycles*. New York: National Bureau of Economic Research.

De Jong, P. 1991. The diffuse Kalman filter. *Annals of Statistics* 19: 1073–1083.

Durbin, J., and S. J. Koopman. 2001. *Time Series Analysis by State Space Methods*. Oxford: Oxford University Press.

Fuller, W. A. 1996. *Introduction to Statistical Time Series*. 2nd ed. New York: Wiley.

Hamilton, J. D. 1994. *Time Series Analysis*. Princeton: Princeton University Press.

Harvey, A. C. 1989. *Forecasting, Structural Time Series Models and the Kalman Filter*. Cambridge: Cambridge University Press.

———. 1993. *Time Series Models*. 2nd ed. Cambridge, MA: MIT Press.

Harvey, A. C., and T. M. Trimbur. 2003. General model-based filters for extracting cycles and trends in economic time series. *The Review of Economics and Statistics* 85: 244–255.

Hipel, K. W., and A. I. McLeod. 1994. *Time Series Modelling of Water Resources and Environmental Systems*. Amsterdam: Elsevier.

Priestley, M. B. 1981. *Spectral Analysis and Time Series*. London: Academic Press.

Trimbur, T. M. 2006. Properties of higher order stochastic cycles. *Journal of Time Series Analysis* 27: 1–17.

Wei, W. W. S. 2006. *Time Series Analysis: Univariate and Multivariate Methods*. 2nd ed. Boston: Pearson.

Also see

[TS] **ucm postestimation** — Postestimation tools for ucm

[TS] **tsset** — Declare data to be time-series data

[TS] **sspace** — State-space models

[TS] **var** — Vector autoregressive models

[TS] **arima** — ARIMA, ARMAX, and other dynamic regression models

[TS] **tsfilter** — Filter a time-series, keeping only selected periodicities

[TS] **tssmooth** — Smooth and forecast univariate time-series data

[U] **20 Estimation and postestimation commands**

Title

ucm postestimation — Postestimation tools for ucm

Description

The following postestimation commands are of special interest after ucm:

Command	Description
estat period	display cycle periods in time units
psdensity	estimate the spectral density

For information about estat period, see below.
For information about psdensity, see [TS] **psdensity**.

The following standard postestimation commands are also available:

Command	Description
estat	AIC, BIC, VCE, and estimation sample summary
estimates	cataloging estimation results
lincom	point estimates, standard errors, testing and inference for linear combinations of coefficients
lrtest	likelihood-ratio test
nlcom	point estimates, standard errors, testing and inference for nonlinear combinations of coefficients
predict	predictions, residuals, influence statistics, and other diagnostic measures
predictnl	point estimates, standard errors, testing, and inference for generalized predictions
test	Wald tests of simple and composite linear hypotheses
testnl	Wald tests of nonlinear hypotheses

See the corresponding entries in the *Base Reference Manual* for details.

Special-interest postestimation commands

estat period transforms an estimated central frequency to an estimated period after ucm.

536

Syntax for predict

predict [*type*] { *stub** | *newvarlist* } [*if*] [*in*] [, *statistic options*]

statistic	Description
Main	
xb	linear prediction using exogenous variables
trend	trend component
seasonal	seasonal component
cycle	cyclical component
residuals	residuals
rstandard	standardized residuals

These statistics are available both in and out of sample; type predict ... if e(sample) ... if wanted only for the estimation sample.

options	Description
Options	
rmse(*stub** | *newvarlist*)	put estimated root mean squared errors of predicted statistics in the new variable
dynamic(*time_constant*)	begin dynamic forecast at specified time
Advanced	
smethod(*method*)	method for predicting unobserved components

method	Description
onestep	predict using past information
smooth	predict using all sample information
filter	predict using past and contemporaneous information

Menu

Statistics > Postestimation > Predictions, residuals, etc.

Options for predict

⌐ Main ⌐

xb, trend, seasonal, cycle, residuals, and rstandard specify the statistic to be predicted.

xb, the default, calculates the linear predictions using the exogenous variables. xb may not be used with the smethod(filter) option.

trend estimates the unobserved trend component.

seasonal estimates the unobserved seasonal component.

cycle estimates the unobserved cyclical component.

residuals calculates the residuals in the equation for the dependent variable. residuals may not be specified with dynamic().

rstandard calculates the standardized residuals, which are the residuals normalized to have unit variances. rstandard may not be specified with the smethod(filter), smethod(smooth), or dynamic() option.

> Options

rmse(*stub** | *newvarlist*) puts the root mean squared errors of the predicted statistic into the specified new variable. Multiple variables are only required for predicting cycles of a model that has more than one cycle. The root mean squared errors measure the variances due to the disturbances but do not account for estimation error. The *stub** syntax is for models with multiple cycles, where you provide the prefix and predict will add a numeric suffix for each predicted cycle.

dynamic(*time_constant*) specifies when predict should start producing dynamic forecasts. The specified *time_constant* must be in the scale of the time variable specified in tsset, and the *time_constant* must be inside a sample for which observations on the dependent variable are available. For example, dynamic(tq(2008q4)) causes dynamic predictions to begin in the fourth quarter of 2008, assuming that your time variable is quarterly; see [D] **datetime**. If the model contains exogenous variables, they must be present for the whole predicted sample. dynamic() may not be specified with the rstandard, residuals, or smethod(smooth) option.

> Advanced

smethod(*method*) specifies the method for predicting the unobserved components. smethod() causes different amounts of information on the dependent variable to be used in predicting the components at each time period.

smethod(onestep), the default, causes predict to estimate the components at each time period using previous information on the dependent variable. The Kalman filter is performed on previous periods, but only the one-step predictions are made for the current period.

smethod(smooth) causes predict to estimate the components at each time period using all the sample data by the Kalman smoother. smethod(smooth) may not be specified with the rstandard option.

smethod(filter) causes predict to estimate the components at each time period using previous and contemporaneous data by the Kalman filter. The Kalman filter is performed on previous periods and the current period. smethod(filter) may not be specified with the xb option.

Syntax for estat period

> estat period [, *options*]

options	Description
Main	
level(*#*)	set confidence level; default is level(95)
cformat(%*fmt*)	numeric format

Menu

Statistics > Postestimation > Reports and statistics

Options for estat period

⌐ Options ⌐

level(#) specifies the confidence level, as a percentage, for confidence intervals. The default is level(95) or as set by set level; see [U] **20.7 Specifying the width of confidence intervals**.

cformat(%*fmt*) sets the display format for the table numeric values. The default is cformat(%9.0g).

Remarks

We assume that you have already read [TS] **ucm**. In this entry, we illustrate some features of predict after using ucm to estimate the parameters of an unobserved-components model.

All predictions after ucm depend on the unobserved components, which are estimated recursively using a Kalman filter. Changing the sample can alter the state estimates, which can change all other predictions.

▷ Example 1

We begin by modeling monthly data on the median duration of employment spells in the United States. We include a stochastic-seasonal component because the data have not been seasonally adjusted.

```
. use http://www.stata-press.com/data/r12/uduration2
(BLS data, not seasonally adjusted)

. ucm duration, seasonal(12) cycle(1) difficult
searching for initial values ...................
(setting technique to bhhh)
Iteration 0:   log likelihood = -409.79512
Iteration 1:   log likelihood =   -403.383
Iteration 2:   log likelihood = -403.37363  (backed up)
Iteration 3:   log likelihood = -403.36891  (backed up)
Iteration 4:   log likelihood = -403.36653  (backed up)
(switching technique to nr)
Iteration 5:   log likelihood = -403.36649  (backed up)
Iteration 6:   log likelihood = -397.87764  (not concave)
Iteration 7:   log likelihood = -396.53184  (not concave)
Iteration 8:   log likelihood = -395.59253  (not concave)
Iteration 9:   log likelihood = -390.32744  (not concave)
Iteration 10:  log likelihood = -389.14967  (not concave)
Iteration 11:  log likelihood =   -388.733  (not concave)
Iteration 12:  log likelihood = -388.50581
Iteration 13:  log likelihood = -388.28068
Iteration 14:  log likelihood = -388.25924
Iteration 15:  log likelihood = -388.25676
Iteration 16:  log likelihood = -388.25675
Refining estimates:
Iteration 0:   log likelihood = -388.25675
Iteration 1:   log likelihood = -388.25675
```

```
Unobserved-components model
Components: random walk, seasonal(12), order 1 cycle
Sample: 1967m7 - 2008m12                    Number of obs   =        498
                                            Wald chi2(2)    =       7.17
Log likelihood = -388.25675                 Prob > chi2     =     0.0277
```

duration	Coef.	OIM Std. Err.	z	P>|z|	[95% Conf. Interval]	
frequency	1.641531	.7250323	2.26	0.024	.2204938	3.062568
damping	.2671232	.1050168	2.54	0.011	.0612939	.4729524
Variance						
level	.1262922	.0221428	5.70	0.000	.0828932	.1696912
seasonal	.0017289	.0009647	1.79	0.037	0	.0036196
cycle1	.0641496	.0211839	3.03	0.001	.0226299	.1056693

```
Note: Model is not stationary.
Note: Tests of variances against zero are one sided, and the two-sided
      confidence intervals are truncated at zero.
```

Below we predict the trend and the seasonal components to get a look at the model fit.

```
. predict strend, trend
. predict season, seasonal
. tsline duration strend, name(trend) nodraw legend(rows(1))
. tsline season, name(season) yline(0,lwidth(vthin)) nodraw
. graph combine trend season, rows(2)
```

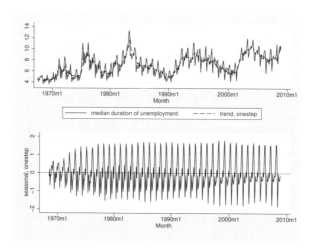

The trend tracks the data well. That the seasonal component appears to change over time indicates that the stochastic-seasonal component might fit better than a deterministic-seasonal component.

◁

▷ Example 2

In this example, we use the model to forecast the median unemployment duration. We use the root mean squared error of the prediction to compute a confidence interval of our dynamic predictions. Recall that the root mean squared error accounts for variances due to the disturbances but not due to the estimation error.

```
. tsappend, add(12)

. predict duration_f, dynamic(tm(2009m1)) rmse(rmse)

. scalar z = invnormal(0.95)

. generate lbound = duration_f - z*rmse if tm>=tm(2008m12)
(497 missing values generated)

. generate ubound = duration_f + z*rmse if tm>=tm(2008m12)
(497 missing values generated)

. label variable lbound "90% forecast interval"

. twoway (tsline duration duration_f if tm>=tm(2006m1))
>        (tsrline lbound ubound if tm>=tm(2008m12)),
>        ysize(2) xtitle("") legend(cols(1))
```

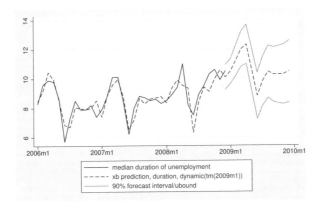

The model forecasts a large temporary increase in the median duration of unemployment.

◁

Methods and formulas

All postestimation commands listed above are implemented as ado-files.

For details on the ucm postestimation methods, see [TS] **sspace postestimation**.

See [TS] **psdensity** for the methods used to estimate the spectral density.

Also see

[TS] **ucm** — Unobserved-components model

[TS] **psdensity** — Parametric spectral density estimation after arima, arfima, and ucm

[TS] **sspace postestimation** — Postestimation tools for sspace

Title

---

var intro — Introduction to vector autoregressive models

Description

Stata has a suite of commands for fitting, forecasting, interpreting, and performing inference on vector autoregressive (VAR) models and structural vector autoregressive (SVAR) models. The suite includes several commands for estimating and interpreting impulse–response functions (IRFs), dynamic-multiplier functions, and forecast-error variance decompositions (FEVDs). The table below describes the available commands.

Fitting a VAR or SVAR

var	[TS] **var**	Fit vector autoregressive models
svar	[TS] **var svar**	Fit structural vector autoregressive models
varbasic	[TS] **varbasic**	Fit a simple VAR and graph IRFs or FEVDs

Model diagnostics and inference

varstable	[TS] **varstable**	Check the stability condition of VAR or SVAR estimates
varsoc	[TS] **varsoc**	Obtain lag-order selection statistics for VARs and VECMs
varwle	[TS] **varwle**	Obtain Wald lag-exclusion statistics after var or svar
vargranger	[TS] **vargranger**	Perform pairwise Granger causality tests after var or svar
varlmar	[TS] **varlmar**	Perform LM test for residual autocorrelation after var or svar
varnorm	[TS] **varnorm**	Test for normally distributed disturbances after var or svar

Forecasting after fitting a VAR or SVAR

fcast compute	[TS] **fcast compute**	Compute dynamic forecasts of dependent variables after var, svar, or vec
fcast graph	[TS] **fcast graph**	Graph forecasts of variables computed by fcast compute

Working with IRFs, dynamic-multiplier functions, and FEVDs

irf	[TS] **irf**	Create and analyze IRFs, dynamic-multiplier functions, and FEVDs

This entry provides an overview of vector autoregressions and structural vector autoregressions. More rigorous treatments can be found in Hamilton (1994), Lütkepohl (2005), and Amisano and Giannini (1997). Stock and Watson (2001) provide an excellent nonmathematical treatment of vector autoregressions and their role in macroeconomics.

Remarks

Remarks are presented under the following headings:

Introduction to VARs
Introduction to SVARs
Short-run SVAR models
Long-run restrictions
IRFs and FEVDs

Introduction to VARs

A VAR is a model in which K variables are specified as linear functions of p of their own lags, p lags of the other $K - 1$ variables, and possibly additional exogenous variables. Algebraically, a p-order VAR model, written VAR(p), with exogenous variables $\mathbf{x}_t$ is given by

$$\mathbf{y}_t = \mathbf{v} + \mathbf{A}_1\mathbf{y}_{t-1} + \cdots + \mathbf{A}_p\mathbf{y}_{t-p} + \mathbf{B}_0\mathbf{x}_t + \mathbf{B}_1\mathbf{x}_{t-1} + \cdots + \mathbf{B}_s\mathbf{x}_{t-s} + \mathbf{u}_t \qquad t \in \{-\infty, \infty\} \quad (1)$$

where

$\mathbf{y}_t = (y_{1t}, \ldots, y_{Kt})'$ is a $K \times 1$ random vector,
$\mathbf{A}_1$ through $\mathbf{A}_p$ are $K \times K$ matrices of parameters,
$\mathbf{x}_t$ is an $M \times 1$ vector of exogenous variables,
$\mathbf{B}_0$ through $\mathbf{B}_s$ are $K \times M$ matrices of coefficients,
$\mathbf{v}$ is a $K \times 1$ vector of parameters, and
$\mathbf{u}_t$ is assumed to be white noise; that is,
$E(\mathbf{u}_t) = \mathbf{0}$,
$E(\mathbf{u}_t\mathbf{u}_t') = \mathbf{\Sigma}$, and
$E(\mathbf{u}_t\mathbf{u}_s') = \mathbf{0}$ for $t \neq s$

There are $K^2 \times p + K \times (M(s + 1) + 1)$ parameters in the equation for $\mathbf{y}_t$, and there are $\{K \times (K + 1)\}/2$ parameters in the covariance matrix $\mathbf{\Sigma}$. One way to reduce the number of parameters is to specify an incomplete VAR, in which some of the $\mathbf{A}$ or $\mathbf{B}$ matrices are set to zero. Another way is to specify linear constraints on some of the coefficients in the VAR.

A VAR can be viewed as the reduced form of a system of dynamic simultaneous equations. Consider the system

$$\mathbf{W}_0\mathbf{y}_t = \mathbf{a} + \mathbf{W}_1\mathbf{y}_{t-1} + \cdots + \mathbf{W}_p\mathbf{y}_{t-p} + \widetilde{\mathbf{W}}_1\mathbf{x}_t + \widetilde{\mathbf{W}}_2\mathbf{x}_{t-2} + \cdots + \widetilde{\mathbf{W}}_s\mathbf{x}_{t-s} + \mathbf{e}_t \quad (2)$$

where $\mathbf{a}$ is a $K \times 1$ vector of parameters, each $\mathbf{W}_i$, $i = 0, \ldots, p$, is a $K \times K$ matrix of parameters, and $\mathbf{e}_t$ is a $K \times 1$ disturbance vector. In the traditional dynamic simultaneous equations approach, sufficient restrictions are placed on the $\mathbf{W}_i$ to obtain identification. Assuming that $\mathbf{W}_0$ is nonsingular, (2) can be rewritten as

$$\begin{aligned}
\mathbf{y}_t = &\mathbf{W}_0^{-1}\mathbf{a} + \mathbf{W}_0^{-1}\mathbf{W}_1\mathbf{y}_{t-1} + \cdots + \mathbf{W}_0^{-1}\mathbf{W}_p\mathbf{y}_{t-p} \\
&+ \mathbf{W}_0^{-1}\widetilde{\mathbf{W}}_1\mathbf{x}_t + \mathbf{W}_0^{-1}\widetilde{\mathbf{W}}_2\mathbf{x}_{t-2} + \cdots + \mathbf{W}_0^{-1}\widetilde{\mathbf{W}}_s\mathbf{x}_{t-s} + \mathbf{W}_0^{-1}\mathbf{e}_t
\end{aligned} \quad (3)$$

which is a VAR with

$$\mathbf{v} = \mathbf{W}_0^{-1}\mathbf{a}$$
$$\mathbf{A}_i = \mathbf{W}_0^{-1}\mathbf{W}_i$$
$$\mathbf{B}_i = \mathbf{W}_0^{-1}\widetilde{\mathbf{W}}_i$$
$$\mathbf{u}_t = \mathbf{W}_0^{-1}\mathbf{e}_t$$

The cross-equation error variance–covariance matrix $\boldsymbol{\Sigma}$ contains all the information about contemporaneous correlations in a VAR and may be the VAR's greatest strength and its greatest weakness. Because no questionable a priori assumptions are imposed, fitting a VAR allows the dataset to speak for itself. However, without imposing some restrictions on the structure of $\boldsymbol{\Sigma}$, we cannot make a causal interpretation of the results.

If we make additional technical assumptions, we can derive another representation of the VAR in (1). If the VAR is stable (see [TS] **varstable**), we can rewrite $\mathbf{y}_t$ as

$$
\mathbf{y}_t = \boldsymbol{\mu} + \sum_{i=0}^{\infty} \mathbf{D}_i \mathbf{x}_{t-i} + \sum_{i=0}^{\infty} \boldsymbol{\Phi}_i \mathbf{u}_{t-i} \tag{4}
$$

where $\boldsymbol{\mu}$ is the $K \times 1$ time-invariant mean of the process and $\mathbf{D}_i$ and $\boldsymbol{\Phi}_i$ are $K \times M$ and $K \times K$ matrices of parameters, respectively. Equation (4) states that the process by which the variables in $\mathbf{y}_t$ fluctuate about their time-invariant means, $\boldsymbol{\mu}$, is completely determined by the parameters in $\mathbf{D}_i$ and $\boldsymbol{\Phi}_i$ and the (infinite) past history of the exogenous variables $\mathbf{x}_t$ and the independent and identically distributed (i.i.d.) shocks or innovations, $\mathbf{u}_{t-1}, \mathbf{u}_{t-2}, \dots$. Equation (4) is known as the vector moving-average representation of the VAR. The $\mathbf{D}_i$ are the dynamic-multiplier functions, or transfer functions. The moving-average coefficients $\boldsymbol{\Phi}_i$ are also known as the simple IRFs at horizon i. The precise relationships between the VAR parameters and the $\mathbf{D}_i$ and $\boldsymbol{\Phi}_i$ are derived in *Methods and formulas* of [TS] **irf create**.

The joint distribution of $\mathbf{y}_t$ is determined by the distributions of $\mathbf{x}_t$ and $\mathbf{u}_t$ and the parameters $\mathbf{v}$, $\mathbf{B}_i$, and $\mathbf{A}_i$. Estimating the parameters in a VAR requires that the variables in $\mathbf{y}_t$ and $\mathbf{x}_t$ be covariance stationary, meaning that their first two moments exist and are time invariant. If the $\mathbf{y}_t$ are not covariance stationary, but their first differences are, a vector error-correction model (VECM) can be used. See [TS] **vec intro** and [TS] **vec** for more information about those models.

If the $\mathbf{u}_t$ form a zero mean, i.i.d. vector process, and $\mathbf{y}_t$ and $\mathbf{x}_t$ are covariance stationary and are not correlated with the $\mathbf{u}_t$, consistent and efficient estimates of the $\mathbf{B}_i$, the $\mathbf{A}_i$, and $\mathbf{v}$ are obtained via seemingly unrelated regression, yielding estimators that are asymptotically normally distributed. When the equations for the variables $\mathbf{y}_t$ have the same set of regressors, equation-by-equation OLS estimates are the conditional maximum likelihood estimates.

Much of the interest in VAR models is focused on the forecasts, IRFs, dynamic-multiplier functions, and the FEVDs, all of which are functions of the estimated parameters. Estimating these functions is straightforward, but their asymptotic standard errors are usually obtained by assuming that $\mathbf{u}_t$ forms a zero mean, i.i.d. Gaussian (normal) vector process. Also, some of the specification tests for VARs have been derived using the likelihood-ratio principle and the stronger Gaussian assumption.

In the absence of contemporaneous exogenous variables, the disturbance variance–covariance matrix contains all the information about contemporaneous correlations among the variables. VARs are sometimes classified into three types by how they account for this contemporaneous correlation. (See Stock and Watson [2001] for one derivation of this taxonomy.) A reduced-form VAR, aside from estimating the variance–covariance matrix of the disturbance, does not try to account for contemporaneous correlations. In a recursive VAR, the K variables are assumed to form a recursive dynamic structural equation model in which the first variable is a function of lagged variables, the second is a function of contemporaneous values of the first variable and lagged values, and so on. In a structural VAR, the theory you are working with places restrictions on the contemporaneous correlations that are not necessarily recursive.

Stata has two commands for fitting reduced-form VARs: var and varbasic. var allows for constraints to be imposed on the coefficients. varbasic allows you to fit a simple VAR quickly without constraints and graph the IRFs.

Because fitting a VAR of the correct order can be important, `varsoc` offers several methods for choosing the lag order p of the VAR to fit. After fitting a VAR, and before proceeding with inference, interpretation, or forecasting, checking that the VAR fits the data is important. `varlmar` can be used to check for autocorrelation in the disturbances. `varwle` performs Wald tests to determine whether certain lags can be excluded. `varnorm` tests the null hypothesis that the disturbances are normally distributed. `varstable` checks the eigenvalue condition for stability, which is needed to interpret the IRFs and IRFs.

Introduction to SVARs

As discussed in [TS] **irf create**, a problem with VAR analysis is that, because $\mathbf{\Sigma}$ is not restricted to be a diagonal matrix, an increase in an innovation to one variable provides information about the innovations to other variables. This implies that no causal interpretation of the simple IRFs is possible: there is no way to determine whether the shock to the first variable caused the shock in the second variable or vice versa.

However, suppose that we had a matrix $\mathbf{P}$ such that $\mathbf{\Sigma} = \mathbf{PP}'$. We can then show that the variables in $\mathbf{P}^{-1}\mathbf{u}_t$ have zero mean and that $E\{\mathbf{P}^{-1}\mathbf{u}_t(\mathbf{P}^{-1}\mathbf{u}_t)'\} = \mathbf{I}_K$. We could rewrite (4) as

$$
\begin{aligned}
\mathbf{y}_t &= \boldsymbol{\mu} + \sum_{s=0}^{\infty} \mathbf{\Phi}_s \mathbf{PP}^{-1}\mathbf{u}_{t-s} \\
&= \boldsymbol{\mu} + \sum_{s=0}^{\infty} \mathbf{\Theta}_s \mathbf{P}^{-1}\mathbf{u}_{t-s} \\
&= \boldsymbol{\mu} + \sum_{s=0}^{\infty} \mathbf{\Theta}_s \mathbf{w}_{t-s}
\end{aligned}
\tag{5}
$$

where $\mathbf{\Theta}_s = \mathbf{\Phi}_s \mathbf{P}$ and $\mathbf{w}_t = \mathbf{P}^{-1}\mathbf{u}_t$. If we had such a $\mathbf{P}$, the $\mathbf{w}_k$ would be mutually orthogonal, and the $\mathbf{\Theta}_s$ would allow the causal interpretation that we seek.

SVAR models provide a framework for estimation of and inference about a broad class of $\mathbf{P}$ matrices. As described in [TS] **irf create**, the estimated $\mathbf{P}$ matrices can then be used to estimate structural IRFs and structural FEVDs. There are two types of SVAR models. Short-run SVAR models identify a $\mathbf{P}$ matrix by placing restrictions on the contemporaneous correlations between the variables. Long-run SVAR models, on the other hand, do so by placing restrictions on the long-term accumulated effects of the innovations.

Short-run SVAR models

A short-run SVAR model without exogenous variables can be written as

$$
\mathbf{A}(\mathbf{I}_K - \mathbf{A}_1 L - \mathbf{A}_2 L^2 - \cdots - \mathbf{A}_p L^p)\mathbf{y}_t = \mathbf{A}\boldsymbol{\epsilon}_t = \mathbf{B}\mathbf{e}_t
\tag{6}
$$

where L is the lag operator; $\mathbf{A}$, $\mathbf{B}$, and $\mathbf{A}_1, \ldots, \mathbf{A}_p$ are $K \times K$ matrices of parameters; $\boldsymbol{\epsilon}_t$ is a $K \times 1$ vector of innovations with $\boldsymbol{\epsilon}_t \sim N(\mathbf{0}, \mathbf{\Sigma})$ and $E[\boldsymbol{\epsilon}_t \boldsymbol{\epsilon}_s'] = \mathbf{0}_K$ for all $s \neq t$; and $\mathbf{e}_t$ is a $K \times 1$ vector of orthogonalized disturbances; that is, $\mathbf{e}_t \sim N(\mathbf{0}, \mathbf{I}_K)$ and $E[\mathbf{e}_t \mathbf{e}_s'] = \mathbf{0}_K$ for all $s \neq t$. These transformations of the innovations allow us to analyze the dynamics of the system in terms of a change to an element of $\mathbf{e}_t$. In a short-run SVAR model, we obtain identification by placing restrictions on $\mathbf{A}$ and $\mathbf{B}$, which are assumed to be nonsingular.

Equation (6) implies that $\mathbf{P}_{\mathrm{sr}} = \mathbf{A}^{-1}\mathbf{B}$, where $\mathbf{P}_{\mathrm{sr}}$ is the $\mathbf{P}$ matrix identified by a particular short-run SVAR model. The latter equality in (6) implies that

$$\mathbf{A}\boldsymbol{\epsilon}_t\boldsymbol{\epsilon}_t'\mathbf{A}' = \mathbf{B}\mathbf{e}_t\mathbf{e}_t'\mathbf{B}'$$

Taking the expectation of both sides yields

$$\boldsymbol{\Sigma} = \mathbf{P}_{\mathrm{sr}}\mathbf{P}_{\mathrm{sr}}'$$

Assuming that the underlying VAR is stable (see [TS] **varstable** for a discussion of stability), we can invert the autoregressive representation of the model in (6) to an infinite-order, moving-average representation of the form

$$\mathbf{y}_t = \boldsymbol{\mu} + \sum_{s=0}^{\infty} \boldsymbol{\Theta}_s^{\mathrm{sr}}\mathbf{e}_{t-s} \tag{7}$$

whereby $\mathbf{y}_t$ is expressed in terms of the mutually orthogonal, unit-variance structural innovations $\mathbf{e}_t$. The $\boldsymbol{\Theta}_s^{\mathrm{sr}}$ contain the structural IRFs at horizon s.

In a short-run SVAR model, the $\mathbf{A}$ and $\mathbf{B}$ matrices model all the information about contemporaneous correlations. The $\mathbf{B}$ matrix also scales the innovations $\mathbf{u}_t$ to have unit variance. This allows the structural IRFs constructed from (7) to be interpreted as the effect on variable i of a one-time unit increase in the structural innovation to variable j after s periods.

$\mathbf{P}_{\mathrm{sr}}$ identifies the structural IRFs by defining a transformation of $\boldsymbol{\Sigma}$, and $\mathbf{P}_{\mathrm{sr}}$ is identified by the restrictions placed on the parameters in $\mathbf{A}$ and $\mathbf{B}$. Because there are only $K(K+1)/2$ free parameters in $\boldsymbol{\Sigma}$, only $K(K+1)/2$ parameters may be estimated in an identified $\mathbf{P}_{\mathrm{sr}}$. Because there are $2K^2$ total parameters in $\mathbf{A}$ and $\mathbf{B}$, the order condition for identification requires that at least $2K^2 - K(K+1)/2$ restrictions be placed on those parameters. Just as in the simultaneous-equations framework, this order condition is necessary but not sufficient. Amisano and Giannini (1997) derive a method to check that an SVAR model is locally identified near some specified values for $\mathbf{A}$ and $\mathbf{B}$.

Before moving on to models with long-run constraints, consider these limitations. We cannot place constraints on the elements of $\mathbf{A}$ in terms of the elements of $\mathbf{B}$, or vice versa. This limitation is imposed by the form of the check for identification derived by Amisano and Giannini (1997). As noted in *Methods and formulas* of [TS] **var svar**, this test requires separate constraint matrices for the parameters in $\mathbf{A}$ and $\mathbf{B}$. Also, we cannot mix short-run and long-run constraints.

Long-run restrictions

A general short-run SVAR has the form

$$\mathbf{A}(\mathbf{I}_K - \mathbf{A}_1 L - \mathbf{A}_2 L^2 - \cdots - \mathbf{A}_p L^p)\mathbf{y}_t = \mathbf{B}\mathbf{e}_t$$

To simplify the notation, let $\bar{\mathbf{A}} = (\mathbf{I}_K - \mathbf{A}_1 L - \mathbf{A}_2 L^2 - \cdots - \mathbf{A}_p L^p)$. The model is assumed to be stable (see [TS] **varstable**), so $\bar{\mathbf{A}}^{-1}$, the matrix of estimated long-run effects of the reduced-form VAR shocks, is well defined. Constraining $\mathbf{A}$ to be an identity matrix allows us to rewrite this equation as

$$\mathbf{y}_t = \bar{\mathbf{A}}^{-1}\mathbf{B}\mathbf{e}_t$$

which implies that $\boldsymbol{\Sigma} = \mathbf{B}\mathbf{B}'$. Thus $\mathbf{C} = \bar{\mathbf{A}}^{-1}\mathbf{B}$ is the matrix of long-run responses to the orthogonalized shocks, and

$$\mathbf{y}_t = \mathbf{C}\mathbf{e}_t$$

In long-run models, the constraints are placed on the elements of $\mathbf{C}$, and the free parameters are estimated. These constraints are often exclusion restrictions. For instance, constraining $\mathbf{C}[1,2]$ to be zero can be interpreted as setting the long-run response of variable 1 to the structural shocks driving variable 2 to be zero.

Stata's `svar` command estimates the parameters of structural VARs. See [TS] **var svar** for more information and examples.

IRFs and FEVDs

IRFs describe how the K endogenous variables react over time to a one-time shock to one of the K disturbances. Because the disturbances may be contemporaneously correlated, these functions do not explain how variable i reacts to a one-time increase in the innovation to variable j after s periods, holding everything else constant. To explain this, we must start with orthogonalized innovations so that the assumption to hold everything else constant is reasonable. Recursive VARs use a Cholesky decomposition to orthogonalize the disturbances and thereby obtain structurally interpretable IRFs. Structural VARs use theory to impose sufficient restrictions, which need not be recursive, to decompose the contemporaneous correlations into orthogonal components.

FEVDs are another tool for interpreting how the orthogonalized innovations affect the K variables over time. The FEVD from j to i gives the fraction of the s-step forecast-error variance of variable i that can be attributed to the jth orthogonalized innovation.

Dynamic–multiplier functions describe how the endogenous variables react over time to a unit change in an exogenous variable. This is a different experiment from that in IRFs and FEVDs because dynamic-multiplier functions consider a change in an exogenous variable instead of a shock to an endogenous variable.

`irf create` estimates IRFs, Cholesky orthogonalized IRFs, dynamic-multiplier functions, and structural IRFs and their standard errors. It also estimates Cholesky and structural FEVDs. The `irf graph`, `irf cgraph`, `irf ograph`, `irf table`, and `irf ctable` commands graph and tabulate these estimates. Stata also has several other commands to manage IRF and FEVD results. See [TS] **irf** for a description of these commands.

`fcast compute` computes dynamic forecasts and their standard errors from VARs. `fcast graph` graphs the forecasts that are generated using `fcast compute`.

VARs allow researchers to investigate whether one variable is useful in predicting another variable. A variable x is said to Granger-cause a variable y if, given the past values of y, past values of x are useful for predicting y. The Stata command `vargranger` performs Wald tests to investigate Granger causality between the variables in a VAR.

References

Amisano, G., and C. Giannini. 1997. *Topics in Structural VAR Econometrics*. 2nd ed. Heidelberg: Springer.

Hamilton, J. D. 1994. *Time Series Analysis*. Princeton: Princeton University Press.

Lütkepohl, H. 2005. *New Introduction to Multiple Time Series Analysis*. New York: Springer.

Stock, J. H., and M. W. Watson. 2001. Vector autoregressions. *Journal of Economic Perspectives* 15: 101–115.

Watson, M. W. 1994. Vector autoregressions and cointegration. In Vol. IV of *Handbook of Econometrics*, ed. R. F. Engle and D. L. McFadden. Amsterdam: Elsevier.

Also see

Title

var — Vector autoregressive models

Syntax

var *depvarlist* $\left[\text{if} \right]$ $\left[\text{in} \right]$ $\left[, \text{options} \right]$

options	Description
Model	
<u>no</u>constant	suppress constant term
<u>lags</u>(*numlist*)	use lags *numlist* in the VAR
<u>ex</u>og(*varlist*)	use exogenous variables *varlist*
Model 2	
<u>constraints</u>(*numlist*)	apply specified linear constraints
<u>nolog</u>	suppress SURE iteration log
<u>iter</u>ate(#)	set maximum number of iterations for SURE; default is iterate(1600)
<u>tol</u>erance(#)	set convergence tolerance of SURE
<u>noisure</u>	use one-step SURE
dfk	make small-sample degrees-of-freedom adjustment
<u>small</u>	report small-sample t and F statistics
<u>nobigf</u>	do not compute parameter vector for coefficients implicitly set to zero
Reporting	
<u>level</u>(#)	set confidence level; default is level(95)
lutstats	report Lütkepohl lag-order selection statistics
<u>nocnsreport</u>	do not display constraints
display_options	control column formats, row spacing, and line width
<u>coeflegend</u>	display legend instead of statistics

You must tsset your data before using var; see [TS] **tsset**.

depvarlist and *varlist* may contain time-series operators; see [U] **11.4.4 Time-series varlists**.

by, rolling, statsby, and xi are allowed; see [U] **11.1.10 Prefix commands**.

coeflegend does not appear in the dialog box.

See [U] **20 Estimation and postestimation commands** for more capabilities of estimation commands.

Menu

Statistics > Multivariate time series > Vector autoregression (VAR)

Description

var fits a multivariate time-series regression of each dependent variable on lags of itself and on lags of all the other dependent variables. var also fits a variant of vector autoregressive (VAR) models known as the VARX model, which also includes exogenous variables. See [TS] **var intro** for a list of commands that are used in conjunction with var.

Options

‎⎍ Model ⎍

noconstant; see [R] **estimation options**.

lags(*numlist*) specifies the lags to be included in the model. The default is lags(1 2). This option takes a *numlist* and not simply an integer for the maximum lag. For example, lags(2) would include only the second lag in the model, whereas lags(1/2) would include both the first and second lags in the model. See [U] **11.1.8 numlist** and [U] **11.4.4 Time-series varlists** for more discussion of numlists and lags.

exog(*varlist*) specifies a list of exogenous variables to be included in the VAR.

‎⎍ Model 2 ⎍

constraints(*numlist*); see [R] **estimation options**.

nolog suppresses the log from the iterated seemingly unrelated regression algorithm. By default, the iteration log is displayed when the coefficients are estimated through iterated seemingly unrelated regression. When the constraints() option is not specified, the estimates are obtained via OLS, and nolog has no effect. For this reason, nolog can be specified only when constraints() is specified. Similarly, nolog cannot be combined with noisure.

iterate(#) specifies an integer that sets the maximum number of iterations when the estimates are obtained through iterated seemingly unrelated regression. By default, the limit is 1,600. When constraints() is not specified, the estimates are obtained using OLS, and iterate() has no effect. For this reason, iterate() can be specified only when constraints() is specified. Similarly, iterate() cannot be combined with noisure.

tolerance(#) specifies a number greater than zero and less than 1 for the convergence tolerance of the iterated seemingly unrelated regression algorithm. By default, the tolerance is 1e-6. When the constraints() option is not specified, the estimates are obtained using OLS, and tolerance() has no effect. For this reason, tolerance() can be specified only when constraints() is specified. Similarly, tolerance() cannot be combined with noisure.

noisure specifies that the estimates in the presence of constraints be obtained through one-step seemingly unrelated regression. By default, var obtains estimates in the presence of constraints through iterated seemingly unrelated regression. When constraints() is not specified, the estimates are obtained using OLS, and noisure has no effect. For this reason, noisure can be specified only when constraints() is specified.

dfk specifies that a small-sample degrees-of-freedom adjustment be used when estimating Σ, the error variance–covariance matrix. Specifically, $1/(T - \overline{m})$ is used instead of the large-sample divisor $1/T$, where $\overline{m}$ is the average number of parameters in the functional form for $\mathbf{y}_t$ over the K equations.

small causes var to report small-sample t and F statistics instead of the large-sample normal and chi-squared statistics.

nobigf requests that var not save the estimated parameter vector that incorporates coefficients that have been implicitly constrained to be zero, such as when some lags have been omitted from a model. e(bf) is used for computing asymptotic standard errors in the postestimation commands irf create and fcast compute; see [TS] **irf create** and [TS] **fcast compute**. Therefore, specifying nobigf implies that the asymptotic standard errors will not be available from irf create and fcast compute. See *Fitting models with some lags excluded*.

[Reporting |

level(#); see [R] **estimation options**.

lutstats specifies that the Lütkepohl (2005) versions of the lag-order selection statistics be reported. See *Methods and formulas* in [TS] **varsoc** for a discussion of these statistics.

nocnsreport; see [R] **estimation options**.

display_options: vsquish, cformat(%*fmt*), pformat(%*fmt*), sformat(%*fmt*), and nolstretch; see [R] **estimation options**.

The following option is available with var but is not shown in the dialog box:

coeflegend; see [R] **estimation options**.

Remarks

Remarks are presented under the following headings:

> *Introduction*
> *Fitting models with some lags excluded*
> *Fitting models with exogenous variables*
> *Fitting models with constraints on the coefficients*

Introduction

A VAR is a model in which K variables are specified as linear functions of p of their own lags, p lags of the other $K - 1$ variables, and possibly exogenous variables. A VAR with p lags is usually denoted a VAR(p). For more information, see [TS] **var intro**.

▷ Example 1: VAR model

To illustrate the basic usage of var, we replicate the example in Lütkepohl (2005, 77–78). The data consists of three variables: the first difference of the natural log of investment, dln_inv; the first difference of the natural log of income, dln_inc; and the first difference of the natural log of consumption, dln_consump. The dataset contains data through the fourth quarter of 1982, though Lütkepohl uses only the observations through the fourth quarter of 1978.

```
. use http://www.stata-press.com/data/r12/lutkepohl2
(Quarterly SA West German macro data, Bil DM, from Lutkepohl 1993 Table E.1)

. tsset
        time variable:  qtr, 1960q1 to 1982q4
                delta:  1 quarter
```

```
. var dln_inv dln_inc dln_consump if qtr<=tq(1978q4), lutstats dfk
Vector autoregression
```

Sample: 1960q4 - 1978q4   No. of obs   =   73
Log likelihood =   606.307   (lutstats) AIC   = -24.63163
FPE   =   2.18e-11   HQIC   = -24.40656
Det(Sigma_ml) =   1.23e-11   SBIC   = -24.06686

Equation	Parms	RMSE	R-sq	chi2	P>chi2
dln_inv	7	.046148	0.1286	9.736909	0.1362
dln_inc	7	.011719	0.1142	8.508289	0.2032
dln_consump	7	.009445	0.2513	22.15096	0.0011

	Coef.	Std. Err.	z	P>|z|	[95% Conf. Interval]	
dln_inv						
dln_inv						
L1.	-.3196318	.1254564	-2.55	0.011	-.5655218	-.0737419
L2.	-.1605508	.1249066	-1.29	0.199	-.4053633	.0842616
dln_inc						
L1.	.1459851	.5556664	0.27	0.789	-.9235013	1.215472
L2.	.1146009	.5345709	0.21	0.830	-.9331388	1.162341
dln_consump						
L1.	.9612288	.6643086	1.45	0.148	-.3407922	2.26325
L2.	.9344001	.6650949	1.40	0.160	-.369162	2.237962
_cons	-.0167221	.0172264	-0.97	0.332	-.0504852	.0170409
dln_inc						
dln_inv						
L1.	.0439309	.0318592	1.38	0.168	-.018512	.1063739
L2.	.0500302	.0317196	1.58	0.115	-.0121391	.1121995
dln_inc						
L1.	-.1527311	.1385702	-1.10	0.270	-.4243237	.1188615
L2.	.0191634	.1357525	0.14	0.888	-.2469067	.2852334
dln_consump						
L1.	.2884992	.168699	1.71	0.087	-.0421448	.6191431
L2.	-.0102	.1688987	-0.06	0.952	-.3412354	.3208353
_cons	.0157672	.0043746	3.60	0.000	.0071932	.0243412
dln_consump						
dln_inv						
L1.	-.002423	.0256763	-0.09	0.925	-.0527476	.0479016
L2.	.0338806	.0255638	1.33	0.185	-.0162235	.0839847
dln_inc						
L1.	.2248134	.1116778	2.01	0.044	.005929	.4436978
L2.	.3549135	.1094069	3.24	0.001	.1404798	.5693471
dln_consump						
L1.	-.2639695	.1359595	-1.94	0.052	-.5304451	.0025062
L2.	-.0222264	.1361204	-0.16	0.870	-.2890175	.2445646
_cons	.0129258	.0035256	3.67	0.000	.0060157	.0198358

The output has two parts: a header and the standard Stata output table for the coefficients, standard errors, and confidence intervals. The header contains summary statistics for each equation in the VAR and statistics used in selecting the lag order of the VAR. Although there are standard formulas for all the lag-order statistics, Lütkepohl (2005) gives different versions of the three information criteria that drop the constant term from the likelihood. To obtain the Lütkepohl (2005) versions, we specified the `lutstats` option. The formulas for the standard and Lütkepohl versions of these statistics are given in *Methods and formulas* of [TS] **varsoc**.

The `dfk` option specifies that the small-sample divisor $1/(T - \overline{m})$ be used in estimating Σ instead of the maximum likelihood (ML) divisor $1/T$, where $\overline{m}$ is the average number of parameters included in each of the K equations. All the lag-order statistics are computed using the ML estimator of Σ. Thus, specifying `dfk` will not change the computed lag-order statistics, but it will change the estimated variance–covariance matrix. Also, when `dfk` is specified, a `dfk`-adjusted log likelihood is computed and saved in `e(ll_dfk)`.

◁

The `lag()` option takes a *numlist* of lags. To specify a model that includes the first and second lags, type

        . var y1 y2 y3, lags(1/2)

not

        . var y1 y2 y3, lags(2)

because the latter specification would fit a model that included only the second lag.

Fitting models with some lags excluded

To fit a model that has only a fourth lag, that is,

$$\mathbf{y}_t = \mathbf{v} + \mathbf{A}_4 \mathbf{y}_{t-4} + \mathbf{u}_t$$

you would specify the `lags(4)` option. Doing so is equivalent to fitting the more general model

$$\mathbf{y}_t = \mathbf{v} + \mathbf{A}_1 \mathbf{y}_{t-1} + \mathbf{A}_2 \mathbf{y}_{t-2} + \mathbf{A}_3 \mathbf{y}_{t-3} + \mathbf{A}_4 \mathbf{y}_{t-4} + \mathbf{u}_t$$

with $\mathbf{A}_1$, $\mathbf{A}_2$, and $\mathbf{A}_3$ constrained to be $\mathbf{0}$. When you fit a model with some lags excluded, `var` estimates the coefficients included in the specification ($\mathbf{A}_4$ here) and saves these estimates in `e(b)`. To obtain the asymptotic standard errors for impulse–response functions and other postestimation statistics, Stata needs the complete set of parameter estimates, including those that are constrained to be zero; `var` stores them in `e(bf)`. Because you can specify models for which the full set of parameter estimates exceeds Stata's limit on the size of matrices, the `nobigf` option specifies that `var` not compute and store `e(bf)`. This means that the asymptotic standard errors of the postestimation functions cannot be obtained, although bootstrap standard errors are still available. Building `e(bf)` can be time consuming, so if you do not need this full matrix, and speed is an issue, use `nobigf`.

Fitting models with exogenous variables

▷ Example 2: VAR model with exogenous variables

We use the exog() option to include exogenous variables in a VAR.

```
. var dln_inc dln_consump if qtr<=tq(1978q4), dfk exog(dln_inv)
Vector autoregression
```

```
Sample:  1960q4 - 1978q4                    No. of obs      =         73
Log likelihood =  478.5663                  AIC             = -12.78264
FPE            = 9.64e-09                    HQIC            = -12.63259
Det(Sigma_ml)  = 6.93e-09                    SBIC            = -12.40612
```

Equation	Parms	RMSE	R-sq	chi2	P>chi2
dln_inc	6	.011917	0.0702	5.059587	0.4087
dln_consump	6	.009197	0.2794	25.97262	0.0001

	Coef.	Std. Err.	z	P>|z|	[95% Conf. Interval]	
dln_inc						
dln_inc						
L1.	-.1343345	.1391074	-0.97	0.334	-.4069801	.1383111
L2.	.0120331	.1380346	0.09	0.931	-.2585097	.2825759
dln_consump						
L1.	.3235342	.1652769	1.96	0.050	-.0004027	.647471
L2.	.0754177	.1648624	0.46	0.647	-.2477066	.398542
dln_inv	.0151546	.0302319	0.50	0.616	-.0440987	.074408
_cons	.0145136	.0043815	3.31	0.001	.0059259	.0231012
dln_consump						
dln_inc						
L1.	.2425719	.1073561	2.26	0.024	.0321578	.452986
L2.	.3487949	.1065281	3.27	0.001	.1400036	.5575862
dln_consump						
L1.	-.3119629	.1275524	-2.45	0.014	-.5619611	-.0619648
L2.	-.0128502	.1272325	-0.10	0.920	-.2622213	.2365209
dln_inv	.0503616	.0233314	2.16	0.031	.0046329	.0960904
_cons	.0131013	.0033814	3.87	0.000	.0064738	.0197288

All the postestimation commands for analyzing VARs work when exogenous variables are included in a model, but the asymptotic standard errors for the h-step-ahead forecasts are not available.

◁

Fitting models with constraints on the coefficients

var permits model specifications that include constraints on the coefficient, though var does not allow for constraints on Σ. See [TS] **var intro** and [TS] **var svar** for ways to constrain Σ.

> Example 3: VAR model with constraints

In the first example, we fit a full VAR(2) to a three-equation model. The coefficients in the equation for dln_inv were jointly insignificant, as were the coefficients in the equation for dln_inc; and many individual coefficients were not significantly different from zero. In this example, we constrain the coefficient on L2.dln_inc in the equation for dln_inv and the coefficient on L2.dln_consump in the equation for dln_inc to be zero.

```
. constraint 1 [dln_inv]L2.dln_inc = 0

. constraint 2 [dln_inc]L2.dln_consump = 0

. var dln_inv dln_inc dln_consump if qtr<=tq(1978q4), lutstats dfk
> constraints(1 2)
Estimating VAR coefficients

Iteration 1:    tolerance =  .00737681
Iteration 2:    tolerance =  3.998e-06
Iteration 3:    tolerance =  2.730e-09

Vector autoregression

Sample:  1960q4 - 1978q4                  No. of obs       =          73
Log likelihood =  606.2804               (lutstats) AIC    = -31.69254
FPE            =  1.77e-14                           HQIC   = -31.46747
Det(Sigma_ml)  =  1.05e-14                           SBIC   = -31.12777

Equation          Parms      RMSE     R-sq      chi2     P>chi2
-------------------------------------------------------------------
dln_inv              6      .043895   0.1280   9.842338   0.0798
dln_inc              6      .011143   0.1141   8.584446   0.1268
dln_consump          7      .008981   0.2512  22.86958    0.0008
-------------------------------------------------------------------

 ( 1)   [dln_inv]L2.dln_inc = 0
 ( 2)   [dln_inc]L2.dln_consump = 0
-------------------------------------------------------------------------------
             |     Coef.   Std. Err.      z    P>|z|    [95% Conf. Interval]
-------------+-----------------------------------------------------------------
dln_inv      |
   dln_inv   |
        L1.  | -.320713    .1247512    -2.57   0.010   -.5652208   -.0762051
        L2.  | -.1607084   .124261     -1.29   0.196   -.4042555    .0828386
             |
   dln_inc   |
        L1.  |  .1195448   .5295669     0.23   0.821   -.9183873   1.157477
        L2.  |  5.66e-19   9.33e-18     0.06   0.952   -1.77e-17   1.89e-17
             |
dln_consump  |
        L1.  |  1.009281   .623501      1.62   0.106   -.2127586   2.231321
        L2.  |  1.008079   .5713486     1.76   0.078   -.1117438   2.127902
             |
      _cons  | -.0162102   .016893     -0.96   0.337   -.0493199    .0168995
-------------------------------------------------------------------------------
```

dln_inc						
dln_inv						
L1.	.0435712	.0309078	1.41	0.159	−.017007	.1041495
L2.	.0496788	.0306455	1.62	0.105	−.0103852	.1097428
dln_inc						
L1.	−.1555119	.1315854	−1.18	0.237	−.4134146	.1023908
L2.	.0122353	.1165811	0.10	0.916	−.2162595	.2407301
dln_consump						
L1.	.29286	.1568345	1.87	0.062	−.01453	.6002501
L2.	−1.53e−18	1.89e−17	−0.08	0.935	−3.85e−17	3.55e−17
_cons	.015689	.003819	4.11	0.000	.0082039	.0231741
dln_consump						
dln_inv						
L1.	−.0026229	.0253538	−0.10	0.918	−.0523154	.0470696
L2.	.0337245	.0252113	1.34	0.181	−.0156888	.0831378
dln_inc						
L1.	.2224798	.1094349	2.03	0.042	.0079912	.4369683
L2.	.3469758	.1006026	3.45	0.001	.1497984	.5441532
dln_consump						
L1.	−.2600227	.1321622	−1.97	0.049	−.519056	−.0009895
L2.	−.0146825	.1117618	−0.13	0.895	−.2337315	.2043666
_cons	.0129149	.003376	3.83	0.000	.0062981	.0195317

None of the free parameter estimates changed by much. Whereas the coefficients in the equation dln_inv are now significant at the 10% level, the coefficients in the equation for dln_inc remain jointly insignificant.

◁

Saved results

var saves the following in e():

Scalars

e(N)	number of observations
e(N_gaps)	number of gaps in sample
e(k)	number of parameters
e(k_eq)	number of equations in e(b)
e(k_dv)	number of dependent variables
e(df_eq)	average number of parameters in an equation
e(df_m)	model degrees of freedom
e(df_r)	residual degrees of freedom (small only)
e(ll)	log likelihood
e(ll_dfk)	dfk adjusted log likelihood (dfk only)
e(obs_#)	number of observations on equation #
e(k_#)	number of parameters in equation #
e(df_m#)	model degrees of freedom for equation #
e(df_r#)	residual degrees of freedom for equation # (small only)
e(r2_#)	R-squared for equation #
e(ll_#)	log likelihood for equation #
e(chi2_#)	x^2 for equation #
e(F_#)	F statistic for equation # (small only)
e(rmse_#)	root mean squared error for equation #
e(aic)	Akaike information criterion
e(hqic)	Hannan–Quinn information criterion
e(sbic)	Schwarz–Bayesian information criterion
e(fpe)	final prediction error
e(mlag)	highest lag in VAR
e(tmin)	first time period in sample
e(tmax)	maximum time
e(detsig)	determinant of e(Sigma)
e(detsig_ml)	determinant of $\widehat{\Sigma}_{ml}$
e(rank)	rank of e(V)

Macros
e(cmd)	var
e(cmdline)	command as typed
e(depvar)	names of dependent variables
e(endog)	names of endogenous variables, if specified
e(exog)	names of exogenous variables, and their lags, if specified
e(exogvars)	names of exogenous variables, if specified
e(eqnames)	names of equations
e(lags)	lags in model
e(exlags)	lags of exogenous variables in model, if specified
e(title)	title in estimation output
e(nocons)	nocons, if noconstant is specified
e(constraints)	constraints, if specified
e(cnslist_var)	list of specified constraints
e(small)	small, if specified
e(lutstats)	lutstats, if specified
e(timevar)	time variable specified in tsset
e(tsfmt)	format for the current time variable
e(dfk)	dfk, if specified
e(properties)	b V
e(predict)	program used to implement predict
e(marginsok)	predictions allowed by margins
e(marginsnotok)	predictions disallowed by margins

Matrices
e(b)	coefficient vector
e(Cns)	constraints matrix
e(Sigma)	$\widehat{\boldsymbol{\Sigma}}$ matrix
e(V)	variance–covariance matrix of the estimators
e(bf)	constrained coefficient vector
e(exlagsm)	matrix mapping lags to exogenous variables
e(G)	Gamma matrix; see *Methods and formulas*

Functions
e(sample)	marks estimation sample

Methods and formulas

var is implemented as an ado-file.

When there are no constraints placed on the coefficients, the VAR(p) is a seemingly unrelated regression model with the same explanatory variables in each equation. As discussed in Lütkepohl (2005) and Greene (2008, 696), performing linear regression on each equation produces the maximum likelihood estimates of the coefficients. The estimated coefficients can then be used to calculate the residuals, which in turn are used to estimate the cross-equation error variance–covariance matrix $\boldsymbol{\Sigma}$.

Per Lütkepohl (2005), we write the VAR(p) with exogenous variables as

$$\mathbf{y}_t = \mathbf{A}\mathbf{Y}_{t-1} + \mathbf{B}_0\mathbf{x}_t + \mathbf{u}_t \tag{5}$$

where

$\mathbf{y}_t$ is the $K \times 1$ vector of endogenous variables,

$\mathbf{A}$ is a $K \times Kp$ matrix of coefficients,

$\mathbf{B}_0$ is a $K \times M$ matrix of coefficients,

x_t is the $M \times 1$ vector of exogenous variables,

u_t is the $K \times 1$ vector of white noise innovations, and

Y_t is the $Kp \times 1$ matrix given by $Y_t = \begin{pmatrix} y_t \\ \vdots \\ y_{t-p+1} \end{pmatrix}$

Although (5) is easier to read, the formulas are much easier to manipulate if it is instead written as

$$Y = BZ + U$$

where

$$\begin{aligned} Y &= (y_1, \ldots, y_T) & Y \text{ is } K \times T \\ B &= (A, B_0) & B \text{ is } K \times (Kp + M) \\ Z &= \begin{pmatrix} Y_0 \ldots, Y_{T-1} \\ x_1 \ldots, x_T \end{pmatrix} & Z \text{ is } (Kp + M) \times T \\ U &= (u_1, \ldots, u_T) & U \text{ is } K \times T \end{aligned}$$

Intercept terms in the model are included in x_t. If there are no exogenous variables and no intercept terms in the model, x_t is empty.

The coefficients are estimated by iterated seemingly unrelated regression. Because the estimation is actually performed by reg3, the methods are documented in [R] **reg3**. See [P] **makecns** for more on estimation with constraints.

Let $\widehat{U}$ be the matrix of residuals that are obtained via $Y - \widehat{B}Z$, where $\widehat{B}$ is the matrix of estimated coefficients. Then the estimator of Σ is

$$\widehat{\Sigma} = \frac{1}{\widetilde{T}} \widehat{U}' \widehat{U}$$

By default, the maximum likelihood divisor of $\widetilde{T} = T$ is used. When dfk is specified, a small-sample degrees-of-freedom adjustment is used; then, $\widetilde{T} = T - \overline{m}$ where $\overline{m}$ is the average number of parameters per equation in the functional form for y_t over the K equations.

small specifies that Wald tests after var be assumed to have F or t distributions instead of chi-squared or standard normal distributions. The standard errors from each equation are computed using the degrees of freedom for the equation.

The "gamma" matrix saved in e(G) referred to in *Saved results* is the $(Kp+1) \times (Kp+1)$ matrix given by

$$\frac{1}{T} \sum_{t=1}^{T} (1, Y_t')(1, Y_t')'$$

The formulas for the lag-order selection criteria and the log likelihood are discussed in [TS] **varsoc**.

Acknowledgment

We thank Christopher F. Baum of Boston College for his helpful comments.

References

Greene, W. H. 2008. *Econometric Analysis.* 6th ed. Upper Saddle River, NJ: Prentice Hall.

Hamilton, J. D. 1994. *Time Series Analysis.* Princeton: Princeton University Press.

Lütkepohl, H. 1993. *Introduction to Multiple Time Series Analysis.* 2nd ed. New York: Springer.

———. 2005. *New Introduction to Multiple Time Series Analysis.* New York: Springer.

Stock, J. H., and M. W. Watson. 2001. Vector autoregressions. *Journal of Economic Perspectives* 15: 101–115.

Watson, M. W. 1994. Vector autoregressions and cointegration. In Vol. IV of *Handbook of Econometrics,* ed. R. F. Engle and D. L. McFadden. Amsterdam: Elsevier.

Also see

[TS] **var postestimation** — Postestimation tools for var

[TS] **tsset** — Declare data to be time-series data

[TS] **dfactor** — Dynamic-factor models

[TS] **mgarch** — Multivariate GARCH models

[TS] **sspace** — State-space models

[TS] **var svar** — Structural vector autoregressive models

[TS] **varbasic** — Fit a simple VAR and graph IRFs or FEVDs

[TS] **vec** — Vector error-correction models

[U] **20 Estimation and postestimation commands**

[TS] **var intro** — Introduction to vector autoregressive models

Title

> **var postestimation** — Postestimation tools for var

Description

The following postestimation commands are of special interest after `var`:

Command	Description
`fcast compute`	obtain dynamic forecasts
`fcast graph`	graph dynamic forecasts obtained from `fcast compute`
`irf`	create and analyze IRFs and FEVDs
`vargranger`	Granger causality tests
`varlmar`	LM test for autocorrelation in residuals
`varnorm`	test for normally distributed residuals
`varsoc`	lag-order selection criteria
`varstable`	check stability condition of estimates
`varwle`	Wald lag-exclusion statistics

For information about these commands, see the corresponding entries in this manual.

The following standard postestimation commands are also available:

Command	Description
`estat`	AIC, BIC, VCE, and estimation sample summary
`estimates`	cataloging estimation results
`lincom`	point estimates, standard errors, testing, and inference for linear combinations of coefficients
`lrtest`	likelihood-ratio test
`margins`	marginal means, predictive margins, marginal effects, and average marginal effects
`marginsplot`	graph the results from margins (profile plots, interaction plots, etc.)
`nlcom`	point estimates, standard errors, testing, and inference for nonlinear combinations of coefficients
`predict`	predictions, residuals, influence statistics, and other diagnostic measures
`predictnl`	point estimates, standard errors, testing, and inference for generalized predictions
`test`	Wald tests of simple and composite linear hypotheses
`testnl`	Wald tests of nonlinear hypotheses

See the corresponding entries in the *Base Reference Manual* for details.

Syntax for predict

predict [*type*] *newvar* [*if*] [*in*] [, *statistic* <u>equation</u>(*eqno* | *eqname*)]

statistic	Description
Main	
xb	linear prediction; the default
stdp	standard error of the linear prediction
<u>r</u>esiduals	residuals

These statistics are available both in and out of sample; type predict ... if e(sample) ... if wanted only for the estimation sample.

Menu

Statistics > Postestimation > Predictions, residuals, etc.

Options for predict

    [Main]

xb, the default, calculates the linear prediction for the specified equation.

stdp calculates the standard error of the linear prediction for the specified equation.

residuals calculates the residuals.

equation(*eqno* | *eqname*) specifies the equation to which you are referring.

> equation() is filled in with one *eqno* or *eqname* for options xb, stdp, and residuals. For example, equation(#1) would mean that the calculation is to be made for the first equation, equation(#2) would mean the second, and so on. You could also refer to the equation by its name; thus, equation(income) would refer to the equation named income and equation(hours), to the equation named hours.

> If you do not specify equation(), the results are the same as if you specified equation(#1).

For more information on using predict after multiple-equation estimation commands, see [R] **predict**.

Remarks

Remarks are presented under the following headings:

> *Model selection and inference*
> *Forecasting*

Model selection and inference

See the following sections for information on model selection and inference after var.

> [TS] **irf** — Create and analyze IRFs, dynamic-multiplier functions, and FEVDs

> [TS] **vargranger** — Perform pairwise Granger causality tests after var or svar

> [TS] **varlmar** — Perform LM test for residual autocorrelation after var or svar

> [TS] **varnorm** — Test for normally distributed disturbances after var or svar

> [TS] **varsoc** — Obtain lag-order selection statistics for VARs and VECMs

> [TS] **varstable** — Check the stability condition of VAR or SVAR estimates

> [TS] **varwle** — Obtain Wald lag-exclusion statistics after var or svar

Forecasting

Two types of forecasts are available after you fit a VAR(p): a one-step-ahead forecast and a dynamic h-step-ahead forecast.

The one-step-ahead forecast produces a prediction of the value of an endogenous variable in the current period by using the estimated coefficients, the past values of the endogenous variables, and any exogenous variables. If you include contemporaneous values of exogenous variables in your model, you must have observations on the exogenous variables that are contemporaneous with the period in which the prediction is being made to compute the prediction. In Stata terms, these one-step-ahead predictions are just the standard linear predictions available after any estimation command. Thus predict, xb eq(*eqno* | *eqname*) produces one-step-ahead forecasts for the specified equation. predict, stdp eq(*eqno* | *eqname*) produces the standard error of the linear prediction for the specified equation. The standard error of the forecast includes an estimate of the variability due to innovations, whereas the standard error of the linear prediction does not.

The dynamic h-step-ahead forecast begins by using the estimated coefficients, the lagged values of the endogenous variables, and any exogenous variables to predict one step ahead for each endogenous variable. Then the one-step-ahead forecast produces two-step-ahead forecasts for each endogenous variable. The process continues for h periods. Because each step uses the predictions of the previous steps, these forecasts are known as dynamic forecasts. See the following sections for information on obtaining forecasts after svar:

[TS] **fcast compute** — Compute dynamic forecasts of dependent variables after var, svar, or vec

[TS] **fcast graph** — Graph forecasts of variables computed by fcast compute

Methods and formulas

All postestimation commands listed above are implemented as ado-files.

Formulas for predict

predict with the xb option provides the one-step-ahead forecast. If exogenous variables are specified, the forecast is conditional on the exogenous x_t variables. Specifying the residuals option causes predict to calculate the errors of the one-step-ahead forecasts. Specifying the stdp option causes predict to calculate the standard errors of the one-step-ahead forecasts.

Also see

[TS] **var** — Vector autoregressive models

[U] **20 Estimation and postestimation commands**

Title

> **var svar** — Structural vector autoregressive models

Syntax

Short-run constraints

svar *depvarlist* $\left[\,if\,\right]$ $\left[\,in\,\right]$, { <u>acon</u>straints(*constraints$_a$*) <u>aeq</u>(*matrix$_{\text{aeq}}$*)

  <u>acns</u>(*matrix$_{\text{acns}}$*) <u>bcon</u>straints(*constraints$_b$*) <u>beq</u>(*matrix$_{\text{beq}}$*) <u>bcns</u>(*matrix$_{\text{bcns}}$*) }

  $\left[\,short\_run\_options\,\right]$

Long-run constraints

svar *depvarlist* $\left[\,if\,\right]$ $\left[\,in\,\right]$, { <u>lrcon</u>straints(*constraints$_{\text{lr}}$*) <u>lreq</u>(*matrix$_{\text{lreq}}$*)

  <u>lrcns</u>(*matrix$_{\text{lrcns}}$*) } $\left[\,long\_run\_options\,\right]$

short_run_options	Description
Model	
<u>nocons</u>tant	suppress constant term
* <u>acon</u>straints(*constraints$_a$*)	apply previously defined *constraints$_a$* to **A**
* <u>aeq</u>(*matrix$_{aeq}$*)	define and apply to **A** equality constraint matrix *matrix$_{aeq}$*
* <u>acns</u>(*matrix$_{acns}$*)	define and apply to **A** cross-parameter constraint matrix *matrix$_{acns}$*
* <u>bcons</u>traints(*constraints$_b$*)	apply previously defined *constraints$_b$* to **B**
* <u>beq</u>(*matrix$_{beq}$*)	define and apply to **B** equality constraint matrix *matrix$_{beq}$*
* <u>bcns</u>(*matrix$_{bcns}$*)	define and apply to **B** cross-parameter constraint *matrix$_{bcns}$*
<u>lags</u>(*numlist*)	use lags *numlist* in the underlying VAR
Model 2	
<u>exog</u>(*varlist$_{exog}$*)	use exogenous variables *varlist*
<u>varc</u>onstraints(*constraints$_v$*)	apply *constraints$_v$* to underlying VAR
<u>noislog</u>	suppress SURE iteration log
<u>isite</u>rate(*#*)	set maximum number of iterations for SURE; default is isiterate(1600)
<u>istol</u>erance(*#*)	set convergence tolerance of SURE
<u>noisure</u>	use one-step SURE
dfk	make small-sample degrees-of-freedom adjustment
<u>small</u>	report small-sample t and F statistics
<u>noiden</u>check	do not check for local identification
<u>nobigf</u>	do not compute parameter vector for coefficients implicitly set to zero
Reporting	
<u>level</u>(*#*)	set confidence level; default is level(95)
<u>full</u>	show constrained parameters in table
var	display underlying var output
<u>lutstats</u>	report Lütkepohl lag-order selection statistics
<u>nocns</u>report	do not display constraints
display_options	control column formats
Maximization	
maximize_options	control the maximization process; seldom used
<u>coefl</u>egend	display legend instead of statistics

* aconstraints(*constraints$_a$*), aeq(*matrix$_{aeq}$*), acns(*matrix$_{acns}$*), bconstraints(*constraints$_b$*), beq(*matrix$_{beq}$*), bcns(*matrix$_{bcns}$*): at least one of these options must be specified.
coeflegend does not appear in the dialog box.

long_run_options	Description
Model	
<u>nocons</u>tant	suppress constant term
* <u>lrc</u>onstraints(*constraints*$_{\text{lr}}$)	apply previously defined *constraints*$_{\text{lr}}$ to **C**
* <u>lre</u>q(*matrix*$_{\text{lreq}}$)	define and apply to **C** equality constraint matrix *matrix*$_{\text{lreq}}$
* <u>lrc</u>ns(*matrix*$_{\text{lrcns}}$)	define and apply to **C** cross-parameter constraint matrix *matrix*$_{\text{lrcns}}$
<u>lags</u>(*numlist*)	use lags *numlist* in the underlying VAR
Model 2	
<u>exog</u>(*varlist*$_{\text{exog}}$)	use exogenous variables *varlist*
<u>varc</u>onstraints(*constraints*$_v$)	apply *constraints*$_v$ to underlying VAR
noislog	suppress SURE iteration log
<u>isiter</u>ate(#)	set maximum number of iterations for SURE; default is isiterate(1600)
<u>istol</u>erance(#)	set convergence tolerance of SURE
<u>nois</u>ure	use one-step SURE
dfk	make small-sample degrees-of-freedom adjustment
<u>sm</u>all	report small-sample t and F statistics
<u>noiden</u>check	do not check for local identification
<u>nobigf</u>	do not compute parameter vector for coefficients implicitly set to zero
Reporting	
<u>level</u>(#)	set confidence level; default is level(95)
<u>full</u>	show constrained parameters in table
var	display underlying var output
lutstats	report Lütkepohl lag-order selection statistics
<u>nocns</u>report	do not display constraints
display_options	control column formats
Maximization	
maximize_options	control the maximization process; seldom used
<u>coefl</u>egend	display legend instead of statistics

* <u>lrc</u>onstraints(*constraints*$_{\text{lr}}$), <u>lre</u>q(*matrix*$_{\text{lreq}}$), <u>lrc</u>ns(*matrix*$_{\text{lrcns}}$): at least one of these options must be specified.

coeflegend does not appear in the dialog box.

You must tsset your data before using svar; see [TS] **tsset**.

depvarlist and *varlist*$_{\text{exog}}$ may contain time-series operators; see [U] **11.4.4 Time-series varlists**.

by, rolling, statsby, and xi are allowed; see [U] **11.1.10 Prefix commands**.

See [U] **20 Estimation and postestimation commands** for more capabilities of estimation commands.

Menu

Statistics > Multivariate time series > Structural vector autoregression (SVAR)

Description

    svar fits a vector autoregressive model subject to short- or long-run constraints you place on the resulting impulse–response functions (IRFs). Economic theory typically motivates the constraints, allowing a causal interpretation of the IRFs to be made. See [TS] **var intro** for a list of commands that are used in conjunction with svar.

Options

       Model

noconstant; see [R] **estimation options**.

aconstraints(*constraints$_a$*), aeq(*matrix$_{\text{aeq}}$*), acns(*matrix$_{\text{acns}}$*)

bconstraints(*constraints$_b$*), beq(*matrix$_{\text{beq}}$*), bcns(*matrix$_{\text{bcns}}$*)

    These options specify the short-run constraints in an SVAR. To specify a short-run SVAR model, you must specify at least one of these options. The first list of options specifies constraints on the parameters of the **A** matrix; the second list specifies constraints on the parameters of the **B** matrix (see *Short-run SVAR models*). If at least one option is selected from the first list and none are selected from the second list, svar sets **B** to the identity matrix. Similarly, if at least one option is selected from the second list and none are selected from the first list, svar sets **A** to the identity matrix.

    None of these options may be specified with any of the options that define long-run constraints.

    aconstraints(*constraints$_a$*) specifies a *numlist* of previously defined Stata constraints that are to be applied to **A** during estimation.

    aeq(*matrix$_{\text{aeq}}$*) specifies a matrix that defines a set of equality constraints. This matrix must be square with dimension equal to the number of equations in the underlying VAR. The elements of this matrix must be *missing* or real numbers. A missing value in the (i, j) element of this matrix specifies that the (i, j) element of **A** is a free parameter. A real number in the (i, j) element of this matrix constrains the (i, j) element of **A** to this real number. For example,

$$\mathbf{A} = \begin{bmatrix} 1 & 0 \\ . & 1.5 \end{bmatrix}$$

    specifies that $\mathbf{A}[1, 1] = 1$, $\mathbf{A}[1, 2] = 0$, $\mathbf{A}[2, 2] = 1.5$, and $\mathbf{A}[2, 1]$ is a free parameter.

acns(*matrix*$_{\text{acns}}$) specifies a matrix that defines a set of exclusion or cross-parameter equality constraints on **A**. This matrix must be square with dimension equal to the number of equations in the underlying VAR. Each element of this matrix must be *missing*, 0, or a positive integer. A missing value in the (i, j) element of this matrix specifies that no constraint be placed on this element of **A**. A zero in the (i, j) element of this matrix constrains the (i, j) element of **A** to be zero. Any strictly positive integers must be in two or more elements of this matrix. A strictly positive integer in the (i, j) element of this matrix constrains the (i, j) element of **A** to be equal to all the other elements of **A** that correspond to elements in this matrix that contain the same integer. For example, consider the matrix

$$\mathbf{A} = \begin{bmatrix} \cdot & 1 \\ 1 & 0 \end{bmatrix}$$

Specifying acns(A) in a two-equation SVAR constrains $\mathbf{A}[2, 1] = \mathbf{A}[1, 2]$ and $\mathbf{A}[2, 2] = 0$ while leaving $\mathbf{A}[1, 1]$ free.

bconstraints(*constraints*$_b$) specifies a *numlist* of previously defined Stata constraints to be applied to **B** during estimation.

beq(*matrix*$_{\text{beq}}$) specifies a matrix that defines a set of equality constraints. This matrix must be square with dimension equal to the number of equations in the underlying VAR. The elements of this matrix must be either *missing* or real numbers. The syntax of implied constraints is analogous to the one described in aeq(), except that it applies to **B** rather than to **A**.

bcns(*matrix*$_{\text{bcns}}$) specifies a matrix that defines a set of exclusion or cross-parameter equality constraints on **B**. This matrix must be square with dimension equal to the number of equations in the underlying VAR. Each element of this matrix must be *missing*, 0, or a positive integer. The format of the implied constraints is the same as the one described in the acns() option above.

lrconstraints(*constraints*$_{\text{lr}}$), lreq(*matrix*$_{\text{lreq}}$), lrcns(*matrix*$_{\text{lrcns}}$)

These options specify the long-run constraints in an SVAR. To specify a long-run SVAR model, you must specify at least one of these options. The list of options specifies constraints on the parameters of the long-run **C** matrix (see *Long-run SVAR models* for the definition of **C**). None of these options may be specified with any of the options that define short-run constraints.

lrconstraints(*constraints*$_{\text{lr}}$) specifies a *numlist* of previously defined Stata constraints to be applied to **C** during estimation.

lreq(*matrix*$_{\text{lreq}}$) specifies a matrix that defines a set of equality constraints on the elements of **C**. This matrix must be square with dimension equal to the number of equations in the underlying VAR. The elements of this matrix must be either *missing* or real numbers. The syntax of implied constraints is analogous to the one described in option aeq(), except that it applies to **C**.

lrcns(*matrix*$_{\text{lrcns}}$) specifies a matrix that defines a set of exclusion or cross-parameter equality constraints on **C**. This matrix must be square with dimension equal to the number of equations in the underlying VAR. Each element of this matrix must be *missing*, 0, or a positive integer. The syntax of the implied constraints is the same as the one described for the acns() option above.

lags(*numlist*) specifies the lags to be included in the underlying VAR model. The default is lags(1 2). This option takes a *numlist* and not simply an integer for the maximum lag. For instance, lags(2) would include only the second lag in the model, whereas lags(1/2) would include both the first and second lags in the model. See [U] **11.1.8 numlist** and [U] **11.4.4 Time-series varlists** for further discussion of *numlists* and lags.

⌐ Model 2 ⌐

exog(*varlist*$_{exog}$) specifies a list of exogenous variables to be included in the underlying VAR.

varconstraints(*constraints*$_v$) specifies a list of constraints to be applied to coefficients in the underlying VAR. Because svar estimates multiple equations, the constraints must specify the equation name for all but the first equation.

noislog prevents svar from displaying the iteration log from the iterated seemingly unrelated regression algorithm. When the varconstraints() option is not specified, the VAR coefficients are estimated via OLS, a noniterative procedure. As a result, noislog may be specified only with varconstraints(). Similarly, noislog may not be combined with noisure.

isiterate(#) sets the maximum number of iterations for the iterated, seemingly unrelated regression algorithm. The default limit is 1,600. When the varconstraints() option is not specified, the VAR coefficients are estimated via OLS, a noniterative procedure. As a result, isiterate() may be specified only with varconstraints(). Similarly, isiterate() may not be combined with noisure.

istolerance(#) specifies the convergence tolerance of the iterated seemingly unrelated regression algorithm. The default tolerance is 1e-6. When the varconstraints() option is not specified, the VAR coefficients are estimated via OLS, a noniterative procedure. As a result, istolerance() may be specified only with varconstraints(). Similarly, istolerance() may not be combined with noisure.

noisure specifies that the VAR coefficients be estimated via one-step seemingly unrelated regression when varconstraints() is specified. By default, svar estimates the coefficients in the VAR via iterated, seemingly unrelated regression when varconstraints() is specified. When the varconstraints() option is not specified, the VAR coefficient estimates are obtained via OLS, a noniterative procedure. As a result, noisure may be specified only with varconstraints().

dfk specifies that a small-sample degrees-of-freedom adjustment be used when estimating Σ, the covariance matrix of the VAR disturbances. Specifically, $1/(T - \overline{m})$ is used instead of the large-sample divisor $1/T$, where $\overline{m}$ is the average number of parameters in the functional form for $\mathbf{y}_t$ over the K equations.

small causes svar to calculate and report small-sample t and F statistics instead of the large-sample normal and chi-squared statistics.

noidencheck requests that the Amisano and Giannini (1997) check for local identification not be performed. This check is local to the starting values used. Because of this dependence on the starting values, you may wish to suppress this check by specifying the noidencheck option. However, be careful in specifying this option. Models that are not structurally identified can still converge, thereby producing meaningless results that only appear to have meaning.

nobigf requests that svar not save the estimated parameter vector that incorporates coefficients that have been implicitly constrained to be zero, such as when some lags have been omitted from a model. e(bf) is used for computing asymptotic standard errors in the postestimation commands irf create and fcast compute. Therefore, specifying nobigf implies that the asymptotic standard errors will not be available from irf create and fcast compute. See *Fitting models with some lags excluded* in [TS] **var**.

⌐ Reporting ⌐

level(#); see [R] **estimation options**.

full shows constrained parameters in table.

var specifies that the output from var also be displayed. By default, the underlying VAR is fit quietly.

lutstats specifies that the Lütkepohl versions of the lag-order selection statistics be reported. See *Methods and formulas* in [TS] **varsoc** for a discussion of these statistics.

nocnsreport; see [R] **estimation options**.

display_options: cformat(%*fmt*), pformat(%*fmt*), and sformat(%*fmt*); see [R] **estimation options**.

___Maximization___

maximize_options: <u>diff</u>icult, <u>tech</u>nique(*algorithm_spec*), <u>iter</u>ate(#), [<u>no</u>]<u>log</u>, <u>trace</u>, gradient, showstep, <u>hess</u>ian, <u>showtol</u>erance, <u>tol</u>erance(#), <u>ltol</u>erance(#), <u>nrtol</u>erance(#), nonrtolerance, and from(*init_specs*); see [R] **maximize**. These options are seldom used.

The following option is available with svar but is not shown in the dialog box:

coeflegend; see [R] **estimation options**.

Remarks

Remarks are presented under the following headings:

>   *Introduction*
>   *Short-run SVAR models*
>   *Long-run SVAR models*

Introduction

This entry assumes that you have already read [TS] **var intro** and [TS] **var**; if not, please do. Here we illustrate how to fit SVARs in Stata subject to short-run and long-run restrictions. For more detailed information on SVARs, see Amisano and Giannini (1997) and Hamilton (1994). For good introductions to VARs, see Lütkepohl (2005), Hamilton (1994), and Stock and Watson (2001).

Short-run SVAR models

A short-run SVAR model without exogenous variables can be written as

$$\mathbf{A}(\mathbf{I}_K - \mathbf{A}_1 L - \mathbf{A}_2 L^2 - \cdots - \mathbf{A}_p L^p)\mathbf{y}_t = \mathbf{A}\epsilon_t = \mathbf{Be}_t$$

where L is the lag operator, $\mathbf{A}$, $\mathbf{B}$, and $\mathbf{A}_1, \ldots, \mathbf{A}_p$ are $K \times K$ matrices of parameters, ϵ_t is a $K \times 1$ vector of innovations with $\epsilon_t \sim N(0, \boldsymbol{\Sigma})$ and $E[\epsilon_t \epsilon_s'] = \mathbf{0}_K$ for all $s \neq t$, and $\mathbf{e}_t$ is a $K \times 1$ vector of orthogonalized disturbances; that is, $\mathbf{e}_t \sim N(0, \mathbf{I}_K)$ and $E[\mathbf{e}_t \mathbf{e}_s'] = \mathbf{0}_K$ for all $s \neq t$. These transformations of the innovations allow us to analyze the dynamics of the system in terms of a change to an element of $\mathbf{e}_t$. In a short-run SVAR model, we obtain identification by placing restrictions on $\mathbf{A}$ and $\mathbf{B}$, which are assumed to be nonsingular.

▷ Example 1: Short-run just-identified SVAR model

Following Sims (1980), the Cholesky decomposition is one method of identifying the impulse–response functions in a VAR; thus, this method corresponds to an SVAR. There are several sets of constraints on $\mathbf{A}$ and $\mathbf{B}$ that are easily manipulated back to the Cholesky decomposition, and the following example illustrates this point.

One way to impose the Cholesky restrictions is to assume an SVAR model of the form

$$\widetilde{\mathbf{A}}(\mathbf{I}_K - \mathbf{A}_1 - \mathbf{A}_2 L^2 - \cdots \mathbf{A}_p L^p)\mathbf{y}_t = \widetilde{\mathbf{B}}\mathbf{e}_t$$

where $\widetilde{\mathbf{A}}$ is a lower triangular matrix with ones on the diagonal and $\widetilde{\mathbf{B}}$ is a diagonal matrix. Because the $\mathbf{P}$ matrix for this model is $\mathbf{P}_{sr} = \widetilde{\mathbf{A}}^{-1}\widetilde{\mathbf{B}}$, its estimate, $\widehat{\mathbf{P}}_{sr}$, obtained by plugging in estimates of $\widetilde{\mathbf{A}}$ and $\widetilde{\mathbf{B}}$, should equal the Cholesky decomposition of $\widehat{\boldsymbol{\Sigma}}$.

To illustrate, we use the German macroeconomic data discussed in Lütkepohl (2005) and used in [TS] **var**. In this example, $\mathbf{y}_t = (\texttt{dln\_inv}, \texttt{dln\_inc}, \texttt{dln\_consump})$, where $\texttt{dln\_inv}$ is the first difference of the log of investment, $\texttt{dln\_inc}$ is the first difference of the log of income, and $\texttt{dln\_consump}$ is the first difference of the log of consumption. Because the first difference of the natural log of a variable can be treated as an approximation of the percent change in that variable, we will refer to these variables as percent changes in $\texttt{inv}$, $\texttt{inc}$, and $\texttt{consump}$, respectively.

We will impose the Cholesky restrictions on this system by applying equality constraints with the constraint matrices

$$\mathbf{A} = \begin{bmatrix} 1 & 0 & 0 \\ . & 1 & 0 \\ . & . & 1 \end{bmatrix} \quad \text{and} \quad \mathbf{B} = \begin{bmatrix} . & 0 & 0 \\ 0 & . & 0 \\ 0 & 0 & . \end{bmatrix}$$

With these structural restrictions, we assume that the percent change in $\texttt{inv}$ is not contemporaneously affected by the percent changes in either $\texttt{inc}$ or $\texttt{consump}$. We also assume that the percent change of $\texttt{inc}$ is affected by contemporaneous changes in $\texttt{inv}$ but not $\texttt{consump}$. Finally, we assume that percent changes in $\texttt{consump}$ are affected by contemporaneous changes in both $\texttt{inv}$ and $\texttt{inc}$.

The following commands fit an SVAR model with these constraints.

```
. use http://www.stata-press.com/data/r12/lutkepohl2
(Quarterly SA West German macro data, Bil DM, from Lutkepohl 1993 Table E.1)

. matrix A = (1,0,0\.,1,0\.,.,1)

. matrix B = (.,0,0\0,.,0\0,0,.)
```

```
. svar dln_inv dln_inc dln_consump if qtr<=tq(1978q4), aeq(A) beq(B)
Estimating short-run parameters
```

(*output omitted*)

```
Structural vector autoregression
( 1)  [a_1_1]_cons = 1
( 2)  [a_1_2]_cons = 0
( 3)  [a_1_3]_cons = 0
( 4)  [a_2_2]_cons = 1
( 5)  [a_2_3]_cons = 0
( 6)  [a_3_3]_cons = 1
( 7)  [b_1_2]_cons = 0
( 8)  [b_1_3]_cons = 0
( 9)  [b_2_1]_cons = 0
(10)  [b_2_3]_cons = 0
(11)  [b_3_1]_cons = 0
(12)  [b_3_2]_cons = 0
```

```
Sample:  1960q4 - 1978q4                    No. of obs      =       73
Exactly identified model                    Log likelihood  =  606.307
```

	Coef.	Std. Err.	z	P>|z|	[95% Conf.	Interval]
/a_1_1	1	(constrained)				
/a_2_1	-.0336288	.0294605	-1.14	0.254	-.0913702	.0241126
/a_3_1	-.0435846	.0194408	-2.24	0.025	-.0816879	-.0054812
/a_1_2	0	(omitted)				
/a_2_2	1	(constrained)				
/a_3_2	-.424774	.0765548	-5.55	0.000	-.5748187	-.2747293
/a_1_3	0	(omitted)				
/a_2_3	0	(omitted)				
/a_3_3	1	(constrained)				
/b_1_1	.0438796	.0036315	12.08	0.000	.036762	.0509972
/b_2_1	0	(omitted)				
/b_3_1	0	(omitted)				
/b_1_2	0	(omitted)				
/b_2_2	.0110449	.0009141	12.08	0.000	.0092534	.0128365
/b_3_2	0	(omitted)				
/b_1_3	0	(omitted)				
/b_2_3	0	(omitted)				
/b_3_3	.0072243	.0005979	12.08	0.000	.0060525	.0083962

The SVAR output has four parts: an iteration log, a display of the constraints imposed, a header with sample and SVAR log-likelihood information, and a table displaying the estimates of the parameters from the $\mathbf{A}$ and $\mathbf{B}$ matrices. From the output above, we can see that the equality constraint matrices supplied to svar imposed the intended constraints and that the SVAR header informs us that the model we fit is just identified. The estimates of a_2_1, a_3_1, and a_3_2 are all negative. Because the off-diagonal elements of the $\mathbf{A}$ matrix contain the negative of the actual contemporaneous effects, the estimated effects are positive, as expected.

The estimates $\widehat{\mathbf{A}}$ and $\widehat{\mathbf{B}}$ are stored in e(A) and e(B), respectively, allowing us to compute the estimated Cholesky decomposition.

```
. matrix Aest = e(A)
. matrix Best = e(B)
. matrix chol_est = inv(Aest)*Best
```

```
. matrix list chol_est
chol_est[3,3]
                  dln_inv     dln_inc  dln_consump
     dln_inv    .04387957           0            0
     dln_inc    .00147562    .01104494           0
 dln_consump    .00253928     .0046916    .00722432
```

svar saves the estimated Σ from the underlying var in e(Sigma). The output below illustrates the computation of the Cholesky decomposition of e(Sigma). It is the same as the output computed from the SVAR estimates.

```
. matrix sig_var = e(Sigma)
. matrix chol_var = cholesky(sig_var)
. matrix list chol_var
chol_var[3,3]
                  dln_inv     dln_inc  dln_consump
     dln_inv    .04387957           0            0
     dln_inc    .00147562    .01104494           0
 dln_consump    .00253928     .0046916    .00722432
```

◁

We might now wonder why we bother obtaining parameter estimates via nonlinear estimation if we can obtain them simply by a transform of the estimates produced by var. When the model is just identified, as in the previous example, the SVAR parameter estimates can be computed via a transform of the VAR estimates. However, when the model is overidentified, such is not the case.

▷ Example 2: Short-run overidentified SVAR model

The Cholesky decomposition example above fit a just-identified model. This example considers an overidentified model. In the previous example, the a_2_1 parameter was not significant, which is consistent with a theory in which changes in our measure of investment affect only changes in income with a lag. We can impose the restriction that a_2_1 is zero and then test this overidentifying restriction. Our **A** and **B** matrices are now

$$\mathbf{A} = \begin{bmatrix} 1 & 0 & 0 \\ 0 & 1 & 0 \\ . & . & 1 \end{bmatrix} \quad \text{and} \quad \mathbf{B} = \begin{bmatrix} . & 0 & 0 \\ 0 & . & 0 \\ 0 & 0 & . \end{bmatrix}$$

The output below contains the commands and results we obtained by fitting this model on the Lütkepohl data.

```
. matrix B = (.,0,0\0,.,0\0,0,.)
. matrix A = (1,0,0\0,1,0\.,.,1)
```

```
. svar dln_inv dln_inc dln_consump if qtr<=tq(1978q4), aeq(A) beq(B)
Estimating short-run parameters
```

(output omitted)

```
Structural vector autoregression
( 1)  [a_1_1]_cons = 1
( 2)  [a_1_2]_cons = 0
( 3)  [a_1_3]_cons = 0
( 4)  [a_2_1]_cons = 0
( 5)  [a_2_2]_cons = 1
( 6)  [a_2_3]_cons = 0
( 7)  [a_3_3]_cons = 1
( 8)  [b_1_2]_cons = 0
( 9)  [b_1_3]_cons = 0
(10)  [b_2_1]_cons = 0
(11)  [b_2_3]_cons = 0
(12)  [b_3_1]_cons = 0
(13)  [b_3_2]_cons = 0
```

```
Sample:  1960q4 - 1978q4                    No. of obs       =        73
Overidentified model                        Log likelihood  =  605.6613
```

	Coef.	Std. Err.	z	P>|z|	[95% Conf. Interval]
/a_1_1	1	(constrained)			
/a_2_1	0	(omitted)			
/a_3_1	-.0435911	.0192696	-2.26	0.024	-.0813589 -.0058233
/a_1_2	0	(omitted)			
/a_2_2	1	(constrained)			
/a_3_2	-.4247741	.0758806	-5.60	0.000	-.5734973 -.2760508
/a_1_3	0	(omitted)			
/a_2_3	0	(omitted)			
/a_3_3	1	(constrained)			
/b_1_1	.0438796	.0036315	12.08	0.000	.036762 .0509972
/b_2_1	0	(omitted)			
/b_3_1	0	(omitted)			
/b_1_2	0	(omitted)			
/b_2_2	.0111431	.0009222	12.08	0.000	.0093356 .0129506
/b_3_2	0	(omitted)			
/b_1_3	0	(omitted)			
/b_2_3	0	(omitted)			
/b_3_3	.0072243	.0005979	12.08	0.000	.0060525 .0083962

```
LR test of identifying restrictions:  chi2( 1)=   1.292  Prob > chi2 = 0.256
```

The footer in this example reports a test of the overidentifying restriction. The null hypothesis of this test is that any overidentifying restrictions are valid. In the case at hand, we cannot reject this null hypothesis at any of the conventional levels.

◁

▷ Example 3: Short-run SVAR model with constraints

svar also allows us to place constraints on the parameters of the underlying VAR. We begin by looking at the underlying VAR for the SVARs that we have used in the previous examples.

```
. var dln_inv dln_inc dln_consump if qtr<=tq(1978q4)
```

Vector autoregression

Sample: 1960q4 - 1978q4			No. of obs	=	73
Log likelihood = 606.307			AIC	=	-16.03581
FPE = 2.18e-11			HQIC	=	-15.77323
Det(Sigma_ml) = 1.23e-11			SBIC	=	-15.37691

Equation	Parms	RMSE	R-sq	chi2	P>chi2
dln_inv	7	.046148	0.1286	10.76961	0.0958
dln_inc	7	.011719	0.1142	9.410683	0.1518
dln_consump	7	.009445	0.2513	24.50031	0.0004

	Coef.	Std. Err.	z	P>|z|	[95% Conf. Interval]	
dln_inv						
dln_inv						
L1.	-.3196318	.1192898	-2.68	0.007	-.5534355	-.0858282
L2.	-.1605508	.118767	-1.35	0.176	-.39333	.0722283
dln_inc						
L1.	.1459851	.5188451	0.28	0.778	-.8709326	1.162903
L2.	.1146009	.508295	0.23	0.822	-.881639	1.110841
dln_consump						
L1.	.9612288	.6316557	1.52	0.128	-.2767936	2.199251
L2.	.9344001	.6324034	1.48	0.140	-.3050877	2.173888
_cons	-.0167221	.0163796	-1.02	0.307	-.0488257	.0153814
dln_inc						
dln_inv						
L1.	.0439309	.0302933	1.45	0.147	-.0154427	.1033046
L2.	.0500302	.0301605	1.66	0.097	-.0090833	.1091437
dln_inc						
L1.	-.1527311	.131759	-1.16	0.246	-.4109741	.1055118
L2.	.0191634	.1290799	0.15	0.882	-.2338285	.2721552
dln_consump						
L1.	.2884992	.1604069	1.80	0.072	-.0258926	.6028909
L2.	-.0102	.1605968	-0.06	0.949	-.3249639	.3045639
_cons	.0157672	.0041596	3.79	0.000	.0076146	.0239198
dln_consump						
dln_inv						
L1.	-.002423	.0244142	-0.10	0.921	-.050274	.045428
L2.	.0338806	.0243072	1.39	0.163	-.0137607	.0815219
dln_inc						
L1.	.2248134	.1061884	2.12	0.034	.0166879	.4329389
L2.	.3549135	.1040292	3.41	0.001	.1510199	.558807
dln_consump						
L1.	-.2639695	.1292766	-2.04	0.041	-.517347	-.010592
L2.	-.0222264	.1294296	-0.17	0.864	-.2759039	.231451
_cons	.0129258	.0033523	3.86	0.000	.0063554	.0194962

The equation-level model tests reported in the header indicate that we cannot reject the null hypotheses that all the coefficients in the first equation are zero, nor can we reject the null that all the coefficients in the second equation are zero at the 5% significance level. We use a combination of theory and the p-values from the output above to place some exclusion restrictions on the underlying VAR(2). Specifically, in the equation for the percent change of inv, we constrain the coefficients on L2.dln_inv, L.dln_inc, L2.dln_inc, and L2.dln_consump to be zero. In the equation for dln_inc, we constrain the coefficients on L2.dln_inv, L2.dln_inc, and L2.dln_consump to be zero. Finally, in the equation for dln_consump, we constrain L.dln_inv and L2.dln_consump to be zero. We then refit the SVAR from the previous example.

```
. constraint 1 [dln_inv]L2.dln_inv = 0
. constraint 2 [dln_inv]L.dln_inc = 0
. constraint 3 [dln_inv]L2.dln_inc = 0
. constraint 4 [dln_inv]L2.dln_consump = 0
. constraint 5 [dln_inc]L2.dln_inv = 0
. constraint 6 [dln_inc]L2.dln_inc = 0
. constraint 7 [dln_inc]L2.dln_consump = 0
. constraint 8 [dln_consump]L.dln_inv = 0
. constraint 9 [dln_consump]L2.dln_consump = 0
. svar dln_inv dln_inc dln_consump if qtr<=tq(1978q4), aeq(A) beq(B) varconst(1/9)
> noislog
Estimating short-run parameters
  (output omitted )
Structural vector autoregression
 ( 1)   [a_1_1]_cons = 1
 ( 2)   [a_1_2]_cons = 0
 ( 3)   [a_1_3]_cons = 0
 ( 4)   [a_2_1]_cons = 0
 ( 5)   [a_2_2]_cons = 1
 ( 6)   [a_2_3]_cons = 0
 ( 7)   [a_3_3]_cons = 1
 ( 8)   [b_1_2]_cons = 0
 ( 9)   [b_1_3]_cons = 0
 (10)   [b_2_1]_cons = 0
 (11)   [b_2_3]_cons = 0
 (12)   [b_3_1]_cons = 0
 (13)   [b_3_2]_cons = 0
```

```
Sample:  1960q4 - 1978q4                      No. of obs     =        73
Overidentified model                          Log likelihood =  601.8591
```

	Coef.	Std. Err.	z	P>|z|	[95% Conf. Interval]
/a_1_1	1	(constrained)				
/a_2_1	0	(omitted)				
/a_3_1	-.0418708	.0187579	-2.23	0.026	-.0786356	-.0051061
/a_1_2	0	(omitted)				
/a_2_2	1	(constrained)				
/a_3_2	-.4255808	.0745298	-5.71	0.000	-.5716565	-.2795051
/a_1_3	0	(omitted)				
/a_2_3	0	(omitted)				
/a_3_3	1	(constrained)				
/b_1_1	.0451851	.0037395	12.08	0.000	.0378557	.0525145
/b_2_1	0	(omitted)				
/b_3_1	0	(omitted)				
/b_1_2	0	(omitted)				
/b_2_2	.0113723	.0009412	12.08	0.000	.0095276	.013217
/b_3_2	0	(omitted)				
/b_1_3	0	(omitted)				
/b_2_3	0	(omitted)				
/b_3_3	.0072417	.0005993	12.08	0.000	.006067	.0084164

```
LR test of identifying restrictions:  chi2( 1)=    .8448  Prob > chi2 = 0.358
```

If we displayed the underlying VAR(2) results by using the var option, we would see that most of the unconstrained coefficients are now significant at the 10% level and that none of the equation-level model statistics fail to reject the null hypothesis at the 10% level. The svar output reveals that the p-value of the overidentification test rose and that the coefficient on a_3_1 is still insignificant at the 1% level but not at the 5% level.

◁

Before moving on to models with long-run constraints, consider these limitations. We cannot place constraints on the elements of $\mathbf{A}$ in terms of the elements of $\mathbf{B}$, or vice versa. This limitation is imposed by the form of the check for identification derived by Amisano and Giannini (1997). As noted in *Methods and formulas*, this test requires separate constraint matrices for the parameters in $\mathbf{A}$ and $\mathbf{B}$. Another limitation is that we cannot mix short-run and long-run constraints.

Long-run SVAR models

As discussed in [TS] **var intro**, a long-run SVAR has the form

$$\mathbf{y}_t = \mathbf{C}e_t$$

In long-run models, the constraints are placed on the elements of $\mathbf{C}$, and the free parameters are estimated. These constraints are often exclusion restrictions. For instance, constraining $\mathbf{C}[1,2]$ to be zero can be interpreted as setting the long-run response of variable 1 to the structural shocks driving variable 2 to be zero.

Similar to the short-run model, the $\mathbf{P}_{\mathrm{lr}}$ matrix such that $\mathbf{P}_{\mathrm{lr}}\mathbf{P}'_{\mathrm{lr}} = \boldsymbol{\Sigma}$ identifies the structural impulse–response functions. $\mathbf{P}_{\mathrm{lr}} = \mathbf{C}$ is identified by the restrictions placed on the parameters in $\mathbf{C}$. There are K^2 parameters in $\mathbf{C}$, and the order condition for identification requires that there be at least $K^2 - K(K+1)/2$ restrictions placed on those parameters. As in the short-run model, this order condition is necessary but not sufficient, so the Amisano and Giannini (1997) check for local identification is performed by default.

▷ Example 4: Long-run SVAR model

Suppose that we have a theory in which unexpected changes to the money supply have no long-run effects on changes in output and, similarly, that unexpected changes in output have no long-run effects on changes in the money supply. The $\mathbf{C}$ matrix implied by this theory is

$$\mathbf{C} = \begin{bmatrix} . & 0 \\ 0 & . \end{bmatrix}$$

```
. use http://www.stata-press.com/data/r12/m1gdp

. matrix lr = (.,0\0,.)

. svar d.ln_m1 d.ln_gdp, lreq(lr)
Estimating long-run parameters

 (output omitted )

Structural vector autoregression
 ( 1)  [c_1_2]_cons = 0
 ( 2)  [c_2_1]_cons = 0
Sample:  1959q4 - 2002q2                      No. of obs       =        171
Overidentified model                         Log likelihood   =   1151.614
```

	Coef.	Std. Err.	z	P>|z|	[95% Conf. Interval]	
/c_1_1	.0301007	.0016277	18.49	0.000	.0269106	.0332909
/c_2_1	0	(omitted)				
/c_1_2	0	(omitted)				
/c_2_2	.0129691	.0007013	18.49	0.000	.0115946	.0143436

```
LR test of identifying restrictions:  chi2( 1)=    .1368  Prob > chi2 = 0.712
```

We have assumed that the underlying VAR has 2 lags; four of the five selection-order criteria computed by varsoc (see [TS] **varsoc**) recommended this choice. The test of the overidentifying restrictions provides no indication that it is not valid.   ◁

Saved results

svar saves the following in e():

Scalars

e(N)	number of observations
e(N_cns)	number of constraints
e(k_eq)	number of equations in e(b)
e(k_dv)	number of dependent variables
e(k_aux)	number of auxiliary parameters
e(ll)	log likelihood from svar
e(ll_#)	log likelihood for equation #
e(N_gaps_var)	number of gaps in the sample
e(k_var)	number of coefficients in VAR
e(k_eq_var)	number of equations in underlying VAR
e(k_dv_var)	number of dependent variables in underlying VAR
e(df_eq_var)	average number of parameters in an equation
e(df_m_var)	model degrees of freedom
e(df_r_var)	if small, residual degrees of freedom
e(obs_#_var)	number of observations on equation #
e(k_#_var)	number of coefficients in equation #
e(df_m#_var)	model degrees of freedom for equation #
e(df_r#_var)	residual degrees of freedom for equation # (small only)
e(r2_#_var)	R-squared for equation #
e(ll_#_var)	log likelihood for equation # VAR
e(chi2_#_var)	χ^2 statistic for equation #
e(F_#_var)	F statistic for equation # (small only)
e(rmse_#_var)	root mean squared error for equation #
e(mlag_var)	highest lag in VAR
e(tparms_var)	number of parameters in all equations
e(aic_var)	Akaike information criterion
e(hqic_var)	Hannan–Quinn information criterion
e(sbic_var)	Schwarz–Bayesian information criterion
e(fpe_var)	final prediction error
e(ll_var)	log likelihood from var
e(detsig_var)	determinant of e(Sigma)
e(detsig_ml_var)	determinant of $\widehat{\Sigma}_{\mathrm{ml}}$
e(tmin)	first time period in the sample
e(tmax)	maximum time
e(chi2_oid)	overidentification test
e(oid_df)	number of overidentifying restrictions
e(rank)	rank of e(V)
e(ic_ml)	number of iterations
e(rc_ml)	return code from ml

Macros
e(cmd)	svar
e(cmdline)	command as typed
e(lrmodel)	long-run model, if specified
e(lags_var)	lags in model
e(depvar_var)	names of dependent variables
e(endog_var)	names of endogenous variables
e(exog_var)	names of exogenous variables, if specified
e(nocons_var)	noconstant, if noconstant specified
e(cns_lr)	long-run constraints
e(cns_a)	cross-parameter equality constraints on **A**
e(cns_b)	cross-parameter equality constraints on **B**
e(dfk_var)	alternate divisor (dfk), if specified
e(eqnames_var)	names of equations
e(lutstats_var)	lutstats, if specified
e(constraints_var)	constraints_var, if there are constraints on VAR
e(small)	small, if specified
e(tsfmt)	format of timevar
e(timevar)	name of timevar
e(title)	title in estimation output
e(properties)	b V
e(predict)	program used to implement predict

Matrices
e(b)	coefficient vector
e(Cns)	constraints matrix
e(Sigma)	$\widehat{\boldsymbol{\Sigma}}$ matrix
e(V)	variance–covariance matrix of the estimators
e(b_var)	coefficient vector of underlying VAR model
e(V_var)	VCE of underlying VAR model
e(bf_var)	full coefficient vector with zeros in dropped lags
e(G_var)	G matrix saved by var; see [TS] **var** *Methods and formulas*
e(aeq)	aeq(*matrix*), if specified
e(acns)	acns(*matrix*), if specified
e(beq)	beq(*matrix*), if specified
e(bcns)	bcns(*matrix*), if specified
e(lreq)	lreq(*matrix*), if specified
e(lrcns)	lrcns(*matrix*), if specified
e(Cns_var)	constraint matrix from var, if varconstraints() is specified
e(A)	estimated A matrix, if a short-run model
e(B)	estimated B matrix
e(C)	estimated C matrix, if a long-run model
e(A1)	estimated $\bar{\text{A}}$ matrix, if a long-run model

Functions
e(sample)	marks estimation sample

Methods and formulas

svar is implemented as an ado-file.

The log-likelihood function for models with short-run constraints is

$$L(\mathbf{A}, \mathbf{B}) = -\frac{NK}{2}\ln(2\pi) + \frac{N}{2}\ln(|\mathbf{W}|^2) - \frac{N}{2}\text{tr}(\mathbf{W}'\mathbf{W}\widehat{\boldsymbol{\Sigma}})$$

where $\mathbf{W} = \mathbf{B}^{-1}\mathbf{A}$.

When there are long-run constraints, because $\mathbf{C} = \bar{\mathbf{A}}^{-1}\mathbf{B}$ and $\mathbf{A} = \mathbf{I}_K$, $\mathbf{W} = \mathbf{B}^{-1} = \mathbf{C}^{-1}\bar{\mathbf{A}}^{-1} = (\bar{\mathbf{A}}\mathbf{C})^{-1}$. Substituting the last term for $\mathbf{W}$ in the short-run log likelihood produces the long-run log likelihood

$$L(\mathbf{C}) = -\frac{NK}{2}\ln(2\pi) + \frac{N}{2}\ln(|\widetilde{\mathbf{W}}|^2) - \frac{N}{2}\operatorname{tr}(\widetilde{\mathbf{W}}'\widetilde{\mathbf{W}}\widehat{\Sigma})$$

where $\widetilde{\mathbf{W}} = (\bar{\mathbf{A}}\mathbf{C})^{-1}$.

For both the short-run and the long-run models, the maximization is performed by the scoring method. See Harvey (1990) for a discussion of this method.

Based on results from Amisano and Giannini (1997), the score vector for the short-run model is

$$\frac{\partial L(\mathbf{A},\mathbf{B})}{\partial[\operatorname{vec}(\mathbf{A}),\operatorname{vec}(\mathbf{B})]} = N\left[\{\operatorname{vec}(\mathbf{W}'^{-1})\}' - \{\operatorname{vec}(\mathbf{W})\}'(\widehat{\Sigma}\otimes\mathbf{I}_K)\right] \times$$
$$\left[(\mathbf{I}_K\otimes\mathbf{B}^{-1}), -(\mathbf{A}'\mathbf{B}'^{-1}\otimes\mathbf{B}^{-1})\right]$$

and the expected information matrix is

$$I\left[\operatorname{vec}(\mathbf{A}),\operatorname{vec}(\mathbf{B})\right] = N\begin{bmatrix}(\mathbf{W}^{-1}\otimes\mathbf{B}'^{-1})\\-(\mathbf{I}_K\otimes\mathbf{B}'^{-1})\end{bmatrix}(\mathbf{I}_{K^2}+\oplus)\left[(\mathbf{W}'^{-1}\otimes\mathbf{B}^{-1}), -(\mathbf{I}_K\otimes\mathbf{B}^{-1})\right]$$

where $\oplus$ is the commutation matrix defined in Magnus and Neudecker (1999, 46–48).

Using results from Amisano and Giannini (1997), we can derive the score vector and the expected information matrix for the case with long-run restrictions. The score vector is

$$\frac{\partial L(\mathbf{C})}{\partial\operatorname{vec}(\mathbf{C})} = N\left[\{\operatorname{vec}(\mathbf{W}'^{-1})\}' - \{\operatorname{vec}(\mathbf{W})\}'(\widehat{\Sigma}\otimes\mathbf{I}_K)\right]\left[-(\bar{\mathbf{A}}'^{-1}\mathbf{C}'^{-1}\otimes\mathbf{C}^{-1})\right]$$

and the expected information matrix is

$$I\left[\operatorname{vec}(\mathbf{C})\right] = N(\mathbf{I}_K\otimes\mathbf{C}'^{-1})(\mathbf{I}_{K^2}+\oplus)(\mathbf{I}_K\otimes\mathbf{C}'^{-1})$$

Checking for identification

This section describes the methods used to check for identification of models with short-run or long-run constraints. Both methods depend on the starting values. By default, svar uses starting values constructed by taking a vector of appropriate dimension and applying the constraints. If there are m parameters in the model, the jth element of the $1\times m$ vector is $1+m/100$. svar also allows the user to provide starting values.

For the short-run case, the model is identified if the matrix

$$\mathbf{V}_{\text{sr}}^* = \begin{bmatrix} \mathbf{N}_K & \mathbf{N}_K \\ \mathbf{N}_K & \mathbf{N}_K \\ \mathbf{R}_a(\mathbf{W}'\otimes\mathbf{B}) & \mathbf{0}_{K^2} \\ \mathbf{0}_{K^2} & \mathbf{R}_a(\mathbf{I}_K\otimes\mathbf{B}) \end{bmatrix}$$

has full column rank of $2K^2$, where $\mathbf{N}_K = (1/2)(\mathbf{I}_{K^2}+\oplus)$, $\mathbf{R}_a$ is the constraint matrix for the parameters in $\mathbf{A}$ (that is, $\mathbf{R}_a\operatorname{vec}(\mathbf{A}) = \mathbf{r}_a$), and $\mathbf{R}_b$ is the constraint matrix for the parameters in $\mathbf{B}$ (that is, $\mathbf{R}_b\operatorname{vec}(\mathbf{B}) = \mathbf{r}_b$).

For the long-run case, based on results from the **C** model in Amisano and Giannini (1997), the model is identified if the matrix

$$\mathbf{V}_{\text{lr}}^{*} = \left[\begin{array}{c} (\mathbf{I} \otimes \mathbf{C}'^{-1})(2\mathbf{N}_K)(\mathbf{I} \otimes \mathbf{C}^{-1}) \\ \mathbf{R}_c \end{array} \right]$$

has full column rank of K^2, where $\mathbf{R}_c$ is the constraint matrix for the parameters in $\mathbf{C}$; that is, $\mathbf{R}_c \text{vec}(\mathbf{C}) = \mathbf{r}_c$.

The test of the overidentifying restrictions is computed as

$$LR = 2(LL_{\text{var}} - LL_{\text{svar}})$$

where LR is the value of the test statistic against the null hypothesis that the overidentifying restrictions are valid, LL_{var} is the log likelihood from the underlying VAR(p) model, and LL_{svar} is the log likelihood from the SVAR model. The test statistic is asymptotically distributed as $\chi^2(q)$, where q is the number of overidentifying restrictions. Amisano and Giannini (1997, 38–39) emphasize that, because this test of the validity of the overidentifying restrictions is an omnibus test, it can be interpreted as a test of the null hypothesis that all the restrictions are valid.

Because constraints might not be independent either by construction or because of the data, the number of restrictions is not necessarily equal to the number of constraints. The rank of e(V) gives the number of parameters that were independently estimated after applying the constraints. The maximum number of parameters that can be estimated in an identified short-run or long-run SVAR is $K(K+1)/2$. This implies that the number of overidentifying restrictions, q, is equal to $K(K+1)/2$ minus the rank of e(V).

The number of overidentifying restrictions is also linked to the order condition for each model. In a short-run SVAR model, there are $2K^2$ parameters. Because no more than $K(K+1)/2$ parameters may be estimated, the order condition for a short-run SVAR model is that at least $2K^2 - K(K+1)/2$ restrictions be placed on the model. Similarly, there are K^2 parameters in long-run SVAR model. Because no more than $K(K+1)/2$ parameters may be estimated, the order condition for a long-run SVAR model is that at least $K^2 - K(K+1)/2$ restrictions be placed on the model.

Acknowledgment

We thank Gianni Amisano, Università degli Studi di Brescia, for his helpful comments.

References

Amisano, G., and C. Giannini. 1997. *Topics in Structural VAR Econometrics.* 2nd ed. Heidelberg: Springer.

Christiano, L. J., M. Eichenbaum, and C. L. Evans. 1999. Monetary policy shocks: What have we learned and to what end? In *Handbook of Macroeconomics: Volume 1A*, ed. J. B. Taylor and M. Woodford. New York: Elsevier.

Hamilton, J. D. 1994. *Time Series Analysis.* Princeton: Princeton University Press.

Harvey, A. C. 1990. *The Econometric Analysis of Time Series.* 2nd ed. Cambridge, MA: MIT Press.

Lütkepohl, H. 1993. *Introduction to Multiple Time Series Analysis.* 2nd ed. New York: Springer.

———. 2005. *New Introduction to Multiple Time Series Analysis.* New York: Springer.

Magnus, J. R., and H. Neudecker. 1999. *Matrix Differential Calculus with Applications in Statistics and Econometrics.* Rev. ed. New York: Wiley.

Rothenberg, T. J. 1971. Identification in parametric models. *Econometrica* 39: 577–591.

Sims, C. A. 1980. Macroeconomics and reality. *Econometrica* 48: 1–48.

Stock, J. H., and M. W. Watson. 2001. Vector autoregressions. *Journal of Economic Perspectives* 15: 101–115.

Watson, M. W. 1994. Vector autoregressions and cointegration. In Vol. IV of *Handbook of Econometrics*, ed. R. F. Engle and D. L. McFadden. Amsterdam: Elsevier.

Also see

[TS] **var svar postestimation** — Postestimation tools for svar

[TS] **tsset** — Declare data to be time-series data

[TS] **var** — Vector autoregressive models

[TS] **varbasic** — Fit a simple VAR and graph IRFs or FEVDs

[TS] **vec** — Vector error-correction models

[U] **20 Estimation and postestimation commands**

[TS] **var intro** — Introduction to vector autoregressive models

Title

var svar postestimation — Postestimation tools for svar

Description

The following postestimation commands are of special interest after svar:

Command	Description
fcast compute	obtain dynamic forecasts
fcast graph	graph dynamic forecasts obtained from fcast compute
irf	create and analyze IRFs and FEVDs
vargranger	Granger causality tests
varlmar	LM test for autocorrelation in residuals
varnorm	test for normally distributed residuals
varsoc	lag-order selection criteria
varstable	check stability condition of estimates
varwle	Wald lag-exclusion statistics

For information about these commands, see the corresponding entries in this manual.

The following standard postestimation commands are also available:

Command	Description
estat	AIC, BIC, VCE, and estimation sample summary
estimates	cataloging estimation results
lincom	point estimates, standard errors, testing, and inference for linear combinations of coefficients
lrtest	likelihood-ratio test
nlcom	point estimates, standard errors, testing, and inference for nonlinear combinations of coefficients
predict	predictions, residuals, influence statistics, and other diagnostic measures
predictnl	point estimates, standard errors, testing, and inference for generalized predictions
test	Wald tests of simple and composite linear hypotheses
testnl	Wald tests of nonlinear hypotheses

See the corresponding entries in the *Base Reference Manual* for details.

Syntax for predict

predict [*type*] *newvar* [*if*] [*in*] [, *statistic* equation(*eqno* | *eqname*)]

statistic	Description
Main	
xb	linear prediction; the default
stdp	standard error of the linear prediction
residuals	residuals

These statistics are available both in and out of sample; type predict ... if e(sample) ... if wanted only for the estimation sample.

584

Menu

Statistics > Postestimation > Predictions, residuals, etc.

Options for predict

⌐Main⌐

xb, the default, calculates the linear prediction for the specified equation.

stdp calculates the standard error of the linear prediction for the specified equation.

residuals calculates the residuals.

equation(*eqno* | *eqname*) specifies the equation to which you are referring.

> equation() is filled in with one *eqno* or *eqname* for options xb, stdp, and residuals. For example, equation(#1) would mean that the calculation is to be made for the first equation, equation(#2) would mean the second, and so on. You could also refer to the equation by its name; thus, equation(income) would refer to the equation named income and equation(hours), to the equation named hours.

> If you do not specify equation(), the results are the same as if you specified equation(#1).

For more information on using predict after multiple-equation estimation commands, see [R] **predict**.

Remarks

Remarks are presented under the following headings:

> *Model selection and inference*
> *Forecasting*

Model selection and inference

See the following sections for information on model selection and inference after var.

> [TS] **irf** — Create and analyze IRFs, dynamic-multiplier functions, and FEVDs
>
> [TS] **vargranger** — Perform pairwise Granger causality tests after var or svar
>
> [TS] **varlmar** — Perform LM test for residual autocorrelation after var or svar
>
> [TS] **varnorm** — Test for normally distributed disturbances after var or svar
>
> [TS] **varsoc** — Obtain lag-order selection statistics for VARs and VECMs
>
> [TS] **varstable** — Check the stability condition of VAR or SVAR estimates
>
> [TS] **varwle** — Obtain Wald lag-exclusion statistics after var or svar

Forecasting

See the following sections for information on obtaining forecasts after svar:

[TS] **fcast compute** — Compute dynamic forecasts of dependent variables after var, svar, or vec

[TS] **fcast graph** — Graph forecasts of variables computed by fcast compute

Methods and formulas

All postestimation commands listed above are implemented as ado-files.

Also see

[TS] **var svar** — Structural vector autoregressive models

[U] **20 Estimation and postestimation commands**

Title

> **varbasic** — Fit a simple VAR and graph IRFs or FEVDs

Syntax

varbasic *depvarlist* $\left[\textit{if}\right]$ $\left[\textit{in}\right]$ $\left[\textit{, options}\right]$

options	Description
Main	
<u>lags</u>(*numlist*)	use lags *numlist* in the model; default is lags(1 2)
<u>irf</u>	produce matrix graph of IRFs
<u>fevd</u>	produce matrix graph of FEVDs
<u>nograph</u>	do not produce a graph
<u>step</u>(*#*)	set forecast horizon *#* for estimating the OIRFs, IRFs, and FEVDs; default is step(8)

You must tsset your data before using varbasic; see [TS] **tsset**.

depvarlist may contain time-series operators; see [U] **11.4.4 Time-series varlists**.

rolling, statsby, and xi are allowed; see [U] **11.1.10 Prefix commands**.

See [U] **20 Estimation and postestimation commands** for more capabilities of estimation commands.

Menu

Statistics > Multivariate time series > Basic VAR

Description

varbasic fits a basic vector autoregressive (VAR) model and graphs the impulse–response functions (IRFs), the orthogonalized impulse–response functions (OIRFs), or the forecast-error variance decompositions (FEVDs).

Options

> ⌐ Main ⌐

lags(*numlist*) specifies the lags to be included in the model. The default is lags(1 2). This option takes a numlist and not simply an integer for the maximum lag. For instance, lags(2) would include only the second lag in the model, whereas lags(1/2) would include both the first and second lags in the model. See [U] **11.1.8 numlist** and [U] **11.4.4 Time-series varlists** for more discussion of numlists and lags.

irf causes varbasic to produce a matrix graph of the IRFs instead of a matrix graph of the OIRFs, which is produced by default.

fevd causes varbasic to produce a matrix graph of the FEVDs instead of a matrix graph of the OIRFs, which is produced by default.

nograph specifies that no graph be produced. The IRFs, OIRFs, and FEVDs are still estimated and saved in the IRF file _varbasic.irf.

step(*#*) specifies the forecast horizon for estimating the IRFs, OIRFs, and FEVDs. The default is eight periods.

Remarks

varbasic simplifies fitting simple VARs and graphing the IRFs, the OIRFs, or the FEVDs. See [TS] **var** and [TS] **var svar** for fitting more advanced VAR models and structural vector autoregressive (SVAR) models. All the postestimation commands discussed in [TS] **var postestimation** work after varbasic.

This entry does not discuss the methods for fitting a VAR or the methods surrounding the IRFs, OIRFs, and FEVDs. See [TS] **var** and [TS] **irf create** for more on these methods. This entry illustrates how to use varbasic to easily obtain results. It also illustrates how varbasic serves as an entry point to further analysis.

▷ Example 1

We fit a three-variable VAR with two lags to the German macro data used by Lütkepohl (2005). The three variables are the first difference of natural log of investment, dln_inv; the first difference of the natural log of income, dln_inc; and the first difference of the natural log of consumption, dln_consump. In addition to fitting the VAR, we want to see the OIRFs. Below we use varbasic to fit a VAR(2) model on the data from the second quarter of 1961 through the fourth quarter of 1978. By default, varbasic produces graphs of the OIRFs.

```
. use http://www.stata-press.com/data/r12/lutkepohl2
(Quarterly SA West German macro data, Bil DM, from Lutkepohl 1993 Table E.1)

. varbasic dln_inv dln_inc dln_consump if qtr<=tq(1978q4)

Vector autoregression

Sample:  1960q4 - 1978q4                    No. of obs    =          73
Log likelihood =   606.307                  AIC           =   -16.03581
FPE            =   2.18e-11                  HQIC          =   -15.77323
Det(Sigma_ml)  =   1.23e-11                  SBIC          =   -15.37691

Equation          Parms      RMSE     R-sq      chi2     P>chi2

dln_inv              7     .046148   0.1286   10.76961   0.0958
dln_inc              7     .011719   0.1142    9.410683  0.1518
dln_consump          7     .009445   0.2513   24.50031   0.0004
```

	Coef.	Std. Err.	z	P>|z|	[95% Conf. Interval]	
dln_inv						
dln_inv						
L1.	-.3196318	.1192898	-2.68	0.007	-.5534355	-.0858282
L2.	-.1605508	.118767	-1.35	0.176	-.39333	.0722283
dln_inc						
L1.	.1459851	.5188451	0.28	0.778	-.8709326	1.162903
L2.	.1146009	.508295	0.23	0.822	-.881639	1.110841
dln_consump						
L1.	.9612288	.6316557	1.52	0.128	-.2767936	2.199251
L2.	.9344001	.6324034	1.48	0.140	-.3050877	2.173888
_cons	-.0167221	.0163796	-1.02	0.307	-.0488257	.0153814
dln_inc						
dln_inv						
L1.	.0439309	.0302933	1.45	0.147	-.0154427	.1033046
L2.	.0500302	.0301605	1.66	0.097	-.0090833	.1091437
dln_inc						
L1.	-.1527311	.131759	-1.16	0.246	-.4109741	.1055118
L2.	.0191634	.1290799	0.15	0.882	-.2338285	.2721552
dln_consump						
L1.	.2884992	.1604069	1.80	0.072	-.0258926	.6028909
L2.	-.0102	.1605968	-0.06	0.949	-.3249639	.3045639
_cons	.0157672	.0041596	3.79	0.000	.0076146	.0239198
dln_consump						
dln_inv						
L1.	-.002423	.0244142	-0.10	0.921	-.050274	.045428
L2.	.0338806	.0243072	1.39	0.163	-.0137607	.0815219
dln_inc						
L1.	.2248134	.1061884	2.12	0.034	.0166879	.4329389
L2.	.3549135	.1040292	3.41	0.001	.1510199	.558807
dln_consump						
L1.	-.2639695	.1292766	-2.04	0.041	-.517347	-.010592
L2.	-.0222264	.1294296	-0.17	0.864	-.2759039	.231451
_cons	.0129258	.0033523	3.86	0.000	.0063554	.0194962

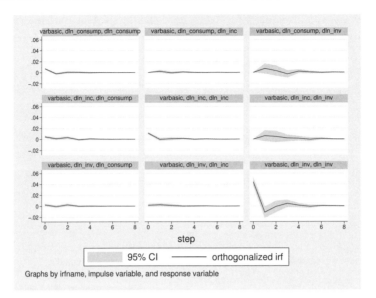

Because we are also interested in looking at the FEVDs, we can use `irf graph` to obtain the graphs. Although the details are available in [TS] **irf** and [TS] **irf graph**, the command below produces what we want after the call to `varbasic`.

```
. irf graph fevd, lstep(1)
```

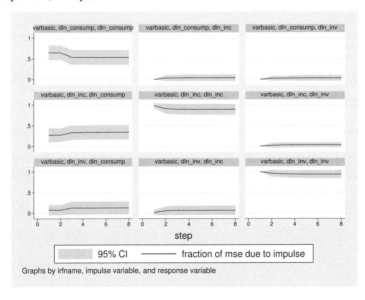

◁

❏ Technical note

Stata stores the estimated IRFs, OIRFs, and FEVDs in a IRF file called `_varbasic.irf` in the current working directory. `varbasic` replaces any `_varbasic.irf` that already exists. Finally, `varbasic` makes `_varbasic.irf` the active IRF file. This means that the graph and table commands `irf graph`, irf cgraph, irf ograph, irf table, and irf ctable will all display results that correspond to the VAR fit by varbasic.

❑

Saved results

See *Saved results* in [TS] **var**.

Methods and formulas

varbasic is implemented as an ado-file.

varbasic uses var and irf graph to obtain its results. See [TS] **var** and [TS] **irf graph** for a discussion of how those commands obtain their results.

References

Lütkepohl, H. 1993. *Introduction to Multiple Time Series Analysis.* 2nd ed. New York: Springer.

——. 2005. *New Introduction to Multiple Time Series Analysis.* New York: Springer.

Also see

[TS] **varbasic postestimation** — Postestimation tools for varbasic

[TS] **tsset** — Declare data to be time-series data

[TS] **var** — Vector autoregressive models

[TS] **var svar** — Structural vector autoregressive models

[U] **20 Estimation and postestimation commands**

[TS] **var intro** — Introduction to vector autoregressive models

Title

> **varbasic postestimation** — Postestimation tools for varbasic

Description

The following postestimation commands are of special interest after `varbasic`:

Command	Description
`fcast compute`	obtain dynamic forecasts
`fcast graph`	graph dynamic forecasts obtained from `fcast compute`
`irf`	create and analyze IRFs and FEVDs
`vargranger`	Granger causality tests
`varlmar`	LM test for autocorrelation in residuals
`varnorm`	test for normally distributed residuals
`varsoc`	lag-order selection criteria
`varstable`	check stability condition of estimates
`varwle`	Wald lag-exclusion statistics

For information about these commands, see the corresponding entries in this manual.

The following standard postestimation commands are also available:

Command	Description
`estat`	AIC, BIC, VCE, and estimation sample summary
`estimates`	cataloging estimation results
`lincom`	point estimates, standard errors, testing, and inference for linear combinations of coefficients
`lrtest`	likelihood-ratio test
`margins`	marginal means, predictive margins, marginal effects, and average marginal effects
`marginsplot`	graph the results from margins (profile plots, interaction plots, etc.)
`nlcom`	point estimates, standard errors, testing, and inference for nonlinear combinations of coefficients
`predict`	predictions, residuals, influence statistics, and other diagnostic measures
`predictnl`	point estimates, standard errors, testing, and inference for generalized predictions
`test`	Wald tests of simple and composite linear hypotheses
`testnl`	Wald tests of nonlinear hypotheses

See the corresponding entries in the *Base Reference Manual* for details.

Syntax for predict

predict [*type*] *newvar* [*if*] [*in*] [, *statistic* equation(*eqno* | *eqname*)]

statistic	Description
Main	
xb	linear prediction; the default
stdp	standard error of the linear prediction
residuals	residuals

These statistics are available both in and out of sample; type predict ... if e(sample) ... if wanted only for the estimation sample.

Menu

Statistics > Postestimation > Predictions, residuals, etc.

Options for predict

⌐ Main ⌐

xb, the default, calculates the linear prediction for the specified equation.

stdp calculates the standard error of the linear prediction for the specified equation.

residuals calculates the residuals.

equation(*eqno* | *eqname*) specifies the equation to which you are referring.

equation() is filled in with one *eqno* or *eqname* for the xb, stdp, and residuals options. For example, equation(#1) would mean that the calculation is to be made for the first equation, equation(#2) would mean the second, and so on. You could also refer to the equation by its name; thus, equation(income) would refer to the equation named income and equation(hours), to the equation named hours.

If you do not specify equation(), the results are the same as if you specified equation(#1).

For more information on using predict after multiple-equation estimation commands, see [R] **predict**.

Remarks

> Example 1

All the postestimation commands discussed in [TS] **var postestimation** work after varbasic. Suppose that we are interested in testing the hypothesis that there is no autocorrelation in the VAR disturbances. Continuing example 1 from [TS] **varbasic**, we now use varlmar to test this hypothesis.

```
. use http://www.stata-press.com/data/r12/lutkepohl2
(Quarterly SA West German macro data, Bil DM, from Lutkepohl 1993 Table E.1)
. varbasic dln_inv dln_inc dln_consump if qtr<=tq(1978q4)
  (output omitted )
. varlmar
```

Lagrange-multiplier test

lag	chi2	df	Prob > chi2
1	5.5871	9	0.78043
2	6.3189	9	0.70763

H0: no autocorrelation at lag order

Because we cannot reject the null hypothesis of no autocorrelation in the residuals, this test does not indicate any model misspecification.

◁

Methods and formulas

All postestimation commands listed above are implemented as ado-files.

Also see

[TS] **varbasic** — Fit a simple VAR and graph IRFs or FEVDs

[U] **20 Estimation and postestimation commands**

Title

> **vargranger** — Perform pairwise Granger causality tests after var or svar

Syntax

vargranger $\left[\, , \underline{est}imates(estname) \; \underline{sep}arator(\#) \,\right]$

vargranger can be used only after var or svar; see [TS] **var** and [TS] **var svar**.

Menu

Statistics > Multivariate time series > VAR diagnostics and tests > Granger causality tests

Description

vargranger performs a set of Granger causality tests for each equation in a VAR, providing a convenient alternative to test; see [R] **test**.

Options

estimates(*estname*) requests that vargranger use the previously obtained set of var or svar estimates saved as *estname*. By default, vargranger uses the active results. See [R] **estimates** for information about saving and restoring estimation results.

separator(#) specifies how often separator lines should be drawn between rows. By default, separator lines appear every K lines, where K is the number of equations in the VAR under analysis. For example, separator(1) would draw a line between each row, separator(2) between every other row, and so on. separator(0) specifies that lines not appear in the table.

Remarks

After fitting a VAR, we may want to know whether one variable "Granger-causes" another (Granger 1969). A variable x is said to Granger-cause a variable y if, given the past values of y, past values of x are useful for predicting y. A common method for testing Granger causality is to regress y on its own lagged values and on lagged values of x and test the null hypothesis that the estimated coefficients on the lagged values of x are jointly zero. Failure to reject the null hypothesis is equivalent to failing to reject the hypothesis that x does not Granger-cause y.

For each equation and each endogenous variable that is not the dependent variable in that equation, vargranger computes and reports Wald tests that the coefficients on all the lags of an endogenous variable are jointly zero. For each equation in a VAR, vargranger tests the hypotheses that each of the other endogenous variables does not Granger-cause the dependent variable in that equation.

Because it may be interesting to investigate these types of hypotheses by using the VAR that underlies an SVAR, vargranger can also produce these tests by using the e() results from an svar. When vargranger uses svar e() results, the hypotheses concern the underlying var estimates.

See [TS] **var** and [TS] **var svar** for information about fitting VARs and SVARs in Stata. See Lütkepohl (2005), Hamilton (1994), and Amisano and Giannini (1997) for information about Granger causality and on VARs and SVARs in general.

▷ Example 1: After var

Here we refit the model with German data described in [TS] **var** and then perform Granger causality tests with vargranger.

```
. use http://www.stata-press.com/data/r12/lutkepohl2
(Quarterly SA West German macro data, Bil DM, from Lutkepohl 1993 Table E.1)
. var dln_inv dln_inc dln_consump if qtr<=tq(1978q4), dfk small
```
(*output omitted*)
```
. vargranger
  Granger causality Wald tests
```

Equation	Excluded	F	df	df_r	Prob > F
dln_inv	dln_inc	.04847	2	66	0.9527
dln_inv	dln_consump	1.5004	2	66	0.2306
dln_inv	ALL	1.5917	4	66	0.1869
dln_inc	dln_inv	1.7683	2	66	0.1786
dln_inc	dln_consump	1.7184	2	66	0.1873
dln_inc	ALL	1.9466	4	66	0.1130
dln_consump	dln_inv	.97147	2	66	0.3839
dln_consump	dln_inc	6.1465	2	66	0.0036
dln_consump	ALL	3.7746	4	66	0.0080

Because the estimates() option was not specified, vargranger used the active e() results. Consider the results of the three tests for the first equation. The first is a Wald test that the coefficients on the two lags of dln_inc that appear in the equation for dln_inv are jointly zero. The null hypothesis that dln_inc does not Granger-cause dln_inv cannot be rejected. Similarly, we cannot reject the null hypothesis that the coefficients on the two lags of dln_consump in the equation for dln_inv are jointly zero, so we cannot reject the hypothesis that dln_consump does not Granger-cause dln_inv. The third test is with respect to the null hypothesis that the coefficients on the two lags of all the other endogenous variables are jointly zero. Because this cannot be rejected, we cannot reject the null hypothesis that dln_inc and dln_consump, jointly, do not Granger-cause dln_inv.

Because we failed to reject most of these null hypotheses, we might be interested in imposing some constraints on the coefficients. See [TS] **var** for more on fitting VAR models with constraints on the coefficients.

◁

▷ Example 2: Using test instead of vargranger

We could have used test to compute these Wald tests, but vargranger saves a great deal of typing. Still, seeing how to use test to obtain the results reported by vargranger is useful.

```
. test [dln_inv]L.dln_inc [dln_inv]L2.dln_inc
 ( 1)  [dln_inv]L.dln_inc = 0
 ( 2)  [dln_inv]L2.dln_inc = 0
       F(  2,    66) =    0.05
            Prob > F =    0.9527
```

```
. test [dln_inv]L.dln_consump [dln_inv]L2.dln_consump, accumulate
 ( 1)  [dln_inv]L.dln_inc = 0
 ( 2)  [dln_inv]L2.dln_inc = 0
 ( 3)  [dln_inv]L.dln_consump = 0
 ( 4)  [dln_inv]L2.dln_consump = 0
       F(  4,    66) =    1.59
            Prob > F =    0.1869
. test [dln_inv]L.dln_inv [dln_inv]L2.dln_inv, accumulate
 ( 1)  [dln_inv]L.dln_inc = 0
 ( 2)  [dln_inv]L2.dln_inc = 0
 ( 3)  [dln_inv]L.dln_consump = 0
 ( 4)  [dln_inv]L2.dln_consump = 0
 ( 5)  [dln_inv]L.dln_inv = 0
 ( 6)  [dln_inv]L2.dln_inv = 0
       F(  6,    66) =    1.62
            Prob > F =    0.1547
```

The first two calls to `test` show how `vargranger` obtains its results. The first test reproduces the first test reported for the `dln_inv` equation. The second test reproduces the ALL entry for the first equation. The third test reproduces the standard F statistic for the `dln_inv` equation, reported in the header of the `var` output in the previous example. The standard F statistic also includes the lags of the dependent variable, as well as any exogenous variables in the equation. This illustrates that the test performed by `vargranger` of the null hypothesis that the coefficients on all the lags of all the other endogenous variables are jointly zero for a particular equation; that is, the All test is not the same as the standard F statistic for that equation.

◁

▷ Example 3: After svar

When `vargranger` is run on `svar` estimates, the null hypotheses are with respect to the underlying `var` estimates. We run `vargranger` after using `svar` to fit an SVAR that has the same underlying VAR as our model in the previous example.

```
. matrix A = (., 0,0 \ ., ., 0\ .,.,.)
. matrix B = I(3)
. svar dln_inv dln_inc dln_consump if qtr<=tq(1978q4), dfk small aeq(A) beq(B)
  (output omitted )
. vargranger
  Granger causality Wald tests
```

Equation	Excluded	F	df	df_r	Prob > F
dln_inv	dln_inc	.04847	2	66	0.9527
dln_inv	dln_consump	1.5004	2	66	0.2306
dln_inv	ALL	1.5917	4	66	0.1869
dln_inc	dln_inv	1.7683	2	66	0.1786
dln_inc	dln_consump	1.7184	2	66	0.1873
dln_inc	ALL	1.9466	4	66	0.1130
dln_consump	dln_inv	.97147	2	66	0.3839
dln_consump	dln_inc	6.1465	2	66	0.0036
dln_consump	ALL	3.7746	4	66	0.0080

As we expected, the `vargranger` results are identical to those in the first example.

◁

Saved results

vargranger saves the following in r():

Matrices
  r(gstats) χ^2, df, and p-values (if e(small)=="")
  r(gstats) F, df, df_r, and p-values (if e(small)!="")

Methods and formulas

vargranger is implemented as an ado-file.

vargranger uses test to obtain Wald statistics of the hypothesis that all coefficients on the lags of variable x are jointly zero in the equation for variable y. vargranger uses the e() results saved by var or svar to determine whether to calculate and report small-sample F statistics or large-sample χ^2 statistics.

> Clive William John Granger (1934–2009) was born in Swansea, Wales, and earned degrees at the University of Nottingham in mathematics and statistics. Joining the staff there, he also worked at Princeton on the spectral analysis of economic time series, before moving in 1973 to the University of California, San Diego. He was awarded the 2003 Nobel Prize in Economics for methods of analyzing economic time series with common trends (cointegration). He was knighted in 2005, thus becoming Sir Clive Granger.

References

Amisano, G., and C. Giannini. 1997. *Topics in Structural VAR Econometrics*. 2nd ed. Heidelberg: Springer.

Granger, C. W. J. 1969. Investigating causal relations by econometric models and cross-spectral methods. *Econometrica* 37: 424–438.

Hamilton, J. D. 1994. *Time Series Analysis*. Princeton: Princeton University Press.

Lütkepohl, H. 1993. *Introduction to Multiple Time Series Analysis*. 2nd ed. New York: Springer.

———. 2005. *New Introduction to Multiple Time Series Analysis*. New York: Springer.

Phillips, P. C. B. 1997. The *ET* Interview: Professor Clive Granger. *Econometric Theory* 13: 253–303.

Also see

[TS] **var** — Vector autoregressive models

[TS] **var svar** — Structural vector autoregressive models

[TS] **varbasic** — Fit a simple VAR and graph IRFs or FEVDs

[TS] **var intro** — Introduction to vector autoregressive models

Title

> **varlmar** — Perform LM test for residual autocorrelation after var or svar

Syntax

varlmar [, *options*]

options	Description
<u>ml</u>ag(*#*)	use *#* for the maximum order of autocorrelation; default is mlag(2)
<u>est</u>imates(*estname*)	use previously saved results *estname*; default is to use active results
<u>sep</u>arator(*#*)	draw separator line after every *#* rows

varlmar can be used only after var or svar; see [TS] **var** and [TS] **var svar**.
You must tsset your data before using varlmar; see [TS] **tsset**.

Menu

Statistics > Multivariate time series > VAR diagnostics and tests > LM test for residual autocorrelation

Description

varlmar implements a Lagrange-multiplier (LM) test for autocorrelation in the residuals of VAR models, which was presented in Johansen (1995).

Options

mlag(*#*) specifies the maximum order of autocorrelation to be tested. The integer specified in mlag() must be greater than 0; the default is 2.

estimates(*estname*) requests that varlmar use the previously obtained set of var or svar estimates saved as *estname*. By default, varlmar uses the active results. See [R] **estimates** for information on saving and restoring estimation results.

separator(*#*) specifies how often separator lines should be drawn between rows. By default, separator lines do not appear. For example, separator(1) would draw a line between each row, separator(2) between every other row, and so on.

Remarks

Most postestimation analyses of VAR models and SVAR models assume that the disturbances are not autocorrelated. varlmar implements the LM test for autocorrelation in the residuals of a VAR model discussed in Johansen (1995, 21–22). The test is performed at lags $j = 1, \ldots, \text{mlag}()$. For each j, the null hypothesis of the test is that there is no autocorrelation at lag j.

varlmar uses the estimation results saved by var or svar. By default, varlmar uses the active estimation results. However, varlmar can use any previously saved var or svar estimation results specified in the estimates() option.

▷ Example 1: After var

Here we refit the model with German data described in [TS] **var** and then call `varlmar`.

```
. use http://www.stata-press.com/data/r12/lutkepohl2
(Quarterly SA West German macro data, Bil DM, from Lutkepohl 1993 Table E.1)
. var dln_inv dln_inc dln_consump if qtr<=tq(1978q4), dfk
(output omitted)
. varlmar, mlag(5)
   Lagrange-multiplier test
```

lag	chi2	df	Prob > chi2
1	5.5871	9	0.78043
2	6.3189	9	0.70763
3	8.4022	9	0.49418
4	11.8742	9	0.22049
5	5.2914	9	0.80821

```
   H0: no autocorrelation at lag order
```

Because we cannot reject the null hypothesis that there is no autocorrelation in the residuals for any of the five orders tested, this test gives no hint of model misspecification. Although we fit the VAR with the `dfk` option to be consistent with the example in [TS] **var**, `varlmar` always uses the ML estimator of Σ. The results obtained from `varlmar` are the same whether or not `dfk` is specified.

◁

▷ Example 2: After svar

When `varlmar` is applied to estimation results produced by `svar`, the sequence of LM tests is applied to the underlying VAR. See [TS] **var svar** for a description of how an SVAR model builds on a VAR. In this example, we fit an SVAR that has an underlying VAR with two lags that is identical to the one fit in the previous example.

```
. matrix A = (.,.,0\0,.,.0\.,.,.)
. matrix B = I(3)
. svar dln_inv dln_inc dln_consump if qtr<=tq(1978q4), dfk aeq(A) beq(B)
(output omitted)
. varlmar, mlag(5)
   Lagrange-multiplier test
```

lag	chi2	df	Prob > chi2
1	5.5871	9	0.78043
2	6.3189	9	0.70763
3	8.4022	9	0.49418
4	11.8742	9	0.22049
5	5.2914	9	0.80821

```
   H0: no autocorrelation at lag order
```

Because the underlying VAR(2) is the same as the previous example (we assure you that this is true), the output from `varlmar` is also the same.

◁

Saved results

varlmar saves the following in r():

Matrices
r(lm) χ^2, df, and p-values

Methods and formulas

varlmar is implemented as an ado-file.

The formula for the LM test statistic at lag j is

$$\text{LM}_s = (T - d - .5) \ln \left(\frac{|\widehat{\boldsymbol{\Sigma}}|}{|\widetilde{\boldsymbol{\Sigma}}_s|} \right)$$

where T is the number of observations in the VAR; d is explained below; $\widehat{\boldsymbol{\Sigma}}$ is the maximum likelihood estimate of $\boldsymbol{\Sigma}$, the variance–covariance matrix of the disturbances from the VAR; and $\widetilde{\boldsymbol{\Sigma}}_s$ is the maximum likelihood estimate of $\boldsymbol{\Sigma}$ from the following augmented VAR.

If there are K equations in the VAR, we can define $\mathbf{e}_t$ to be a $K \times 1$ vector of residuals. After we create the K new variables e1, e2, ..., eK containing the residuals from the K equations, we can augment the original VAR with lags of these K new variables. For each lag s, we form an augmented regression in which the new residual variables are lagged s times. Per the method of Davidson and MacKinnon (1993, 358), the missing values from these s lags are replaced with zeros. $\widetilde{\boldsymbol{\Sigma}}_s$ is the maximum likelihood estimate of $\boldsymbol{\Sigma}$ from this augmented VAR, and d is the number of coefficients estimated in the augmented VAR. See [TS] **var** for a discussion of the maximum likelihood estimate of $\boldsymbol{\Sigma}$ in a VAR.

The asymptotic distribution of LM_s is χ^2 with K^2 degrees of freedom.

References

Davidson, R., and J. G. MacKinnon. 1993. *Estimation and Inference in Econometrics*. New York: Oxford University Press.

Johansen, S. 1995. *Likelihood-Based Inference in Cointegrated Vector Autoregressive Models*. Oxford: Oxford University Press.

Also see

[TS] **var** — Vector autoregressive models

[TS] **var svar** — Structural vector autoregressive models

[TS] **varbasic** — Fit a simple VAR and graph IRFs or FEVDs

[TS] **var intro** — Introduction to vector autoregressive models

Title

> **varnorm** — Test for normally distributed disturbances after var or svar

Syntax

varnorm [, *options*]

options	Description
<u>j</u>bera	report Jarque–Bera statistic; default is to report all three statistics
<u>s</u>kewness	report skewness statistic; default is to report all three statistics
<u>k</u>urtosis	report kurtosis statistic; default is to report all three statistics
<u>est</u>imates(*estname*)	use previously saved results *estname*; default is to use active results
<u>cho</u>lesky	use Cholesky decomposition
<u>separator</u>(#)	draw separator line after every # rows

varnorm can be used only after var or svar; see [TS] **var** and [TS] **var svar**.
You must **tsset** your data before using varnorm; see [TS] **tsset**.

Menu

Statistics > Multivariate time series > VAR diagnostics and tests > Test for normally distributed disturbances

Description

varnorm computes and reports a series of statistics against the null hypothesis that the disturbances in a VAR are normally distributed. For each equation, and for all equations jointly, up to three statistics may be computed: a skewness statistic, a kurtosis statistic, and the Jarque–Bera statistic. By default, all three statistics are reported.

Options

jbera requests that the Jarque–Bera statistic and any other explicitly requested statistic be reported. By default, the Jarque–Bera, skewness, and kurtosis statistics are reported.

skewness requests that the skewness statistic and any other explicitly requested statistic be reported. By default, the Jarque–Bera, skewness, and kurtosis statistics are reported.

kurtosis requests that the kurtosis statistic and any other explicitly requested statistic be reported. By default, the Jarque–Bera, skewness, and kurtosis statistics are reported.

estimates(*estname*) specifies that varnorm use the previously obtained set of var or svar estimates saved as *estname*. By default, varnorm uses the active results. See [R] **estimates** for information on saving and restoring estimation results.

cholesky specifies that varnorm use the Cholesky decomposition of the estimated variance–covariance matrix of the disturbances, $\widehat{\Sigma}$, to orthogonalize the residuals when varnorm is applied to svar results. By default, when varnorm is applied to svar results, it uses the estimated structural decomposition $\widehat{A}^{-1}\widehat{B}$ on $\widehat{C}$ to orthogonalize the residuals. When applied to var e() results, varnorm always uses the Cholesky decomposition of $\widehat{\Sigma}$. For this reason, the cholesky option may not be specified when using var results.

separator(#) specifies how often separator lines should be drawn between rows. By default, separator lines do not appear. For example, separator(1) would draw a line between each row, separator(2) between every other row, and so on.

Remarks

Some of the postestimation statistics for VAR and SVAR assume that the K disturbances have a K-dimensional multivariate normal distribution. varnorm uses the estimation results produced by var or svar to produce a series of statistics against the null hypothesis that the K disturbances in the VAR are normally distributed.

Per the notation in Lütkepohl (2005), call the skewness statistic $\widehat{\lambda}_1$, the kurtosis statistic $\widehat{\lambda}_2$, and the Jarque–Bera statistic $\widehat{\lambda}_3$. The Jarque–Bera statistic is a combination of the other two statistics. The single-equation results are from tests against the null hypothesis that the disturbance for that particular equation is normally distributed. The results for all the equations are from tests against the null hypothesis that the K disturbances follow a K-dimensional multivariate normal distribution. Failure to reject the null hypothesis indicates a lack of model misspecification.

▷ Example 1: After var

We refit the model with German data described in [TS] **var** and then call varnorm.

```
. use http://www.stata-press.com/data/r12/lutkepohl2
(Quarterly SA West German macro data, Bil DM, from Lutkepohl 1993 Table E.1)
. var dln_inv dln_inc dln_consump if qtr<=tq(1978q4), dfk
(output omitted )
. varnorm
```

Jarque-Bera test

Equation	chi2	df	Prob > chi2
dln_inv	2.821	2	0.24397
dln_inc	3.450	2	0.17817
dln_consump	1.566	2	0.45702
ALL	7.838	6	0.25025

Skewness test

Equation	Skewness	chi2	df	Prob > chi2
dln_inv	.11935	0.173	1	0.67718
dln_inc	-.38316	1.786	1	0.18139
dln_consump	-.31275	1.190	1	0.27532
ALL		3.150	3	0.36913

Kurtosis test

Equation	Kurtosis	chi2	df	Prob > chi2
dln_inv	3.9331	2.648	1	0.10367
dln_inc	3.7396	1.664	1	0.19710
dln_consump	2.6484	0.376	1	0.53973
ALL		4.688	3	0.19613

dfk estimator used in computations

In this example, neither the single-equation Jarque–Bera statistics nor the joint Jarque–Bera statistic come close to rejecting the null hypothesis.

The skewness and kurtosis results have similar structures.

The Jarque–Bera results use the sum of the skewness and kurtosis statistics. The skewness and kurtosis results are based on the skewness and kurtosis coefficients, respectively. See *Methods and formulas*.

◁

▷ Example 2: After svar

The test statistics are computed on the orthogonalized VAR residuals; see *Methods and formulas*. When varnorm is applied to var results, varnorm uses a Cholesky decomposition of the estimated variance–covariance matrix of the disturbances, $\widehat{\Sigma}$, to orthogonalize the residuals.

By default, when varnorm is applied to svar estimation results, it uses the estimated structural decomposition $\widehat{A}^{-1}\widehat{B}$ on $\widehat{C}$ to orthogonalize the residuals of the underlying VAR. Alternatively, when varnorm is applied to svar results and the cholesky option is specified, varnorm uses the Cholesky decomposition of $\widehat{\Sigma}$ to orthogonalize the residuals of the underlying VAR.

We fit an SVAR that is based on an underlying VAR with two lags that is the same as the one fit in the previous example. We impose a structural decomposition that is the same as the Cholesky decomposition, as illustrated in [TS] **var svar**.

```
. matrix a = (.,0,0\.,.,0\.,.,.)
. matrix b = I(3)
. svar dln_inv dln_inc dln_consump if qtr<=tq(1978q4), dfk aeq(a) beq(b)
(output omitted)
. varnorm
```

Jarque-Bera test

Equation	chi2	df	Prob > chi2
dln_inv	2.821	2	0.24397
dln_inc	3.450	2	0.17817
dln_consump	1.566	2	0.45702
ALL	7.838	6	0.25025

Skewness test

Equation	Skewness	chi2	df	Prob > chi2
dln_inv	.11935	0.173	1	0.67718
dln_inc	-.38316	1.786	1	0.18139
dln_consump	-.31275	1.190	1	0.27532
ALL		3.150	3	0.36913

Kurtosis test

Equation	Kurtosis	chi2	df	Prob > chi2
dln_inv	3.9331	2.648	1	0.10367
dln_inc	3.7396	1.664	1	0.19710
dln_consump	2.6484	0.376	1	0.53973
ALL		4.688	3	0.19613

dfk estimator used in computations

Because the estimated structural decomposition is the same as the Cholesky decomposition, the varnorm results are the same as those from the previous example.

◁

❏ Technical note

The statistics computed by varnorm depend on $\widehat{\Sigma}$, the estimated variance–covariance matrix of the disturbances. var uses the maximum likelihood estimator of this matrix by default, but the dfk option produces an estimator that uses a small-sample correction. Thus specifying dfk in the call to var or svar will affect the test results produced by varnorm.

❏

Saved results

varnorm saves the following in r():

Macros
  r(dfk)   dfk, if specified
Matrices
  r(kurtosis)   kurtosis test, df, and p-values
  r(skewness)   skewness test, df, and p-values
  r(jb)   Jarque–Bera test, df, and p-values

Methods and formulas

varnorm is implemented as an ado-file.

varnorm is based on the derivations found in Lütkepohl (2005, 174–181). Let $\widehat{\mathbf{u}}_t$ be the $K \times 1$ vector of residuals from the K equations in a previously fitted VAR or the residuals from the K equations of the VAR underlying a previously fitted SVAR. Similarly, let $\widehat{\Sigma}$ be the estimated covariance matrix of the disturbances. (Note that $\widehat{\Sigma}$ depends on whether the dfk option was specified.) The skewness, kurtosis, and Jarque–Bera statistics must be computed using the orthogonalized residuals.

Because

$$\widehat{\Sigma} = \widehat{\mathbf{P}}\widehat{\mathbf{P}}'$$

implies that

$$\widehat{\mathbf{P}}^{-1}\widehat{\Sigma}\widehat{\mathbf{P}}^{-1\prime} = \mathbf{I}_K$$

premultiplying $\widehat{\mathbf{u}}_t$ by $\widehat{\mathbf{P}}$ is one way of performing the orthogonalization. When varnorm is applied to var results, $\widehat{\mathbf{P}}$ is defined to be the Cholesky decomposition of $\widehat{\Sigma}$. When varnorm is applied to svar results, $\widehat{\mathbf{P}}$ is set, by default, to the estimated structural decomposition; that is, $\widehat{\mathbf{P}} = \widehat{\mathbf{A}}^{-1}\widehat{\mathbf{B}}$, where $\widehat{\mathbf{A}}$ and $\widehat{\mathbf{B}}$ are the svar estimates of the $\mathbf{A}$ and $\mathbf{B}$ matrices, or $\widehat{\mathbf{C}}$, where $\widehat{\mathbf{C}}$ is the long-run SVAR estimation of $\mathbf{C}$. (See [TS] **var svar** for more on the origin and estimation of the $\mathbf{A}$ and $\mathbf{B}$ matrices.) When varnorm is applied to svar results and the cholesky option is specified, $\widehat{\mathbf{P}}$ is set to the Cholesky decomposition of $\widehat{\Sigma}$.

Define $\widehat{\mathbf{w}}_t$ to be the orthogonalized VAR residuals given by

$$\widehat{\mathbf{w}}_t = (\widehat{w}_{1t}, \ldots, \widehat{w}_{Kt})' = \widehat{\mathbf{P}}^{-1}\widehat{\mathbf{u}}_t$$

The $K \times 1$ vectors of skewness and kurtosis coefficients are then computed using the orthogonalized residuals by

$$\widehat{\mathbf{b}}_1 = (\widehat{b}_{11}, \ldots, \widehat{b}_{K1})'; \qquad \widehat{b}_{k1} = \frac{1}{T} \sum_{i=1}^{T} \widehat{w}_{kt}^3$$

$$\widehat{\mathbf{b}}_2 = (\widehat{b}_{12}, \ldots, \widehat{b}_{K2})'; \qquad \widehat{b}_{k2} = \frac{1}{T} \sum_{i=1}^{T} \widehat{w}_{kt}^4$$

Under the null hypothesis of multivariate Gaussian disturbances,

$$\widehat{\lambda}_1 = \frac{T\widehat{\mathbf{b}}_1'\widehat{\mathbf{b}}_1}{6} \xrightarrow{d} \chi^2(K)$$

$$\widehat{\lambda}_2 = \frac{T(\widehat{\mathbf{b}}_2 - 3)'(\widehat{\mathbf{b}}_2 - 3)}{24} \xrightarrow{d} \chi^2(K)$$

and

$$\widehat{\lambda}_3 = \widehat{\lambda}_1 + \widehat{\lambda}_2 \xrightarrow{d} \chi^2(2K)$$

$\widehat{\lambda}_1$ is the skewness statistic, $\widehat{\lambda}_2$ is the kurtosis statistic, and $\widehat{\lambda}_3$ is the Jarque–Bera statistic.

$\widehat{\lambda}_1$, $\widehat{\lambda}_2$, and $\widehat{\lambda}_3$ are for tests of the null hypothesis that the $K \times 1$ vector of disturbances follows a multivariate normal distribution. The corresponding statistics against the null hypothesis that the disturbances from the kth equation come from a univariate normal distribution are

$$\widehat{\lambda}_{1k} = \frac{T\widehat{b}_{k1}^2}{6} \xrightarrow{d} \chi^2(1)$$

$$\widehat{\lambda}_{2k} = \frac{T(\widehat{b}_{k2}^2 - 3)^2}{24} \xrightarrow{d} \chi^2(1)$$

and

$$\widehat{\lambda}_{3k} = \widehat{\lambda}_1 + \widehat{\lambda}_2 \xrightarrow{d} \chi^2(2)$$

References

Hamilton, J. D. 1994. *Time Series Analysis*. Princeton: Princeton University Press.

Jarque, C. M., and A. K. Bera. 1987. A test for normality of observations and regression residuals. *International Statistical Review* 2: 163–172.

Lütkepohl, H. 1993. *Introduction to Multiple Time Series Analysis*. 2nd ed. New York: Springer.

———. 2005. *New Introduction to Multiple Time Series Analysis*. New York: Springer.

Also see

[TS] **var** — Vector autoregressive models

[TS] **var svar** — Structural vector autoregressive models

[TS] **varbasic** — Fit a simple VAR and graph IRFs or FEVDs

[TS] **var intro** — Introduction to vector autoregressive models

Title

> **varsoc** — Obtain lag-order selection statistics for VARs and VECMs

Syntax

Preestimation syntax

varsoc *depvarlist* $\begin{bmatrix} if \end{bmatrix}$ $\begin{bmatrix} in \end{bmatrix}$ $\begin{bmatrix} , & preestimation\_options \end{bmatrix}$

Postestimation syntax

varsoc $\begin{bmatrix} , & \underline{est}imates(estname) \end{bmatrix}$

preestimation_options	Description
Main	
<u>max</u>lag(#)	set maximum lag order to #; default is maxlag(4)
<u>exog</u>(varlist)	use *varlist* as exogenous variables
<u>constr</u>aints(constraints)	apply constraints to exogenous variables
<u>nocons</u>tant	suppress constant term
<u>lutstats</u>	use Lütkepohl's version of information criteria
<u>level</u>(#)	set confidence level; default is level(95)
<u>separator</u>(#)	draw separator line after every # rows

You must tsset your data before using varsoc; see [TS] **tsset**.
by is allowed with the preestimation version of varsoc; see [U] **11.1.10 Prefix commands**.

Menu

Preestimation for VARs

Statistics > Multivariate time series > VAR diagnostics and tests > Lag-order selection statistics (preestimation)

Postestimation for VARs

Statistics > Multivariate time series > VAR diagnostics and tests > Lag-order selection statistics (postestimation)

Preestimation for VECMs

Statistics > Multivariate time series > VEC diagnostics and tests > Lag-order selection statistics (preestimation)

Postestimation for VECMs

Statistics > Multivariate time series > VEC diagnostics and tests > Lag-order selection statistics (postestimation)

Description

varsoc reports the final prediction error (FPE), Akaike's information criterion (AIC), Schwarz's Bayesian information criterion (SBIC), and the Hannan and Quinn information criterion (HQIC) lag-order selection statistics for a series of vector autoregressions of order 1, ..., maxlag(). A sequence of likelihood-ratio test statistics for all the full VARs of order less than or equal to the highest lag order is also reported. In the postestimation version, the maximum lag and estimation options are based on the model just fit or the model specified in estimates(*estname*).

607

The preestimation version of varsoc can also be used to select the lag order for a vector error-correction model (VECM). As shown by Nielsen (2001), the lag-order selection statistics discussed here can be used in the presence of I(1) variables.

Preestimation options

⌐ Main ⌐

maxlag(#) specifies the maximum lag order for which the statistics are to be obtained.

exog(*varlist*) specifies exogenous variables to include in the VARs fit by varsoc.

constraints(*constraints*) specifies a list of constraints on the exogenous variables to be applied. Do not specify constraints on the lags of the endogenous variables because specifying one would mean that at least one of the VAR models considered by varsoc will not contain the lag specified in the constraint. Use var directly to obtain selection-order criteria with constraints on lags of the endogenous variables.

noconstant suppresses the constant terms from the model. By default, constant terms are included.

lutstats specifies that the Lütkepohl (2005) versions of the lag-order selection statistics be reported. See *Methods and formulas* for a discussion of these statistics.

level(#) specifies the confidence level, as a percentage, that is used to identify the first likelihood-ratio test that rejects the null hypothesis that the additional parameters from adding a lag are jointly zero. The default is level(95) or as set by set level; see [U] **20.7 Specifying the width of confidence intervals**.

separator(#) specifies how often separator lines should be drawn between rows. By default, separator lines do not appear. For example, separator(1) would draw a line between each row, separator(2) between every other row, and so on.

Postestimation option

estimates(*estname*) specifies the name of a previously stored set of var or svar estimates. When no *depvarlist* is specified, varsoc uses the *postestimation syntax* and uses the currently active estimation results or the results specified in estimates(*estname*). See [R] **estimates** for information about saving and restoring estimation results.

Remarks

Many selection-order statistics have been developed to assist researchers in fitting a VAR of the correct order. Several of these selection-order statistics appear in the [TS] **var** output. The varsoc command computes these statistics over a range of lags p while maintaining a common sample and option specification.

varsoc can be used as a preestimation or a postestimation command. When it is used as a preestimation command, a *depvarlist* is required, and the default maximum lag is 4. When it is used as a postestimation command, varsoc uses the model specification stored in *estname* or the previously fitted model.

varsoc computes four information criteria as well as a sequence of likelihood ratio (LR) tests. The information criteria include the FPE, AIC, the HQIC, and SBIC.

For a given lag p, the LR test compares a VAR with p lags with one with $p - 1$ lags. The null hypothesis is that all the coefficients on the pth lags of the endogenous variables are zero. To use this sequence of LR tests to select a lag order, we start by looking at the results of the test for the model with the most lags, which is at the bottom of the table. Proceeding up the table, the first test that rejects the null hypothesis is the lag order selected by this process. See Lütkepohl (2005, 143–144) for more information on this procedure. An '*' appears next to the LR statistic indicating the optimal lag.

For the remaining statistics, the lag with the smallest value is the order selected by that criterion. An '*' indicates the optimal lag. Strictly speaking, the FPE is not an information criterion, though we include it in this discussion because, as with an information criterion, we select the lag length corresponding to the lowest value; and, naturally, we want to minimize the prediction error. The AIC measures the discrepancy between the given model and the true model, which, of course, we want to minimize. Amemiya (1985) provides an intuitive discussion of the arguments in Akaike (1973). The SBIC and the HQIC can be interpreted similarly to the AIC, though the SBIC and the HQIC have a theoretical advantage over the AIC and the FPE. As Lütkepohl (2005, 148–152) demonstrates, choosing p to minimize the SBIC or the HQIC provides consistent estimates of the true lag order, p. In contrast, minimizing the AIC or the FPE will overestimate the true lag order with positive probability, even with an infinite sample size.

▷ Example 1: Preestimation

Here we use varsoc as a preestimation command.

```
. use http://www.stata-press.com/data/r12/lutkepohl2
(Quarterly SA West German macro data, Bil DM, from Lutkepohl 1993 Table E.1)
. varsoc dln_inv dln_inc dln_consump if qtr<=tq(1978q4), lutstats
   Selection-order criteria (lutstats)
   Sample:  1961q2 - 1978q4                     Number of obs      =        71
```

lag	LL	LR	df	p	FPE	AIC	HQIC	SBIC
0	564.784				2.7e-11	-24.423	-24.423*	-24.423*
1	576.409	23.249	9	0.006	2.5e-11	-24.497	-24.3829	-24.2102
2	588.859	24.901*	9	0.003	2.3e-11*	-24.5942*	-24.3661	-24.0205
3	591.237	4.7566	9	0.855	2.7e-11	-24.4076	-24.0655	-23.5472
4	598.457	14.438	9	0.108	2.9e-11	-24.3575	-23.9012	-23.2102

```
   Endogenous:  dln_inv dln_inc dln_consump
   Exogenous:   _cons
```

The sample used begins in 1961q2 because all the VARs are fit to the sample defined by any if or in conditions and the available data for the maximum lag specified. The default maximum number of lags is four. Because we specified the lutstats option, the table contains the Lütkepohl (2005) versions of the information criteria, which differ from the standard definitions in that they drop the constant term from the log likelihood. In this example, the likelihood-ratio tests selected a model with two lags. AIC and FPE have also both chosen a model with two lags, whereas SBIC and HQIC have both selected a model with zero lags.

◁

▷ Example 2: Postestimation

`varsoc` works as a postestimation command when no dependent variables are specified.

```
. var dln_inc dln_consump if qtr<=tq(1978q4), lutstats exog(l.dln_inv)
(output omitted)
. varsoc
```

```
  Selection-order criteria (lutstats)
  Sample:  1960q4 - 1978q4                     Number of obs    =        73
```

lag	LL	LR	df	p	FPE	AIC	HQIC	SBIC
0	460.646				1.3e-08	-18.2962	-18.2962	-18.2962*
1	467.606	13.919	4	0.008	1.2e-08	-18.3773	-18.3273	-18.2518
2	477.087	18.962*	4	0.001	1.0e-08*	-18.5275*	-18.4274*	-18.2764

```
  Endogenous:  dln_inc dln_consump
  Exogenous:   L.dln_inv  _cons
```

Because we included one lag of `dln_inv` in our original model, `varsoc` did likewise with each model it fit.

◁

Based on the work of Tsay (1984), Paulsen (1984), and Nielsen (2001), these lag-order selection criteria can be used to determine the lag length of the VAR underlying a VECM. See [TS] **vec intro** for an example in which we use `varsoc` to choose the lag order for a VECM.

Saved results

`varsoc` saves the following in `r()`:

Scalars
r(N)	number of observations	r(mlag)	maximum lag order
r(tmax)	last time period in sample	r(N_gaps)	the number of gaps in
r(tmin)	first time period in sample		the sample

Macros
r(endog)	names of endogenous variables	r(exog)	names of exogenous variables
r(lutstats)	lutstats, if specified	r(rmlutstats)	rmlutstats, if specified
r(cns#)	the #th constraint		

Matrices
r(stats)	LL, LR, FPE, AIC, HQIC, SBIC, and p-values

Methods and formulas

`varsoc` is implemented as an ado-file.

As shown by Hamilton (1994, 295–296), the log likelihood for a VAR(p) is

$$\text{LL} = \left(\frac{T}{2}\right)\left\{\ln\left(|\widehat{\boldsymbol{\Sigma}}^{-1}|\right) - K\ln(2\pi) - K\right\}$$

where T is the number of observations, K is the number of equations, and $\widehat{\Sigma}$ is the maximum likelihood estimate of $E[\mathbf{u}_t\mathbf{u}_t']$, where $\mathbf{u}_t$ is the $K \times 1$ vector of disturbances. Because

$$\ln\left(\left|\widehat{\Sigma}^{-1}\right|\right) = -\ln\left(\left|\widehat{\Sigma}\right|\right)$$

the log likelihood can be rewritten as

$$\text{LL} = -\left(\frac{T}{2}\right)\left\{\ln\left(\left|\widehat{\Sigma}\right|\right) + K\ln(2\pi) + K\right\}$$

Letting $\text{LL}(j)$ be the value of the log likelihood with j lags yields the LR statistic for lag order j as

$$\text{LR}(j) = 2\left\{\text{LL}(j) - \text{LL}(j-1)\right\}$$

Model-order statistics

The formula for the FPE given in Lütkepohl (2005, 147) is

$$\text{FPE} = |\Sigma_u|\left(\frac{T + Kp + 1}{T - Kp - 1}\right)^K$$

This formula, however, assumes that there is a constant in the model and that none of the variables are dropped because of collinearity. To deal with these problems, the FPE is implemented as

$$\text{FPE} = |\Sigma_u|\left(\frac{T + \overline{m}}{T - \overline{m}}\right)^K$$

where $\overline{m}$ is the average number of parameters over the K equations. This implementation accounts for variables dropped because of collinearity.

By default, the AIC, SBIC, and HQIC are computed according to their standard definitions, which include the constant term from the log likelihood. That is,

$$\text{AIC} = -2\left(\frac{\text{LL}}{T}\right) + \frac{2t_p}{T}$$

$$\text{SBIC} = -2\left(\frac{\text{LL}}{T}\right) + \frac{\ln(T)}{T}t_p$$

$$\text{HQIC} = -2\left(\frac{\text{LL}}{T}\right) + \frac{2\ln\left\{\ln(T)\right\}}{T}t_p$$

where t_p is the total number of parameters in the model and LL is the log likelihood.

Lutstats

Lütkepohl (2005) advocates dropping the constant term from the log likelihood because it does not affect inference. The Lütkepohl versions of the information criteria are

$$\text{AIC} = \ln\big(|\boldsymbol{\Sigma}_u|\big) + \frac{2pK^2}{T}$$

$$\text{SBIC} = \ln\big(|\boldsymbol{\Sigma}_u|\big) + \frac{\ln(T)}{T}pK^2$$

$$\text{HQIC} = \ln\big(|\boldsymbol{\Sigma}_u|\big) + \frac{2\ln\{\ln(T)\}}{T}pK^2$$

References

Akaike, H. 1973. Information theory and an extension of the maximum likelihood principle. In *Second International Symposium on Information Theory*, ed. B. N. Petrov and F. Csaki, 267–281. Budapest: Akailseoniai–Kiudo.

Amemiya, T. 1985. *Advanced Econometrics*. Cambridge, MA: Harvard University Press.

Hamilton, J. D. 1994. *Time Series Analysis*. Princeton: Princeton University Press.

Lütkepohl, H. 1993. *Introduction to Multiple Time Series Analysis*. 2nd ed. New York: Springer.

——. 2005. *New Introduction to Multiple Time Series Analysis*. New York: Springer.

Nielsen, B. 2001. Order determination in general vector autoregressions. Working paper, Department of Economics, University of Oxford and Nuffield College. http://ideas.repec.org/p/nuf/econwp/0110.html.

Paulsen, J. 1984. Order determination of multivariate autoregressive time series with unit roots. *Journal of Time Series Analysis* 5: 115–127.

Tsay, R. S. 1984. Order selection in nonstationary autoregressive models. *Annals of Statistics* 12: 1425–1433.

Also see

Title

> **varstable** — Check the stability condition of VAR or SVAR estimates

Syntax

varstable [, *options*]

options	Description
Main	
est̲imates(*estname*)	use previously saved results *estname*; default is to use active results
amat(*matrix_name*)	save the companion matrix as *matrix_name*
graph	graph eigenvalues of the companion matrix
dlabel	label eigenvalues with the distance from the unit circle
modlabel	label eigenvalues with the modulus
marker_options	change look of markers (color, size, etc.)
rlopts(*cline_options*)	affect rendition of reference unit circle
nogrid	suppress polar grid circles
pgrid([...])	specify radii and appearance of polar grid circles; see *Options* for details
Add plots	
addplot(*plot*)	add other plots to the generated graph
Y axis, X axis, Titles, Legend, Overall	
twoway_options	any options other than by() documented in [G-3] *twoway_options*

varstable can be used only after var or svar; see [TS] **var** and [TS] **var svar**.

Menu

Statistics > Multivariate time series > VAR diagnostics and tests > Check stability condition of VAR estimates

Description

varstable checks the eigenvalue stability condition after estimating the parameters of a vector autoregression using var or svar.

Options

> Main

estimates(*estname*) requests that varstable use the previously obtained set of var estimates saved as *estname*. By default, varstable uses the active estimation results. See [R] **estimates** for information about saving and restoring estimation results.

amat(*matrix_name*) specifies a valid Stata matrix name by which the companion matrix $\mathbf{A}$ can be saved (see *Methods and formulas* for the definition of the matrix $\mathbf{A}$). The default is not to save the $\mathbf{A}$ matrix.

graph causes `varstable` to draw a graph of the eigenvalues of the companion matrix.

dlabel labels each eigenvalue with its distance from the unit circle. dlabel cannot be specified with modlabel.

modlabel labels the eigenvalues with their moduli. modlabel cannot be specified with dlabel.

marker_options specify the look of markers. This look includes the marker symbol, the marker size, and its color and outline; see [G-3] *marker_options*.

rlopts(*cline_options*) affect the rendition of the reference unit circle; see [G-3] *cline_options*.

nogrid suppresses the polar grid circles.

pgrid([*numlist*] [, *line_options*]) determines the radii and appearance of the polar grid circles. By default, the graph includes nine polar grid circles with radii .1, .2, ..., .9 that have the grid line style. The *numlist* specifies the radii for the polar grid circles. The *line_options* determine the appearance of the polar grid circles; see [G-3] *line_options*. Because the pgrid() option can be repeated, circles with different radii can have distinct appearances.

⌐ Add plots ⌐

addplot(*plot*) adds specified plots to the generated graph. See [G-3] *addplot_option*.

⌐ Y axis, X axis, Titles, Legend, Overall ⌐

twoway_options are any of the options documented in [G-3] *twoway_options*, except by(). These include options for titling the graph (see [G-3] *title_options*) and for saving the graph to disk (see [G-3] *saving_option*).

Remarks

Inference after var and svar requires that variables be covariance stationary. The variables in $\mathbf{y}_t$ are covariance stationary if their first two moments exist and are independent of time. More explicitly, a variable y_t is covariance stationary if

1. $E[y_t]$ is finite and independent of t.

2. $\text{Var}[y_t]$ is finite and independent of t

3. $\text{Cov}[y_t, y_s]$ is a finite function of $|t - s|$ but not of t or s alone.

Interpretation of VAR models, however, requires that an even stricter stability condition be met. If a VAR is stable, it is invertible and has an infinite-order vector moving-average representation. If the VAR is stable, impulse–response functions and forecast-error variance decompositions have known interpretations.

Lütkepohl (2005) and Hamilton (1994) both show that if the modulus of each eigenvalue of the matrix $\mathbf{A}$ is strictly less than one, the estimated VAR is stable (see *Methods and formulas* for the definition of the matrix $\mathbf{A}$).

▷ Example 1

After fitting a VAR with var, we can use `varstable` to check the stability condition. Using the same VAR model that was used in [TS] **var**, we demonstrate the use of `varstable`.

```
. use http://www.stata-press.com/data/r12/lutkepohl2
(Quarterly SA West German macro data, Bil DM, from Lutkepohl 1993 Table E.1)
. var dln_inv dln_inc dln_consump if qtr>=tq(1961q2) & qtr<=tq(1978q4)
(output omitted)
```

```
. varstable, graph
  Eigenvalue stability condition
```

Eigenvalue	Modulus
.5456253	.545625
−.3785754 + .3853982*i*	.540232
−.3785754 − .3853982*i*	.540232
−.0643276 + .4595944*i*	.464074
−.0643276 − .4595944*i*	.464074
−.3698058	.369806

```
All the eigenvalues lie inside the unit circle.
VAR satisfies stability condition.
```

Because the modulus of each eigenvalue is strictly less than 1, the estimates satisfy the eigenvalue stability condition.

Specifying the `graph` option produced a graph of the eigenvalues with the real components on the x axis and the complex components on the y axis. The graph below indicates visually that these eigenvalues are well inside the unit circle.

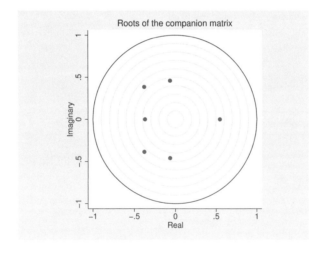

◁

> Example 2

This example illustrates two other features of the `varstable` command. First, `varstable` can check the stability of the estimates of the VAR underlying an SVAR fit by `var svar`. Second, `varstable` can check the stability of any previously stored `var` or `var svar` estimates.

We begin by refitting the previous VAR and storing the results as `var1`. Because this is the same VAR that was fit in the previous example, the stability results should be identical.

```
. var dln_inv dln_inc dln_consump if qtr>=tq(1961q2) & qtr<=tq(1978q4)
  (output omitted)
. estimates store var1
```

Now we use `svar` to fit an SVAR with a different underlying VAR and check the estimates of that underlying VAR for stability.

```
. matrix A = (.,0\.,.)
. matrix B = I(2)
. svar d.ln_inc d.ln_consump, aeq(A) beq(B)
 (output omitted )
. varstable
  Eigenvalue stability condition
```

Eigenvalue		Modulus
.548711		.548711
−.2979493 +	.4328013i	.525443
−.2979493 −	.4328013i	.525443
−.3570825		.357082

```
  All the eigenvalues lie inside the unit circle.
  VAR satisfies stability condition.
```

The `estimates()` option allows us to check the stability of the `var` results stored as `var1`.

```
. varstable, est(var1)
  Eigenvalue stability condition
```

Eigenvalue		Modulus
.5456253		.545625
−.3785754 +	.3853982i	.540232
−.3785754 −	.3853982i	.540232
−.0643276 +	.4595944i	.464074
−.0643276 −	.4595944i	.464074
−.3698058		.369806

```
  All the eigenvalues lie inside the unit circle.
  VAR satisfies stability condition.
```

The results are identical to those obtained in the previous example, confirming that we were checking the results in `var1`.

◁

Saved results

`varstable` saves the following in `r()`:

Matrices
r(Re)	real part of the eigenvalues of A
r(Im)	imaginary part of the eigenvalues of A
r(Modulus)	modulus of the eigenvalues of A

Methods and formulas

`varstable` is implemented as an ado-file.

`varstable` forms the companion matrix

$$\mathbf{A} = \begin{pmatrix} \mathbf{A}_1 & \mathbf{A}_2 & \ldots & \mathbf{A}_{p-1} & \mathbf{A}_p \\ \mathbf{I} & \mathbf{0} & \ldots & \mathbf{0} & \mathbf{0} \\ \mathbf{0} & \mathbf{I} & \ldots & \mathbf{0} & \mathbf{0} \\ \vdots & \vdots & \ddots & \vdots & \vdots \\ \mathbf{0} & \mathbf{0} & \ldots & \mathbf{I} & \mathbf{0} \end{pmatrix}$$

and obtains its eigenvalues by using `matrix eigenvalues`. The modulus of the complex eigenvalue $r + ci$ is $\sqrt{r^2 + c^2}$. As shown by Lütkepohl (2005) and Hamilton (1994), the VAR is stable if the modulus of each eigenvalue of $\mathbf{A}$ is strictly less than 1.

References

Hamilton, J. D. 1994. *Time Series Analysis*. Princeton: Princeton University Press.

Lütkepohl, H. 1993. *Introduction to Multiple Time Series Analysis*. 2nd ed. New York: Springer.

——. 2005. *New Introduction to Multiple Time Series Analysis*. New York: Springer.

Also see

[TS] **var** — Vector autoregressive models

[TS] **var svar** — Structural vector autoregressive models

[TS] **varbasic** — Fit a simple VAR and graph IRFs or FEVDs

[TS] **var intro** — Introduction to vector autoregressive models

Title

> **varwle** — Obtain Wald lag-exclusion statistics after var or svar

Syntax

varwle $\big[$, <u>est</u>imates(*estname*) <u>separator</u>(#) $\big]$

varwle can be used only after var or svar; see [TS] **var** and [TS] **var svar**.

Menu

Statistics > Multivariate time series > VAR diagnostics and tests > Wald lag-exclusion statistics

Description

varwle reports Wald tests the hypothesis that the endogenous variables at a given lag are jointly zero for each equation and for all equations jointly.

Options

<u>est</u>imates(*estname*) requests that varwle use the previously obtained set of var or svar estimates saved as *estname*. By default, varwle uses the active estimation results. See [R] **estimates** for information about saving and restoring estimation results.

<u>separator</u>(#) specifies how often separator lines should be drawn between rows. By default, separator lines do not appear. For example, separator(1) would draw a line between each row, separator(2) between every other row, and so on.

Remarks

After fitting a VAR, one hypothesis of interest is that all the endogenous variables at a given lag are jointly zero. varwle reports Wald tests of this hypothesis for each equation and for all equations jointly. varwle uses the estimation results from a previously fitted var or svar. By default, varwle uses the active estimation results, but you may also use a stored set of estimates by specifying the estimates() option.

If the VAR was fit with the small option, varwle also presents small-sample F statistics; otherwise, varwle presents large-sample chi-squared statistics.

▷ Example 1: After var

We analyze the model with the German data described in [TS] **var** using varwle.

```
. use http://www.stata-press.com/data/r12/lutkepohl2
(Quarterly SA West German macro data, Bil DM, from Lutkepohl 1993 Table E.1)
. var dln_inv dln_inc dln_consump if qtr<=tq(1978q4), dfk small
  (output omitted )
. varwle
```

Equation: dln_inv

lag	F	df	df_r	Prob > F
1	2.64902	3	66	0.0560
2	1.25799	3	66	0.2960

Equation: dln_inc

lag	F	df	df_r	Prob > F
1	2.19276	3	66	0.0971
2	.907499	3	66	0.4423

Equation: dln_consump

lag	F	df	df_r	Prob > F
1	1.80804	3	66	0.1543
2	5.57645	3	66	0.0018

Equation: All

lag	F	df	df_r	Prob > F
1	3.78884	9	66	0.0007
2	2.96811	9	66	0.0050

Because the VAR was fit with the `dfk` and `small` options, `varwle` used the small-sample estimator of $\widehat{\Sigma}$ in constructing the VCE, producing an F statistic. The first two equations appear to have a different lag structure from that of the third. In the first two equations, we cannot reject the null hypothesis that all three endogenous variables have zero coefficients at the second lag. The hypothesis that all three endogenous variables have zero coefficients at the first lag can be rejected at the 10% level for both of the first two equations. In contrast, in the third equation, the coefficients on the second lag of the endogenous variables are jointly significant, but not those on the first lag. However, we strongly reject the hypothesis that the coefficients on the first lag of the endogenous variables are zero in all three equations jointly. Similarly, we can also strongly reject the hypothesis that the coefficients on the second lag of the endogenous variables are zero in all three equations jointly.

If we believe these results strongly enough, we might want to refit the original VAR, placing some constraints on the coefficients. See [TS] **var** for details on how to fit VAR models with constraints.

◁

▷ Example 2: After svar

Here we fit a simple SVAR and then run `varwle`:

```
. matrix a = (.,0\.,.)
. matrix b = I(2)
```

```
. svar dln_inc dln_consump, aeq(a) beq(b)
Estimating short-run parameters

Iteration 0:   log likelihood = -159.21683
Iteration 1:   log likelihood =  490.92264
Iteration 2:   log likelihood =  528.66126
Iteration 3:   log likelihood =  573.96363
Iteration 4:   log likelihood =  578.05136
Iteration 5:   log likelihood =  578.27633
Iteration 6:   log likelihood =  578.27699
Iteration 7:   log likelihood =  578.27699

Structural vector autoregression

 ( 1)  [a_1_2]_cons = 0
 ( 2)  [b_1_1]_cons = 1
 ( 3)  [b_1_2]_cons = 0
 ( 4)  [b_2_1]_cons = 0
 ( 5)  [b_2_2]_cons = 1
```

Sample: 1960q4 - 1982q4		No. of obs	=	89
Exactly identified model		Log likelihood	=	578.277

	Coef.	Std. Err.	z	P>|z|	[95% Conf. Interval]	
/a_1_1	89.72411	6.725107	13.34	0.000	76.54315	102.9051
/a_2_1	-64.73622	10.67698	-6.06	0.000	-85.66271	-43.80973
/a_1_2	0	.	.	.	.	.
/a_2_2	126.2964	9.466318	13.34	0.000	107.7428	144.8501
/b_1_1	1	.	.	.	.	.
/b_2_1	0	.	.	.	.	.
/b_1_2	0	.	.	.	.	.
/b_2_2	1	.	.	.	.	.

The output table from var svar gives information about the estimates of the parameters in the **A** and **B** matrices in the structural VAR. But, as discussed in [TS] **var svar**, an SVAR model builds on an underlying VAR. When varwle uses the estimation results produced by svar, it performs Wald lag-exclusion tests on the underlying VAR model. Next we run varwle on these svar results.

```
. varwle
```

Equation: dln_inc

lag	chi2	df	Prob > chi2
1	6.88775	2	0.032
2	1.873546	2	0.392

Equation: dln_consump

lag	chi2	df	Prob > chi2
1	9.938547	2	0.007
2	13.89996	2	0.001

Equation: All

lag	chi2	df	Prob > chi2
1	34.54276	4	0.000
2	19.44093	4	0.001

Now we fit the underlying VAR with two lags and apply `varwle` to these results.

```
. var dln_inc dln_consump
(output omitted )

. varwle
```

    Equation: dln_inc

lag	chi2	df	Prob > chi2
1	6.88775	2	0.032
2	1.873546	2	0.392

    Equation: dln_consump

lag	chi2	df	Prob > chi2
1	9.938547	2	0.007
2	13.89996	2	0.001

    Equation: All

lag	chi2	df	Prob > chi2
1	34.54276	4	0.000
2	19.44093	4	0.001

Because `varwle` produces the same results in these two cases, we can conclude that when `varwle` is applied to `svar` results, it performs Wald lag-exclusion tests on the underlying VAR.

◁

Saved results

`varwle` saves the following in `r()`:

Matrices
if e(small)==""

r(chi2)	χ^2 test statistics
r(df)	degrees of freedom
r(p)	p-values

if e(small)!=""

r(F)	F test statistics
r(df_r)	numerator degrees of freedom
r(df)	denominator degree of freedom
r(p)	p-values

Methods and formulas

`varwle` is implemented as an ado-file.

`varwle` uses `test` to obtain Wald statistics of the hypotheses that all the endogenous variables at a given lag are jointly zero for each equation and for all equations jointly. Like the `test` command, `varwle` uses estimation results saved by `var` or `var svar` to determine whether to calculate and report small-sample F statistics or large-sample chi-squared statistics.

Abraham Wald (1902–1950) was born in Cluj, in what is now Romania. He studied mathematics at the University of Vienna, publishing at first on geometry, but then became interested in economics and econometrics. He moved to the United States in 1938 and later joined the faculty at Columbia. His major contributions to statistics include work in decision theory, optimal sequential sampling, large-sample distributions of likelihood-ratio tests, and nonparametric inference. Wald died in a plane crash in India.

References

Amisano, G., and C. Giannini. 1997. *Topics in Structural VAR Econometrics*. 2nd ed. Heidelberg: Springer.

Hamilton, J. D. 1994. *Time Series Analysis*. Princeton: Princeton University Press.

Lütkepohl, H. 1993. *Introduction to Multiple Time Series Analysis*. 2nd ed. New York: Springer.

Mangel, M., and F. J. Samaniego. 1984. Abraham Wald's work on aircraft survivability. *Journal of the American Statistical Association* 79: 259–267.

Wolfowitz, J. 1952. Abraham Wald, 1902–1950. *Annals of Mathematical Statistics* 23: 1–13 (and other reports in same issue).

Also see

[TS] **var** — Vector autoregressive models

[TS] **var svar** — Structural vector autoregressive models

[TS] **varbasic** — Fit a simple VAR and graph IRFs or FEVDs

[TS] **var intro** — Introduction to vector autoregressive models

Title

> **vec intro** — Introduction to vector error-correction models

Description

Stata has a suite of commands for fitting, forecasting, interpreting, and performing inference on vector error-correction models (VECMs) with cointegrating variables. After fitting a VECM, the irf commands can be used to obtain impulse–response functions (IRFs) and forecast-error variance decompositions (FEVDs). The table below describes the available commands.

Fitting a VECM

vec	[TS] **vec**	Fit vector error-correction models

Model diagnostics and inference

vecrank	[TS] **vecrank**	Estimate the cointegrating rank of a VECM
veclmar	[TS] **veclmar**	Perform LM test for residual autocorrelation after vec
vecnorm	[TS] **vecnorm**	Test for normally distributed disturbances after vec
vecstable	[TS] **vecstable**	Check the stability condition of VECM estimates
varsoc	[TS] **varsoc**	Obtain lag-order selection statistics for VARs and VECMs

Forecasting from a VECM

fcast compute	[TS] **fcast compute**	Compute dynamic forecasts of dependent variables after var, svar, or vec
fcast graph	[TS] **fcast graph**	Graph forecasts of variables computed by fcast compute

Working with IRFs and FEVDs

irf	[TS] **irf**	Create and analyze IRFs and FEVDs

This manual entry provides an overview of the commands for VECMs; provides an introduction to integration, cointegration, estimation, inference, and interpretation of VECM models; and gives an example of how to use Stata's vec commands.

Remarks

vec estimates the parameters of cointegrating VECMs. You may specify any of the five trend specifications in Johansen (1995, sec. 5.7). By default, identification is obtained via the Johansen normalization, but vec allows you to obtain identification by placing your own constraints on the parameters of the cointegrating vectors. You may also put more restrictions on the adjustment coefficients.

vecrank is the command for determining the number of cointegrating equations. vecrank implements Johansen's multiple trace test procedure, the maximum eigenvalue test, and a method based on minimizing either of two different information criteria.

Because Nielsen (2001) has shown that the methods implemented in `varsoc` can be used to choose the order of the autoregressive process, no separate `vec` command is needed; you can simply use `varsoc`. `veclmar` tests that the residuals have no serial correlation, and `vecnorm` tests that they are normally distributed.

All the `irf` routines described in [TS] **irf** are available for estimating, interpreting, and managing estimated IRFs and FEVDs for VECMs.

Remarks are presented under the following headings:

> *Introduction to cointegrating VECMs*
> > *What is cointegration?*
> > *The multivariate VECM specification*
> > *Trends in the Johansen VECM framework*
> *VECM estimation in Stata*
> > *Selecting the number of lags*
> > *Testing for cointegration*
> > *Fitting a VECM*
> > *Fitting VECMs with Johansen's normalization*
> > *Postestimation specification testing*
> > *Impulse–response functions for VECMs*
> > *Forecasting with VECMs*

Introduction to cointegrating VECMs

This section provides a brief introduction to integration, cointegration, and cointegrated vector error-correction models. For more details about these topics, see Hamilton (1994), Johansen (1995), Lütkepohl (2005), and Watson (1994).

What is cointegration?

Standard regression techniques, such as ordinary least squares (OLS), require that the variables be covariance stationary. A variable is covariance stationary if its mean and all its autocovariances are finite and do not change over time. Cointegration analysis provides a framework for estimation, inference, and interpretation when the variables are not covariance stationary.

Instead of being covariance stationary, many economic time series appear to be "first-difference stationary". This means that the level of a time series is not stationary but its first difference is. First-difference stationary processes are also known as integrated processes of order 1, or I(1) processes. Covariance-stationary processes are I(0). In general, a process whose dth difference is stationary is an integrated process of order d, or I(d).

The canonical example of a first-difference stationary process is the random walk. This is a variable x_t that can be written as

$$x_t = x_{t-1} + \epsilon_t \tag{1}$$

where the ϵ_t are independently and identically distributed (i.i.d.) with mean zero and a finite variance σ^2. Although $E[x_t] = 0$ for all t, $\text{Var}[x_t] = T\sigma^2$ is not time invariant, so x_t is not covariance stationary. Because $\Delta x_t = x_t - x_{t-1} = \epsilon_t$ and ϵ_t is covariance stationary, x_t is first-difference stationary.

These concepts are important because, although conventional estimators are well behaved when applied to covariance-stationary data, they have nonstandard asymptotic distributions and different rates of convergence when applied to I(1) processes. To illustrate, consider several variants of the model

$$y_t = a + bx_t + e_t \tag{2}$$

Throughout the discussion, we maintain the assumption that $E[e_t] = 0$.

If both y_t and x_t are covariance-stationary processes, e_t must also be covariance stationary. As long as $E[x_t e_t] = 0$, we can consistently estimate the parameters a and b by using OLS. Furthermore, the distribution of the OLS estimator converges to a normal distribution centered at the true value as the sample size grows.

If y_t and x_t are independent random walks and $b = 0$, there is no relationship between y_t and x_t, and (2) is called a spurious regression. Granger and Newbold (1974) performed Monte Carlo experiments and showed that the usual t statistics from OLS regression provide spurious results: given a large enough dataset, we can almost always reject the null hypothesis of the test that $b = 0$ even though b is in fact zero. Here the OLS estimator does not converge to any well-defined population parameter.

Phillips (1986) later provided the asymptotic theory that explained the Granger and Newbold (1974) results. He showed that the random walks y_t and x_t are first-difference stationary processes and that the OLS estimator does not have its usual asymptotic properties when the variables are first-difference stationary.

Because Δy_t and Δx_t are covariance stationary, a simple regression of Δy_t on Δx_t appears to be a viable alternative. However, if y_t and x_t cointegrate, as defined below, the simple regression of Δy_t on Δx_t is misspecified.

If y_t and x_t are I(1) and $b \neq 0$, e_t could be either I(0) or I(1). Phillips and Durlauf (1986) have derived the asymptotic theory for the OLS estimator when e_t is I(1), though it has not been widely used in applied work. More interesting is the case in which $e_t = y_t - a - bx_t$ is I(0). y_t and x_t are then said to be cointegrated. Two variables are cointegrated if each is an I(1) process but a linear combination of them is an I(0) process.

It is not possible for y_t to be a random walk and x_t and e_t to be covariance stationary. As Granger (1981) pointed out, because a random walk cannot be equal to a covariance-stationary process, the equation does not "balance". An equation balances when the processes on each side of the equal sign are of the same order of integration. Before attacking any applied problem with integrated variables, make sure that the equation balances before proceeding.

An example from Engle and Granger (1987) provides more intuition. Redefine y_t and x_t to be

$$y_t + \beta x_t = \epsilon_t, \qquad \epsilon_t = \epsilon_{t-1} + \xi_t \tag{3}$$
$$y_t + \alpha x_t = \nu_t, \qquad \nu_t = \rho \nu_{t-1} + \zeta_t, \quad |\rho| < 1 \tag{4}$$

where ξ_t and ζ_t are i.i.d. disturbances over time that are correlated with each other. Because ϵ_t is I(1), (3) and (4) imply that both x_t and y_t are I(1). The condition that $|\rho| < 1$ implies that ν_t and $y_t + \alpha x_t$ are I(0). Thus y_t and x_t cointegrate, and $(1, \alpha)$ is the cointegrating vector.

Using a bit of algebra, we can rewrite (3) and (4) as

$$\Delta y_t = \beta \delta z_{t-1} + \eta_{1t} \tag{5}$$
$$\Delta x_t = -\delta z_{t-1} + \eta_{2t} \tag{6}$$

where $\delta = (1-\rho)/(\alpha-\beta)$, $z_t = y_t + \alpha x_t$, and η_{1t} and η_{2t} are distinct, stationary, linear combinations of ξ_t and ζ_t. This representation is known as the vector error-correction model (VECM). One can think of $z_t = 0$ as being the point at which y_t and x_t are in equilibrium. The coefficients on z_{t-1} describe how y_t and x_t adjust to z_{t-1} being nonzero, or out of equilibrium. z_t is the "error" in the system, and (5) and (6) describe how system adjusts or corrects back to the equilibrium. As ρ goes to 1, the system degenerates into a pair of correlated random walks. The VECM parameterization highlights this point, because $\delta \to 0$ as $\rho \to 1$.

If we knew α, we would know z_t, and we could work with the stationary system of (5) and . Although knowing α seems silly, we can conduct much of the analysis as if we knew α because there is an estimator for the cointegrating parameter α that converges to its true value at a faster rate than the estimator for the adjustment parameters β and δ.

The definition of a bivariate cointegrating relation requires simply that there exist a linear combination of the I(1) variables that is I(0). If y_t and x_t are I(1) and there are two finite real numbers $a \neq 0$ and $b \neq 0$, such that $ay_t + bx_t$ is I(0), then y_t and x_t are cointegrated. Although there are two parameters, a and b, only one will be identifiable because if $ay_t + bx_t$ is I(0), so is $cay_t + cbx_t$ for any finite, nonzero, real number c. Obtaining identification in the bivariate case is relatively simple. The coefficient on y_t in (4) is unity. This natural construction of the model placed the necessary identification restriction on the cointegrating vector. As we discuss below, identification in the multivariate case is more involved.

If $\mathbf{y}_t$ is a $K \times 1$ vector of I(1) variables and there exists a vector β, such that $\beta \mathbf{y}_t$ is a vector of I(0) variables, then $\mathbf{y}_t$ is said to be cointegrating of order (1,0) with cointegrating vector β. We say that the parameters in β are the parameters in the cointegrating equation. For a vector of length K, there may be at most $K - 1$ distinct cointegrating vectors. Engle and Granger (1987) provide a more general definition of cointegration, but this one is sufficient for our purposes.

The multivariate VECM specification

In practice, most empirical applications analyze multivariate systems, so the rest of our discussion focuses on that case. Consider a VAR with p lags

$$\mathbf{y}_t = \mathbf{v} + \mathbf{A}_1\mathbf{y}_{t-1} + \mathbf{A}_2\mathbf{y}_{t-2} + \cdots + \mathbf{A}_p\mathbf{y}_{t-p} + \boldsymbol{\epsilon}_t \tag{7}$$

where $\mathbf{y}_t$ is a $K \times 1$ vector of variables, $\mathbf{v}$ is a $K \times 1$ vector of parameters, $\mathbf{A}_1$–$\mathbf{A}_p$ are $K \times K$ matrices of parameters, and $\boldsymbol{\epsilon}_t$ is a $K \times 1$ vector of disturbances. $\boldsymbol{\epsilon}_t$ has mean $\mathbf{0}$, has covariance matrix $\boldsymbol{\Sigma}$, and is i.i.d. normal over time. Any VAR(p) can be rewritten as a VECM. Using some algebra, we can rewrite (7) in VECM form as

$$\Delta\mathbf{y}_t = \mathbf{v} + \mathbf{\Pi}\mathbf{y}_{t-1} + \sum_{i=1}^{p-1} \mathbf{\Gamma}_i \Delta\mathbf{y}_{t-i} + \boldsymbol{\epsilon}_t \tag{8}$$

where $\mathbf{\Pi} = \sum_{j=1}^{j=p} \mathbf{A}_j - \mathbf{I}_k$ and $\mathbf{\Gamma}_i = -\sum_{j=i+1}^{j=p} \mathbf{A}_j$. The $\mathbf{v}$ and $\boldsymbol{\epsilon}_t$ in (7) and (8) are identical.

Engle and Granger (1987) show that if the variables $\mathbf{y}_t$ are I(1) the matrix $\mathbf{\Pi}$ in (8) has rank $0 \leq r < K$, where r is the number of linearly independent cointegrating vectors. If the variables cointegrate, $0 < r < K$ and (8) shows that a VAR in first differences is misspecified because it omits the lagged level term $\mathbf{\Pi}\mathbf{y}_{t-1}$.

Assume that $\mathbf{\Pi}$ has reduced rank $0 < r < K$ so that it can be expressed as $\mathbf{\Pi} = \alpha\beta'$, where α and β are both $r \times K$ matrices of rank r. Without further restrictions, the cointegrating vectors are not identified: the parameters (α, β) are indistinguishable from the parameters $(\alpha\mathbf{Q}, \beta\mathbf{Q}^{-1'})$ for any $r \times r$ nonsingular matrix $\mathbf{Q}$. Because only the rank of $\mathbf{\Pi}$ is identified, the VECM is said to identify the rank of the cointegrating space, or equivalently, the number of cointegrating vectors. In practice, the estimation of the parameters of a VECM requires at least r^2 identification restrictions. Stata's vec command can apply the conventional Johansen restrictions discussed below or use constraints that the user supplies.

The VECM in (8) also nests two important special cases. If the variables in $\mathbf{y}_t$ are I(1) but not cointegrated, $\mathbf{\Pi}$ is a matrix of zeros and thus has rank 0. If all the variables are I(0), $\mathbf{\Pi}$ has full rank K.

There are several different frameworks for estimation and inference in cointegrating systems. Although the methods in Stata are based on the maximum likelihood (ML) methods developed by Johansen (1988, 1991, 1995), other useful frameworks have been developed by Park and Phillips (1988, 1989); Sims, Stock, and Watson (1990); Stock (1987); and Stock and Watson (1988); among others. The ML framework developed by Johansen was independently developed by Ahn and Reinsel (1990). Maddala and Kim (1998) and Watson (1994) survey all these methods. The cointegration methods in Stata are based on Johansen's maximum likelihood framework because it has been found to be particularly useful in several comparative studies, including Gonzalo (1994) and Hubrich, Lütkepohl, and Saikkonen (2001).

Trends in the Johansen VECM framework

Deterministic trends in a cointegrating VECM can stem from two distinct sources; the mean of the cointegrating relationship and the mean of the differenced series. Allowing for a constant and a linear trend and assuming that there are r cointegrating relations, we can rewrite the VECM in (8) as

$$\Delta\mathbf{y}_t = \alpha\beta'\mathbf{y}_{t-1} + \sum_{i=1}^{p-1}\Gamma_i\Delta\mathbf{y}_{t-i} + \mathbf{v} + \delta t + \epsilon_t \tag{9}$$

where δ is a $K \times 1$ vector of parameters. Because (9) models the differences of the data, the constant implies a linear time trend in the levels, and the time trend δt implies a quadratic time trend in the levels of the data. Often we may want to include a constant or a linear time trend for the differences without allowing for the higher-order trend that is implied for the levels of the data. VECMs exploit the properties of the matrix α to achieve this flexibility.

Because α is a $K \times r$ rank matrix, we can rewrite the deterministic components in (9) as

$$\mathbf{v} = \alpha\mu + \gamma \tag{10a}$$

$$\delta t = \alpha\rho t + \tau t \tag{10b}$$

where μ and ρ are $r \times 1$ vectors of parameters and γ and τ are $K \times 1$ vectors of parameters. γ is orthogonal to $\alpha\mu$, and τ is orthogonal to $\alpha\rho$; that is, $\gamma'\alpha\mu = 0$ and $\tau'\alpha\rho = 0$, allowing us to rewrite (9) as

$$\Delta\mathbf{y}_t = \alpha(\beta'\mathbf{y}_{t-1} + \mu + \rho t) + \sum_{i=1}^{p-1}\Gamma_i\Delta\mathbf{y}_{t-i} + \gamma + \tau t + \epsilon_t \tag{11}$$

Placing restrictions on the trend terms in (11) yields five cases.

CASE 1: Unrestricted trend

> If no restrictions are placed on the trend parameters, (11) implies that there are quadratic trends in the levels of the variables and that the cointegrating equations are stationary around time trends (trend stationary).

CASE 2: Restricted trend, $\tau = \mathbf{0}$

> By setting $\tau = \mathbf{0}$, we assume that the trends in the levels of the data are linear but not quadratic. This specification allows the cointegrating equations to be trend stationary.

CASE 3: Unrestricted constant, $\tau = \mathbf{0}$ and $\rho = \mathbf{0}$

> By setting $\tau = \mathbf{0}$ and $\rho = \mathbf{0}$, we exclude the possibility that the levels of the data have quadratic trends, and we restrict the cointegrating equations to be stationary around constant means. Because γ is not restricted to zero, this specification still puts a linear time trend in the levels of the data.

CASE 4: Restricted constant, $\tau = 0$, $\rho = 0$, and $\gamma = 0$

> By adding the restriction that $\gamma = 0$, we assume there are no linear time trends in the levels of the data. This specification allows the cointegrating equations to be stationary around a constant mean, but it allows no other trends or constant terms.

CASE 5: No trend, $\tau = 0$, $\rho = 0$, $\gamma = 0$, and $\mu = 0$

> This specification assumes that there are no nonzero means or trends. It also assumes that the cointegrating equations are stationary with means of zero and that the differences and the levels of the data have means of zero.

This flexibility does come at a price. Below we discuss testing procedures for determining the number of cointegrating equations. The asymptotic distribution of the LR for hypotheses about r changes with the trend specification, so we must first specify a trend specification. A combination of theory and graphical analysis will aid in specifying the trend before proceeding with the analysis.

VECM estimation in Stata

We provide an overview of the vec commands in Stata through an extended example. We have monthly data on the average selling prices of houses in four cities in Texas: Austin, Dallas, Houston, and San Antonio. In the dataset, these average housing prices are contained in the variables austin, dallas, houston, and sa. The series begin in January of 1990 and go through December 2003, for a total of 168 observations. The following graph depicts our data.

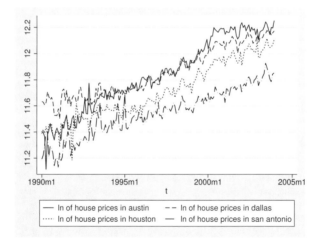

The plots on the graph indicate that all the series are trending and potential I(1) processes. In a competitive market, the current and past prices contain all the information available, so tomorrow's price will be a random walk from today's price. Some researchers may opt to use [TS] **dfgls** to investigate the presence of a unit root in each series, but the test for cointegration we use includes the case in which all the variables are stationary, so we defer formal testing until we test for cointegration. The time trends in the data appear to be approximately linear, so we will specify trend(constant) when modeling these series, which is the default with vec.

The next graph shows just Dallas' and Houston's data, so we can more carefully examine their relationship.

Except for the crash at the end of 1991, housing prices in Dallas and Houston appear closely related. Although average prices in the two cities will differ because of resource variations and other factors, if the housing markets become too dissimilar, people and businesses will migrate, bringing the average housing prices back toward each other. We therefore expect the series of average housing prices in Houston to be cointegrated with the series of average housing prices in Dallas.

Selecting the number of lags

To test for cointegration or fit cointegrating VECMs, we must specify how many lags to include. Building on the work of Tsay (1984) and Paulsen (1984), Nielsen (2001) has shown that the methods implemented in varsoc can be used to determine the lag order for a VAR model with I(1) variables. As can be seen from (9), the order of the corresponding VECM is always one less than the VAR. vec makes this adjustment automatically, so we will always refer to the order of the underlying VAR. The output below uses varsoc to determine the lag order of the VAR of the average housing prices in Dallas and Houston.

```
. use http://www.stata-press.com/data/r12/txhprice
. varsoc dallas houston
   Selection-order criteria
   Sample:  1990m5 - 2003m12                Number of obs     =       164
```

lag	LL	LR	df	p	FPE	AIC	HQIC	SBIC
0	299.525				.000091	-3.62835	-3.61301	-3.59055
1	577.483	555.92	4	0.000	3.2e-06	-6.9693	-6.92326	-6.85589
2	590.978	26.991*	4	0.000	2.9e-06*	-7.0851*	-7.00837*	-6.89608*
3	593.437	4.918	4	0.296	2.9e-06	-7.06631	-6.95888	-6.80168
4	596.364	5.8532	4	0.210	3.0e-06	-7.05322	-6.9151	-6.71299

```
   Endogenous:  dallas houston
   Exogenous:   _cons
```

We will use two lags for this bivariate model because the Hannan–Quinn information criterion (HQIC) method, Schwarz Bayesian information criterion (SBIC) method, and sequential likelihood-ratio (LR) test all chose two lags, as indicated by the "*" in the output.

The reader can verify that when all four cities' data are used, the LR test selects three lags, the HQIC method selects two lags, and the SBIC method selects one lag. We will use three lags in our four-variable model.

Testing for cointegration

The tests for cointegration implemented in `vecrank` are based on Johansen's method. If the log likelihood of the unconstrained model that includes the cointegrating equations is significantly different from the log likelihood of the constrained model that does not include the cointegrating equations, we reject the null hypothesis of no cointegration.

Here we use `vecrank` to determine the number of cointegrating equations:

```
. vecrank dallas houston
                    Johansen tests for cointegration
Trend: constant                                  Number of obs =      166
Sample:  1990m3 - 2003m12                                  Lags =        2
                                                         5%
maximum                                    trace    critical
   rank    parms       LL     eigenvalue  statistic   value
      0        6  576.26444          .      46.8252    15.41
      1        9  599.58781    0.24498       0.1785*    3.76
      2       10  599.67706    0.00107
```

Besides presenting information about the sample size and time span, the header indicates that test statistics are based on a model with two lags and a constant trend. The body of the table presents test statistics and their critical values of the null hypotheses of no cointegration (line 1) and one or fewer cointegrating equations (line 2). The eigenvalue shown on the last line is used to compute the trace statistic in the line above it. Johansen's testing procedure starts with the test for zero cointegrating equations (a maximum rank of zero) and then accepts the first null hypothesis that is not rejected.

In the output above, we strongly reject the null hypothesis of no cointegration and fail to reject the null hypothesis of at most one cointegrating equation. Thus we accept the null hypothesis that there is one cointegrating equation in the bivariate model.

Using all four series and a model with three lags, we find that there are two cointegrating relationships.

```
. vecrank austin dallas houston sa, lag(3)
                    Johansen tests for cointegration
Trend: constant                                  Number of obs =      165
Sample:  1990m4 - 2003m12                                  Lags =        3
                                                         5%
maximum                                    trace    critical
   rank    parms       LL     eigenvalue  statistic   value
      0       36  1107.7833          .     101.6070    47.21
      1       43  1137.7484    0.30456      41.6768    29.68
      2       48  1153.6435    0.17524       9.8865*   15.41
      3       51  1158.4191    0.05624       0.3354     3.76
      4       52  1158.5868    0.00203
```

Fitting a VECM

`vec` estimates the parameters of cointegrating VECMs. There are four types of parameters of interest:

1. The parameters in the cointegrating equations β

2. The adjustment coefficients α

3. The short-run coefficients

4. Some standard functions of β and α that have useful interpretations

Although all four types are discussed in [TS] **vec**, here we discuss only types 1–3 and how they appear in the output of vec.

Having determined that there is a cointegrating equation between the Dallas and Houston series, we now want to estimate the parameters of a bivariate cointegrating VECM for these two series by using vec.

```
. vec dallas houston

Vector error-correction model

Sample:  1990m3 - 2003m12                       No. of obs   =        166
                                                AIC          = -7.115516
Log likelihood =  599.5878                      HQIC         =  -7.04703
Det(Sigma_ml)  =  2.50e-06                      SBIC         = -6.946794

Equation          Parms      RMSE     R-sq      chi2     P>chi2

D_dallas             4     .038546   0.1692   32.98959   0.0000
D_houston            4     .045348   0.3737   96.66399   0.0000
```

	Coef.	Std. Err.	z	P>|z|	[95% Conf. Interval]	
D_dallas						
_ce1						
L1.	-.3038799	.0908504	-3.34	0.001	-.4819434	-.1258165
dallas						
LD.	-.1647304	.0879356	-1.87	0.061	-.337081	.0076202
houston						
LD.	-.0998368	.0650838	-1.53	0.125	-.2273988	.0277251
_cons	.0056128	.0030341	1.85	0.064	-.0003339	.0115595
D_houston						
_ce1						
L1.	.5027143	.1068838	4.70	0.000	.2932258	.7122028
dallas						
LD.	-.0619653	.1034547	-0.60	0.549	-.2647327	.1408022
houston						
LD.	-.3328437	.07657	-4.35	0.000	-.4829181	-.1827693
_cons	.0033928	.0035695	0.95	0.342	-.0036034	.010389

```
Cointegrating equations

Equation          Parms      chi2     P>chi2

_ce1                 1     1640.088   0.0000

Identification:  beta is exactly identified
                 Johansen normalization restriction imposed
```

beta	Coef.	Std. Err.	z	P>|z|	[95% Conf. Interval]	
_ce1						
dallas	1	.	.	.	.	.
houston	-.8675936	.0214231	-40.50	0.000	-.9095821	-.825605
_cons	-1.688897	.	.	.	.	.

The header contains information about the sample, the fit of each equation, and overall model fit statistics. The first estimation table contains the estimates of the short-run parameters, along with their standard errors, z statistics, and confidence intervals. The two coefficients on L._ce1 are the parameters in the adjustment matrix α for this model. The second estimation table contains the estimated parameters of the cointegrating vector for this model, along with their standard errors, z statistics, and confidence intervals.

Using our previous notation, we have estimated

$$\widehat{\alpha} = (-0.304, 0.503) \qquad \widehat{\beta} = (1, -0.868) \qquad \widehat{\mathbf{v}} = (0.0056, 0.0034)$$

and

$$\widehat{\Gamma} = \begin{pmatrix} -0.165 & -0.0998 \\ -0.062 & -0.333 \end{pmatrix}$$

Overall, the output indicates that the model fits well. The coefficient on houston in the cointegrating equation is statistically significant, as are the adjustment parameters. The adjustment parameters in this bivariate example are easy to interpret, and we can see that the estimates have the correct signs and imply rapid adjustment toward equilibrium. When the predictions from the cointegrating equation are positive, dallas is above its equilibrium value because the coefficient on dallas in the cointegrating equation is positive. The estimate of the coefficient [D_dallas]L._ce1 is $-.3$. Thus when the average housing price in Dallas is too high, it quickly falls back toward the Houston level. The estimated coefficient [D_houston]L._ce1 of .5 implies that when the average housing price in Dallas is too high, the average price in Houston quickly adjusts toward the Dallas level at the same time that the Dallas prices are adjusting.

Fitting VECMs with Johansen's normalization

As discussed by Johansen (1995), if there are r cointegrating equations, then at least r^2 restrictions are required to identify the free parameters in β. Johansen proposed a default identification scheme that has become the conventional method of identifying models in the absence of theoretically justified restrictions. Johansen's identification scheme is

$$\beta' = (\mathbf{I}_r, \widetilde{\beta}')$$

where $\mathbf{I}_r$ is the $r \times r$ identity matrix and $\widetilde{\beta}$ is an $(K - r) \times r$ matrix of identified parameters. vec applies Johansen's normalization by default.

To illustrate, we fit a VECM with two cointegrating equations and three lags on all four series. We are interested only in the estimates of the parameters in the cointegrating equations, so we can specify the noetable option to suppress the estimation table for the adjustment and short-run parameters.

```
. vec austin dallas houston sa, lags(3) rank(2) noetable
```

Vector error-correction model

Sample: 1990m4 – 2003m12	No. of obs	=	165
	AIC	=	-13.40174
Log likelihood = 1153.644	HQIC	=	-13.03496
Det(Sigma_ml) = 9.93e-12	SBIC	=	-12.49819

Cointegrating equations

Equation	Parms	chi2	P>chi2
_ce1	2	586.3044	0.0000
_ce2	2	2169.826	0.0000

Identification: beta is exactly identified

Johansen normalization restrictions imposed

beta	Coef.	Std. Err.	z	P>|z|	[95% Conf. Interval]	
_ce1						
austin	1	.	.	.	.	.
dallas	-2.37e-17	.	.	.	.	.
houston	-.2623782	.1893625	-1.39	0.166	-.6335219	.1087655
sa	-1.241805	.229643	-5.41	0.000	-1.691897	-.7917128
_cons	5.577099	.	.	.	.	.
_ce2						
austin	-1.58e-17	.	.	.	.	.
dallas	1	.	.	.	.	.
houston	-1.095652	.0669898	-16.36	0.000	-1.22695	-.9643545
sa	.2883986	.0812396	3.55	0.000	.1291718	.4476253
_cons	-2.351372	.	.	.	.	.

The Johansen identification scheme has placed four constraints on the parameters in β: [_ce1]austin=1, [_ce1]dallas=0, [_ce2]austin=0, and [_ce2]dallas=1. (The computational method used imposes zero restrictions that are numerical rather than exact. The values $-3.48e-17$ and $-1.26e-17$ are indistinguishable from zero.) We interpret the results of the first equation as indicating the existence of an equilibrium relationship between the average housing price in Austin and the average prices of houses in Houston and San Antonio.

The Johansen normalization restricted the coefficient on dallas to be unity in the second cointegrating equation, but we could instead constrain the coefficient on houston. Both sets of restrictions define just-identified models, so fitting the model with the latter set of restrictions will yield the same maximized log likelihood. To impose the alternative set of constraints, we use the constraint command.

```
. constraint 1 [_ce1]austin = 1
. constraint 2 [_ce1]dallas = 0
. constraint 3 [_ce2]austin = 0
. constraint 4 [_ce2]houston = 1
```

```
. vec austin dallas houston sa, lags(3) rank(2) noetable bconstraints(1/4)

Iteration 1:      log likelihood = 1148.8745
  (output omitted )
Iteration 25:     log likelihood = 1153.6435

Vector error-correction model
```

Sample: 1990m4 - 2003m12		No. of obs	=	165
		AIC	=	-13.40174
Log likelihood = 1153.644		HQIC	=	-13.03496
Det(Sigma_ml) = 9.93e-12		SBIC	=	-12.49819

Cointegrating equations

Equation	Parms	chi2	P>chi2
_ce1	2	586.3392	0.0000
_ce2	2	3455.469	0.0000

```
Identification:  beta is exactly identified
 ( 1)  [_ce1]austin = 1
 ( 2)  [_ce1]dallas = 0
 ( 3)  [_ce2]austin = 0
 ( 4)  [_ce2]houston = 1
```

beta	Coef.	Std. Err.	z	P>|z|	[95% Conf. Interval]	
_ce1						
austin	1	.	.	.	.	.
dallas	0	(omitted)				
houston	-.2623784	.1876727	-1.40	0.162	-.6302102	.1054534
sa	-1.241805	.2277537	-5.45	0.000	-1.688194	-.7954157
_cons	5.577099	.	.	.	.	.
_ce2						
austin	0	(omitted)				
dallas	-.9126985	.0595804	-15.32	0.000	-1.029474	-.7959231
houston	1	.	.	.	.	.
sa	-.2632209	.0628791	-4.19	0.000	-.3864617	-.1399802
_cons	2.146094	.	.	.	.	.

Only the estimates of the parameters in the second cointegrating equation have changed, and the new estimates are simply the old estimates divided by -1.095652 because the new constraints are just an alternative normalization of the same just-identified model. With the new normalization, we can interpret the estimates of the parameters in the second cointegrating equation as indicating an equilibrium relationship between the average house price in Houston and the average prices of houses in Dallas and San Antonio.

Postestimation specification testing

Inference on the parameters in α depends crucially on the stationarity of the cointegrating equations, so we should check the specification of the model. As a first check, we can predict the cointegrating equations and graph them over time.

```
. predict ce1, ce equ(#1)
. predict ce2, ce equ(#2)
```

. twoway line ce1 t

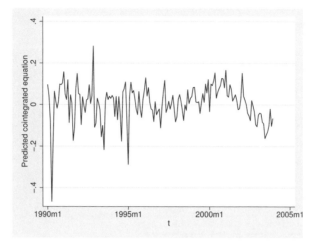

. twoway line ce2 t

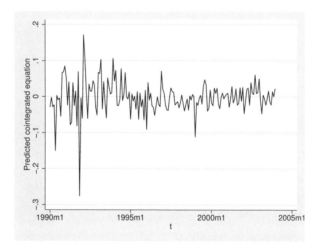

Although the large shocks apparent in the graph of the levels have clear effects on the predictions from the cointegrating equations, our only concern is the negative trend in the first cointegrating equation since the end of 2000. The graph of the levels shows that something put a significant brake on the growth of housing prices after 2000 and that the growth of housing prices in San Antonio slowed during 2000 but then recuperated while Austin maintained slower growth. We suspect that this indicates that the end of the high-tech boom affected Austin more severely than San Antonio. This difference is what causes the trend in the first cointegrating equation. Although we could try to account for this effect with a more formal analysis, we will proceed as if the cointegrating equations are stationary.

We can use `vecstable` to check whether we have correctly specified the number of cointegrating equations. As discussed in [TS] **vecstable**, the companion matrix of a VECM with K endogenous variables and r cointegrating equations has $K - r$ unit eigenvalues. If the process is stable, the moduli of the remaining r eigenvalues are strictly less than one. Because there is no general distribution theory for the moduli of the eigenvalues, ascertaining whether the moduli are too close to one can be difficult.

```
. vecstable, graph
```

Eigenvalue stability condition

Eigenvalue			Modulus
1			1
1			1
-.6698661			.669866
.3740191	+	.4475996i	.583297
.3740191	-	.4475996i	.583297
-.386377	+	.395972i	.553246
-.386377	-	.395972i	.553246
.540117			.540117
-.0749239	+	.5274203i	.532715
-.0749239	-	.5274203i	.532715
-.2023955			.202395
.09923966			.09924

The VECM specification imposes 2 unit moduli.

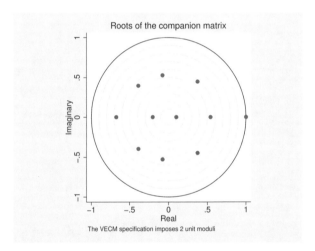

Because we specified the graph option, vecstable plotted the eigenvalues of the companion matrix. The graph of the eigenvalues shows that none of the remaining eigenvalues appears close to the unit circle. The stability check does not indicate that our model is misspecified.

Here we use veclmar to test for serial correlation in the residuals.

```
. veclmar, mlag(4)
```

Lagrange-multiplier test

lag	chi2	df	Prob > chi2
1	56.8757	16	0.00000
2	31.1970	16	0.01270
3	30.6818	16	0.01477
4	14.6493	16	0.55046

H0: no autocorrelation at lag order

The results clearly indicate serial correlation in the residuals. The results in Gonzalo (1994) indicate that underspecifying the number of lags in a VECM can significantly increase the finite-sample bias in the parameter estimates and lead to serial correlation. For this reason, we refit the model with five lags instead of three.

```
. vec austin dallas houston sa, lags(5) rank(2) noetable bconstraints(1/4)

Iteration 1:     log likelihood = 1200.5402
  (output omitted)
Iteration 20:    log likelihood = 1203.9465

Vector error-correction model

Sample:  1990m6 - 2003m12                      No. of obs    =        163
                                               AIC           = -13.79075
Log likelihood =  1203.946                     HQIC          =  -13.1743
Det(Sigma_ml)  =  4.51e-12                      SBIC          = -12.27235

Cointegrating equations

Equation           Parms      chi2     P>chi2
-------------------------------------------------
_ce1                  2     498.4682    0.0000
_ce2                  2     4125.926    0.0000
-------------------------------------------------

Identification:  beta is exactly identified
 ( 1)   [_ce1]austin = 1
 ( 2)   [_ce1]dallas = 0
 ( 3)   [_ce2]austin = 0
 ( 4)   [_ce2]houston = 1
```

beta	Coef.	Std. Err.	z	P>|z|	[95% Conf. Interval]
_ce1						
austin	1	.	.	.	.	.
dallas	0	(omitted)				
houston	-.6525574	.2047061	-3.19	0.001	-1.053774	-.2513407
sa	-.6960166	.2494167	-2.79	0.005	-1.184864	-.2071688
_cons	3.846275	.	.	.	.	.
_ce2						
austin	0	(omitted)				
dallas	-.932048	.0564332	-16.52	0.000	-1.042655	-.8214409
houston	1	.	.	.	.	.
sa	-.2363915	.0599348	-3.94	0.000	-.3538615	-.1189215
_cons	2.065719	.	.	.	.	.

Comparing these results with those from the previous model reveals that

1. there is now evidence that the coefficient [_ce1]houston is not equal to zero,

2. the two sets of estimated coefficients for the first cointegrating equation are different, and

3. the two sets of estimated coefficients for the second cointegrating equation are similar.

The assumption that the errors are independently, identically, and normally distributed with zero mean and finite variance allows us to derive the likelihood function. If the errors do not come from a normal distribution but are just independently and identically distributed with zero mean and finite variance, the parameter estimates are still consistent, but they are not efficient.

We use `vecnorm` to test the null hypothesis that the errors are normally distributed.

```
. qui vec austin dallas houston sa, lags(5) rank(2) bconstraints(1/4)
. vecnorm
```

Jarque-Bera test

Equation	chi2	df	Prob > chi2
D_austin	74.324	2	0.00000
D_dallas	3.501	2	0.17370
D_houston	245.032	2	0.00000
D_sa	8.426	2	0.01481
ALL	331.283	8	0.00000

Skewness test

Equation	Skewness	chi2	df	Prob > chi2
D_austin	.60265	9.867	1	0.00168
D_dallas	.09996	0.271	1	0.60236
D_houston	-1.0444	29.635	1	0.00000
D_sa	.38019	3.927	1	0.04752
ALL		43.699	4	0.00000

Kurtosis test

Equation	Kurtosis	chi2	df	Prob > chi2
D_austin	6.0807	64.458	1	0.00000
D_dallas	3.6896	3.229	1	0.07232
D_houston	8.6316	215.397	1	0.00000
D_sa	3.8139	4.499	1	0.03392
ALL		287.583	4	0.00000

The results indicate that we can strongly reject the null hypothesis of normally distributed errors. Most of the errors are both skewed and kurtotic.

Impulse–response functions for VECMs

With a model that we now consider acceptably well specified, we can use the `irf` commands to estimate and interpret the IRFs. Whereas IRFs from a stationary VAR die out over time, IRFs from a cointegrating VECM do not always die out. Because each variable in a stationary VAR has a time-invariant mean and finite, time-invariant variance, the effect of a shock to any one of these variables must die out so that the variable can revert to its mean. In contrast, the I(1) variables modeled in a cointegrating VECM are not mean reverting, and the unit moduli in the companion matrix imply that the effects of some shocks will not die out over time.

These two possibilities gave rise to new terms. When the effect of a shock dies out over time, the shock is said to be transitory. When the effect of a shock does not die out over time, the shock is said to be permanent.

Below we use `irf create` to estimate the IRFs and `irf graph` to graph two of the orthogonalized IRFs.

```
. irf create vec1, set(vecintro, replace) step(24)
(file vecintro.irf created)
(file vecintro.irf now active)
(file vecintro.irf updated)

. irf graph oirf, impulse(austin dallas) response(sa) yline(0)
```

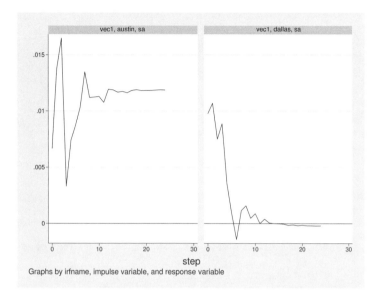

Graphs by irfname, impulse variable, and response variable

The graphs indicate that an orthogonalized shock to the average housing price in Austin has a permanent effect on the average housing price in San Antonio but that an orthogonalized shock to the average price of housing in Dallas has a transitory effect. According to this model, unexpected shocks that are local to the Austin housing market will have a permanent effect on the housing market in San Antonio, but unexpected shocks that are local to the Dallas housing market will have only a transitory effect on the housing market in San Antonio.

Forecasting with VECMs

Cointegrating VECMs are also used to produce forecasts of both the first-differenced variables and the levels of the variables. Comparing the variances of the forecast errors of stationary VARs with those from a cointegrating VECM reveals a fundamental difference between the two models. Whereas the variances of the forecast errors for a stationary VAR converge to a constant as the prediction horizon grows, the variances of the forecast errors for the levels of a cointegrating VECM diverge with the forecast horizon. (See sec. 6.5 of Lütkepohl [2005] for more about this result.) Because all the variables in the model for the first differences are stationary, the forecast errors for the dynamic forecasts of the first differences remain finite. In contrast, the forecast errors for the dynamic forecasts of the levels diverge to infinity.

We use `fcast compute` to obtain dynamic forecasts of the levels and `fcast graph` to graph these dynamic forecasts, along with their asymptotic confidence intervals.

```
. tsset
        time variable:  t, 1990m1 to 2003m12
                delta:  1 month
. fcast compute m1_, step(24)
. fcast graph m1_austin m1_dallas m1_houston m1_sa
```

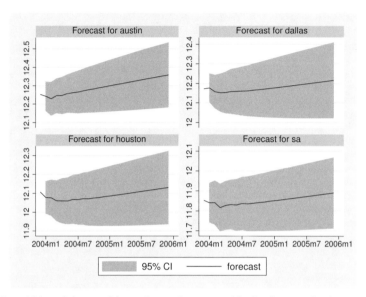

As expected, the widths of the confidence intervals grow with the forecast horizon.

References

Ahn, S. K., and G. C. Reinsel. 1990. Estimation for partially nonstationary multivariate autoregressive models. *Journal of the American Statistical Association* 85: 813–823.

Engle, R. F., and C. W. J. Granger. 1987. Co-integration and error correction: Representation, estimation, and testing. *Econometrica* 55: 251–276.

Gonzalo, J. 1994. Five alternative methods of estimating long-run equilibrium relationships. *Journal of Econometrics* 60: 203–233.

Granger, C. W. J. 1981. Some properties of time series data and their use in econometric model specification. *Journal of Econometrics* 16: 121–130.

Granger, C. W. J., and P. Newbold. 1974. Spurious regressions in econometrics. *Journal of Econometrics* 2: 111–120.

Hamilton, J. D. 1994. *Time Series Analysis*. Princeton: Princeton University Press.

Hubrich, K., H. Lütkepohl, and P. Saikkonen. 2001. A review of systems cointegration tests. *Econometric Reviews* 20: 247–318.

Johansen, S. 1988. Statistical analysis of cointegration vectors. *Journal of Economic Dynamics and Control* 12: 231–254.

———. 1991. Estimation and hypothesis testing of cointegration vectors in Gaussian vector autoregressive models. *Econometrica* 59: 1551–1580.

———. 1995. *Likelihood-Based Inference in Cointegrated Vector Autoregressive Models*. Oxford: Oxford University Press.

Lütkepohl, H. 2005. *New Introduction to Multiple Time Series Analysis*. New York: Springer.

Maddala, G. S., and I.-M. Kim. 1998. *Unit Roots, Cointegration, and Structural Change*. Cambridge: Cambridge University Press.

Nielsen, B. 2001. Order determination in general vector autoregressions. Working paper, Department of Economics, University of Oxford and Nuffield College. http://ideas.repec.org/p/nuf/econwp/0110.html.

Park, J. Y., and P. C. B. Phillips. 1988. Statistical inference in regressions with integrated processes: Part I. *Econometric Theory* 4: 468–497.

———. 1989. Statistical inference in regressions with integrated processes: Part II. *Econometric Theory* 5: 95–131.

Paulsen, J. 1984. Order determination of multivariate autoregressive time series with unit roots. *Journal of Time Series Analysis* 5: 115–127.

Phillips, P. C. B. 1986. Understanding spurious regressions in econometrics. *Journal of Econometrics* 33: 311–340.

Phillips, P. C. B., and S. N. Durlauf. 1986. Multiple time series regressions with integrated processes. *Review of Economic Studies* 53: 473–495.

Sims, C. A., J. H. Stock, and M. W. Watson. 1990. Inference in linear time series models with some unit roots. *Econometrica* 58: 113–144.

Stock, J. H. 1987. Asymptotic properties of least squares estimators of cointegrating vectors. *Econometrica* 55: 1035–1056.

Stock, J. H., and M. W. Watson. 1988. Testing for common trends. *Journal of the American Statistical Association* 83: 1097–1107.

Tsay, R. S. 1984. Order selection in nonstationary autoregressive models. *Annals of Statistics* 12: 1425–1433.

Watson, M. W. 1994. Vector autoregressions and cointegration. In Vol. IV of *Handbook of Econometrics*, ed. R. F. Engle and D. L. McFadden. Amsterdam: Elsevier.

Also see

Title

> **vec** — Vector error-correction models

Syntax

> vec *varlist* $\left[\,if\,\right]$ $\left[\,in\,\right]$ $\left[\,,\ options\,\right]$

options	Description
Model	
<u>r</u>ank(*#*)	use *#* cointegrating equations; default is rank(1)
<u>lag</u>s(*#*)	use *#* for the maximum lag in underlying VAR model
trend(<u>c</u>onstant)	include an unrestricted constant in model; the default
trend(<u>rc</u>onstant)	include a restricted constant in model
trend(<u>tre</u>nd)	include a linear trend in the cointegrating equations and a quadratic trend in the undifferenced data
trend(<u>rt</u>rend)	include a restricted trend in model
trend(<u>n</u>one)	do not include a trend or a constant
<u>bc</u>onstraints(*constraints*$_{\rm bc}$)	place *constraints*$_{\rm bc}$ on cointegrating vectors
<u>ac</u>onstraints(*constraints*$_{\rm ac}$)	place *constraints*$_{\rm ac}$ on adjustment parameters
Adv. model	
<u>si</u>ndicators(*varlist*$_{\rm si}$)	include normalized seasonal indicator variables *varlist*$_{\rm si}$
noreduce	do not perform checks and corrections for collinearity among lags of dependent variables
Reporting	
<u>l</u>evel(*#*)	set confidence level; default is level(95)
<u>nobt</u>able	do not report parameters in the cointegrating equations
<u>noid</u>test	do not report the likelihood-ratio test of overidentifying restrictions
<u>al</u>pha	report adjustment parameters in separate table
pi	report parameters in $\Pi = \alpha\beta'$
<u>nop</u>table	do not report elements of Π matrix
<u>ma</u>i	report parameters in the moving-average impact matrix
<u>noe</u>table	do not report adjustment and short-run parameters
dforce	force reporting of short-run, beta, and alpha parameters when the parameters in beta are not identified; advanced option
<u>nocns</u>report	do not display constraints
display_options	control column formats, row spacing, and line width
Maximization	
maximize_options	control the maximization process; seldom used
<u>coefl</u>egend	display legend instead of statistics

vec does not allow gaps in the data.

You must tsset your data before using vec; see [TS] **tsset**.

varlist must contain at least two variables and may contain time-series operators; see [U] **11.4.4 Time-series varlists**.

by, rolling, statsby, and xi are allowed; see [U] **11.1.10 Prefix commands**.

coeflegend does not appear in the dialog box.

See [U] **20 Estimation and postestimation commands** for more capabilities of estimation commands.

Menu

Statistics > Multivariate time series > Vector error-correction model (VECM)

Description

vec fits a type of vector autoregression in which some of the variables are cointegrated by using Johansen's (1995) maximum likelihood method. Constraints may be placed on the parameters in the cointegrating equations or on the adjustment terms. See [TS] **vec intro** for a list of commands that are used in conjunction with vec.

Options

⌐ Model ⌐

rank(*#*) specifies the number of cointegrating equations; rank(1) is the default.

lags(*#*) specifies the maximum lag to be included in the underlying VAR model. The maximum lag in a VECM is one smaller than the maximum lag in the corresponding VAR in levels; the number of lags must be greater than zero but small enough so that the degrees of freedom used up by the model are fewer than the number of observations. The default is lags(2).

trend(*trend_spec*) specifies which of Johansen's five trend specifications to include in the model. These specifications are discussed in *Specification of constants and trends* below. The default is trend(constant).

bconstraints(*constraints*$_{bc}$) specifies the constraints to be placed on the parameters of the cointegrating equations. When no constraints are placed on the adjustment parameters—that is, when the aconstraints() option is not specified—the default is to place the constraints defined by Johansen's normalization on the parameters of the cointegrating equations. When constraints are placed on the adjustment parameters, the default is not to place constraints on the parameters in the cointegrating equations.

aconstraints(*constraints*$_{ac}$) specifies the constraints to be placed on the adjustment parameters. By default, no constraints are placed on the adjustment parameters.

⌐ Adv. model ⌐

sindicators(*varlist*$_{si}$) specifies the normalized seasonal indicator variables to include in the model. The indicator variables specified in this option must be normalized as discussed in Johansen (1995). If the indicators are not properly normalized, the estimator of the cointegrating vector does not converge to the asymptotic distribution derived by Johansen (1995). More details about how these variables are handled are provided in *Methods and formulas*. sindicators() cannot be specified with trend(none) or with trend(rconstant).

noreduce causes vec to skip the checks and corrections for collinearity among the lags of the dependent variables. By default, vec checks to see whether the current lag specification causes some of the regressions performed by vec to contain perfectly collinear variables; if so, it reduces the maximum lag until the perfect collinearity is removed.

⌐ Reporting └

level(#); see [R] **estimation options**.

nobtable suppresses the estimation table for the parameters in the cointegrating equations. By default, vec displays the estimation table for the parameters in the cointegrating equations.

noidtest suppresses the likelihood-ratio test of the overidentifying restrictions, which is reported by default when the model is overidentified.

alpha displays a separate estimation table for the adjustment parameters, which is not displayed by default.

pi displays a separate estimation table for the parameters in $\Pi = \alpha\beta'$, which is not displayed by default.

noptable suppresses the estimation table for the elements of the Π matrix, which is displayed by default when the parameters in the cointegrating equations are not identified.

mai displays a separate estimation table for the parameters in the moving-average impact matrix, which is not displayed by default.

noetable suppresses the main estimation table that contains information about the estimated adjustment parameters and the short-run parameters, which is displayed by default.

dforce displays the estimation tables for the short-run parameters and α and β—if the last two are requested—when the parameters in β are not identified. By default, when the specified constraints do not identify the parameters in the cointegrating equations, estimation tables are displayed only for Π and the MAI.

nocnsreport; see [R] **estimation options**.

display_options: vsquish, cformat(%*fmt*), pformat(%*fmt*), sformat(%*fmt*), and nolstretch; see [R] **estimation options**.

⌐ Maximization └

maximize_options: iterate(#), nolog, trace, toltrace, tolerance(#), ltolerance(#), afrom(*matrix_a*), and bfrom(*matrix_b*); see [R] **maximize**.

toltrace displays the relative differences for the log likelihood and the coefficient vector at every iteration. This option cannot be specified if no constraints are defined or if nolog is specified.

afrom(*matrix_a*) specifies a $1 \times (K * r)$ row vector with starting values for the adjustment parameters, where K is the number of endogenous variables and r is the number of cointegrating equations specified in the rank() option. The starting values should be ordered as they are reported in e(alpha). This option cannot be specified if no constraints are defined.

bfrom(*matrix_b*) specifies a $1 \times (m_1 * r)$ row vector with starting values for the parameters of the cointegrating equations, where m_1 is the number of variables in the trend-augmented system and r is the number of cointegrating equations specified in the rank() option. (See *Methods and formulas* for more details about m_1.) The starting values should be ordered as they are reported in e(betavec). As discussed in *Methods and formulas*, for some trend specifications, e(beta) contains parameter estimates that are not obtained directly from the optimization algorithm. bfrom() should specify only starting values for the parameters reported in e(betavec). This option cannot be specified if no constraints are defined.

The following option is available with vec but is not shown in the dialog box:

coeflegend; see [R] **estimation options**.

Remarks

Remarks are presented under the following headings:

> *Introduction*
> *Specification of constants and trends*
> *Collinearity*

Introduction

VECMs are used to model the stationary relationships between multiple time series that contain unit roots. vec implements Johansen's approach for estimating the parameters of a VECM.

[TS] **vec intro** reviews the basics of integration and cointegration and highlights why we need special methods for modeling the relationships between processes that contain unit roots. This manual entry assumes familiarity with the material in [TS] **vec intro** and provides examples illustrating how to use the vec command. See Johansen (1995) and Hamilton (1994) for more in-depth introductions to cointegration analysis.

▷ Example 1

This example uses annual data on the average per-capita disposable personal income in the eight U.S. Bureau of Economic Analysis (BEA) regions of the United States. We use data from 1948–2002 in logarithms. Unit-root tests on these series fail to reject the null hypothesis that per-capita disposable income in each region contains a unit root. Because capital and labor can move easily between the different regions of the United States, we would expect that no one series will diverge from all the remaining series and that cointegrating relationships exist.

Below we graph the natural logs of average disposal income in the New England and the Southeast regions.

```
. use http://www.stata-press.com/data/r12/rdinc
. line ln_ne ln_se year
```

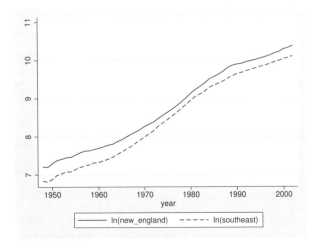

The graph indicates a differential between the two series that shrinks between 1960 and about 1980 and then grows until it stabilizes around 1990. We next estimate the parameters of a bivariate VECM with one cointegrating relationship.

```
. vec ln_ne ln_se

Vector error-correction model

Sample:  1950 - 2002                          No. of obs    =         53
                                              AIC           = -11.00462
Log likelihood =  300.6224                    HQIC          = -10.87595
Det(Sigma_ml)  =  4.06e-08                    SBIC          = -10.67004

Equation          Parms      RMSE     R-sq     chi2     P>chi2

D_ln_ne              4     .017896   0.9313   664.4668   0.0000
D_ln_se              4     .018723   0.9292   642.7179   0.0000
```

	Coef.	Std. Err.	z	P>|z|	[95% Conf.	Interval]
D_ln_ne						
_ce1						
L1.	-.4337524	.0721365	-6.01	0.000	-.5751373	-.2923675
ln_ne						
LD.	.7168658	.1889085	3.79	0.000	.3466119	1.08712
ln_se						
LD.	-.6748754	.2117975	-3.19	0.001	-1.089991	-.2597599
_cons	-.0019846	.0080291	-0.25	0.805	-.0177214	.0137521
D_ln_se						
_ce1						
L1.	-.3543935	.0754725	-4.70	0.000	-.5023168	-.2064701
ln_ne						
LD.	.3366786	.1976448	1.70	0.088	-.050698	.7240553
ln_se						
LD.	-.1605811	.2215922	-0.72	0.469	-.5948939	.2737317
_cons	.002429	.0084004	0.29	0.772	-.0140355	.0188936

```
Cointegrating equations

Equation          Parms     chi2     P>chi2

_ce1                1    29805.02    0.0000

Identification:   beta is exactly identified
                  Johansen normalization restriction imposed
```

beta	Coef.	Std. Err.	z	P>|z|	[95% Conf.	Interval]
_ce1						
ln_ne	1	.	.	.	.	.
ln_se	-.9433708	.0054643	-172.64	0.000	-.9540807	-.9326609
_cons	-.8964065	.	.	.	.	.

The default output has three parts. The header provides information about the sample, the model fit, and the identification of the parameters in the cointegrating equation. The main estimation table contains the estimates of the short-run parameters, along with their standard errors and confidence intervals. The second estimation table reports the estimates of the parameters in the cointegrating equation, along with their standard errors and confidence intervals.

The results indicate strong support for a cointegrating equation such that

$$\texttt{ln\_ne} - .943\,\texttt{ln\_se} - .896$$

should be a stationary series. Identification of the parameters in the cointegrating equation is achieved by constraining some of them to be fixed, and fixed parameters do not have standard errors. In this example, the coefficient on ln_ne has been normalized to 1, so its standard error is missing. As discussed in *Methods and formulas*, the constant term in the cointegrating equation is not directly estimated in this trend specification but rather is backed out from other estimates. Not all the elements of the VCE that correspond to this parameter are readily available, so the standard error for the _cons parameter is missing.

To get a better idea of how our model fits, we predict the cointegrating equation and graph it over time:

```
. predict ce, ce
. line ce year
```

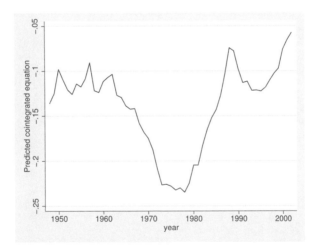

Although the predicted cointegrating equation has the right appearance for the time before the mid-1960s, afterward the predicted cointegrating equation does not look like a stationary series. A better model would account for the trends in the size of the differential.

◁

As discussed in [TS] **vec intro**, simply normalizing one of the coefficients to be one is sufficient to identify the parameters of the single cointegrating vector. When there is more than one cointegrating equation, more restrictions are required.

▷ Example 2

We have data on monthly unemployment rates in Indiana, Illinois, Kentucky, and Missouri from January 1978 through December 2003. We suspect that factor mobility will keep the unemployment rates in equilibrium. The following graph plots the data.

```
. use http://www.stata-press.com/data/r12/urates, clear
. line missouri indiana kentucky illinois t
```

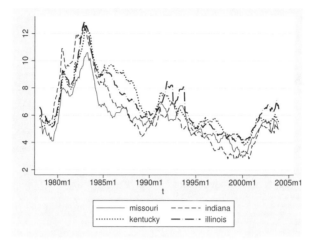

The graph shows that although the series do appear to move together, the relationship is not as clear as in the previous example. There are periods when Indiana has the highest rate and others when Indiana has the lowest rate. Although the Kentucky rate moves closely with the other series for most of the sample, there is a period in the mid-1980s when the unemployment rate in Kentucky does not fall at the same rate as the other series.

We will model the series with two cointegrating equations and no linear or quadratic time trends in the original series. Because we are focusing on the cointegrating vectors, we use the `noetable` option to suppress displaying the short-run estimation table.

```
. vec missouri indiana kentucky illinois, trend(rconstant) rank(2) lags(4)
> noetable
```

Vector error-correction model

Sample: 1978m5 - 2003m12	No. of obs	=	308
	AIC	=	-2.306048
Log likelihood = 417.1314	HQIC	=	-2.005818
Det(Sigma_ml) = 7.83e-07	SBIC	=	-1.555184

Cointegrating equations

Equation	Parms	chi2	P>chi2
_ce1	2	133.3885	0.0000
_ce2	2	195.6324	0.0000

Identification: beta is exactly identified
                 Johansen normalization restrictions imposed

beta	Coef.	Std. Err.	z	P>|z|	[95% Conf. Interval]	
_ce1						
missouri	1	.	.	.	.	.
indiana	-2.19e-18	.	.	.	.	.
kentucky	.3493902	.2005537	1.74	0.081	-.0436879	.7424683
illinois	-1.135152	.2069063	-5.49	0.000	-1.540681	-.7296235
_cons	-.3880707	.4974323	-0.78	0.435	-1.36302	.5868787
_ce2						
missouri	1.09e-16	.	.	.	.	.
indiana	1	.	.	.	.	.
kentucky	.2059473	.2718678	0.76	0.449	-.3269038	.7387985
illinois	-1.51962	.2804792	-5.42	0.000	-2.069349	-.9698907
_cons	2.92857	.6743122	4.34	0.000	1.606942	4.250197

Except for the coefficients on `kentucky` in the two cointegrating equations and the constant term in the first, all the parameters are significant at the 5% level. We can refit the model with the Johansen normalization and the overidentifying constraint that the coefficient on `kentucky` in the second cointegrating equation is zero.

```
. constraint 1 [_ce1]missouri = 1
. constraint 2 [_ce1]indiana  = 0
. constraint 3 [_ce2]missouri = 0
. constraint 4 [_ce2]indiana  = 1
. constraint 5 [_ce2]kentucky = 0
```

```
. vec missouri indiana kentucky illinois, trend(rconstant) rank(2) lags(4)
> noetable bconstraints(1/5)

Iteration 1:      log likelihood = 416.97177
  (output omitted )
Iteration 20:     log likelihood = 416.9744

Vector error-correction model

Sample:  1978m5 - 2003m12                      No. of obs     =         308
                                               AIC            =   -2.311522
Log likelihood =   416.9744                    HQIC           =   -2.016134
Det(Sigma_ml)  =   7.84e-07                     SBIC           =   -1.572769

Cointegrating equations

Equation            Parms      chi2     P>chi2

_ce1                   2     145.233    0.0000
_ce2                   1    209.9344    0.0000


Identification:  beta is overidentified

 ( 1)  [_ce1]missouri = 1
 ( 2)  [_ce1]indiana = 0
 ( 3)  [_ce2]missouri = 0
 ( 4)  [_ce2]indiana = 1
 ( 5)  [_ce2]kentucky = 0
```

beta	Coef.	Std. Err.	z	P>|z|	[95% Conf. Interval]	
_ce1						
missouri	1	.	.	.	.	.
indiana	0	(omitted)				
kentucky	.2521685	.1649653	1.53	0.126	-.0711576	.5754946
illinois	-1.037453	.1734165	-5.98	0.000	-1.377343	-.6975626
_cons	-.3891102	.4726968	-0.82	0.410	-1.315579	.5373586
_ce2						
missouri	0	(omitted)				
indiana	1	.	.	.	.	.
kentucky	0	(omitted)				
illinois	-1.314265	.0907071	-14.49	0.000	-1.492048	-1.136483
_cons	2.937016	.6448924	4.55	0.000	1.67305	4.200982

```
LR test of identifying restrictions:  chi2(  1) =    .3139  Prob > chi2 = 0.575
```

The test of the overidentifying restriction does not reject the null hypothesis that the restriction is valid, and the *p*-value on the coefficient on kentucky in the first cointegrating equation indicates that it is not significant. We will leave the variable in the model and attribute the lack of significance to whatever caused the kentucky series to temporarily rise above the others from 1985 until 1990, though we could instead consider removing kentucky from the model.

Next, we look at the estimates of the adjustment parameters. In the output below, we replay the previous results. We specify the alpha option so that vec will display an estimation table for the estimates of the adjustment parameters, and we specify nobtable to suppress the table for the parameters of the cointegrating equations because we have already looked at those.

```
. vec, alpha nobtable noetable
Vector error-correction model
```

```
Sample:  1978m5 - 2003m12                          No. of obs    =        308
                                                   AIC           = -2.311522
Log likelihood =   416.9744                        HQIC          = -2.016134
Det(Sigma_ml)  = 7.84e-07                          SBIC          = -1.572769
```
Adjustment parameters

Equation	Parms	chi2	P>chi2
D_missouri	2	19.39607	0.0001
D_indiana	2	6.426086	0.0402
D_kentucky	2	8.524901	0.0141
D_illinois	2	22.32893	0.0000

alpha	Coef.	Std. Err.	z	P>|z|	[95% Conf. Interval]	
D_missouri						
_ce1						
L1.	−.0683152	.0185763	−3.68	0.000	−.1047242	−.0319063
_ce2						
L1.	.0405613	.0112417	3.61	0.000	.018528	.0625946
D_indiana						
_ce1						
L1.	−.0342096	.0220955	−1.55	0.122	−.0775159	.0090967
_ce2						
L1.	.0325804	.0133713	2.44	0.015	.0063732	.0587877
D_kentucky						
_ce1						
L1.	−.0482012	.0231633	−2.08	0.037	−.0936004	−.0028021
_ce2						
L1.	.0374395	.0140175	2.67	0.008	.0099657	.0649133
D_illinois						
_ce1						
L1.	.0138224	.0227041	0.61	0.543	−.0306768	.0583215
_ce2						
L1.	.0567664	.0137396	4.13	0.000	.0298373	.0836955

LR test of identifying restrictions: chi2(1) =   .3139 Prob > chi2 = 0.575

All the coefficients are significant at the 5% level, except those on Indiana and Illinois in the first cointegrating equation. From an economic perspective, the issue is whether the unemployment rates in Indiana and Illinois adjust when the first cointegrating equation is out of equilibrium. We could impose restrictions on one or both of those parameters and refit the model, or we could just decide to use the current results.

◁

❏ Technical note

vec can be used to fit models in which the parameters in β are not identified, in which case only the parameters in Π and the moving-average impact matrix $\mathbf{C}$ are identified. When the parameters in β are not identified, the values of $\widehat{\beta}$ and $\widehat{\alpha}$ can vary depending on the starting values. However, the estimates of Π and $\mathbf{C}$ are identified and have known asymptotic distributions. This method is valid because these additional normalization restrictions impose no restriction on Π or $\mathbf{C}$.

❏

Specification of constants and trends

As discussed in [TS] **vec intro**, allowing for a constant term and linear time trend allow us to write the VECM as

$$\Delta \mathbf{y}_t = \boldsymbol{\alpha}(\beta \mathbf{y}_{t-1} + \boldsymbol{\mu} + \boldsymbol{\rho} t) + \sum_{i=1}^{p-1} \boldsymbol{\Gamma}_i \Delta \mathbf{y}_{t-i} + \boldsymbol{\gamma} + \boldsymbol{\tau} t + \epsilon_t$$

Five different trend specifications are available:

Option in trend()	Parameter restrictions	Johansen (1995) notation
trend	none	$H(r)$
rtrend	$\tau = 0$	$H^*(r)$
constant	$\rho = 0$, and $\tau = 0$	$H_1(r)$
rconstant	$\rho = 0$, $\gamma = 0$ and $\tau = 0$	$H_1^*(r)$
none	$\mu = 0$, $\rho = 0$, $\gamma = 0$, and $\tau = 0$	$H_2(r)$

trend(trend) allows for a linear trend in the cointegrating equations and a quadratic trend in the undifferenced data. A linear trend in the cointegrating equations implies that the cointegrating equations are assumed to be trend stationary.

trend(rtrend) defines a restricted trend model that excludes linear trends in the differenced data but allows for linear trends in the cointegrating equations. As in the previous case, a linear trend in a cointegrating equation implies that the cointegrating equation is trend stationary.

trend(constant) defines a model with an unrestricted constant. This allows for a linear trend in the undifferenced data and cointegrating equations that are stationary around a nonzero mean. This is the default.

trend(rconstant) defines a model with a restricted constant in which there is no linear or quadratic trend in the undifferenced data. A nonzero μ allows for the cointegrating equations to be stationary around nonzero means, which provide the only intercepts for differenced data. Seasonal indicators are not allowed with this specification.

trend(none) defines a model that does not include a trend or a constant. When there is no trend or constant, the cointegrating equations are restricted to being stationary with zero means. Also, after adjusting for the effects of lagged endogenous variables, the differenced data are modeled as having mean zero. Seasonal indicators are not allowed with this specification.

❏ Technical note

vec uses a switching algorithm developed by Boswijk (1995) to maximize the log-likelihood function when constraints are placed on the parameters. The starting values affect both the ability of the algorithm to find a maximum and its speed in finding that maximum. By default, vec uses the parameter estimates that correspond to Johansen's normalization. Sometimes, other starting values will cause the algorithm to find a maximum faster.

To specify starting values for the parameters in α, we specify a $1 \times (K * r)$ matrix in the afrom() option. Specifying starting values for the parameters in β is slightly more complicated. As explained in *Methods and formulas*, specifying trend(constant), trend(rtrend), or trend(trend) causes some of the estimates of the trend parameters appearing in $\widehat{\beta}$ to be "backed out". The switching algorithm estimates only the parameters of the cointegrating equations whose estimates are saved in e(betavec). For this reason, only the parameters saved in e(betavec) can have their initial values set via bfrom().

The table below describes which trend parameters in the cointegrating equations are estimated by the switching algorithm for each of the five specifications.

Trend specification	Trend parameters in cointegrating equations	Trend parameter estimated via switching algorithm
none	none	none
rconstant	_cons	_cons
constant	_cons	none
rtrend	_cons, _trend	_trend
trend	_cons, _trend	none

❏

Collinearity

As expected, collinearity among variables causes some parameters to be unidentified numerically. If vec encounters perfect collinearity among the dependent variables, it exits with an error.

In contrast, if vec encounters perfect collinearity that appears to be due to too many lags in the model, vec displays a warning message and reduces the maximum lag included in the model in an effort to find a model with fewer lags in which all the parameters are identified by the data. Specifying the noreduce option causes vec to skip over these additional checks and corrections for collinearity. Thus the noreduce option can be used to force the estimation to proceed when not all the parameters are identified by the data. When some parameters are not identified because of collinearity, the results cannot be interpreted but can be used to find the source of the collinearity.

Saved results

vec saves the following in e():

Scalars
e(N)	number of observations
e(k_rank)	number of unconstrained parameters
e(k_eq)	number of equations in e(b)
e(k_dv)	number of dependent variables
e(k_ce)	number of cointegrating equations
e(n_lags)	number of lags
e(df_m)	model degrees of freedom
e(ll)	log likelihood
e(chi2_res)	value of test of overidentifying restrictions
e(df_lr)	degrees of freedom of the test of overidentifying restrictions
e(beta_iden)	1 if the parameters in β are identified and 0 otherwise
e(beta_icnt)	number of independent restrictions placed on β
e(k_#)	number of variables in equation #
e(df_m#)	model degrees of freedom in equation #
e(r2_#)	R^2 of equation #
e(chi2_#)	$\chi2$ statistic for equation #
e(rmse_#)	RMSE of equation #
e(aic)	value of AIC
e(hqic)	value of HQIC
e(sbic)	value of SBIC
e(tmin)	minimum time
e(tmax)	maximum time
e(detsig_ml)	determinant of the estimated covariance matrix
e(rank)	rank of e(V)
e(converge)	1 if the switching algorithm converged, 0 if it did not converge

Macros
e(cmd)	vec
e(cmdline)	command as typed
e(trend)	trend specified
e(tsfmt)	format of the time variable
e(tvar)	variable denoting time within groups
e(endog)	endogenous variables
e(covariates)	list of covariates
e(eqnames)	equation names
e(cenames)	names of cointegrating equations
e(reduce_opt)	noreduce, if noreduce is specified
e(reduce_lags)	list of maximum lags to which the model has been reduced
e(title)	title in estimation output
e(aconstraints)	constraints placed on α
e(bconstraints)	constraints placed on β
e(sindicators)	sindicators, if specified
e(properties)	b V
e(predict)	program used to implement predict
e(marginsok)	predictions allowed by margins
e(marginsnotok)	predictions disallowed by margins

Matrices

e(b)	estimates of short-run parameters
e(V)	VCE of short-run parameter estimates
e(beta)	estimates of $\boldsymbol{\beta}$
e(V_beta)	VCE of $\widehat{\boldsymbol{\beta}}$
e(betavec)	directly obtained estimates of $\boldsymbol{\beta}$
e(pi)	estimates of $\widehat{\boldsymbol{\Pi}}$
e(V_pi)	VCE of $\widehat{\boldsymbol{\Pi}}$
e(alpha)	estimates of $\boldsymbol{\alpha}$
e(V_alpha)	VCE of $\widehat{\boldsymbol{\alpha}}$
e(omega)	estimates of $\widehat{\boldsymbol{\Omega}}$
e(mai)	estimates of $\mathbf{C}$
e(V_mai)	VCE of $\widehat{\mathbf{C}}$

Functions

e(sample)	marks estimation sample

Methods and formulas

vec is implemented as an ado-file. *Methods and formulas* are presented under the following headings:

> *General specification of the VECM*
> *The log-likelihood function*
>> *Unrestricted trend*
>> *Restricted trend*
>> *Unrestricted constant*
>> *Restricted constant*
>> *No trend*
> *Estimation with Johansen identification*
> *Estimation with constraints:* β *identified*
> *Estimation with constraints:* β *not identified*
> *Formulas for the information criteria*
> *Formulas for predict*

General specification of the VECM

vec estimates the parameters of a VECM that can be written as

$$\Delta \mathbf{y}_t = \boldsymbol{\alpha}\boldsymbol{\beta}'\mathbf{y}_{t-1} + \sum_{i=1}^{p-1} \boldsymbol{\Gamma}_i \Delta \mathbf{y}_{t-i} + \mathbf{v} + \boldsymbol{\delta}t + \mathbf{w}_1 s_1 + \cdots + \mathbf{w}_m s_m + \boldsymbol{\epsilon}_t \qquad (1)$$

where

$\mathbf{y}_t$ is a $K \times 1$ vector of endogenous variables,

$\boldsymbol{\alpha}$ is a $K \times r$ matrix of parameters,

$\boldsymbol{\beta}$ is a $K \times r$ matrix of parameters,

$\boldsymbol{\Gamma}_1, \ldots, \boldsymbol{\Gamma}_{p-1}$ are $K \times K$ matrices of parameters,

$\mathbf{v}$ is a $K \times 1$ vector of parameters,

$\boldsymbol{\delta}$ is a $K \times 1$ vector of trend coefficients,

t is a linear time trend,

$s_1, \ldots, s_m$ are orthogonalized seasonal indicators specified in the `sindicators()` option, and

$\mathbf{w}_1, \ldots, \mathbf{w}_m$ are $K \times 1$ vectors of coefficients on the orthogonalized seasonal indicators.

There are two types of deterministic elements in (1): the trend, $\mathbf{v} + \delta t$, and the orthogonalized seasonal terms, $\mathbf{w}_1 s_1 + \cdots + \mathbf{w}_m s_m$. Johansen (1995, chap. 11) shows that inference about the number of cointegrating equations is based on nonstandard distributions and that the addition of any term that generalizes the deterministic specification in (1) changes the asymptotic distributions of the statistics used for inference on the number of cointegrating equations and the asymptotic distribution of the ML estimator of the cointegrating equations. In fact, Johansen (1995, 84) notes that including event indicators causes the statistics used for inference on the number of cointegrating equations to have asymptotic distributions that must be computed case by case. For this reason, event indicators may not be specified in the present version of `vec`.

If seasonal indicators are included in the model, they cannot be collinear with a constant term. If they are collinear with a constant term, one of the indicator variables is omitted.

As discussed in *Specification of constants and trends*, we can reparameterize the model as

$$\Delta \mathbf{y}_t = \boldsymbol{\alpha}(\boldsymbol{\beta} \mathbf{y}_{t-1} + \boldsymbol{\mu} + \boldsymbol{\rho} t) + \sum_{i=1}^{p-1} \boldsymbol{\Gamma}_i \Delta \mathbf{y}_{t-i} + \boldsymbol{\gamma} + \boldsymbol{\tau} t + \boldsymbol{\epsilon}_t \tag{2}$$

The log-likelihood function

We can maximize the log-likelihood function much more easily by writing it in concentrated form. In fact, as discussed below, in the simple case with the Johansen normalization on $\boldsymbol{\beta}$ and no constraints on $\boldsymbol{\alpha}$, concentrating the log-likelihood function produces an analytical solution for the parameter estimates.

To concentrate the log likelihood, rewrite (2) as

$$\mathbf{Z}_{0t} = \boldsymbol{\alpha}\widetilde{\boldsymbol{\beta}}' \mathbf{Z}_{1t} + \boldsymbol{\Psi}\mathbf{Z}_{2t} + \boldsymbol{\epsilon}_t \tag{3}$$

where $\mathbf{Z}_{0t}$ is a $K \times 1$ vector of variables $\Delta \mathbf{y}_t$, $\boldsymbol{\alpha}$ is the $K \times r$ matrix of adjustment coefficients, and $\boldsymbol{\epsilon}_t$ is a $K \times 1$ vector of independently and identically distributed normal vectors with mean 0 and contemporaneous covariance matrix $\boldsymbol{\Omega}$. $\mathbf{Z}_{1t}$, $\mathbf{Z}_{2t}$, $\widetilde{\boldsymbol{\beta}}$, and $\boldsymbol{\Psi}$ depend on the trend specification and are defined below.

The log-likelihood function for the model in (3) is

$$L = -\frac{1}{2}\Big\{ TK \ln(2\pi) + T \ln(|\boldsymbol{\Omega}|)$$

$$+ \sum_{t=1}^{T} (\mathbf{Z}_{0t} - \boldsymbol{\alpha}\widetilde{\boldsymbol{\beta}}' \mathbf{Z}_{1t} - \boldsymbol{\Psi}\mathbf{Z}_{2t})' \boldsymbol{\Omega}^{-1} (\mathbf{Z}_{0t} - \boldsymbol{\alpha}\widetilde{\boldsymbol{\beta}}' \mathbf{Z}_{1t} - \boldsymbol{\Psi}\mathbf{Z}_{2t}) \Big\} \tag{4}$$

with the constraints that $\boldsymbol{\alpha}$ and $\widetilde{\boldsymbol{\beta}}$ have rank r.

Johansen (1995, chap. 6), building on Anderson (1951), shows how the $\boldsymbol{\Psi}$ parameters can be expressed as analytic functions of $\boldsymbol{\alpha}$, $\widetilde{\boldsymbol{\beta}}$, and the data, yielding the concentrated log-likelihood function

$$L_c = -\frac{1}{2}\Big\{ TK \ln(2\pi) + T \ln(|\boldsymbol{\Omega}|)$$

$$+ \sum_{t=1}^{T} (\mathbf{R}_{0t} - \boldsymbol{\alpha}\widetilde{\boldsymbol{\beta}}' \mathbf{R}_{1t})' \boldsymbol{\Omega}^{-1} (\mathbf{R}_{0t} - \boldsymbol{\alpha}\widetilde{\boldsymbol{\beta}}' \mathbf{R}_{1t}) \Big\} \tag{5}$$

where

$$\mathbf{M}_{ij} = T^{-1} \sum_{t=1}^{T} \mathbf{Z}_{it}\mathbf{Z}'_{jt}, \qquad i, j \in \{0, 1, 2\};$$

$$\mathbf{R}_{0t} = \mathbf{Z}_{0t} - \mathbf{M}_{02}\mathbf{M}_{22}^{-1}\mathbf{Z}_{2t}; \text{ and}$$

$$\mathbf{R}_{1t} = \mathbf{Z}_{1t} - \mathbf{M}_{12}\mathbf{M}_{22}^{-1}\mathbf{Z}_{2t}.$$

The definitions of $\mathbf{Z}_{1t}$, $\mathbf{Z}_{2t}$, $\widetilde{\boldsymbol{\beta}}$, and $\boldsymbol{\Psi}$ change with the trend specifications, although some of their components stay the same.

Unrestricted trend

When the trend in the VECM is unrestricted, we can define the variables in (3) directly in terms of the variables in (1):

$\mathbf{Z}_{1t} = \mathbf{y}_{t-1}$ is $K \times 1$

$\mathbf{Z}_{2t} = (\Delta\mathbf{y}'_{t-1}, \ldots, \Delta\mathbf{y}'_{t-p+1}, 1, t, s_1, \ldots, s_m)'$ is $\{K(p-1) + 2 + m\} \times 1$;

$\boldsymbol{\Psi} = (\boldsymbol{\Gamma}_1, \ldots, \boldsymbol{\Gamma}_{p-1}, \mathbf{v}, \boldsymbol{\delta}, \mathbf{w}_1, \ldots, \mathbf{w}_m)$ is $K \times \{K(p-1) + 2 + m\}$

$\widetilde{\boldsymbol{\beta}} = \boldsymbol{\beta}$ is the $K \times r$ matrix composed of the r cointegrating vectors.

In the unrestricted trend specification, $m_1 = K$, $m_2 = K(p-1) + 2 + m$, and there are $n_{\text{parms}} = Kr + Kr + K\{K(p-1) + 2 + m\}$ parameters in (3).

Restricted trend

When there is a restricted trend in the VECM in (2), $\boldsymbol{\tau} = 0$, but the intercept $\mathbf{v} = \boldsymbol{\alpha}\boldsymbol{\mu} + \boldsymbol{\gamma}$ is unrestricted. The VECM with the restricted trend can be written as

$$\Delta\mathbf{y}_t = \boldsymbol{\alpha}(\boldsymbol{\beta}', \boldsymbol{\rho}) \begin{pmatrix} \mathbf{y}_{t-1} \\ t \end{pmatrix} + \sum_{i=1}^{p-1} \boldsymbol{\Gamma}_i \Delta\mathbf{y}_{t-i} + \mathbf{v} + \mathbf{w}_1 s_1 + \cdots + \mathbf{w}_m s_m + \epsilon_t$$

This equation can be written in the form of (3) by defining

$\mathbf{Z}_{1t} = (\mathbf{y}'_{t-1}, t)'$ is $(K + 1) \times 1$

$\mathbf{Z}_{2t} = (\Delta\mathbf{y}'_{t-1}, \ldots, \Delta\mathbf{y}'_{t-p+1}, 1, s_1, \ldots, s_m)'$ is $\{K(p-1) + 1 + m\} \times 1$

$\boldsymbol{\Psi} = (\boldsymbol{\Gamma}_1, \ldots, \boldsymbol{\Gamma}_{p-1}, \mathbf{v}, \mathbf{w}_1, \ldots, \mathbf{w}_m)$ is $K \times \{K(p-1) + 1 + m\}$

$\widetilde{\boldsymbol{\beta}} = (\boldsymbol{\beta}', \boldsymbol{\rho})'$ is the $(K + 1) \times r$ matrix composed of the r cointegrating vectors and the r trend coefficients $\boldsymbol{\rho}$

In the restricted trend specification, $m_1 = K + 1$, $m_2 = \{K(p-1) + 1 + m\}$, and there are $n_{\text{parms}} = Kr + (K+1)r + K\{K(p-1) + 1 + m\}$ parameters in (3).

Unrestricted constant

An unrestricted constant in the VECM in (2) is equivalent to setting $\boldsymbol{\delta} = 0$ in (1), which can be written in the form of (3) by defining

$\mathbf{Z}_{1t} = \mathbf{y}_{t-1}$ is $(K \times 1)$

$\mathbf{Z}_{2t} = (\Delta\mathbf{y}'_{t-1}, \ldots, \Delta\mathbf{y}'_{t-p+1}, 1, s_1, \ldots, s_m)'$ is $\{K(p-1) + 1 + m\} \times 1$;

$\boldsymbol{\Psi} = (\boldsymbol{\Gamma}_1, \ldots, \boldsymbol{\Gamma}_{p-1}, \mathbf{v}, \mathbf{w}_1, \ldots, \mathbf{w}_m)$ is $K \times \{K(p-1) + 1 + m\}$

$\widetilde{\boldsymbol{\beta}} = \boldsymbol{\beta}$ is the $K \times r$ matrix composed of the r cointegrating vectors

In the unrestricted constant specification, $m_1 = K$, $m_2 = \{K(p-1) + 1 + m\}$, and there are $n_{\text{parms}} = Kr + Kr + K\{K(p-1) + 1 + m\}$ parameters in (3).

Restricted constant

When there is a restricted constant in the VECM in (2), it can be written in the form of (3) by defining

$\mathbf{Z}_{1t} = \left(\mathbf{y}'_{t-1}, 1\right)'$ is $(K+1) \times 1$

$\mathbf{Z}_{2t} = (\Delta \mathbf{y}'_{t-1}, \ldots, \Delta \mathbf{y}'_{t-p+1})'$ is $K(p-1) \times 1$

$\boldsymbol{\Psi} = (\boldsymbol{\Gamma}_1, \ldots, \boldsymbol{\Gamma}_{p-1})$ is $K \times K(p-1)$

$\widetilde{\boldsymbol{\beta}} = \left(\boldsymbol{\beta}', \boldsymbol{\mu}\right)'$ is the $(K+1) \times r$ matrix composed of the r cointegrating vectors and the r constants in the cointegrating relations.

In the restricted trend specification, $m_1 = K + 1$, $m_2 = K(p-1)$, and there are $n_{\text{parms}} = Kr + (K+1)r + K\{K(p-1)\}$ parameters in (3).

No trend

When there is no trend in the VECM in (2), it can be written in the form of (3) by defining

$\mathbf{Z}_{1t} = \mathbf{y}_{t-1}$ is $K \times 1$

$\mathbf{Z}_{2t} = (\Delta \mathbf{y}'_{t-1}, \ldots, \Delta \mathbf{y}'_{t-p+1})'$ is $K(p-1) + m \times 1$

$\boldsymbol{\Psi} = (\boldsymbol{\Gamma}_1, \ldots, \boldsymbol{\Gamma}_{p-1})$ is $K \times K(p-1)$

$\widetilde{\boldsymbol{\beta}} = \boldsymbol{\beta}$ is $K \times r$ matrix of r cointegrating vectors

In the no-trend specification, $m_1 = K$, $m_2 = K(p-1)$, and there are $n_{\text{parms}} = Kr + Kr + K\{K(p-1)\}$ parameters in (3).

Estimation with Johansen identification

Not all the parameters in $\boldsymbol{\alpha}$ and $\widetilde{\boldsymbol{\beta}}$ are identified. Consider the simple case in which $\widetilde{\boldsymbol{\beta}}$ is $K \times r$ and let $\mathbf{Q}$ be a nonsingular $r \times r$ matrix. Then

$$\boldsymbol{\alpha}\widetilde{\boldsymbol{\beta}}' = \boldsymbol{\alpha}\mathbf{Q}\mathbf{Q}^{-1}\widetilde{\boldsymbol{\beta}}' = \boldsymbol{\alpha}\mathbf{Q}(\widetilde{\boldsymbol{\beta}}\mathbf{Q}'^{-1})' = \dot{\boldsymbol{\alpha}}\dot{\boldsymbol{\beta}}'$$

Substituting $\dot{\boldsymbol{\alpha}}\dot{\boldsymbol{\beta}}'$ into the log likelihood in (5) for $\boldsymbol{\alpha}\widetilde{\boldsymbol{\beta}}'$ would not change the value of the log likelihood, so some a priori identification restrictions must be found to identify $\boldsymbol{\alpha}$ and $\widetilde{\boldsymbol{\beta}}$. As discussed in Johansen (1995, chap. 5 and 6) and Boswijk (1995), if the restrictions exactly identify or overidentify $\widetilde{\boldsymbol{\beta}}$, the estimates of the unconstrained parameters in $\widetilde{\boldsymbol{\beta}}$ will be superconsistent, meaning that the estimates of the free parameters in $\widetilde{\boldsymbol{\beta}}$ will converge at a faster rate than estimates of the short-run parameters in $\boldsymbol{\alpha}$ and $\boldsymbol{\Gamma}_i$. This allows the distribution of the estimator of the short-run parameters to be derived conditional on the estimated $\widetilde{\boldsymbol{\beta}}$.

Johansen (1995, chap. 6) has proposed a normalization method for use when theory does not provide sufficient a priori restrictions to identify the cointegrating vector. This method has become widely adopted by researchers. Johansen's identification scheme is

$$\widetilde{\boldsymbol{\beta}}' = (\mathbf{I}_r, \breve{\boldsymbol{\beta}}') \tag{6}$$

where $\mathbf{I}_r$ is the $r \times r$ identity matrix and $\breve{\beta}$ is a $(m_1 - r) \times r$ matrix of identified parameters.

Johansen's identification method places r^2 linearly independent constraints on the parameters in $\tilde{\beta}$, thereby defining an exactly identified model. The total number of freely estimated parameters is $n_{\mathrm{parms}} - r^2 = \{K + m_2 + (K + m_1 - r)r\}$, and the degrees of freedom d is calculated as the integer part of $(n_{\mathrm{parms}} - r^2)/K$.

When only the rank and the Johansen identification restrictions are placed on the model, we can further manipulate the log likelihood in (5) to obtain analytic formulas for the parameters in $\tilde{\beta}$, α, and Ω. For a given value of $\tilde{\beta}$, α and Ω can be found by regressing $\mathbf{R}_{0t}$ on $\tilde{\beta}' \mathbf{R}_{1t}$. This allows a further simplification of the problem in which

$$\alpha(\tilde{\beta}) = \mathbf{S}_{01} \tilde{\beta} (\tilde{\beta}' \mathbf{S}_{11} \tilde{\beta})^{-1}$$

$$\Omega(\tilde{\beta}) = \mathbf{S}_{00} - \mathbf{S}_{01} \tilde{\beta} (\tilde{\beta}' \mathbf{S}_{11} \tilde{\beta})^{-1} \tilde{\beta}' \mathbf{S}_{10}$$

$$\mathbf{S}_{ij} = (1/T) \sum_{t=1}^{T} R_{it} R_{jt}' \qquad i, j \in \{0, 1\}$$

Johansen (1995) shows that by inserting these solutions into equation (5), $\widehat{\beta}$ is given by the r eigenvectors $\mathbf{v}_1, \ldots, \mathbf{v}_r$ corresponding to the r largest eigenvalues $\lambda_1, \ldots, \lambda_r$ that solve the generalized eigenvalue problem

$$|\lambda_i \mathbf{S}_{11} - \mathbf{S}_{10} \mathbf{S}_{00}^{-1} \mathbf{S}_{01}| = 0 \tag{7}$$

The eigenvectors corresponding to $\lambda_1, \ldots, \lambda_r$ that solve (7) are the unidentified parameter estimates. To impose the identification restrictions in (6), we normalize the eigenvectors such that

$$\lambda_i \mathbf{S}_{11} \mathbf{v}_i = \mathbf{S}_{01} \mathbf{S}_{00}^{-1} \mathbf{S}_{01} \mathbf{v}_i \tag{8}$$

and

$$\mathbf{v}_i' \mathbf{S}_{11} \mathbf{v}_j = \begin{cases} 1 & \text{if } i = j \\ 0 & \text{otherwise} \end{cases} \tag{9}$$

At the optimum the log-likelihood function with the Johansen identification restrictions can be expressed in terms of T, K, $\mathbf{S}_{00}$, and the r largest eigenvalues

$$L_c = -\frac{1}{2} T \left\{ K \ln(2\pi) + K + \ln(|\mathbf{S}_{00}|) + \sum_{i=1}^{r} \ln(1 - \widehat{\lambda}_i) \right\}$$

where the $\widehat{\lambda}_i$ are the eigenvalues that solve (7), (8), and (9).

Using the normalized $\widehat{\beta}$, we can then obtain the estimates

$$\widehat{\alpha} = \mathbf{S}_{01} \widehat{\beta} (\widehat{\beta}' S_{11} \widehat{\beta})^{-1} \tag{10}$$

and

$$\widehat{\Omega} = \mathbf{S}_{00} - \widehat{\alpha} \widehat{\beta}' \mathbf{S}_{10}$$

Let $\widehat{\beta}_y$ be a $K \times r$ matrix that contains the estimates of the parameters in β in (1). $\widehat{\beta}_y$ differs from $\widehat{\beta}$ in that any trend parameter estimates are omitted from $\widehat{\beta}$. We can then use $\widehat{\beta}_y$ to obtain predicted values for the r nondemeaned cointegrating equations

$$\widehat{\widetilde{\mathbf{E}}}_t = \widehat{\beta}_y' \mathbf{y}_t$$

The r series in $\widehat{\widetilde{E}}_t$ are called the predicted, nondemeaned cointegrating equations because they still contain the terms μ and ρ. We want to work with the predicted, demeaned cointegrating equations. Thus we need estimates of μ and ρ. In the `trend(rconstant)` specification, the algorithm directly produces the estimator $\widehat{\mu}$. Similarly, in the `trend(rtrend)` specification, the algorithm directly produces the estimator $\widehat{\rho}$. In the remaining cases, to back out estimates of μ and ρ, we need estimates of $\mathbf{v}$ and $\boldsymbol{\delta}$, which we can obtain by estimating the parameters of the following VAR:

$$\Delta \mathbf{y}_t = \boldsymbol{\alpha} \widehat{\widetilde{\mathbf{E}}}_{t-1} + \sum_{i=1}^{p-1} \boldsymbol{\Gamma}_i \Delta \mathbf{y}_{t-i} + \mathbf{v} + \boldsymbol{\delta} t + \mathbf{w}_1 s_1 + \cdots + \mathbf{w}_m s_m + \boldsymbol{\epsilon}_t \tag{11}$$

Depending on the trend specification, we use $\widehat{\boldsymbol{\alpha}}$ to back out the estimates of

$$\widehat{\mu} = (\widehat{\boldsymbol{\alpha}}' \widehat{\boldsymbol{\alpha}})^{-1} \widehat{\boldsymbol{\alpha}}' \widehat{\mathbf{v}} \tag{12}$$

$$\widehat{\rho} = (\widehat{\boldsymbol{\alpha}}' \widehat{\boldsymbol{\alpha}})^{-1} \widehat{\boldsymbol{\alpha}}' \widehat{\boldsymbol{\delta}} \tag{13}$$

if they are not already in $\widehat{\beta}$ and are included in the trend specification.

We then augment $\widehat{\beta}_y$ to

$$\widehat{\beta}'_f = (\widehat{\beta}'_y, \widehat{\mu}, \widehat{\rho})$$

where the estimates of $\widehat{\mu}$ and $\widehat{\rho}$ are either obtained from $\widehat{\beta}$ or backed out using (12) and (13). We next use $\widehat{\beta}_f$ to obtain the r predicted, demeaned cointegrating equations, $\widehat{\mathbf{E}}_t$, via

$$\widehat{\mathbf{E}}_t = \widehat{\beta}'_f \left(\mathbf{y}'_t, 1, t \right)'$$

We last obtain estimates of all the short-run parameters from the VAR:

$$\Delta \mathbf{y}_t = \boldsymbol{\alpha} \widehat{\mathbf{E}}_{t-1} + \sum_{i=1}^{p-1} \boldsymbol{\Gamma}_i \Delta \mathbf{y}_{t-i} + \boldsymbol{\gamma} + \boldsymbol{\tau} t + \mathbf{w}_1 s_1 + \cdots + \mathbf{w}_m s_m + \boldsymbol{\epsilon}_t \tag{14}$$

Because the estimator $\widehat{\beta}_f$ converges in probability to its true value at a rate faster than $T^{-\frac{1}{2}}$, we can take our estimated $\widehat{\mathbf{E}}_{t-1}$ as given data in (14). This allows us to estimate the variance–covariance (VCE) matrix of the estimates of the parameters in (14) by using the standard VAR VCE estimator. Equation (11) can be used to obtain consistent estimates of all the parameters and of the VCE of all the parameters, except $\mathbf{v}$ and $\boldsymbol{\delta}$. The standard VAR VCE of $\widehat{\mathbf{v}}$ and $\widehat{\boldsymbol{\delta}}$ is incorrect because these estimates converge at a faster rate. This is why it is important to use the predicted, demeaned cointegrating equations, $\widehat{\mathbf{E}}_{t-1}$, when estimating the short-run parameters and trend terms. In keeping with the cointegration literature, `vec` makes a small-sample adjustment to the VCE estimator so that the divisor is $(T - d)$ instead of T, where d represents the degrees of freedom of the model. d is calculated as the integer part of n_{parms}/K, where n_{parms} is the total number of freely estimated parameters in the model.

In the `trend(rconstant)` specification, the estimation procedure directly estimates μ. For `trend(constant)`, `trend(rtrend)`, and `trend(trend)`, the estimates of μ are backed out using (12). In the `trend(rtrend)` specification, the estimation procedure directly estimates ρ. In the `trend(trend)` specification, the estimates of ρ are backed out using (13). Because the elements of the estimated VCE are readily available only when the estimates are obtained directly, when the trend parameter estimates are backed out, their elements in the VCE for $\widehat{\beta}_f$ are missing.

Under the Johansen identification restrictions, vec obtains $\widehat{\beta}$, the estimates of the parameters in the $r \times m_1$ matrix $\widetilde{\beta}'$ in (5). The VCE of $\text{vec}(\widehat{\beta})$ is $rm_1 \times rm_1$. Per Johansen (1995), the asymptotic distribution of $\widehat{\beta}$ is mixed Gaussian, and its VCE is consistently estimated by

$$\left(\frac{1}{T-d}\right)(\mathbf{I}_r \otimes \mathbf{H}_J)\left\{(\widehat{\alpha}'\mathbf{\Omega}^{-1}\widehat{\alpha}) \otimes (\mathbf{H}_J'\mathbf{S}_{11}\mathbf{H}_J)\right\}^{-1}(\mathbf{I}_r \otimes \mathbf{H}_J)' \tag{15}$$

where $\mathbf{H}_J$ is the $m_1 \times (m_1 - r)$ matrix given by $\mathbf{H}_J = (\mathbf{0}'_{r \times (m_1-r)}, \mathbf{I}_{m_1-r})'$. The VCE reported in e(V_beta) is the estimated VCE in (15) augmented with missing values to account for any backed-out estimates of μ or ρ.

The parameter estimates $\widehat{\alpha}$ can be found either as a function of $\widehat{\beta}$, using (10) or from the VAR in (14). The estimated VCE of $\widehat{\alpha}$ reported in e(V_alpha) is given by

$$\frac{1}{(T-d)}\widehat{\mathbf{\Omega}} \otimes \widehat{\mathbf{\Sigma}}_B$$

where $\widehat{\mathbf{\Sigma}}_B = (\widehat{\beta}'\mathbf{S}_{11}\widehat{\beta})^{-1}$.

As we would expect, the estimator of $\Pi = \alpha\beta'$ is

$$\widehat{\Pi} = \widehat{\alpha}\widehat{\beta}'$$

and its estimated VCE is given by

$$\frac{1}{(T-d)}\widehat{\mathbf{\Omega}} \otimes (\widehat{\beta}\widehat{\mathbf{\Sigma}}_B\widehat{\beta}')$$

The moving-average impact matrix $\mathbf{C}$ is estimated by

$$\widehat{\mathbf{C}} = \widehat{\beta}_\perp(\widehat{\alpha}_\perp\widehat{\mathbf{\Gamma}}\widehat{\beta}_\perp)^{-1}\widehat{\alpha}_\perp'$$

where $\widehat{\beta}_\perp$ is the orthogonal complement of $\widehat{\beta}_y$, $\widehat{\alpha}_\perp$ is the orthogonal complement of $\widehat{\alpha}$, and $\widehat{\mathbf{\Gamma}} = \mathbf{I}_K - \sum_{i=1}^{p-1}\mathbf{\Gamma}_i$. The orthogonal complement of a $K \times r$ matrix $\mathbf{Q}$ that has rank r is a matrix $\mathbf{Q}_\perp$ of rank $K - r$, such that $\mathbf{Q}'\mathbf{Q}_\perp = \mathbf{0}$. Although this operation is not uniquely defined, the results used by vec do not depend on the method of obtaining the orthogonal complement. vec uses the following method: the orthogonal complement of $\mathbf{Q}$ is given by the r eigenvectors with the highest eigenvalues from the matrix $\mathbf{Q}'(\mathbf{Q}'\mathbf{Q})^{-1}\mathbf{Q}'$.

Per Johansen (1995, chap. 13) and Drukker (2004), the VCE of $\widehat{\mathbf{C}}$ is estimated by

$$\frac{T-d}{T}\widehat{\mathbf{S}}_q\widehat{\mathbf{V}}_{\widehat{\nu}}\widehat{\mathbf{S}}_q' \tag{16}$$

where

$$\widehat{\mathbf{S}}_q = \widehat{\mathbf{C}} \otimes \widehat{\xi}$$

$$\widehat{\xi} = \begin{cases} (\widehat{\xi}_1, \widehat{\xi}_2) & \text{if } p > 1 \\ \widehat{\xi}_1 & \text{if } p = 1 \end{cases}$$

$$\widehat{\xi}_1 = (\widehat{\mathbf{C}}'\widehat{\mathbf{\Gamma}}' - \mathbf{I}_K)\bar{\alpha}$$

$$\bar{\alpha} = \widehat{\alpha}(\widehat{\alpha}'\widehat{\alpha})^{-1}$$

$$\widehat{\xi}_2 = \iota_{p-1} \otimes \widehat{\mathbf{C}}$$

ι_{p-1} is a $(p-1) \times 1$ vector of ones

$\widehat{\mathbf{V}}_{\widehat{\nu}}$ is the estimated VCE of $\widehat{\nu} = (\widehat{\alpha}, \widehat{\mathbf{\Gamma}}_1, \dots \widehat{\mathbf{\Gamma}}_{p-1})$

Estimation with constraints: β identified

vec can also fit models in which the adjustment parameters are subject to homogeneous linear constraints and the cointegrating vectors are subject to general linear restrictions. Mathematically, vec allows for constraints of the form

$$\mathbf{R}'_{\alpha}\mathrm{vec}(\boldsymbol{\alpha}) = \mathbf{0} \tag{17}$$

where $\mathbf{R}_{\alpha}$ is a known $Kr \times n_{\alpha}$ constraint matrix, and

$$\mathbf{R}'_{\widetilde{\beta}}\mathrm{vec}(\widetilde{\beta}) = \mathbf{b} \tag{18}$$

where $\mathbf{R}_{\widetilde{\beta}}$ is a known $m_1 r \times n_{\beta}$ constraint matrix and $\mathbf{b}$ is a known $n_{\beta} \times 1$ vector of constants. Although (17) and (18) are intuitive, they can be rewritten in a form to facilitate computation. Specifically, (17) can be written as

$$\mathrm{vec}(\boldsymbol{\alpha}') = \mathbf{G}\mathbf{a} \tag{19}$$

where $\mathbf{G}$ is $Kr \times n_{\alpha}$ and $\mathbf{a}$ is $n_{\alpha} \times 1$. Equation (18) can be rewritten as

$$\mathrm{vec}(\widetilde{\beta}) = \mathbf{H}\mathbf{b} + \mathbf{h}_0 \tag{20}$$

where $\mathbf{H}$ is a known $n_1 r \times n_{\beta}$ matrix, $\mathbf{b}$ is an $n_{\beta} \times 1$ matrix of parameters, and $\mathbf{h}_0$ is a known $n_1 r \times 1$ matrix. See [P] **makecns** for a discussion of the different ways of specifying the constraints.

When constraints are specified via the aconstraints() and bconstraints() options, the Boswijk (1995) rank method determines whether the parameters in $\widetilde{\beta}$ are underidentified, exactly identified, or overidentified.

Boswijk (1995) uses the Rothenberg (1971) method to determine whether the parameters in $\widetilde{\beta}$ are identified. Thus the parameters in $\widetilde{\beta}$ are exactly identified if $\rho_{\beta} = r^2$, and the parameters in $\widetilde{\beta}$ are overidentified if $\rho_{\beta} > r^2$, where

$$\rho_{\beta} = \mathrm{rank}\left\{\mathbf{R}_{\widetilde{\beta}}(\mathbf{I}_r \otimes \ddot{\beta})\right\}$$

and $\ddot{\beta}$ is a full-rank matrix with the same dimensions as $\widetilde{\beta}$. The computed ρ_{β} is saved in e(beta_icnt).

Similarly, the number of freely estimated parameters in $\boldsymbol{\alpha}$ and $\widetilde{\beta}$ is given by ρ_{jacob}, where

$$\rho_{\mathrm{jacob}} = \mathrm{rank}\left\{(\widehat{\boldsymbol{\alpha}} \otimes \mathbf{I}_{m_1})\mathbf{H}, (\mathbf{I}_K \otimes \widehat{\beta})\mathbf{G}\right\}$$

Using ρ_{jacob}, we can calculate several other parameter counts of interest. In particular, the degrees of freedom of the overidentifying test are given by $(K + m_1 - r)r - \rho_{\mathrm{jacob}}$, and the number of freely estimated parameters in the model is $n_{\mathrm{parms}} = Km_2 + \rho_{\mathrm{jacob}}$.

Although the problem of maximizing the log-likelihood function in (4), subject to the constraints in (17) and (18), could be handled by the algorithms in [R] **ml**, the switching algorithm of Boswijk (1995) has proven to be more convergent. For this reason, vec uses the Boswijk (1995) switching algorithm to perform the optimization.

Given starting values $(\widehat{\mathbf{b}}_0, \widehat{\mathbf{a}}_0, \widehat{\boldsymbol{\Omega}}_0)$, the algorithm iteratively updates the estimates until convergence is achieved, as follows:

$\widehat{\boldsymbol{\alpha}}_j$ is constructed from (19) and $\widehat{\mathbf{a}}_j$

$\widehat{\boldsymbol{\beta}}_j$ is constructed from (20) and $\widehat{\mathbf{b}}_j$

$$\widehat{\mathbf{b}}_{j+1} = \{\mathbf{H}'(\widehat{\boldsymbol{\alpha}}_j'\widehat{\boldsymbol{\Omega}}_j^{-1}\widehat{\boldsymbol{\alpha}}_j \otimes \mathbf{S}_{11})\mathbf{H}\}^{-1}\mathbf{H}'(\widehat{\boldsymbol{\alpha}}_j\widehat{\boldsymbol{\Omega}}_j^{-1} \otimes \mathbf{S}_{11})\{\text{vec}(\widehat{\mathbf{P}}) - (\widehat{\boldsymbol{\alpha}}_j \otimes \mathbf{I}_{n_{Z1}})\mathbf{h}_0\}$$

$$\widehat{\mathbf{a}}_{j+1} = \{\mathbf{G}(\widehat{\boldsymbol{\Omega}}_j^{-1} \otimes \widehat{\boldsymbol{\beta}}_j\mathbf{S}_{11}\widehat{\boldsymbol{\beta}}_j)\mathbf{G}\}^{-1}\mathbf{G}'(\widehat{\boldsymbol{\Omega}}_j^{-1} \otimes \widehat{\boldsymbol{\beta}}_j\mathbf{S}_{11})\text{vec}(\widehat{\mathbf{P}})$$

$$\widehat{\boldsymbol{\Omega}}_{j+1} = \mathbf{S}_{00} - \mathbf{S}_{01}\widehat{\boldsymbol{\beta}}_j\widehat{\boldsymbol{\alpha}}_j' - \widehat{\boldsymbol{\alpha}}_j\widehat{\boldsymbol{\beta}}_j'\mathbf{S}_{10} + \widehat{\boldsymbol{\alpha}}_j\widehat{\boldsymbol{\beta}}_j'\mathbf{S}_{11}\widehat{\boldsymbol{\beta}}_j\widehat{\boldsymbol{\alpha}}_j'$$

The estimated VCE of $\widehat{\boldsymbol{\beta}}$ is given by

$$\frac{1}{(T-d)}\mathbf{H}\{\mathbf{H}'(\mathbf{W} \otimes \mathbf{S}_{11})\mathbf{H}\}^{-1}\mathbf{H}'$$

where $\mathbf{W}$ is $\widehat{\boldsymbol{\alpha}}'\widehat{\boldsymbol{\Omega}}^{-1}\widehat{\boldsymbol{\alpha}}$. As in the case without constraints, the estimated VCE of $\widehat{\boldsymbol{\alpha}}$ can be obtained either from the VCE of the short-run parameters, as described below, or via the formula

$$\widehat{V}_{\widehat{\boldsymbol{\alpha}}} = \frac{1}{(T-d)}\mathbf{G}\left[\mathbf{G}'\left\{\widehat{\boldsymbol{\Omega}}^{-1} \otimes (\widehat{\boldsymbol{\beta}}'\mathbf{S}_{11}\widehat{\boldsymbol{\beta}})\mathbf{G}\right\}^{-1}\right]\mathbf{G}'$$

Boswijk (1995) notes that, as long as the parameters of the cointegrating equations are exactly identified or overidentified, the constrained ML estimator produces superconsistent estimates of $\widetilde{\boldsymbol{\beta}}$. This implies that the method of estimating the short-run parameters described above applies in the presence of constraints, as well, albeit with a caveat: when there are constraints placed on $\boldsymbol{\alpha}$, the VARs must be estimated subject to these constraints.

With these estimates and the estimated VCE of the short-run parameter matrix $\widehat{\mathbf{V}}_{\widehat{\nu}}$, Drukker (2004) shows that the estimated VCE for $\widehat{\boldsymbol{\Pi}}$ is given by

$$(\widehat{\boldsymbol{\beta}} \otimes \mathbf{I}_K)\widehat{V}_{\widehat{\boldsymbol{\alpha}}}(\widehat{\boldsymbol{\beta}} \otimes \mathbf{I}_K)'$$

Drukker (2004) also shows that the estimated VCE of $\widehat{\mathbf{C}}$ can be obtained from (16) with the extension that $\widehat{V}_{\widehat{\nu}}$ is the estimated VCE of $\widehat{\nu}$ that takes into account any constraints on $\widehat{\boldsymbol{\alpha}}$.

Estimation with constraints: β not identified

When the parameters in β are not identified, only the parameters in $\boldsymbol{\Pi} = \boldsymbol{\alpha}\boldsymbol{\beta}$ and $\mathbf{C}$ are identified. The estimates of $\boldsymbol{\Pi}$ and $\mathbf{C}$ would not change if more identification restrictions were imposed to achieve exact identification. Thus the VCE matrices for $\widehat{\boldsymbol{\Pi}}$ and $\widehat{\mathbf{C}}$ can be derived as if the model exactly identified β.

Formulas for the information criteria

The AIC, SBIC, and HQIC are calculated according to their standard definitions, which include the constant term from the log likelihood; that is,

$$\text{AIC} = -2\left(\frac{L}{T}\right) + \frac{2n_{\text{parms}}}{T}$$

$$\text{SBIC} = -2\left(\frac{L}{T}\right) + \frac{\ln(T)}{T}n_{\text{parms}}$$

$$\text{HQIC} = -2\left(\frac{L}{T}\right) + \frac{2\ln\{\ln(T)\}}{T}n_{\text{parms}}$$

where n_{parms} is the total number of parameters in the model and L is the value of the log likelihood at the optimum.

Formulas for predict

`xb`, `residuals` and `stdp` are standard and are documented in [R] **predict**. `ce` causes predict to compute $\widehat{E}_t = \widehat{\beta}_f \mathbf{y}_t$ for the requested cointegrating equation.

`levels` causes `predict` to compute the predictions for the levels of the data. Let $\widehat{y}_t^d$ be the predicted value of Δy_t. Because the computations are performed for a given equation, y_t is a scalar. Using $\widehat{y}_t^d$, we can predict the level by $\widehat{y}_t = \widehat{y}_t^d + y_{t-1}$.

Because the residuals from the VECM for the differences and the residuals from the corresponding VAR in levels are identical, there is no need for an option for predicting the residuals in levels.

References

Anderson, T. W. 1951. Estimating linear restrictions on regression coefficients for multivariate normal distributions. *Annals of Mathematical Statistics* 22: 327–351.

Boswijk, H. P. 1995. Identifiability of cointegrated systems. Discussion Paper #95-78, Tinbergen Institute. http://www1.fee.uva.nl/pp/bin/258fulltext.pdf.

Boswijk, H. P., and J. A. Doornik. 2004. Identifying, estimating and testing restricted cointegrating systems: An overview. *Statistica Neerlandica* 58: 440–465.

Drukker, D. M. 2004. Some further results on estimation and inference in the presence of constraints on alpha in a cointegrating VECM. Working paper, StataCorp.

Engle, R. F., and C. W. J. Granger. 1987. Co-integration and error correction: Representation, estimation, and testing. *Econometrica* 55: 251–276.

Hamilton, J. D. 1994. *Time Series Analysis*. Princeton: Princeton University Press.

Johansen, S. 1988. Statistical analysis of cointegration vectors. *Journal of Economic Dynamics and Control* 12: 231–254.

——. 1991. Estimation and hypothesis testing of cointegration vectors in Gaussian vector autoregressive models. *Econometrica* 59: 1551–1580.

——. 1995. *Likelihood-Based Inference in Cointegrated Vector Autoregressive Models*. Oxford: Oxford University Press.

Maddala, G. S., and I.-M. Kim. 1998. *Unit Roots, Cointegration, and Structural Change*. Cambridge: Cambridge University Press.

Park, J. Y., and P. C. B. Phillips. 1988. Statistical inference in regressions with integrated processes: Part I. *Econometric Theory* 4: 468–497.

——. 1989. Statistical inference in regressions with integrated processes: Part II. *Econometric Theory* 5: 95–131.

Phillips, P. C. B. 1986. Understanding spurious regressions in econometrics. *Journal of Econometrics* 33: 311–340.

Phillips, P. C. B., and S. N. Durlauf. 1986. Multiple time series regressions with integrated processes. *Review of Economic Studies* 53: 473–495.

Rothenberg, T. J. 1971. Identification in parametric models. *Econometrica* 39: 577–591.

Sims, C. A., J. H. Stock, and M. W. Watson. 1990. Inference in linear time series models with some unit roots. *Econometrica* 58: 113–144.

Stock, J. H. 1987. Asymptotic properties of least squares estimators of cointegrating vectors. *Econometrica* 55: 1035–1056.

Stock, J. H., and M. W. Watson. 1988. Testing for common trends. *Journal of the American Statistical Association* 83: 1097–1107.

Watson, M. W. 1994. Vector autoregressions and cointegration. In Vol. IV of *Handbook of Econometrics*, ed. R. F. Engle and D. L. McFadden. Amsterdam: Elsevier.

Also see

[TS] **vec postestimation** — Postestimation tools for vec

[TS] **tsset** — Declare data to be time-series data

[TS] **var** — Vector autoregressive models

[TS] **var svar** — Structural vector autoregressive models

[U] **20 Estimation and postestimation commands**

[TS] **vec intro** — Introduction to vector error-correction models

Title

---

vec postestimation — Postestimation tools for vec

Description

The following postestimation commands are of special interest after `vec`:

Command	Description
`fcast compute`	obtain dynamic forecasts
`fcast graph`	graph dynamic forecasts obtained from `fcast compute`
`irf`	create and analyze IRFs and FEVDs
`veclmar`	LM test for autocorrelation in residuals
`vecnorm`	test for normally distributed residuals
`vecstable`	check stability condition of estimates

For information about these commands, see the corresponding entries in this manual.

The following standard postestimation commands are also available:

Command	Description
`estat`	AIC, BIC, VCE, and estimation sample summary
`estimates`	cataloging estimation results
`lincom`	point estimates, standard errors, testing, and inference for linear combinations of coefficients
`lrtest`	likelihood-ratio test
`margins`	marginal means, predictive margins, marginal effects, and average marginal effects
`marginsplot`	graph the results from margins (profile plots, interaction plots, etc.)
`nlcom`	point estimates, standard errors, testing, and inference for nonlinear combinations of coefficients
`predict`	predictions, residuals, influence statistics, and other diagnostic measures
`predictnl`	point estimates, standard errors, testing, and inference for generalized predictions
`test`	Wald tests of simple and composite linear hypotheses
`testnl`	Wald tests of nonlinear hypotheses

See the corresponding entries in the *Base Reference Manual* for details.

Syntax for predict

predict [*type*] *newvar* [*if*] [*in*] [, *statistic* <u>eq</u>uation(*eqno* | *eqname*)]

statistic	Description
Main	
xb	linear prediction; the default
stdp	standard error of the linear prediction
<u>r</u>esiduals	residuals
ce	the predicted value of specified cointegrating equation
<u>l</u>evels	one-step prediction of the level of the endogenous variable
<u>u</u>sece(*varlist*$_{ce}$)	compute the predictions using previously predicted cointegrating equations

These statistics are available both in and out of sample; type predict ... if e(sample) ... if wanted only for the estimation sample.

Menu

Statistics > Postestimation > Predictions, residuals, etc.

Options for predict

> Main

xb, the default, calculates the linear prediction for the specified equation. The form of the VECM implies that these fitted values are the one-step predictions for the first-differenced variables.

stdp calculates the standard error of the linear prediction for the specified equation.

residuals calculates the residuals from the specified equation of the VECM.

ce calculates the predicted value of the specified cointegrating equation.

levels calculates the one-step prediction of the level of the endogenous variable in the requested equation.

usece(*varlist*$_{ce}$) specifies that previously predicted cointegrating equations saved under the names in *varlist*$_{ce}$ be used to compute the predictions. The number of variables in the *varlist*$_{ce}$ must equal the number of cointegrating equations specified in the model.

equation(*eqno* | *eqname*) specifies to which equation you are referring.

equation() is filled in with one *eqno* or *eqname* for xb, residuals, stdp, ce, and levels options. equation(#1) would mean that the calculation is to be made for the first equation, equation(#2) would mean the second, and so on. You could also refer to the equation by its name. equation(D_income) would refer to the equation named D_income and equation(_ce1), to the first cointegrating equation, which is named _ce1 by vec.

If you do not specify equation(), the results are as if you specified equation(#1).

For more information on using predict after multiple-equation estimation commands, see [R] **predict**.

Remarks

Remarks are presented under the following headings:

>   *Model selection and inference*
>   *Forecasting*

Model selection and inference

See the following sections for information on model selection and inference after vec.

[TS] **irf** — Create and analyze IRFs, dynamic-multiplier functions, and FEVDs

[TS] **varsoc** — Obtain lag-order selection statistics for VARs and VECMs

[TS] **veclmar** — Perform LM test for residual autocorrelation after vec

[TS] **vecnorm** — Test for normally distributed disturbances after vec

[TS] **vecrank** — Estimate the cointegrating rank of a VECM

[TS] **vecstable** — Check the stability condition of VECM estimates

Forecasting

See the following sections for information on obtaining forecasts after vec:

[TS] **fcast compute** — Compute dynamic forecasts of dependent variables after var, svar, or vec

[TS] **fcast graph** — Graph forecasts of variables computed by fcast compute

Methods and formulas

All postestimation commands listed above are implemented as ado-files.

Also see

[TS] **vec** — Vector error-correction models

[U] **20 Estimation and postestimation commands**

[TS] **vec intro** — Introduction to vector error-correction models

Title

> **veclmar** — Perform LM test for residual autocorrelation after vec

Syntax

> veclmar [, *options*]

options	Description
<u>ml</u>ag(#)	use # for the maximum order of autocorrelation; default is mlag(2)
<u>est</u>imates(*estname*)	use previously saved results *estname*; default is to use active results
<u>se</u>parator(#)	draw separator line after every # rows

veclmar can be used only after vec; see [TS] **vec**.

You must tsset your data before using veclmar; see [TS] **tsset**.

Menu

Statistics > Multivariate time series > VEC diagnostics and tests > LM test for residual autocorrelation

Description

veclmar implements a Lagrange-multiplier (LM) test for autocorrelation in the residuals of vector error-correction models (VECMs).

Options

mlag(#) specifies the maximum order of autocorrelation to be tested. The integer specified in mlag() must be greater than 0; the default is 2.

estimates(*estname*) requests that veclmar use the previously obtained set of vec estimates saved as *estname*. By default, veclmar uses the active results. See [R] **estimates** for information on saving and restoring estimation results.

separator(#) specifies how many rows should appear in the table between separator lines. By default, separator lines do not appear. For example, separator(1) would draw a line between each row, separator(2) between every other row, and so on.

Remarks

Estimation, inference, and postestimation analysis of VECMs is predicated on the errors' not being autocorrelated. veclmar implements the LM test for autocorrelation in the residuals of a VECM discussed in Johansen (1995, 21–22). The test is performed at lags $j = 1, \ldots, \texttt{mlag}()$. For each j, the null hypothesis of the test is that there is no autocorrelation at lag j.

▷ Example 1

We fit a VECM using the regional income data described in [TS] **vec** and then call `veclmar` to test for autocorrelation.

```
. use http://www.stata-press.com/data/r12/rdinc
. vec ln_ne ln_se
(output omitted )
. veclmar, mlag(4)
  Lagrange-multiplier test
```

lag	chi2	df	Prob > chi2
1	8.9586	4	0.06214
2	4.9809	4	0.28926
3	4.8519	4	0.30284
4	0.3270	4	0.98801

```
  H0: no autocorrelation at lag order
```

At the 5% level, we cannot reject the null hypothesis that there is no autocorrelation in the residuals for any of the orders tested. Thus this test finds no evidence of model misspecification.

◁

Saved results

`veclmar` saves the following in `r()`:

Matrices
    `r(lm)` χ^2, df, and p-values

Methods and formulas

`veclmar` is implemented as an ado-file.

Consider a VECM without any trend:

$$\Delta \mathbf{y}_t = \boldsymbol{\alpha}\boldsymbol{\beta}\mathbf{y}_{t-1} + \sum_{i=1}^{p-1} \boldsymbol{\Gamma}_i \Delta \mathbf{y}_{t-i} + \epsilon_t$$

As discussed in [TS] **vec**, as long as the parameters in the cointegrating vectors, $\boldsymbol{\beta}$, are exactly identified or overidentified, the estimates of these parameters are superconsistent. This implies that the $r \times 1$ vector of estimated cointegrating relations

$$\widehat{\mathbf{E}}_t = \widehat{\boldsymbol{\beta}}\mathbf{y}_t \tag{1}$$

can be used as data with standard estimation and inference methods. When the parameters of the cointegrating equations are not identified, (1) does not provide consistent estimates of $\widehat{\mathbf{E}}_t$; in these cases, `veclmar` exits with an error message.

The VECM above can be rewritten as

$$\Delta \mathbf{y}_t = \alpha \widehat{\mathbf{E}}_t + \sum_{i=1}^{p-1} \Gamma_i \Delta \mathbf{y}_{t-i} + \epsilon_t$$

which is just a VAR with $p - 1$ lags where the endogenous variables have been first-differenced and is augmented with the exogenous variables $\widehat{\mathbf{E}}$. veclmar fits this VAR and then calls varlmar to compute the LM test for autocorrelation.

The above discussion assumes no trend and implicitly ignores constraints on the parameters in α. As discussed in vec, the other four trend specifications considered by Johansen (1995, sec. 5.7) complicate the estimation of the free parameters in β but do not alter the basic result that the $\widehat{\mathbf{E}}_t$ can be used as data in the subsequent VAR. Similarly, constraints on the parameters in α imply that the subsequent VAR must be estimated with these constraints applied, but $\widehat{\mathbf{E}}_t$ can still be used as data in the VAR.

See [TS] **varlmar** for more information on the Johansen LM test.

Reference

Johansen, S. 1995. *Likelihood-Based Inference in Cointegrated Vector Autoregressive Models.* Oxford: Oxford University Press.

Also see

[TS] **vec** — Vector error-correction models

[TS] **varlmar** — Perform LM test for residual autocorrelation after var or svar

[TS] **vec intro** — Introduction to vector error-correction models

Title

vecnorm — Test for normally distributed disturbances after vec

Syntax

vecnorm [, *options*]

options	Description
jbera	report Jarque–Bera statistic; default is to report all three statistics
skewness	report skewness statistic; default is to report all three statistics
kurtosis	report kurtosis statistic; default is to report all three statistics
estimates(*estname*)	use previously saved results *estname*; default is to use active results
dfk	make small-sample adjustment when computing the estimated variance–covariance matrix of the disturbances
separator(*#*)	draw separator line after every # rows

vecnorm can be used only after vec; see [TS] **vec**.

Menu

Statistics > Multivariate time series > VEC diagnostics and tests > Test for normally distributed disturbances

Description

vecnorm computes and reports a series of statistics against the null hypothesis that the disturbances in a VECM are normally distributed.

Options

jbera requests that the Jarque–Bera statistic and any other explicitly requested statistic be reported. By default, the Jarque–Bera, skewness, and kurtosis statistics are reported.

skewness requests that the skewness statistic and any other explicitly requested statistic be reported. By default, the Jarque–Bera, skewness, and kurtosis statistics are reported.

kurtosis requests that the kurtosis statistic and any other explicitly requested statistic be reported. By default, the Jarque–Bera, skewness, and kurtosis statistics are reported.

estimates(*estname*) requests that vecnorm use the previously obtained set of vec estimates saved as *estname*. By default, vecnorm uses the active results. See [R] **estimates** for information about saving and restoring estimation results.

dfk requests that a small-sample adjustment be made when computing the estimated variance–covariance matrix of the disturbances.

separator(*#*) specifies how many rows should appear in the table between separator lines. By default, separator lines do not appear. For example, separator(1) would draw a line between each row, separator(2) between every other row, and so on.

Remarks

vecnorm computes a series of test statistics of the null hypothesis that the disturbances in a VECM are normally distributed. For each equation and all equations jointly, up to three statistics may be computed: a skewness statistic, a kurtosis statistic, and the Jarque–Bera statistic. By default, all three statistics are reported; if you specify only one statistic, the others are not reported. The Jarque–Bera statistic tests skewness and kurtosis jointly. The single-equation results are against the null hypothesis that the disturbance for that particular equation is normally distributed. The results for all the equations are against the null that all K disturbances have a K-dimensional multivariate normal distribution. Failure to reject the null hypothesis indicates lack of model misspecification.

As noted by Johansen (1995, 141), the log likelihood for the VECM is derived assuming the errors are independently and identically distributed (i.i.d.) normal, though many of the asymptotic properties can be derived under the weaker assumption that the errors are merely i.i.d. Many researchers still prefer to test for normality. vecnorm uses the results from vec to produce a series of statistics against the null hypothesis that the K disturbances in the VECM are normally distributed.

▷ Example 1

This example uses vecnorm to test for normality after estimating the parameters of a VECM using the regional income data.

```
. use http://www.stata-press.com/data/r12/rdinc
. vec ln_ne ln_se
(output omitted )
. vecnorm
```

Jarque-Bera test

Equation	chi2	df	Prob > chi2
D_ln_ne	0.094	2	0.95417
D_ln_se	0.586	2	0.74608
ALL	0.680	4	0.95381

Skewness test

Equation	Skewness	chi2	df	Prob > chi2
D_ln_ne	.05982	0.032	1	0.85890
D_ln_se	.243	0.522	1	0.47016
ALL		0.553	2	0.75835

Kurtosis test

Equation	Kurtosis	chi2	df	Prob > chi2
D_ln_ne	3.1679	0.062	1	0.80302
D_ln_se	2.8294	0.064	1	0.79992
ALL		0.126	2	0.93873

The Jarque–Bera results present test statistics for each equation and for all equations jointly against the null hypothesis of normality. For the individual equations, the null hypothesis is that the disturbance term in that equation has a univariate normal distribution. For all equations jointly, the null hypothesis is that the K disturbances come from a K-dimensional normal distribution. In this example, the single-equation and overall Jarque–Bera statistics do not reject the null of normality.

The single-equation skewness test statistics are of the null hypotheses that the disturbance term in each equation has zero skewness, which is the skewness of a normally distributed variable. The row marked ALL shows the results for a test that the disturbances in all equations jointly have zero skewness. The skewness results shown above do not suggest nonnormality.

The kurtosis of a normally distributed variable is three, and the kurtosis statistics presented in the table test the null hypothesis that the disturbance terms have kurtosis consistent with normality. The results in this example do not reject the null hypothesis.

◁

The statistics computed by vecnorm are based on the estimated variance–covariance matrix of the disturbances. vec saves the ML estimate of this matrix, which vecnorm uses by default. Specifying the dfk option instructs vecnorm to make a small-sample adjustment to the variance–covariance matrix before computing the test statistics.

Saved results

vecnorm saves the following in r():

Macros
 r(dfk)   dfk, if specified

Matrices
 r(jb)   Jarque–Bera χ^2, df, and p-values
 r(skewness)   skewness χ^2, df, and p-values
 r(kurtosis)   kurtosis χ^2, df, and p-values

Methods and formulas

vecnorm is implemented as an ado-file.

As discussed in *Methods and formulas* of [TS] **vec**, a cointegrating VECM can be rewritten as a VAR in first differences that includes the predicted cointegrating equations as exogenous variables. vecnorm computes the tests discussed in [TS] **varnorm** for the corresponding augmented VAR in first differences. See *Methods and formulas* of [TS] **veclmar** for more information on this approach.

When the parameters of the cointegrating equations are not identified, the consistent estimates of the cointegrating equations are not available, and, in these cases, vecnorm exits with an error message.

References

Hamilton, J. D. 1994. *Time Series Analysis*. Princeton: Princeton University Press.

Jarque, C. M., and A. K. Bera. 1987. A test for normality of observations and regression residuals. *International Statistical Review* 2: 163–172.

Johansen, S. 1995. *Likelihood-Based Inference in Cointegrated Vector Autoregressive Models*. Oxford: Oxford University Press.

Lütkepohl, H. 2005. *New Introduction to Multiple Time Series Analysis*. New York: Springer.

Also see

[TS] **vec** — Vector error-correction models

[TS] **varnorm** — Test for normally distributed disturbances after var or svar

[TS] **vec intro** — Introduction to vector error-correction models

Title

> **vecrank** — Estimate the cointegrating rank of a VECM

Syntax

vecrank *depvar* $[if]$ $[in]$ $[$, *options* $]$

options	Description
Model	
<u>lags</u>(#)	use # for the maximum lag in underlying VAR model
<u>trend(c</u>onstant)	include an unrestricted constant in model; the default
<u>trend(rc</u>onstant)	include a restricted constant in model
<u>trend(t</u>rend)	include a linear trend in the cointegrating equations and a quadratic trend in the undifferenced data
<u>trend(rt</u>rend)	include a restricted trend in model
<u>trend(n</u>one)	do not include a trend or a constant
Adv. model	
<u>s</u>indicators(*varlist*$_{si}$)	include normalized seasonal indicator variables *varlist*$_{si}$
noreduce	do not perform checks and corrections for collinearity among lags of dependent variables
Reporting	
<u>notrace</u>	do not report the trace statistic
<u>max</u>	report maximum-eigenvalue statistic
<u>ic</u>	report information criteria
level99	report 1% critical values instead of 5% critical values
levela	report both 1% and 5% critical values

You must tsset your data before using vecrank; see [TS] **tsset**.

depvar may contain time-series operators; see [U] **11.4.4 Time-series varlists**.

by, rolling, and statsby are allowed; see [U] **11.1.10 Prefix commands**.

vecrank does not allow gaps in the data.

Menu

Statistics > Multivariate time series > Cointegrating rank of a VECM

Description

vecrank produces statistics used to determine the number of cointegrating equations in a vector error-correction model (VECM).

Options

lags(*#*) specifies the number of lags in the VAR representation of the model. The VECM will include one fewer lag of the first differences. The number of lags must be greater than zero but small enough so that the degrees of freedom used by the model are less than the number of observations.

trend(*trend_spec*) specifies one of five trend specifications to include in the model. See [TS] **vec intro** and [TS] **vec** for descriptions. The default is trend(constant).

sindicators(*varlist*$_{si}$) specifies normalized seasonal indicator variables to be included in the model. The indicator variables specified in this option must be normalized as discussed in Johansen (1995, 84). If the indicators are not properly normalized, the likelihood-ratio–based tests for the number of cointegrating equations do not converge to the asymptotic distributions derived by Johansen. For details, see *Methods and formulas* of [TS] **vec**. sindicators() cannot be specified with trend(none) or trend(rconstant)

noreduce causes vecrank to skip the checks and corrections for collinearity among the lags of the dependent variables. By default, vecrank checks whether the current lag specification causes some of the regressions performed by vecrank to contain perfectly collinear variables and reduces the maximum lag until the perfect collinearity is removed. See *Collinearity* in [TS] **vec** for more information.

notrace requests that the output for the trace statistic not be displayed. The default is to display the trace statistic.

max requests that the output for the maximum-eigenvalue statistic be displayed. The default is to not display this output.

ic causes the output for the information criteria to be displayed. The default is to not display this output.

level99 causes the 1% critical values to be displayed instead of the default 5% critical values.

levela causes both the 1% and the 5% critical values to be displayed.

Remarks

Remarks are presented under the following headings:

> *Introduction*
> *The trace statistic*
> *The maximum-eigenvalue statistic*
> *Minimizing an information criterion*

Introduction

Before estimating the parameters of a VECM models, you must choose the number of lags in the underlying VAR, the trend specification, and the number of cointegrating equations. vecrank offers several ways of determining the number of cointegrating vectors conditional on a trend specification and lag order.

vecrank implements three types of methods for determining r, the number of cointegrating equations in a VECM. The first is Johansen's "trace" statistic method. The second is his "maximum eigenvalue" statistic method. The third method chooses r to minimize an information criterion.

All three methods are based on Johansen's maximum likelihood (ML) estimator of the parameters of a cointegrating VECM. The basic VECM is

$$\Delta \mathbf{y}_t = \boldsymbol{\alpha}\boldsymbol{\beta}'\mathbf{y}_{t-1} + \sum_{t=1}^{p-1} \boldsymbol{\Gamma}_i \Delta \mathbf{y}_{t-i} + \epsilon_t$$

where $\mathbf{y}$ is a $(K \times 1)$ vector of I(1) variables, $\boldsymbol{\alpha}$ and $\boldsymbol{\beta}$ are $(K \times r)$ parameter matrices with rank $r < K$, $\boldsymbol{\Gamma}_1, \ldots, \boldsymbol{\Gamma}_{p-1}$ are $(K \times K)$ matrices of parameters, and ϵ_t is a $(K \times 1)$ vector of normally distributed errors that is serially uncorrelated but has contemporaneous covariance matrix $\boldsymbol{\Omega}$.

Building on the work of Anderson (1951), Johansen (1995) derives an ML estimator for the parameters and two likelihood-ratio (LR) tests for inference on r. These LR tests are known as the trace statistic and the maximum-eigenvalue statistic because the log likelihood can be written as the log of the determinant of a matrix plus a simple function of the eigenvalues of another matrix.

Let $\lambda_1, \ldots, \lambda_K$ be the K eigenvalues used in computing the log likelihood at the optimum. Furthermore, assume that these eigenvalues are sorted from the largest λ_1 to the smallest λ_K. If there are $r < K$ cointegrating equations, $\boldsymbol{\alpha}$ and $\boldsymbol{\beta}$ have rank r and the eigenvalues $\lambda_{r+1}, \ldots, \lambda_K$ are zero.

The trace statistic

The null hypothesis of the trace statistic is that there are no more than r cointegrating relations. Restricting the number of cointegrating equations to be r or less implies that the remaining $K - r$ eigenvalues are zero. Johansen (1995, chap. 11 and 12) derives the distribution of the trace statistic

$$-T \sum_{i=r+1}^{K} \ln(1 - \widehat{\lambda}_i)$$

where T is the number of observations and the $\widehat{\lambda}_i$ are the estimated eigenvalues. For any given value of r, large values of the trace statistic are evidence against the null hypothesis that there are r or fewer cointegrating relations in the VECM.

One of the problems in determining the number of cointegrating equations is that the process involves more than one statistical test. Johansen (1995, chap. 6, 11, and 12) derives a method based on the trace statistic that has nominal coverage despite evaluating multiple tests. This method can be interpreted as being an estimator $\widehat{r}$ of the true number of cointegrating equations r_0. The method starts testing at $r = 0$ and accepts as $\widehat{r}$ the first value of r for which the trace statistic fails to reject the null.

▷ Example 1

We have quarterly data on the natural logs of aggregate consumption, investment, and GDP in the United States from the first quarter of 1959 through the fourth quarter of 1982. As discussed in King et al. (1991), the balanced-growth hypothesis in economics implies that we would expect to find two cointegrating equations among these three variables. In the output below, we use vecrank to determine the number of cointegrating equations using Johansen's multiple-trace test method.

```
. use http://www.stata-press.com/data/r12/balance2
(macro data for VECM/balance study)

. vecrank y i c, lags(5)
```

```
                    Johansen tests for cointegration
Trend: constant                              Number of obs =      91
Sample:   1960q2 - 1982q4                         Lags =          5
```

					5%
maximum				trace	critical
rank	parms	LL	eigenvalue	statistic	value
0	39	1231.1041	.	46.1492	29.68
1	44	1245.3882	0.26943	17.5810	15.41
2	47	1252.5055	0.14480	3.3465*	3.76
3	48	1254.1787	0.03611		

The header produces information about the sample, the trend specification, and the number of lags included in the model. The main table contains a separate row for each possible value of r, the number of cointegrating equations. When $r = 3$, all three variables in this model are stationary.

In this example, because the trace statistic at $r = 0$ of 46.1492 exceeds its critical value of 29.68, we reject the null hypothesis of no cointegrating equations. Similarly, because the trace statistic at $r = 1$ of 17.581 exceeds its critical value of 15.41, we reject the null hypothesis that there is one or fewer cointegrating equation. In contrast, because the trace statistic at $r = 2$ of 3.3465 is less than its critical value of 3.76, we cannot reject the null hypothesis that there are two or fewer cointegrating equations. Because Johansen's method for estimating r is to accept as $\hat{r}$ the first r for which the null hypothesis is not rejected, we accept $r = 2$ as our estimate of the number of cointegrating equations between these three variables. The "*" by the trace statistic at $r = 2$ indicates that this is the value of r selected by Johansen's multiple-trace test procedure. The eigenvalue shown in the last line of output computes the trace statistic in the preceding line.

◁

▷ Example 2

In the previous example, we used the default 5% critical values. We can estimate r with 1% critical values instead by specifying the level99 option.

```
. vecrank y i c, lags(5) level99
```

```
                    Johansen tests for cointegration
Trend: constant                              Number of obs =      91
Sample:   1960q2 - 1982q4                         Lags =          5
```

					1%
maximum				trace	critical
rank	parms	LL	eigenvalue	statistic	value
0	39	1231.1041	.	46.1492	35.65
1	44	1245.3882	0.26943	17.5810*	20.04
2	47	1252.5055	0.14480	3.3465	6.65
3	48	1254.1787	0.03611		

The output indicates that switching from the 5% to the 1% level changes the resulting estimate from $r = 2$ to $r = 1$.

◁

The maximum-eigenvalue statistic

The alternative hypothesis of the trace statistic is that the number of cointegrating equations is strictly larger than the number r assumed under the null hypothesis. Instead, we could assume a given r under the null hypothesis and test this against the alternative that there are $r + 1$ cointegrating equations. Johansen (1995, chap. 6, 11, and 12) derives an LR test of the null of r cointegrating relations against the alternative of $r + 1$ cointegrating relations. Because the part of the log likelihood that changes with r is a simple function of the eigenvalues of a $(K \times K)$ matrix, this test is known as the maximum-eigenvalue statistic. This method is used less often than the trace statistic method because no solution to the multiple-testing problem has yet been found.

▷ Example 3

In the output below, we reexamine the balanced-growth hypothesis. We use the `levela` option to obtain both the 5% and 1% critical values, and we use the `notrace` option to suppress the table of trace statistics.

```
. vecrank y i c, lags(5) max levela notrace
```

                    Johansen tests for cointegration
    Trend: constant   Number of obs =   91
    Sample: 1960q2 - 1982q4   Lags =   5

maximum rank	parms	LL	eigenvalue	max statistic	5% critical value	1% critical value
0	39	1231.1041		28.5682	20.97	25.52
1	44	1245.3882	0.26943	14.2346	14.07	18.63
2	47	1252.5055	0.14480	3.3465	3.76	6.65
3	48	1254.1787	0.03611			

We can reject $r = 1$ in favor of $r = 2$ at the 5% level but not at the 1% level. As with the trace statistic method, whether we choose to specify one or two cointegrating equations in our VECM will depend on the significance level we use here.

◁

Minimizing an information criterion

Many multiple-testing problems in the time-series literature have been solved by defining an estimator that minimizes an information criterion with known asymptotic properties. Selecting the lag length in an autoregressive model is probably the best-known example. Gonzalo and Pitarakis (1998) and Aznar and Salvador (2002) have shown that this approach can be applied to determining the number of cointegrating equations in a VECM. As in the lag-length selection problem, choosing the number of cointegrating equations that minimizes either the Schwarz Bayesian information criterion (SBIC) or the Hannan and Quinn information criterion (HQIC) provides a consistent estimator of the number of cointegrating equations.

▷ Example 4

We use these information-criteria methods to estimate the number of cointegrating equations in our balanced-growth data.

```
. vecrank y i c, lags(5) ic notrace
                     Johansen tests for cointegration
Trend: constant                                  Number of obs =       91
Sample:   1960q2 - 1982q4                               Lags =          5
```

maximum rank	parms	LL	eigenvalue	SBIC	HQIC	AIC
0	39	1231.1041		-25.12401	-25.76596	-26.20009
1	44	1245.3882	0.26943	-25.19009	-25.91435	-26.40414
2	47	1252.5055	0.14480	-25.19781*	-25.97144*	-26.49463
3	48	1254.1787	0.03611	-25.18501	-25.97511	-26.50942

Both the SBIC and the HQIC estimators suggest that there are two cointegrating equations in the balanced-growth data.

◁

Saved results

vecrank saves the following in e():

Scalars
e(N)	number of observations
e(k_eq)	number of equations in e(b)
e(k_dv)	number of dependent variables
e(tmin)	minimum time
e(tmax)	maximum time
e(n_lags)	number of lags
e(k_ce95)	number of cointegrating equations chosen by multiple trace tests with level(95)
e(k_ce99)	number of cointegrating equations chosen by multiple trace tests with level(99)
e(k_cesbic)	number of cointegrating equations chosen by minimizing SBIC
e(k_cehqic)	number of cointegrating equations chosen by minimizing HQIC

Macros
e(cmd)	vecrank
e(cmdline)	command as typed
e(trend)	trend specified
e(reduced_lags)	list of maximum lags to which the model has been reduced
e(reduce_opt)	noreduce, if noreduce is specified
e(tsfmt)	format for current time variable

Matrices
e(max)	vector of maximum-eigenvalue statistics
e(trace)	vector of trace statistics
e(lambda)	vector of eigenvalues
e(k_rank)	vector of numbers of unconstrained parameters
e(hqic)	vector of HQIC values
e(sbic)	vector of SBIC values
e(aic)	vector of AIC values
e(ll)	vector of log-likelihood values

Methods and formulas

vecrank is implemented as an ado-file.

As shown in *Methods and formulas* of [TS] **vec**, given a lag, trend, and seasonal specification when there are $0 \le r \le K$ cointegrating equations, the log likelihood with the Johansen identification restrictions can be written as

$$L = -\frac{1}{2}T\left[K\left\{\ln\left(2\pi\right) + 1\right\} + \ln\left(|S_{00}|\right) + \sum_{i=1}^{r}\ln\left(1 - \widehat{\lambda}_i\right)\right] \tag{1}$$

where the $(K \times K)$ matrix S_{00} and the eigenvalues $\widehat{\lambda}_i$ are defined in *Methods and formulas* of [TS] **vec**.

The trace statistic compares the null hypothesis that there are r or fewer cointegrating relations with the alternative hypothesis that there are more than r cointegrating equations. Under the alternative hypothesis, the log likelihood is

$$L_A = -\frac{1}{2}T\left[K\left\{\ln\left(2\pi\right) + 1\right\} + \ln\left(|S_{00}|\right) + \sum_{i=1}^{K}\ln\left(1 - \widehat{\lambda}_i\right)\right] \tag{2}$$

Thus the LR test that compares the unrestricted model in (2) with the restricted model in (1) is given by

$$LR_{\text{trace}} = -T\sum_{i=r+1}^{K}\ln\left(1 - \widehat{\lambda}_i\right)$$

As discussed by Johansen (1995), the trace statistic has a nonstandard distribution under the null hypothesis because the null hypothesis places restrictions on the coefficients on $\mathbf{y}_{t-1}$, which is assumed to have $K - r$ random-walk components. vecrank reports the Osterwald-Lenum (1992) critical values.

The maximum-eigenvalue statistic compares the null model containing r cointegrating relations with the alternative model that has $r + 1$ cointegrating relations. Thus using these two values for r in (1) and a few lines of algebra implies that the LR test of this hypothesis is

$$LR_{\max} = -T\ln\left(1 - \widehat{\lambda}_{r+1}\right)$$

As for the trace statistic, because this test involves restrictions on the coefficients on a vector of I(1) variables, the test statistic's distribution will be nonstandard. vecrank reports the Osterwald-Lenum (1992) critical values.

The formulas for the AIC, SBIC, and HQIC are given in *Methods and formulas* of [TS] **vec**.

Søren Johansen (1939–) earned degrees in mathematical statistics at the University of Copenhagen, where he is now based. In addition to making contributions to mathematical statistics, probability theory, and medical statistics, he has worked mostly in econometrics—in particular, on the theory of cointegration.

References

Anderson, T. W. 1951. Estimating linear restrictions on regression coefficients for multivariate normal distributions. *Annals of Mathematical Statistics* 22: 327–351.

Aznar, A., and M. Salvador. 2002. Selecting the rank of the cointegration space and the form of the intercept using an information criterion. *Econometric Theory* 18: 926–947.

Engle, R. F., and C. W. J. Granger. 1987. Co-integration and error correction: Representation, estimation, and testing. *Econometrica* 55: 251–276.

Gonzalo, J., and J.-Y. Pitarakis. 1998. Specification via model selection in vector error correction models. *Economics Letters* 60: 321–328.

Hamilton, J. D. 1994. *Time Series Analysis*. Princeton: Princeton University Press.

Hubrich, K., H. Lütkepohl, and P. Saikkonen. 2001. A review of systems cointegration tests. *Econometric Reviews* 20: 247–318.

Johansen, S. 1988. Statistical analysis of cointegration vectors. *Journal of Economic Dynamics and Control* 12: 231–254.

———. 1991. Estimation and hypothesis testing of cointegration vectors in Gaussian vector autoregressive models. *Econometrica* 59: 1551–1580.

———. 1995. *Likelihood-Based Inference in Cointegrated Vector Autoregressive Models*. Oxford: Oxford University Press.

King, R. G., C. I. Plosser, J. H. Stock, and M. W. Watson. 1991. Stochastic trends and economic fluctuations. *American Economic Review* 81: 819–840.

Lütkepohl, H. 2005. *New Introduction to Multiple Time Series Analysis*. New York: Springer.

Maddala, G. S., and I.-M. Kim. 1998. *Unit Roots, Cointegration, and Structural Change*. Cambridge: Cambridge University Press.

Osterwald-Lenum, M. 1992. A note with quantiles of the asymptotic distribution of the maximum likelihood cointegration rank test statistics. *Oxford Bulletin of Economics and Statistics* 54: 461–472.

Park, J. Y., and P. C. B. Phillips. 1988. Statistical inference in regressions with integrated processes: Part I. *Econometric Theory* 4: 468–497.

———. 1989. Statistical inference in regressions with integrated processes: Part II. *Econometric Theory* 5: 95–131.

Phillips, P. C. B. 1986. Understanding spurious regressions in econometrics. *Journal of Econometrics* 33: 311–340.

Phillips, P. C. B., and S. N. Durlauf. 1986. Multiple time series regressions with integrated processes. *Review of Economic Studies* 53: 473–495.

Sims, C. A., J. H. Stock, and M. W. Watson. 1990. Inference in linear time series models with some unit roots. *Econometrica* 58: 113–144.

Stock, J. H. 1987. Asymptotic properties of least squares estimators of cointegrating vectors. *Econometrica* 55: 1035–1056.

Stock, J. H., and M. W. Watson. 1988. Testing for common trends. *Journal of the American Statistical Association* 83: 1097–1107.

Watson, M. W. 1994. Vector autoregressions and cointegration. In Vol. IV of *Handbook of Econometrics*, ed. R. F. Engle and D. L. McFadden. Amsterdam: Elsevier.

Also see

[TS] **tsset** — Declare data to be time-series data

[TS] **vec** — Vector error-correction models

[TS] **vec intro** — Introduction to vector error-correction models

Title

vecstable — Check the stability condition of VECM estimates

Syntax

vecstable [, *options*]

options	Description
Main	
est̲imates(*estname*)	use previously saved results *estname*; default is to use active results
amat(*matrix_name*)	save the companion matrix as *matrix_name*
graph	graph eigenvalues of the companion matrix
dlabel	label eigenvalues with the distance from the unit circle
modlabel	label eigenvalues with the modulus
marker_options	change look of markers (color, size, etc.)
rl̲opts(*cline_options*)	affect rendition of reference unit circle
nogrid	suppress polar grid circles
pgrid([...])	specify radii and appearance of polar grid circles; see *Options* for details
Add plots	
addplot(*plot*)	add other plots to the generated graph
Y axis, X axis, Titles, Legend, Overall	
twoway_options	any options other than by() documented in [G-3] ***twoway_options***

vecstable can be used only after vec; see [TS] **vec**.

Menu

Statistics > Multivariate time series > VEC diagnostics and tests > Check stability condition of VEC estimates

Description

vecstable checks the eigenvalue stability condition in a vector error-correction model (VECM) fit using vec.

Options

<u>Main</u>

estimates(*estname*) requests that vecstable use the previously obtained set of vec estimates saved as *estname* By default, vecstable uses the active results. See [R] **estimates** for information about saving and restoring estimation results.

amat(*matrix_name*) specifies a valid Stata matrix name by which the companion matrix can be saved. The companion matrix is referred to as the **A** matrix in Lütkepohl (2005) and [TS] **varstable**. The default is not to save the companion matrix.

684

graph causes vecstable to draw a graph of the eigenvalues of the companion matrix.

dlabel labels the eigenvalues with their distances from the unit circle. dlabel cannot be specified with modlabel.

modlabel labels the eigenvalues with their moduli. modlabel cannot be specified with dlabel.

marker_options specify the look of markers. This look includes the marker symbol, the marker size, and its color and outline; see [G-3] ***marker_options***.

rlopts(*cline_options*) affects the rendition of the reference unit circle; see [G-3] ***cline_options***.

nogrid suppresses the polar grid circles.

pgrid([*numlist*] [, *line_options*]) [pgrid([*numlist*] [, *line_options*])] ...

   pgrid([*numlist*] [, *line_options*])] determines the radii and appearance of the polar grid circles. By default, the graph includes nine polar grid circles with radii .1, .2, ..., .9 that have the grid linestyle. The *numlist* specifies the radii for the polar grid circles. The *line_options* determine the appearance of the polar grid circles; see [G-3] ***line_options***. Because the pgrid() option can be repeated, circles with different radii can have distinct appearances.

⌐ Add plots ⌐

addplot(*plot*) adds specified plots to the generated graph; see [G-3] ***addplot_option***.

⌐ Y axis, X axis, Titles, Legend, Overall ⌐

twoway_options are any of the options documented in [G-3] ***twoway_options***, excluding by(). These include options for titling the graph (see [G-3] ***title_options***) and for saving the graph to disk (see [G-3] ***saving_option***).

Remarks

Inference after vec requires that the cointegrating equations be stationary and that the number of cointegrating equations be correctly specified. Although the methods implemented in vecrank identify the number of stationary cointegrating equations, they assume that the individual variables are I(1). vecstable provides indicators of whether the number of cointegrating equations is misspecified or whether the cointegrating equations, which are assumed to be stationary, are not stationary.

vecstable is analogous to varstable. vecstable uses the coefficient estimates from the previously fitted VECM to back out estimates of the coefficients of the corresponding VAR and then compute the eigenvalues of the companion matrix. See [TS] **varstable** for details about how the companion matrix is formed and about how to interpret the resulting eigenvalues for covariance-stationary VAR models.

If a VECM has K endogenous variables and r cointegrating vectors, there will be $K - r$ unit moduli in the companion matrix. If any of the remaining moduli computed by vecrank are too close to one, either the cointegrating equations are not stationary or there is another common trend and the rank() specified in the vec command is too high. Unfortunately, there is no general distribution theory that allows you to determine whether an estimated root is too close to one for all the cases that commonly arise in practice.

▷ Example 1

In example 1 of [TS] **vec**, we estimated the parameters of a bivariate VECM of the natural logs of the average disposable incomes in two of the economic regions created by the U.S. Bureau of Economic Analysis. In that example, we concluded that the predicted cointegrating equation was probably not stationary. Here we continue that example by refitting that model and using vecstable to analyze the eigenvalues of the companion matrix of the corresponding VAR.

```
. use http://www.stata-press.com/data/r12/rdinc
. vec ln_ne ln_se
  (output omitted)
. vecstable
```

Eigenvalue stability condition

Eigenvalue		Modulus
1		1
.9477854		.947785
.2545357 +	.2312756i	.343914
.2545357 −	.2312756i	.343914

The VECM specification imposes a unit modulus.

The output contains a table showing the eigenvalues of the companion matrix and their associated moduli. The table shows that one of the roots is 1. The table footer reminds us that the specified VECM imposes one unit modulus on the companion matrix.

The output indicates that there is a real root at about 0.95. Although there is no distribution theory to measure how close this root is to one, per other discussions in the literature (for example, Johansen [1995, 137–138]), we conclude that the root of 0.95 supports our earlier analysis, in which we concluded that the predicted cointegrating equation is probably not stationary.

If we had included the graph option with vecstable, the following graph would have been displayed:

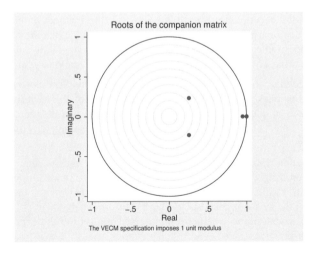

The graph plots the eigenvalues of the companion matrix with the real component on the x axis and the imaginary component on the y axis. Although the information is the same as in the table, the graph shows visually how close the root with modulus 0.95 is to the unit circle.

◁

Saved results

vecstable saves the following in r():

Scalars
    r(unitmod)   number of unit moduli imposed on the companion matrix

Matrices
    r(Re)   real part of the eigenvalues of A
    r(Im)   imaginary part of the eigenvalues of A
    r(Modulus)   moduli of the eigenvalues of A

where A is the companion matrix of the VAR that corresponds to the VECM.

Methods and formulas

vecstable is implemented as an ado-file.

vecstable uses the formulas given in *Methods and formulas* of [TS] **irf create** to obtain estimates of the parameters in the corresponding VAR from the vec estimates. With these estimates, the calculations are identical to those discussed in [TS] **varstable**. In particular, the derivation of the companion matrix, A, from the VAR point estimates is given in [TS] **varstable**.

References

Hamilton, J. D. 1994. *Time Series Analysis*. Princeton: Princeton University Press.

Johansen, S. 1995. *Likelihood-Based Inference in Cointegrated Vector Autoregressive Models*. Oxford: Oxford University Press.

Lütkepohl, H. 2005. *New Introduction to Multiple Time Series Analysis*. New York: Springer.

Also see

[TS] **vec** — Vector error-correction models

[TS] **vec intro** — Introduction to vector error-correction models

Title

> **wntestb** — Bartlett's periodogram-based test for white noise

Syntax

> wntestb *varname* [*if*] [*in*] [, *options*]

options	Description
Main	
<u>table</u>	display a table instead of graphical output
<u>level</u>(*#*)	set confidence level; default is level(95)
Plot	
marker_options	change look of markers (color, size, etc.)
marker_label_options	add marker labels; change look or position
cline_options	add connecting lines; change look
Add plots	
addplot(*plot*)	add other plots to the generated graph
Y axis, X axis, Titles, Legend, Overall	
twoway_options	any options other than by() documented in [G-3] ***twoway_options***

You must tsset your data before using wntestb; see [TS] **tsset**. In addition, the time series must be dense (nonmissing with no gaps in the time variable) in the specified sample.

varname may contain time-series operators; see [U] **11.4.4 Time-series varlists**.

Menu

Statistics > Time series > Tests > Bartlett's periodogram-based white-noise test

Description

wntestb performs Bartlett's periodogram-based test for white noise. The result is presented graphically by default but optionally may be presented as text in a table.

Options

Main

table displays the test results as a table instead of as the default graph.

level(*#*) specifies the confidence level, as a percentage, for the confidence bands included on the graph. The default is level(95) or as set by set level; see [U] **20.7 Specifying the width of confidence intervals**.

688

Plot

marker_options specify the look of markers. This look includes the marker symbol, the marker size, and its color and outline; see [G-3] ***marker_options***.

marker_label_options specify if and how the markers are to be labeled; see [G-3] ***marker_label_options***.

cline_options specify if the points are to be connected with lines and the rendition of those lines; see [G-3] ***cline_options***.

Add plots

addplot(*plot*) adds specified plots to the generated graph; see [G-3] ***addplot_option***.

Y axis, X axis, Titles, Legend, Overall

twoway_options are any of the options documented in [G-3] ***twoway_options***, excluding by(). These include options for titling the graph (see [G-3] ***title_options***) and for saving the graph to disk (see [G-3] ***saving_option***).

Remarks

Bartlett's test is a test of the null hypothesis that the data come from a white-noise process of uncorrelated random variables having a constant mean and a constant variance.

For a discussion of this test, see Bartlett (1955, 92–94), Newton (1988, 172), or Newton (1996).

▷ Example 1

In this example, we generate two time series and show the graphical and statistical tests that can be obtained from this command. The first time series is a white-noise process, and the second is a white-noise process with an embedded deterministic cosine curve.

```
. drop _all
. set seed 12393
. set obs 100
obs was 0, now 100
. generate x1 = rnormal()
. generate x2 = rnormal() + cos(2*_pi*(_n-1)/10)
. generate time = _n
. tsset time
        time variable:  time, 1 to 100
                delta:  1 unit
```

We can then submit the white-noise data to the wntestb command by typing

. wntestb x1

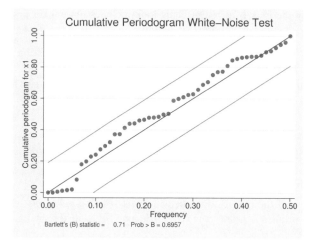

We can see in the graph that the values never appear outside the confidence bands. The test statistic has a p-value of 0.91, so we conclude that the process is not different from white noise. If we had wanted only the statistic without the plot, we could have used the `table` option.

Turning our attention to the other series (x2), we type

. wntestb x2

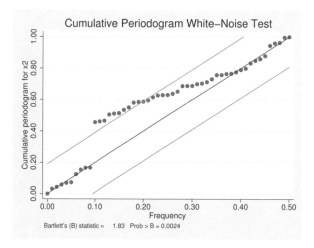

Here the process does appear outside of the bands. In fact, it steps out of the bands at a frequency of 0.1 (exactly as we synthesized this process). We also have confirmation from the test statistic, at a p-value of 0.001, that the process is significantly different from white noise.

◁

Saved results

wntestb saves the following in r():

Scalars
r(stat)	Bartlett's statistic	r(p)	probability value

Methods and formulas

wntestb is implemented as an ado-file.

If $x(1), \ldots, x(T)$ is a realization from a white-noise process with variance σ^2, the spectral distribution would be given by $F(\omega) = \sigma^2 \omega$ for $\omega \in [0, 1]$, and we would expect the cumulative periodogram (see [TS] **cumsp**) of the data to be close to the points $S_k = k/q$ for $q = \lfloor n/2 \rfloor + 1, k = 1, \ldots, q$. $\lfloor n/2 \rfloor$ is the greatest integer less than or equal to $n/2$.

Except for $\omega = 0$ and $\omega = .5$, the random variables $2\widehat{f}(\omega_k)/\sigma^2$ are asymptotically independently and identically distributed as χ_2^2. Because χ_2^2 is the same as twice a random variable distributed exponentially with mean 1, the cumulative periodogram has approximately the same distribution as the ordered values from a uniform (on the unit interval) distribution. Feller (1948) shows that this results in

$$\lim_{q \to \infty} \Pr\left(\max_{1 \le k \le q} \sqrt{q}\left|U_k - \frac{k}{q}\right| \le a\right) = \sum_{j=-\infty}^{\infty} (-1)^j e^{-2a^2 j^2} = G(a)$$

where U_k is the ordered uniform quantile. The Bartlett statistic is computed as

$$B = \max_{1 \le k \le q} \sqrt{\frac{n}{2}} \left|\widehat{F}_k - \frac{k}{q}\right|$$

where $\widehat{F}_k$ is the cumulative periodogram defined in terms of the sample spectral density $\widehat{f}$ (see [TS] **pergram**) as

$$\widehat{F}_k = \frac{\sum_{j=1}^{k} \widehat{f}(\omega_j)}{\sum_{j=1}^{q} \widehat{f}(\omega_j)}$$

The associated p-value for the Bartlett statistic and the confidence bands on the graph are computed as $1 - G(B)$ using Feller's result.

Maurice Stevenson Bartlett (1910–2002) was a British statistician. Apart from a short period in industry, he spent his career teaching and researching at the universities of Cambridge, Manchester, London (University College), and Oxford. His many contributions include work on the statistical analysis of multivariate data (especially factor analysis) and time series and on stochastic models of population growth, epidemics, and spatial processes.

Acknowledgment

wntestb is based on the wntestf command by H. Joseph Newton (1996), Department of Statistics, Texas A&M University.

References

Bartlett, M. S. 1955. *An Introduction to Stochastic Processes with Special Reference to Methods and Applications*. Cambridge: Cambridge University Press.

Feller, W. 1948. On the Kolmogorov–Smirnov limit theorems for empirical distributions. *Annals of Mathematical Statistics* 19: 177–189.

Gani, J. 2002. Professor M. S. Bartlett FRS, 1910–2002. *Statistician* 51: 399–402.

Newton, H. J. 1988. *TIMESLAB: A Time Series Analysis Laboratory*. Belmont, CA: Wadsworth.

———. 1996. sts12: A periodogram-based test for white noise. *Stata Technical Bulletin* 34: 36–39. Reprinted in *Stata Technical Bulletin Reprints*, vol. 6, pp. 203–207. College Station, TX: Stata Press.

Olkin, I. 1989. A conversation with Maurice Bartlett. *Statistical Science* 4: 151–163.

Also see

[TS] **tsset** — Declare data to be time-series data

[TS] **corrgram** — Tabulate and graph autocorrelations

[TS] **cumsp** — Cumulative spectral distribution

[TS] **pergram** — Periodogram

[TS] **wntestq** — Portmanteau (Q) test for white noise

Title

wntestq — Portmanteau (Q) test for white noise

Syntax

wntestq *varname* $\left[\textit{if}\right]$ $\left[\textit{in}\right]$ $\left[\,\text{, }\underline{\text{lags}}(\#)\,\right]$

You must tsset your data before using wntestq; see [TS] **tsset**. Also the time series must be dense (nonmissing with no gaps in the time variable) in the specified sample.

varname may contain time-series operators; see [U] **11.4.4 Time-series varlists**.

Menu

Statistics > Time series > Tests > Portmanteau white-noise test

Description

wntestq performs the portmanteau (or Q) test for white noise.

Option

lags(#) specifies the number of autocorrelations to calculate. The default is to use $\min(\lfloor n/2\rfloor - 2, 40)$, where $\lfloor n/2\rfloor$ is the greatest integer less than or equal to $n/2$.

Remarks

Box and Pierce (1970) developed a portmanteau test of white noise that was refined by Ljung and Box (1978). See also Diggle (1990, sec. 2.5).

▷ Example 1

In the example shown in [TS] **wntestb**, we generated two time series. One (x1) was a white-noise process, and the other (x2) was a white-noise process with an embedded cosine curve. Here we compare the output of the two tests.

```
. drop _all
. set seed 12393
. set obs 100
obs was 0, now 100
. generate x1 = rnormal()
. generate x2 = rnormal() + cos(2*_pi*(_n-1)/10)
. generate time = _n
. tsset time
        time variable:  time, 1 to 100
                delta:  1 unit
```

693

```
. wntestb x1, table
```

Cumulative periodogram white-noise test

Bartlett's (B) statistic	=	0.7093
Prob > B	=	0.6957

```
. wntestq x1
```

Portmanteau test for white noise

Portmanteau (Q) statistic =		32.6863
Prob > chi2(40)	=	0.7875

```
. wntestb x2, table
```

Cumulative periodogram white-noise test

Bartlett's (B) statistic	=	1.8323
Prob > B	=	0.0024

```
. wntestq x2
```

Portmanteau test for white noise

Portmanteau (Q) statistic =		129.4436
Prob > chi2(40)	=	0.0000

This example shows that both tests agree. For the first process, the Bartlett and portmanteau tests result in nonsignificant test statistics: a p-value of 0.9053 for wntestb and one of 0.9407 for wntestq.

For the second process, each test has a significant result to 0.0010.

◁

Saved results

wntestq saves the following in r():

Scalars

r(stat)	Q statistic	r(p)	probability value
r(df)	degrees of freedom		

Methods and formulas

wntestq is implemented as an ado-file.

The portmanteau test relies on the fact that if $x(1), \ldots, x(n)$ is a realization from a white-noise process. Then

$$Q = n(n+2) \sum_{j=1}^{m} \frac{1}{n-j} \, \widehat{\rho}^2(j) \longrightarrow \chi_m^2$$

where m is the number of autocorrelations calculated (equal to the number of lags specified) and $\longrightarrow$ indicates convergence in distribution to a χ^2 distribution with m degrees of freedom. $\widehat{\rho}_j$ is the estimated autocorrelation for lag j; see [TS] **corrgram** for details.

References

Box, G. E. P., and D. A. Pierce. 1970. Distribution of residual autocorrelations in autoregressive-integrated moving average time series models. *Journal of the American Statistical Association* 65: 1509–1526.

Diggle, P. J. 1990. *Time Series: A Biostatistical Introduction.* Oxford: Oxford University Press.

Ljung, G. M., and G. E. P. Box. 1978. On a measure of lack of fit in time series models. *Biometrika* 65: 297–303.

Sperling, R., and C. F. Baum. 2001. sts19: Multivariate portmanteau (Q) test for white noise. *Stata Technical Bulletin* 60: 39–41. Reprinted in *Stata Technical Bulletin Reprints*, vol. 10, pp. 373–375. College Station, TX: Stata Press.

Also see

[TS] **tsset** — Declare data to be time-series data

[TS] **corrgram** — Tabulate and graph autocorrelations

[TS] **cumsp** — Cumulative spectral distribution

[TS] **pergram** — Periodogram

[TS] **wntestb** — Bartlett's periodogram-based test for white noise

Title

> **xcorr** — Cross-correlogram for bivariate time series

Syntax

> xcorr *varname*$_1$ *varname*$_2$ $\left[\,if\,\right]$ $\left[\,in\,\right]$ $\left[\,,\ options\,\right]$

options	Description
Main	
<u>gen</u>erate(*newvar*)	create *newvar* containing cross-correlation values
<u>tab</u>le	display a table instead of graphical output
noplot	do not include the character-based plot in tabular output
<u>lag</u>s(*#*)	include *#* lags and leads in graph
Plot	
base(*#*)	value to drop to; default is 0
marker_options	change look of markers (color, size, etc.)
marker_label_options	add marker labels; change look or position
line_options	change look of dropped lines
Add plots	
addplot(*plot*)	add other plots to the generated graph
Y axis, X axis, Titles, Legend, Overall	
twoway_options	any options other than by() documented in [G-3] **twoway_options**

You must tsset your data before using xcorr; see [TS] **tsset**.

varname$_1$ and *varname*$_2$ may contain time-series operators; see [U] **11.4.4 Time-series varlists**.

Menu

Statistics > Time series > Graphs > Cross-correlogram for bivariate time series

Description

xcorr plots the sample cross-correlation function.

Options

> ⌐ Main ⌐

generate(*newvar*) specifies a new variable to contain the cross-correlation values.

table requests that the results be presented as a table rather than the default graph.

noplot requests that the table not include the character-based plot of the cross-correlations.

lags(*#*) indicates the number of lags and leads to include in the graph. The default is to use $\min(\lfloor n/2 \rfloor - 2, 20)$.

Plot

base(*#*) specifies the value from which the lines should extend. The default is base(0).

marker_options, *marker_label_options*, and *line_options* affect the rendition of the plotted cross-correlations.

   marker_options specify the look of markers. This look includes the marker symbol, the marker size, and its color and outline; see [G-3] ***marker_options***.

   marker_label_options specify if and how the markers are to be labeled; see [G-3] ***marker_label_options***.

   line_options specify the look of the dropped lines, including pattern, width, and color; see [G-3] ***line_options***.

Add plots

addplot(*plot*) provides a way to add other plots to the generated graph. See [G-3] ***addplot_option***.

Y axis, X axis, Titles, Legend, Overall

twoway_options are any of the options documented in [G-3] ***twoway_options***, excluding by(). These include options for titling the graph (see [G-3] ***title_options***) and for saving the graph to disk (see [G-3] ***saving_option***).

Remarks

▷ Example 1

We have a bivariate time series (Box, Jenkins, and Reinsel 2008, Series J) on the input and output of a gas furnace, where 296 paired observations on the input (gas rate) and output (% CO_2) were recorded every 9 seconds. The cross-correlation function is given by

```
. use http://www.stata-press.com/data/r12/furnace
(TIMESLAB: Gas furnace)

. xcorr input output, xline(5) lags(40)
```

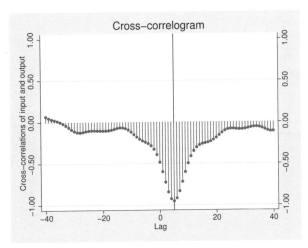

We included a vertical line at lag 5, because there is a well-defined peak at this value. This peak indicates that the output lags the input by five periods. Further, the fact that the correlations are negative indicates that as input (coded gas rate) is increased, output (% CO_2) decreases.

We may obtain the table of autocorrelations and the character-based plot of the cross-correlations (analogous to the univariate time-series command corrgram) by specifying the table option.

```
. xcorr input output, table
                        -1        0        1
     LAG       CORR    [Cross-correlation]

     -20     -0.1033               |
     -19     -0.1027               |
     -18     -0.0998               |
     -17     -0.0932               |
     -16     -0.0832               |
     -15     -0.0727               |
     -14     -0.0660               |
     -13     -0.0662               |
     -12     -0.0751               |
     -11     -0.0927               |
     -10     -0.1180               |
      -9     -0.1484              -|
      -8     -0.1793              -|
      -7     -0.2059              -|
      -6     -0.2266              -|
      -5     -0.2429              -|
      -4     -0.2604             --|
      -3     -0.2865             --|
      -2     -0.3287             --|
      -1     -0.3936            ---|
       0     -0.4845            ---|
       1     -0.5985           ----|
       2     -0.7251          -----|
       3     -0.8429         ------|
       4     -0.9246        -------|
       5     -0.9503        -------|
       6     -0.9146        -------|
       7     -0.8294         ------|
       8     -0.7166          -----|
       9     -0.5998           ----|
      10     -0.4952            ---|
      11     -0.4107            ---|
      12     -0.3479             --|
      13     -0.3049             --|
      14     -0.2779             --|
      15     -0.2632             --|
      16     -0.2548             --|
      17     -0.2463              -|
      18     -0.2332              -|
      19     -0.2135              -|
      20     -0.1869              -|
```

Once again, the well-defined peak is apparent in the plot.

◁

Methods and formulas

`xcorr` is implemented as an ado-file.

The cross-covariance function of lag k for time series x_1 and x_2 is given by

$$\mathrm{Cov}\Big\{x_1(t), x_2(t+k)\Big\} = R_{12}(k)$$

This function is not symmetric about lag zero; that is,

$$R_{12}(k) \neq R_{12}(-k)$$

We define the cross-correlation function as

$$\rho_{ij}(k) = \mathrm{Corr}\Big\{x_i(t), x_j(t+k)\Big\} = \frac{R_{ij}(k)}{\sqrt{R_{ii}(0)R_{jj}(0)}}$$

where ρ_{11} and ρ_{22} are the autocorrelation functions for x_1 and x_2, respectively. The sequence $\rho_{12}(k)$ is the cross-correlation function and is drawn for lags $k \in (-Q, -Q+1, \ldots, -1, 0, 1, \ldots, Q-1, Q)$.

If $\rho_{12}(k) = 0$ for all lags, x_1 and x_2 are not cross-correlated.

References

Box, G. E. P., G. M. Jenkins, and G. C. Reinsel. 2008. *Time Series Analysis: Forecasting and Control*. 4th ed. Hoboken, NJ: Wiley.

Hamilton, J. D. 1994. *Time Series Analysis*. Princeton: Princeton University Press.

Newton, H. J. 1988. *TIMESLAB: A Time Series Analysis Laboratory*. Belmont, CA: Wadsworth.

Also see

Glossary

ARCH model. An autoregressive conditional heteroskedasticity (ARCH) model is a regression model in which the conditional variance is modeled as an autoregressive (AR) process. The ARCH(m) model is

$$y_t = \mathbf{x}_t \boldsymbol{\beta} + \epsilon_t$$
$$E(\epsilon_t^2 | \epsilon_{t-1}^2, \epsilon_{t-2}^2, \ldots) = \alpha_0 + \alpha_1 \epsilon_{t-1}^2 + \cdots + \alpha_m \epsilon_{t-m}^2$$

where ϵ_t is a white-noise error term. The equation for y_t represents the conditional mean of the process, and the equation for $E(\epsilon_t^2 | \epsilon_{t-1}^2, \epsilon_{t-2}^2, \ldots)$ specifies the conditional variance as an autoregressive function of its past realizations. Although the conditional variance changes over time, the unconditional variance is time invariant because y_t is a stationary process. Modeling the conditional variance as an AR process raises the implied unconditional variance, making this model particularly appealing to researchers modeling fat-tailed data, such as financial data.

ARFIMA model. An autoregressive fractionally integrated moving-average (ARFIMA) model is a time-series model suitable for use with long-memory processes. ARFIMA models generalize autoregressive integrated moving-average (ARIMA) models by allowing the differencing parameter to be a real number in $(-0.5, 0.5)$ instead of requiring it to be an integer.

ARIMA model. An autoregressive integrated moving-average (ARIMA) model is a time-series model suitable for use with integrated processes. In an ARIMA(p, d, q) model, the data is differenced d times to obtain a stationary series, and then an ARMA(p, q) model is fit to this differenced data. ARIMA models that include exogenous explanatory variables are known as ARMAX models.

ARMA model. An autoregressive moving-average (ARMA) model is a time-series model in which the current period's realization is the sum of an autoregressive (AR) process and a moving-average (MA) process. An ARMA(p, q) model includes p AR terms and q MA terms. ARMA models with just a few lags are often able to fit data, as well as pure AR or MA models with many more lags.

ARMAX model. An ARMAX model is a time-series model in which the current period's realization is an ARMA process plus a linear function of a set of a exogenous variables. Equivalently, an ARMAX model is a linear regression model in which the error term is specified to follow an ARMA process.

autocorrelation function. The autocorrelation function (ACF) expresses the correlation between periods t and $t - k$ of a time series as function of the time t and the lag k. For a stationary time series, the ACF does not depend on t and is symmetric about $k = 0$, meaning that the correlation between periods t and $t - k$ is equal to the correlation between periods t and $t + k$.

autoregressive process. An autoregressive process is a time-series model in which the current value of a variable is a linear function of its own past values and a white-noise error term. A first-order autoregressive process, denoted as an AR(1) process, is $y_t = \rho y_{t-1} + \epsilon_t$. An AR$(p)$ model contains p lagged values of the dependent variable.

band-pass filter. Time-series filters are designed to pass or block stochastic cycles at specified frequencies. Band-pass filters, such as those implemented in `tsfilter bk` and `tsfilter cf`, pass through stochastic cycles in the specified range of frequencies and block all other stochastic cycles.

Cholesky ordering. Cholesky ordering is a method used to orthogonalize the error term in a VAR or VECM to impose a recursive structure on the dynamic model, so that the resulting impulse–response functions can be given a causal interpretation. The method is so named because it uses the Cholesky decomposition of the error covariance matrix.

Cochrane–Orcutt estimator. This estimation is a linear regression estimator that can be used when the error term exhibits first-order autocorrelation. An initial estimate of the autocorrelation parameter ρ is obtained from OLS residuals, and then OLS is performed on the transformed data $\widetilde{y}_t = y_t - \rho y_{t-1}$ and $\widetilde{\mathbf{x}}_t = \mathbf{x}_t - \rho \mathbf{x}_{t-1}$.

cointegrating vector. A cointegrating vector specifies a stationary linear combination of nonstationary variables. Specifically, if each of the variables $x_1, x_2, \ldots, x_k$ is integrated of order one and there exists a set of parameters $\beta_1, \beta_2, \ldots, \beta_k$ such that $z_t = \beta_1 x_1 + \beta_2 x_2 + \cdots + \beta_k x_k$ is a stationary process, the variables $x_1, x_2, \ldots, x_k$ are said to be cointegrated, and the vector β is known as a cointegrating vector.

conditional variance. Although the conditional variance is simply the variance of a conditional distribution, in time-series analysis the conditional variance is often modeled as an autoregressive process, giving rise to ARCH models.

correlogram. A correlogram is a table or graph showing the sample autocorrelations or partial autocorrelations of a time series.

covariance stationarity. A process is covariance stationary if the mean of the process is finite and independent of t, the unconditional variance of the process is finite and independent of t, and the covariance between periods t and $t - s$ is finite and depends on $t - s$ but not on t or s themselves. Covariance-stationary processes are also known as weakly stationary processes.

cross-correlation function. The cross-correlation function expresses the correlation between one series at time t and another series at time $t - k$ as a function of the time t and lag k. If both series are stationary, the function does not depend on t. The function is not symmetric about $k = 0$: $\rho_{12}(k) \neq \rho_{12}(-k)$.

cyclical component. A cyclical component is a part of a time series that is a periodic function of time. Deterministic functions of time are deterministic cyclical components, and random functions of time are stochastic cyclical components. For example, fixed seasonal effects are deterministic cyclical components and random seasonal effects are stochastic seasonal components.

Random coefficients on time inside of periodic functions form an especially useful class of stochastic cyclical components; see [TS] **ucm**.

deterministic trend. A deterministic trend is a deterministic function of time that specifies the long-run tendency of a time series.

difference operator. The difference operator Δ denotes the change in the value of a variable from period $t - 1$ to period t. Formally, $\Delta y_t = y_t - y_{t-1}$, and $\Delta^2 y_t = \Delta(y_t - y_{t-1}) = (y_t - y_{t-1}) - (y_{t-1} - y_{t-2}) = y_t - 2y_{t-1} + y_{t-2}$.

drift. Drift is the constant term in a unit-root process. In

$$y_t = \alpha + y_{t-1} + \epsilon_t$$

α is the drift when ϵ_t is a stationary, zero-mean process.

dynamic forecast. A dynamic forecast is one in which the current period's forecast is calculated using forecasted values for prior periods.

dynamic-multiplier function. A dynamic-multiplier function measures the effect of a shock to an exogenous variable on an endogenous variable. The kth dynamic-multiplier function of variable i on variable j measures the effect on variable j in period $t + k$ in response to a one-unit shock to variable i in period t, holding everything else constant.

endogenous variable. An endogenous variable is a regressor that is correlated with the unobservable error term. Equivalently, an endogenous variable is one whose values are determined by the equilibrium or outcome of a structural model.

exogenous variable. An exogenous variable is one that is correlated with none of the unobservable error terms in the model. Equivalently, an exogenous variable is one whose values change independently of the other variables in a structural model.

exponential smoothing. Exponential smoothing is a method of smoothing a time series in which the smoothed value at period t is equal to a fraction α of the series value at time t plus a fraction $1 - \alpha$ of the previous period's smoothed value. The fraction α is known as the smoothing parameter.

forecast-error variance decomposition. Forecast-error variance decompositions measure the fraction of the error in forecasting variable i after h periods that is attributable to the orthogonalized shocks to variable j.

forward operator. The forward operator F denotes the value of a variable at time $t + 1$. Formally, $Fy_t = y_{t+1}$, and $F^2 y_t = Fy_{t+1} = y_{t+2}$.

frequency-domain analysis. Frequency-domain analysis is analysis of time-series data by considering its frequency properties. The spectral density and distribution functions are key components of frequency-domain analysis, so it is often called spectral analysis. In Stata, the `cumsp` and `pergram` commands are used to analyze the sample spectral distribution and density functions, respectively. `psdensity` estimates the spectral density or the spectral distribution function after estimating the parameters of a parametric model using `arfima`, `arima`, or `ucm`.

gain (of a linear filter). The gain of a linear filter scales the spectral density of the unfiltered series into the spectral density of the filtered series for each frequency. Specifically, at each frequency, multiplying the spectral density of the unfiltered series by the square of the gain of a linear filter yields the spectral density of the filtered series. If the gain at a particular frequency is 1, the filtered and unfiltered spectral densities are the same at that frequency and the corresponding stochastic cycles are passed through perfectly. If the gain at a particular frequency is 0, the filter removes all the corresponding stochastic cycles from the unfiltered series.

GARCH model. A generalized autoregressive conditional heteroskedasticity (GARCH) model is a regression model in which the conditional variance is modeled as an ARMA process. The GARCH(m, k) model is

$$y_t = \mathbf{x}_t \boldsymbol{\beta} + \epsilon_t$$
$$\sigma_t^2 = \gamma_0 + \gamma_1 \epsilon_{t-1}^2 + \cdots + \gamma_m \epsilon_{t-m}^2 + \delta_1 \sigma_{t-1}^2 + \cdots + \delta_k \sigma_{t-k}^2$$

where the equation for y_t represents the conditional mean of the process and σ_t represents the conditional variance. See [TS] **arch** or Hamilton (1994, chap. 21) for details on how the conditional variance equation can be viewed as an ARMA process. GARCH models are often used because the ARMA specification often allows the conditional variance to be modeled with fewer parameters than are required by a pure ARCH model. Many extensions to the basic GARCH model exist; see [TS] **arch** for those that are implemented in Stata. See also *ARCH model*.

generalized least-squares estimator. A generalized least-squares (GLS) estimator is used to estimate the parameters of a regression function when the error term is heteroskedastic or autocorrelated. In the linear case, GLS is sometimes described as "OLS on transformed data" because the GLS estimator can be implemented by applying an appropriate transformation to the dataset and then using OLS.

Granger causality. The variable x is said to Granger-cause variable y if, given the past values of y, past values of x are useful for predicting y.

high-pass filter. Time-series filters are designed to pass or block stochastic cycles at specified frequencies. High-pass filters, such as those implemented in `tsfilter bw` and `tsfilter hp`, pass through stochastic cycles above the cutoff frequency and block all other stochastic cycles.

Holt–Winters smoothing. A set of methods for smoothing time-series data that assume that the value of a time series at time t can be approximated as the sum of a mean term that drifts over time, as well as a time trend whose strength also drifts over time. Variations of the basic method allow for seasonal patterns in data, as well.

impulse–response function. An impulse–response function (IRF) measures the effect of a shock to an endogenous variable on itself or another endogenous variable. The kth impulse–response function of variable i on variable j measures the effect on variable j in period $t + k$ in response to a one-unit shock to variable i in period t, holding everything else constant.

independent and identically distributed. A series of observations is independently and identically distributed (i.i.d.) if each observation is an independent realization from the same underlying distribution. In some contexts, the definition is relaxed to mean only that the observations are independent and have identical means and variances; see Davidson and MacKinnon (1993, 42).

integrated process. A nonstationary process is integrated of order d, written I(d), if the process must be differenced d times to produce a stationary series. An I(1) process y_t is one in which Δy_t is stationary.

Kalman filter. The Kalman filter is a recursive procedure for predicting the state vector in a state-space model.

lag operator. The lag operator L denotes the value of a variable at time $t - 1$. Formally, $Ly_t = y_{t-1}$, and $L^2 y_t = Ly_{t-1} = y_{t-2}$.

linear filter. A linear filter is a sequence of weights used to compute a weighted average of a time series at each time period. More formally, a linear filter $\alpha(L)$ is

$$\alpha(L) = \alpha_0 + \alpha_1 L + \alpha_2 L^2 + \cdots = \sum_{\tau=0}^{\infty} \alpha_\tau L^\tau$$

where L is the lag operator. Applying the linear filter $\alpha(L)$ to the time series x_t yields a sequence of weighted averages of x_t:

$$\alpha(L)x_t = \sum_{\tau=0}^{\infty} \alpha_\tau L^\tau x_{t-\tau}$$

long-memory process. A long-memory process is a stationary process whose autocorrelations decay at a slower rate than a short-memory process. ARFIMA models are typically used to represent long-memory processes, and ARMA models are typically used to represent short-memory processes.

moving-average process. A moving-average process is a time-series process in which the current value of a variable is modeled as a weighted average of current and past realizations of a white-noise process and, optionally, a time-invariant constant. By convention, the weight on the current realization of the white-noise process is equal to one, and the weights on the past realizations are known as the moving-average (MA) coefficients. A first-order moving-average process, denoted as an MA(1) process, is $y_t = \theta \epsilon_{t-1} + \epsilon_t$.

multivariate GARCH models. Multivariate GARCH models are multivariate time-series models in which the conditional covariance matrix of the errors depends on its own past and its past shocks. The acute trade-off between parsimony and flexibility has given rise to a plethora of models; see [TS] **mgarch**.

Newey–West covariance matrix. The Newey–West covariance matrix is a member of the class of heteroskedasticity- and autocorrelation-consistent (HAC) covariance matrix estimators used with time-series data that produces covariance estimates that are robust to both arbitrary heteroskedasticity and autocorrelation up to a prespecified lag.

orthogonalized impulse–response function. An orthogonalized impulse–response function (OIRF) measures the effect of an orthogonalized shock to an endogenous variable on itself or another endogenous variable. An orthogonalized shock is one that affects one variable at time t but no other variables. See [TS] **irf create** for a discussion of the difference between IRFs and OIRFs.

partial autocorrelation function. The partial autocorrelation function (PACF) expresses the correlation between periods t and $t-k$ of a time series as a function of the time t and lag k, after controlling for the effects of intervening lags. For a stationary time series, the PACF does not depend on t. The PACF is not symmetric about $k=0$: the partial autocorrelation between y_t and y_{t-k} is not equal to the partial autocorrelation between y_t and y_{t+k}.

periodogram. A periodogram is a graph of the spectral density function of a time series as a function of frequency. The `pergram` command first standardizes the amplitude of the density by the sample variance of the time series, and then plots the logarithm of that standardized density. Peaks in the periodogram represent cyclical behavior in the data.

phase function. The phase function of a linear filter specifies how the filter changes the relative importance of the random components at different frequencies in the frequency domain.

portmanteau statistic. The portmanteau, or Q, statistic is used to test for white noise and is calculated using the first m autocorrelations of the series, where m is chosen by the user. Under the null hypothesis that the series is a white-noise process, the portmanteau statistic has a χ^2 distribution with m degrees of freedom.

Prais–Winsten estimator. A Prais–Winsten estimator is a linear regression estimator that is used when the error term exhibits first-order autocorrelation; see also *Cochrane–Orcutt estimator*. Here the first observation in the dataset is transformed as $\widetilde{y}_1 = \sqrt{1-\rho^2}\, y_1$ and $\widetilde{\mathbf{x}}_1 = \sqrt{1-\rho^2}\, \mathbf{x}_1$, so that the first observation is not lost. The Prais–Winsten estimator is a generalized least-squares estimator.

priming values. Priming values are the initial, preestimation values used to begin a recursive process.

random walk. A random walk is a time-series process in which the current period's realization is equal to the previous period's realization plus a white-noise error term: $y_t = y_{t-1} + \epsilon_t$. A *random walk with drift* also contains a nonzero time-invariant constant: $y_t = \delta + y_{t-1} + \epsilon_t$. The constant term δ is known as the drift parameter. An important property of random-walk processes is that the best predictor of the value at time $t+1$ is the value at time t plus the value of the drift parameter.

recursive regression analysis. A recursive regression analysis involves performing a regression at time t by using all available observations from some starting time t_0 through time t, performing another regression at time $t+1$ by using all observations from time t_0 through time $t+1$, and so on. Unlike a rolling regression analysis, the first period used for all regressions is held fixed.

rolling regression analysis. A rolling, or moving window, regression analysis involves performing regressions for each period by using the most recent m periods' data, where m is known as the window size. At time t the regression is fit using observations for times $t-19$ through time t; at time $t+1$ the regression is fit using the observations for time $t-18$ through $t+1$; and so on.

seasonal difference operator. The period-s seasonal difference operator Δ_s denotes the difference in the value of a variable at time t and time $t-s$. Formally, $\Delta_s y_t = y_t - y_{t-s}$, and $\Delta_s^2 y_t = \Delta_s(y_t - y_{t-s}) = (y_t - y_{t-s}) - (y_{t-s} - y_{t-2s}) = y_t - 2y_{t-s} + y_{t-2s}$.

serial correlation. Serial correlation refers to regression errors that are correlated over time. If a regression model does not contained lagged dependent variables as regressors, the OLS estimates are consistent in the presence of mild serial correlation, but the covariance matrix is incorrect. When the model includes lagged dependent variables and the residuals are serially correlated, the

OLS estimates are biased and inconsistent. See, for example, Davidson and MacKinnon (1993, chap. 10) for more information.

serial correlation tests. Because OLS estimates are at least inefficient and potentially biased in the presence of serial correlation, econometricians have developed many tests to detect it. Popular ones include the Durbin–Watson (1950, 1951, 1971) test, the Breusch–Pagan (1980) test, and Durbin's (1970) alternative test. See [R] **regress postestimation time series**.

smoothing. Smoothing a time series refers to the process of extracting an overall trend in the data. The motivation behind smoothing is the belief that a time series exhibits a trend component as well as an irregular component and that the analyst is interested only in the trend component. Some smoothers also account for seasonal or other cyclical patterns.

spectral analysis. See *frequency-domain analysis*.

spectral density function. The spectral density function is the derivative of the spectral distribution function. Intuitively, the spectral density function $f(\omega)$ indicates the amount of variance in a time series that is attributable to sinusoidal components with frequency ω. See also *spectral distribution function*. The spectral density function is sometimes called the *spectrum*.

spectral distribution function. The (normalized) spectral distribution function $F(\omega)$ of a process describes the proportion of variance that can be explained by sinusoids with frequencies in the range $(0, \omega)$, where $0 \leq \omega \leq \pi$. The spectral distribution and density functions used in frequency-domain analysis are closely related to the autocorrelation function used in time-domain analysis; see Chatfield (2004, chap. 6) and Wei (2006, chap. 12).

spectrum. See *spectral density function*.

state-space model. A state-space model describes the relationship between an observed time series and an unobservable state vector that represents the "state" of the world. The measurement equation expresses the observed series as a function of the state vector, and the transition equation describes how the unobserved state vector evolves over time. By defining the parameters of the measurement and transition equations appropriately, one can write a wide variety of time-series models in the state-space form.

steady-state equilibrium. The steady-state equilibrium is the predicted value of a variable in a dynamic model, ignoring the effects of past shocks, or, equivalently, the value of a variable, assuming that the effects of past shocks have fully died out and no longer affect the variable of interest.

stochastic trend. A stochastic trend is a nonstationary random process. Unit-root process and random coefficients on time are two common stochastic trends. See [TS] **ucm** for examples and discussions of more commonly applied stochastic trends.

strict stationarity. A process is strictly stationary if the joint distribution of $y_1, \ldots, y_k$ is the same as the joint distribution of $y_{1+\tau}, \ldots, y_{k+\tau}$ for all k and τ. Intuitively, shifting the origin of the series by τ units has no effect on the joint distributions.

structural model. In time-series analysis, a structural model is one that describes the relationship among a set of variables, based on underlying theoretical considerations. Structural models may contain both endogenous and exogenous variables.

SVAR. A structural vector autoregressive (SVAR) model is a type of VAR in which short- or long-run constraints are placed on the resulting impulse–response functions. The constraints are usually motivated by economic theory and therefore allow causal interpretations of the IRFs to be made.

time-domain analysis. Time-domain analysis is analysis of data viewed as a sequence of observations observed over time. The autocorrelation function, linear regression, ARCH models, and ARIMA models are common tools used in time-domain analysis.

trend. The trend specifies the long-run behavior in a time series. The trend can be deterministic or stochastic. Many economic, biological, health, and social time series have long-run tendencies to increase or decrease. Before the 1980s, most time-series analysis specified the long-run tendencies as deterministic functions of time. Since the 1980s, the stochastic trends implied by unit-root processes have become a standard part of the toolkit.

unit-root process. A unit-root process is one that is integrated of order one, meaning that the process is nonstationary but that first-differencing the process produces a stationary series. The simplest example of a unit-root process is the random walk. See Hamilton (1994, chap. 15) for a discussion of when general ARMA processes may contain a unit root.

unit-root tests. Whether a process has a unit root has both important statistical and economic ramifications, so a variety of tests have been developed to test for them. Among the earliest tests proposed is the one by Dickey and Fuller (1979), though most researchers now use an improved variant called the augmented Dickey–Fuller test instead of the original version. Other common unit-root tests implemented in Stata include the DF–GLS test of Elliott, Rothenberg, and Stock (1996) and the Phillips–Perron (1988) test. See [TS] **dfuller**, [TS] **dfgls**, and [TS] **pperron**.

Variants of unit-root tests suitable for panel data have also been developed; see [XT] **xtunitroot**.

VAR. A vector autoregressive (VAR) model is a multivariate regression technique in which each dependent variable is regressed on lags of itself and on lags of all the other dependent variables in the model. Occasionally, exogenous variables are also included in the model.

VECM. A vector error-correction model (VECM) is a type of VAR that is used with variables that are cointegrated. Although first-differencing variables that are integrated of order one makes them stationary, fitting a VAR to such first-differenced variables results in misspecification error if the variables are cointegrated. See *The multivariate VECM specification* in [TS] **vec intro** for more on this point.

white noise. A variable u_t represents a white-noise process if the mean of u_t is zero, the variance of u_t is σ^2, and the covariance between u_t and u_s is zero for all $s \neq t$. Gaussian white noise refers to white noise in which u_t is normally distributed.

Yule–Walker equations. The Yule–Walker equations are a set of difference equations that describe the relationship among the autocovariances and autocorrelations of an autoregressive moving-average (ARMA) process.

References

Breusch, T. S., and A. R. Pagan. 1980. The Lagrange multiplier test and its applications to model specification in econometrics. *Review of Economic Studies* 47: 239–253.

Chatfield, C. 2004. *The Analysis of Time Series: An Introduction.* 6th ed. Boca Raton, FL: Chapman & Hall/CRC.

Davidson, R., and J. G. MacKinnon. 1993. *Estimation and Inference in Econometrics.* New York: Oxford University Press.

Dickey, D. A., and W. A. Fuller. 1979. Distribution of the estimators for autoregressive time series with a unit root. *Journal of the American Statistical Association* 74: 427–431.

Durbin, J. 1970. Testing for serial correlation in least-squares regressions when some of the regressors are lagged dependent variables. *Econometrica* 38: 410–421.

Durbin, J., and G. S. Watson. 1950. Testing for serial correlation in least squares regression. I. *Biometrika* 37: 409–428.

———. 1951. Testing for serial correlation in least squares regression. II. *Biometrika* 38: 159–177.

———. 1971. Testing for serial correlation in least squares regression. III. *Biometrika* 58: 1–19.

Elliott, G., T. J. Rothenberg, and J. H. Stock. 1996. Efficient tests for an autoregressive unit root. *Econometrica* 64: 813–836.

Hamilton, J. D. 1994. *Time Series Analysis.* Princeton: Princeton University Press.

Phillips, P. C. B., and P. Perron. 1988. Testing for a unit root in time series regression. *Biometrika* 75: 335–346.

Wei, W. W. S. 2006. *Time Series Analysis: Univariate and Multivariate Methods.* 2nd ed. Boston: Pearson.

Subject and author index

This is the subject and author index for the *Time-Series Reference Manual*. Readers interested in topics other than time series should see the combined subject index (and the combined author index) in the *Quick Reference and Index*.

Semicolons set off the most important entries from the rest. Sometimes no entry will be set off with semicolons, meaning that all entries are equally important.

A

Abraham, B., [TS] **tssmooth**, [TS] **tssmooth dexponential**, [TS] **tssmooth exponential**, [TS] **tssmooth hwinters**, [TS] **tssmooth shwinters**

ac command, [TS] **corrgram**

add, irf subcommand, [TS] **irf add**

Adkins, L. C., [TS] **arch**

Ahn, S. K., [TS] **vec intro**

Aielli, G. P., [TS] **mgarch**, [TS] **mgarch dcc**

Akaike, H., [TS] **varsoc**

Amemiya, T., [TS] **varsoc**

Amisano, G., [TS] **irf create**, [TS] **var intro**, [TS] **var svar**, [TS] **vargranger**, [TS] **varwle**

An, S., [TS] **arfima**

Anderson, B. D. O., [TS] **sspace**

Anderson, T. W., [TS] **vec**, [TS] **vecrank**

Ansley, C. F., [TS] **arima**

A-PARCH, [TS] **arch**

AR, [TS] **arch**, [TS] **arfima**, [TS] **arima**, [TS] **dfactor**, [TS] **sspace**, [TS] **ucm**

ARCH, *also see* multivariate GARCH
  effects, [TS] **arch**
  model, [TS] **arch**, [TS] **Glossary**
  postestimation, [TS] **arch postestimation**

arch command, [TS] **arch**, [TS] **arch postestimation**

ARFIMA
  model, [TS] **arfima**, [TS] **Glossary**
  postestimation, [TS] **arfima postestimation**, [TS] **psdensity**

arfima command, [TS] **arfima**, [TS] **arfima postestimation**

ARIMA
  model, [TS] **arima**, [TS] **Glossary**
  postestimation, [TS] **arima postestimation**, [TS] **psdensity**

arima command, [TS] **arima**, [TS] **arima postestimation**

ARMA, [TS] **arch**, [TS] **arfima**, [TS] **arima**, [TS] **sspace**, [TS] **ucm**, [TS] **Glossary**

ARMAX, [TS] **arfima**, [TS] **arima**, [TS] **dfactor**, [TS] **sspace**, [TS] **ucm**, [TS] **Glossary**

autocorrelation, [TS] **arch**, [TS] **arfima**, [TS] **arima**, [TS] **corrgram**, [TS] **dfactor**, [TS] **newey**, [TS] **prais**, [TS] **psdensity**, [TS] **sspace**, [TS] **ucm**, [TS] **var**, [TS] **varlmar**, [TS] **Glossary**

autoregressive
  conditional heteroskedasticity, *see* ARCH
  fractionally integrated moving-average model, *see* ARFIMA
  integrated moving average, *see* ARIMA
  model, [TS] **dfactor**, [TS] **psdensity**, [TS] **sspace**, [TS] **ucm**
  moving average, *see* ARMA
  process, [TS] **Glossary**

Aznar, A., [TS] **vecrank**

B

Baillie, R. T., [TS] **arfima**

band-pass filters, [TS] **tsfilter bk**, [TS] **tsfilter cf**, [TS] **Glossary**

Bartlett, M. S., [TS] **wntestb**

Bartlett's
  bands, [TS] **corrgram**
  periodogram test, [TS] **wntestb**

Baum, C. F., [TS] **time series**, [TS] **arch**, [TS] **arima**, [TS] **dfgls**, [TS] **rolling**, [TS] **tsset**, [TS] **var**, [TS] **wntestq**

Bauwens, L., [TS] **mgarch**

Baxter–King filter, [TS] **tsfilter**, [TS] **tsfilter bk**

Baxter, M., [TS] **tsfilter**, [TS] **tsfilter bk**, [TS] **tsfilter cf**

Becketti, S., [TS] **corrgram**

Bera, A. K., [TS] **arch**, [TS] **varnorm**, [TS] **vecnorm**

Beran, J., [TS] **arfima**, [TS] **arfima postestimation**

Berkes, I., [TS] **mgarch**

Berndt, E. K., [TS] **arch**, [TS] **arima**

Bianchi, G., [TS] **tsfilter**, [TS] **tsfilter bw**

bk, tsfilter subcommand, [TS] **tsfilter bk**

Black, F., [TS] **arch**

block exogeneity, [TS] **vargranger**

Bloomfield, P., [TS] **arfima**

Bollerslev, T., [TS] **arch**, [TS] **arima**, [TS] **mgarch**, [TS] **mgarch ccc**, [TS] **mgarch dvech**

Boswijk, H. P., [TS] **vec**

Bowerman, B. L., [TS] **tssmooth**, [TS] **tssmooth dexponential**, [TS] **tssmooth exponential**, [TS] **tssmooth hwinters**, [TS] **tssmooth shwinters**

Box, G. E. P., [TS] **arfima**, [TS] **arima**, [TS] **corrgram**, [TS] **cumsp**, [TS] **dfuller**, [TS] **pergram**, [TS] **pperron**, [TS] **psdensity**, [TS] **wntestq**, [TS] **xcorr**

Breusch, T. S., [TS] **Glossary**

Brockwell, P. J., [TS] **corrgram**, [TS] **sspace**

Burns, A. F., [TS] **tsfilter**, [TS] **tsfilter bk**, [TS] **tsfilter bw**, [TS] **tsfilter cf**, [TS] **tsfilter hp**, [TS] **ucm**

business days, [TS] **intro**

Butterworth filter, [TS] **tsfilter**, [TS] **tsfilter bw**

Butterworth, S., [TS] **tsfilter**, [TS] **tsfilter bw**

bw, tsfilter subcommand, [TS] **tsfilter bw**

K

Kalman
> filter, [TS] **arima**, [TS] **dfactor**, [TS] **dfactor postestimation**, [TS] **sspace**, [TS] **sspace postestimation**, [TS] **ucm**, [TS] **ucm postestimation**, [TS] **Glossary**
> forecast, [TS] **dfactor postestimation**, [TS] **sspace postestimation**, [TS] **ucm postestimation**
> smoothing, [TS] **dfactor postestimation**, [TS] **sspace postestimation**, [TS] **ucm postestimation**

Kalman, R. E., [TS] **arima**

Kim, I.-M., [TS] **vec intro**, [TS] **vec**, [TS] **vecrank**

King, M. L., [TS] **prais**

King, R. G., [TS] **tsfilter**, [TS] **tsfilter bk**, [TS] **tsfilter cf**, [TS] **tsfilter hp**, [TS] **vecrank**

Kmenta, J., [TS] **arch**, [TS] **prais**, [TS] **rolling**

Koehler, A. B., [TS] **tssmooth**, [TS] **tssmooth dexponential**, [TS] **tssmooth exponential**, [TS] **tssmooth hwinters**, [TS] **tssmooth shwinters**

Kohn, R., [TS] **arima**

Koopman, S. J., [TS] **ucm**

Kroner, K. F., [TS] **arch**

kurtosis, [TS] **varnorm**, [TS] **vecnorm**

L

lag-exclusion statistics, [TS] **varwle**

lag operator, [TS] **Glossary**

lag-order selection statistics, [TS] **var**, [TS] **var intro**, [TS] **var svar**, [TS] **vec intro**, [TS] **varsoc**

Lagrange-multiplier test, [TS] **varlmar**, [TS] **veclmar**

Lai, K. S., [TS] **dfgls**

Laurent, S., [TS] **mgarch**

leap seconds, [TS] **tsset**

Ledolter, J., [TS] **tssmooth**, [TS] **tssmooth dexponential**, [TS] **tssmooth exponential**, [TS] **tssmooth hwinters**, [TS] **tssmooth shwinters**

Lee, T.-C., [TS] **arch**, [TS] **prais**

Leser, C. E. V., [TS] **tsfilter**, [TS] **tsfilter hp**

Lieberman, O., [TS] **mgarch**

Lilien, D. M., [TS] **arch**

Lim, G. C., [TS] **arch**

linear
> filter, [TS] **Glossary**
> regression, [TS] **newey**, [TS] **prais**, [TS] **var intro**, [TS] **var**, [TS] **var svar**, [TS] **varbasic**

Ling, S., [TS] **mgarch**

Ljung, G. M., [TS] **wntestq**

long-memory process, [TS] **arfima**, [TS] **Glossary**

Lu, J. Y., [TS] **prais**

Lütkepohl, H., [TS] **time series**, [TS] **arch**, [TS] **dfactor**, [TS] **fcast compute**, [TS] **irf**, [TS] **irf create**, [TS] **mgarch dvech**, [TS] **prais**, [TS] **sspace**, [TS] **sspace postestimation**, [TS] **var intro**, [TS] **var**, [TS] **var svar**,

Lütkepohl, H., *continued*
> [TS] **varbasic**, [TS] **vargranger**, [TS] **varnorm**, [TS] **varsoc**, [TS] **varstable**, [TS] **varwle**, [TS] **vec intro**, [TS] **vecnorm**, [TS] **vecrank**, [TS] **vecstable**

M

MA, [TS] **arch**, [TS] **arfima**, [TS] **arima**, [TS] **sspace**, [TS] **ucm**

ma, tssmooth subcommand, [TS] **tssmooth ma**

MacKinnon, J. G., [TS] **arch**, [TS] **arima**, [TS] **dfuller**, [TS] **pperron**, [TS] **prais**, [TS] **sspace**, [TS] **varlmar**, [TS] **Glossary**

Maddala, G. S., [TS] **vec intro**, [TS] **vec**, [TS] **vecrank**

Magnus, J. R., [TS] **var svar**

Mandelbrot, B., [TS] **arch**

Mangel, M., [TS] **varwle**

Maravall, A., [TS] **tsfilter hp**

McAleer, M., [TS] **mgarch**

McCullough, B. D., [TS] **corrgram**

McDowell, A. W., [TS] **arima**

McLeod, A. I., [TS] **arima**, [TS] **ucm**

Meiselman, D., [TS] **arima**

MGARCH, *see* multivariate GARCH

mgarch
> ccc command, [TS] **mgarch ccc**
> dcc command, [TS] **mgarch dcc**
> dvech command, [TS] **mgarch dvech**
> vcc command, [TS] **mgarch vcc**

Miller, J. I., [TS] **sspace**

Mitchell, W. C., [TS] **tsfilter**, [TS] **tsfilter bk**, [TS] **tsfilter bw**, [TS] **tsfilter cf**, [TS] **tsfilter hp**, [TS] **ucm**

Monfort, A., [TS] **arima**, [TS] **mgarch ccc**, [TS] **mgarch dcc**, [TS] **mgarch vcc**

Montgomery, D. C., [TS] **tssmooth**, [TS] **tssmooth dexponential**, [TS] **tssmooth exponential**, [TS] **tssmooth hwinters**, [TS] **tssmooth shwinters**

Moore, J. B., [TS] **sspace**

moving average
> model, *see* MA
> process, [TS] **Glossary**
> smoother, [TS] **tssmooth**, [TS] **tssmooth ma**

multiplicative heteroskedasticity, [TS] **arch**

multivariate GARCH, [TS] **mgarch**, [TS] **Glossary**
> model,
>> constant conditional correlation, [TS] **mgarch ccc**
>> diagonal vech, [TS] **mgarch dvech**
>> dynamic conditional correlation, [TS] **mgarch dcc**
>> varying conditional correlation, [TS] **mgarch vcc**
> postestimation,
>> after ccc model, [TS] **mgarch ccc postestimation**
>> after dcc model, [TS] **mgarch dcc postestimation**
>> after dvech model, [TS] **mgarch dvech postestimation**